Core Curriculum

Introductory Craft Skills

Trainee Guide
Fourth Edition

Prentice Hall
Boston Columbus Indianapolis New York San Francisco Upper Saddle River
Amsterdam Cape Town Dubai London Madrid Milan Munich Paris Montreal Toronto
Delhi Mexico City Sao Paulo Sydney Hong Kong Seoul Singapore Taipei Tokyo

National Center for Construction Education and Research
President: Don Whyte
Director of Product Development: Daniele Stacey
Core Curriculum Project Manager: Daniele Stacey
Production Manager: Tim Davis
Quality Assurance Coordinator: Debie Ness
Desktop Publishing Coordinator: James McKay
Editors: Rob Richardson, Matt Tischler

Writing and development services provided by MetaMedia Training International, Inc.
Project Manager: Donna Lynn
Assistant Project Manager: Kerrie Ogden
Sr. Instructional Designer: Gina DeVitis
Sr. Instructional Designer: Mark Colone
Art Director: Daniel Govar
Graphic Artist/DTP: Kimberly Graham

Pearson Education, Inc.
Editor in Chief: Vernon R. Anthony
Senior Product Manager: Lori Cowen
Product Development Editor: Janet Ryerson
Managing Editor: JoEllen Gohr
Project Manager: Steve Robb
AV Project Manager: Janet Portisch
Operations Supervisor: Laura Weaver
Art Director: Diane Y. Ernsberger
Text and Cover Designer: Kristina D. Holmes
Cover Photo: Tim Davis/NCCER
Director of Marketing: David Gesell
Executive Marketing Manager: Derril Trakalo
Senior Marketing Coordinator: Alicia Wozniak
Full Service Project Management: DeAnn Montoya, S4Carlisle Publishing Services
Composition: S4Carlisle Publishing Services
Printer: Courier/Kendallville, Inc.
Cover Printer: Lehigh-Phoenix Color/Hagerstown
Text Fonts: Palatino and Univers

Credits and acknowledgments borrowed from other sources and reproduced, with permission, in this textbook appear on pages FC.1 and FC.2.

The cover art is from a photograph of a tile mosaic, titled "Tools of the Trade," which depicts the tools of craftsmanship in stained glass. This 8-foot by 8-foot mosaic was designed by Mary and Mike McIntyre and hangs in the Charles Perry Construction Yard at the M. E. Rinker, Sr. School of Building Construction at the University of Florida in Gainesville. The National Center for Construction Education and Research is an affiliate of the University of Florida.

---

10 9 8 7 6 5

Prentice Hall
is an imprint of

www.pearsonhighered.com

Perfect bound: ISBN 10: 0-13-608637-3
ISBN 13: 978-0-13-608637-6
Loose leaf: ISBN 10: 0-13-608638-1
ISBN 13: 978-0-13-608638-3
Case bound: ISBN 13: 0-13-608636-5
ISBN 13: 978-0-13-608636-9

Preface

To the Trainee

Welcome to the world of construction! You are joining the eight million Americans who have chosen a career in this lucrative field. Construction is one of the nation's largest industries, offering excellent opportunities for high earnings, career advancement, and business ownership.

Work in construction offers a great variety of career opportunities. People with many different talents and educational backgrounds—skilled craftspersons, managers, supervisors, and superintendents—find job opportunities in construction and related fields. As you will learn throughout your training, many other industries depend upon the work you will do in construction. From houses and office buildings to factories, roads, and bridges—*everything* begins with construction.

New with *Core Curriculum: Introductory Craft Skills*

NCCER and Pearson are pleased to present the fourth edition of *Core Curriculum: Introductory Craft Skills.* This full-color textbook now includes nine modules for building foundation skills in construction. To help entry-level craftworkers step up their awareness of materials handling techniques and equipment, NCCER brings you "Introduction to Materials Handling."

We are also excited to provide a revised "Basic Safety" module that now aligns to OSHA's 10-hour program. This means that instructors who are OSHA-500 certified are able to issue 10-hour OSHA cards to their students who successfully complete the module. Combined with an NCCER credential, the OSHA 10-hour card will show employers a credible and valuable training record.

We keep math "real" for students in this edition of "Introduction to Construction Math" by adding more application rather than theory-related exercises. By keeping math "real," it makes the language of math much easier to understand.

Interested in the evolving world of green building? Then you'll be pleased to notice NCCER *greening* the curriculum with *Going Green* features. These features highlight building products, practices, and projects that are energy efficient, sustainable, and earth-friendly.

Finally, NCCER is pleased to debut a new "Introduction to Construction Drawings" module that features drawings from LEED-Gold Certified Rinker Hall at the University of Florida. The M.E. Rinker, Sr. School of Building Construction—an NCCER partner—is one of the top-ranked building construction programs in the country.

We invite you to visit the NCCER website at **www.nccer.org** for the latest releases, training information, newsletter, and much more. You can also reference the Contren® product catalog online at **www.nccer.org**. Your feedback is welcome. You may e-mail your comments to **curriculum@nccer.org** or send general comments and inquiries to **info@nccer.org**.

Contren® Learning Series

The National Center for Construction Education and Research (NCCER) is a not-for-profit 501(c)(3) education foundation established in 1995 by the world's largest and most progressive construction companies and national construction associations. It was founded to address the severe workforce shortage facing the industry and to develop a standardized training process and curricula. Today, NCCER is supported by hundreds of leading construction and maintenance companies, manufacturers, and national associations. The Contren® Learning Series was developed by NCCER in partnership with Pearson Education, Inc., the world's largest educational publisher.

Some features of NCCER's Contren® Learning Series are as follows:

- An industry-proven record of success
- Curricula developed by the industry for the industry
- National standardization providing portability of learned job skills and educational credits
- Compliance with Apprenticeship, Training, Employer, and Labor Services (ATELS) requirements for related classroom training (CFR 29:29)
- Well-illustrated, up-to-date, and practical information

NCCER also maintains a National Registry that provides transcripts, certificates, and wallet cards to individuals who have successfully completed modules of NCCER's Contren® Learning Series. Training programs must be delivered by an NCCER Accredited Training Sponsor in order to receive these credentials.

Special Features

In an effort to provide a comprehensive user-friendly training resource, we have incorporated many different features for your use. Whether you are a visual or hands-on learner, this book will provide you with the proper tools to get started in the construction industry.

Introduction Page

This page is found at the beginning of each module and lists the Objectives, Key Trade Terms, Required Trainee Materials, Prerequisites, and Course Map for that module. The Objectives list the skills and knowledge you will need in order to complete the module successfully. The list of Key Trade Terms identifies important terms you will need to know by the end of the module. Required Trainee Materials list the materials and supplies needed for the module. The Prerequisites for the module are listed and illustrated in the Course Map. The Course Map also gives a visual overview of the entire course and a suggested learning sequence for you to follow.

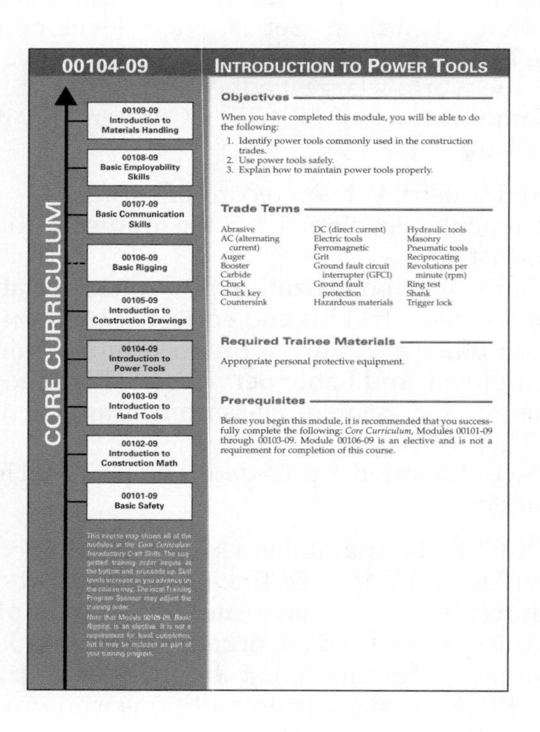

Notes, Cautions, and Warnings

Safety features are set off from the main text in highlighted boxes and organized into three categories based on the potential danger of the issue being addressed. Notes simply provide additional information on the topic area. Cautions alert you of a danger that does not present potential injury but may cause damage to equipment. Warnings stress a potentially dangerous situation that may cause injury to you or a co-worker.

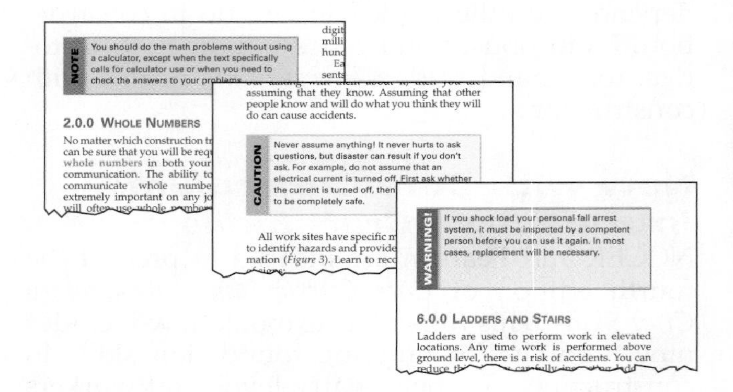

On-Site

The On-Site features offer technical hints and tips from the construction industry. These often include nice-to-know information that you will find helpful. On-Sites also present real-life scenarios similar to those you might encounter on the job site.

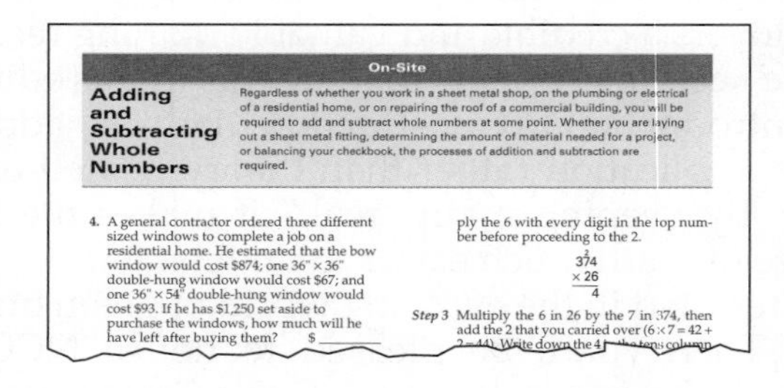

Going Green

Going Green looks at ways to preserve the environment, save energy, and make good choices regarding the health of the planet. Through the introduction of new construction practices and products, you will see how the "greening of America" has already taken root.

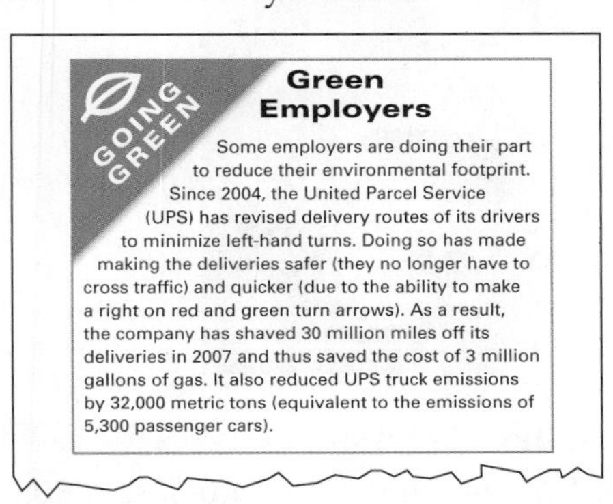
GOING GREEN

Green Employers

Some employers are doing their part to reduce their environmental footprint. Since 2004, the United Parcel Service (UPS) has revised delivery routes of its drivers to minimize left-hand turns. Doing so has made making the deliveries safer (they no longer have to cross traffic) and quicker (due to the ability to make a right on red and green turn arrows). As a result, the company has shaved 30 million miles off its deliveries in 2007 and thus saved the cost of 3 million gallons of gas. It also reduced UPS truck emissions by 32,000 metric tons (equivalent to the emissions of 5,300 passenger cars).

Did You Know?

The Did You Know? features introduce historical tidbits or modern information about the construction industry. Interesting and sometimes surprising facts about construction are also presented.

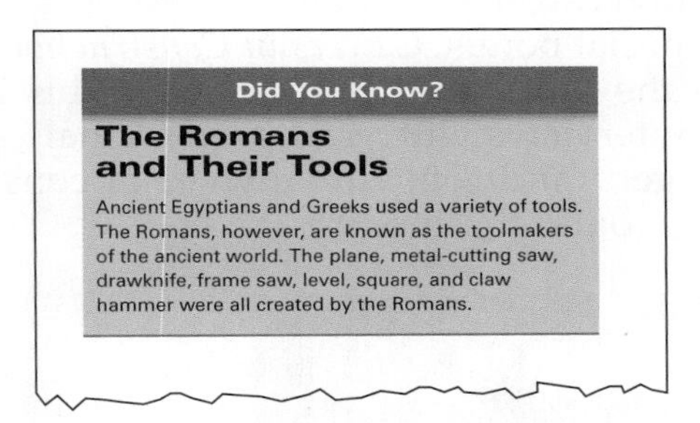
Did You Know?

The Romans and Their Tools

Ancient Egyptians and Greeks used a variety of tools. The Romans, however, are known as the toolmakers of the ancient world. The plane, metal-cutting saw, drawknife, frame saw, level, square, and claw hammer were all created by the Romans.

Color Illustrations and Photographs

Full-color illustrations and photographs are used throughout each module to provide vivid detail. These figures highlight important concepts from the text and provide clarity for complex instructions. Each figure is denoted in the text in *italic type* for easy reference.

Figure 4 Horseplay at work can be dangerous.

Review Questions

Review Questions are provided to reinforce the knowledge you have gained. This makes them a useful tool for measuring what you have learned.

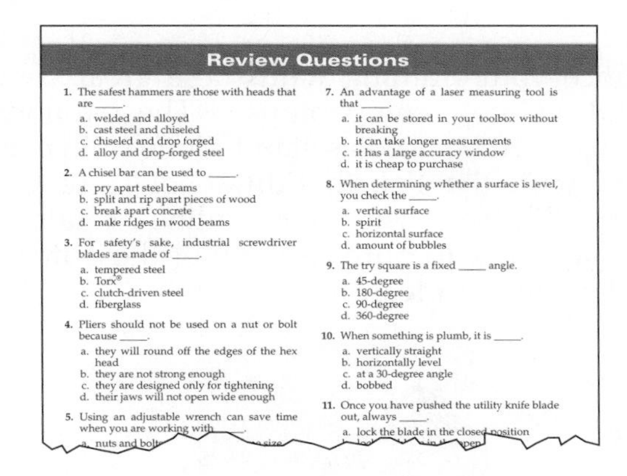
Review Questions

1. The safest hammers are those with heads that are _____.
   a. welded and alloyed
   b. cast steel and chiseled
   c. chiseled and drop forged
   d. alloy and drop-forged steel
2. A chisel bar can be used to _____.
   a. pry apart steel beams
   b. split and rip apart pieces of wood
   c. break apart concrete
   d. make ridges in wood beams
3. For safety's sake, industrial screwdriver blades are made of _____.
   a. tempered steel
   b. Torx®
   c. clutch-driven steel
   d. fiberglass
4. Pliers should not be used on a nut or bolt because _____.
   a. they will round off the edges of the hex head
   b. they are not strong enough
   c. they are designed only for tightening
   d. their jaws will not open wide enough
5. Using an adjustable wrench can save time when you are working with _____.
   a. nuts and bolt
7. An advantage of a laser measuring tool is that _____.
   a. it can be stored in your toolbox without breaking
   b. it can take longer measurements
   c. it has a large accuracy window
   d. it is cheap to purchase
8. When determining whether a surface is level, you check the _____.
   a. vertical surface
   b. spirit
   c. horizontal surface
   d. amount of bubbles
9. The try square is a fixed _____ angle.
   a. 45-degree
   b. 180-degree
   c. 90-degree
   d. 360-degree
10. When something is plumb, it is _____.
    a. vertically straight
    b. horizontally level
    c. at a 30-degree angle
    d. bobbed
11. Once you have pushed the utility knife blade out, always _____.
    a. lock the blade in the closed position

Trade Terms

Each module presents a list of Trade Terms that are discussed within the text, defined in the Glossary at the end of the module, and reinforced with a Trade Terms Quiz. These terms are denoted in the text with **green bold type** upon their first occurrence. To make searches for key information easier, a comprehensive Glossary of Trade Terms from all modules is found at the back of this book.

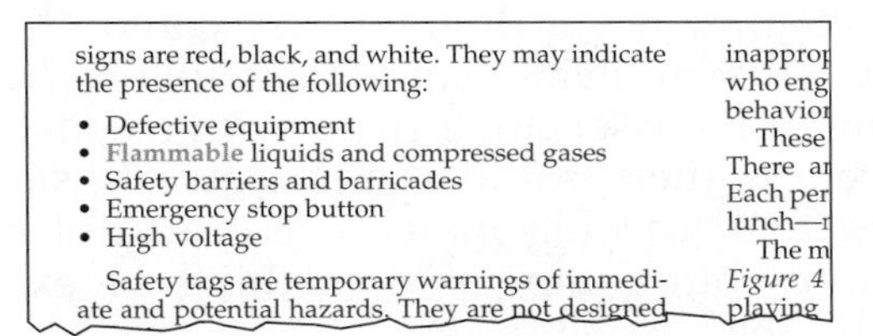
signs are red, black, and white. They may indicate the presence of the following:

- Defective equipment
- **Flammable** liquids and compressed gases
- Safety barriers and barricades
- Emergency stop button
- High voltage

Safety tags are temporary warnings of immediate and potential hazards. They are not designed

Cornerstone of Craftsmanship

Cornerstones of Craftsmanship share the experiences of and advice from successful craftspersons.

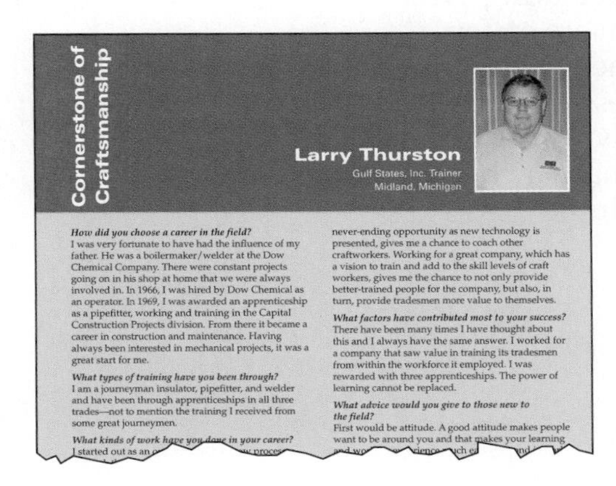
Cornerstone of Craftsmanship

Larry Thurston
Gulf States, Inc. Trainer
Midland, Michigan

How did you choose a career in the field?
I was very fortunate to have had the influence of my father. He was a boilermaker/welder at the Dow Chemical Company. There were constant projects going on in his shop at home that we were always involved in. In 1966, I was hired by Dow Chemical as an operator. In 1969, I was awarded an apprenticeship as a pipefitter, working and training in the Capital Construction Projects division. From there it became a career in construction and maintenance. Having always been interested in mechanical projects, it was a great start for me.

What types of training have you been through?
I am a journeyman insulator, pipefitter, and welder and have been through apprenticeships in all three trades—not to mention the training I received from some great journeymen.

What kinds of work have you done in your career?
I started out as an

never-ending opportunity as new technology is presented, gives me a chance to coach other craftworkers. Working for a great company, which has a vision to train and add to the skill levels of craft workers, gives me the chance to not only provide better-trained people for the company, but also, in turn, provide tradesmen more value to themselves.

What factors have contributed most to your success?
There have been many times I have thought about this and I always have the same answer. I worked for a company that saw value in training its tradesmen from within the workforce it employed. I was rewarded with three apprenticeships. The power of learning cannot be replaced.

What advice would you give to those new to the field?
First would be attitude. A good attitude makes people want to be around you and that makes your learning

Core Curriculum Companions

Enhance your training with these great supplemental Core companions. The following resources can be used as stand-alone or in combination with the Fourth Edition of Core Curriculum. Please visit our online catalog at www.nccer.org to purchase any of these items to supplement your learning.

Applied Construction Math

Paperback Trainee Guide: ISBN 0-13-227298-9

Applied Construction Math: A Novel Approach features a story that students can relate to and math skills they never thought they could. Its innovative style motivates students to follow the lessons taught by associating math with events that will occur in their real-lives. Students will stop thinking that learning math is as dreadful as doing chores but rather as something as exciting as building a new house!

Thirteen chapters teach basic math skills including:

- Division
- Decimals/Percentages
- Reading Measurements
- Calculating Area
- Powers of Ten
- Linear Measure, Angles, Volumes,
- Pressure and Slopes
- Solving for Unknowns
- Square Inches, Feet, and Yards
- Volume

Careers in Construction

Paperback Trainee Guide: ISBN 0-13-228605-X

NCCER's *Careers in Construction* showcases the world of construction and career opportunities available to high school students and anyone interested in pursuing a construction career. The guide includes a variety of pictures and illustrations as it reviews the pride and excitement of a construction career.

As a special bonus, *Careers in Construction* also includes the Build Your Future DVD. This DVD features interviews with craft professionals, project managers, and company owners at construction sites around the country.

Tools for Success: *Critical Skills for the Construction Industry 3/E*

Paperback Trainee Guide: ISBN 0-13-610649-8

The *Tools for Success* workbook includes classroom activities to help students navigate their way through intangible workplace issues such as conflict resolution, diversity, problem-solving, professionalism, and proper communications techniques.

Your Role in the Green Environment

Paperback Trainee Guide: ISBN 0-13-606596-1

Geared to entry-level craft workers or to anyone wishing to learn more about green building, this module provides fundamental instruction in the green environment, green construction practices, and green building rating systems.

This module and built-in study guide have been endorsed by Green Advantage® to assist students and practitioners in their preparations for the Green Advantage® Commercial/Residential certification exam. NCCER is also a United States Green Building Council Education Provider, and, as such, is committed to enhancing the ongoing development of building industry professionals.

One Industry. One Training Program. One Online Solution.

Contren Connect®: *An Interactive Online Course*

Core Trainee Access: ISBN 0-13-610656-0

Ideal for blended or distance education, Contren® Connect is a unique web-based supplement in the form of an electronic book that provides a range of visual, auditory, and interactive elements to enhance your training. It can be used in a variety of settings such as self-study, blended/distance education, or in the traditional classroom environment! It's the perfect way to review content from a class you may have missed or to practice at your own pace.

Features:

- **Online Lectures** – Each ebook module features a written summary of key content accompanied by an optional Audio Summary so if you need a refresher, this tool is always available.
- **Video Presentations** – Throughout, you'll find dynamic video presentations that demonstrate difficult skills and concepts! Of special note are the safe/unsafe scenarios shot on a live construction site testing your knowledge of the four 'high hazards' presented in the Basic Safety module.
- **Personalization Tools** – With the "highlighter" and "notes" options you can easily personalize your own Contren® Connect ebook to keep track of important information or create your own study guide.
- **Review Quizzes** – Short multiple-choice concept check quizzes at the end of each module section act as the ideal study tool and provide immediate feedback. Additionally, you'll find fill-in-the-blank trade terms quizzes, applied math questions, and comprehension questions at the end of each module.
- **Active Figures** – Interactive exercises bring key concepts to life, including animation in the Introduction to Construction Drawings module that will help you make the mental transition from a flat, 2-dimensional plan to a 3-dimensional finished structure.

Visit www.contrenconnect.com to view a demo.

Contren® Connect is available with:

Core Curriculum
Carpentry Levels 1-4
Construction Technology
Electrical Levels 1-4
Electronic Systems Technician Levels 1-4
HVAC Levels 1-4
Plumbing Levels 1-4
Your Role in the Green Environment

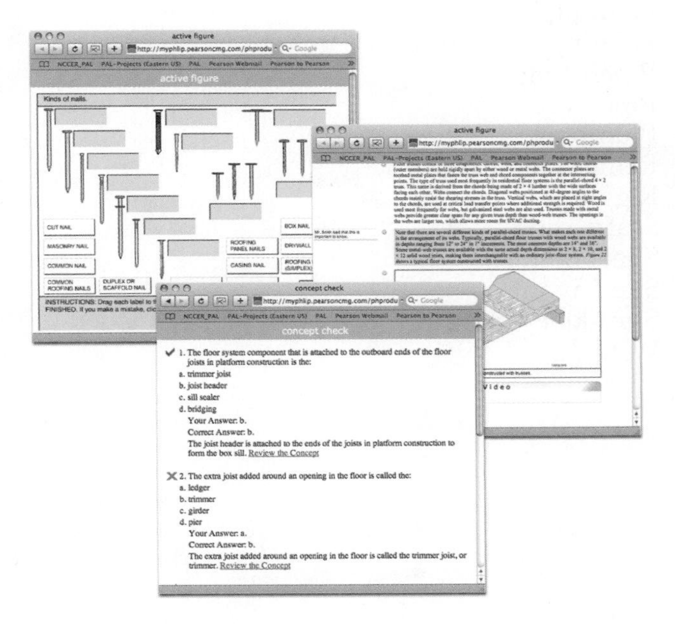

Contren® Curricula

NCCER's training programs comprise over 50 construction, maintenance, and pipeline areas and include skills assessments, safety training, and management education.

Boilermaking
Cabinetmaking
Carpentry
Concrete Finishing
Construction Craft Laborer
Construction Technology
Core Curriculum: Introductory Craft Skills
Drywall
Electrical
Electronic Systems Technician
Heating, Ventilating, and Air Conditioning
Heavy Equipment Operations
Highway/Heavy Construction
Hydroblasting
Industrial Maintenance Electrical and Instrumentation Technician
Industrial Maintenance Mechanic
Instrumentation
Insulating
Ironworking
Masonry
Millwright
Mobile Crane Operations
Painting
Painting, Industrial
Pipefitting
Pipelayer
Plumbing
Reinforcing Ironwork
Rigging
Scaffolding
Sheet Metal
Site Layout
Sprinkler Fitting
Welding

Pipeline

Control Center Operations, Liquid
Corrosion Control
Electrical and Instrumentation
Field Operations, Liquid
Field Operations, Gas
Maintenance
Mechanical

Safety

Field Safety
Safety Orientation
Safety Technology

Management

Introductory Skills for the Crew Leader
Project Management
Project Supervision

Spanish Translations

Acabado de Concreto
Aislamiento
Albañilería
Andamios
Carpintería de Formas
Currículo Básico: Habilidades Introductorias del Oficio
Electricidad
Herreria de Refuerzo
Instalación de Rociadores Nivel Uno
Instalación de Tubería Industrial
Orientación de Seguridad
Seguridad de Campo

Supplemental Titles

Applied Construction Math
Careers in Construction
Tool for Success
Your Role in the Green Environment

Acknowledgments

This curriculum was revised as a result of the farsightedness and leadership of the following sponsors:

ABC of Metro Washington
Alfred State College
Associated Training Services
BE&K
Corbin's Electric
Crossland Construction
Gulf States, Inc.
Ivey Mechanical Company
NC Department of Public Instruction
River Valley Technical Center
Swanson Center for Youth

This curriculum would not exist were it not for the dedication and unselfish energy of those volunteers who served on the Authoring Team. A sincere thanks is extended to the following:

Salvatore Benevegna
Joe Beyer
Frank Branham
Jonathan Byrd
Tim Eldridge
Bob Fitzgerald
Tim Grattan
Erin M. Hunter
Jon Jones
Peter Klapperich
Rick Klepin
Robert Moffett
Thomas G. Murphy
Jason C. Roberts
Aaron Thompson
D. Larry Thurston
Robert F. Tilley
Marcel Veronneau

A final note: This book is the result of a collaborative effort involving the production, editorial, and development staff at Prentice Hall and the National Center for Construction Education and Research. Thanks to all of the dedicated people involved in the many stages of this project.

NCCER Partnering Associations

American Fire Sprinkler Association
Associated Builders and Contractors, Inc.
Associated General Contractors of America
Association for Career and Technical Education
Association for Skilled and Technical Sciences
BOCES
Carolinas AGC, Inc.
Carolinas Electrical Contractors Association
Center for the Improvement of Construction Management and Processes
Construction Industry Institute
Construction Users Roundtable
Design Build Institute of America
Green Advantage
IMTI of NY and IMTI of CT
Merit Contractors Association of Canada
Metal Building Manufacturers Association
NACE International
National Association of Minority Contractors
National Association of Women in Construction
National Insulation Association
National Ready Mixed Concrete Association
National Systems Contractors Association
National Technical Honor Society
National Utility Contractors Association
NAWIC Education Foundation
North American Crane Bureau
North American Technician Excellence
Painting & Decorating Contractors of America
Portland Cement Association
SafeTek USA
SkillsUSA
Steel Erectors Association of America
Texas Gulf Coast Chapter, ABC
Tri-Counties Multi Trade Centers
University of Florida
U.S. Army Corps of Engineers
Women Construction Owners & Executives, USA

Contents

00101-09 Basic Safety 1.i

Complies with OSHA-10 training requirements. Explains the safety obligations of workers, supervisors, and managers to ensure a safe workplace. Discusses the causes and results of accidents and the impact of accident costs. Reviews the role of company policies and OSHA regulations. Introduces common job-site hazards and identifies proper protections. Defines safe work procedures, proper use of personal protective equipment, and working with hazardous chemicals. Identifies other potential construction hazards, including hazardous material exposures, welding and cutting hazards, and confined spaces. **(12.5 hours)**

00102-09 Introduction to Construction Math 2.i

Reviews basic mathematical functions such as adding, subtracting, dividing, and multiplying whole numbers, fractions, and decimals, and explains their applications to the construction trades. Explains how to use and read various length measurement tools, including standard and metric rulers and tape measures, and the architect's and engineer's scales. Explains decimal-fraction conversions and the metric system, using practical examples. Also reviews basic geometry as applied to common shapes and forms. **(10 hours)**

00103-09 Introduction to Hand Tools 3.i

Introduces trainees to hand tools that are widely used in the construction industry, such as hammers, saws, levels, pullers, and clamps. Explains the specific applications of each tool and shows how to use them properly. Also discusses important safety and maintenance issues related to hand tools. **(10 hours)**

00104-09 Introduction to Power Tools 4.i

Provides detailed descriptions of commonly used power tools, such as drills, saws, grinders, and sanders. Reviews applications, proper use, safety, and maintenance. Many illustrations show power tools used in on-the-job settings. **(10 hours)**

00105-09 Introduction to Construction Drawings 5.i

Familiarizes trainees with basic terms for construction drawings, components, and symbols. Explains the different types of drawings (civil, architectural, structural, mechanical, plumbing/piping, electrical, and fire protection) and instructs trainees on how to interpret and use drawing dimensions. Four oversized drawings are included. **(10 hours)**

00106-09 Basic Rigging . . . 6.i

Explains how ropes, chains, hoists, loaders, and cranes are used to move material and equipment from one location to another on a job site. Describes inspection techniques and load-handling safety practices. Also reviews American National Standards Institute (ANSI) hand signals. **(15 Elective hours)**

00107-09 Basic Communication Skills . . 7.i

Provides trainees with techniques for communicating effectively with co-workers and supervisors. Includes practical examples that emphasize the importance of verbal and written information and instructions on the job. Also discusses effective telephone and email communication skills. **(7.5 hours)**

00108-09 Basic Employability Skills 8.i

Identifies the roles of individuals and companies in the construction industry. Introduces trainees to critical thinking and problem-solving skills and computer systems and their industry applications. Also reviews effective relationship skills, effective self-presentation, and key workplace issues such as sexual harassment, stress, and substance abuse. **(7.5 hours)**

00109-09 Introduction to Materials Handling 9.i

Recognizes hazards associated with materials handling and explains proper materials handling techniques and procedures. Also introduces materials handling equipment, and identifies appropriate equipment for common job-site tasks. **(5 hours)**

Glossary of Trade Terms G.1

Figure Credits FC.1

Index I.1

Basic Safety

00101-09

CORE CURRICULUM

00109-09 Introduction to Materials Handling

00108-09 Basic Employability Skills

00107-09 Basic Communication Skills

00106-09 Basic Rigging

00105-09 Introduction to Construction Drawings

00104-09 Introduction to Power Tools

00103-09 Introduction to Hand Tools

00102-09 Introduction to Construction Math

00101-09 Basic Safety

This course map shows all of the modules in the *Core Curriculum: Introductory Craft Skills*. The suggested training order begins at the bottom and proceeds up. Skill levels increase as you advance on the course map. The local Training Program Sponsor may adjust the training order.

Note that Module 00106-09, *Basic Rigging,* is an elective. It is not a requirement for level completion, but it may be included as part of your training program.

Objectives

When you have completed this module, you will be able to do the following:

1. Explain the idea of a safety culture and its importance in the construction crafts.
2. Identify causes of accidents and the impact of accident costs.
3. Explain the role of OSHA in job-site safety.
4. Explain OSHA's General Duty Clause and *1926 CFR Subpart C*.
5. Recognize hazard recognition and risk assessment techniques.
6. Explain fall protection, ladder, stair, and scaffold procedures and requirements.
7. Identify struck-by hazards and demonstrate safe working procedures and requirements.
8. Identify caught-in-between hazards and demonstrate safe working procedures and requirements.
9. Define safe work procedures to use around electrical hazards.
10. Demonstrate the use and care of appropriate personal protective equipment (PPE).
11. Explain the importance of hazard communications (HazCom) and Material Safety Data Sheets (MSDSs).
12. Identify other construction hazards on your job site, including hazardous material exposures, environmental elements, welding and cutting hazards, confined spaces, and fires.

Trade Terms

Apparatus
Arc
Arc welding
Combustible
Competent person
Concealed receptacle
Confined space
Cross-bracing
Dross
Electrical distribution panel
Excavation
Experience modification rate (EMR)
Extension ladder
Flammable
Flash
Flashback
Flash burn
Flash goggles
Flash point
Foot-candle
Ground
Ground fault circuit interrupter (GFCI)
Guarded
Hand line
Hazard Communication Standard (HazCom)
Lanyard
Lockout/tagout
Management system
Material safety data sheet (MSDS)
Maximum allowable slope
Maximum intended load
Mid-rail
Occupational Safety and Health Administration (OSHA)
Permit-required confined space
Personal protective equipment (PPE)
Planked
Proximity work
Qualified person
Respirator
Safety culture
Scaffold
Shoring
Signaler
Six-foot rule
Slag
Spall
Stepladder
Straight ladder
Switch enclosure
Toeboard
Top rail
Trench
Welding shield
Wind sock

Prerequisites

There are no prerequisites for this module.

Contents

Topics to be presented in this module include:

1.0.0 Introduction 1.1
2.0.0 Importance of Safety 1.1
2.1.0 Safety Culture 1.1
3.0.0 Accidents: Causes and Results 1.1
3.1.0 Accident Costs 1.2
3.2.0 What Causes Accidents? 1.3
3.2.1 Failure to Communicate 1.3
3.2.2 Poor Work Habits 1.5
3.2.3 Alcohol and Drug Abuse 1.5
3.2.4 Lack of Skill 1.7
3.2.5 Intentional Acts 1.8
3.2.6 Unsafe Acts 1.8
3.2.7 Rationalizing Risk 1.8
3.2.8 Unsafe Conditions 1.8
3.2.9 Management System Failure 1.9
3.3.0 Housekeeping 1.9
3.4.0 Company Safety Policies and OSHA Regulations 1.9
3.4.1 The Code of Federal Regulations 1.9
3.4.2 The General Duty Clause 1.10
3.4.3 Employee Rights and Responsibilities 1.10
3.4.4 Inspections 1.11
3.4.5 Violations 1.11
3.4.6 Compliance 1.11
3.4.7 Record Keeping 1.12
3.5.0 Reporting Injuries, Accidents, and Incidents 1.12
3.6.0 The Four High-Hazard Areas 1.13
3.7.0 Evacuation Procedures 1.13
4.0.0 Hazard Recognition, Evaluation, and Control 1.13
4.1.0 Hazard Recognition 1.13
4.2.0 Job Safety Analysis (JSA) and Task Safety Analysis (TSA) 1.14
4.3.0 Risk Assessment 1.15
5.0.0 Elevated Work and Fall Protection 1.17
5.1.0 Fall Hazards 1.17
5.2.0 Walking and Working Surfaces 1.17
5.3.0 Unprotected Sides, Wall Openings, and Floor Holes 1.17
5.4.0 Personal Fall Arrest Systems 1.19
5.4.1 PFAS Inspection 1.20
5.4.2 Donning a Harness 1.20
6.0.0 Ladders and Stairs 1.21
6.1.0 Straight Ladders 1.22
6.1.1 Inspecting Straight Ladders 1.23
6.1.2 Using Straight Ladders 1.23

6.2.0 Extension Ladders 1.24
6.2.1 Inspecting Extension Ladders 1.25
6.2.2 Using Extension Ladders 1.26
6.3.0 Stepladders 1.26
6.3.1 Inspecting Stepladders 1.26
6.3.2 Using Stepladders 1.26
6.4.0 Stairways 1.28
6.4.1 Stairway Maintenance and Housekeeping 1.28
7.0.0 Scaffolds 1.28
7.1.0 Types of Scaffolds 1.28
7.1.1 Manufactured Scaffolds 1.28
7.1.2 Rolling Scaffolds 1.28
7.2.0 Inspecting Scaffolds 1.28
7.3.0 Using Scaffolds 1.31
8.0.0 Struck-By Hazards 1.31
8.1.0 Vehicle and Roadway Hazards 1.31
8.2.0 Falling Objects 1.32
8.3.0 Flying Objects 1.32
9.0.0 Caught-In-Between Hazards 1.33
9.1.0 Trenching and Excavation 1.33
9.1.1 Cave-Ins 1.33
9.1.2 Inspections 1.34
9.1.3 Protective Systems 1.34
9.1.4 Spoil Pile and Material Hazards 1.36
9.1.5 Access and Egress 1.37
9.1.6 Emergency Response 1.37
9.2.0 Tool and Machine Guarding 1.37
10.0.0 Electrical Hazards 1.38
10.1.0 Basics of Electricity 1.38
10.2.0 Electrical Safety Guidelines 1.39
10.3.0 Electrical Power Systems 1.42
10.3.1 Assured Equipment Grounding Conductor Programs 1.42
10.3.2 Ground Fault Circuit Interrupters 1.43
10.4.0 Lockout/Tagout 1.43
10.5.0 Working Near Energized Electrical Equipment 1.44
10.6.0 If Someone Is Shocked 1.44
11.0.0 Personal Protective Equipment 1.45
11.1.0 Personal Protective Equipment Needs 1.45
11.2.0 Personal Protective Equipment Use and Care 1.45
11.3.0 Clothing and Jewelry 1.45
11.4.0 Hard Hat 1.45
11.5.0 Eye and Face Protection 1.45
11.6.0 Gloves 1.46
11.7.0 Leg Protection 1.48
11.8.0 Foot Protection 1.48
11.9.0 Skin Protection 1.48
11.10.0 Hearing Protection 1.48
11.11.0 Respiratory Protection 1.49
11.11.1 Respirator Requirements 1.50
11.11.2 Selecting Respirators 1.51
11.11.3 Testing Respirators 1.51
11.11.4 Inspecting Respirators 1.52
11.11.5 Maintaining Respirators 1.52

12.0.0 Hazard Communication Standard 1.52
12.1.0 Material Safety Data Sheets 1.52
12.2.0 Your Responsibilities Under HazCom 1.52
13.0.0 Other Job-Site Hazards 1.56
13.1.0 Job-Site Exposures 1.56
13.1.1 Lead 1.57
13.1.2 Asbestos 1.57
13.1.3 Silica 1.57
13.1.4 Bloodborne Pathogens. 1.57
13.1.5 Chemical Splashes. 1.58
13.2.0 Proximity Work 1.58
13.2.1 Pressurized or High-Temperature Systems 1.59
13.3.0 Heat Stress 1.59
13.3.1 Heat Cramps 1.59
13.3.2 Heat Exhaustion. 1.60
13.3.3 Heat Stroke 1.60
13.4.0 Cold Stress 1.60
13.4.1 Frostbite. 1.61
13.4.2 Hypothermia 1.61
13.5.0 Welding and Cutting Hazards 1.62
13.5.1 Flame Cutting. 1.64
13.5.2 Hoses and Regulators 1.64
13.5.3 Work Area 1.66
13.6.0 Confined Spaces 1.66
13.7.0 Construction Ergonomics 1.67
13.8.0 Fire Hazards 1.69
13.8.1 How Fires Start 1.70
13.8.2 Fire Prevention. 1.70
13.8.3 Types of Combustibles. 1.70
13.8.4 Firefighting 1.71

1.0.0 Introduction

When you take a job, you have a safety obligation to your employer, co-workers, family, and yourself. In exchange for your wages and benefits, you agree to work safely. You are also obligated to make sure anyone you work with is working safely. Your employer is likewise obligated to maintain a safe workplace for all employees. The ultimate responsibility for on-the-job safety, however, rests with you. Safety is part of your job. In this module, you will learn to ensure your safety and that of the people you work with by adhering to the following rules:

- Follow safe work practices and procedures.
- Inspect safety equipment before use.
- Use safety equipment properly.

To take full advantage of the wide variety of training, job, and career opportunities the construction industry offers, you must first understand the importance of safety. Successful completion of this module will be your first step toward achieving this goal. Subsequent modules offer more detailed explanations of safety procedures and opportunities to practice them.

2.0.0 Importance of Safety

On a typical job site, there are often many workers from many trades in one place. These workers are all performing different tasks and operations. As a result, the job site is constantly changing and hazards are continually emerging. These hazards can jeopardize your safety. Your employer will make every effort to plan safety into each job and to provide a safe and healthful job site. Ultimately, your safety is in your own hands. Throughout the workday, you may perform tasks that may be repeated with little conscious thought. This routine work can dull alertness and increase the chances of an accident. Safety consciousness is a vital part of your work. Safety training is conducted to make you aware that dangers exist all around you every day. The time you spend learning and practicing safety procedures can save your life and the lives of others.

Did You Know?

Safety training is required for all activities. Never operate tools, machinery, or equipment without prior training. Always refer to the manufacturer's instructions.

2.1.0 Safety Culture

Your boss might say, "I want my company to have a perfect safety record." What does that mean? A safety record is more than the number of days a company has worked without an accident. Safety is a learned behavior and attitude. Safety is a way of working that must be incorporated into the company as a culture.

A **safety culture** is created when the whole company sees the value of a safe work environment. Creating and maintaining a safety culture is an ongoing process that includes a sound safety structure and attitude, and relates to both organizations and individuals. Everybody in the company, from management to laborers, must be responsible for safety every day they come to work.

There are many benefits to having a safety culture. Companies with strong safety cultures usually have:

- Fewer at-risk behaviors
- Lower accident rates
- Less turnover
- Lower absenteeism
- Higher productivity

A strong safety culture can also lower your company's **experience modification rate (EMR)**, which leads to winning more bids and keeping workers employed. Contractors with high EMRs are sometimes excluded from bidding. Factors that contribute to a strong safety culture include:

- Perceiving safety as a core value
- Strong leadership
- Establishing and enforcing high standards of expectation and performance
- The involvement of all employees
- Effective communication and commonly understood and agreed-upon goals
- Using the workplace as a learning environment
- Encouraging workers to have a questioning attitude
- Good organizational learning and responsiveness to change
- Providing timely response to safety issues and concerns
- Continually monitoring performance

3.0.0 Accidents: Causes and Results

No person is immune to an accident. Accidents can happen to anyone at any time, in any place. Both poor behavior and poor working conditions can cause accidents. You can help prevent accidents by using safe work habits, understanding what causes accidents, and learning how to prevent them.

The lessons you will learn in this module will help you work safely. You will be able to spot and avoid hazardous conditions on the job site. By following safety procedures and being aware of the need for safety, you will help keep your workplace free from accidents and protect yourself and your co-workers from injury or even death.

An accident is defined as an unplanned event that may or may not result in personal injury or property damage. Accidents are often categorized by their severity and impact, as follows:

- *Near-miss* – An unplanned event or occurrence in which no one was injured and no damage to property occurred, but during which either could have happened. Near-miss incidents are warnings which should not be overlooked or taken lightly.
- *Property damage* – An unplanned event that resulted in damage to tools, materials, or equipment, but no injuries.
- *Minor injuries* – Personnel may have received minor cuts, bruises, or strains, but the injured workers returned to full duty on their next regularly scheduled work shift.
- *Serious or disabling injuries* – Personnel received injuries that resulted in temporary or permanent disability. Included in this category would be lost time accidents, restricted duty or motion cases, and those which resulted in permanent partial or permanent total disability.
- *Fatalities*.

Studies have shown that for every serious or disabling injury, there were 10 injuries of a less serious nature and 30 property damage accidents (*Figure 1*). A further study showed that 600 near-miss incidents occurred for every serious or disabling injury.

3.1.0 Accident Costs

When an accident happens, everyone loses—the injured worker, the employer, and the insurance company. Accidents cost billions of dollars each year and cause much needless suffering. The National Safety Council estimates that the organized safety movement has saved more than 4.2 million lives since it began in 1913. This section examines why accidents happen and how you can help prevent them.

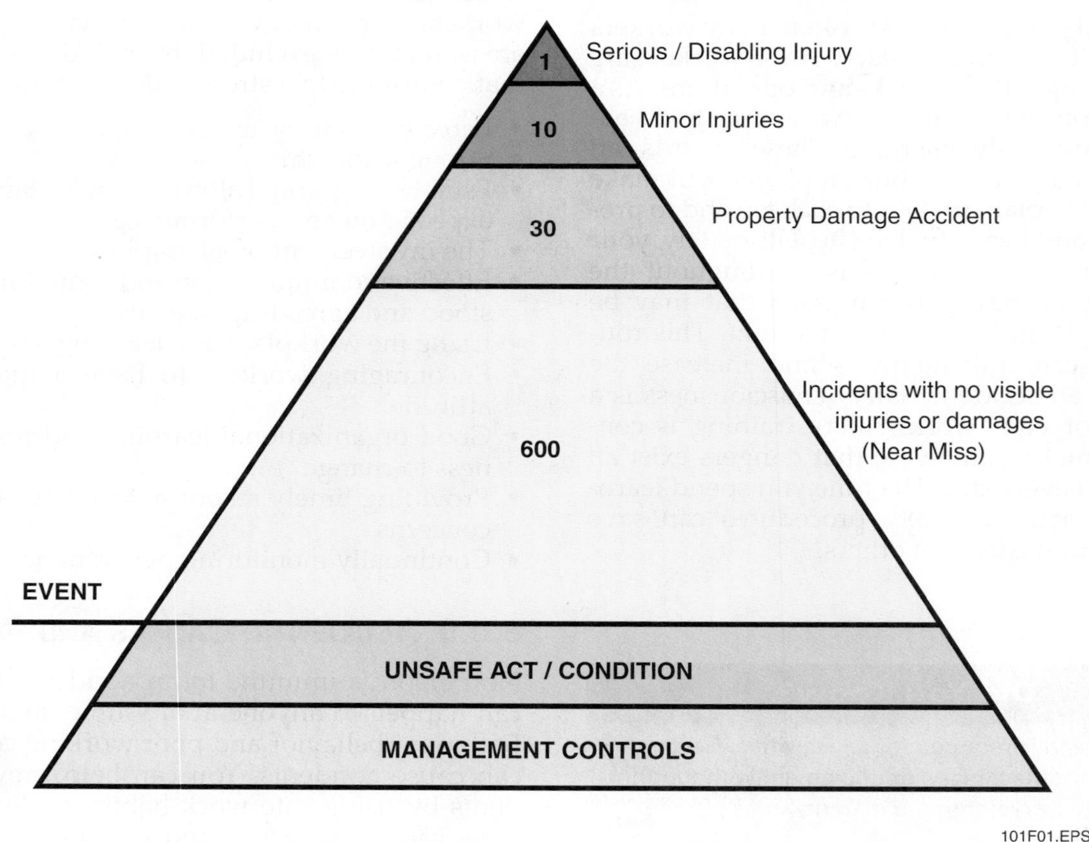

Figure 1 Accident ratio study.

Accident costs are often classified as direct and indirect. Direct costs include medical costs and other workers' compensation insurance benefits, as well as liability and property damage insurance payments. Of these, claims under workers' compensation, the insurance that covers workers on the job, are the most substantial of direct costs.

Indirect or hidden costs can be compared to the hidden nine-tenths of an iceberg, with the tip of the iceberg representing the direct or insured costs (*Figure 2*). Studies have shown that the hidden costs of accidents can and usually do exceed the direct costs of accidents from two to seven times. These hidden expenses include the costs associated with:

- Training replacement workers
- Accident investigation and corrective measures
- Scheduling delays
- Lost productivity
- Repairing damaged equipment and property
- Absenteeism

Many contract awards are based, in part, on a company's safety record. Therefore, accidents can also result in the loss of future jobs, which affects the company's financial position. This can mean layoffs, hiring freezes, or inability to purchase new equipment or tools. In this way, an accident indirectly affects everyone on the job site.

3.2.0 What Causes Accidents?

You may already know some of the main causes of accidents. They include the following:

- Failure to communicate
- Poor work habits
- Alcohol or drug abuse
- Lack of skill
- Intentional acts
- Unsafe acts
- Rationalizing risks
- Unsafe conditions
- **Management system** failure

3.2.1 Failure to Communicate

Many accidents happen because of a lack of communication. For example, you may learn how to do things one way on one job, but what happens when you go to a new job site? You need to communicate with the people at the new job site to find out whether they do things the way you have learned to do them. If you do not communicate clearly, accidents can happen. Remember that different people, companies, and job sites do things in different ways.

> **NOTE**
> Toolbox talks are one way to effectively keep all workers aware and informed of safety issues and guidelines. Toolbox talks are short, 5- to 10-minute meetings that review specific health and safety topics.

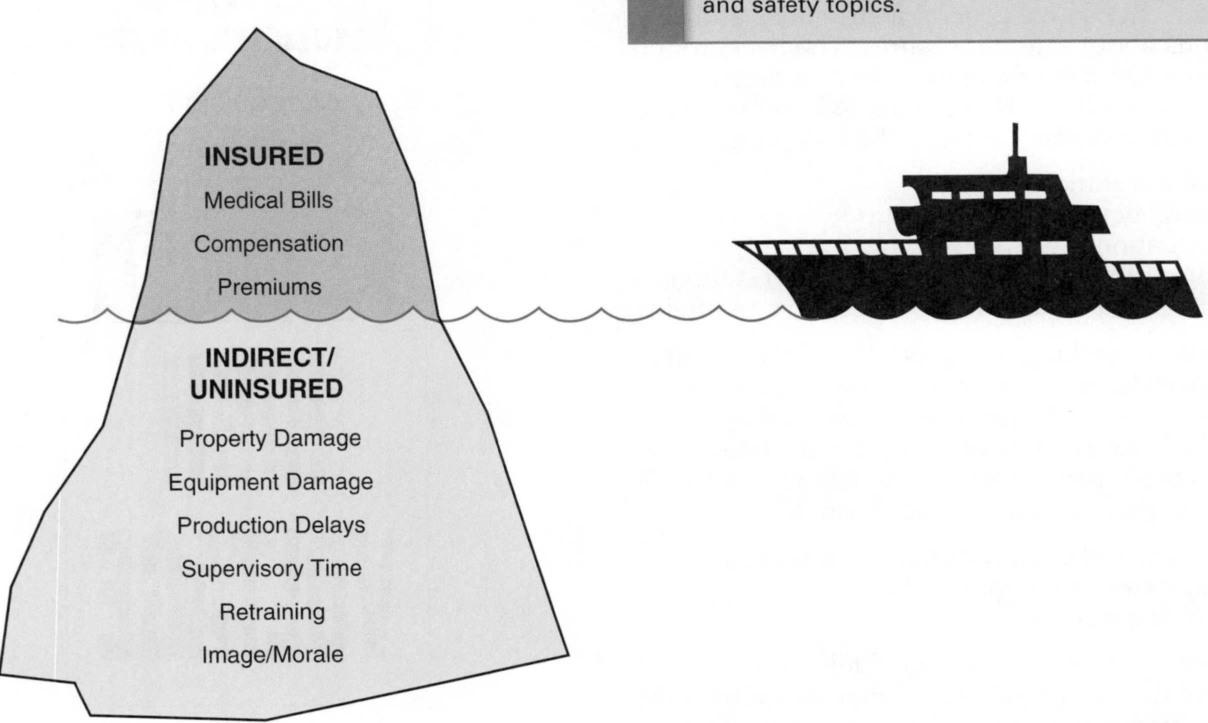

Figure 2 Hidden costs of accidents.

If you think that people know something without talking with them about it, then you are assuming that they know. Assuming that other people know and will do what you think they will do can cause accidents.

> **CAUTION**
>
> Never assume anything! It never hurts to ask questions, but disaster can result if you don't ask. For example, do not assume that an electrical current is turned off. First ask whether the current is turned off, then check it yourself to be completely safe.

All work sites have specific markings and signs to identify hazards and provide emergency information (*Figure 3*). Learn to recognize these types of signs:

- Informational
- Safety
- Caution
- Danger
- Temporary warnings

Informational markings or signs provide general information. These signs are blue. The following are considered informational signs:

- No Admittance
- No Trespassing
- For Employees Only

Safety signs give general instructions and suggestions about safety measures. The background on these signs is white; most have a green panel with white letters. These signs tell you where to find such important areas as the following:

- First-aid stations
- Emergency eye-wash stations
- Evacuation routes
- **Material Safety Data Sheet (MSDS)** stations
- Exits (usually have white letters on a red field)

Caution markings or signs tell you about potential hazards or warn against unsafe acts. When you see a caution sign, protect yourself against a possible hazard. Caution signs are yellow and have a black panel with yellow letters. They may give you the following information:

- Hearing and eye protection are required
- **Respirators** are required
- Smoking is not allowed

Danger markings or signs tell you that an immediate hazard exists and that you must take certain precautions to avoid an accident. Danger

INFORMATION SIGN

SAFETY SIGN

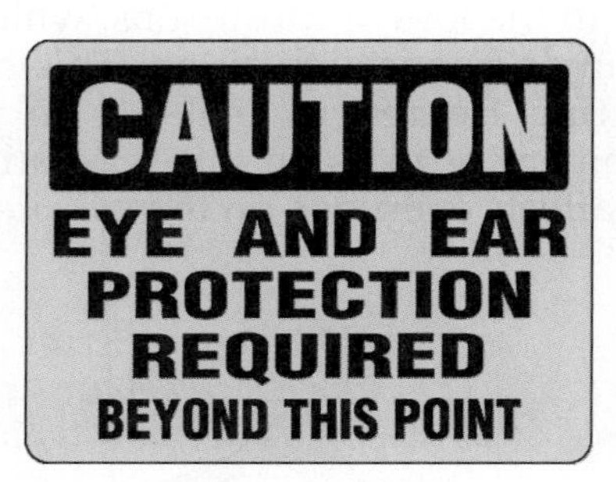

CAUTION SIGN

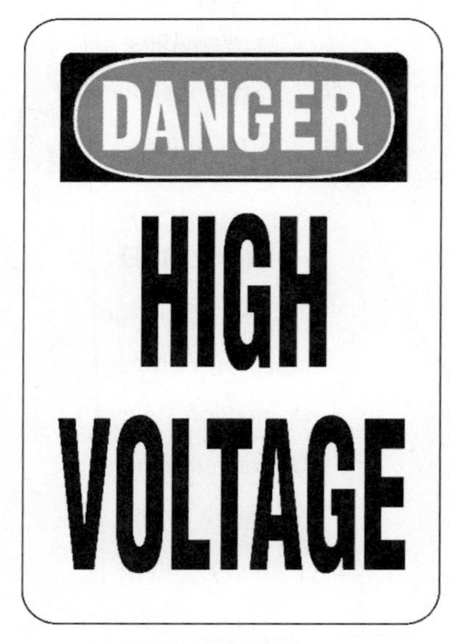

DANGER SIGN

101F03.EPS

Figure 3 Communication tags/signs.

signs are red, black, and white. They may indicate the presence of the following:

- Defective equipment
- **Flammable** liquids and compressed gases
- Safety barriers and barricades
- Emergency stop button
- High voltage

Safety tags are temporary warnings of immediate and potential hazards. They are not designed to replace signs or to serve as permanent means of protection. Learn to recognize the standard accident prevention signs and tags (*Table 1*).

3.2.2 Poor Work Habits

Poor work habits can cause serious accidents. Examples of poor work habits are procrastination, carelessness, and horseplay. Procrastination, or putting things off, is a common cause of accidents. For example, delaying the repair, inspection, or cleaning of equipment and tools can cause accidents. If you try to push machines and equipment beyond their operating capacities, you risk injuring yourself and your co-workers.

Machines, power tools, and even a pair of pliers can hurt you if you don't use them safely. It is your responsibility to be careful. Tools and machines don't know the difference between wood or steel and flesh and bone.

Work habits and work attitudes are closely related. If you resist taking orders, you may also resist listening to warnings. If you let yourself be easily distracted, you won't be able to concentrate. If you aren't concentrating, you could cause an accident.

Your safety is affected not only by how you do your work, but also by how you act on the job site. This is why most companies have strict policies for employee behavior. Horseplay and other inappropriate behavior are forbidden. Workers who engage in horseplay and other inappropriate behavior on the job site will be fired.

These strict policies are for your protection. There are many hazards on construction sites. Each person's behavior—at work, on a break, or at lunch—must follow the principles of safety.

The man pouring the liquid on his co-worker in *Figure 4* may look as if he's just having fun by playing a prank on his co-worker. In fact, what he's doing could cause his co-worker serious, even fatal, injury. If you horse around on the job, play pranks, or don't concentrate on what you are doing, you are showing a poor work attitude that can lead to a serious accident.

3.2.3 Alcohol and Drug Abuse

Alcohol and drug abuse costs the construction industry millions of dollars a year in accidents, lost time, and lost productivity. The true cost of alcohol and drug abuse is much more than just money, of course. Substance abuse can cost lives. Just as drunk driving kills thousands of people on our highways every year, alcohol and drug abuse kills on the construction site. Examine the person in *Figure 5*. Would you want to be like him or be working near him?

On-Site

Dull blades cause more accidents than sharp ones. If you do not keep your cutting tools sharpened, they won't cut very easily. When you have a hard time cutting, you exert more force on the tool. When that happens, something is bound to slip. And when something slips, you can get cut.

Table 1 Tags and Signs

Basic Stock (background)	Safety Colors (ink)	Message(s)
White	Red panel with white or gray letters	Do Not Operate Do Not Start
White	Black square with a red oval and white letters	Danger Unsafe Do Not Use
Yellow	Black square with yellow letters	Caution
White	Black square with white letters	Out of Order Do Not Use
Yellow	Red/magenta (purple) panel with black letters and a radiation symbol	Radiation Hazard
White	Fluorescent orange square with black letters and a biohazard symbol	Biological Hazard

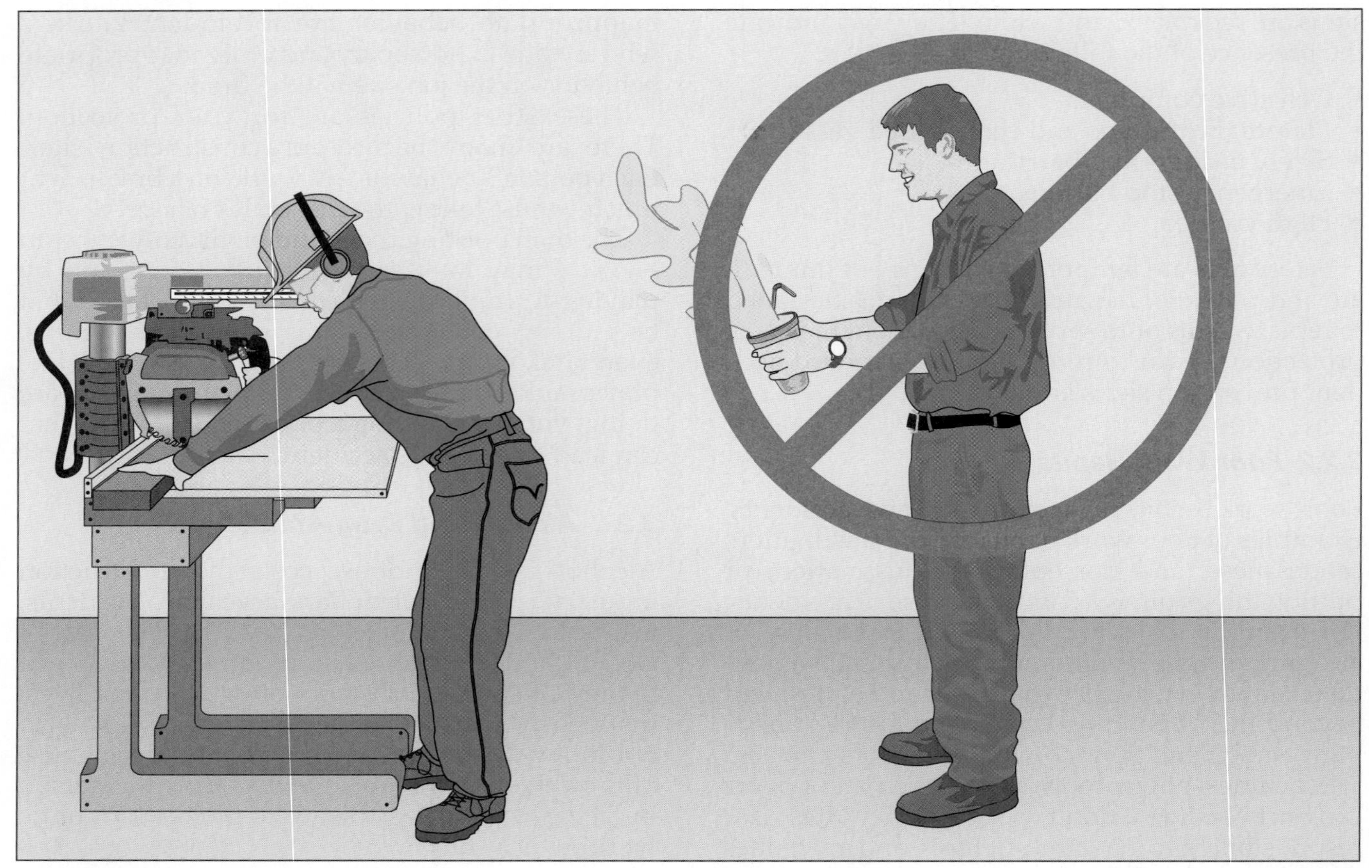

101F04.EPS

Figure 4 Horseplay at work can be dangerous.

Think About It

Look at the man who's trying to play a prank on his co-worker in *Figure 4.* Think of ways in which this work situation could be safer.

- No hard hat
- No safety goggles
- No gloves
- Breaking shop rules

Using alcohol or drugs creates a risk of injury for everyone on a job site. Many states have laws that prevent workers from collecting insurance benefits if they are injured while under the influence of alcohol or illegal drugs.

Would you trust your life to a crane operator who was high on drugs? Would you bet your life on the responses of a co-worker using alcohol or drugs? Alcohol and drug abuse have no place in the construction industry. A person on a construction site who is under the influence of alcohol or drugs is an accident waiting to happen—possibly a fatal accident.

People who work while using alcohol or drugs are at risk of accident or injury, and their

Did You Know?

Stress creates a chemical change in your body. Although stress may heighten your hearing, vision, energy, and strength, long-term stress can harm your health.

Not all stress is job-related; some stress develops from the pressures of dealing with family and friends and daily living. In the end, your ability to handle and manage your stress determines whether stress hurts or helps you. Use common sense when you are dealing with stressful situations. For example, consider the following:

- Keep daily occurrences in perspective. Not everything is worth getting upset, angry, or anxious about.
- When you have a particularly difficult workday scheduled, get plenty of rest the night before.
- Manage your time. The feeling of always being behind creates a lot of stress. Waiting until the last minute to finish an important task adds unnecessary stress.
- Talk to your supervisor. Your supervisor may understand what is causing your stress and may be able to suggest ways to manage it better.

Figure 5 Identify the safety hazards.

co-workers are at risk as well. That's why your employer probably has a formal substance abuse policy. You should know that policy and follow it for your own safety.

You don't have to be abusing illegal drugs such as marijuana, cocaine, or heroin to create a job hazard. Many prescription and over-the-counter drugs, taken for legitimate reasons, can affect your ability to work safely. Amphetamines, barbiturates, and antihistamines are only a few of the legal drugs that can affect your ability to work or operate machinery safely.

CAUTION

If your doctor prescribes any medication that you think might affect your job performance, ask about its effects. Your safety and the safety of your co-workers depend on everyone being alert on the job.

Do yourself and the people you work with a big favor. Be aware of and follow your employer's substance abuse policy. Avoid any substances that can affect your job performance. The life you save could be your own.

Think About It

If the man in *Figure 5* doesn't kill himself first, he will almost certainly kill someone else eventually. How many violations can you identify?

- Consuming alcohol on the job site
- Not following proper safety procedures for operating a motorized vehicle near hazardous materials
- Not using both hands to drive the vehicle
- Not wearing a seat/safety belt
- Not wearing a hard hat
- Not wearing safety glasses or goggles

3.2.4 Lack of Skill

You should learn and practice new skills under careful supervision. Never perform new tasks alone until you've been checked out by a supervisor.

Lack of skill can cause accidents quickly. For example, suppose you are told to cut some 2 × 8 boards with a circular saw, but you aren't skilled with that tool. A basic rule of circular saw operation is never to cut without a properly functioning guard. Because you haven't been trained, you don't know this. You find that the guard on the saw is slowing you down. So you jam the guard open with a small block of wood. The result could be a serious accident. Proper training can prevent this type of accident.

CAUTION

Never operate a power tool until you have been trained to use it. You can greatly reduce the chances of accidents by learning the safety rules for each task you perform.

3.2.5 Intentional Acts

When someone purposely causes an accident, it is called an intentional act. Sometimes an angry or dissatisfied employee may purposely create a situation that leads to property damage or personal injury. If someone you are working with threatens to get even or pay back someone, let your supervisor know at once.

3.2.6 Unsafe Acts

An unsafe act is a change from an accepted, normal, or correct procedure that usually causes an accident. It can be any conduct that causes unnecessary exposure to a job-site hazard or that makes an activity less safe than usual. Here are examples of unsafe acts:

- Failing to use **personal protective equipment (PPE)**
- Failing to warn co-workers of hazards
- Lifting improperly
- Loading or placing equipment or supplies improperly
- Making safety devices (such as saw guards) inoperable
- Operating equipment at improper speeds
- Operating equipment without authority
- Servicing equipment in motion
- Taking an improper working position
- Using defective equipment
- Using equipment improperly

3.2.7 Rationalizing Risk

Everybody takes risks every day. When you get in your car to drive to work, you know there is a risk of being involved in an accident. Yet when you drive using all the safety practices you have learned, you know that there is a good chance that you will arrive at your destination safely. Driving is an appropriate risk because you have some control over your own safety and that of others.

Some risks are not appropriate. On the job, you must never take risks that endanger yourself or others just because you can make an excuse for doing so. This is called rationalizing risk. Rationalizing risk means ignoring safety warnings and practices. For example, because you are late for work, you might decide to run a red light. By trying to save time, you could cause a serious accident.

The following are common examples of rationalized risks on the job:

- Crossing boundaries because no activity is in sight
- Not wearing gloves because it will take only a minute to make a cut
- Removing your hard hat because you are hot and you cannot see anyone working overhead
- Not tying off your fall protection because you only have to lean over by about a foot

Think about the job before you do it. If you think that it is unsafe, then it is unsafe. Stop working until the job can be done safely. Bring your concerns to the attention of your supervisor. Your health and safety, and that of your co-workers, make it worth taking extra care.

3.2.8 Unsafe Conditions

An unsafe condition is a physical state that is different from the acceptable, normal, or correct condition found on the job site. It usually causes an accident. It can be anything that reduces the degree of safety normally present. The following are some examples of unsafe conditions:

- Congested workplace
- Defective tools, equipment, or supplies
- Excessive noise
- Fire and explosive hazards
- Hazardous atmospheric conditions (such as gases, dusts, fumes, and vapors)
- Inadequate supports or guards
- Inadequate warning systems
- Poor housekeeping
- Poor lighting
- Poor ventilation
- Radiation exposure
- Unguarded moving parts such as pulleys, drive chains, and belts

On-Site

Most workers who die from falls were wearing harnesses but had failed to tie off properly. Always follow the manufacturer's instructions when wearing a harness. Know and follow your company's safety procedures when working on roofs, ladders, and other elevated locations.

3.2.9 Management System Failure

Sometimes the cause of an accident is failure of the management system. The management system should be designed to prevent or correct the acts and conditions that can cause accidents. If the management system did not do these things, that system failure may have caused the accident.

What traits could mean the difference between a management system that fails and one that succeeds? A company implementing a good management system will:

- Put safety policies and procedures in writing
- Distribute written safety policies and procedures to each employee
- Review safety policies and procedures periodically
- Enforce all safety policies and procedures fairly and consistently
- Evaluate supplies, equipment, and services to see whether they are safe
- Provide regular, periodic safety training for employees

3.3.0 Housekeeping

In construction, housekeeping means keeping your work area clean and free of scraps or spills. It also means being orderly and organized. You must store your materials and supplies safely and label them properly. Arranging your tools and equipment to permit safe, efficient work practices and easy cleaning is also important.

If the work site is indoors, make sure it is well-lit and ventilated. Don't allow aisles and exits to be blocked by materials and equipment. Make sure that flammable liquids are stored in safety cans. Oily rags must be placed only in approved, self-closing metal containers.

Remember that the major goal of housekeeping is to prevent accidents. Good housekeeping reduces the chances for slips, fires, explosions, and falling objects. Here are some good housekeeping rules:

- Remove from work areas all scrap material and lumber with nails protruding.
- Clean up spills to prevent falls.
- Remove all **combustible** scrap materials regularly.
- Make sure you have containers for the collection and separation of refuse. Containers for flammable or harmful refuse must be covered.
- Dispose of wastes often.
- Store all tools and equipment when you're finished using them.

Another term for good housekeeping is pride of workmanship. If you take pride in what you are doing, you won't let trash build up around you. The saying "A place for everything and everything in its place" is the right idea on the job site.

3.4.0 Company Safety Policies and OSHA Regulations

The mission of the **Occupational Safety and Health Administration (OSHA)** is to save lives, prevent injuries, and protect the health of America's workers. To accomplish this, federal and state governments work in partnership with the 111 million working men and women and their 7 million employers who are covered by the Occupational Safety and Health Act (OSH Act) of 1970.

Nearly every worker in the nation comes under OSHA's jurisdiction. There are some exceptions, such as miners, transportation workers, many public employees, and the self-employed.

3.4.1 The Code of Federal Regulations

The *Code of Federal Regulations (CFR) Part 1910* covers the OSHA standards for general industry. *CFR Part 1926* covers the OSHA standards for the construction industry. Either or both may apply to you, depending on where you are working and what you are doing. If a job-site condition is covered in the CFR book, then that standard must be used. However, if a more stringent requirement is listed in *CFR 1910*, it should also be met. Check with your supervisor to find out which standards apply to your job.

29 CFR 1926 is divided into 26 lettered subparts (A through Z). As you progress in task-specific training, you will learn about all the subparts applicable to your work. *Subpart C of 29 CFR 1926* applies to all construction and maintenance work. It outlines the general safety and health provisions for the construction industry. It covers the following topics:

- Safety training and education
- Injury reporting and recording
- First aid and medical attention
- Housekeeping
- Illumination
- Sanitation
- PPE
- Standards incorporated by reference
- Definitions
- Access to employee exposure and medical records
- Means of egress
- Employee emergency action plans

See *Figure 6* to identify parts, sections, paragraphs, and subparagraphs of an OSHA standard.

An OSHA Standard reference may look like this:

29 CFR 1926.501 (a)(1)(i)(A)

and breaks down like this:

29	=	Title (Labor)
CFR	=	Code of Federal Regulations
1926	=	Part (Construction)
.501	=	Section
(a)	=	Paragraph
(1)	=	Subparagraph
(i)	=	Subparagraph
(A)	=	Subparagraph

101F06.EPS

Figure 6 Reading OSHA standards.

On-Site

All of OSHA's safety requirements in the *Code of Federal Regulations* apply to residential as well as commercial construction. In the past, OSHA enforced safety only at commercial sites. The increasing rate of accidents at residential sites led OSHA to enforce safety guidelines for the building of houses and townhouses. Today, however, OSHA still focuses its enforcement efforts on commercial construction.

3.4.2 The General Duty Clause

If a standard does not specifically address a hazard, the general duty clause must be invoked. Failing to adhere to the general duty clause can result in heavy fines for your employer. The general duty clause reads as follows:

> In practice, OSHA, court precedent, and the review commission have established that if the following elements are present, a general duty clause citation may be issued:

- The employers failed to keep the workplace free of a hazard to which employees of that employer were exposed.
- The hazard was recognized. (Examples might include: through your safety personnel, employees, organization, trade organization, or industry customs.)
- The hazard was causing or was likely to cause death or serious physical harm.
- There was a feasible and useful method to correct the hazard.

3.4.3 Employee Rights and Responsibilities

While it is the employer's responsibility to keep workers safe by complying with the General Duty Clause and all other OSHA regulations, workers have certain rights and responsibilities on the job site as well. First and foremost, workers must follow their employers' safety rules. While workers cannot be cited or fined by OSHA, they can be disciplined for violating their employer's safety rules. Workers must also wear the provided personal protective equipment. Workers should also inform their foreman about health and safety concerns on the job.

Section 11(c) of the OSH Act prohibits employers from disciplining or discriminating against any worker for practicing their rights under OSHA, including filing a complaint. You have the right to file a complaint if you do not think that your employer is protecting your health and safety at work. You may submit a written request to OSHA asking for an inspection of your worksite. Workers who file a complaint have the right to have their names withheld from their employers, and OSHA will not reveal this information.

Workers who would like an on-site inspection must submit a written request. You have the following rights when job site inspection is conducted:

- You must be informed of imminent dangers. An OSHA inspector must tell you if you are exposed to an imminent danger. An imminent danger is one that could cause death or serious injury now or in the near future. The inspector will also ask your employer to stop any dangerous activity.
- You have the right to accompany the OSHA inspector in the walk-around inspection. Walk-around activities include all opening and closing conferences related to the conduct of the inspection.
- You have the right to be told about citations issued at your workplace. Notices of OSHA citations must be posted in the workplace near the site where the violation occurred and must remain posted for three days or until the hazard is corrected, whichever is longer.

After an inspection has been performed, OSHA will give the employer a date by which any hazards cited must be fixed. Employers can appeal these dates, and appeals must be filed within 15 days of the citation. Workers have the right to meet privately with the OSHA inspector to discuss the results of the inspection.

If you have been discriminated against for asserting your OSHA rights, you have the right to file a complaint with the OSHA area office within 30 days of the incident. Make sure you file your complaint as soon as possible, as the time limit is strictly enforced.

You also have the right to see and copy any medical records about you that the employer has obtained. Your employer is required by OSHA *29 CFR 1926.33* and OSHA *29 CFR1910.1020* to maintain your medical records for 30 years after you leave employment. If you are employed for less than one year, the employer can maintain your records or give them to you when you leave the job.

3.4.4 Inspections

OSHA conducts six types of inspections to determine if employers are in compliance with standards:

- *Imminent danger inspections* – OSHA's top priority for inspection, conducted when workers face an immediate risk of death or serious physical harm.
- *Catastrophe inspections* – Performed after an accident that requires hospitalization of three or more workers. Employers are required to report fatalities and catastrophes to OSHA within eight hours.
- *Worker complaint and referral inspections* – Conducted due to complaints by workers or a worker representative, or a referral from a recognized professional.
- *Programmed inspection* – Aimed at high-risk areas based on OSHA's targeting and priority methods.
- *Follow-up inspection* – Completed after citations to assure employer has corrected violations.
- *Monitoring inspection* – Used for long-term abatement follow-up or to assure compliance with variances.

Before beginning an inspection, OSHA staff must be able to determine from the complaint that there are reasonable grounds to believe that a violation of an OSHA standard or a safety or health hazard exists. If OSHA has information indicating the employer is aware of the hazard and is correcting it, the agency may not conduct an inspection after obtaining the necessary documentation from the employer.

Complaint inspections are typically limited to the hazards listed in the complaint, although other violations in plain sight may be cited as well. The inspector may decide to expand the inspection based on professional judgment or conversations with workers.

Complaints are not necessarily inspected in first-come, first-served order. OSHA ranks complaints based on the severity of the alleged hazard and the number of workers exposed. That is why lower-priority complaints can often be handled more quickly using the phone/fax method than through on-site inspections.

Inspections are typically performed by conducting a walk-around. During a walk-around inspection, the inspector typically does the following:

- Observes conditions of the job site.
- Talks to workers.
- Inspects records.
- Examines posted hazard warnings and signs.
- Points out hazards and suggests ways to reduce or eliminate them.

After the walk-around inspection, there is typically a closing conference held between the inspector and the site contractor or company managers. During this conference, inspectors discuss their findings, citing specific violations and suggested abatement methods. Inspectors may also conduct interviews with the employers, workers, and representatives at this point.

3.4.5 Violations

Employers who violate OSHA regulations can be fined. The fines are not always high, but they can harm a company's reputation for safety. Fines for serious safety violations can cost up to $7,000. Fines for each violation that was done willfully can cost up to $70,000. In 2002, more than 78,000 fines were levied at a cost of $70,000 per violation.

3.4.6 Compliance

Just as employers are responsible to OSHA for compliance, employees must comply with their company's safety policies and rules. Employers are required to identify hazards and potential hazards within the workplace and eliminate them, control them, or provide protection from them. This can only be done through the combined efforts of the employer and employees. Employers must provide written programs and training on hazards, and employees must follow the procedures. You, as the employee, must read and understand the OSHA poster at your job site explaining your rights and responsibilities. If you are unsure where the OSHA poster is, ask your supervisor.

To help employers provide a safe workplace, OSHA requires companies to provide a **competent person** to ensure the safety of the employees. In *OSHA 29 CFR 1926*, OSHA defines a competent person as follows:

> Competent person means one who is capable of identifying existing and predictable hazards in the surroundings or working conditions which

are unsanitary, hazardous, or dangerous to employees, and who has authorization to take prompt corrective measures to eliminate them.

In comparison, *OSHA 29 CFR 1926* defines a **qualified person** as follows:

> Someone who, by possession of a recognized degree, certificate, or professional standing, or who by extensive knowledge, training, and experience, has successfully demonstrated his ability to solve or resolve problems relating to the subject matter, work, or the project.

In other words, a competent person is experienced and knowledgeable about the specific operation and has the authority from the employer to correct the problem or shut down the operation until it is safe. A qualified person has the knowledge and experience to handle problems. A competent person is not necessarily a qualified person.

These terms will be an important part of your career. It is important for you to know who the competent person is on your job site. OSHA requires a competent person for many of the tasks you may be assigned to perform, such as **confined space** entry, ladder use, and trenching. Different individuals may be assigned as a competent person for different tasks, according to their expertise. To ensure safety for you and your co-workers, work closely with your competent person and supervisor.

3.4.7 Record Keeping

Accurate record keeping is a particularly important part of OSHA compliance. OSHA 29 *CFR 1904* outlines the recording and reporting information for all occupational injuries and illnesses. Its purpose is to set the guidelines so that employers know when and how to report and record all workplace fatalities, injuries, and illnesses. The rules vary between companies, so it's important to revisit *CFR 1904* frequently when you participate in any record keeping or reporting activities.

On-Site

For more information on OSHA, visit the OSHA website at www.osha.gov. To report an emergency, fatality, or imminent life-threatening situation, call OSHA's toll free line immediately:

1-800-321-OSHA (6742) or TTY – 1-877-889-5627

To file a complaint online, go to http://www.osha.gov/pls/osha7/eComplaintForm.html.

3.5.0 Reporting Injuries, Accidents, and Incidents

There are three categories of on-the-job events: injuries, accidents, and incidents. An injury is anything that requires treatment, even minor first aid. An accident is anything that causes an injury or property damage. An incident is anything that could have caused an injury or damage but, because it was caught in time, did not.

You must report all on-the-job injuries, accidents, or incidents, no matter how minor, to your supervisor (*Figure 7*). Some workers think they will get in trouble if they report minor injuries. That's not true. Small injuries, like cuts and scrapes, can later become big problems because of infection and other complications.

Many employers are required to maintain a log of significant work-related injuries and illnesses using *OSHA Form 300*. Employee names can be kept confidential in certain circumstances. A summary of these injuries must be posted at certain intervals, although employers do not need to submit it to OSHA unless requested. Employers can calculate the total number of injuries and illnesses and compare it with the average national rates for similar companies.

101F07.EPS

Figure 7 All accidents, injuries, or incidents must be reported to your supervisor.

By analyzing accidents, companies and OSHA can improve safety policies and procedures. By reporting an accident, you can help keep similar accidents from happening in the future.

3.6.0 The Four High-Hazard Areas

Construction has four leading causes of death. These are often referred to as the four high-hazard areas, and include falls, struck-by hazards, caught-in or caught-between hazards, and electrical hazards. Of all construction fatalities, 82% fall into one of these four categories (*Figure 8*).

Here are explanations of the four leading hazard groups:

- Falls from elevation are accidents involving failure of, failure to provide, or failure to use appropriate fall protection.
- Struck-by accidents involve unsafe operation of equipment, machinery, and vehicles, as well as improper handling of materials, such as through unsafe rigging operations.
- Caught-in or caught-between accidents involve unsafe operation of equipment, machinery, and vehicles, as well as improper safety procedures at **trench** sites and in other confined spaces.
- Electrical shock accidents involve contact with overhead wires, use of defective tools, failure to disconnect power source before repairs, or improper **ground** fault protection.

3.7.0 Evacuation Procedures

In many work environments, specific evacuation procedures are needed. These procedures go into effect when dangerous situations arise, such as fires, chemical spills, and gas leaks. In an emergency, you must know the evacuation procedures. You must also know the signal (usually a horn or siren) that tells workers to evacuate.

When you hear the evacuation signal, follow the evacuation procedures exactly. That usually means taking a certain route to a designated assembly area and telling the person in charge that you are there. If hazardous materials are released into the air, you may have to look at the **wind sock** to see which way the wind is blowing. Different evacuation routes are planned for different wind directions. Taking the right route will keep you from being exposed to the hazardous material.

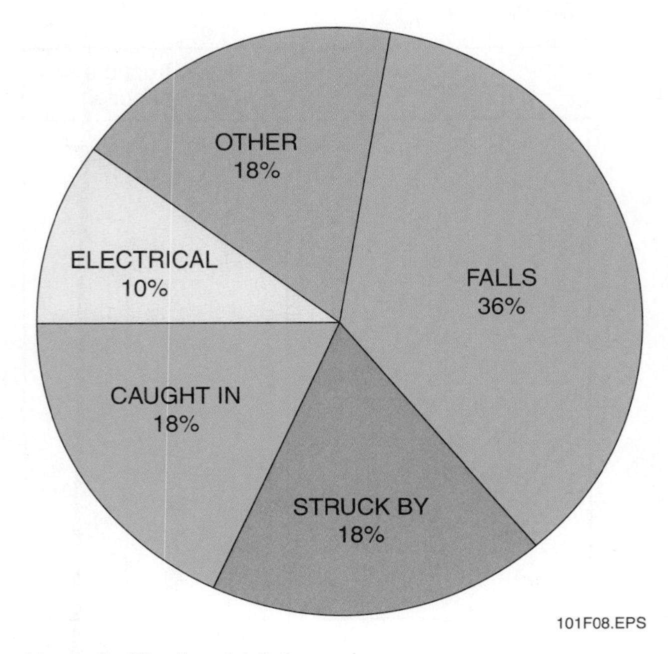

Figure 8 The four high-hazard areas.

4.0.0 Hazard Recognition, Evaluation, and Control

The process of hazard recognition, evaluation, and control is the foundation of an effective safety program. When hazards are identified and assessed, they can be addressed quickly, reducing the hazard potential. Simply put, the more aware you are of your surroundings and the dangers in them, the less likely you are to be involved in an accident.

4.1.0 Hazard Recognition

Accident/incident types and energy sources are considered potential hazard indicators. The best approach in determining if a situation or equipment is potentially hazardous is to ask yourself these questions:

- How can this situation or equipment cause harm?
- What types of energy sources are present that can cause an accident?
- What is the magnitude of the energy?
- What could go wrong to release the energy?
- How can the energy be eliminated or controlled?
- Will I be exposed to any hazardous materials?

Before you can fully answer these questions, you need to know the different types of accidents that can happen and the energy sources behind the accidents. Some of the different types of accidents that can cause injuries include the following:

- Falls on the same elevations or falls from elevations
- Being caught in, on, or between equipment
- Being struck by falling objects
- Contact with acid, electricity, heat, cold, radiation, pressurized liquid, gas, or toxic substances
- Being cut by tools or equipment
- Exposure to high noise levels
- Repetitive motion or excessive vibration

> **NOTE**
> Did you spot the four high-hazard areas in this list? Remember, these four types of accidents cause 82% of all construction fatalities: falls, struck-by, caught-in, and electrical.

When equipment is the cause of an accident, it is usually because there was an uncontrolled release of energy. The different types of energy sources that can be released include the following:

- Mechanical
- Pneumatic
- Hydraulic
- Electrical
- Chemical
- Thermal
- Radioactive
- Gravitational
- Stored energy

There are a number of ways to recognize hazards and potential hazards on a job site. Some techniques are more complicated than others. In order to be effective, they all must answer this question: What could go wrong with this situation or operation? No matter what hazard recognition technique you use, answering that question in advance will save lives and prevent equipment damage.

4.2.0 Job Safety Analysis (JSA) and Task Safety Analysis (TSA)

Performing a job safety analysis (JSA), also known as job hazard analysis (JHA), is one approach to hazard recognition. Another common technique is performing a task safety analysis (TSA), also called a task hazard analysis (THA).

In a JSA, the task at hand is broken down into its individual parts or steps and then each step is analyzed for its potential hazards. Once a hazard is identified, certain actions or procedures are recommended that will correct that hazard. For example, during a JSA, it is determined that using a come-along to install a pump motor in a tight space would be safer than having a worker do it manually. By using the come-along, the chance that the worker's hand would get crushed during installation is reduced. Using the job safety analysis saved the worker from injury. *Figure 9* shows an example of a form used to conduct a job safety analysis.

JOB SAFETY ANALYSIS FORM

Job Title: Date of Analysis:

Job Location: Conducted by:

PPE: Staffing:

Tools, Materials and Equipment: Duration:

Step	Hazards	Quality Concern	Environmental Concern	New Procedure or Protection

101F09.EPS

Figure 9 Job safety analysis form.

JSAs can also be used as pre-planning tools. This helps to ensure that safety is planned into the job. You may be asked to take part in a JSA during job planning. When JSAs are used as pre-planning tools, they contain the following information:

- Tools, materials, and equipment needs
- Staffing or manpower requirements
- Duration of the job
- Quality concerns

Task safety analysis is similar to job safety analysis in that both require workers to identify potential hazards and needed safeguards associated with a job they are about to do. The difference is the form used to report the hazards. During a TSA, a pre-printed, fill-in-the-blank checklist, like the one shown in *Figure 10*, is often used to document any hazard found during analysis. Before work begins, the first-line supervisor or team leader should discuss the conclusions found during the TSA with the crew. Some companies require workers to sign the completed TSA forms or checklists before they start work. This helps companies document and ensure that workers have been told of potential hazards and safety procedures.

4.3.0 Risk Assessment

Whether an action is considered safe is often a matter of evaluating risk. Risk is a measure of the probability, consequences, and exposure related to an event. Probability is the chance that a given event will occur. Consequences are the results of an action, condition, or event. Exposure is the amount of time and/or the degree to which someone or something is exposed to an unsafe condition, material, or environment.

A safe operation is one in which there is an acceptable level of risk. This means there is a low probability of an accident and that the consequences and exposure risk are all acceptable. For example, climbing a ladder has risk that is considered to be acceptable if the proper ladder is being used as intended, if it is set up correctly, and if it is in good condition. The probability of an accident, exposure to danger, and potential consequences are all low. If any one of these conditions were different, climbing the ladder would have an unacceptable level of risk.

TASK SAFETY ANALYSIS CHECKLIST			
Name: Location:			
Signature: Date:			
	Yes	No	N/A
1. Have underground utilities been located prior to excavation?			
2. In areas where there are known or suspected unexploded ordnances, has the area been cleared by qualified explosive ordnance disposal (EOD) personnel?			
3. Are excavations, the adjacent areas, and protective systems inspected and documented daily?			
4. Are excavations over 5 feet in depth adequately protected by shoring, trench box or sloping?			
5. When excavations are undercut, is the overhanging material safely supported?			
6. Have methods been taken to control the accumulation of water in excavations?			
7. Are employees protected from falling materials (loose rock or soil)?			
8. Are substantial stop logs or barricades installed where vehicles or equipment are used or allowed adjacent to an excavation?			
9. Have steps been taken to prevent the public, workers or equipment from falling into excavations?			
10. Are all wells, calyx holes, pits, shafts, etc. barricaded or covered?			
11. Are walkways provided where employees or equipment are required or permitted to cross over excavations?			
12. Where employees are required to enter excavations, is access/egress provided every 25 feet laterally?			
13. For excavations less than 20 feet, is the maximum slope 1½ horizontal to 1 vertical?			
14. Are support systems drawn from manufacturer's tabulated data in accordance with all manufacturer's specifications?			
15. Are copies of the tabulated data maintained at the job site?			
16. Are members of support systems securely connected together?			
17. Are shields installed in a manner to restrict lateral or other hazardous movement?			
Comments:			

101F10.EPS

Figure 10 TSA checklist.

5.0.0 Elevated Work and Fall Protection

Falls from elevated areas are one of the leading causes of fatalities among construction workers. Falls from elevated heights account for approximately ⅓ of all deaths in the construction trade. Approximately 85% of those workers injured lose time from work; approximately 33% require hospitalization; some never return to the job.

While the risk of falls is high in construction, there is much you can do to safeguard yourself. Using the appropriate personal protective equipment, following proper safety procedures, practicing good housekeeping habits, and staying alert at all times will help you stay safe when working at an elevation.

5.1.0 Fall Hazards

Falls are classified into two groups: falls from an elevation and falls from the same level. Falls from an elevation can happen during work from **scaffolding**, work platforms, decking, concrete forms, ladders, stairs, and work near **excavations**. Falls from elevation often result in death. Falls on the same level are usually caused by tripping or slipping. Sharp edges and pointed objects, such as exposed rebar, could cut and otherwise harm a worker. Head injuries are also common results.

OSHA Subpart M requires fall protection for platforms or work surfaces with unprotected sides or edges that are six feet or higher than the ground or level below it. This is commonly referred to as the **six-foot rule**. However, some state OSHA regulations and company policies may require fall protection for heights less than six feet.

5.2.0 Walking and Working Surfaces

Slips, trips, and falls on walking and working surfaces cause 15% of all accidental deaths in the construction industry. Some accidents occur due to environmental conditions, such as snow, ice, or wet surfaces. Others happen because of poor housekeeping and careless behavior, such as leaving tools, materials, and equipment out and unattended. You can avoid slips, trips, and falls by being aware of your surroundings and following the rules on your site. Remember these general walking and working surface guidelines to avoid accidents:

- Keep all walking and working areas clean and dry. If you see a spill or ice patch, clean it up, or barricade the area until it can be properly attended to.
- Keep all walking and working surfaces clear of clutter and debris.
- Install cables, extension cords, and hoses so that they will not become tripping hazards.
- Do not run on scaffolding, work platforms, decking, roofs, or other elevated work areas.

5.3.0 Unprotected Sides, Wall Openings, and Floor Holes

Any opening in a wall or floor is a safety hazard. There are two types of protection for these openings: (1) they can be **guarded** or (2) they can be covered. Cover any hole in the floor when possible. When it is not practical to cover the hole, use barricades. If the bottom edge of a wall opening is fewer than 39 inches above the floor and would allow someone to fall six feet or more, then place guards around the opening.

The types of barriers and barricades used vary from job site to job site. There may also be different procedures for when and how barricades are put up. Learn and follow the policies at your job site.

Several different types of guards are commonly used:

- Railings are used across wall openings or as a barrier around floor openings to prevent falls (*Figure 11A*).
- Warning barricades alert workers to hazards but provide no real protection (*Figure 11B*). Typical warning barricades are made of plastic tape or rope strung from wire or between posts. The tape or rope is color-coded as follows:
  - Red means danger. No one may enter an area with a red warning barricade. A red barricade is used when there is danger from falling objects or when a load is suspended over an area.
  - Yellow means caution. You may enter an area with a yellow barricade, but be sure you know what the hazard is, and be careful. Yellow barricades are used around wet areas or areas containing loose dust. Yellow with black lettering warns of physical hazards such as bumping into something, stumbling, or falling.
  - Yellow and purple means radiation warning. No one may pass a yellow and purple barricade without authorization, training, and the appropriate personal protective equipment. These barricades are often used where piping welds are being X-rayed.
- Protective barricades give both a visual warning and protection from injury (*Figure 11C*). They can be wooden posts and rails, posts and chain, or steel cable. People should not be able to get past protective barricades.

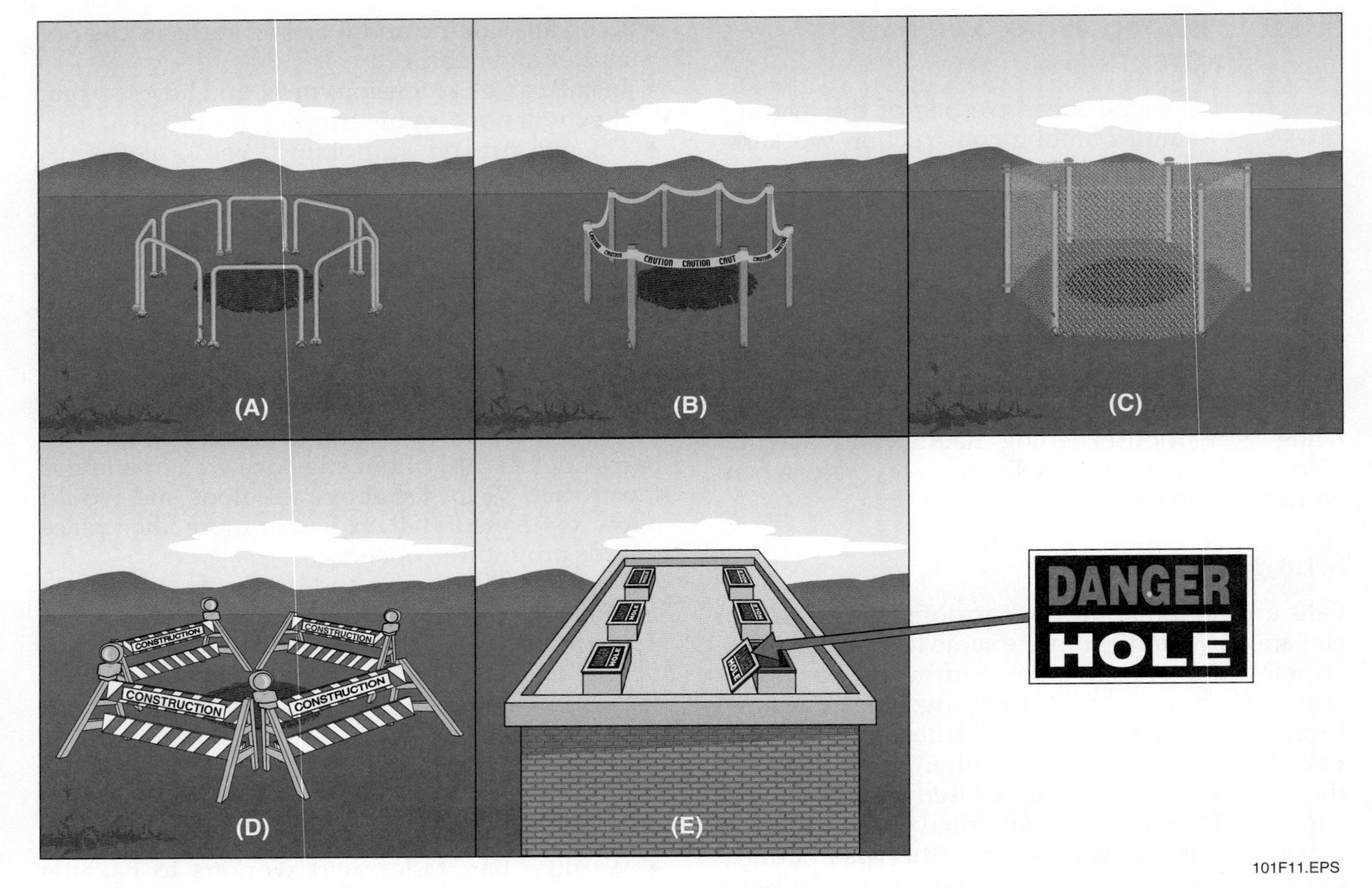

Figure 11 Common types of barriers and barricades.

- Blinking lights are placed on barricades so they can be seen at night (*Figure 11D*).
- Hole covers are used to cover open holes in a floor or in the ground (*Figure 11E*).

> **WARNING!** Never remove a barricade unless you have been authorized to do so. Follow your employer's procedures for putting up and removing barricades.

Follow these guidelines when working near unprotected sides, floor holes, and wall openings:

- Mark hole or cover with a sign or marking that reads "Warning: Temporary Cover – Do Not Remove Unless Authorized" (*Figure 12*), or otherwise identify them.
- Hole covers must be cleated, wired, or otherwise secured to prevent slipping sideways or horizontally beyond the hole.

Case History

- A worker taking measurements was killed when he fell backward from an unguarded balcony to the concrete below.
- A roofer handling a piece of fiberboard backed up and tripped over a 7½-inch parapet. He fell more than 50 feet to the ground level and died of severe head injuries.
- Two connectors were erecting light-weight steel I-beams on the third floor of a 12-story building, 54 feet above the ground. One worker removed a choker sling from a beam and then attempted to place the sling onto a lower hook on a series of stringers. While the crawler tower crane was booming away from the steel, the wind moved the stringer into the beam the worker was standing on. The beam moved while the worker was trying to disengage the hook, causing him to lose his balance and fall to his death.

WARNING

TEMPORARY COVER
DO NOT REMOVE
UNLESS AUTHORIZED

101F12.EPS

Figure 12 Sign to mark a hole cover.

- Covers must extend adequately beyond the edge of the hole.
- Hole covers must be strong enough to support twice the weight of anything that may be placed on top of them. Use ¾-inch plywood as a hole cover, provided that one dimension of the opening is less than 18 inches; otherwise, 2-inch lumber is required.
- Never store material or equipment on a hole cover.
- Guard all stairway floor openings, with the exception of the entrance, with standard railing and **toeboards**.
- Guard all wall openings from which there is a drop of more than 6 feet and for which the bottom of the opening is less than 39 inches above the working surface.
- Guard all open-sided floors and platforms 6 feet or more above adjacent floor or ground level, using a standard railing or the equivalent.

5.4.0 Personal Fall Arrest Systems

One of the main pieces of personal protective equipment used to prevent falls and protect workers who do fall is the personal fall arrest system (PFAS). The components of a personal fall arrest system include:

- Body harness (*Figure 13*)
- **Lanyards** (*Figure 14*)
- Lifeline
- Connecting devices
- Anchor points

Safety harnesses should be worn when working under the following conditions:

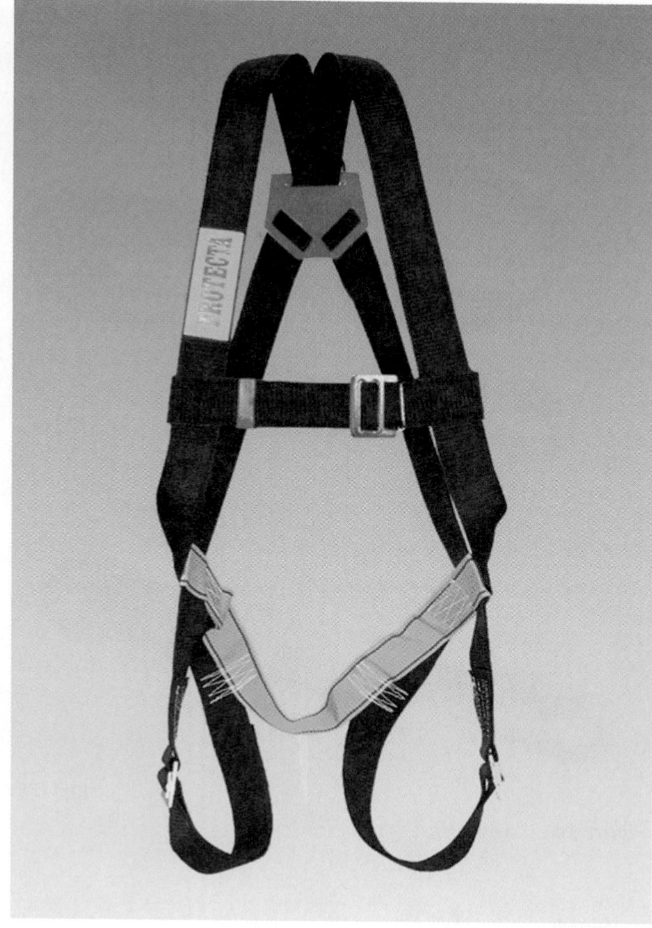

101F13.EPS

Figure 13 Typical full-body safety harness.

- In any area more than 6 feet above the ground or on a solid work surface that is not protected by a guardrail or safety net.
- When assembling and disassembling scaffolding, on any scaffold with incomplete handrail or decking, and on any suspended scaffold, platform, or stage over 10 feet above the ground or lower level.
- Around floor openings and when removing floor planks from the last panel in a temporary floor.
- On all sloping roofs and any flat roofs without handrails within 6 feet of the edge and around roof openings.
- When placing and tying reinforcing steel at heights of 6 feet or greater and in areas exposed to protruding reinforcing steel or any impalement hazard.
- On extendable boom lifts.
- When required by your company's safety rules or policies.

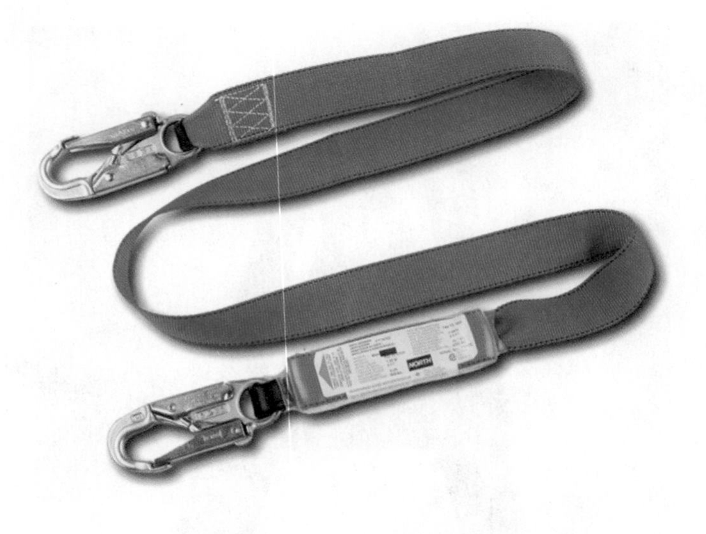

101F14.EPS

Figure 14 Lanyards.

> **CAUTION**
>
> The safety harness and lanyard are parts of a personal fall protection system. Workers must know how to properly inspect, don, and maintain their system.

> **WARNING!**
>
> Never use a safety harness and lanyard for anything except their intended purpose. Always follow the manufacturer's instructions for hooking up a safety harness or lanyard.

5.4.1 PFAS Inspection

Treat a safety harness as if your life depends on it, because it does! To maintain their service life and high performance, all belts and harnesses should be inspected frequently. Damage to fall arrest systems includes burns, hardening due to chemical contact, and excessive wear. When inspecting a harness, check that the buckles and D-ring are not bent or deeply scratched. Check the harness for any cuts or rough spots.

> **WARNING!**
>
> Always inspect your lanyard, harness, and any other fall protection gear prior to use. Never use damaged equipment.

You should also have your personal fall arrest system inspected monthly by a competent person. The competent person has the authority to impose prompt corrective measures to eliminate any hazards. If there is any question about a defect that is found, no matter how small it may seem to be, err on the side of caution. Take the fall arrest component(s) out of service for testing or replacement.

5.4.2 Donning a Harness

Safety harnesses are extra-heavy-duty harnesses that buckle around your body. They have leg, shoulder, chest, and pelvic straps.

Safety harnesses have a D-ring attached to one end of a short section of rope called a lanyard. The D-ring or support point on a safety harness should be placed to the rear, between your shoulder blades.The other end of the lanyard should be attached to a strong anchor point located above the work area. (A qualified person will tell you what a strong anchor point is.) The lanyard should be long enough to let you work but short enough to keep you from falling more than 6 feet.

> **WARNING!**
>
> Lanyards must be secured to a suitable anchorage capable of supporting 5,000 pounds (or two times the intended load), and tied off as short as possible. (Maximum fall is 6 feet.)

If you do not know how to use a safety harness correctly, ask for instruction. Improper use leaves you at risk of falling and sustaining serious injury. Workers should plan their PFAS fall protection in advance. PFAS come in different sizes and should be ordered to specifically fit each individual worker.

> **WARNING!**
>
> Always use a safety harness and lanyard properly. Over 70% of reported job-site accidents are caused by improper use of the lanyard and harness.

You must understand each job site's fall protection program, the height requirements for that site, and the emergency action plan to rescue fallen co-workers prior to the start of work. Some operations require dual lanyards to allow for 100% tie-off. A dual lanyard is an excellent method to enhance safe work operation. When welding, cutting, or burning, it is preferable to use a lanyard with a wire rope center. Always refrain from securing to electrical anchor points, such as conduit or electric cables.

In order to ensure that you use the appropriate lanyard with your harness, you must know the height of the elevated work surface, length of your lanyard, and the maximum length of the lanyard's shock absorber. For example, if you are using a 6-foot long, shock-absorbing lanyard, its manufacturer's data may show that its shock absorber will expand to 3.5 feet. To represent the height of an average worker, add 6 feet to that, plus an extra 3-foot safety factor, and you will find that you may work safely above heights of 18.5 feet (*Figure 15*). The math for this example would look like this:

$$6 + 3.5 + 6 + 3 = 18.5$$

> **WARNING!** Before using a shock-absorbing lanyard or a self-retracting lifeline, calculate your fall distance and select the proper equipment to meet estimated fall clearance. Failure to select proper equipment and calculate fall distance may result in serious personal injury or death.

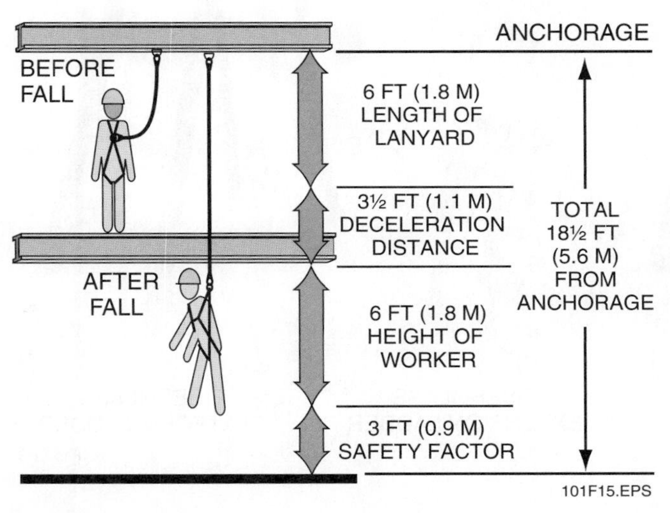

Figure 15 Calculating potential fall distance.

> **WARNING!** If you shock load your personal fall arrest system, it must be inspected by a competent person before you can use it again. In most cases, replacement will be necessary.

6.0.0 Ladders and Stairs

Ladders are used to perform work in elevated locations. Any time work is performed above ground level, there is a risk of accidents. You can reduce this risk by carefully inspecting ladders before you use them and by using them properly.

Overloading means exceeding the **maximum intended load** of a ladder. Overloading can cause ladder failure, which means that the ladder could buckle, break, or topple, among other possibilities. The maximum intended load is the total weight of all people, equipment, tools, materials, loads that are being carried, and other loads that the ladder can hold at any one time. Check the manufacturer's specifications to determine the maximum intended load. Ladders are usually given a duty rating that indicates their load capacity, as shown in *Table 2*.

Case History

A worker was climbing a 10-foot ladder to access a landing, which was 9 feet above the adjacent floor. The ladder slid down, and the worker fell to the floor, sustaining fatal injuries. Although the ladder had slip-resistant feet, it was not secured, and the railings did not extend 3 feet above the landing.

Table 2 Ladder Duty Ratings and Load Capacities

Duty Ratings	Load Capacities
Type IAA	375 lbs., extra-heavy duty/professional use
Type IA	300 lbs., extra-heavy duty/professional use
Type I	250 lbs., heavy duty/industrial use
Type II	225 lbs., medium duty/commercial use
Type III	200 lbs., light duty/household use

There are different types of ladders to use for different jobs (*Figure 16*). Selecting the right ladder for the job at hand is important to complete a job as safely and efficiently as possible. Ladder types include portable **straight ladders**, **extension ladders**, and **stepladders**.

> **WARNING!** Use ladders for their intended uses only. Ladders are not interchangeable, and incorrect use can result in injury or damage.

> **WARNING!** When you use a ladder, be sure to maintain three-point contact with the ladder when ascending or descending. Three-point contact means that either two feet and one hand or one foot and two hands are always touching the ladder.

6.1.0 Straight Ladders

Straight ladders consist of two rails, rungs between the rails, and safety feet on the bottom of the rails (*Figure 17*). The straight ladders used in construction are made of aluminum, wood, or fiberglass.

Metal ladders conduct electricity and should never be used around electrical equipment. Any portable metal ladder must have "Danger. Do Not Use Around Electrical Installations" stenciled on the rails in two-inch, red letters. Ladders made of dry wood or fiberglass, neither of which conducts electricity, should be used around electrical equipment. Check that any wooden ladder is, in fact, completely dry before using it; even a small amount of water will conduct electricity.

Different types of ladders are intended for use in specific situations. Aluminum ladders are corrosion-resistant and can be used where they might be exposed to the elements. They are also lightweight and can be used where they must frequently be lifted and moved. Fiberglass ladders are very durable, so they are useful where some amount of rough treatment is unavoidable. Wooden ladders, which are heavier and sturdier

Figure 16 Different types of ladders.

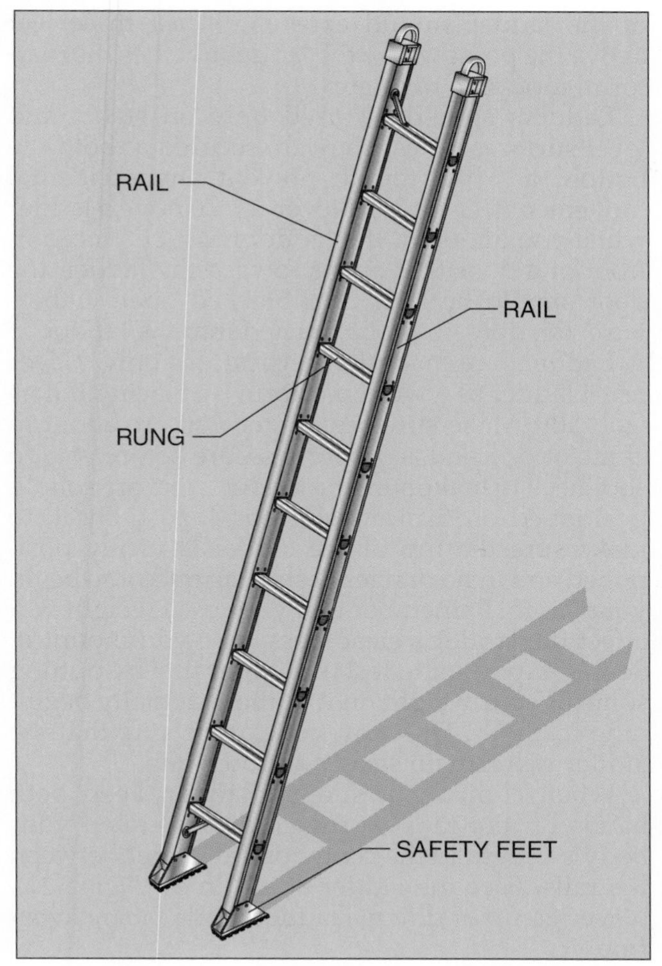

Figure 17 Portable straight ladder.

than fiberglass or aluminum ladders, can be used where heavy loads must be moved up and down. Both fiberglass and aluminum are easier to clean than wood.

6.1.1 Inspecting Straight Ladders

Always inspect a ladder before you use it. Check the rails and rungs for cracks or other damage. Also, check for loose rungs. If you find any damage, do not use the ladder. Check the entire ladder for loose nails, screws, brackets, or other hardware. If you find any hardware problems, tighten the loose parts or have the ladder repaired before you use it. OSHA requires regular inspections of all ladders and an inspection just before each use.

> **CAUTION**
> Wooden ladders should never be painted. The paint could hide cracks in the rungs or rails. Clear varnish, shellac, or a preservative oil finish will protect the wood without hiding defects.

Figure 18 shows the safety feet attached to a straight ladder. Make sure the feet are securely attached and that they are not damaged or worn down. Do not use a ladder if its safety feet are not in good working order.

6.1.2 Using Straight Ladders

It is very important to place a straight ladder at the proper angle before using it. A ladder placed at an improper angle will be unstable and could cause you to fall. *Figure 19* shows a properly positioned straight ladder.

The distance between the foot of a ladder and the base of the structure it is leaning against must be one-fourth of the distance between the ground and the point where the ladder touches the structure. For example, if the height of the wall shown in *Figure 19* is 16 feet, the base of the ladder should be 4 feet from the base of the wall. If you are going to step off a ladder onto a platform or roof, the top

Figure 18 Ladder safety feet.

of the ladder should extend at least three feet above the point where the ladder touches the platform, roof, side rails, etc.

Ladders should be used only on stable and level surfaces unless they are secured at both the bottom and the top to prevent any accidental movement (*Figure 20*). Never try to move a ladder while you are on it. If a ladder must be placed in front of a door that opens toward the ladder, the door should be locked or blocked open. Otherwise, the door could be opened into the ladder.

Ladders are made for vertical use only. Never use a ladder as a work platform by placing it horizontally. Make sure the ladder you are about to climb or descend is properly secure before you do so. Check to make sure the ladder's feet are solidly positioned on firm, level ground. Also check to make sure the top of the ladder is firmly positioned and in no danger of shifting once you begin your climb. Remember that your own weight will affect the ladder's steadiness once you mount it. So it is important to test the ladder first by putting some of your weight on it without actually beginning to climb. This way, you can be sure that the ladder will remain steady as you climb.

When climbing a straight ladder, keep both hands on the rails or rungs. Always keep your body's weight in the center of the ladder between the rails. Face the ladder at all times (*Figure 21*). Never go up or down a ladder while facing away from it.

To carry a tool while you are on the ladder, use a **hand line** or tagline attached to the tool. Climb the ladder and then pull up the tool. Don't carry tools in your hands while you are climbing a ladder.

Figure 19 Proper positioning of a straight ladder.

6.2.0 Extension Ladders

An extension ladder is actually two straight ladders. They are connected so you can adjust the overlap between them and change the length of the ladder as needed (*Figure 22*).

On-Site

Three-Point Contact

When climbing or working from a ladder, you run the risk of falling. An important rule to safeguard yourself against a fall is to maintain three-point contact with the ladder at all times. This means that you either have two hands and one foot or two feet and one hand touching the ladder constantly. Maintaining three-point contact with the ladder will help prevent falling.

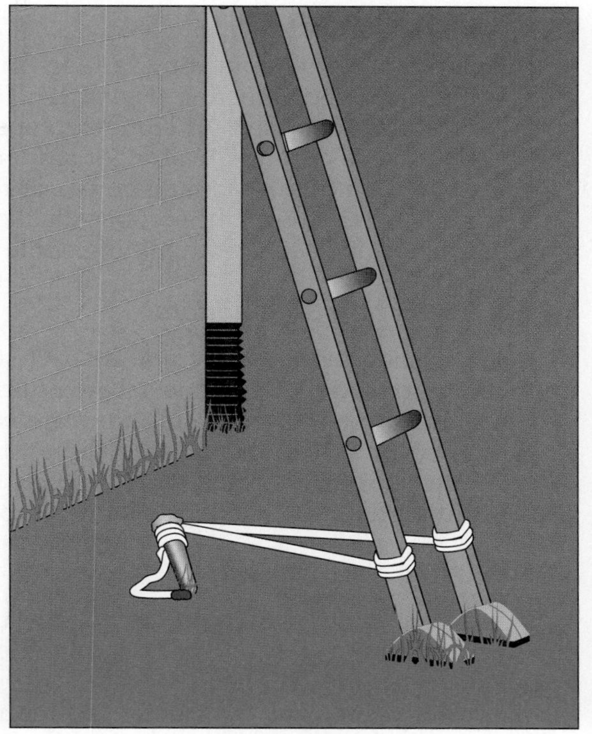

Figure 20 Securing a ladder.

Figure 21 Moving up or down a ladder.

6.2.1 Inspecting Extension Ladders

The same rules for inspecting straight ladders apply to extension ladders. In addition, you should inspect the rope that is used to raise and lower the movable section of the ladder. If the rope is frayed or has worn spots, it should be replaced before the ladder is used.

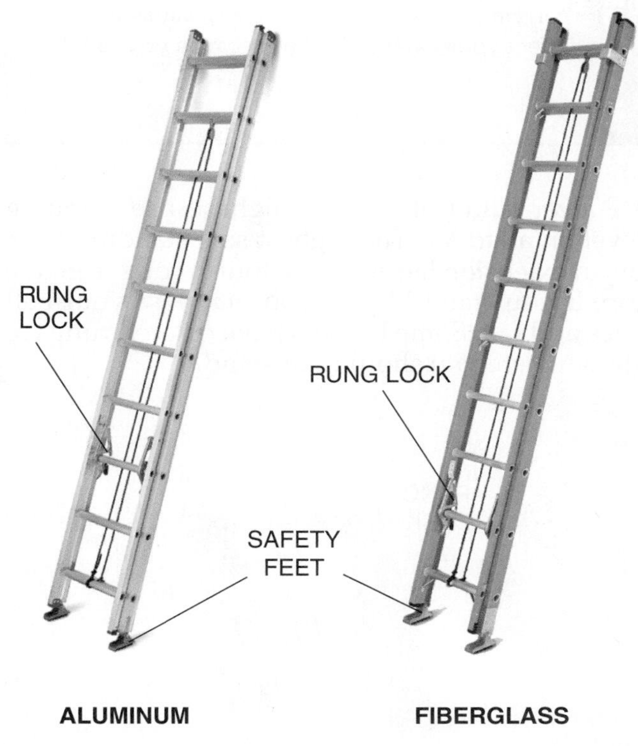

Figure 22 Typical extension ladders.

The rung locks (*Figure 23*) support the entire weight of the movable section and the person climbing the ladder. Inspect them for damage before each use. If they are damaged, they should be repaired or replaced before the ladder is used.

6.2.2 Using Extension Ladders

Extension ladders are positioned and secured following the same rules as straight ladders. When you adjust the length of an extension ladder, always reposition the movable section from the bottom, not the top, so you can make sure the rung locks are properly engaged after you make the adjustment. Check to make sure the section locking mechanism is fully hooked over the desired rung. Also check to make sure that all ropes used for raising and lowering the extension are clear and untangled.

WARNING!

Extension ladders have a built-in extension stop mechanism. Do not remove it. If the mechanism is removed, it could cause the ladder to collapse under a load.

WARNING!

Haul materials up on a line rather than hand-carrying them up an extension ladder. Avoid carrying anything on a ladder, because it will affect your balance and may cause you to fall.

Never stand above the highest safe standing level on a ladder. The highest safe standing level on an extension ladder is the fourth rung from the top. If you stand higher, you may lose your balance and fall. Some ladders have colored rungs to show where you should not stand.

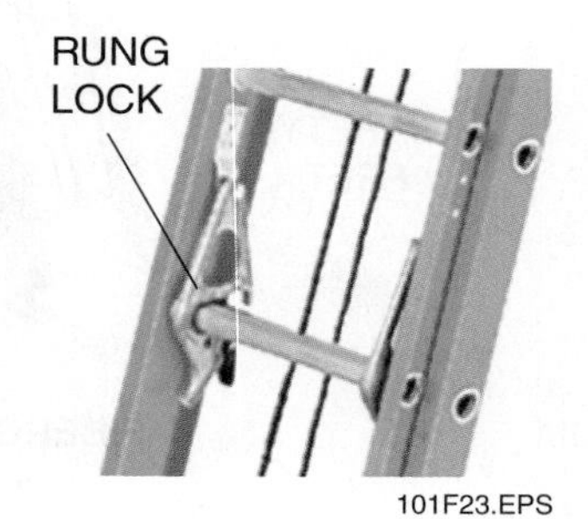

Figure 23 Rung locks.

6.3.0 Stepladders

Stepladders are self-supporting ladders made of two sections hinged at the top (*Figure 24*). The section of a stepladder used for climbing consists of rails and rungs like those on straight ladders. The other section consists of rails and braces. Spreaders are hinged arms between the sections that keep the ladder stable and keep it from folding while in use.

6.3.1 Inspecting Stepladders

Inspect stepladders the way you inspect straight and extension ladders. Pay special attention to the hinges and spreaders to be sure they are in good repair. Also, be sure the rungs are clean. A stepladder's rungs are usually flat, so oil, grease, or dirt can build up on them and make them slippery.

6.3.2 Using Stepladders

When you position a stepladder, be sure that all four feet are on a hard, even surface. Otherwise, the ladder can rock from side to side or corner to corner when you climb it. With the ladder in position, be sure the spreaders are locked in the fully open position.

Never stand on the top step or the top of a stepladder. Putting your weight this high will make the ladder unstable. The top of the ladder is made to support the hinges, not to be used as a step. And, although the rear braces may look like rungs, they are not designed to support your weight. Never use the braces for climbing. And never climb the back of a stepladder. (For certain jobs, however, there are specially designed two-person ladders with steps on both sides.) *Figure 25* shows common dos and don'ts for using ladders.

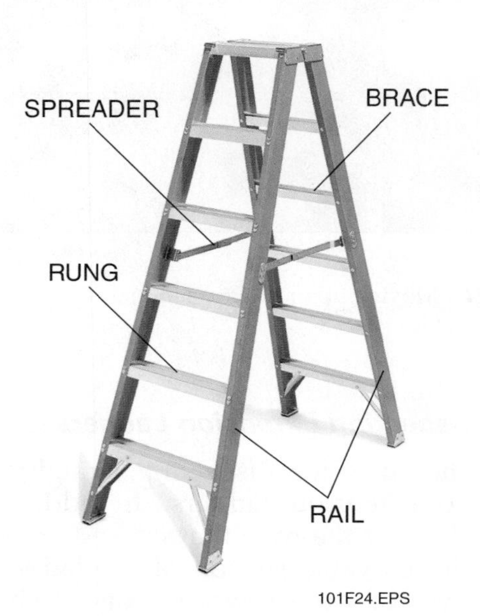

Figure 24 Typical fiberglass stepladder.

DOs

- Be sure your ladder has been properly set up and is used in accordance with safety instructions and warnings.
- Wear shoes with non-slip soles.

- Keep your body centered on the ladder. Hold the ladder with one hand while working with the other. Never let your belt buckle pass beyond either ladder rail.

- Move materials with extreme caution. Be careful pushing or pulling anything while on a ladder. You may lose your balance or tip the ladder.

- Get help with a ladder that is too heavy to handle alone. If possible, have another person hold the ladder when you are working on it.

- Climb facing the ladder. Center your body between the rails. Maintain a firm grip.
- Always move one step at a time, firmly setting one foot before moving the other.

- Haul materials up on a line rather than carry them up an extension ladder.
- Use extra caution when carrying anything on a ladder.

Read ladder labels for additional information.

DON'Ts

- DON'T stand above the highest safe standing level.
- DON'T stand above the second step from the top of a stepladder and the 4th rung from the top of an extension ladder. A person standing higher may lose their balance and fall.

- DON'T climb a closed stepladder. It may slip out from under you.
- DON'T climb on the back of a stepladder. It is not designed to hold a person.

- DON'T stand or sit on a step-ladder top or pail shelf. They are not designed to carry your weight.
- DON'T climb a ladder if you are not physically and mentally up to the task.

- DON'T exceed the Duty Rating, which is the maximum load capacity of the ladder. Do not permit more than one person on a single-sided stepladder or on any extension ladder.

- DON'T place the base of an extension ladder too close to the building as it may tip over backward.
- DON'T place the base of an extension ladder too far away from the building, as it may slip out at the bottom. **Please refer to the 4 to 1 Ratio Box.**

- DON'T over-reach, lean to one side, or try to move a ladder while on it. You could lose your balance or tip the ladder. **Climb down and then reposition the ladder closer to your work!**

4 TO 1 Ratio

Place an extension ladder at a 75-1/2° angle. The set-back ("S") needs to be 1 ft. for each 4 ft. of length ("L") to the upper support point.

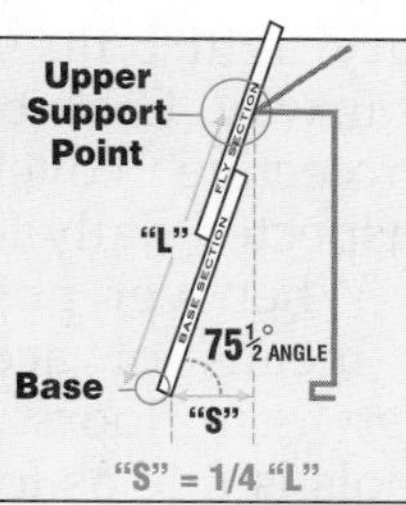

101F25.EPS

Figure 25 Ladder safety dos and don'ts.

WARNING!

Never stand on a step with your knees higher than the top of a stepladder. You need to be able to hold on to the ladder with your hand. Keep your body centered between the side rails.

6.4.0 Stairways

Stairways are also routinely used on construction sites where there is a break in elevation of 19 inches or more, and no ramp, runway, sloped embankment, or personnel hoist is provided. Observe the following regulations, based on OSHA standards, when using stairways on a job site:

- Stairways having four or more risers or rising more than 30 inches (whichever is less) must be equipped with at least one handrail and one stair railing system along each unprotected side.
- Winding and spiral stairways must be equipped with a handrail offset sufficiently to prevent walking on those portions of the stairways where the tread width is less than 6 inches.
- Stair railings must be not less than 36 inches from the upper surface of the stair railing system to the surface of the tread, in line with the face of the riser at the forward edge of the tread.

6.4.1 Stairway Maintenance and Housekeeping

To reduce the likelihood of slips, trips, or falls, keep stairways clean and clear of debris. Do not store any tools or materials on stairways, and clean up liquid spills, rain water, or mud immediately.

Stairways must be adequately lit. OSHA construction standards require five **foot-candles** of light in stairways. A foot-candle is a unit of measure of the intensity of light falling on a surface, equal to one lumen per square foot and originally defined with reference to a standardized candle burning at one foot from a given surface. Adequate lighting can sometimes be a problem because permanent lighting is usually installed after stairway construction is completed. The amount of light can be measured with a standard light meter. If the lighting is inadequate, a temporary light bulb string should be installed in the stairway. Each bulb should be equipped with a protective cover and the string should be inspected daily for burned out or broken bulbs.

Whenever possible, avoid using stairways as a means of access for transporting materials between floors. Carrying small materials and tools is fine as long as the materials do not block your vision. Going up or down a stairway while carrying large items is physically demanding and increases the chance of injuries and falls. Use the building elevator or crane service to transport large materials from one floor to another.

7.0.0 SCAFFOLDS

Scaffolds provide safe elevated work platforms for people and materials. They are designed and built to comply with high safety standards, but normal wear and tear or accidentally putting too much weight on them can weaken them and make them unsafe. That's why it is important to inspect every part of a scaffold before each use.

CAUTION

Only a competent person has the authority to supervise setting up, moving, and taking down scaffolding. Only a competent person can approve the use of scaffolding on the job site after inspecting the scaffolding.

7.1.0 Types of Scaffolds

Two basic types of scaffolds—self-supporting scaffolds and suspended scaffolds—are used in the construction industry. The rules for safe use apply to both of them.

7.1.1 Manufactured Scaffolds

Manufactured scaffolds are made of painted steel, stainless steel, or aluminum (*Figure 26*). They are stronger and more fire-resistant than wooden scaffolds. They are supplied in ready-made, individual units, which are assembled on site.

7.1.2 Rolling Scaffolds

A rolling scaffold has wheels on its legs so that it can be easily moved (*Figure 27*). The scaffold wheels have brakes so the scaffold will not move while workers are standing on it.

WARNING!

Never unlock the wheel brakes of a rolling scaffold while anyone is on it. People on a moving scaffold can lose their balance and fall.

7.2.0 Inspecting Scaffolds

Any scaffold that is assembled on the job site should be tagged. These tags indicate whether the scaffold meets OSHA standards and is safe to use.

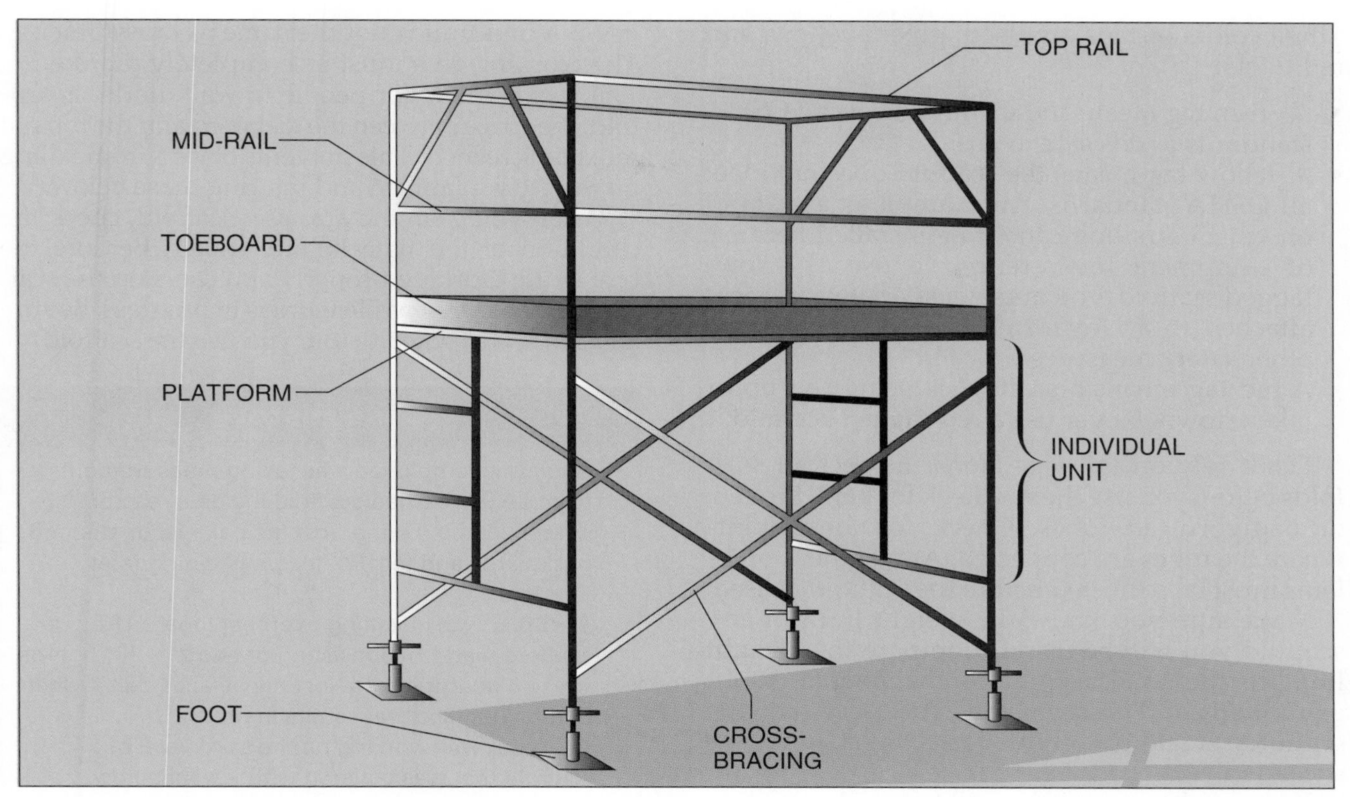

Figure 26 Typical manufactured scaffold.

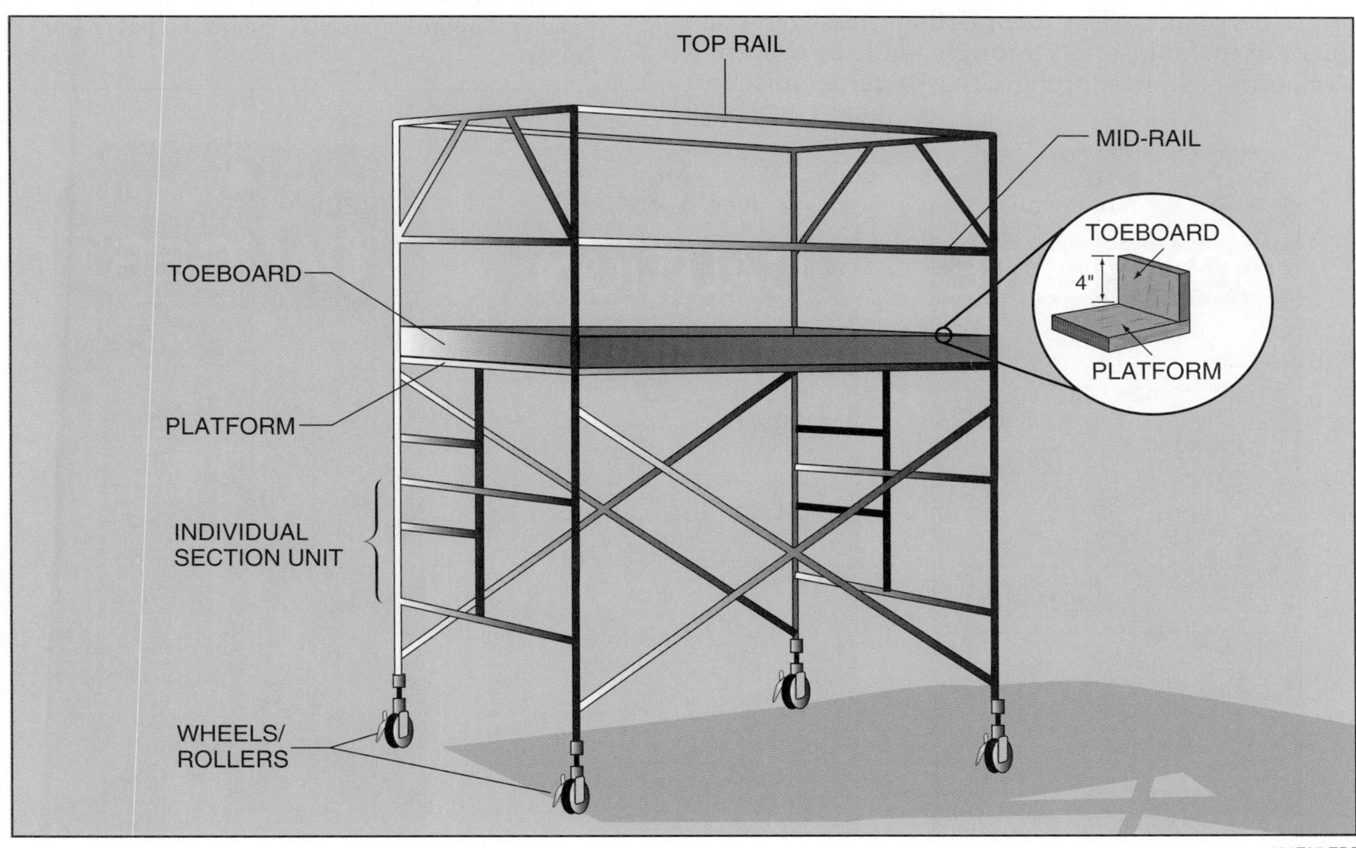

Figure 27 Typical rolling scaffold.

Three colors of tags are used: green, yellow, and red (*Figure 28*).

- A green tag means the scaffold meets all OSHA standards and is safe to use.
- A yellow tag means the scaffold does not meet all OSHA standards. An example is a scaffold on which a railing cannot be installed because of equipment interference. To use a yellow-tagged scaffold, you must wear a safety harness attached to a lanyard. You may have to take other safety measures as well.
- A red tag means a scaffold is being put up or taken down. Never use a red-tagged scaffold.

Don't rely on the tags alone. Inspect all scaffolds before you use them. Check for bent, broken, or badly rusted tubes. Check for loose joints where the tubes are connected. Any of these problems must be corrected before the scaffold is used.

Make sure you know the weight limit of any scaffold you will be using. Compare this weight limit to the total weight of the people, tools, equipment, and material you expect to put on the scaffold. Scaffold weight limits must never be exceeded.

If a scaffold is more than 10 feet high, check to see that it is equipped with **top rails**, **mid-rails**, and toeboards, or use the PFAS. All connections must be pinned. That means they must have a piece of metal inserted through a hole to prevent connections from slipping. **Cross-bracing** must be used. A handrail is not the same as cross-bracing. The working area must be completely **planked**.

If it is possible for people to walk under a scaffold, the space between the toeboard and the top rail must be screened. This prevents objects from falling off the work platform and injuring those below.

When you examine a rolling scaffold, check the condition of the wheels and brakes. Be sure the brakes are working properly and can stop the scaffold from moving while work is in progress. Be sure all brakes are locked before you use the scaffold.

Case History

- A contract employee was taking measurements from an unguarded scaffold inside a reactor vessel when he either lost his balance or stepped backwards and fell 14½ feet, sustaining fatal injuries.
- A worker was installing overhead boards from a scaffold platform consisting of two 2" × 10" boards with no guardrails. He lost his balance, fell 7½ feet to the floor, and was fatally injured.
- A laborer was working on the third level of a scaffold that was covered with ice and snow. Planking on the scaffold was inadequate; there was no guardrail and no access ladder for the various scaffold levels. The worker slipped and fell head-first approximately 20 feet to the pavement below.

COMPLETE
ERECTED FOR
DATE
COMPLETE
HANDRAILS
TOEBOARDS
DECK
LADDER
LIFE LINE
SCAFFOLD OVER
LOCATION
CHARGE #
ERECTION FOREMAN
SIGNATURE
DATE
LOCATION
SIGNATURE

CAUTION
ERECTED FOR
DATE
INCOMPLETE
HANDRAILS
TOEBOARDS
DECK
LADDER
LIFE LINE
SCAFFOLD OVER
REASON INCOMPLETE
LOCATION
CHARGE #
ERECTION FOREMAN
SIGNATURE
DATE
LOCATION
SIGNATURE

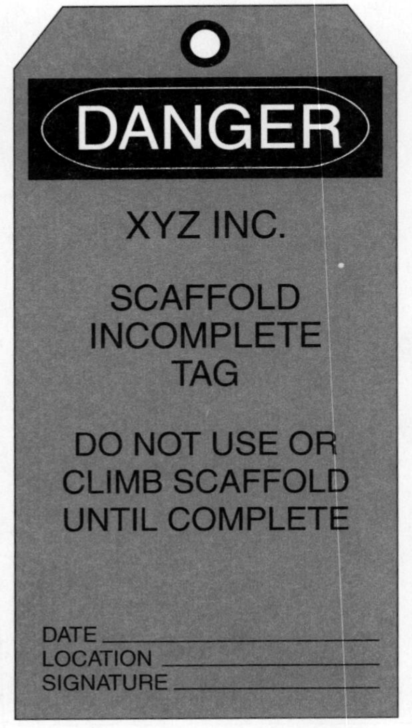

101F28.EPS

Figure 28 Typical scaffold tags.

7.3.0 Using Scaffolds

Be sure that a competent person inspects the scaffold before you use it. There should be firm footing under each leg of a scaffold before you put any weight on it. If you are working on loose or soft soil, you can put matting under the scaffold's legs or wheels.

When you move a rolling scaffold, always follow these steps:

Step 1 Unlock the wheel brakes.

Step 2 Move the scaffold.

Step 3 Re-lock the wheel brakes.

Step 4 Get back on the scaffold.

WARNING!

Keep scaffolds a minimum of three feet from power lines up to 300 volts and a minimum of 10 feet from power lines above 300 volts in accordance with OSHA guidelines. Refer to *OSHA 29 CFR 1926.*

On-Site

Falls from scaffolding and falls from ladders account for approximately 27% of total fall deaths, according to data collected by the U.S. Department of Labor. Prevent accidents by following OSHA and company guidelines.

8.0.0 Struck-By Hazards

Struck-by accidents rank along with caught-in accidents as the second highest cause of construction-related deaths. In fact, struck-by accidents accounted for 18% of construction fatalities between 2003 and 2004. Approximately 75% of struck-by fatalities involve heavy equipment such as trucks or cranes. The primary causes of struck-by accidents include the following hazards: vehicle and equipment strikes, falling objects, and flying objects.

8.1.0 Vehicle and Roadway Hazards

The most common causes of accidents involving highway workers on foot are hazards such as being run over by equipment, especially backing equipment, or by equipment tip-over. The most common cause of death for equipment operators is equipment rollover. If vehicle safety practices are not observed at your site, you risk being pinned between construction vehicles and walls, struck by swinging backhoes, crushed beneath overturned vehicles, or other similar accidents. If you work near public roadways, you risk being struck by passenger or commercial vehicles. When working near moving vehicles and equipment, follow these guidelines:

- Stay alert at all times and keep a safe distance from vehicles and equipment.
- Maintain eye contact with vehicle or equipment operators to ensure that they see you.
- Never get into blind spots of operators.
- Keep off equipment unless authorized.
- Wear reflective or high-visibility vests or other suitable garments (*Figure 29*).
- Never stand between pieces of equipment unless they are secured.
- Never stand under loads handled by lifting or digging equipment, or near vehicles being loaded or unloaded.

Operators must also use caution when driving vehicles. The operator of any vehicle is responsible for the safety of passengers and the protection

101F29.EPS

Figure 29 Worker wearing reflective gear.

Did You Know?

One in four struck-by vehicle deaths involve construction workers, more than any other occupation.

Case History

A contractor was operating a backhoe when an employee attempted to walk between the swinging superstructure of the backhoe and a concrete wall. As the employee approached from the operator's blind side, the superstructure hit the victim, crushing him against the wall. Employees had not been trained in safe work practices, and no barricades had been erected to prevent employee access to a hazardous area.

of the load. Follow these safety guidelines when you operate vehicles on a job site:

- Always wear a seat belt.
- Be sure that each person in the vehicle has a firmly secured seat and seat belt.
- Obey all speed limits. Reduce speed in crowded areas.
- Look to the rear and sound the horn before backing up. If your rear vision is blocked, get a **signaler** to direct you.
- Every vehicle must have a backup alarm. Make sure the backup alarm works.
- Always turn off the engine when you are fueling.
- Turn off the engine and set the brakes before you leave the vehicle.
- Never stay on or in a truck that is being loaded by excavating equipment.
- Keep windshields, rearview mirrors, and lights clean and functional.
- Carry road flares, fire extinguishers, and other standard safety equipment at all times.

WARNING!

Driving a vehicle indoors without good ventilation can make you sick or kill you because of the carbon monoxide given off by the exhaust. Carbon monoxide is especially dangerous because you cannot see, smell, or taste it. Make sure there is good ventilation before you operate any motorized vehicle indoors.

8.2.0 Falling Objects

You are at risk from falling objects when you are beneath machinery and equipment, such as cranes and scaffolds, where overhead work is being performed, or when working around stacked materials. To protect against struck-by injuries from falling objects, always wear an approved hard hat when working under potential falling object hazards. Many employers require workers to wear hard hats at all times on construction sites.

When working near machinery and equipment such as cranes, never stand or work beneath the load. Barricade hazard areas where rigging equipment is in use, and post warning signs to inform other workers of falling object hazards. Inspect cranes and rigging components before use and do not exceed the rated load capacity.

When performing overhead work, be sure all tools, materials, and equipment are secured to prevent them from falling on people below. Use protective measures such as toeboards, debris nets, catch platforms, or canopies to catch or deflect falling objects.

Many workers are hurt or killed by falling stacks of material. Do not stack materials higher than 4:1 height-to-base ratio. Secure all loads by stacking, blocking, and interlocking them. Be aware of changing weather conditions, such as wind, that may lift and shift loads.

8.3.0 Flying Objects

There is a danger from flying objects when power tools or activities such as pushing, pulling, or prying cause objects to become airborne. Chipping, grinding, brushing, or hammering are all examples of job tasks that may cause flying object hazards. Tools that move at amazing speed, like pneumatic and powder-actuated tools, can be very dangerous. Injuries from flying objects can range from minor abrasions to concussions, blindness, or death.

Case History

A worker was standing under a suspended scaffold that was hoisting a workman and three sections of ladder. Sections of that ladder became unlatched and fell 50 feet, striking the worker in the skull. The worker was not wearing any head protection and died from his injuries.

Case History

A carpenter was attempting to anchor a plywood form in preparation for pouring a concrete wall, using a powder-actuated tool. The nail passed through the hollow wall, traveled 72 feet, and struck an apprentice in the head, killing him. The tool operator had never been trained in the proper use of the tool, and none of the employees in the area, including the victim, were wearing PPE.

To protect against flying object hazards, follow these guidelines:

- Use eye protection, such as safety glasses, goggles, or face shields where machines or tools may cause flying particles.
- Inspect tools and machines to ensure that protective guards are in place and in good condition.
- Make sure you are trained in the proper operation of pneumatic and powder-actuated tools.

9.0.0 Caught-In-Between Hazards

Congested work sites, heavy equipment, and multiple trades can contribute to caught-in-between hazards. The primary causes of caught-in-between fatalities include trench/excavation collapse, rotating equipment, and unguarded parts.

9.1.0 Trenching and Excavation

In many construction jobs you will need to work in and around trenches and excavations. Trenching and excavation require specialized training. You must know and follow appropriate safety procedures before entering or working around an excavation.

An excavation is any man-made cut, cavity, trench, or depression formed by the mechanical or hand removal of earth or soil. Sometimes the terms *excavation* and *trench* are used interchangeably, but there is a difference. A trench is an excavation that is deeper than it is wide, and usually not wider than 15 feet. Nearly all trenches are dangerous if not protected. Because trenches are narrow, workers can easily become trapped. Some excavations, especially those on large job sites, may only be dangerous near the slopes.

Did You Know?

The fatality rate for excavation work is 112% higher than the rate for general construction.

Did You Know?

One cubic yard of earth weighs approximately 2,700 pounds—that's the weight of a small car!

Hazards involved with trench and excavation work include:

- Cave-ins
- Water accumulation
- Falling objects
- Collapse of nearby structures
- Hazardous atmospheres produced by toxic gases in the soil

WARNING!

Just 2 to 3 feet of soil can put enough pressure on your lungs to prevent you from breathing. In as little as 4 to 6 minutes without oxygen, you can sustain considerable brain damage.

9.1.1 Cave-Ins

Cave-ins are the most common and deadly hazard in excavation work. When dirt is removed from an excavation, the surrounding soil becomes unstable and gravity forces it to collapse. Cave-ins occur when soil or rock falls or slides into an excavation. Cave-ins can entrap, bury, or otherwise injure or immobilize a person. Most cave-ins occur in trenches 5 to 15 feet deep, and happen suddenly with little or no warning. On average, about 1,000 trench collapses occur each year in the United States.

The four main hazards that cause cave-ins to occur are:

- Failure to properly or routinely inspect excavations
- Lack of protective systems
- Excessive weight from spoil piles or machinery
- No safe means of egress/access

Soil conditions can change and shift, so they must be constantly evaluated. There are certain

Did You Know?

Each year in the United States, more than 100 people are killed and many more are seriously injured in cave-in accidents.The chances of a trapped worker being killed can be as high as 50%.

factors that could change the surroundings of the site, making a cave-in more likely. These factors include the following:

- Changes in weather conditions, such as freezing, thawing, or sudden heavy rain
- An excavation dug in unstable or previously disturbed soil
- Excessive vibration around the excavation
- Water accumulation in an excavation

Sometimes there are visible warning signs around the excavation that can be spotted before a cave-in occurs. Being aware of the warning signs increases your chances of getting out before a collapse occurs. Visible warning signs of a potential cave-in include the following:

- Ground settlement or narrow cracks in the sidewalls, slopes, or surface next to the excavation
- Flakes, pebbles, or clumps of soil separating and falling into the excavation
- Changes or bulges in the wall slope

If you notice any of these signs, get out of the excavation immediately, and get your co-workers out, too.

Soil type is a key factor in determining the type of protective system necessary to ensure that the trench will be safe. Solid rock is the most stable, while sandy soil is the least stable. The four types of soil classifications are shown in *Table 3*.

To be safe, treat soil as if it is Type C soil unless proven otherwise. It is better to over-prepare for a stronger soil than to not prepare enough for a weaker one.

9.1.2 Inspections

A competent person will inspect excavations daily and decide whether cave-ins or failures of protective systems could occur, and whether there are any other hazardous conditions present. The competent person must conduct inspections before any work begins, as needed throughout the shift, and after every rainstorm or other hazard-increasing incident.

When conducting an inspection, the competent person performs a visual analysis and looks at the soil and trench, the surrounding area, and the excavation site in general. During the visual analysis, the competent person looks for the following:

- Cracks and **spalls** in the sides of the opened excavation and adjacent surface area.
- Existing utility or other underground structures and previously disturbed soil in and around the excavation.
- Sources of vibration that may affect the stability of the excavation face.
- Surface water, water seeping from the sides of the excavation, or the location of the level of the water table.
- Layered systems in the excavation (two or more distinct types of soil).

If a competent person finds indications of protective systems failure, hazardous atmospheres, or a possible cave-in, workers must be removed from the hazardous area and may not return until necessary precautions have been taken. Always ask the competent person on site or your immediate supervisor if you have questions about proper safety practices.

9.1.3 Protective Systems

Once visual and manual tests are performed and the soil type is determined, then a protective system must be chosen. Protective systems are required in nearly all excavations. Protective systems protect workers from:

- Cave-ins
- Materials that could fall or roll into an excavation
- Collapse of adjacent structures

There are various types of trench protective systems to meet each type of soil condition. Selecting a protective system for an excavation depends on soil conditions, the depth of the trench, and the environmental conditions surrounding the site.

Table 3 Soil Types

Name	Type/Characteristics
Solid Rock	Excavation walls stay vertical as long as the excavation is open.
Type A Soil	Fine-grained, cohesive: clay, hardpan, and caliche. Particles too small to see with the naked eye.
Type B Soil	Angular rock, silt, and similar soil.
Type C Soil	Coarse-grained, granular: sand, gravel, and loamy sand. Particles are visible to the naked eye.

Case History

A worker was in a trench installing forms for concrete footers when it caved in, causing fatal injuries. The trench, which was 7½ feet deep, was in loose, sandy (Type C) soil, and no inspection was conducted prior to the start of the shift.

A registered professional engineer must design protective systems for excavations deeper than 20 feet. There are two basic systems of trench protection: sloping and benching systems and support systems.

Sloping and benching are forms of trench protection that cut away and slant the excavation face. A sloping system is a method in which the sides of an excavation are cut back to a safe angle using relatively smooth inclines (*Figure 30*). A benching system is similar to a sloping system, but instead of smooth inclines, the sides of the trench wall are cut back using a series of steps (*Figure 31*).

Both of these systems call for a safe angle called the **maximum allowable slope** (MAS). Maximum allowable slope means the steepest incline of an excavation face that is acceptable for the most favorable site conditions as protection against cave-ins, and is expressed as the ratio of horizontal distance to vertical rise. The maximum allowable slope for sloping or benching systems depends on the type of soil that is being dug. The maximum allowable slopes for A, B, and C soil types are as follows:

- *Type A soil* – ¾ to 1, or 53° (*Figure 32*)
- *Type B soil* – 1 to 1, or 45° (*Figure 33*)
- *Type C soil* – 1½ to 1, or 34° (*Figure 34*)

The maximum allowable slope varies for layered soils. A simple rule for maximum allowable slope is the more granular the soil, the more gradual the incline. It is also important to know that benching systems cannot be used in Type C soils. They will not hold the bench properly and could collapse.

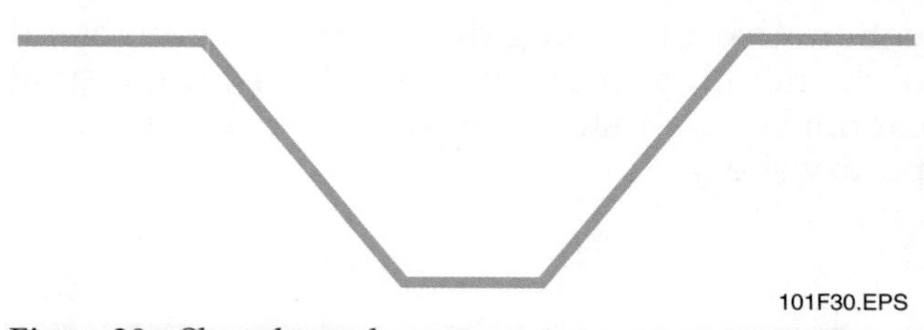

Figure 30 Sloped trench.

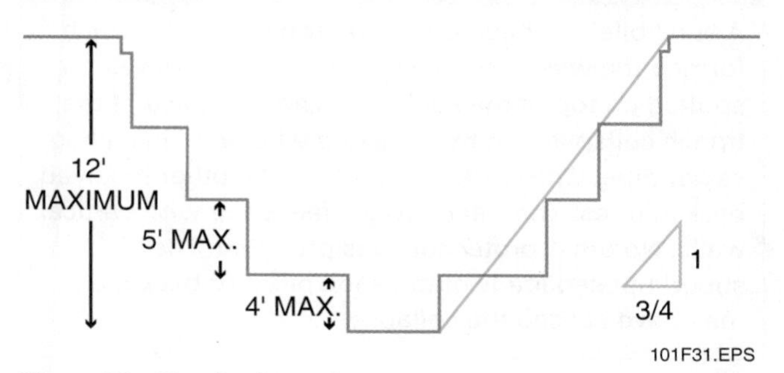

Figure 31 Benched trench.

Many times trenches are in narrow places, so sloping and benching are not options. In these situations, support systems like **shoring** or shielding must be utilized. Shoring structures are typically made of metal or wood and are used to support the sides of a trench and prevent soil from caving in. They consist of plating held firmly in place with expandable braces (*Figure 35*). There are many types of shoring systems. Some of them are easy to install, and others require experience and engineering. Shielding structures, or trench boxes, are placed inside trenches or excavations, and are strong enough to protect workers in the event of a cave-in, so long as the workers are within the confines of the box (*Figure 36*). Regardless of what type of system is used, if the excavation is more than 20 feet deep, the entire excavation protective system has to be designed by a registered professional engineer.

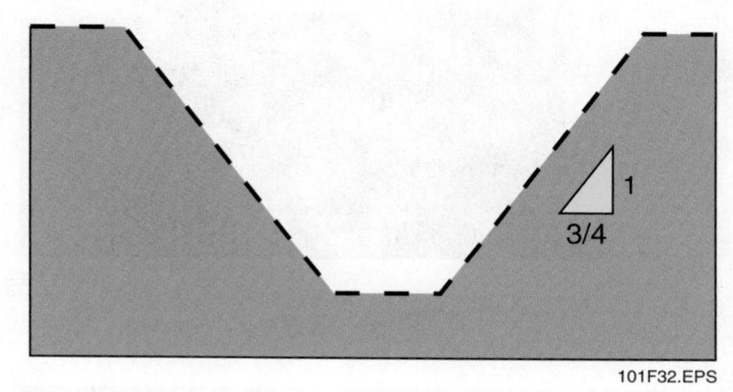

Figure 32 Type A MAS.

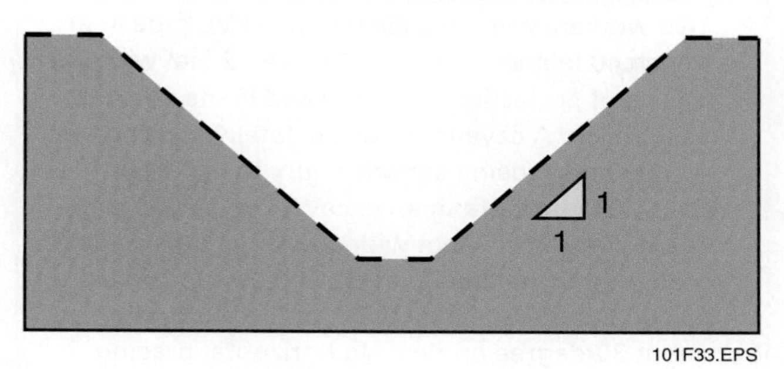

Figure 33 Type B MAS.

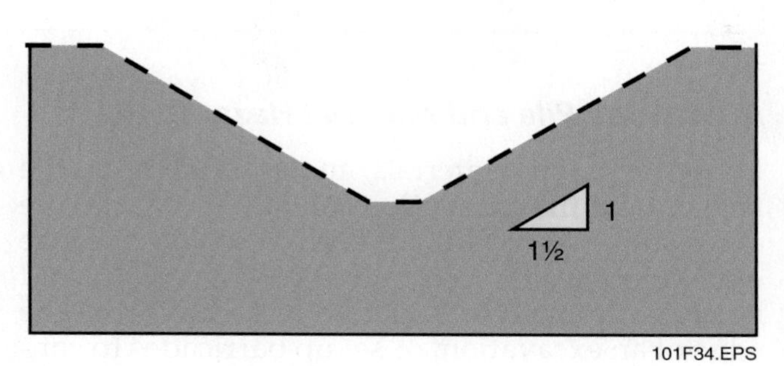

Figure 34 Type C MAS.

101F35.EPS

Figure 35 Shoring structure.

101F36.EPS

Figure 36 Shielding structure.

Case History

- Two workers were installing 6-inch PVC pipe in a trench 40 feet long, 9 feet deep, and 2 feet wide. No means of protection was provided in the vertical wall trench. A cave-in occurred, fatally injuring one worker and causing serious injury to the other.
- Four workers were in an excavation 32 feet long, 7 feet deep, and 9 feet wide, boring a hole under a road. Eight-foot steel plates used as shoring were placed against the side walls of the excavation at about 30-degree angles. No horizontal bracing was used. One of the plates tipped over, crushing one worker.

9.1.4 Spoil Pile and Material Hazards

Loose rock, soil, materials, and equipment on the face or near the excavation can fall or roll into the excavation, or overload and possibly collapse excavation walls. Keep all spoil, materials, and heavy equipment at least two feet away from the edge of an excavation, or set up barricades to contain falling material (*Figure 37*). Scale the excavation face to remove loose material, and place spoils so that rainwater runs away from the excavation. Use a retaining device strong enough and high enough to resist expected loads. If the spoil cannot be safely stored on site, remove it to a temporary site.

Case History

A spoil pile had been placed on top of a curb, which formed the west face of a trench. A backhoe was spotted on top of the spoil pile. The west face of the trench collapsed on two workers who were installing sewer pipe. One worker was killed; the other received back injuries. The trench was 8 feet deep with vertical walls. No other protection was provided. The superimposed loads of the spoil pile and backhoe may have caused the collapse.

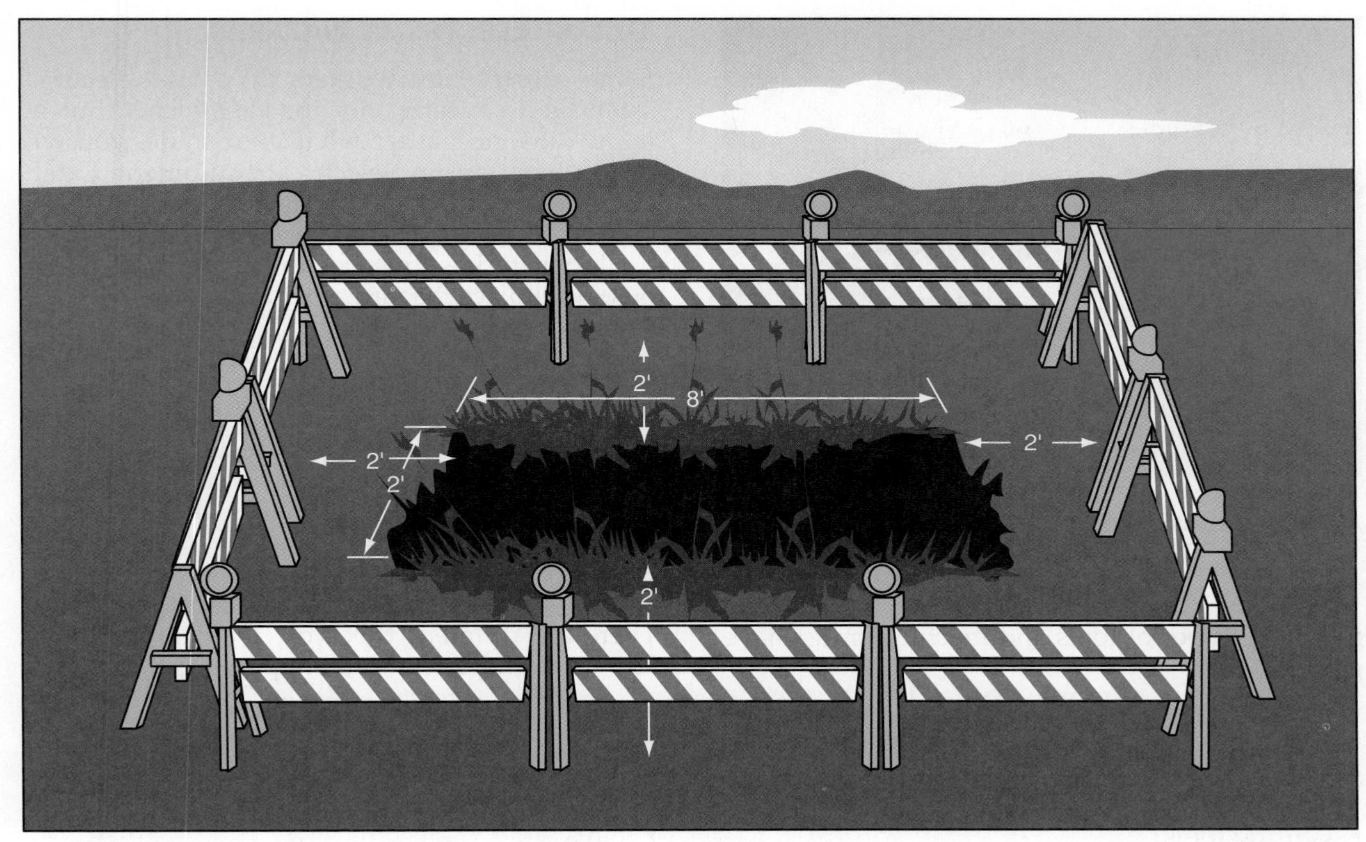

Figure 37 Barricade around a trench.

9.1.5 Access and Egress

When working in a trench, there has to be a safe means of entry and exit for workers, such as a stairway, ladder, or ramp. There must be an exit every 25 feet for every trench over 4 feet deep. Lifting equipment such as loader buckets and backhoe shovels are not safe means for entering or exiting a trench. Once you are in the trench and before you begin work, take a moment to look around and find the nearest ladder so that you can plan your exit, if necessary.

Case History

Two workers were laying sewer pipe in a 15-foot deep trench that was not shored or sloped properly. The only way to get out was by climbing the backfill. While exiting the trench, one of the workers was trapped by a small cave-in. The second worker tried to free him, but a second cave-in occurred, trapping the second worker at the waist. The second cave-in caused the death of the first worker, while the second worker sustained a hip injury.

9.1.6 Emergency Response

Your company should have an emergency action plan that must be communicated to every worker. If you're not sure what the emergency action plan is for your site, don't be afraid to ask questions. Your knowledge could help prevent serious injury or even death, and your supervisor wants you and everybody working with you to be as safe as possible.

Most importantly, try to prevent emergencies before they happen. When in doubt, get out! If you notice potentially dangerous conditions while working in an excavation, get yourself and your co-workers out of danger immediately, and inform the competent person of your concerns.

9.2.0 Tool and Machine Guarding

Almost all tools and machines used in construction are equipped with guards that protect workers from rotating parts (*Figure 38*). All tools and machines that could cause potential harm to workers should have a guard shielding the hazard. The

101F38.EPS

Figure 38 Machine with guard.

following types of tools and machines must have guards:

- Grinding tools
- Shearing tools
- Presses
- Punches
- Cutting tools
- Rolling machines
- Tools or machines with pinch points
- Tools or machines with sharp edges

Machine guards should prevent moving parts of the machine from coming into contact with your arms, hands, or any other part of the body, while allowing you to use the machine comfortably and efficiently without hindering work. Some workers find machine guards to be aggravating and try to remove them from machines. Guards should be secure and should not be easily removed. They should be maintained in good condition, made of durable material, and bolted or screwed to the machine so that tools are necessary for their removal. Follow these guidelines for using and caring for tool and machine guards:

- Do not remove a guard from a tool or machine except for cleaning purposes or to change a blade or perform other service.
- When you are finished with cleaning or maintenance, replace the guard immediately.
- Do not use any material to wedge a guard open.
- Guards and attachments are designed for the specific tool or machine you are using. Only use attachments that are specifically designed for that tool or machine.

10.0.0 Electrical Hazards

Some construction workers think that electrical safety is a concern only for electricians. But on many jobs, no matter what your trade, you will use or work around electrical equipment. Extension cords, power tools, portable lights, and many other pieces of equipment use electricity.

Electricity has long been recognized as a serious workplace hazard, exposing employees to such dangers as electric shock, electrocution, burns, fires, and explosions. During the decade 1992–2002, the Bureau of Labor Statistics reported that 5% of all workplace fatalities, almost 3,400 deaths, were the direct result of electrocution. The construction industry accounted for 47% of these fatalities. What makes these statistics even more tragic is that most of these fatalities could have easily been avoided.

Not all electrical accidents result in death. There are different types of electrical accidents. Any of the following can happen:

- Burns
- Electric shock
- Explosions
- Falls caused by electric shock
- Fires

From 1992 to 2002, over 47,000 nonfatal injuries occurred, including 18,000 burns and 29,000 electrical shocks.

10.1.0 Basics of Electricity

Electricity can be described as the flow of electrons through a conductor. This flow of electrons is called electrical current. Some materials—such as silver, copper, steel, and aluminum—are excellent conductors, allowing electrical current to flow easily through them. The human body, especially when it is wet, is also a good conductor.

To create an electrical current, a path must be provided in a circular route, or a circuit. If the circuit is interrupted, the electrical current will complete its circular route by flowing along the path

Did You Know?

Electric shocks or burns are a major cause of accidents in the construction industry. According to the Bureau of Labor Statistics, electrocution is the fourth leading cause of death among construction workers.

of least resistance. This means that it will flow into and through any conductor that is touching it. If it cannot complete its circuit, the electrical current will go to the ground. This means that it will find the path of least resistance that allows it to flow as directly as possible into the earth. All of this takes place almost instantly.

If the human body comes in contact with an electrically energized conductor and is in contact with the ground at the same time, the human body becomes the path of least resistance for the electricity. The electricity flows through the body in less than the blink of an eye without warning. That's why safety precautions are so important when working with and around electrical currents. When a person's body conducts electrical current and the amount of that current is high enough, the person can be electrocuted (killed by electric shock). *Table 4* shows the effects of different amounts of electrical current on the human body and lists some common tools that operate using those currents.

WARNING!

Less than one amp of electrical current can kill. Always take precautions when working around electricity.

Here's an example. A craftsperson is operating a portable power drill while standing on damp ground. The power cord inside the drill has become frayed, and the electric wire inside the cord touches the metal drill frame. Three amps of current pass from the wire through the frame, then through the worker's body and into the ground. *Table 4* shows that this worker will probably die.

10.2.0 Electrical Safety Guidelines

OSHA and your company have specific policies and procedures to keep the workplace safe from electrical hazards. You can do many things to reduce the chance of an electrical accident. If you ever have any questions about electrical safety on the job site, ask your supervisor. Here are some basic job-site electrical safety guidelines:

- Use three-wire extension cords and protect them from damage. Never fasten them with staples, hang them from nails, or suspend them from wires. Never use damaged cords.
- Use three-wire cords for portable power tools and make sure they are properly connected (*Figure 39A*). The three-wire system is one of the most common safety grounding systems used to protect you from accidental electrical shock. The third wire is connected to a ground. If the insulation in a tool fails, the current will pass to ground through the third wire—not through your body.
- Double-insulated cords are also effective in preventing shocks when using power tools (*Figure 39B*). *Figure 39* shows the double-insulated symbol that can be found on double-insulated tools.

NOTE

It is becoming more common to use double-insulated tools because they are safer than relying on a three-wire cord alone.

Table 4 Effects of Electrical Current on the Human Body

Current	Common Item/Tool	Reaction to Current
0.001 amps	Watch battery	Faint tingle.
0.005 amps	9-volt battery	Slight shock.
0.006–0.025 amps (women) 0.009–0.030 amps (men)	Christmas tree light bulb	Painful shock. Muscular control is lost.
0.050–0.9 amps	Small electric radio	Extreme pain. Breathing stops; severe muscular contractions occur. Death may result.
1.0–9.9 amps	Jigsaw (4 amps); Sawsall® or Port-a-Band® saw (6 amps); portable drill (3–8 amps)	Ventricular fibrillation and nerve damage occur. Death may result.
10 amps and above	ShopVac® (15-gallon); circular saw	Heart stops beating; severe burns occur. Death may result.

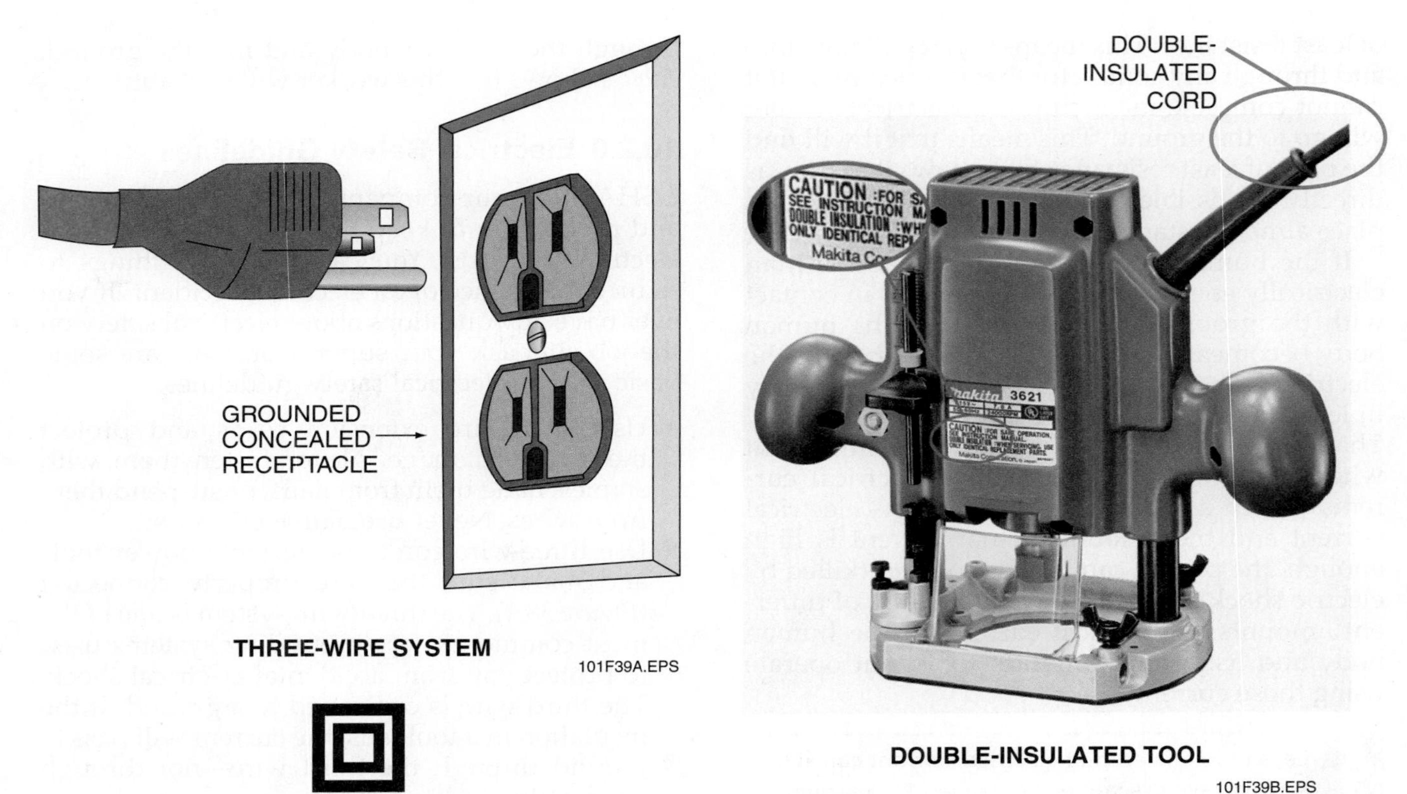

Figure 39 Three-wire system, double-insulated tool, and double insulated symbol.

- Use a **ground fault circuit interrupter (GFCI)** (*Figure 40A*), a portable GFCI (*Figures 40B* and *40C*), or an assured grounding program with every tool.

CAUTION

All tools used in construction must be ground-fault protected. This helps ensure the safety of workers.

- Make sure that panels, switches, outlets, and plugs are grounded.
- Never use bare electrical wire.
- Never use metal ladders near any source of electricity.
- Never wear a metal hard hat.
- Always inspect electrical power tools before you use them.
- Never operate any piece of electrical equipment that has a danger tag or lockout device attached to it.

WARNING!

All work on electrical equipment should be done with circuits de-energized, locked out, and confirmed. All conductors, buses, and connections should be considered energized unless proven otherwise.

- Never use worn or frayed cables (*Figure 41*).
- Make sure all light bulbs have protective guards to prevent accidental contact (*Figure 42*).
- Do not hang temporary lights by their power cords unless they are specifically designed for this use.
- Check the cable and ground prong. Check for cuts in the cords and make sure the cords are clean of grease.
- Use approved **concealed receptacles** for plugs. If different voltages or types of current are used in the same area, the receptacles should be designed so that the plugs are not interchangeable.
- Any repairs to cords must be performed by a qualified person.
- Always make sure all tools are grounded before use.

101F40A.EPS

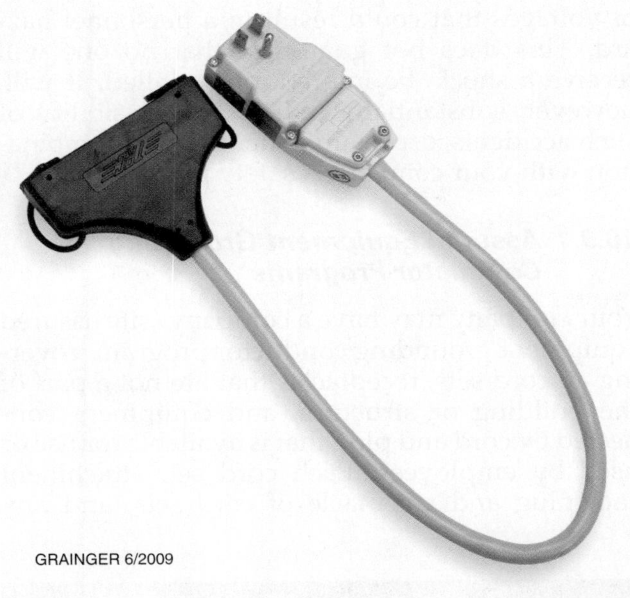

Figure 40 GFCI and portable GFCI.

101F40B.EPS

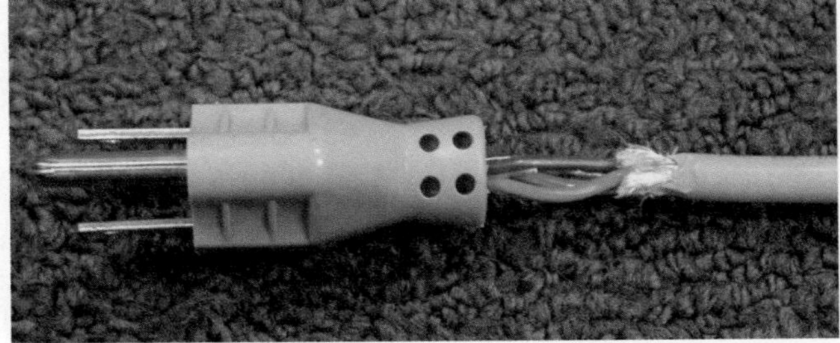

101F41A.EPS

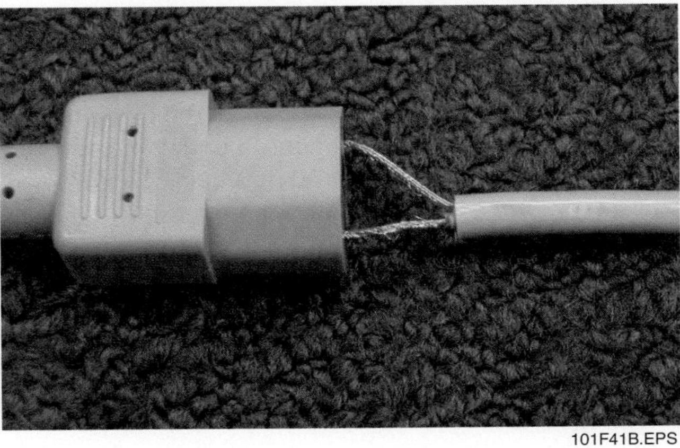

101F41B.EPS

Figure 41 Never use damaged cords.

Case History

A fan connected to a 120-volt electrical system with an extension cord provided ventilation for a worker performing a chipping operation from an aluminum stepladder. The insulation on the extension cord was cut through and exposed bare, energized conductors which made contact with the ladder. The ground wire was not attached on the male end of the cord's plug. When the energized conductor made contact with the ladder, the path to the ground included the worker's body, resulting in death.

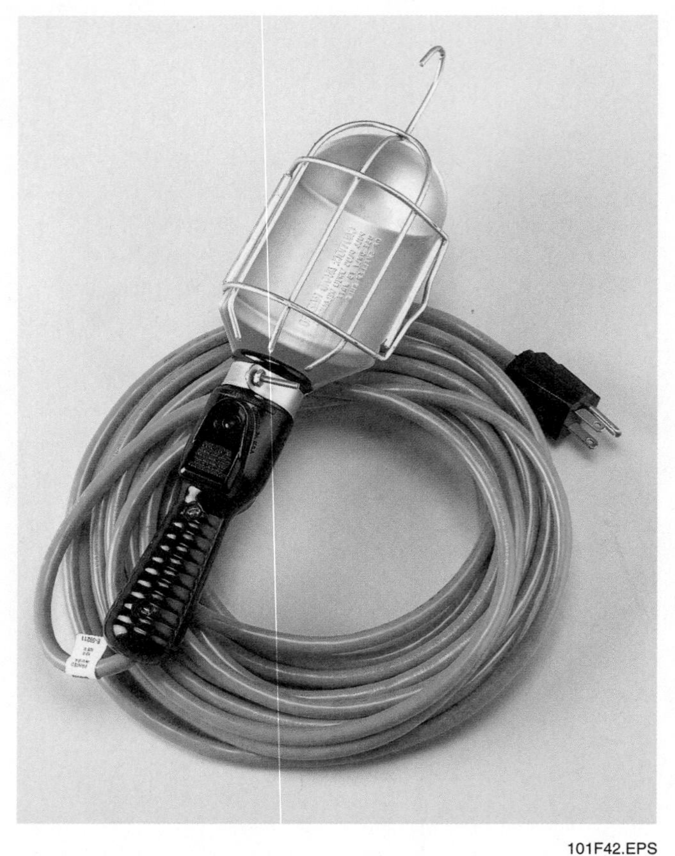

101F42.EPS

Figure 42 Work light with protective guard.

10.3.0 Electrical Power Systems

Grounding is a method of protecting employees from electric shock; however, it is normally a secondary protective measure. The term *ground* refers to a conductive body, usually the earth. A ground is a conductive connection, whether intentional or accidental, by which an electric circuit or equipment is connected to earth or the ground plane. By grounding a tool or electrical system, a low-resistance path to the earth is intentionally created. When properly done, this path offers sufficiently low resistance and has sufficient current-carrying capacity to prevent the buildup of voltages that could result in a personnel hazard. This does not guarantee that no one will receive a shock, be injured, or be killed. It will, however, substantially reduce the possibility of such accidents, especially when used in combination with your company's safety program.

10.3.1 Assured Equipment Grounding Conductor Programs

Your company may have a company/site assured equipment grounding conductor program, covering all cord sets, receptacles that are not a part of the building or structure, and equipment connected by cord and plug that is available for use or used by employees. Each cord set, attachment cap, plug and receptacle of cord sets, and any

On-Site

Electrical Cord Safety

Electrical cords are frequently seen on construction sites, yet they are often overlooked. Use the following safety guidelines to ensure your safety and the safety of other workers.

- Every electrical cord should have an Underwriters Laboratory (UL) label attached to it. Check the UL label for specific wattage. Do not plug more than the specified number of watts into an electrical cord.
- A cord set not marked for outdoor use is to be used indoors only. Check the UL label on the cord for an outdoor marking.
- Do not remove, bend, or modify any metal prongs or pins of an electrical cord.
- Extension cords used with portable tools and equipment must be three-wire type and designated for hard or extra-hard use. Check the UL label for the cord's use designation.
- Avoid overheating an electrical cord. Make sure the cord is uncoiled, and that it does not run under any covering materials, such as tarps, insulation rolls, or lumber.
- Do not run a cord through doorways or through holes in ceilings, walls, and floors, which might pinch the cord. Also, check to see that there are no sharp corners along the cord's path. Any of these situations will lead to cord damage.
- Extension cords are a tripping hazard. They should never be left unattended and should always be put away when not in use.

equipment connected by cord and plug, except cord sets and receptacles that are fixed and not exposed to damage, must be visually inspected before each day's use. Check for external defects, such as deformed or missing pins or insulation damage, and for indication of possible internal damage. If your employer has such a program, you must become familiar with it. Each company or employer should have a written description of the program, including the specific procedures adopted by the employer, available at the job site for inspection by any employee. There must be one or more designated competent persons to implement the program.

The following tests must be performed on all cord sets, receptacles that are not part of the permanent building or structure, and cord- and plug-connected equipment required to be grounded:

- All equipment grounding conductors must be tested for continuity and must be electronically continuous.
- Each receptacle and attachment cap or plug must be tested for correct attachment of the equipment grounding conductor. The equipment grounding conductor must be connected to its proper terminal.

Remember to perform all required electrical equipment tests before the first use and before equipment is returned to service following any repairs. Equipment should also be checked after any incident that can be reasonably suspected to have caused damage (for example, when a cord set is run over) or at least every three months, with the exception of those cord sets and receptacles that are fixed and not exposed to damage, which must be tested at least every six months.

The employer must not make available or permit employees to use any equipment that has not met these requirements. Tests performed as required in this section must be recorded. This test record must identify each receptacle, cord set, and cord- and plug-connected equipment that passed the test and must indicate the last date it was tested or the interval for which it was tested. This record will be kept by means of logs, color coding, other coding, or other effective means and must be made available on the job site for inspection by any affected employee.

Not every company has an assured equipment grounding program. Some use ground-fault circuit interrupters instead. You must know which of these systems your company uses. If you do not know, ask your supervisor.

10.3.2 Ground Fault Circuit Interrupters

A good method of protection from accidental electrocution is the use of a ground fault circuit interrupter. The GFCI is designed to shut off electrical power within as little as 1⁄40 of a second. It works by comparing the amount of current going to electrical equipment against the amount of current returning from the equipment along the circuit connectors. If the current difference exceeds five milliamperes, the GFCI interrupts the current quickly enough to prevent electrocution. The GFCI is used in high-risk areas such as wet locations and construction sites. A GFCI must be used on all receptacles that are not part of the building's permanent wiring, such as temporary power and extension cords. It should be a standard piece of equipment for all construction employees.

10.4.0 Lockout/Tagout

A **lockout/tagout** system safeguards workers from hazardous energy while they work with machines and equipment. A lockout/tagout system protects workers from hazards such as the following:

- Acids
- Air pressure
- Chemicals
- Electricity
- Flammable liquids
- High temperatures
- Hydraulics
- Machinery
- Steam
- Other forms of energy

When people are working on or around any of these hazards, mechanical and other systems are shut down, drained, or de-energized. Tags and locks are placed on each switch, circuit breaker,

Case History

A journeyman HVAC worker was installing metal ductwork using a double-insulated drill connected to a drop-light cord. Power was supplied through two extension cords from a nearby residence. The individual's perspiration-soaked clothing/body contacted exposed conductors on one of the cords, causing an electrocution. No GFCIs were used. Additionally, the ground prongs were missing from the two cords.

valve, or other component to make sure that motors aren't started, valves aren't opened or closed, and no other changes are made that would endanger workers. Lockouts and tagouts protect workers from all possible sources of energy, including electrical, mechanical, hydraulic, thermal, pneumatic (air), and high temperature.

Generally, each lock has its own key, and the person who puts the lock on keeps the key. That person is the only one who can remove the lock. Tags have the words DANGER or DO NOT REMOVE (*Figure 43*).

Follow these rules for a safe lockout/tagout system:

- Never operate any device, valve, switch, or piece of equipment that has a lock or a tag attached to it.
- Use only tags that have been approved for your job site.
- If a device, valve, switch, or piece of equipment is locked out, make sure the proper tag is attached.
- Lock out and tag all electrical systems when they are not in use.
- Lock out and tag pipelines containing acids, explosive fluids, or high-pressure steam during maintenance or repair.
- Tag motorized vehicles and equipment when they are being repaired and before anyone starts work. Also, disconnect or disable the starting devices.

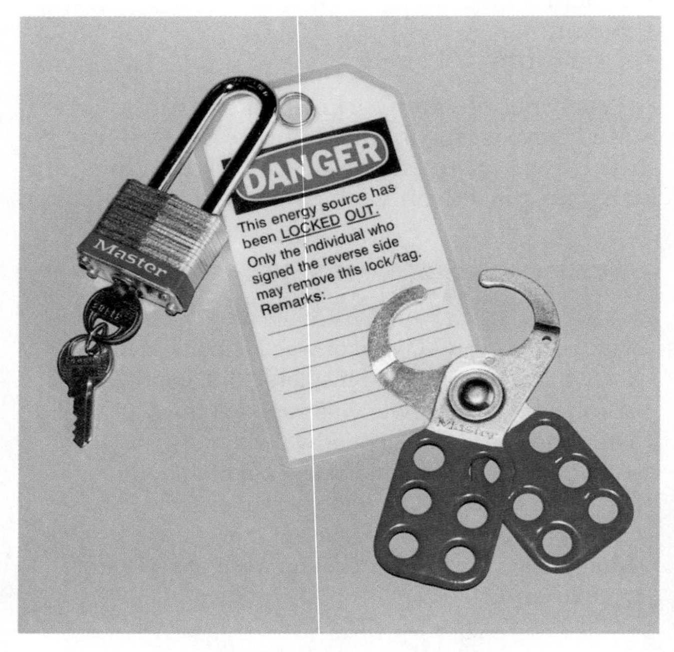

Figure 43 Lockout/tagout safety tags.

The exact procedures for lockout/tagout may vary at different companies and job sites. Ask your supervisor to explain the lockout/tagout procedure on your job site. You must know and follow this procedure. This is for your safety and the safety of your co-workers. If you have any questions about lockout/tagout procedures, ask your supervisor.

10.5.0 Working Near Energized Electrical Equipment

No matter what your trade, your job may include working near exposed electrical equipment or conductors. This is one example of **proximity work**. Often, **electrical distribution panels**, **switch enclosures**, and other equipment must be left open during construction. This leaves the wires and components in them exposed. Some or all of the wires and components may be energized. Working near exposed electrical equipment can be safe, but only if you keep a safe working distance.

Regulations and company policies tell you the minimum safe working distances from exposed conductors. The safe working distance ranges from a few inches to several feet, depending on the voltage. The higher the voltage, the greater the safe working distance.

You must learn the safe working distance for each situation. Make sure you never get any part of your body or any tool you are using closer to exposed conductors than that distance. You can get information on safe working distances from your instructor, your supervisor, company safety policies, and regulatory documents.

10.6.0 If Someone Is Shocked

If you are there when someone gets an electrical shock, you can save a life by taking immediate action. Here's what to do:

Step 1 Immediately disconnect the circuit.

Step 2 If you can't disconnect the circuit, do not try to separate the victim from the circuit. If you touch a person who is being electrocuted, the current will flow through you, too.

> **WARNING!**
> Do not touch the victim or the electrical source with your hand, foot, or any other part of your body or with any object or material. You could become another victim.

Step 3 Once the circuit is disconnected, give first aid and call an ambulance. If you cannot disconnect the circuit, call an ambulance.

11.0.0 Personal Protective Equipment

You are responsible for wearing appropriate personal protective equipment, or PPE, on the job. When worn correctly, PPE is designed to protect you from injury. You must keep it in good condition and use it when you need to. Many workers are injured on the job because they are not using PPE.

11.1.0 Personal Protective Equipment Needs

You will not see all the potentially dangerous conditions just by looking around a job site. It's important to stop and consider what type of accidents could happen on any job that you are about to do. Using common sense and knowing how to use PPE will greatly reduce your chance of getting hurt.

11.2.0 Personal Protective Equipment Use and Care

Remember, while PPE is the last line of defense against personal injury, using it properly and taking care of it are the first steps toward protecting yourself on the job. The best protective equipment is of no use to you unless you do the following four things:

- Regularly inspect it.
- Properly care for it.
- Use properly when it is needed.
- Never alter or modify it in any way.

The sections that follow describe protective equipment commonly used on construction sites and tell how to use and care for each piece of equipment. Be sure to wear the equipment according to the manufacturer's specifications.

11.3.0 Clothing and Jewelry

Your clothing must comply with good general work and safety practices. Do not wear clothing or jewelry that could get caught in machinery or otherwise cause an accident, such as loose clothing, baggy shirts, or dragging pants. You must wear a shirt at all times; some tasks will require long-sleeved shirts. Your shirt should always be tucked in unless you are performing welding.

Did You Know?

Hard hats used to be made of metal. However, metal conducts electricity, so most hard hats are now made of reinforced plastic or fiberglass.

11.4.0 Hard Hat

Figure 44 shows a typical hard hat. The outer shell of the hat can protect your head from a hard blow. The webbing inside the hat keeps space between the shell and your head. Adjust the headband so that the webbing fits your head and there is at least one inch of space between your head and the shell.

Do not alter your hard hat in any way. Inspect your hard hat every time you use it. If there are any cracks or dents in the shell, or if the webbing straps are worn or torn, get a new hard hat. Wash the webbing and headband with soapy water as often as needed to keep them clean. Wear the hard hat only as the manufacturer recommends. Never wear anything under your hard hat.

11.5.0 Eye and Face Protection

Wear eye protection (*Figure 45*) wherever there is even the slightest chance of an eye injury. Eye and face protection must meet the requirements specified in American National Standards Institute (ANSI) *Standard Z87.1-1968* and should be used during the following tasks:

- Grinding and chipping
- Using power saws and other tools/equipment that can throw out solid material
- Working with molten lead, tar pots, and other molten materials
- Working with chemicals, acids, and corrosive liquids
- Arc welding

101F44.EPS

Figure 44 Typical hard hat.

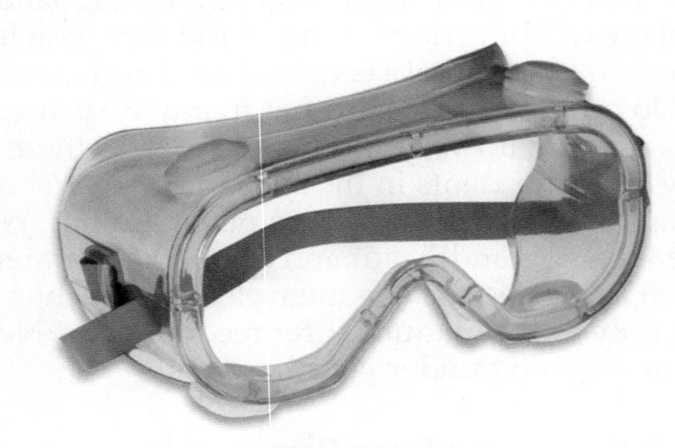

Figure 45 Typical safety glasses, goggles, and full face shields.

On the job site, areas where there are potential eye hazards are usually identified, but you should always be on the lookout for possible hazards.

> **CAUTION**
>
> Standard eyeglasses are not adequate protection in a hazardous work environment.

Regular safety glasses will protect you from falling objects or from objects flying toward your face. You can add side shields for protection from the sides. In some cases, you may need a face shield. Safety goggles give your eyes the best protection from all directions.

> **NOTE**
>
> Handle safety glasses and goggles with care. If they get scratched, replace them. The scratches will interfere with your vision. Clean the lenses regularly with lens tissues or a soft cloth.

Welders must use tinted goggles or welding hoods. The tinted lenses protect the eyes from the bright welding arc or flame. Welders should use filter lenses of not less than a No. 10 shade. Welder's helpers and all employees working in the vicinity of arc welding should not look directly at the welding process and should also use approved eye protection. Gas welding and burning requires the use of a filter lens of not less than a No. 4 shade.

Follow these general precautions for eye care:

- Always report all eye injuries and suspected foreign material in your eye to your supervisor immediately. Do not try to remove foreign matter yourself.
- Keep your hands away from your eyes.
- Keep materials out of your eyes by regularly clearing debris from your hard hat brim, the top of your goggles, and your face shield.
- Flood your eyes with water if you feel something in them. Never rub them, as this can make the problem worse.
- Know the location of eyewash stations and how to use them.

11.6.0 Gloves

On many construction jobs, you must wear heavy-duty gloves to protect your hands (*Figure 46*). Construction work gloves are usually made of cloth, canvas, or leather. Never wear gloves around rotating or moving equipment. They can easily get caught in the equipment.

Gloves help prevent cuts and scrapes when you handle sharp or rough materials. Heat-resistant gloves are sometimes used for hot materials. Electricians use special rubber-insulated gloves when they work on or around live circuits.

Figure 46 Work gloves.

WARNING!

Only specially trained employees are allowed to use dielectrically tested rubber gloves to work on energized equipment. Never attempt this work without proper training and authorization.

Replace gloves when they become worn, torn, or soaked with oil or chemicals.

CAUTION

Keep the inside of gloves free of chemicals.

Electrician's rubber-insulated gloves should be tested regularly to make sure they will protect the wearer. After visually inspecting rubber-insulated gloves, other defects may be observed by applying the following air test (*Figure 47*):

Step 1 Stretch the glove.

Step 2 To trap air inside, either twirl the glove around quickly, or roll it down from the glove gauntlet.

Step 3 Trap the air by squeezing the gauntlet with one hand. Use the other hand to squeeze the palm, fingers, and thumb to check for weaknesses and defects.

Step 4 Hold the glove up to your ear to try to detect any escaping air.

Step 5 If the glove does not pass this inspection, it must be turned in for disposal.

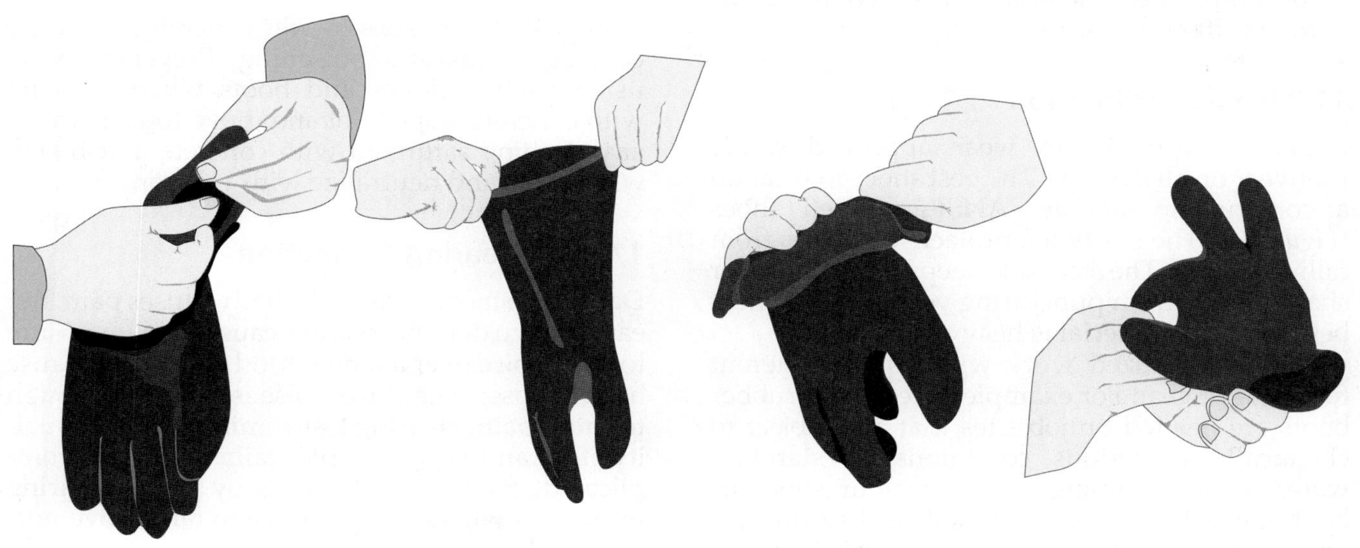

Figure 47 Glove inspection.

11.7.0 Leg Protection

In construction, your legs are at risk for strains, sprains, cuts, bruises, crushing, amputation injuries, and more. Following the appropriate work procedures and using the correct PPE will greatly reduce your chances of injury.

Overalls or pants should not have loose, torn, or dragging fabric that can become caught on objects or pose a tripping hazard. Ideally, you should wear pants without cuffs; this reduces the risk of being caught on something or having a hot or burning object caught in your cuff. Never wear shorts on the job site. Besides the greater risk of injuries, shorts expose your skin to the sun and a number of chemicals and substances on the job site that can cause skin irritations.

WARNING!

Always tape your pants into rubber boot tops when working in concrete or with chemicals.

Remember these general guidelines relating to leg protection:

- Never carry pointed tools, such as scissors or shears, in your pants pockets. Use a canvas or leather tool sheath with all sharp ends pointing down.
- When using certain special equipment, such as chainsaws and brush hooks, use shin guards.
- Consider stability when stepping into or onto locations where materials are stored. Materials may shift and pinch your legs and/or feet.

11.8.0 Foot Protection

Generally, you should wear approved safety footwear on all job sites. The best shoes to wear on a construction site are ANSI-approved shoes (*Figure 48*). The safety toe protects your toes from falling objects. The steel sole keeps nails and other sharp objects from puncturing your feet. The next best footwear material is heavy leather.

Some specialized work will require different footwear or gear. For example, safety-toed rubber boots are needed on job sites that are subject to chemically hazardous conditions or standing water. When climbing a ladder, your shoes or boots must have a well-defined heel to prevent your feet from slipping off the rungs. You will need foot guards when using jackhammers and similar equipment.

101F48.EPS

Figure 48 Safety shoe.

Never wear canvas shoes or sandals on a construction site. They do not provide adequate protection. Always replace boots or shoes when the sole tread becomes worn or the shoes have holes, even if the holes are on top. Don't wear oil-soaked shoes when you are welding, because of the risk of fire.

11.9.0 Skin Protection

Skin is susceptible to dermatitis (skin irritation) caused by exposure to chemicals that strip oils from the skin. Avoid this condition by using gloves and avoiding exposure to chemicals. Use skin creams if you notice your skin drying out, and notify your supervisor immediately.

CAUTION

Always protect your skin from burns, the sun, arc welding, and chemical exposure.

Repeated exposure to wet concrete can cause concrete burns and poisoning. Prevent this by using rubber gloves and boots when working with concrete. Tape boot and glove tops. Remove any clothing saturated with concrete. Flush skin with water and neutralize with vinegar.

11.10.0 Hearing Protection

Damage to most parts of the body causes pain. But ear damage does not always cause pain. Exposure to loud noise over a long period of time can cause hearing loss, even if the noise is not loud enough to cause pain. Hearing loss diminishes your quality of life and makes simple, daily tasks more complicated. Save your hearing by using hearing protection whenever you have to talk above normal levels.

> **CAUTION**
>
> If you have to raise your voice to be heard by someone who is less than two feet away, you need to wear hearing protection.

Most construction companies follow OSHA rules in deciding when hearing protection must be used. One type of hearing protection is specially designed earplugs that fit into your ears and filter out noise (*Figure 49*). Clean earplugs regularly with soap and water to prevent ear infection.

Another type of hearing protection is earmuffs, which are large padded covers for the entire ear (*Figure 50*). You must adjust the headband on earmuffs for a snug fit. If the noise level is very high, you may need to wear both earplugs and earmuffs.

Noise-induced hearing loss can be prevented by using noise control measures and personal protective devices. *Table 5* shows the recommended maximum length of exposure to sound levels rated 90 decibels and higher. When noise levels exceed those outlined in *Table 5*, an effective hearing conservation program is required. A company-appointed program administrator will oversee this program. If you have questions about the hearing conservation program on your site, see your supervisor or the program administrator.

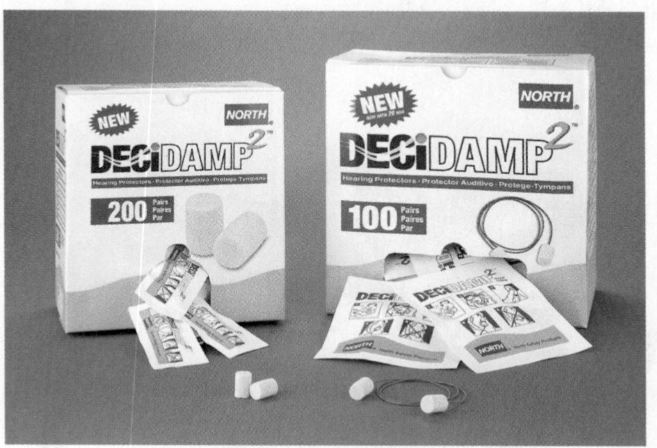

101F49.EPS

Figure 49 Earplugs for hearing protection.

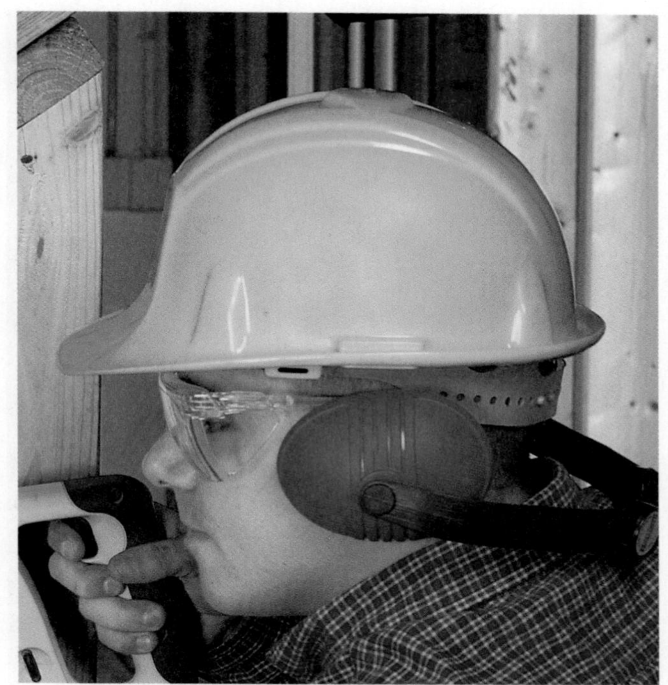
101F50.EPS

Figure 50 Earmuffs for hearing protection.

Table 5 Maximum Noise Levels

Sound Level (decibels)	Maximum Hours of Continuous Exposure per Day	Examples
90	8	Power lawn mower
92	6	Belt sander
95	4	Tractor
97	3	Hand drill
100	2	Chain saw
102	1.5	Impact wrench
105	1	Spray painter
110	0.5	Power shovel
115	0.25 or less	Hammer drill

11.11.0 Respiratory Protection

Wherever there is danger of an inhalation hazard, you must use a respirator. Federal law specifies which type of respirator to use for different types of hazards. There are four general types of respirators (*Figure 51*):

- Self-contained breathing **apparatus** (SCBA)
- Supplied air mask
- Full facepiece mask with chemical canister (gas mask)
- Half mask or mouthpiece with mechanical filter

A SCBA carries its own air supply in a compressed air tank. It is used where there is not enough oxygen or where there are dangerous gases or fumes in the air.

A supplied air mask uses a remote compressor or air tank to provide breathable air in oxygen-deficient atmospheres. Supplied air masks can be used under the same conditions as SCBAs.

(A) SELF-CONTAINED BREATHING APPARATUS (SCBA)

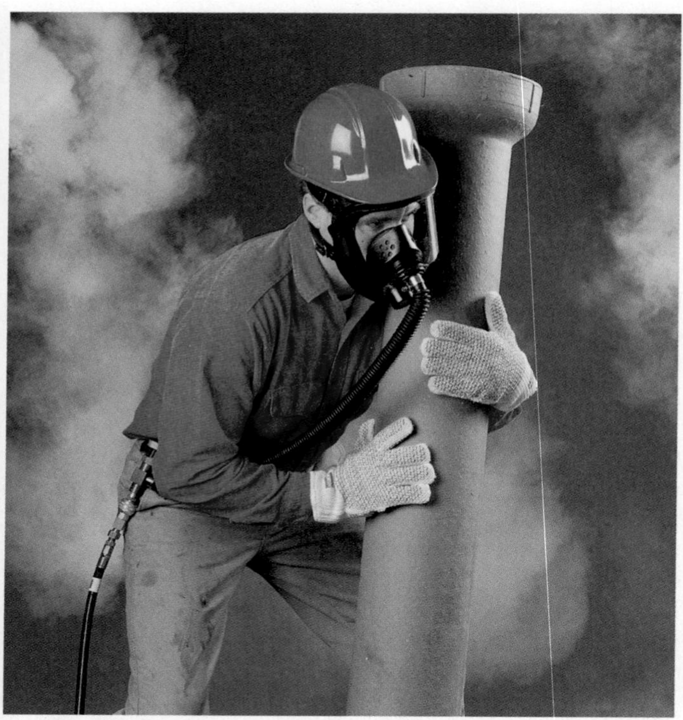

(B) SUPPLIED AIR MASK

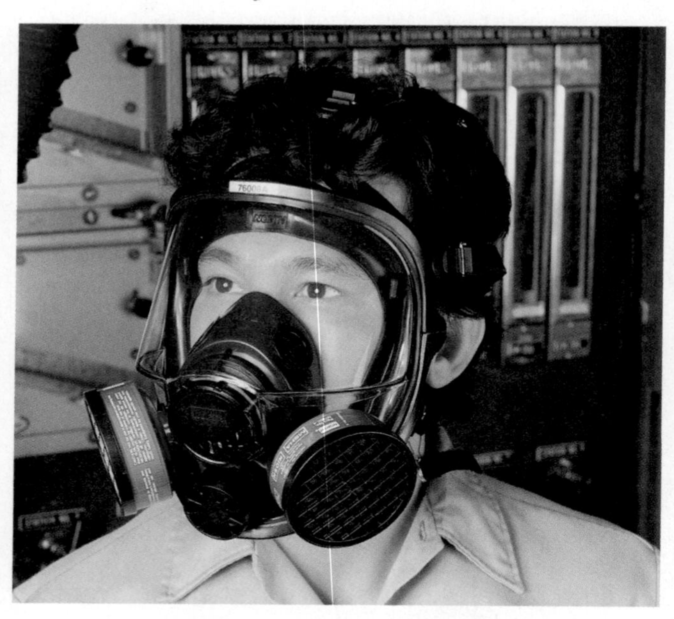

(C) FULL FACEPIECE MASK

(D) HALF MASK

101F51.EPS

Figure 51 Examples of respirators.

A full facepiece mask with chemical canisters is used to protect against brief exposure to dangerous gases or fumes. A half mask or mouthpiece with a mechanical filter is used in areas where you might inhale dust or other solid particles.

11.11.1 Respirator Requirements

If you need to use a respirator, your employer must provide you with the appropriate training to select, test, wear, and maintain this equipment. Your employer is also responsible for having you medically evaluated to ensure you are fit enough to wear respiratory protection equipment without being harmed and tested to ensure a proper fit and a good seal.

All respirators generally place a burden on the employee. Self-contained breathing apparatuses are heavy and can be burdensome to carry, negative pressure respirators restrict breathing, and other respirators may cause claustrophobia. These conditions may adversely affect the health of some employees who wear respirators. Employees must be medically evaluated by a physician or other licensed health care professional to determine under what conditions they may safely wear respirators.

WARNING!

When a respirator is required, a personal monitoring device is usually also required. This device samples the air to measure the concentration of hazardous chemicals.

11.11.2 Selecting Respirators

Follow company and OSHA procedures when choosing the type of respirator for a particular job. Also be sure that it is safe for you to wear a respirator. Under *OSHA's Respiratory Protection Standard,* workers must fill out a questionnaire to determine any potential problems in wearing a respirator. Depending on the answers, a medical exam may be required. When a respirator is not required, workers may voluntarily use a dust or particle mask for general protection. These masks do not require fit testing or a medical examination.

Always use the appropriate respiratory protective device for the hazardous material involved and the extent and nature of the work to be performed. Before using a respirator, you must determine the type and concentration level of the contaminant, and whether the respirator can be properly fitted on your face.

CAUTION

All respirator instructions, warnings, and use limitations contained on each package must be read and understood by the wearer before use.

11.11.3 Testing Respirators

It is very important to check a respirator carefully for damage and for proper fit. A leaking facepiece can be as dangerous as not wearing a respirator at all. All respirators must be fitted properly, and their facepiece-to-face seal must be checked each time the respirator is used. When conditions prevent a good seal, the respirator cannot be worn. The following conditions will interfere with the respirator's seal:

- Facial hair (such as sideburns or beards)
- Skullcaps that project under the facepiece
- Temple bars on glasses (especially when wearing full-face respirators)
- Absence of upper, lower, or all teeth
- Absence of dentures
- Gum and tobacco chewing

To obtain the best protection from your respirator, you must perform positive and negative fit checks each time you wear it. These fit checks must be repeated until you have obtained a good face seal. To perform a positive fit check, do the following:

Step 1 Adjust the facepiece for the best fit; then adjust the head and neck straps to ensure good fit and comfort.

Step 2 Block the exhalation valve with your hand or other material.

Step 3 Breathe out into the mask.

Step 4 Check for air leakage around the edges of the facepiece.

Step 5 If the facepiece puffs out slightly for a few seconds, a good face seal has been obtained.

To perform a negative fit check, do the following:

Step 1 Block the inhalation valve with your hand or other material.

Step 2 Attempt to inhale.

Step 3 Check for air leakage around the edges of the facepiece.

Step 4 If the facepiece caves in slightly for a few seconds, a good face seal has been obtained.

On-Site

OSHA no longer bans the wearing of contact lenses with respirators (*29 CFR 1910.134 [g][1][ii]*). After sponsoring research and studies on the issue, OSHA recently concluded that wearing contact lenses with respirators does not pose an increased risk to the wearer's safety.

If you wear contact lenses, practice wearing a respirator with your contact lenses to see whether you have any problems. That way, you will identify any problems before you use the respirator under hazardous conditions.

11.11.4 Inspecting Respirators

Respirators should be inspected before and after each use and checked at least monthly, even if they are not used. Check the respirator function, tightness of connections, and the condition of various parts including:

- Facepiece
- Head straps
- Valves
- Connecting tube
- Threads
- Cartridges (if applicable)

Look for cracks, tears, holes, and excessive dirt on all parts, and check the elastic parts for inflexibility and signs of deterioration.

11.11.5 Maintaining Respirators

A respirator must be clean and in good condition, with all parts in place, for it to give you the proper protection. Respirators used by only one person should be cleaned after each day of use, or more often if necessary. Those used by more than one person should be cleaned and disinfected (made germ-free) after each use. Follow these guidelines for cleaning your respirator:

- Remove the cartridge and filter, hand wash the respirator using mild soap and a soft brush, and let it air dry overnight. Do not clean respirators with solvents.
- Sanitize your respirator each week. Remove the cartridge and filter, then soak the respirator in sanitizing solution for at least two minutes. Thoroughly rinse with warm water, and let it air dry overnight.
- Store your respirator in its resealable plastic bag. Do not store respirators face down. This will cause distortion to the facepiece.

12.0.0 HAZARD COMMUNICATION STANDARD

OSHA has a rule that affects every worker in the construction industry. It is called the *Hazard Communication Standard (HazCom)*. You may have heard it called the *Right to Know* requirement. It requires all contractors to educate their employees about the hazardous chemicals they may be exposed to on the job site. Employees must be taught how to work safely around these materials.

Many people think that there are very few hazardous chemicals on construction job sites. That isn't true. In the OSHA standard, the term *hazardous chemical* applies to paint, concrete, and even wood dust, as well as other substances.

12.1.0 Material Safety Data Sheets

A Material Safety Data Sheet (MSDS) must accompany every shipment of a hazardous substance and must be available to you on the job site. Use the MSDS to manage, use, and dispose of hazardous materials safely. *Figure 52* shows a typical MSDS.

The information found on an MSDS includes the following:

- Identity of the substance
- Exposure limits
- Physical and chemical characteristics of the substance
- Type of hazard the substance presents
- Precautions for safe handling and use
- Reactivity of the substance
- Specific control measures
- Emergency first-aid procedures
- Manufacturer contact information

12.2.0 Your Responsibilities Under HazCom

As an employee, you have the following responsibilities under HazCom:

- Know where MSDSs are on your job site.
- Report any hazards you spot on the job site to your supervisor.
- Know the physical and health hazards of any hazardous materials on your job site, and know and practice the precautions needed to protect yourself from these hazards.
- Know what to do in an emergency.
- Know the location and content of your employer's written hazard communication program.

The final responsibility for your safety rests with you. Your employer must provide you with information about hazards, but you must know this information and follow safety rules.

Material Safety Data Sheet

1 - Chemical Product and Company Identification

Manufacturer: WD-40 Company	**Chemical Name: Organic Mixture**
Address: 1061 Cudahy Place (92110) P.O. Box 80607 San Diego, California, USA 92138 –0607	**Trade Name: WD-40 Aerosol** **Product Use: Cleaner, Lubricant, Penetrant**
Telephone: 1-800-448-9340 Emergency only: 1-888-324-7596 (PROZAR) Information: 1-888-324-7596	**MSDS Date Of Preparation: 5/16/07**

2 – Hazards Identification

Emergency Overview:
DANGER! Harmful or fatal if swallowed. Flammable aerosol. Contents under pressure. Avoid eye contact. Use with adequate ventilation. Keep away from heat, sparks and all other sources of ignition.

Symptoms of Overexposure:
Inhalation: High concentrations may cause nasal and respiratory irritation and central nervous system effects such as headache, dizziness and nausea. Intentional abuse may be harmful or fatal.
Skin Contact: Prolonged and/or repeated contact may produce mild irritation and defatting with possible dermatitis.
Eye Contact: Contact may be mildly irritating to eyes. May cause redness and tearing.
Ingestion: This product has low oral toxicity. Swallowing may cause gastrointestinal irritation, nausea, vomiting and diarrhea. The liquid contents are an aspiration hazard. If swallowed, can enter the lungs and may cause chemical pneumonitis.
Chronic Effects: None expected.
Medical Conditions Aggravated by Exposure: Preexisting eye, skin and respiratory conditions may be aggravated by exposure.

Suspected Cancer Agent:
Yes No X

3 - Composition/Information on Ingredients

Ingredient	CAS #	Weight Percent
Aliphatic Hydrocarbon	64742-47-8 64742-48-9 64742-88-7	45-50
Petroleum Base Oil	64742-65-0	15-25
LVP Aliphatic Hydrocarbon	64742-47-8	12-18
Carbon Dioxide	124-38-9	2-3
Non-Hazardous Ingredients	Mixture	<10

4 – First Aid Measures

Ingestion (Swallowed): Aspiration Hazard. DO NOT induce vomiting. Call physician, poison control center or the WD-40 Safety Hotline at 1-888-324-7596 immediately.
Eye Contact: Flush thoroughly with water. Get medical attention if irritation persists.
Skin Contact: Wash with soap and water. If irritation develops and persists, get medical attention.

Page 1 of 4

101F52A.EPS

Figure 52 WD-40 MSDS. (1 of 4)

Inhalation (Breathing): If irritation is experienced, move to fresh air. Get medical attention if irritation or other symptoms develop and persist.

5 – Fire Fighting Measures

Extinguishing Media: Use water fog, dry chemical, carbon dioxide or foam. Do not use water jet or flooding amounts of water. Burning product will float on the surface and spread fire.
Special Fire Fighting Procedures: Firefighters should always wear positive pressure self-contained breathing apparatus and full protective clothing. Cool fire-exposed containers with water. Use shielding to protect against bursting containers.
Unusual Fire and Explosion Hazards: Contents under pressure. Aerosol containers may burst under fire conditions. Vapors are heavier than air and may travel along surfaces to remote ignition sources and flash back.

6 – Accidental Release Measures

Wear appropriate protective clothing (see Section 8). Eliminate all sources of ignition and ventilate area. Leaking cans should be placed in a plastic bag or open pail until the pressure has dissipated. Contain and collect liquid with an inert absorbent and place in a container for disposal. Clean spill area thoroughly. Report spills to authorities as required.

7 – Handling and Storage

Handling: Avoid contact with eyes. Avoid prolonged contact with skin. Avoid breathing vapors or aerosols. Use with adequate ventilation. Keep away from heat, sparks, hot surfaces and open flames. Wash thoroughly with soap and water after handling. Do not puncture or incinerate containers. Keep can away from electrical current or battery terminals. Electrical arcing can cause burn-through (puncture) which may result in flash fire, causing serious injury. Keep out of the reach of children.
Storage: Do not store above 120°F or in direct sunlight. U.F.C (NFPA 30B) Level 3 Aerosol.

8 – Exposure Controls/Personal Protection

Chemical	Occupational Exposure Limits
Aliphatic Hydrocarbon	100 ppm TWA (ACGIH) 1200 mg/m3 TWA (manufacturer recommended)
Petroleum Base Oil	5 mg/m3 TWA (OSHA/ACGIH)
LVP Aliphatic Hydrocarbon	1200 mg/m3 TWA (manufacturer recommended)
Carbon Dioxide	5000 ppm TWA (OSHA/ACGIH), 30,000 ppm STEL (ACGIH)
Non-Hazardous Ingredients	None Established

The Following Controls are Recommended for Normal Consumer Use of this Product
Engineering Controls: Use in a well-ventilated area.
Personal Protection:
Eye Protection: Avoid eye contact. Safety glasses or goggles recommended.
Skin Protection: Avoid prolonged skin contact. Chemical resistant gloves recommended for operations where skin contact is likely.
Respiratory Protection: None needed for normal use with adequate ventilation.

For Bulk Processing or Workplace Use the Following Controls are Recommended
Engineering Controls: Use adequate general and local exhaust ventilation to maintain exposure levels below that occupational exposure limits.
Personal Protection:
Eye Protection: Safety goggles recommended where eye contact is possible.
Skin Protection: Wear chemical resistant gloves.
Respiratory Protection: None required if ventilation is adequate. If the occupational exposure limits are exceeded, wear a NIOSH approved respirator. Respirator selection and use should be

Page 2 of 4

101F52B.EPS

Figure 52 WD-40 MSDS. (2 of 4)

based on contaminant type, form and concentration. Follow OSHA 1910.134, ANSI Z88.2 and good Industrial Hygiene practice.
Work/Hygiene Practices: Wash with soap and water after handling.

9 – Physical and Chemical Properties

Boiling Point:	323°F (minimum)	Specific Gravity:	0.817 @ 72°F
Solubility in Water:	Insoluble	pH:	Not Applicable
Vapor Pressure:	110 PSI @ 70°F	Vapor Density:	Greater than 1
Percent Volatile:	74%	VOC:	412 grams/liter (49.5%)
Coefficient of Water/Oil Distribution:	Not Determined	Appearance/Odor	Light amber liquid/mild odor
Flash Point:	131°F (concentrate) Tag Closed Cup	Flammable Limits: (Solvent Portion)	LEL: 1.1% UE:: 8.9%

10 – Stability and Reactivity

Stability: Stable
Hazardous Polymerization: Will not occur.
Conditions to Avoid: Avoid heat, sparks, flames and other sources of ignition. Do not puncture or incinerate containers.
Incompatibilities: Strong oxidizing agents.
Hazardous Decomposition Products: Carbon monoxide and carbon dioxide.

11 – Toxicological Information

The oral toxicity of this product is estimated to be greater than 5,000 mg/kg based on an assessment of the ingredients. This product is not classified as toxic by established criteria. It is an aspiration hazard.
None of the components of this product is listed as a carcinogen or suspected carcinogen or is considered a reproductive hazard.

12 – Ecological Information

No data is currently available.

13 - Disposal Considerations

If this product becomes a waste, it would be expected to meet the criteria of a RCRA ignitable hazardous waste (D001). However, it is the responsibility of the generator to determine at the time of disposal the proper classification and method of disposal. Dispose in accordance with federal, state, and local regulations.

14 – Transportation Information_

DOT Surface Shipping Description: Consumer Commodity, ORM-D
IMDG Shipping Description: Aerosols, 2, UN1950

15 – Regulatory Information

U.S. Federal Regulations:
CERCLA 103 Reportable Quantity: This product is not subject to CERCLA reporting requirements, however, oil spills are reportable to the National Response Center under the Clean Water Act and many states have more stringent release reporting requirements. Report spills required under federal, state and local regulations.
SARA TITLE III:
Hazard Category For Section 311/312: Acute Health, Fire Hazard, Sudden Release of Pressure

Page 3 of 4

101F52C.EPS

Figure 52 WD-40 MSDS. (3 of 4)

Section 313 Toxic Chemicals: This product contains the following chemicals subject to SARA Title III Section 313 Reporting requirements: None
Section 302 Extremely Hazardous Substances (TPQ): None
EPA Toxic Substances Control Act (TSCA) Status: All of the components of this product are listed on the TSCA inventory
Canadian Environmental Protection Act: All of the ingredients are listed on the Canadian Domestic Substances List or exempt from notification
Canadian WHMIS Classification: Class B-5 (Flammable Aerosol)
This MSDS has been prepared according to the criteria of the Controlled Products Regulation (CPR) and the MSDS contains all of the information required by the CPR.

16 – Other Information:

HMIS Hazard Rating:
Health – 1 (slight hazard), Fire Hazard – 4 (severe hazard), Reactivity – 0 (minimal hazard)

SIGNATURE: [signature] TITLE: Director of Global Quality Assurance

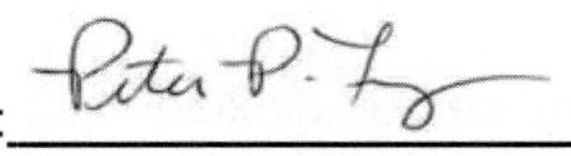

REVISION DATE: Revision Date: May 2007 SUPERSEDES: December 2004

Page 4 of 4

101F52D.EPS

Figure 52 WD-40 MSDS. (4 of 4)

13.0.0 Other Job-Site Hazards

It's impossible to list all the hazards that can exist on a construction job site. This section describes some of the more common hazards and explains how to deal with them.

For your safety, you must know the specific hazards where you are working and how to prevent accidents and injuries. If you have questions specific to your job site, ask your supervisor.

13.1.0 Job-Site Exposures

You will frequently hear the term *exposure* in discussions about safety hazards and risk analysis. Exposure refers to contact with a chemical, biological, or physical hazard. Exposure can be chronic or acute. Chronic exposure is long-term and repeated; it may be mild or severe. Acute exposure is short-term and intense.

According to the HazCom standard, your employer must inform you of any hazards to which you might be exposed. In order to better protect you, your employer may require pre-employment and periodic medical examinations to ensure that your health is not being negatively affected by chemical exposure during work. Periodic exams are usually conducted annually, and generally involve target organ testing. Because most medical tests cannot directly check for specific chemical exposure, blood tests are used to check the status of the particular organs (called target organs) known to be most affected by the chemicals to which you have been exposed. The results of these tests are compared to pre-employment test results and any prior annual or other period exams you may have had. Employers also frequently request a final screening when you leave their company.

You can be exposed to chemicals and other hazardous materials in a variety of ways. Routes of exposure include breathing (inhalation), eating or drinking (ingestion), and skin contact (absorption). OSHA has specific guidelines with respect to chemical exposure. There are strict regulations about exposure, including permissible exposure limits (PELs). A PEL is the maximum concentration of a substance that a worker can be exposed to in an eight-hour shift.

There are various types of hazardous materials you may come in contact with during various types of construction work. Some of these hazardous materials include:

- Lead
- Asbestos
- Silica
- Bloodborne pathogens
- Chemicals

Taking the time to understand each of these hazards will help you stay healthy and safe. Working around these hazards requires special training and personal protective equipment. Never handle any of these materials without proper training, authorization, and PPE.

13.1.1 Lead

Lead occurs naturally in the earth's crust and is spread throughout the environment through human activities, such as burning fossil fuels and mining. As an element, lead is indestructible—it does not break down. Once it is released into the environment, it can move from one medium to another. For example, lead in dust can be carried long distances, dissolve in water, and find its way into soil where it can remain for years. Lead is difficult to detect because it has no distinctive taste or smell.

Lead has many useful properties. It is soft and easily shaped, durable, resistant to some chemicals, and fairly common. Because lead is so versatile, it is used in the production of piping, batteries, and casting metals.

However, lead is a toxic metal that can cause serious health problems. You can be exposed to lead by breathing air, drinking water, eating food, and swallowing or touching dust or dirt that contains lead.

You may encounter lead-based paints during demolition or renovation of buildings constructed prior to 1978, when lead-based paint was banned. Dust created from sanding lead-based paints is hazardous. Protective clothing and equipment must always be used when lead levels are above the permissible exposure limit. If you are unsure whether lead is present, consult your supervisor.

All waste contaminated with lead is considered hazardous. Those who handle hazardous waste must have special training. Never handle hazardous materials or waste without proper training, authorization, and PPE.

13.1.2 Asbestos

Asbestos is a hazardous material that can be harmful to your lungs. Prolonged exposure can cause lung cancer, asbestosis (scarring of the lung tissue), and a cancer called mesothelioma. It may take more than 20 years for these diseases to develop. Smoking significantly increases the risk of lung disease. If you smoke, tar from cigarettes will stick to the asbestos fibers in the lungs, making it more difficult for your body to get rid of the asbestos.

Sources of asbestos include older insulation and other building materials, such as floor tiles, pipe insulation drywall compounds, reinforced plaster, and some roofing materials. The U.S. banned production of most asbestos products in the 1970s, meaning that asbestos-containing materials (ACMs) are generally found in structures built before 1980. During renovation or maintenance operations, asbestos may be dislodged and become airborne. Although you will not feel the effects immediately, this exposure can be a serious health hazard. Never handle asbestos without proper training, authorization, and PPE.

You must have special training before performing any job that may deal with asbestos, including, but not limited to, the following:

- Demolition
- Removal
- Alteration
- Repair
- Maintenance
- Installation
- Cleanup
- Transportation
- Disposal
- Storage

13.1.3 Silica

Silica is a mineral found in concrete, masonry, and rock. During construction, silica may be found in a dust form. Prolonged exposure to silica dust could cause silicosis and lung cancer. Silicosis is an incurable, and sometimes fatal, lung disease. The time it takes for silicosis to develop varies depending on how long the exposure lasted and how much silica the person was exposed to.

When you are working in an area where silica dust is present, you must use the appropriate respiratory protection. Never work around silica dust without proper training, authorization, and PPE.

13.1.4 Bloodborne Pathogens

Bloodborne pathogens are another health hazard you may encounter on the job site. Sometimes you will hear them referred to as bloodborne infectious diseases. They are those diseases that can be transmitted by contact with an infectious person's blood, the most common being HIV (the virus that causes AIDS) and hepatitis B and C.

You could be exposed to another person's blood on the job site when administering first aid to an injured person or being involved in a multi-victim accident. In order to safeguard yourself, you must know the universal precautions to prevent exposure and follow them closely:

- Always use appropriate gloves when administering first aid.
- If you come in contact with someone's blood, immediately wash the affected areas with soap and water.
- Notify your supervisor of the contact right away.

You may need to seek medical attention for precautionary screening, particularly if that person has any bloodborne infectious diseases.

13.1.5 Chemical Splashes

No matter how careful you are on the job site, there is always the potential for injury when working with hazardous chemicals. If you are splashed with any chemicals, get to the nearest shower, remove the clothing around the affected area, rinse yourself for at least 15 minutes, and then seek immediate medical attention. If your eyes are splashed, use the nearest eyewash station and seek immediate medical attention.

WARNING!

Chemical splashes are serious medical emergencies. Before working with any chemicals, ensure that you know where the nearest shower and eyewash station are, and that they are functioning properly.

Remember to wear appropriate PPE on the job site at all times. This will help guard against your skin and eyes being splashed with chemicals. If you are not sure of the appropriate PPE needed on your job site or for a certain task, consult your supervisor before beginning work.

CAUTION

Never handle any chemicals without proper training and PPE. Review the MSDS for each chemical, and strictly follow MSDS first aid instructions if you suspect exposure.

13.2.0 Proximity Work

Work that is done near a hazard but not in direct contact with it is called proximity work. Proximity work requires extra caution and awareness. The hazard may be hot piping, energized electrical equipment, or running motors or machinery (*Figure 53*). You must do your work so that you do not come into contact with the nearby hazard.

You may need to put up barricades to prevent accidental contact. Lifting and rigging operations may have to be done in a way that minimizes the risk of dropping things on the hazard. A monitor may watch you and alert you if you are in danger of touching the hazard while you work.

Energized electrical equipment is very hazardous. Regulations and policies will tell you the minimum safe working distance from energized electrical conductors.

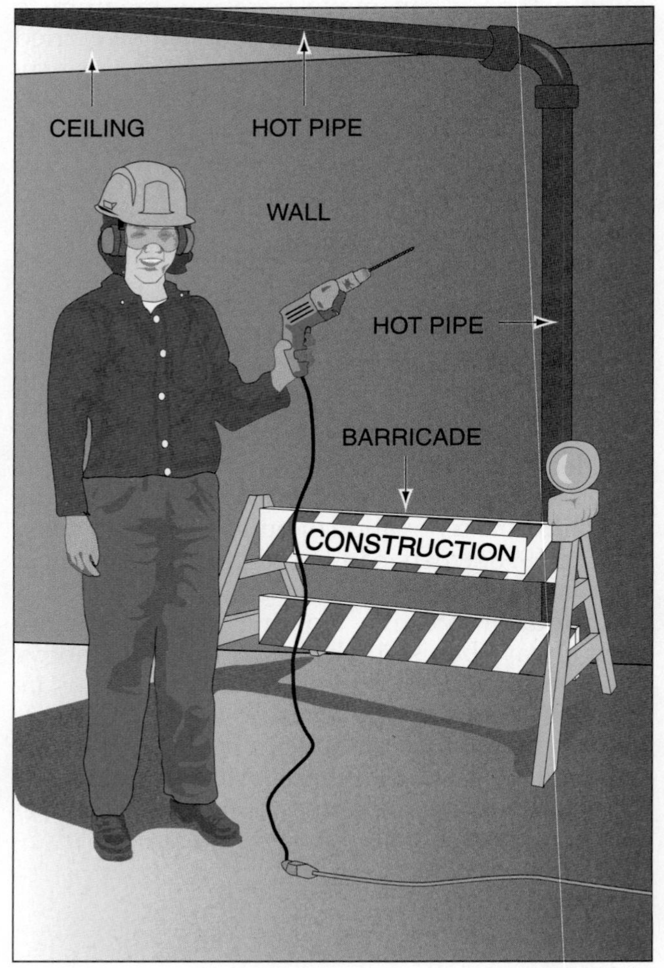

Figure 53 Proximity work.

13.2.1 Pressurized or High-Temperature Systems

In many construction jobs, you must work close to tanks, piping systems, and pumps that contain pressurized or high-temperature fluids. Be aware that touching a container of high-temperature fluid can cause burns (*Figure 54*). Many industrial processes involve fluids that are as hot as several thousand degrees. Also, if a container holding pressurized fluids is damaged, it may leak and spray dangerous fluids.

Any work around pressurized or high-temperature systems is proximity work. Barricades, a monitor, or both may be needed for safety (*Figure 55*).

13.3.0 Heat Stress

Heat stress occurs when abnormally hot air and/or high humidity or extremely heavy exertion prevents your body from cooling itself fast enough. When this happens, you may suffer a heat stroke, heat exhaustion, or heat cramps. To prevent heat stress, take the following precautions:

- Drink plenty of water.
- Avoid alcoholic or caffeinated drinks.
- Do not overexert yourself.
- When possible, perform the most strenuous work during cooler parts of the day.
- Wear lightweight, light-colored clothing.
- Wear loose-fitting cotton clothing if it does not create a hazard.
- Keep your head covered and face shaded.
- Take frequent, short breaks.
- Rest in the shade whenever possible.

> **NOTE**
> Drink one to one-and-a-half gallons of water per day when working in hot conditions. Start by having one or two glasses of water before beginning work, and then drink one-half to one glass every 15 to 20 minutes in hot weather during heavy work.

Figure 54 Avoid touching high-temperature components.

13.3.1 Heat Cramps

Heat cramps are muscular pains and spasms caused by heavy exertion. Any muscles can be affected, but most often it is the muscles you've been using most. Loss of water and electrolytes from heavy sweating causes these cramps. Symptoms of heat cramps include the following:

- Painful muscle spasms and cramping
- Pale, sweaty skin
- Normal body temperature
- Abdominal pain
- Nausea

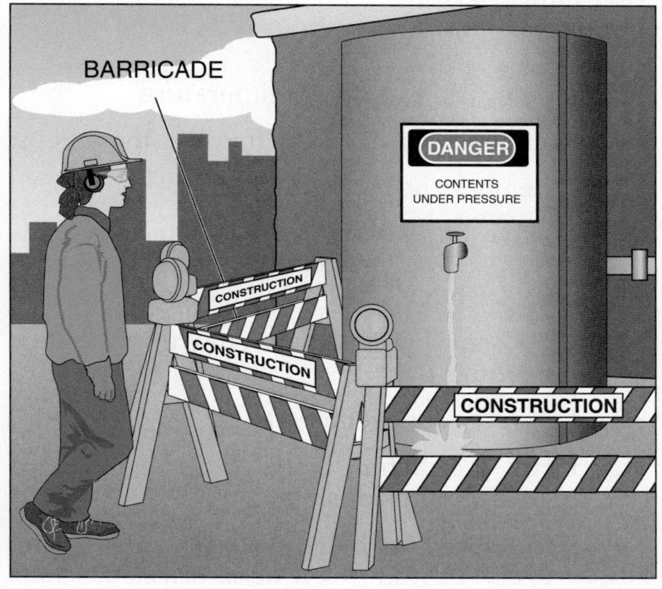

Figure 55 Work safely near pressurized or high-temperature systems.

If you experience heat cramps, take the following steps:

Step 1 Sit or lie down in a cool area.

Step 2 Drink half a glass of water every 15 minutes.

Step 3 Gently stretch and massage cramped muscles.

13.3.2 Heat Exhaustion

Heat exhaustion typically occurs when people exercise heavily or work in a warm, humid place where body fluids are lost through heavy sweating. When it's humid, sweat does not evaporate fast enough to cool the body properly.

> **WARNING!** Heat exhaustion can escalate to heat stroke, a potentially fatal condition. If you suspect you or a co-worker is suffering from heat exhaustion, take immediate action.

Symptoms of heat exhaustion include the following:

- Cool, pale, and moist skin
- Heavy sweating
- Headache, nausea, vomiting
- Dilated pupils
- Dizziness
- Possible fainting
- Fast, weak pulse
- Slight elevation in body temperature

First aid for heat exhaustion involves these steps:

Step 1 Remove the victim from heat.

Step 2 Have the victim lie down, and raise his legs six to eight inches. If the victim is nauseous, have him lie down on his side.

Step 3 Loosen clothing and remove any heavy clothing.

Step 4 Apply cool, wet towels.

Step 5 Fan the victim, but stop if he develops goose bumps or shivers.

Step 6 Get the victim half a glass of water to drink every 15 minutes if he is fully conscious and can tolerate it.

Step 7 If the victim's condition does not improve within a few minutes, call emergency medical services (911).

13.3.3 Heat Stroke

Heat stroke is life threatening. The body's temperature-control system, which produces sweat to cool the body, stops working. The body temperature can rise so high that brain damage and death may result if the body is not cooled quickly.

> **WARNING!** If you suspect someone has heat stroke, call emergency medical services immediately.

The symptoms of heat stroke include the following:

- Hot, dry, or spotted skin
- Extremely high body temperature
- Very small pupils
- Mental confusion
- Headache
- Vision impairment
- Convulsions
- Loss of consciousness

First aid for heat stroke involves these steps:

Step 1 Call emergency medical services (911) immediately.

Step 2 Remove the victim from heat.

Step 3 Have the victim lie down on his back. If the victim is nauseous, have him lie down on his side.

Step 4 Move all nearby objects, as heat stroke may cause convulsions or seizures.

Step 5 Cool the victim by fanning, spraying with cool water mist, covering with a wet sheet, or wiping with a wet cloth.

Step 6 If the victim is alert enough to do so and is not nauseous, give small amounts of cool water (a cup every 15 minutes).

Step 7 Place ice packs under the armpits and the groin.

13.4.0 Cold Stress

When your body temperature drops even a few degrees below normal, which is about 98.6°F, you can begin to shiver uncontrollably and become weak, drowsy, disoriented, unconscious, or even fatally ill. This loss of body heat is known as cold stress, or hypothermia.

Construction workers who work outdoors during the winter need to learn how to protect against loss of body heat. The following guidelines can help you keep your body warm and avoid the dangerous consequences of hypothermia, frostbite, and overexposure to the cold.

> **WARNING!**
> Always seek immediate medical attention if you suspect hypothermia or frostbite.

Outdoors, indoors, in mild weather, or in cold, it pays to dress in layers. Layering your clothes allows you to adjust what you're wearing to suit the temperature conditions. In cold weather, wear cotton, polypropylene, or lightweight wool next to your skin, and wool layers over your undergarments. For outdoor activities, choose clothing made of waterproof, wind-resistant fabrics such as nylon. Since a great deal of body heat is lost through the head, always wear a hat for added protection.

Water chills your body far more rapidly than air or wind. Always take along an extra set of clothing whenever you are working outdoors. If clothes become damp, change to dry clothes to prevent your body temperature from dropping in cold weather. Wear waterproof boots in damp or snowy weather, and always pack raingear even if the forecast calls for sunny skies.

13.4.1 Frostbite

Frostbite is a dangerous condition that can have lifelong effects on your body. It usually affects the hands, fingers, feet, toes, ears, and nose. Symptoms of frostbite include a pale, waxy-white skin color and hard, numb skin.

Remember the following when providing first aid for frostbite:

- Never rub the affected area. This can damage the skin and tissue.
- After the affected area has been warmed, it may become puffy and blister.
- The affected area may have a burning feeling or numbness. When normal feeling, movement, and skin color have returned, the affected areas should be dried and wrapped to keep it warm.
- If there is a chance the affected area may get cold again, do not warm the skin. If the skin is warmed and then becomes cold again, it will cause severe tissue damage.

Follow these steps when providing first aid for frostbite:

Step 1 Stay with the victim, and move him to a warm, dry area.

Step 2 Remove any wet or tight clothing that may cut blood flow to the affected area.

Step 3 Gently place the affected area in warm (105°F) water, and monitor the water temperature to slowly warm the tissue. Do not pour warm water directly on the affected area because it will warm the tissue too fast, causing tissue damage. Warming takes about 25 minutes.

Step 4 Seek medical attention as soon as possible.

13.4.2 Hypothermia

Hypothermia is a serious, potentially fatal condition. You do not have to be in below-freezing temperatures to be at risk for hypothermia. Hypothermia can happen on land or in water.

The effects of hypothermia can be gradual and often go unnoticed until it's too late. If you know you'll be working outdoors for an extended period of time, work with a buddy. At the very least, let someone know where you'll be and what time you expect to return. Ask your buddy to check you for overexposure to the cold, and do the same for your buddy. Check for shivering, slurred speech, mental confusion, drowsiness, and weakness. If anyone shows any of these signs, they should be moved indoors as soon as possible to warm up.

> **WARNING!**
> Call for emergency medical assistance if a co-worker exhibits uncontrolled shivering, slurred speech, clumsy movements, fatigue, and confused behavior.

Symptoms of hypothermia include the following:

- A drop in body temperature
- Fatigue or drowsiness
- Uncontrollable shivering
- Slurred speech
- Clumsy movements
- Irritable, irrational, confused behavior

Follow these steps when providing first aid for hypothermia on land:

Step 1 Call for emergency medical assistance (911).

Step 2 Stay with the victim, and move him to a warm, dry area.

Step 3 Remove any wet clothing and replace with warm, dry clothing, or wrap the victim in blankets.

Step 4 Have the victim drink warm, sweet drinks, if he is alert enough to do so. Avoid drinks with caffeine and alcohol.

Step 5 Have the victim move his arms and legs to create muscle heat. If he is unable to do this, place warm bottles or hot packs around the armpits, groin, neck, and head. Do not rub the victim's body or place him in a warm bath. This may cause a heart attack.

Follow these steps if you experience hypothermia in water:

Step 1 Immediately call for emergency help and medical assistance (911). Body heat is lost up to 25 times faster in water than on land.

Step 2 Do not remove any clothing. Close up and tighten all clothing to create a layer of trapped water close to the body. This provides insulation that slows heat loss. Keep your head out of the water and put on a hat or hood if possible.

Step 3 Get out of the water as quickly as possible, or climb onto a floating object. Do not attempt to swim unless you can reach a floating object or another person. Swimming or other physical activity uses the body's heat and reduces survival time by about 50%.

Step 4 If getting out of the water is not possible, wait quietly, and conserve your body heat by folding your arms across your chest, keeping your thighs together, bending your knees and crossing your ankles. If another person is in the water, huddle together with your chests pressed tightly together.

13.5.0 Welding and Cutting Hazards

Even if you're not welding, you can be injured when you are around a welding operation. The oxygen and acetylene used in gas welding are very dangerous. The cylinders containing oxygen and acetylene must be transported, stored, and handled very carefully. Always follow these safety guidelines:

- Keep the work area clean and free from potentially hazardous items such as combustible materials and grease or petroleum products.

> **WARNING!**
> Keep oxygen away from sources of flame and combustible materials, especially substances containing oil, grease, or other petroleum products. Compressed oxygen mixed with oil or grease will explode. Never use petroleum-based products around fittings that serve compressed oxygen lines.

- Use great caution when you handle compressed gas cylinders.
- Store cylinders in an upright position where they will not be struck, and where they will be away from corrosives. Separate oxygen and fuel cylinders by 20 feet or a 5-foot high ½-hour fire-rated barrier.
- Secure cylinders so they cannot tip over or fall.

> **WARNING!**
> Do not remove the protective cap unless a cylinder is secured. If the cylinder falls over and the nozzle breaks off, the cylinder will shoot off like a rocket, injuring or killing anyone in its path.

- Never look at an arc welding operation without wearing the proper eye protection. The arc will burn your eyes.

> **WARNING!**
> In an arc welding operation, even a reflected arc can harm your eyes. It is extremely important to follow proper safety procedures at and around all welding operations. Serious eye injury or even blindness can result from unsafe conditions.

- If you are welding, use the proper PPE (*Figure 56*), including the following:

  Full face shield with proper lens
  Earplugs to prevent flying sparks from entering your ears
  All-leather, gauntlet-type welder's gloves
  High-top leather boots to prevent **slag** from dropping inside your boots
  Cuffless trousers that cover your ankles and boot tops
  A respirator, if necessary

Figure 56 Personal protective equipment for welding.

CAUTION

When welding on construction job sites, a hard hat with a full face shield may be required.

- If you are welding and other workers are in the area around your work, set up **welding shields**. Make sure everyone wears **flash goggles**. These goggles protect the eyes from the **flash**, which is the sudden bright light associated with starting a welding operation.
- A welder must be protected when the welding shield on the welder's headgear is down, because the shield restricts the welder's field of vision. A helper or monitor must watch the welder and the surrounding area in case of a fire or similar emergency, or rope off the area to prevent collisions and keep other workers away from the area.
- Welded material is hot! Mark it with a sign and stay clear for a while after the welding has been completed.

Pay special attention to the safety guidelines about never looking at the arc without proper eye protection. Even a brief exposure to the ultraviolet light from arc welding can cause a **flash burn** and damage your eyes badly. You may not notice the symptoms until some time after the exposure. Here are some symptoms of flash burns to the eye:

- Headache
- Feeling of sand in your eyes
- Red or weeping eyes
- Trouble opening your eyes
- Impaired vision
- Swollen eyes

On-Site

Post a fire watch when you are welding or cutting. One person other than the welding or cutting operator must constantly scan the work area for fires. Fire watch personnel should have ready access to fire extinguishers and alarms and know how to use them. Welding and cutting operations should never be performed without a fire watch. The area where welding is done must be monitored afterwards until there is no longer a risk of fire.

If you think you may have a flash burn to your eyes, seek medical help at once.

CAUTION

Never wear contact lenses while you are welding. The ultraviolet rays may dry out the moisture beneath the contact lens, causing it to stick to your eye.

13.5.1 Flame Cutting

Many of the safety guidelines for welding apply to flame cutting as well. Cutting is not dangerous as long as you follow safety precautions. Here are some of the precautions:

- Wear appropriate PPE.
- Never open the valve of an acetylene cylinder near an open flame.
- Store oxygen cylinders separately from fuel gas cylinders.
- Store acetylene cylinders in an upright position.
- Always use a friction striker to light a cutting torch.

WARNING!

Never cut galvanized metal without proper ventilation. The zinc oxide fumes given off as the galvanized material is cut are hazardous. Also, use a respirator when you are cutting galvanized material.

CAUTION

The cutting process results in oxides that mix with molten iron and produce dross. The dross is blown from the cut by the jet of cutting oxygen. Hot dross can cause severe injury or can start fires on contact with flammable materials.

Before and during welding and cutting operations, you must follow certain safety procedures. *Figure 57* shows oxyacetylene welding/cutting equipment. As the operator, you must check the hoses, regulators, and work area.

13.5.2 Hoses and Regulators

Always use the proper hose. The fuel gas hose is usually red (sometimes black) and has a left-hand threaded nut for connecting to the torch. The oxygen hose is green and has a right-hand threaded nut for connecting to the torch.

Hoses with leaks, burns, worn places, or other defects that make them unfit for service must be repaired or replaced. Check that hoses are not taped up to cover leaks (*Figure 58*). When inspecting hoses, look for charred sections close to the torch. These may have been caused by **flashback**, which is the result of a welding flame flaring up and charring the hose near the torch connection. Flashback is caused by improperly mixed fuel.

CAUTION

If the torch goes out and begins to hiss, shut off the gas supply to the torch immediately. Otherwise, a flashback could occur.

WARNING!

Never relight a torch from hot metal. Doing so could cause an explosion.

New hoses contain talc and loose bits of rubber. These materials must be removed from the hoses before the torch is connected. If they are not removed, they will clog the torch needle valves. Common industry practice is to use compressed air to blow these materials out of the hose. Always make sure that the regulator valve is turned down to minimal pressure before using compressed air to clean a hose.

CAUTION

Never point a compressed air hose toward anyone. Flying debris and particles of dirt may cause serious injury.

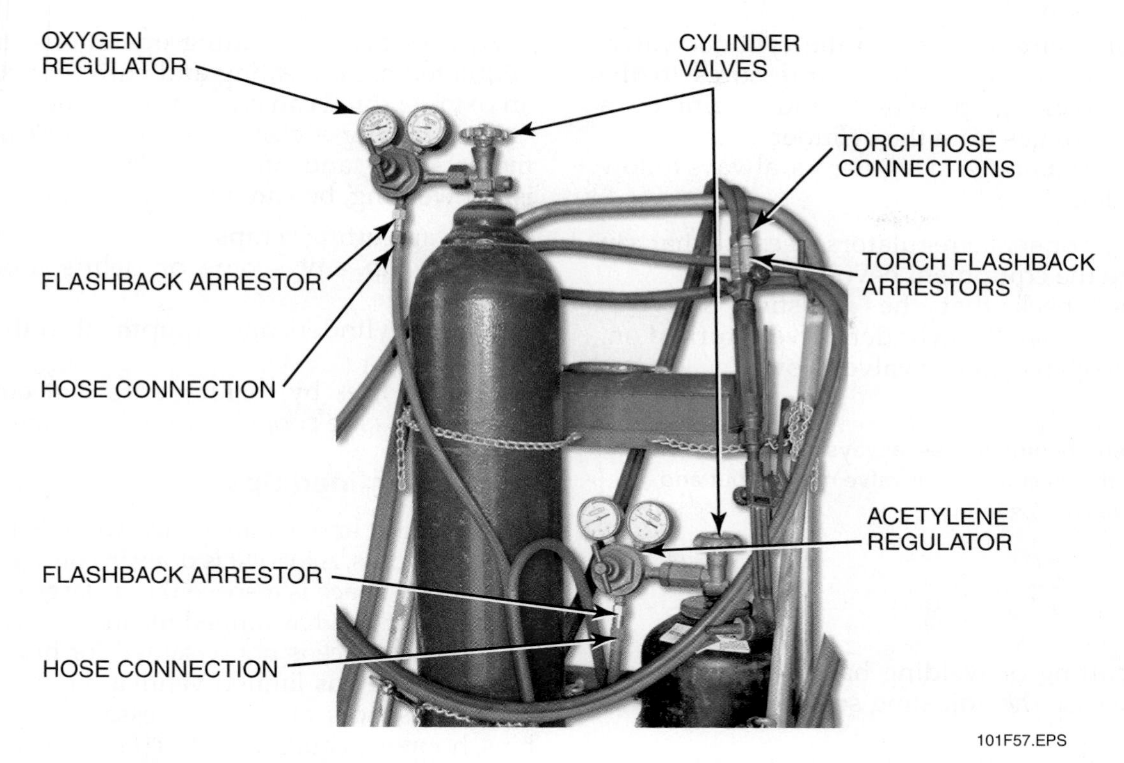

Figure 57 Typical oxyacetylene welding/cutting outfit.

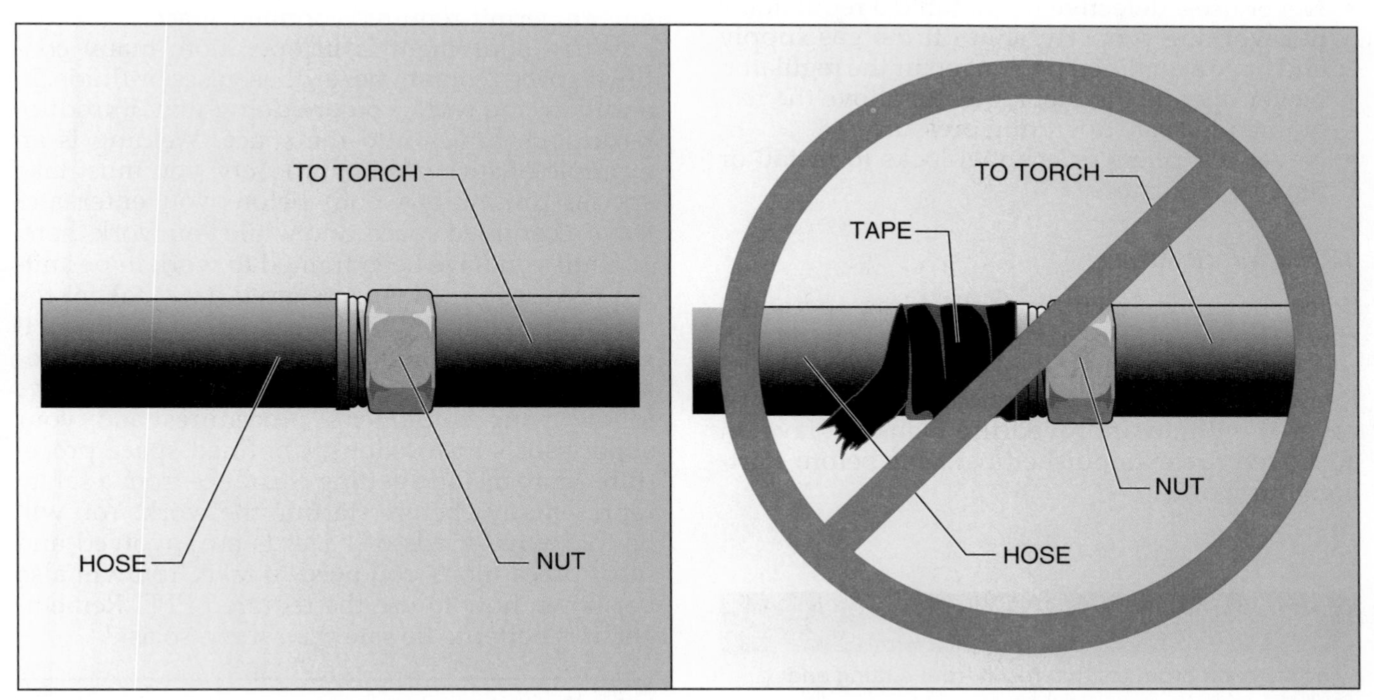

Figure 58 Proper and improper hose connection.

Regulators are attached to the cylinder valve. They lower the high cylinder pressures to the required working pressures and maintain a steady flow of gas from the cylinder.

To prevent damage to regulators, always follow these guidelines:

- Never jar or shake regulators, because that can damage the equipment beyond repair.
- Always check that the adjusting screw is released before the cylinder valve is turned on.
- Always open cylinder valves slowly.

> **CAUTION**
>
> When opening valves, always stand to the side. Dirt that is stuck in the valve may fly out and cause serious injury.

- Once cutting or welding has been completed, fully release the adjusting screw to relieve line pressure.
- Never use oil to lubricate a regulator, because that can cause an explosion.
- Never operate fuel regulators on oxygen cylinders or oxygen regulators on fuel gas cylinders.
- Never use a defective regulator. If a regulator is not working properly, shut off the gas supply and have a qualified person repair the regulator.
- Never operate the fuel regulator above the recommended safe operating pressure.
- Never use pliers or channel locks to install or remove regulators.

13.5.3 Work Area

Before beginning a cutting or welding operation, check the area for fire hazards. Cutting sparks can fly 30 feet or more and can fall several floors. Remove any flammable material in the area or cover it with an approved fire blanket. Have an approved fire extinguisher available before starting your work.

> **On-Site**
>
> The slag and products that result from cutting and welding operations can start fires and cause severe injuries. Always wear appropriate personal protective equipment, including gloves and eye protection, when cutting or welding. Do not wear clothes made of polyester when welding or cutting. Observe the safety instructions of both the manufacturer and your shop.

Always perform cutting operations in a well-ventilated area. Heating and cutting metals with an oxyfuel torch can create toxic fumes.

Maintaining a clean and neat work area promotes safety and efficiency. When you are finished welding, be sure to do the following:

- Pick up cutting scraps.
- Sweep up any scraps or debris around the work area.
- Return cylinders and equipment to the proper places.
- Prevent fires by making sure that cut metals and **dross** are cooled before disposing of them.

13.6.0 Confined Spaces

Construction and maintenance work isn't always done outdoors. A lot of it is done in confined spaces. A confined space is a space that is large enough to work in but that has limited means of entry or exit. A confined space is not designed for human occupancy, and it has limited ventilation. Examples of confined spaces are tanks, vessels, silos, storage bins, hoppers, vaults, and pits (*Figure 59*).

A **permit-required confined space** is a type of confined space that has been evaluated by a qualified person and found to have actual or potential hazards. You must have written authorization to enter a permit-required confined space.

When equipment is in operation, many confined spaces contain hazardous gases or fluids. In addition, the work you are doing may introduce hazardous fumes into the space. Welding is an example of such work. For safety, you must take special precautions both before you enter and leave a confined space, and while you work there.

Until you have been trained to work in permit-required confined spaces and have taken the needed precautions, you must stay out of them. If you aren't sure whether a confined space requires a permit, ask your supervisor. You must always follow your employer's procedures and your supervisor's instructions. Confined space procedures may include getting clearance from a safety representative before starting the work. You will be told what kinds of hazards are involved and what precautions you need to take. You will also be shown how to use the required PPE. Remember, it is better to be safe than sorry, so ask!

> **WARNING!**
>
> Without proper training, no employee is allowed to enter a permit-required or non-permit-required confined space. Employers are required to have programs to control entry to and hazards in both types of confined spaces.

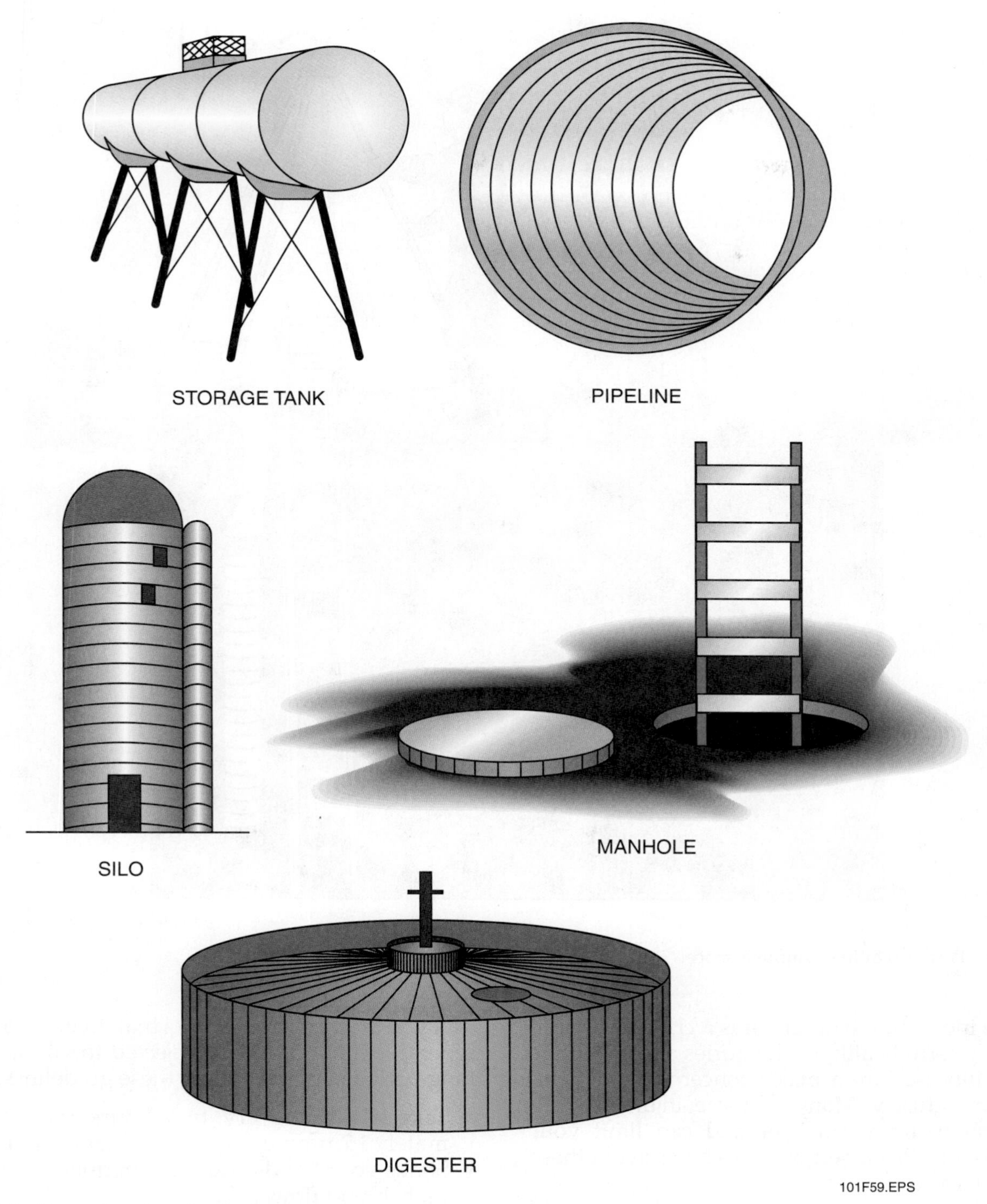

Figure 59 Examples of confined spaces.

Did You Know?

About 100 people die in confined spaces each year. Most of these deaths are caused by inhaling poisonous substances or being engulfed by a substance such as grain. The remainder of deaths are caused by lack of oxygen.

Never work alone in a confined space. OSHA requires an attendant to remain outside a permit-required confined space. The attendant monitors entry, work, and exit (*Figure 60*).

13.7.0 Construction Ergonomics

Ergonomics is the study of the physical impact that work-related movements, motions, and postures

Figure 60 Permit-required confined space.

have on individual workers. It is a critical factor in your long-term health. Back injuries and repetitive motion illnesses are a major concern for the construction industry. Many of these injuries or illnesses show up years later and can limit your quality of life. Proper ergonomics helps avoid these conditions.

Stretching muscles prior to work can help reduce injuries and improve performance. Construction work can be as strenuous as many athletic events. No true athletes would participate in an event without warming up first, and it should be this way in construction.

If you are performing tasks that involve constant repetitive motion or are subjected to vibration for long periods of time, follow these guidelines:

- Take a break every two hours for approximately 15 minutes to get your blood circulating.
- Shake your hands and arms frequently to stimulate blood flow.
- Rotate with co-workers as often as possible.
- Use anti-vibration gloves when needed.

To reduce back injuries when lifting, follow these guidelines (*Figure 61*):

- Determine the weight of the load prior to lifting.
- Plan your lift; know where it is to be unloaded and if there are any tripping hazards.
- Make sure you have firm footing.
- Bend your knees.

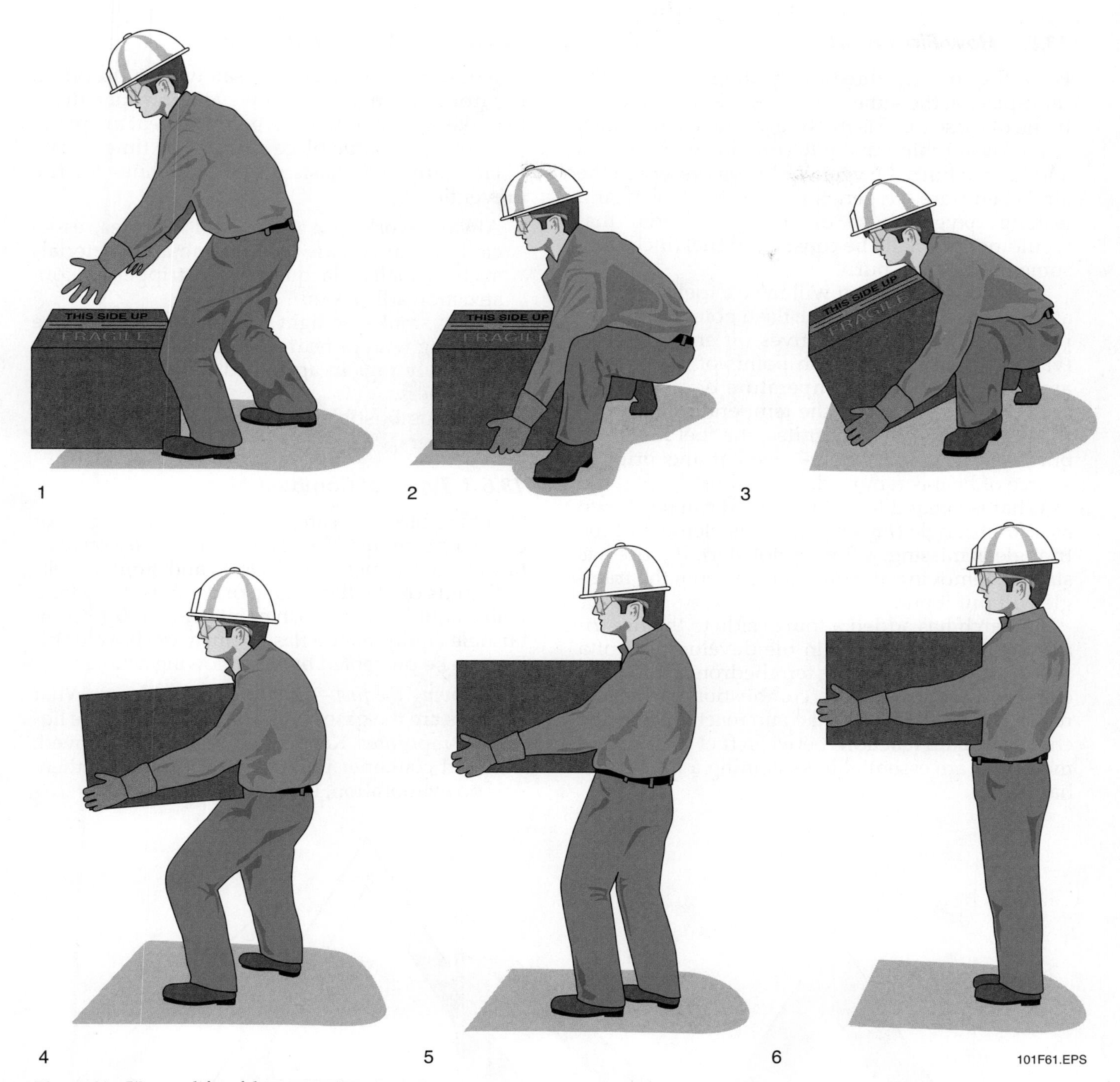

Figure 61 How to lift safely.

- Get a good grip.
- Lift with your legs, keep your back straight, and keep your head up.
- Keep the load close to your body.
- Never turn or twist until you are standing straight, then pivot your feet and body.
- Ask for assistance when lifting heavy loads or break the load down.
- Use mechanical devices when available.

13.8.0 Fire Hazards

Fire is always a hazard on construction job sites. Many of the materials used in construction are flammable. In addition, welding, grinding, and many other construction activities create heat or sparks that can cause a fire. Fire safety involves two elements: fire prevention and fire fighting.

13.8.1 How Fires Start

For a fire to start, three things are needed in the same place at the same time: fuel, heat, and oxygen. If one of these three is missing, a fire will not start.

Fuel is anything that will combine with oxygen and heat to burn. Oxygen is always present in the air. When pure oxygen is present, such as near a leaking oxygen hose or fitting, material that would not normally be considered fuel (including some metals) will burn.

Heat is anything that will raise a fuel's temperature to the **flash point**. The flash point is the temperature at which a fuel gives off enough gases (vapors) to burn. The flash points of many fuels are quite low—room temperature or less. When the burning gases raise the temperature of a fuel to the point at which it ignites, the fuel itself will burn—and keep burning—even if the original source of heat is removed.

What is needed for a fire to start can be shown as a fire triangle (*Figure 62*). If one element of the triangle is missing, a fire cannot start. If a fire has started, removing any one element from the triangle will put it out.

Research has added a fourth side to the fire triangle concept, resulting in the development of a new model called the fire tetrahedron. The fourth element involved in the combustion process is referred to as the chemical chain reaction. Specific chemical chain reactions between fuel and oxygen molecules are essential to sustaining a fire once it has begun.

13.8.2 Fire Prevention

The best way to ensure fire safety is to prevent a fire from starting. The best way to prevent a fire is to make sure that fuel, oxygen, and heat are never present in the same place at the same time.

Here are some basic safety guidelines for fire prevention:

- Always work in a well-ventilated area, especially when you are using flammable materials such as shellac, lacquer, paint stripper, or construction adhesives.
- Never smoke or light matches when you are working with or near flammable materials.
- Keep oily rags in approved, self-closing metal containers.
- Store combustible materials only in approved containers.

13.8.3 Types of Combustibles

Combustibles are categorized as liquid, gas, or ordinary combustibles. The term *ordinary combustibles* means paper, wood, cloth, and similar fuels.

Liquids can be flammable or combustible. Flammable liquids have a flash point below 100°F. Combustible liquids have a flash point at or above 100°F. Fire can be prevented by the following actions:

- *Removing the fuel* – Liquid does not burn. What burns are the gases (vapors) given off as the liquid evaporates. Keeping liquids in an approved, sealed container prevents evaporation. If there is no evaporation, there is no fuel to burn.

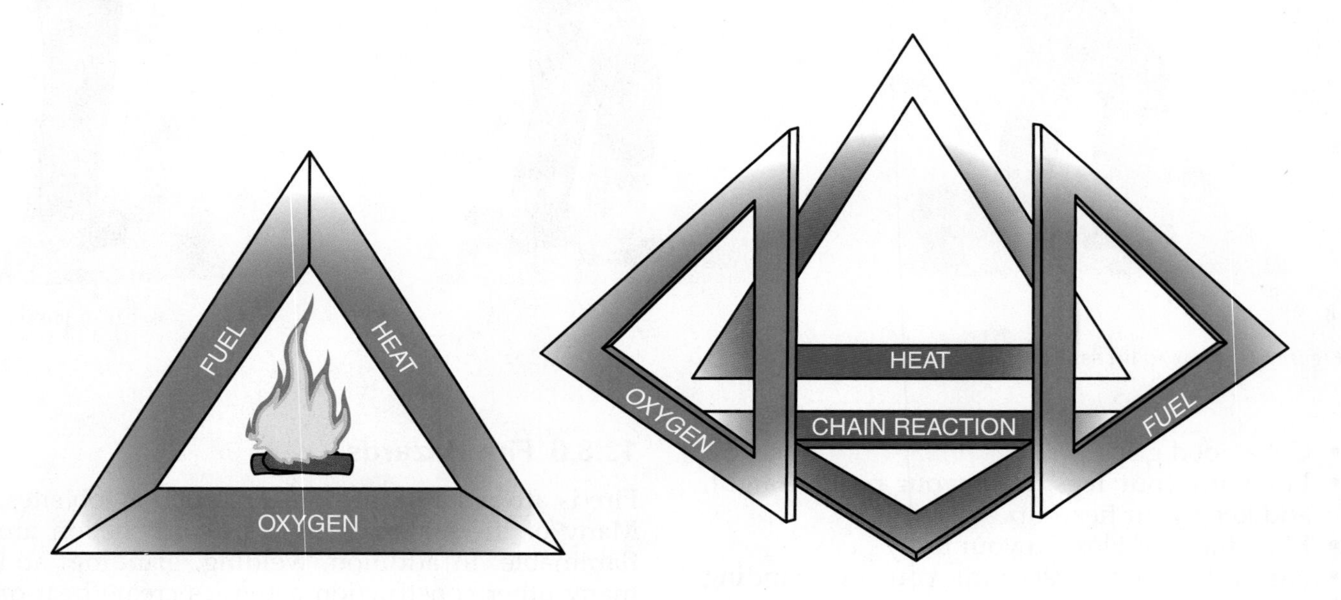

Figure 62 Basic fire requirements.

On-Site

Prevention and Preparation Are the Keys to Fire Safety

Any fire in the workplace can cause serious injury or property damage. When chemicals are involved, the risks are even greater. Prevention is the key to eliminating the hazards of fire where you work. Preparation is the key to controlling any fires that do start. Take the following precautions to make sure you are safe from fire in your workplace:

- Keep work areas clean and clutter-free.
- Know how to handle and store chemicals.
- Know what you are expected to do in case of a fire emergency.
- Should a fire start, call for professional help immediately. Don't let a fire get out of control.
- Know what chemicals you work with. You might have to tell firefighters at a chemical fire what kinds of hazardous substances are involved.
- Make sure you are familiar with your company's emergency action plan for fires.
- Use caution when using power tools near flammable substances.

- *Removing the heat* – If the liquid is stored or used away from a heat source, it will not be able to ignite.
- *Removing the oxygen* – The vapor from a liquid will not burn if oxygen is not present. Keeping safety containers tightly sealed prevents oxygen from coming into contact with the fuel.

Flammable gases used on construction sites include acetylene, hydrogen, ethane, and propane (liquid propane gas, or LPG). To save space, these gases are compressed so that a large amount is stored in a small cylinder or bottle. As long as the gas is kept in the cylinder, oxygen cannot get to it and start a fire. The cylinders should be stored away from sources of heat. If oxygen is allowed to escape and mix with a flammable gas, the resulting mixture will explode under certain conditions.

WARNING!

Never use grease or oil on the fittings of oxygen bottles and hoses. Never allow greasy or oily rags to come near any part of an oxygen system. Oil and pressurized oxygen form a very dangerous mixture that can ignite at low temperatures.

The easiest way to prevent fire in ordinary combustibles is to keep a neat, clean work area. If there are no scraps of paper, cloth, or wood lying around, there will be no fuel for starting a fire. Establish and maintain good housekeeping habits. Use approved storage cabinets and containers for all waste and other ordinary combustibles.

13.8.4 Firefighting

You are not expected to be an expert firefighter. But you may have to deal with a fire to protect your safety and the safety of others. You need to know the locations of firefighting equipment on your job site. You also need to know which equipment to use on different types of fires. However, only qualified personnel are authorized to fight fires.

Most companies tell new employees where fire extinguishers are kept. If you have not been told, be sure to ask. Also, ask how to report fires. The telephone number of the nearest fire department should be clearly posted in your work area. If your company has a company fire brigade, learn how to contact them. Learn your company's fire safety procedures. Know what type of extinguisher to use for different kinds of fires and how to use them. Make sure all extinguishers are fully charged. Never remove the tag from an extinguisher—it shows the date the extinguisher was last serviced and inspected (*Figure 63*).

Four classes of fuels can be involved in fires (*Table 6*). You've already learned about liquids, gases, and ordinary combustibles. Another fuel is metal. Each class of fuel requires a different method of firefighting and a different type of extinguisher. The label on a fire extinguisher clearly shows the class of fire on which it can be used (*Figure 64*).

When you check the extinguishers in your work area, you will see that some are rated for more than one class of fire. You can use an extinguisher that has the three codes A, B, and C on it to fight a class A, B, or C fire. But remember, if the extinguisher has only one code letter, do not use it on any other class of fire, even in an emergency. You could make the fire worse and put yourself in great danger.

DO NOT REMOVE

FOR CITY, STATE AND FIRE INSURANCE INSPECTION

FULL WT ____

D.O.T. CERT. #

LICENSE NO. ____

SERVICED BY ____

- AFFF/LD. STRM
- CARBON DIOXIDE
- PRES. WATER
- HALON 1211
- CO2 SYSTEM
- HALON SYSTEM
- ABC DRY CHEM
- STD. DRY CHEM
- PK DRY CHEM
- DRY CHEM SYS
- WET CHEM SYS

NATIONAL ASSOCIATION OF FIRE EQUIPMENT DISTRIBUTORS ® NAFED

1999/2000/2001

VOID 1 YR FROM MO PUNCHED SYSTEMS 6 MOS

SERVICED	NEW	RECHARGED

DEC NOV OCT SEPT AUG JULY JUNE MAY APR MAR FEB JAN

____ (Model No.)

____ (Mfr.)

OWNER'S I.D. NO. (if used) ____

REMARKS ____

Dry and Wet Chemical Fixed Temperature-Sensing Element Data

Year Manufactured ____

Date Installed ____

MONTHLY INSPECTION RECORD

DATE	BY	DATE	BY

PRINTED IN U.S.A.

101F63.EPS

Figure 63 Fire extinguisher tag.

(A)

101F64A.EPS

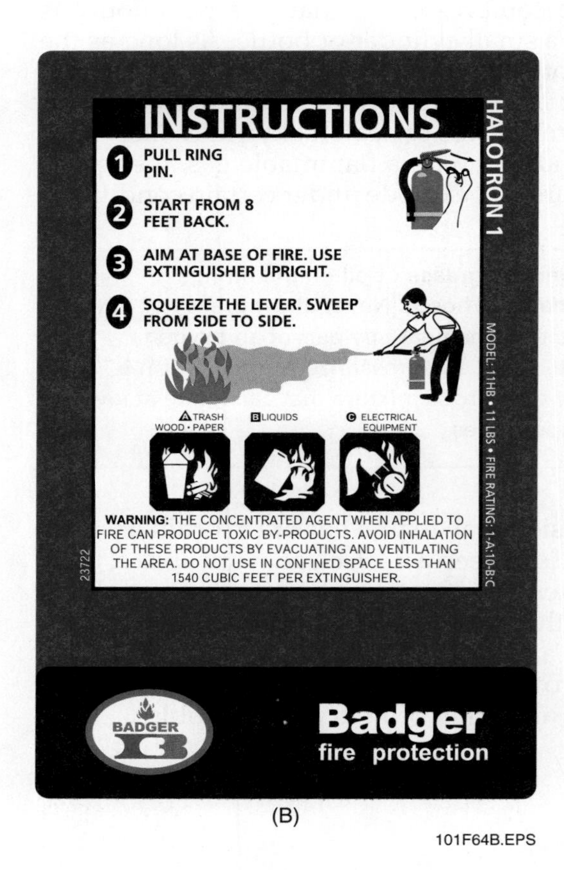

(B)

101F64B.EPS

Figure 64 Typical fire extinguisher labels. (1 of 2)

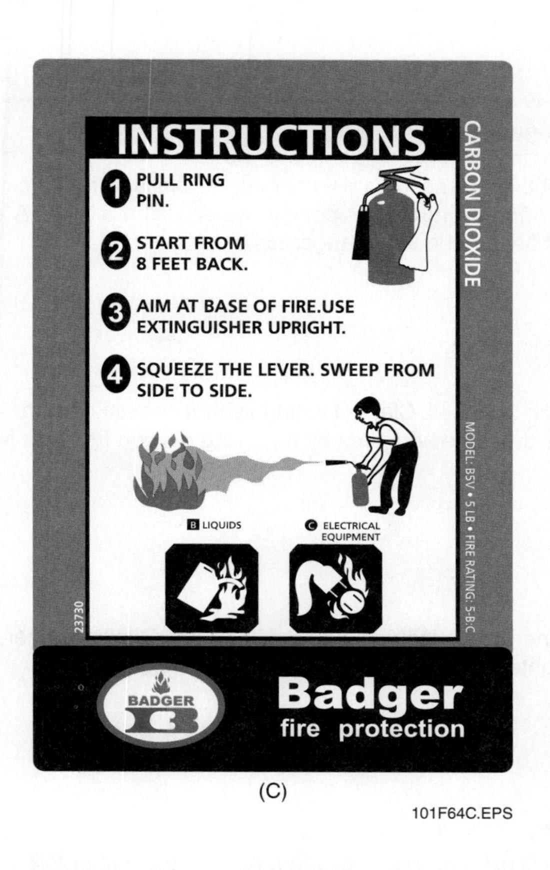

(C)

101F64C.EPS

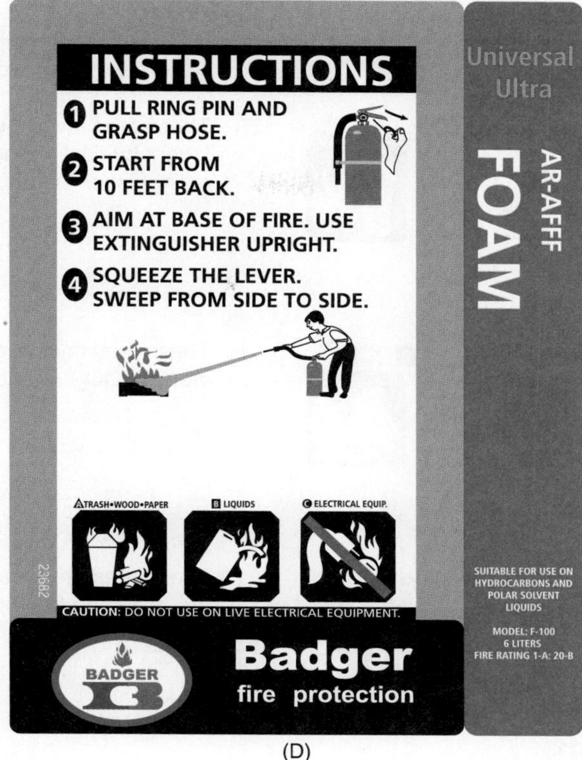

(D)

101F64D.EPS

COMBUSTIBLE METALS  **FIRE EXTINGUISHER**

1. HOLD UPRIGHT - PULL RING PIN.
2. HOLD NOZZLE OVER FIRE.
3. SQUEEZE LEVER AND COVER ALL BURNING METAL
4. REAPPLY AGENT TO HOT SPOTS.

CAUTION: FIRE MAY RE-IGNITE, ALLOW METAL TO COOL BEFORE CLEANUP.

(E)

101F64E.EPS

Figure 64 Typical fire extinguisher labels. (2 of 2)

On-Site

How to Use a Fire Extinguisher

Step 1 Hold the extinguisher upright.

Step 2 Pull the pin, breaking the plastic seal.

Step 3 Stand back 8 to 10 feet from the fire. Standing any closer may cause burning objects to scatter, spreading the fire.

Step 4 Aim at the base of the fire.

Step 5 Keep the extinguisher upright. Squeeze the handles together to discharge. Sweep from side to side.

Step 6 Move closer as the fire is extinguished (watch for scattering burning material).

Step 7 When the fire is out, watch for re-ignition.

Table 6 Classes of Fires

Class	Materials and Proper Fire Extinguisher
Class A fires  ORDINARY COMBUSTIBLES 	These fires involve ordinary combustibles such as wood or paper. Class A fires are fought by cooling the fuel. Class A fire extinguishers contain water. Using a Class A extinguisher on any other type of fire can be very dangerous.
Class B fires  B FLAMMABLE LIQUIDS 	These fires involve grease, liquids, or gases. Class B extinguishers contain carbon dioxide (CO_2) or another material that smothers fires by removing oxygen from the fire.
Class C fires  C ELECTRICAL EQUIPMENT 	These fires are near or involve energized electrical equipment. Class C extinguishers are designed to protect the firefighter from electrical shock. Class C extinguishers smother fires.
Class D fires 	These fires involve metals. Class D extinguishers contain a powder that either forms a crust around the burning metal or gives off gases that prevent oxygen from reaching the fire. Some metals will keep burning even though they have been coated with powder from a Class D extinguisher. The best way to fight these fires is to keep using the extinguisher so the fire will not spread to other fuels.

Review Questions

1. The time you spend learning and practicing _____ can save your life and the lives of others.
   a. tricks of the trade
   b. shortcuts
   c. safety procedures
   d. basic construction skills

2. In construction, keeping your work area clean and free of scraps or spills is referred to as _____.
   a. managing
   b. organizing
   c. housekeeping
   d. stacking and storing

3. The federal agency whose mission is to save lives, prevent injuries, and protect the health of America's workers is the _____.
   a. National Institute for Occupational Safety and Health (NIOSH)
   b. Occupational Safety and Health Administration (OSHA)
   c. Environmental Protection Agency (EPA)
   d. Centers for Disease Control and Prevention (CDC)

4. A(n) _____ inspection is conducted after an accident that requires hospitalization of three or more workers.
   a. imminent danger
   b. catastrophe
   c. worker complaint
   d. monitoring

5. _____ must be reported to your supervisor.
   a. Only major injuries
   b. Only incidents and major injuries
   c. All injuries, accidents, and incidents
   d. Only incidents in which a death occurred

6. The four leading causes of death in the construction industry include electrical accidents, struck-by accidents, caught-in or caught-between accidents, and _____.
   a. vehicular accidents
   b. falls
   c. radiation exposure
   d. chemical burns

7. The process of hazard recognition, evaluation, and control is the foundation of an effective _____ program.
   a. safety
   b. work
   c. training
   d. job mentoring

8. People should not be able to get past _____ barricades.
   a. yellow
   b. protective
   c. blinking light
   d. taped

9. _____ are PPE used to prevent falls and protect workers who do fall.
   a. Tie-offs
   b. Ropes and pulleys
   c. Personal fall arrest systems
   d. Anchors

10. Metal ladders should *not* be used near _____.
    a. stairways
    b. scaffolding
    c. electrical equipment
    d. windows

11. If you lean a straight ladder against the top of a 16-foot wall, the base of the ladder should be _____ feet from the base of the wall.
    a. 3
    b. 4
    c. 5
    d. 6

12. On construction sites where there is a break in elevation of 19 inches or more, and no ramp, runway, sloped embankment, or personnel hoist is provided, _____ may be used.
    a. scaffolds
    b. stairways
    c. hoist ropes
    d. elevators

13. OSHA construction standards require _____ foot-candles of light in stairways.
    a. three
    b. twelve
    c. ten
    d. five

14. The two basic types of scaffolds are _____.
    a. manufactured and rolling scaffolds
    b. fixed and portable scaffolds
    c. metal and wooden scaffolds
    d. assembled and deliverable scaffolds

15. The most common cause of death for equipment operators is _____.
    a. hit and run
    b. equipment rollover
    c. brake malfunction
    d. head-on collisions

16. To reduce the risk of workers being hurt or killed by falling materials, do not stack materials higher than a _____ height-to-base ratio.
    a. 2:1
    b. 10:1
    c. 8:1
    d. 4:1

17. A(n) _____ is a(n) _____ that is deeper than it is wide, and usually not wider than 15 feet.
    a. excavation; trench
    b. trench; hole
    c. slope; trench face
    d. trench; excavation

18. Most cave-ins occur in trenches 5 to _____ feet deep and happen suddenly with little or no warning.
    a. 10
    b. 15
    c. 20
    d. 25

19. In every trench over four feet deep, there must be an exit every _____.
    a. 10 feet
    b. 12 feet
    c. 18 feet
    d. 25 feet

20. Observing proper safety precautions when working with and around electrical current is important because the human body _____.
    a. resists the electricity's path
    b. can conduct electrical current
    c. won't conduct electricity
    d. doesn't offer electricity a circular route

21. The minimum safe working distance from exposed electrical conductors _____.
    a. depends on the voltage
    b. is a few inches
    c. is one foot
    d. is unlimited

22. Which of the following provide the best eye protection?
    a. Welding hoods
    b. Face shields
    c. Safety goggles
    d. Strap-on glasses

23. The best shoes to wear on a construction site are _____-approved.
    a. OSHA
    b. ANSI
    c. NIOSH
    d. EPA

24. A _____ has its own clean air supply.
    a. half mask
    b. mouthpiece with mechanical filter
    c. self-contained breathing apparatus
    d. full facepiece mask

25. HazCom classifies all paint, concrete, and wood dust as _____ materials.
    a. hazardous
    b. common
    c. inexpensive
    d. nonhazardous

26. Under HazCom, if you spot a hazard on your job site you must _____.
    a. report it to your supervisor
    b. leave immediately
    c. notify your co-workers
    d. correct the problem

27. Routes of exposure to hazardous materials include inhalation, ingestion, and _____.
    a. osmosis
    b. absorption
    c. radiation
    d. proximity

28. Work that is performed near a hazard but not in direct contact with it is called _____ work.
    a. close call
    b. near miss
    c. proximity
    d. barricade

29. A confined space _____.
    a. has a limited amount of ventilation
    b. has no means of entry
    c. is too small to work in
    d. may be entered by untrained employees

30. _____ must be present in the same place at the same time for a fire to occur.
    a. Oxygen, carbon dioxide, and heat
    b. Oxygen, heat, and fuel
    c. Hydrogen, oxygen, and wood
    d. Grease, liquid, and heat

SUMMARY

Although the typical construction site has many hazards, it does not have to be a dangerous place to work. Your employer has programs to deal with potential hazards. Basic rules and regulations help protect you and your co-workers from unnecessary risks.

This module has presented many of the basic guidelines you must follow to ensure your safety and the safety of your co-workers. These guidelines fall into the following categories:

- Following safe work practices and procedures
- Inspecting safety equipment before use
- Using safety equipment properly

The basic approach to safety is to eliminate hazards in the equipment and the workplace; to learn the rules and procedures for working safely with and around the remaining hazards; and to apply those rules and procedures. The information covered here offers you the groundwork for a safe, productive, and rewarding construction career.

Trade Terms Quiz

Fill in the blank with the correct key term that you learned from your study of this module.

1. ____________ is a formal procedure for taking equipment out of service and ensuring it cannot be operated until a qualified person has returned it to service.
2. ____________ is the process of joining metal parts by fusion.
3. Because ____________ scrap materials catch fire and burn easily, remove them regularly from the work area.
4. The cutting process results in oxides that mix with molten iron and produce ____________.
5. A(n) ____________ is any man-made cut, cavity, trench, or depression in an earth surface, formed by removing earth.
6. A(n) ____________ identifies unsanitary, hazardous, or dangerous working conditions and has the authority to correct or eliminate them.
7. A(n) ____________ is large enough to work in but has limited means of entry or exit; sometimes a permit is required to work in it.
8. Store ____________ liquids in safety cans to avoid the risk of fire.
9. If a scaffold is more than 10 feet high, ____________ — or pieces of wood or metal placed diagonally from the bottom of one rail to the top of another rail that add support to a structure—must be used.
10. The ____________ houses the circuits that distribute electricity throughout a structure.
11. A(n) ____________ is basically two straight ladders that are connected so the length of the ladder can be changed.
12. Wear ____________ to protect your eyes from the ____________, which is the sudden, bright light that occurs when you start up a welding operation.
13. To save lives, prevent injuries, and protect the health of America's workers, ____________ publishes rules and regulations that employees and employers must follow.
14. If fuel is improperly mixed, it can cause a(n) ____________.
15. An opening in a wall or floor is a safety hazard and must be either covered or ____________.
16. Even a brief exposure to the ultraviolet light from arc welding can damage your eyes, causing a(n) ____________.
17. ____________ is an OSHA rule requiring all contractors to educate their employees about the hazardous chemicals they may be exposed to on the job site.
18. The temperature at which a fuel gives off enough gases to burn is called the ____________.
19. A(n) ____________ is the conducting connection between electrical equipment or an electrical circuit and the earth.
20. When climbing a ladder or scaffold, use a tagline or ____________ to pull up your tools.
21. SCBA is an example of a(n) ____________, or an assembly of machines used to do a particular job.
22. If a work area is ____________, that means it has pieces of material at least 2 inches thick and 6 inches wide used as flooring, decking, or scaffolding.
23. When doing work more than 6 feet above the ground, you must wear a safety harness with a(n) ____________ that is attached to a strong anchor point.
24. A good ____________ helps prevent or correct conditions that can cause accidents.
25. To prevent a cave-in, follow OSHA regulations for ____________ up a trench.
26. Refer to the ____________ to learn about how to handle hazardous substances.
27. Overloading, which means exceeding the ____________ of a ladder, can cause ladder failure.
28. If you are operating a vehicle on a job site and cannot see to your rear, get a(n) ____________ to direct you.
29. Before working in a(n) ____________ you must be trained, obtain written authorization, and take the necessary precautions.

30. A(n) ____________ is a narrow excavation made below the surface of the ground that is generally deeper than it is wide.
31. The ____________ on a scaffold is placed halfway between the toeboard and the top rail.
32. ____________ for welding includes a face shield, ear plugs, and gloves.
33. A(n) ____________ has proven his or her extensive knowledge, training, and experience and has successfully demonstrated the ability to solve problems relating to the work.
34. A(n) ____________ provides clean air for breathing.
35. Manufactured and rolling are the two basic types of ____________.
36. When doing ____________, you must be careful not to come into contact with the nearby hazard.
37. A(n) ____________ is a self-supporting ladder made of two sections hinged at the top.
38. Use only approved ____________ for plugs.
39. A(n) ____________ is nonadjustable and consists of two rails, rungs between the rails, and safety feet on the bottom of the rails.
40. A vertical barrier called a(n) ____________ is used at floor level on scaffolds to prevent materials from falling.
41. A(n) ____________ tells you which way the wind is blowing.
42. A horizontal board called a(n) ____________ is used at top-level on all open sides of scaffolding and platforms.
43. Never look at the ____________ caused by welding without proper eye protection.
44. When you are welding and other workers are in the area, set up a protective screen called a(n) ____________.
45. A(n) ____________ houses electrical switches used to regulate and distribute electricity in a building.
46. ____________ is the waste material from welding operations.
47. A(n) ____________ is a unit of measure of the intensity of light falling on a surface, equal to one lumen per square foot and originally defined with reference to a standardized candle burning at one foot from a given surface.
48. The ____________ rule is a general construction rule that states that fall protection is required any time you are working six feet above a lower level.
49. A chip or fragment of rock or soil that has broken off from the main mass is called a(n) ____________.
50. A device that interrupts and de-energizes an electrical circuit to protect a person from electrocution is called a(n) ____________.

Trade Terms

Apparatus
Arc
Arc welding
Combustible
Competent person
Concealed receptacle
Confined space
Cross-bracing
Dross
Electrical distribution panel
Excavation
Experience modification rate (EMR)
Extension ladder
Flammable
Flash
Flashback
Flash burn
Flash goggles
Flash point
Foot-candle
Ground
Ground fault circuit interrupter
Guarded
Hand line
Hazard Communication Standard (HazCom)
Lanyard
Lockout/tagout
Management system
Material Safety Data Sheet (MSDS)
Maximum allowable slope
Maximum intended load
Mid-rail
Occupational Safety and Health Administration (OSHA)
Permit-required confined space
Personal protective equipment (PPE)
Planked
Proximity work
Qualified person
Respirator
Safety culture
Scaffold
Shoring
Signaler
Six-foot rule
Slag
Spall
Stepladder
Straight ladder
Switch enclosure
Toeboard
Top rail
Trench
Welding shield
Wind sock

Cornerstone of Craftsmanship

Robert F. Tilley, Jr.

SafeTek USA Owner and CEO

How did you choose a career in the field?
My grandfather once told me, "Do what you enjoy, and you will never have to work a day in your life!" Wise words from a wise man. I just never figured I would end up in this industry.

I decided early on that sitting behind a desk all day and having someone on my back all the time was not the job for me. I wanted freedom and to be able to feel a sense of accomplishment. Growing up, I wanted to be a park ranger. Hiking through the mountains and teaching people about nature seemed exciting to me, until I got older and found out the pay wasn't so great. I decided that a job as a police officer would work, so I went to college, graduated from the police academy and became a police officer. It was fun and exciting, but I wasn't too thrilled that the most I would be making after 20 years of service was $60,000 a year. I decided I should find something where I could still have freedom, and have the potential to make a lot of money. That is where I found construction!

What types of training have you been through?
Since I am in the field of construction safety, I have had a lot of training and certifications in safety. I have also complemented that with public speaking, math, employee management, and writing. My certifications are as a confined space instructor, trenching and excavation instructor, scaffolding instructor, forklift instructor, OSHA 500, and NCCER construction site safety master, just to name a few!

What kinds of work have you done in your career?
I have had the opportunity to save lives! I have helped design temporary and permanent fall protection safety systems that are used by businesses large and small, and also by the U.S. military. I also regularly consult with business and government regarding safety compliance issues. I have had the opportunity to assist in developing software that is revolutionizing safety compliance, making it easy for employers to follow OSHA regulations.

Tell us about your present job and what you like about it.
I am the CEO and owner of SafeTek USA. The thing I like most about it is actually being able to help people, and the freedom that comes from being in the construction industry. I have the opportunity to be outside quite a bit, and it is wonderful. Construction is a dangerous business, and fixing construction hazards is big business!

What factors have contributed most to your success?
Hard work, putting others first, and always making things right! Whether you own your own company, are a manager, or a construction technician doing the actual work, just a few things will make you stand out above the rest. Following these principles has helped me greatly. Being on time is late; being early is on time. Put others before yourself as much as reasonable and help them as much as you can, because if you help someone succeed, they just may help you. And if you make a mistake, forget something, mess something up, no matter how big it is, bring it to the attention of your boss or the person in the company it affects, instead of waiting for them to find out. That is called taking responsibility for yourself and your actions; it is hard to find nowadays and will put you ahead of everyone else!

What advice would you give to those new to the field?
The best advice I have for someone new to the field is don't settle for the job you have, but don't expect riches to come tomorrow. Set realistic goals for yourself and where you want to be in a year or five years and work toward those goals. It is all too often easy to become consumed with day-to-day life at home or at the job. The goals are not your focus and in five years you are not even close to where you want to be. Put in the work, and make yourself stand out among the rest.

Tell us some interesting career-related facts or accomplishments:
I would say my biggest accomplishment is starting and running a successful business. I am also a technical writer for various national magazines and publications, and I help develop the curriculum for NCCER.

Cornerstone of Craftsmanship

Bob Fitzgerald

BE&K Industrial Services (a KBR Company)
Area Manager - HSE Execution

How did you choose a career in the field?
After many years working as an emergency medical technician (EMT) in the public and private sectors, I was hired as a medic on a construction site. What sparked my interest in safety and health was the notion of not just treating the injured, but actually being able to do something to prevent the incident from happening in the first place. When working in the medical field, you have to really care and have compassion for people. I think that is one of the major attributes of a successful safety professional.

What types of training have you been through?
I have taken all sorts of safety and health classes through the years, including trenching, shoring, electrical, fall protection, respiratory protection, and more. I was privileged to attend the NCCER Safety Academy at Clemson University early in my safety career and I think that was a turning point in my philosophy and thinking about safety management. Since then I have earned my bachelor's of science and a master's degree in safety and health. I also hold the Certified Safety Professional and Construction Health and Safety Technician designation.

What kinds of work have you done in your career?
I have had two jobs in my professional career: EMT and Safety Professional.

Tell us about your present job and what you like about it.
I serve as an Area Manager – HSE Execution for BE&K Industrial Services (a KBR Company). I am headquartered at the BE&K offices in Birmingham, Alabama, supporting our projects and our employees all across the United States and globally. I enjoy helping people and making a difference in their lives. Maybe something I say or do will help them go home at the end of the day safe to their families.

What factors have contributed most to your success?
You really have to care about people. My early experiences in the medical field helped me to have empathy for people and to see the pain and suffering that could happen if things go wrong. An important success factor is that someone did not have to suffer from an unsafe situation that I may have had a chance to correct or influence.

What advice would you give to those new to the field?
Realize that the value for safety and health is fundamental in all crafts or whatever role you perform in construction or industrial work. Watch out for each other and don't be afraid to speak up if you see an unsafe situation or action. Take the risk of intervening to help someone. Learn all you can about safety and put it to use.

Tell us some interesting career-related facts or accomplishments:
In 1995, I had the privilege to work with a wonderful team of construction professionals and craftworkers to earn the OSHA Voluntary Protection Program (VPP) STAR designation at a site in Stevenson, Alabama. This particular construction site was the first in the state of Alabama to earn the OSHA STAR. The effort that went into achieving OSHA STAR still echoes today in folks that I meet that worked on that site. They have instilled a culture of safety and a lifelong safety commitment. It was, and is, great!

Trade Terms Introduced in this Module

Apparatus: An assembly of machines used together to do a particular job.

Arc: The flow of electrical current through a gas (such as air) from one pole to another pole.

Arc welding: The joining of metal parts by fusion, in which the necessary heat is produced by means of an electric arc.

Combustible: Capable of easily igniting and rapidly burning; used to describe a fuel with a flash point at or above 100°F.

Competent person: A person who is capable of identifying existing and predictable hazards in the surroundings or working conditions which are unsanitary, hazardous, or dangerous to employees, and who has authorization to take prompt corrective measures to eliminate them.

Concealed receptacle: The electrical outlet that is placed inside the structural elements of a building, such as inside the walls. The face of the receptacle is flush with the finished wall surface and covered with a plate.

Confined space: A work area large enough for a person to work, but arranged in such a way that an employee must physically enter the space to perform work. A confined space has a limited or restricted means of entry and exit. It is not designed for continuous work. Tanks, vessels, silos, pits, vaults, and hoppers are examples of confined spaces. See also *permit-required confined space.*

Cross-bracing: Braces (metal or wood) placed diagonally from the bottom of one rail to the top of another rail that add support to a structure.

Dross: Waste material resulting from cutting using a thermal process.

Electrical distribution panel: Part of the electrical distribution system that brings electricity from the street source (power poles and transformers) through the service lines to the electrical meter mounted on the outside of the building and to the panel inside the building. The panel houses the circuits that distribute electricity throughout the structure.

Excavation: Any man-made cut, cavity, trench, or depression in an earth surface, formed by removing earth. It can be made for anything from basements to highways. Also see *trench.*

Experience modification rate (EMR): A rate computation to determine surcharge or credit to workers' compensation premium based on a company's previous accident experience.

Extension ladder: A ladder made of two straight ladders that are connected so that the overall length can be adjusted.

Flammable: Capable of easily igniting and rapidly burning; used to describe a fuel with a flash point below 100°F.

Flash: A sudden bright light associated with starting up a welding torch.

Flashback: A welding flame that flares up and chars the hose at or near the torch connection. It is caused by improperly mixed fuel.

Flash burn: The damage that can be done to eyes after even brief exposure to ultraviolet light from arc welding. A flash burn requires medical attention.

Flash goggles: Eye protective equipment worn during welding operations.

Flash point: The temperature at which fuel gives off enough gases (vapors) to burn.

Foot-candle: A unit of measure of the intensity of light falling on a surface, equal to one lumen per square foot and originally defined with reference to a standardized candle burning at one foot from a given surface.

Ground: The conducting connection between electrical equipment or an electrical circuit and the earth.

Ground fault circuit interrupter (GFCI): A device that interrupts and de-energizes an electrical circuit to protect a person from electrocution.

Guarded: Enclosed, fenced, covered, or otherwise protected by barriers, rails, covers, or platforms to prevent dangerous contact.

Hand line: A line attached to a tool or object so a worker can pull it up after climbing a ladder or scaffold.

Hazard Communication Standard (HazCom): The Occupational Safety and Health Administration standard that requires contractors to educate employees about hazardous chemicals on the job site and how to work with them safely.

Lanyard: A short section of rope or strap, one end of which is attached to a worker's safety harness and the other to a strong anchor point above the work area.

Lockout/tagout: A formal procedure for taking equipment out of service and ensuring that it cannot be operated until a qualified person has removed the lockout or tagout device (such as a lock or warning tag).

Management system: The organization of a company's management, including reporting procedures, supervisory responsibility, and administration.

Material safety data sheet (MSDS): A document that must accompany any hazardous substance. The MSDS identifies the substance and gives the exposure limits, the physical and chemical characteristics, the kind of hazard it presents, precautions for safe handling and use, and specific control measures.

Maximum allowable slope: The steepest incline of an excavation face that is acceptable for the most favorable site conditions as protection against cave-ins, expressed as the ratio of horizontal distance to vertical rise.

Maximum intended load: The total weight of all people, equipment, tools, materials, and loads that a ladder can hold at one time.

Mid-rail: Mid-level, horizontal board required on all open sides of scaffolding and platforms that are more than 14 inches from the face of the structure and more than 10 feet above the ground. It is placed halfway between the toeboard and the top rail.

Occupational Safety and Health Administration (OSHA): An agency of the U.S. Department of Labor. Also refers to the Occupational Safety and Health Act of 1970, a law that applies to more than more than 111 million workers and 7 million job sites in the country.

Permit-required confined space: A confined space that has been evaluated and found to have actual or potential hazards, such as a toxic atmosphere or other serious safety or health hazard. Workers need written authorization to enter a permit-required confined space. Also see *confined space*.

Personal protective equipment (PPE): Equipment or clothing designed to prevent or reduce injuries.

Planked: Having pieces of material 2 or more inches thick and 6 or more inches wide used as flooring, decking, or scaffolding.

Proximity work: Work done near a hazard but not actually in contact with it.

Qualified person: A person who, by possession of a recognized degree, certificate, or professional standing, or by extensive knowledge, training, and experience, has demonstrated the ability to solve or prevent problems relating to a certain subject, work, or project.

Respirator: A device that provides clean, filtered air for breathing, no matter what is in the surrounding air.

Safety culture: The culture created when the whole company sees the value of a safe work environment.

Scaffold: An elevated platform for workers and materials.

Shoring: Using pieces of timber, usually in a diagonal position, to hold a wall in place temporarily.

Signaler: A person who is responsible for directing a vehicle when the driver's vision is blocked in any way.

Six-foot rule: A rule stating that platforms or work surfaces with unprotected sides or edges that are six feet or higher than the ground or level below it require fall protection.

Slag: Waste material from welding operations.

Spall: A chip or fragment of rock or soil that has broken off from the main mass.

Stepladder: A self-supporting ladder consisting of two elements hinged at the top.

Straight ladder: A nonadjustable ladder.

Switch enclosure: A box that houses electrical switches used to regulate and distribute electricity in a building.

Toeboard: A vertical barrier at floor level attached along exposed edges of a platform, runway, or ramp to prevent materials and people from falling.

Top rail: A top-level, horizontal board required on all open sides of scaffolding and platforms that are more than 14 inches from the face of the structure and more than 10 feet above the ground.

Trench: A narrow excavation made below the surface of the ground that is generally deeper than it is wide, with a maximum width of 15 feet. Also see *excavation*.

Welding shield: (1) A protective screen set up around a welding operation designed to safeguard workers not directly involved in that operation. (2) A shield that provides eye and face protection for welders by either connecting to helmet-like headgear or attaching directly to a hard hat; also called a welding helmet.

Wind sock: A cloth cone open at both ends mounted in a high place to show which direction the wind is blowing.

CONTREN® LEARNING SERIES – USER UPDATE

NCCER makes every effort to keep these textbooks up-to-date and free of technical errors. We appreciate your help in this process. If you have an idea for improving this textbook, or if you find an error, a typographical mistake, or an inaccuracy in NCCER's Contren® textbooks, please write us, using this form or a photocopy. Be sure to include the exact module number, page number, a detailed description, and the correction, if applicable. Your input will be brought to the attention of the Technical Review Committee. Thank you for your assistance.

Instructors – If you found that additional materials were necessary in order to teach this module effectively, please let us know so that we may include them in the Equipment/Materials list in the Annotated Instructor's Guide.

Write: Product Development and Revision
National Center for Construction Education and Research
3600 NW 43rd St., Bldg. G, Gainesville, FL 32606

Fax: 352-334-0932

E-mail: curriculum@nccer.org

Craft ______ Module Name ______

Copyright Date ______ Module Number ______ Page Number(s) ______

Description ______

(Optional) Correction ______

(Optional) Your Name and Address ______

Introduction to Construction Math

00102-09

CORE CURRICULUM

00109-09
Introduction to Materials Handling

00108-09
Basic Employability Skills

00107-09
Basic Communication Skills

00106-09
Basic Rigging

00105-09
Introduction to Construction Drawings

00104-09
Introduction to Power Tools

00103-09
Introduction to Hand Tools

00102-09
Introduction to Construction Math

00101-09
Basic Safety

This course map shows all of the modules in the *Core Curriculum: Introductory Craft Skills*. The suggested training order begins at the bottom and proceeds up. Skill levels increase as you advance on the course map. The local Training Program Sponsor may adjust the training order.

Note that Module 00106-09, *Basic Rigging,* is an elective. It is not a requirement for level completion, but it may be included as part of your training program.

Objectives

When you have completed this module, you will be able to do the following:

1. Add, subtract, multiply, and divide whole numbers, with and without a calculator.
2. Use a standard ruler, a metric ruler, and a measuring tape to measure.
3. Add, subtract, multiply, and divide fractions.
4. Add, subtract, multiply, and divide decimals, with and without a calculator.
5. Convert decimals to percentages and percentages to decimals.
6. Convert fractions to decimals and decimals to fractions.
7. Explain what the metric system is and how it is important in the construction trade.
8. Recognize and use metric units of length, weight, volume, and temperature.
9. Recognize some of the basic shapes used in the construction industry and apply basic geometry to measure them.

Trade Terms

Acute angle
Adjacent angle
Angle
Architect's scale
Area
Bisect
Borrow
Carry
Circle
Circumference
Convert
Cubic
Decimal
Degree
Denominator
Diagonal
Diameter
Difference
Digit
Engineer's scale
English ruler
Equilateral triangle
Equivalent fractions
Formula
Fraction
Improper fraction
Invert
Isosceles triangle
Machinist's rule
Meter
Metric ruler
Metric scale
Mixed number
Negative numbers
Numerator
Obtuse angle
Opposite angle
Percentage
Perimeter
Pi
Place value
Positive numbers
Product
Quotient
Radius
Rectangle
Remainder
Right angle
Right triangle
Scalene triangle
Square
Standard ruler
Straight angle
Sum
Triangle
Vertex
Volume
Whole numbers

Required Trainee Materials

1. Sharpened pencils and paper
2. Standard ruler (with 1⁄16-inch markings)
3. Metric ruler (with centimeters [cm] and millimeters [mm])
4. Tape measure
5. Calculator
6. Architect's scale

Prerequisites

Before you begin this module, it is recommended that you successfully complete the following: *Core Curriculum: Introductory Craft Skills,* Module 00101-09. Module 00106-09 is an elective and is not necessary to receive successful completion of this course.

Contents

Topics to be presented in this module include:

1.0.0 Introduction 2.1
2.0.0 Whole Numbers 2.1
2.1.0 Parts of a Whole Number 2.1
2.1.1 Study Problems: Parts of a Whole Number 2.2
2.2.0 Adding Whole Numbers 2.2
2.3.0 Subtracting Whole Numbers 2.3
2.4.0 Study Problems: Adding and Subtracting Whole Numbers 2.3
2.5.0 Multiplying Simple Whole Numbers 2.4
2.6.0 Dividing Whole Numbers 2.5
2.7.0 Study Problems: Multiplying and Dividing Whole Numbers 2.7
2.8.0 Practical Applications 2.7
3.0.0 Working with Length Measurements 2.8
3.1.0 Reading a Standard Ruler (English Ruler) 2.9
3.1.1 Study Problems: Reading a Standard Ruler 2.10
3.2.0 The Metric Ruler 2.10
3.2.1 Reading a Metric Ruler 2.10
3.2.2 Study Problems: Reading a Metric Ruler 2.11
3.3.0 The Measuring Tape (Standard and Metric) 2.11
3.3.1 The Standard Tape Measure 2.11
3.3.2 Using a Tape Measure 2.13
3.3.3 Study Problems: Reading a Metric and English Tape Measure 2.13
3.4.0 Other Types of Scales 2.14
3.4.1 The Architect's Scale 2.14
3.4.2 Reading an Architect's Scale 2.15
3.4.3 Metric Scale (Metric Architect's Scale) 2.15
3.4.4 Reading a Metric Scale 2.16
3.4.5 The Engineer's Scale 2.16
3.4.6 Reading an Engineer's Scale 2.16
3.4.7 Study Problems: Reading Other Types of Scales 2.17
3.5.0 Converting Measurements 2.18
3.5.1 Study Problems: Converting Measurements 2.19
4.0.0 What Are Fractions? 2.20
4.1.0 Finding Equivalent Fractions 2.20
4.1.1 Study Problems: Finding Equivalent Fractions 2.21
4.2.0 Reducing Fractions to Their Lowest Terms 2.22
4.2.1 Study Problems: Reducing Fractions to Their Lowest Terms . . 2.22
4.3.0 Comparing Fractions and Finding Lowest Common Denominators 2.22
4.3.1 Study Problems: Finding the Lowest Common Denominator . 2.23
4.4.0 Adding Fractions 2.23
4.4.1 Study Problems: Adding Fractions 2.23
4.5.0 Subtracting Fractions 2.24
4.5.1 Study Problems: Subtracting Fractions 2.24
4.5.2 Subtracting a Fraction from a Whole Number 2.24
4.5.3 Study Problems: Subtracting a Fraction from a Whole Number 2.24
4.6.0 Multiplying Fractions 2.25
4.6.1 Study Problems: Multiplying Fractions 2.25

4.7.0 Dividing Fractions . . . 2.25
4.7.1 Study Problems: Dividing Fractions . . . 2.25
4.8.0 Practical Applications with Fractions . . . 2.26
5.0.0 Decimals . . . 2.26
5.1.0 Comparing Whole Numbers with Decimals . . . 2.27
5.1.1 Study Problems: Comparing Whole Numbers with Decimals . . . 2.27
5.2.0 Comparing Decimals with Decimals . . . 2.29
5.2.1 Study Problems: Comparing Decimals . . . 2.29
5.3.0 Adding and Subtracting Decimals . . . 2.29
5.3.1 Study Problems: Adding and Subtracting Decimals . . . 2.30
5.4.0 Multiplying Decimals . . . 2.30
5.4.1 Study Problems: Multiplying Decimals . . . 2.31
5.5.0 Dividing with Decimals . . . 2.31
5.5.1 Dividing with a Decimal in the Number Being Divided . 2.32
5.5.2 Study Problems: Dividing with Decimals, Part 1 . . . 2.32
5.5.3 Dividing with a Decimal in the Number You Are Dividing By . 2.32
5.5.4 Study Problems: Dividing with Decimals, Part 2 . . . 2.32
5.5.5 Dividing with Decimals in Both Numbers . . . 2.33
5.5.6 Study Problems: Dividing with Decimals, Part 3 . . . 2.33
5.6.0 Rounding Decimals . . . 2.33
5.6.1 Study Problems: Rounding Decimals . . . 2.34
5.7.0 Using the Calculator to Add, Subtract, Multiply, and Divide Decimals . . . 2.34
5.7.1 Study Problems: Using Decimals on the Calculator . . . 2.35
5.8.0 Practical Application . . . 2.35
6.0.0 Conversion Process . . . 2.37
6.1.0 Converting Decimals to Percentages and Percentages to Decimals . . . 2.37
6.1.1 Study Problems: Converting Decimals to Percentages and Percentages to Decimals . . . 2.37
6.2.0 Converting Fractions to Decimals . . . 2.38
6.2.1 Study Problems: Converting Fractions to Decimals . . . 2.39
6.3.0 Converting Decimals to Fractions . . . 2.39
6.3.1 Study Problems: Converting Decimals to Fractions . . . 2.39
6.4.0 Converting Inches to Decimal Equivalents in Feet . . . 2.39
6.4.1 Study Problems: Converting Inches to Decimals . . . 2.40
6.5.0 Practical Applications . . . 2.40
7.0.0 Introduction to Construction Geometry . . . 2.41
7.1.0 Angles . . . 2.41
7.2.0 Shapes . . . 2.42
7.2.1 Rectangle . . . 2.44
7.2.2 Square . . . 2.44
7.2.3 Triangle . . . 2.44
7.2.4 Circle . . . 2.45
7.3.0 Area of Shapes . . . 2.46
7.3.1 Study Problems: Calculating the Area of a Shape . . . 2.47
7.4.0 Volume of Shapes . . . 2.48
7.4.1 Study Problems: Calculating the Volume of Shapes . . . 2.51
7.5.0 Practical Applications . . . 2.51
APPENDIX A Multiplication Table . . . 2.65
APPENDIX B Inches Converted to Decimals of a Foot . . . 2.66
APPENDIX C Conversion Tables . . . 2.68
APPENDIX D Area and Volume Formulas . . . 2.69
APPENDIX E Using a Calculator . . . 2.70

1.0.0 INTRODUCTION

In the construction trades, workers use math all the time. Plumbers use math to calculate pipe length, read plans, and lay out fixtures. Carpenters use math to lay out floor systems and frame walls and ceilings. When you measure a length of material, fill a container with a specified amount of liquid, or calculate the dimensions of a room, you are using mathematical operations.

This module explains the theories behind some basic mathematical procedures and provides the opportunity to apply those procedures to practice tasks related to construction activities.

> **NOTE**
> You should do the math problems without using a calculator, except when the text specifically calls for calculator use or when you need to check the answers to your problems.

2.0.0 WHOLE NUMBERS

No matter which construction trade you enter, you can be sure that you will be required to work with **whole numbers** in both your written and oral communication. The ability to read, write, and communicate whole numbers accurately is extremely important on any job site. Carpenters will often use whole numbers during the early phases of planning to quickly estimate the square feet of drywall or linear feet of baseboard necessary to finish a room. Sheet metal workers will use whole numbers during the planning phase for installing an air handling system and to estimate, with relative accuracy, trunk line dimensions, lengths, and the **cubic** feet of air they will be required to handle.

Be aware that any mistake made when communicating with any type of number can result in wasted time, effort, and materials—all of which negatively affect the bottom line of the company's budget for that project.

> **Did You Know?**
>
> The following are whole numbers:
>
> 1 5 67 335 2,654
>
> The following are *not* whole numbers:
>
> ½ ¾ 7⅛ 0.45 4.25

In this section, you will learn how to work with whole numbers. Whole numbers are complete units without **fractions** or **decimals**. Your work during this section will be with whole numbers only. Later, you will work with fractions and decimals.

2.1.0 Parts of a Whole Number

Look at the parts of a whole number. The whole number shown in *Figure 1* has seven **digits**. A digit is any of the numerical symbols from 0 to 9. The seven digits in *Figure 1* are the numbers shown across the top of that figure. If you read this seven-digit whole number out loud, you would say "five million, three hundred sixteen thousand, two hundred forty-seven."

Each of this whole number's seven digits represents a **place value**. Each digit has a value that depends on its place, or location, in the whole number. In this whole number, for example, the place value of the 5 is five million, and the place value of the 2 is two hundred.

Other important points to keep in mind about whole numbers include the following:

- Numbers larger than zero are called **positive (+) numbers** (such as 1, 2, 3 . . .).
- Numbers less than zero are called **negative (–) numbers** (such as –1, –2, –3 . . .).
- Zero (0) is neither positive nor negative.
- Except for zero, all numbers without a minus sign in front of them are positive.
- Some whole numbers may contain the digit zero. For example, the whole number 7,093 has a zero in the hundreds place. When you read that number out loud, you would say "seven thousand ninety-three."

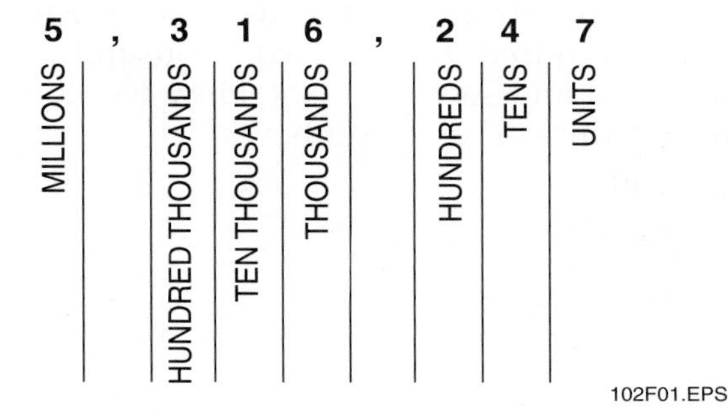

Figure 1 Place values.

2.1.1 *Study Problems: Parts of a Whole Number*

1. Look at this description of a number.

   Digit in the units place: 4
   Digit in the tens place: 6
   Digit in the hundreds place: 9
   Digit in the thousands place: 3

   This number would be written as _____.

   a. 3,964
   b. 4,693
   c. 30,964
   d. 39,064

2. In the number 25,718, the numeral 5 is in the _____ place.

   a. tens
   b. thousands
   c. units
   d. hundreds

3. An estimate for a project house requires four thousand, six hundred ninety-three square feet of carpet to complete the first floor. How would you record this amount as a whole number?

   a. 463
   b. 4,693
   c. 40,693
   d. 46,093

4. The supervisor estimates that a commercial building will require sixteen thousand, five hundred linear feet of PVC piping to complete all of the restroom facilities. How would you record this measurement as a number?

   a. 1,650
   b. 16,500
   c. 160,500
   d. 16,000,500

5. The supervisor estimates that the total cost to install the building's HVAC system will be three hundred twenty-two thousand, nine hundred and seven dollars. How would you record this cost as a number?

   a. 3,297
   b. 300,297
   c. 322,907
   d. 322,000,907

2.2.0 Adding Whole Numbers

To add means to combine the values of two or more numbers together into one **sum** or total. To add whole numbers, use the following steps:

Step 1 Line up the digits in the top number and the bottom number by place value columns. Position the greater number on the top.

$$\begin{array}{r} 723 \\ +\ 84 \\ \hline \end{array}$$

Step 2 Beginning at the right side, add the 3 and the 4 in the units column.

$$\begin{array}{r} 723 \\ +\ 84 \\ \hline 7 \end{array}$$

Step 3 Continue to add the digits in each column to the left. In this example, when you add the 2 and the 8 in the tens column you get 10. This requires you to **carry** a 1 over to the hundreds column. To do this, place the 0 in the tens column and carry the 1 to the top of the hundreds column.

$$\begin{array}{r} {}^{1} \\ 723 \\ +\ 84 \\ \hline 07 \end{array}$$

Step 4 Add the 7 in the hundreds column to the 1 that you carried over to get a sum of 807.

$$\begin{array}{r} {}^{1} \\ 723 \\ +\ 84 \\ \hline 807 \end{array}$$

Note that when working with larger numbers, you may need to carry over more than once to calculate the sum.

Did You Know?

Numeral Systems

People in ancient Egypt, Babylon, Greece, and Rome developed different numeral systems or ways of writing numbers. Some of these early systems were very complex and difficult to use. A new numeral system came into use around 750 B.C. Originally developed by the Hindus in India, the system was spread by Arab traders. The Hindu-Arabic system uses only ten symbols—1, 2, 3, 4, 5, 6, 7, 8, 9, 0—and is still in use today. These ten symbols (also called numerals or digits) can be combined to write any number.

2.3.0 Subtracting Whole Numbers

To subtract means to take away a given amount of one number from the total amount of a second number to find the **difference**. To subtract whole numbers, use the following steps:

Step 1 Line up the digits in the top number and the bottom number by place value columns. Position the larger number on the top.

$$\begin{array}{r} 12,766 \\ -\ 1,483 \\ \hline \end{array}$$

Step 2 Subtract the 3 from the 6 in the units column to get 3. As you work your way left into the tens column, you will see that you are unable to subtract 8 from 6. This will require you to **borrow** a 1 from the hundreds column. To borrow, cross out the 7 in the hundreds column, change it to a 6, and carry a 1 over to the 6 in the tens column making it 16 instead of 6. Subtract 8 from 16 to get 8.

$$\begin{array}{r} {\scriptstyle 6\ 1\ \ } \\ 12,\not{7}66 \\ -1,483 \\ \hline 83 \end{array}$$

Step 3 Continue to work your way left, completing each subtraction for the remaining place values.

$$\begin{array}{r} {\scriptstyle 6\ 1\ \ } \\ 12,\not{7}66 \\ -1,483 \\ \hline 11,283 \end{array}$$

Note that when working with larger numbers, you may need to borrow more than once to calculate the difference.

$$\begin{array}{r} {\scriptstyle 3\ 12\ 10\ \ \ } \\ 9\not{4}\not{3},\not{1}53 \\ -436,372 \\ \hline 506,781 \end{array}$$

2.4.0 Study Problems: Adding and Subtracting Whole Numbers

Use addition and subtraction to solve the following problems. Read each question carefully to determine the appropriate procedure. Be sure to show all of your work.

1. In calculating a bid for a roof restoration, the contractor estimates that he will need $847 for lumber, $456 for roofing shingles, and $169 for hardware. What is his total cost for the bid? $ _________

2. A plumbing contractor allotted $4,265 in his bank account to complete three residential jobs. If he estimates Job 1 to cost $1,032, Job 2 to cost $943, and Job 3 to cost $1,341, how much money will he have left in the account for unexpected change orders? $ _________

3. An HVAC contracting company sent out three work crews to complete three installations over the past week. If Crew One worked 10-, 9-, 11-, 12-, and 9-hour days, Crew Two worked 9-, 12-, 12-, 9-, and 9-hour days, and Crew Three worked 12-, 12-, 10-, 9-, and 11-hour days, how many total hours did the three crews work for the week? _________ hrs

On-Site

Adding It Up!

It is your first day on the job. Your supervisor asks you to distribute materials to three job sites. She asks you to take the following to each site:

Job 1	Job 2	Job 3
4 boxes of 10d nails	10 boxes of 10d nails	13 boxes of 10d nails
6 boxes of drywall screws	9 boxes of drywall screws	11 boxes of drywall screws
18 rolls of R-19 insulation	23 rolls of R-19 insulation	5 rolls of R-19 insulation
12 bags of mortar mix	4 bags of mortar mix	14 bags of mortar mix

How much of each material do you load on the truck?

_______ boxes of 10d nails

_______ boxes of drywall screws

_______ rolls of insulation

_______ bags of mortar mix

On-Site

Adding and Subtracting Whole Numbers

Regardless of whether you work in a sheet metal shop, on the plumbing or electrical of a residential home, or on repairing the roof of a commercial building, you will be required to add and subtract whole numbers at some point. Whether you are laying out a sheet metal fitting, determining the amount of material needed for a project, or balancing your checkbook, the processes of addition and subtraction are required.

4. A general contractor ordered three different sized windows to complete a job on a residential home. He estimated that the bow window would cost $874; one 36" × 36" double-hung window would cost $67; and one 36" × 54" double-hung window would cost $93. If he has $1,250 set aside to purchase the windows, how much will he have left after buying them? $ ________

5. An electrical contractor has $24,296 in his business's bank account. If at the end of the week he deposits $3,428 in payments made, then pays out $2,263 in wages, how much money will he have left in his account? $ ________

2.5.0 Multiplying Simple Whole Numbers

Multiplication is a quick way to add the same number many times. For example, it would be easier to multiply 8 times 4 than to add 4 eight times to get a **product** of 32. To multiply whole numbers, use the following steps:

Step 1 Line up the digits in the top number and the bottom number by place value columns. Position the greater number on the top.

```
 374
× 26
```

Step 2 Start with the units column (the right column) and multiply the 6 in 26 by the 4 in 374 (6 × 4 = 24). Write down the 4 in the units column and carry the 2 to the tens column. Unlike addition, you will multiply the 6 with every digit in the top number before proceeding to the 2.

```
  2
 374
× 26
   4
```

Step 3 Multiply the 6 in 26 by the 7 in 374, then add the 2 that you carried over (6 × 7 = 42 + 2 = 44). Write down the 4 in the tens column and carry the 4 to the hundreds column.

```
 42
 374
× 26
  44
```

Step 4 Multiply the 6 in 26 by the 3 in 374, then add the 4 that you carried over (6 × 3 = 18 + 4 = 22). Write down a 2 in the hundreds column and a 2 in the thousands column.

```
  42
  374
×  26
2,244
```

Step 5 You will then need to multiply each digit in 374 by the 2 in 26. Begin by multiplying by the 4 in 374 (2 × 4 = 8). Write the 8 down under the tens column. Notice that you do not write the 8 down in the units column because the 2 in the number 26 is in the tens column. A zero (0) can be placed in the units column to help keep the columns aligned.

```
  374
×  26
2,244
   80
```

Step 6 Multiply the 2 in 26 by the 7 in 374 (2 × 7 = 14). Write down a 4 in the hundreds column and carry the 1.

```
   1
  374
×  26
2,244
  480
```

Step 7 Multiply the 2 in 26 by the 3 in 374, then add the 1 that you carried over ($2 \times 3 = 6 + 1 = 7$). Write a 7 in the thousands column.

```
    1
  374
×  26
2,244
+7,480
```

Step 8 Add the two products to get a final product of 9,724.

```
    1
  374
×  26
2,244
+7,480
9,724
```

2.6.0 Dividing Whole Numbers

Division is the opposite of multiplication. Instead of adding a number several times ($5 + 5 + 5 = 15$; $5 \times 3 = 15$), when you divide you subtract a number several times to find a **quotient**. To divide whole numbers, use the following steps:

Step 1 Begin by setting up the division problem using the division bar as shown below. Position the number you are dividing by on the left side of the division bar and position the number you are dividing into on the right side of the division bar.

```
24)2,638
```

Step 2 Unlike addition, subtraction, and multiplication, where you start the procedure in the units column, with division you start with the place value that is farthest to the left. In this example, you divide 24 into the 26 of 2,638 because 24 cannot go into 2. It does, however, go into 26 one time, so write a 1 above the 6 and write 24 under the 26 in 2,638. Then subtract the two numbers to get 2. Remember to use zeros (0) to help keep your place values straight.

```
    0,1??
24)2,638
   -2 4
    0 2
```

Step 3 Bring down the next number in 2,638 (the 3) and place it next to the 2. Next, determine if 24 can go into 23. The answer is no, so put a 0 above the 3 on the answer line. 24 times 0 is 0, so write zeros in the appropriate place value columns and subtract the two numbers to get 23.

```
    0,10?
24)2,638
   -2 4↓
    0 23
   -0 00
    0 23
```

Step 4 Bring down the next number in 2,638 (the 8) and place it next to the 3. Next, determine if 24 can go into 238. Yes it can, but how many times? To figure this, think: 24 is close to 25. 238 is close to 250. How many times can 25 go into 250? 10 times. 10 times 24 is 240, so that is too high. Try 9; 9 times 24 is 216, so it works. Write a 9 in the answer line next to the 0. Write 216 below the 238 and subtract to get a **remainder** of 22.

The answer is written as 109 r22.

```
    0 10?            0 109  r22
24)2,638         24)2,638
   -2 4↓            -2 4↓
    0 23             0 23
   -0 00↓           -0 00↓
    0 238            0 238
                    -0 216
                        22
```

On-Site

Multiplying and Dividing Whole Numbers

Just like adding and subtracting whole numbers, regardless of which craft you choose, you will be required to multiply and divide whole numbers at some point. Whether you are estimating the number of drywall sheets required to finish a room or determining the number of electrical outlets necessary to wire a three-story building, the processes of multiplication and division are required.

On-Site

Application of Multiplication and Division

Let's take a look at this example of how these processes are used on a job site.

For 2 feet O.C. (on center) ceiling joists or trusses, the nail spacing for a drywall ceiling application is commonly 7 inches O.C. on ends and double-nail 12 inches O.C. in the field. When using this pattern, it takes approximately 56 drywall nails for each 4' × 12' sheet. If you use this nail pattern, approximately how many drywall nails will you need to hang the ceiling drywall in Bedroom 4?

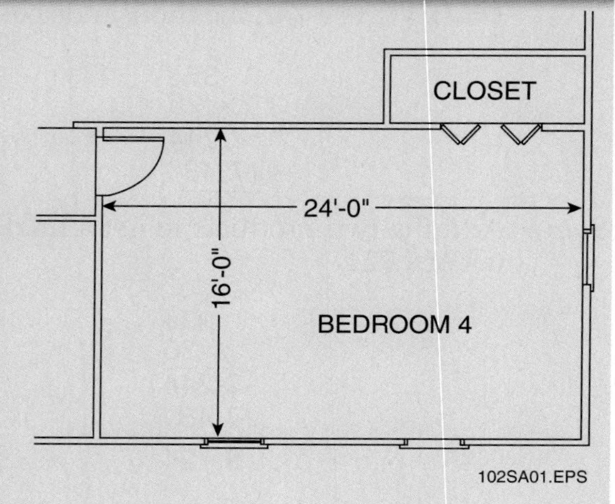

_____________ nails

Solution

Step 1 Calculate the square feet of ceiling space using multiplication. Multiply the room length by the room width.

$$\begin{array}{r} \overset{2}{2}4 \\ \times 16 \\ \hline 144 \\ 24 \\ \hline 384 \end{array}$$ square feet of ceiling

Step 2 Using multiplication, calculate the total number of square feet covered by each drywall sheet. Multiply the sheet length by the sheet width.

$$\begin{array}{r} 12 \\ \times\ 4 \\ \hline 48 \end{array}$$ square feet covered by each drywall sheet

Step 3 Using division, determine how many drywall sheets are required to cover the ceiling. Divide the ceiling square feet by the square feet covered by one drywall sheet.

$$\begin{array}{r} 008 \\ 48\overline{)384} \\ \underline{384} \\ 000 \end{array}$$ sheets of drywall

Step 4 Using multiplication, determine the number of drywall nails required per drywall sheet. Multiply the number of nails required per sheet by the total number of sheets required.

$$\begin{array}{r} \overset{4}{5}6 \\ \times\ 8 \\ \hline 448 \end{array}$$ nails required

Step 5 448 drywall nails are required to hang the ceiling drywall in Bedroom 4.

2.7.0 Study Problems: Multiplying and Dividing Whole Numbers

Use multiplication and division to solve the following problems. Read each question carefully to determine the appropriate procedure. Be aware that addition or subtraction may also be required. Be sure to show all of your work.

1. Your supervisor sends you to the truck for 180 feet of electrical wire. When you get there, you find that the coils of wire come in 15-foot lengths. How many coils of wire will you need to bring back?__________
2. If the following amounts of lumber need to be delivered to 2 different staging areas at 4 different job sites, how many total boards of each size will you need?
   1. 65: 2 × 4s __________
   2. 45: 2 × 8s __________
   3. 25: 2 × 10s __________
3. If one plumbing job requires 135 feet of plastic pipe, and a second job requires 90 feet, how many sections of pipe will you need if it comes in 20-foot sections?__________

   Will there be any pipe left over?__________
4. If a crane rental company charges $383 a day, $1,224 a week (5-day week), and $3,381 a month:

   (a) How much would it cost to rent the crane for 3 days?__________

   (b) How much would it cost to rent the crane for 12 days?__________

   (c) How much would it cost for one month and 17 days?__________
5. If a house is 34 feet wide by 56 feet long, and you need parallel lengths of rebar to run through the footing around its **perimeter**, how many 20-foot lengths of rebar will you need?____________________

Did You Know?

Except for the division bar (which actually has no official name but has been referred to as the obelus), every other part of a math problem has an official name.

Addition

3 addend
+4 addend
7 sum

Subtraction

5 minuend
−2 subtrahend
3 difference

Multiplication

4 multiplicand
×3 multiplier
12 product

Division

3 quotient
divisor 5)15 dividend

2.8.0 Practical Applications

The practical applications below illustrate how a tradesperson may use the processes of addition, subtraction, multiplication, and division on a job site. Use the information provided to solve each problem. Be sure to show all of your work.

1. Find the perimeter (distance around) each of the foundations in *Figures 2 and 3.*

   a. Foundation A perimeter = _____ feet
   b. Foundation B perimeter = _____ feet

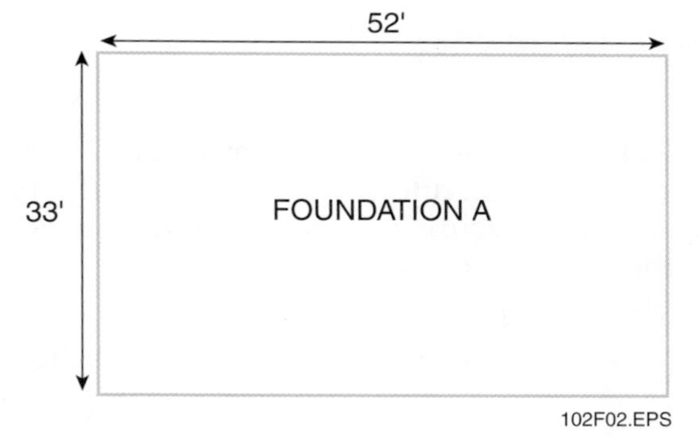

Figure 2 Foundation A.

Did You Know?

An 8" × 8" × 16" concrete block is actually 15⅝" × 7⅝" × 7⅝" in dimension. It is referred to as an 8" × 8" × 16" block because of the addition of the ⅜-inch mortar joint between the blocks. This measurement has been adopted as the trade standard for joints. Adding the ⅜-inch to the block's actual dimensions causes the block's measurements to come out in even inches, rather than as fractions. This holds true for other sizes and types of blocks.

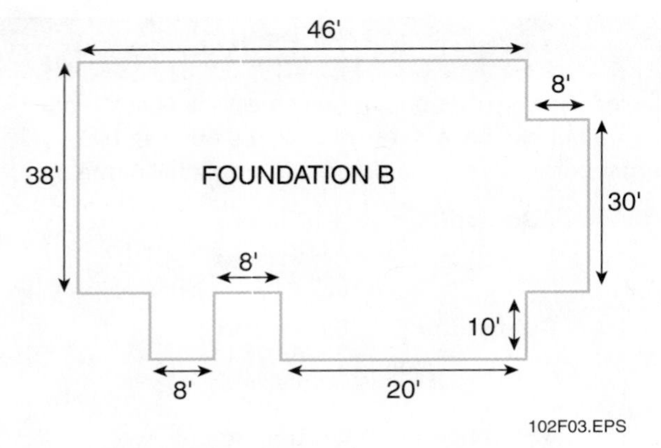

Figure 3 Foundation B.

2. How many of the concrete blocks shown in *Figure 4* are required to erect the 8-foot high basement walls for Foundation A in the previous question? _____

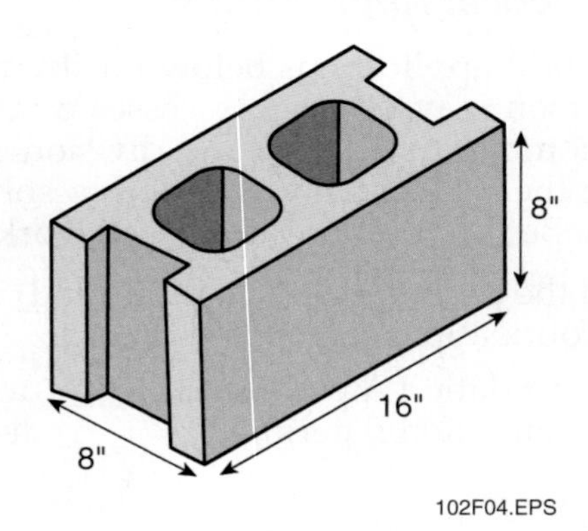

Figure 4 Concrete block.

3. Determine the total number of full plywood sheets required to cover one side of the roof shown in *Figure 5*, if the plywood comes in the following dimensions:
   a. 4' × 8' sheets; requires _____ full sheets
   b. 4' × 12' sheets; requires _____ full sheets

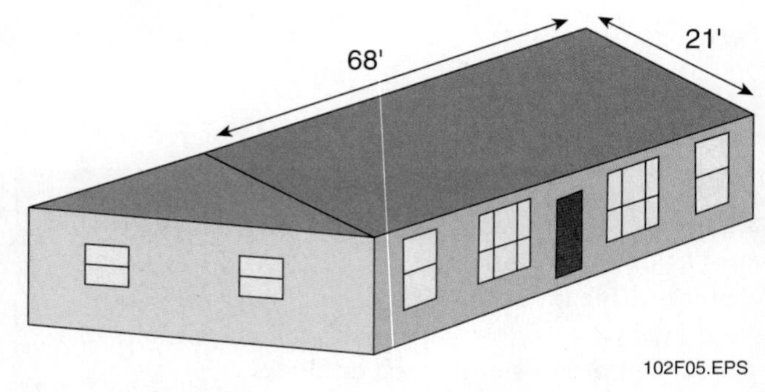

Figure 5 Roof.

4. Determine the total number of whole roofing underlayment rolls required to cover both sides of the roof shown in *Figure 5* if the rolls come in the following dimensions:
   a. 250' × 4' rolls; requires _____ rolls
   b. 432' × 3' rolls; requires _____ rolls
5. Determine the total number of drip edge strips required to go around the perimeter of the roof shown in *Figure 5* if the drip edge comes in the following dimensions:
   a. 10 foot strips; requires _____ strips
   b. 12 foot strips; requires _____ strips

3.0.0 Working with Length Measurements

In the construction trade, you will need to use a ruler or measuring tape to measure various objects. There are two primary rulers you will see on the job: the **standard ruler** (also known as the **English ruler**) and the **metric ruler** (*Figure 6*). You may also use a standard or metric measuring tape. Other types of measurement tools include the **architect's scale** and the **engineer's scale** (also referred to as a civil scale), both of which are used by draftsmen when creating a set of drawings. In this section, you will learn how to use these measuring tools, and then you will practice taking measurements.

NOTE

Always remember to double-check your measurements before cutting a piece of material. Keep in mind the old saying, "Measure twice, cut once."

Did You Know?

A 2 × 4 board used in the field is not actually 2 inches by 4 inches in dimension. The only time the board is truly 2 inches by 4 inches is when it is initially rough-cut from the log. After the board has been dried and planed, it is reduced to a finished size of 1½ inches by 3½ inches. Take a look at the true dimensions of some other common board sizes.

Nominal Size	Actual Measure	Metric Measure
1" × 10"	¾" × 9¼"	19 × 235 mm
2" × 6"	1½" × 5½"	38 × 140 mm
2" × 8"	1½" × 7¼"	38 × 184 mm
2" × 10"	1½" × 9¼"	38 × 235 mm
4" × 4"	3½" × 3½"	89 × 89 mm

On-Site

Number of Courses

When erecting block walls, you will often hear the phrase, "The number of courses." The number of courses refers to the number of times the blocks will need to be stacked in order to establish the correct height of the wall. If you are told to erect a 15-course wall using 8" × 8" × 16" concrete blocks, you would be building a wall that is 10 feet high.

Block height = 8"
Number of courses = 15
12 inches = 1 foot

Therefore:

8" × 15 = 120"
120" ÷ 12" = 10 feet

CAUTION

Most rulers similar to the examples shown in *Figure 6*, are designed with extra material at the ends in order to maintain accurate measurements in the event that one of the ends is damaged. When using this type of ruler, make sure that you start your measurement at the first marked line and not at the absolute end of the ruler.

3.1.0 Reading a Standard Ruler (English Ruler)

The standard ruler is divided into whole inches and then halves, fourths, eighths, and sixteenths. Some standard rulers are divided into thirty-seconds and some into sixty-fourths. These represent fractions of an inch. In this section, you will work with a standard ruler and standard fractions. In *Figure 6* the ruler is marked with both $\frac{1}{16}$- and $\frac{1}{8}$-inch increments. Standard rulers are usually designed with two sets of measurements; however, the distances on the standard ruler in *Figure 7* are marked only in $\frac{1}{16}$-inch increments.

In *Figure 7* the increment between the $\frac{1}{8}$-inch and $\frac{1}{4}$-inch increments would be labeled as $\frac{3}{16}$ of an inch. Similarly, the increment immediately after the $\frac{3}{4}$-inch increment would be labeled as $\frac{13}{16}$ of an inch.

THE STANDARD, OR ENGLISH, RULER

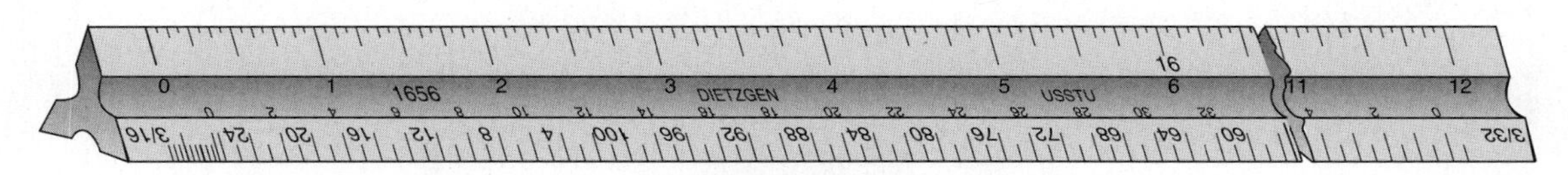

ARCHITECT'S SCALE

THE METRIC RULER

102F06.EPS

Figure 6 Types of measurement tools.

3.1.1 Study Problems: Reading a Standard Ruler

Use the information from *Figure 7* to help you identify the marked lengths numbered 1 through 10 in *Figures 8* and *9*. Label each length in the space provided.

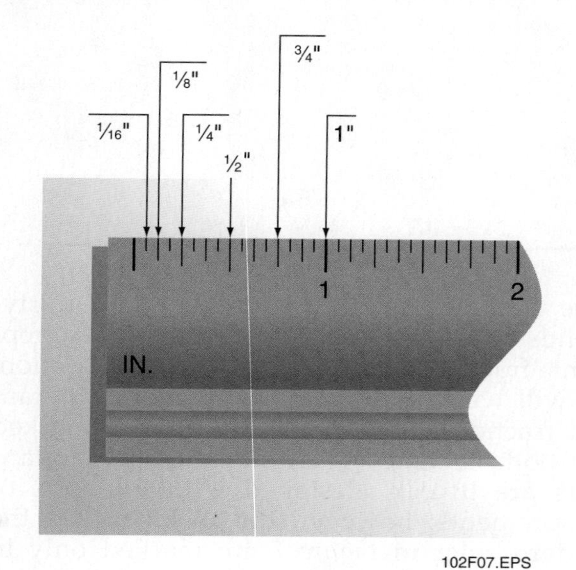

Figure 7 Standard ruler showing 1/16-inch increments.

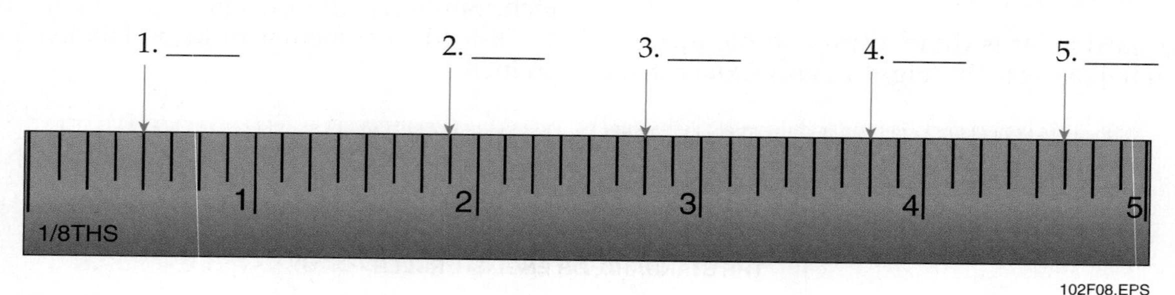

Figure 8 1/8-inch ruler.

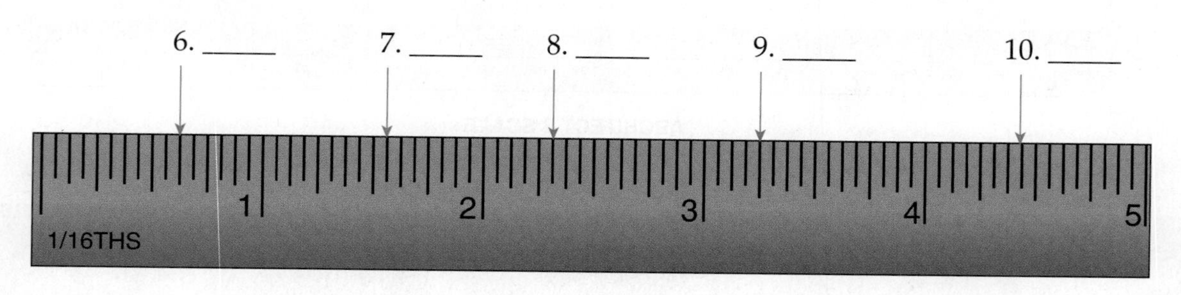

Figure 9 1/16-inch ruler.

3.2.0 The Metric Ruler

Each year, more corporations and industries are **converting** to the metric system of measurements to run their day-to-day operations. There is also a big shift in the construction trades toward the use of the metric system, with the potential for a full conversion in the next several years. Like our standard (English) system, the metric system is also used to measure weight, length, temperature, and **volume**. A common length used in the metric system is the **meter**, which is approximately 39.37 inches. National code books are a good example of where you can see both English and metric measurements.

3.2.1 Reading a Metric Ruler

Metric ruler increments (*Figure 10*) are divided into centimeters and millimeters. The larger lines with numbers printed next to them are centimeters, and the smaller lines are millimeters. The metric system is known as a base ten system, since each millimeter increment is 1/10 of a centimeter. For example, if you measure 7 marks after 2 centimeters (*Figure 10*), it is 2.7 (two point seven) centimeters or 27 millimeters.

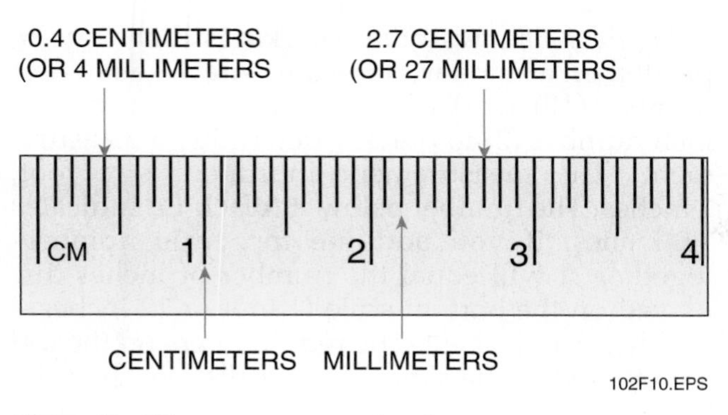

Figure 10 Distances on a metric ruler.

A measurement less than 1 centimeter, for example, six marks before the 1-centimeter mark (*Figure 10*), would be recorded as 0.4 (point four) centimeters or 4 millimeters.

3.2.2 Study Problems: Reading a Metric Ruler

Identify the marked lengths in *Figure 11*. Record the lengths for increments 1, 2, and 3 in centimeters, and the increments for 4 and 5 in millimeters.

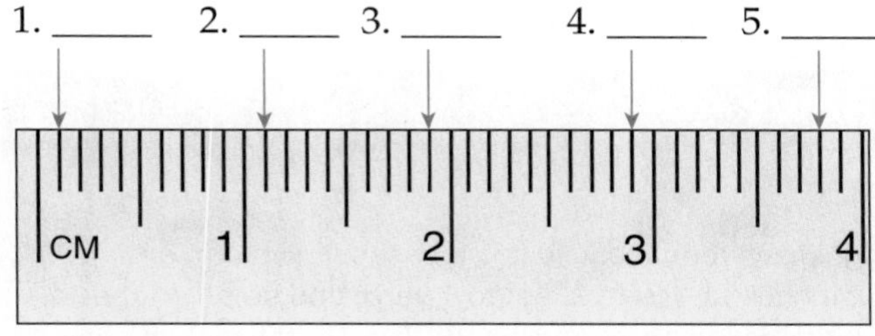

Figure 11 Metric ruler study problem.

3.3.0 The Measuring Tape (Standard and Metric)

The measuring tape's blades show either standard markings, metric markings, or both (*Figure 12*). If the blade only has one set of markings, they will be printed exactly the same on both sides of the blade so that measurements can be taken accurately from either side.

Common lengths for English or metric measuring tapes are 16 feet (5 meters) and 25 feet (7.5 meters). These are also common lengths for measuring tapes with only one set of markings on the blade.

3.3.1 The Standard Tape Measure

The standard measuring tape is marked similarly to the 16ths-inch standard ruler covered earlier in this section with a few additional markings. Along with the ½-inch, ¼-inch, ⅛-inch, and ¹⁄₁₆-inch markings, the standard measuring tape also features additional markings that make the task of wall framing easier.

As shown in *Figure 13A,* every 24 inches is marked with a contrasting black background. 24-inch spacing on center is used most commonly for nonbearing walls. The markings in *Figure 13B* are used for the 16-inch on center spacing most

Figure 12 Measuring tape showing English and metric measurements.

GOING GREEN

Recycling Lumber

A recent waste characterization study found that 22 percent of our nation's solid waste comes from construction and demolition activities (136 million tons per year), and that wood represents about 25 percent of the waste.

Even though only a small percentage of salvaged wood can be reused directly as lumber, almost all of it can be recycled to make other wood products, such as particle board, strand board, or larger dimensional wood for nonstructural applications. For a company, recycling lumber can save them money in disposal fees and can also create some revenue from the sales of the salvaged materials.

In terms of the environment, recycling and reusing salvaged lumber (if applied on a national level) can significantly reduce the timbering of many types of trees, especially pine, which is commonly used for 2 × 4s. Besides lumber you can also salvage and reuse or recycle other materials such as drywall, metals (metal is most commonly recycled), masonry, carpet, pipe, rocks, dirt, plastic, or green waste related to land development.

Before incorporating a recycling program into your company's policies, you must first check with your local governing body for information on a lumber grading service that is certified by the American Lumber Standards (ALS).

commonly used for loadbearing walls. These are highlighted with a red background. The distinction 19.2 inches in *Figure 13C* is marked with a small black diamond; this spacing is an alternate and less-commonly used spacing scheme.

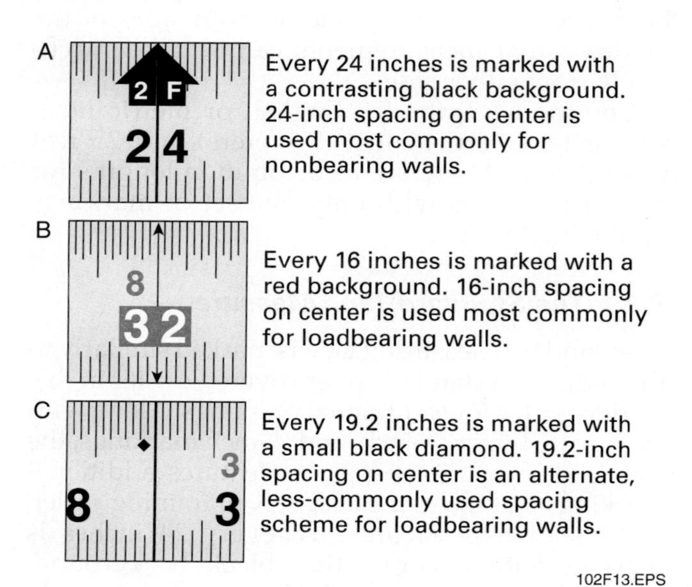

Figure 13 Wall framing markings on a tape measure.

Figure 14 shows other markings that may be printed on a standard tape measure. The red foot number (1F) works in conjunction with the red inch number (2) to quickly determine a measurement. The measurement indicated is 1 foot, 2 inches. The number below it (black 14) indicates 14 inches. If you add the top scale numbers together, it will equal the number of inches displayed on the bottom scale (1 foot or 12 inches + 2 inches = 14 inches). The red numbers on the top are marked in the same way for the length of the blade.

Figure 14 Other markings on a standard tape measure.

Did You Know?

Metric System Facts, Part One

It is believed that the metric system is easier to use than the English system because it is a base ten system, and because it does not require you to make conversions such as 12 inches to 1 foot, 3 feet to 1 yard, and so on. In the metric system, you only need to remember a list of prefixes that are based on powers of 10. These prefixes are shown in the table below.

To make a conversion in the metric system, simply move the decimal point left or right to the appropriate power of 10 (deci, deka, etc.), and add zeros if necessary. For example, if you have 100 centiliters and want to convert to hectoliters, simply move the decimal four places to the left, adding zeros (or divide by 10,000). You will see that 100 centiliters converts to 0.01 hectoliters.

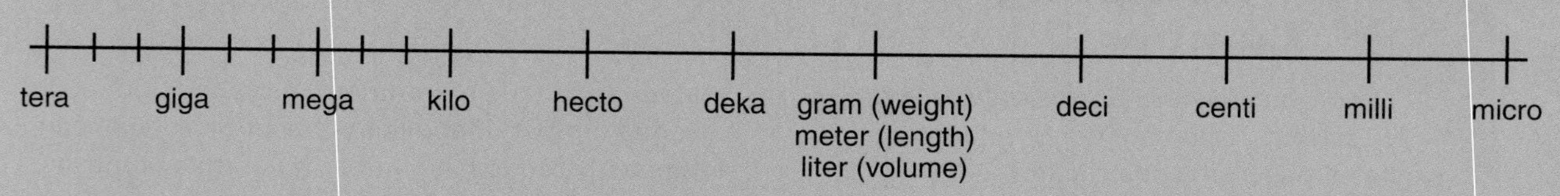

$10^0 = 1$ **BASE UNIT:** GRAM (WEIGHT), METER (LENGTH), LITER (VOLUME) $10^0 = 1$

MULTIPLIER	PREFIX	MEANING	PREFIX	MEANING	MULTIPLIER
$10^1 = 10$	deka-	ten	deci-	tenth	$10^{-1} = 0.1$
$10^2 = 100$	hecto-	hundred	centi-	hundredth	$10^{-2} = 0.01$
$10^3 = 1,000$	kilo-	thousand	milli-	thousandth	$10^{-3} = 0.001$
$10^6 = 1,000,000$	mega-	million	micro-	millionth	$10^{-6} = 0.000001$
$10^9 = 1,000,000,000$	giga-	billion	nano-	billionth	$10^{-9} = 0.000000001$
$10^{12} = 1,000,000,000,000$	tera-	trillion	pico-	trillionth	$10^{-12} = 0.000000000001$

102SA02.EPS

3.3.2 Using a Tape Measure

Step 1 Pull the tape straight out of its case with one hand.

Step 2 Hook the tape to one end of the object being measured using the metal hook on the end of the tape. If you are unable to hook onto the material, it may be necessary to have another worker hold the end of the tape. Be aware that the hook end is designed to slide a small amounnt to compensate for its own thickness. If the end does not slide freely, your measurements can be off by as much as 1⁄16 of an inch. A bent or damaged hook can have the same effect.

Step 3 Extend the tape by pulling the case to the desired length.

Step 4 When the desired length is attained, slide the thumb lock on the case down to hold the tape in place. Not all tape measures have a thumb lock, and some tape measures will lock automatically. On automatic locking tape measures, a thumb hold release is used to retract the tape. Never let the tape retract rapidly back into the case, because this can cause damage to the tool or injury.

Did You Know?

Standard Measure

Determining the distance to a neighbor's farm may have been the earliest measurement of interest to people. Imagine that you wanted to measure the distance to a favorite fishing pond. You could walk—pace off—the distance and count the number of steps you took. You might find that it was 70 paces between your home and the pond. You could then pace off the distance to another pond. By comparing the paces of one distance to the other, you could tell which distance was longer and by how much. Pace became a unit of measure.

Feet, arms, hands, and fingers were useful for measuring all sorts of things. In fact, the body was such a common basis for measuring that we still have traces of that system in our measurement standards. Lengths are measured in feet. Horses are said to be so many hands high. An inch roughly matches the width of someone's thumb.

Complications arise when you have to decide whose legs, arms, or hands to use for measuring. If you wanted to mark off a piece of land, you might choose a tall person with long legs in hopes of getting more than your money's worth. If you owed several lengths of rope, you would want someone with short arms to measure the quantity and possibly save you some rope.

The solution is to arrive at a measurement that everyone agrees on. Legend says that the foot measurement used in France was the length of Charlemagne's foot. The yard is supposed to have been the distance from King Henry I of England's nose to the fingertips of his outstretched arm. Kings were not going to travel around making measurements for people, so the solution was to transfer the measurement to a stick. The distance marked on the stick became the standard measurement, and the government could send out duplicate sticks to each of the towns and cities as secondary standards. Civil officials could then check local merchants' measurements using the secondary standards.

During the Middle Ages, associations of craftspeople enforced strict adherence to the established standards. The success and reputation of the craft depended on providing accurately measured goods. Violators who used faulty measurements were often punished.

3.3.3 Study Problems: Reading a Metric and English Tape Measure

Identify the marked lengths numbered 1 through 5 in *Figures 15* and *16*.

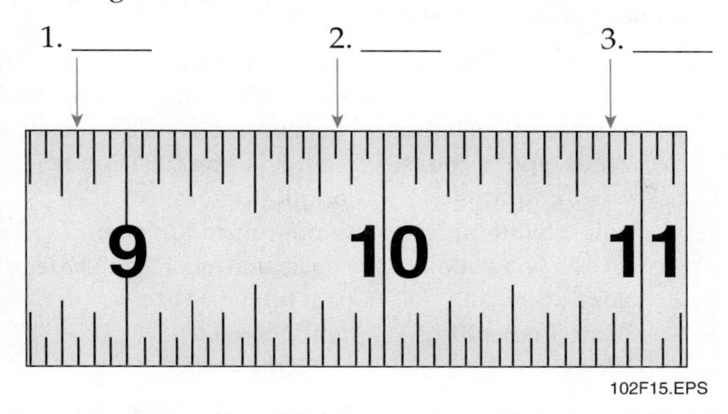

Figure 15 English measuring tape.

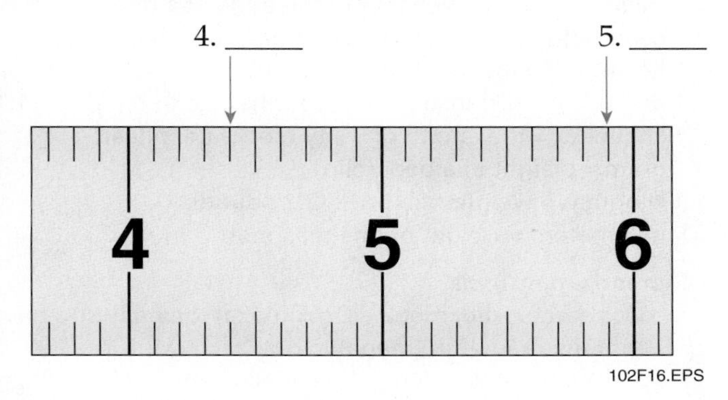

Figure 16 Metric measuring tape.

3.4.0 Other Types of Scales

Standard and metric rulers and measuring tapes are usually used in a shop or in the field for a variety of measuring tasks. However, there are other types of rulers called scales, which are used by draftsmen to produce drawings. These include the architect's scales, the **metric scale** (metric architect's scale), and the engineer's scale. Knowing how these scales work will help you understand the information contained in a set of drawings.

3.4.1 The Architect's Scale

The architect's scale is often used to create construction drawings. An architect's scale translates the large measurements of real structures (rooms, walls, doors, windows, duct, etc.) into smaller measurements for drawings. Architect's scales are available in several types, but the most common include the triangular and flat scales. A flat scale is shown in *Figure 17*. The triangular architect's scale is most commonly used because it can combine up

Did You Know?

Metric System Facts, Part Two

Ease of Use

The metric system is very easy to use. The basic unit of length is a meter. Anything longer than 1 meter is a multiple of the meter. For example, kilo- is a prefix meaning thousand. So a kilometer is 1,000 meters. The prefix mega- means one million. So a megameter is 1,000,000 meters (although you rarely hear this term used).

Anything shorter than a meter is a part of a meter and can be calculated using division. The most common smaller units of a meter used in measuring are the centimeter and the millimeter. The centi- prefix means one hundredth ($\frac{1}{100}$) and milli- means one thousandth ($\frac{1}{1,000}$). A centimeter is $\frac{1}{100}$ of a meter. A millimeter is $\frac{1}{1,000}$ of a meter.

All the units of measure in the metric system differ by multiples of 10. This makes it easy to convert measurements within the metric system.

Centimeter = 1×0.01
Millimeter = 1×0.001
Meter = 1
Kilometer = $1 \times 1,000$
Megameter = $1 \times 1,000,000$

To convert 1 kilometer to meters, multiply by 1,000. Thus, 1 kilometer equals 1,000 meters ($1 \times 1,000$). To convert 1 meter to centimeters, divide by 0.01. Thus, 1 meter equals 100 centimeters ($1 \div 0.01 = 100$).

The name of each metric measurement tells you two things:

- What type of measurement it is (the basic unit):
  weight (grams) length (meters)
  volume (liters) temperature (Celsius)
- Its size (in relation to the basic unit, such as the meter):
  deka (da) = 10 deci (d) = 0.1
  hecto (h) = 100 centi (c) = 0.01
  kilo (k) = 1,000 milli (m) = 0.001

Metric Examples Visualized (estimates)

1 millimeter = the thickness of the edge of a dime
1 centimeter = the width of a standard paperclip
1 decimeter = the length of a crayon
1 meter = the distance from a door handle to the floor (1.1 yard)
1 kilometer = the length of 6 city blocks (.6 miles)
1 gram = weight of a paperclip
1 kilogram = weight of a brick (2.2 pounds)
5 millimeters = weight of one teaspoon

Memorization Tools

- deca, hecto, kilo, mega – Deck (deka) the halls and have a heck of a (hecto) good time, but don't keel (kilo) over from the mega-fun (mega)!
- deci, centi, milli, micro – Desi (deci) sent (centi) Milli (milli) to Micronesia (micro).

to twelve different scales on one tool. Each side of the triangular form has two faces. Two scales are combined on each face. Architect's scales are available in 6- and 12-inch lengths.

3.4.2 Reading an Architect's Scale

Each scale on an architect's scale is designated to a different fraction of an inch that equals a foot. These fraction designations appear on the right and left corners of each scale. You read an architect's scale from left to right or right to left, depending on which scale you are reading.

Look at the point of measurement in *Figure 17*. It represents 57 feet when read from the left on the ⅛-inch scale and equals 18 feet when read from the right on the ¼-inch scale.

Now look at *Figure 18*. Using the ⅜-inch scale and reading from the right, you can determine that the section of duct is 7-feet, 6-inches long. Notice how the 0 point on an architect's scale is not at the extreme end of the measuring line. This is because numbers to the right of the 0 represent fractions of one foot (or inches).

3.4.3 Metric Scale (Metric Architect's Scale)

Similar to the architect's scale is the metric scale, sometimes referred to as a metric architect's scale (*Figure 19*). Common lengths for metric scales are 30 and 60 millimeters. Like the architect's scale, a metric scale can have a number of scales on it and is used to generate drawings.

Did You Know?

Standard measuring tapes can also be marked with 1/32-inch increments. These measuring tapes are used mainly in engineering where precise measurements are required. However, some 1/16-inch measuring tapes can be marked with 1/32-inch increments for the first few feet, as shown in the illustration.

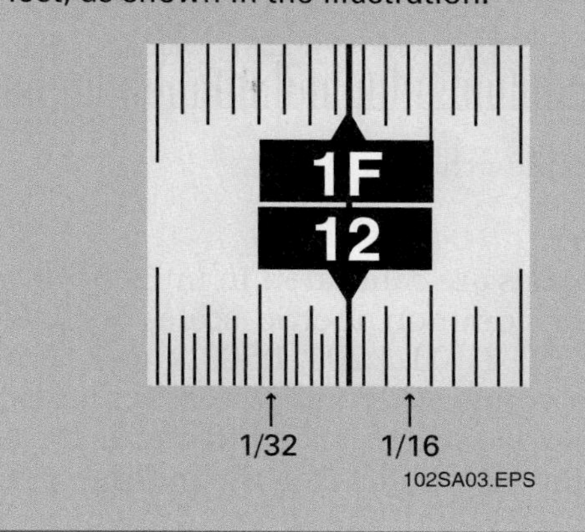

Did You Know?

When communicating measurements verbally on the job site that include ¼- or ¾-inch increments, you will hear them referred to as a quarter inch and three quarters of an inch, not as one-fourth inch and three-fourths inch. For example, a measurement of 5'-3¾" would be said as five feet and three and three-quarters of an inch, not as five feet and three and three-fourths of an inch.

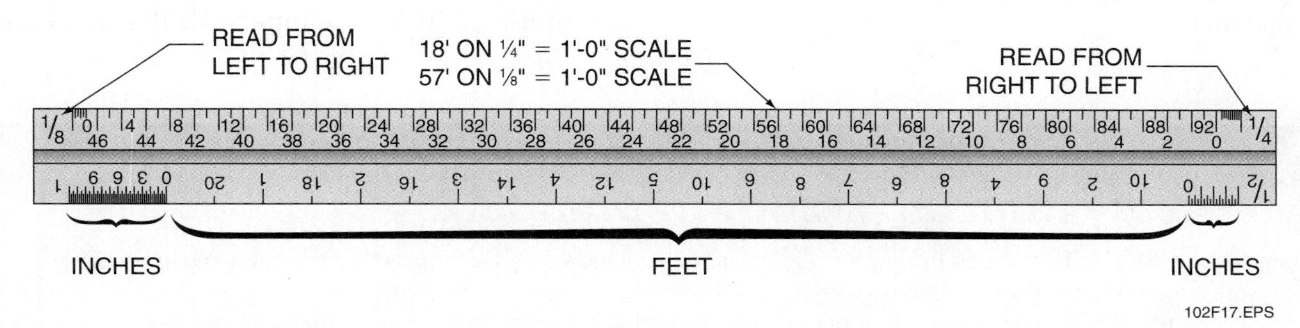

Figure 17 The architect's scale.

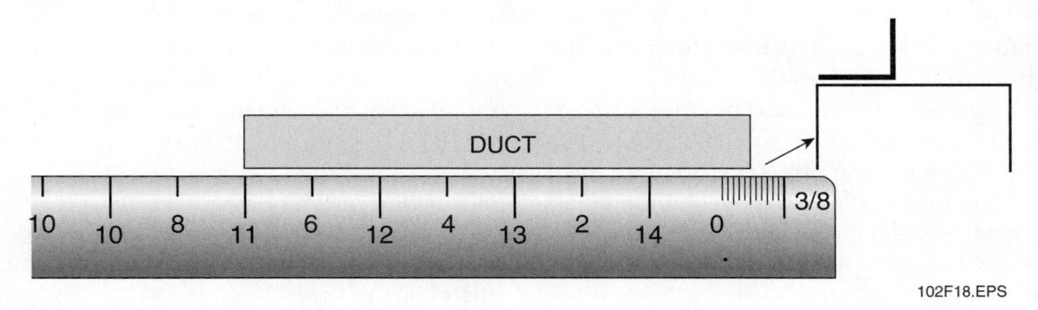

Figure 18 Measuring a section of duct with an architect's scale.

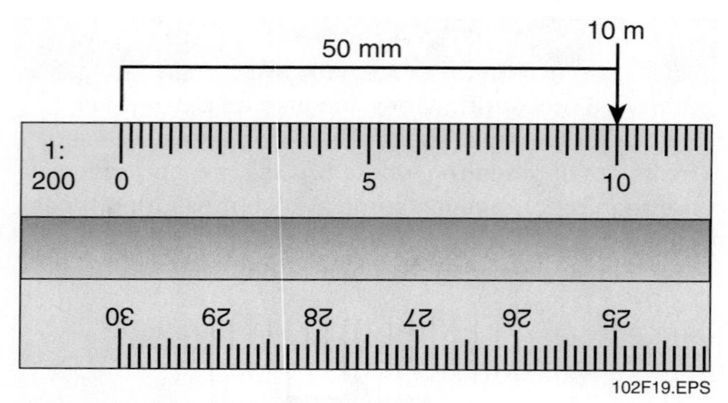

Figure 19 Metric architect's scale.

Metric scales are calibrated in units of 10. Some of the most common metric scales include 1:5, 1:10, 1:20, 1:50, 1:100, 1:200, 1:500, and 1:1,000.

The two common length measurements used with the metric scale on architectural drawings are the meter and millimeter, the millimeter being $\frac{1}{1000}$ of a meter. On drawings drawn to scale between 1:1 and 1:100, the millimeter is typically used. The millimeter symbol (mm) will not be shown, but there should be a note on the drawing indicating that all dimensions are given in millimeters unless otherwise noted.

On drawings with scales between 1:200 (*Figure 19*) and 1:2,000, the meter is generally used. Again, the meter symbol (m) will not be shown, but the drawing will have a note indicating that all dimensions are in meters unless otherwise noted. Land distances shown on site and plot plans, expressed in metric units, are typically given in meters or kilometers (1,000 meters).

3.4.4 Reading a Metric Scale

Reading a metric scale is easy once you identify the scale to use and the length that the scale increments represent on the drawing. Here's how it works.

The unit of length in *Figure 19* is the meter and the object on the drawing starts at 0 and extends out to the 10 on the scale. At a 1:200 ratio this object would be 10 meters long. This is because every millimeter on the scale represents 200 millimeters on the object. If you have 50 millimeters from 0 to 10 and multiply the 50 by 200, you will come up with 10,000 mm (50 mm × 200 = 10,000 mm). Because there are 1000 mm in one meter, if you divide 10,000 by 10, and because the unit being used is the meter, you get 10 meters. This same process is also used to determine lengths for other ratios on the scale.

3.4.5 The Engineer's Scale

The engineer's scale (*Figure 20*) is used mainly for land measurements on site plans, which means the scale must accommodate very large measurements. Each engineer's scale is set up as multiples of 10 and the measurements are taken in decimals. This is different from the architect's scale in that a unit is represented by a portion of an inch. The most common engineer's scales are 10, 20, 30, 40, 50, and 60, which can all be combined on the triangular engineer's scale.

3.4.6 Reading an Engineer's Scale

For each scale, the measurements can represent various units derived from that scale number and a multiple of 10. For example on a 10 scale, 1 inch

Did You Know?

Metric System as a Modern Standard

As scientific thought developed during the 1500s and later, scholars and scientists had trouble explaining their measurements to one another. The measuring standards varied among countries and sometimes even within one country. By the 1700s, scientists were debating how to establish a uniform system for measurement.

In France in the 1790s, scientists created a standard length called a meter (based on the Latin word for measure). This became the basis for the metric system that is still in use throughout most of the world. Not until the 1970s did both the United States and Canada begin to switch over to the metric system. There is still resistance to the metric system in the United States, even though virtually every other country in the world uses it.

The original international standard bar measuring a meter was a platinum-iridium bar kept in the International Bureau of Weights and Measures near Paris, France. The bar was made from a platinum-iridium mixture because it would not rust or change over time. That ensured the accuracy of the standard.

In more modern times, scientists have found that a natural standard measure was more accurate than anything made out of metal. In 1983 the speed of light was calculated as exactly 299,792,458 meters per second. So the distance that light travels in $\frac{1}{299,792,458}$ second is the definition of the length of a meter. This may be a more precise measure than you will need on the construction site, but it illustrates the accuracy of modern techniques.

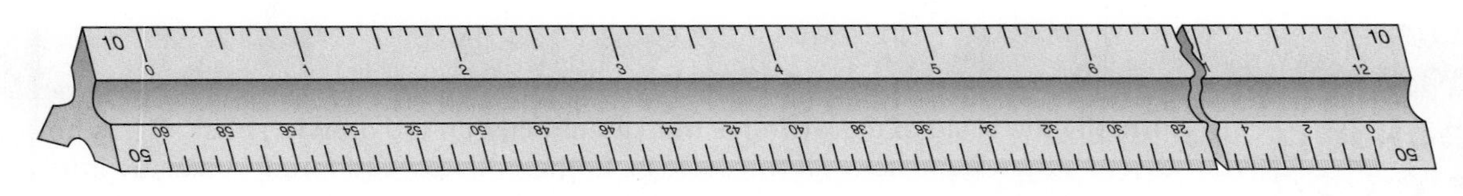

Figure 20 The engineer's scale.

can represent 1 foot (10 × 0.10), 10 feet (10 × 1.0), 100 feet (10 × 10.0), and so on. On a 50 scale, an inch can be 5 feet (50 × 0.10), 50 feet (50 × 1.0), or 500 feet (50 × 10.0), and so on. This same process can also be used to determine lengths on the scales.

3.4.7 Study Problems: Reading Other Types of Scales

1. Determine the lengths (in feet and inches) of the sections of duct using the architect's scales shown in *Figure 21*:

   a. ______

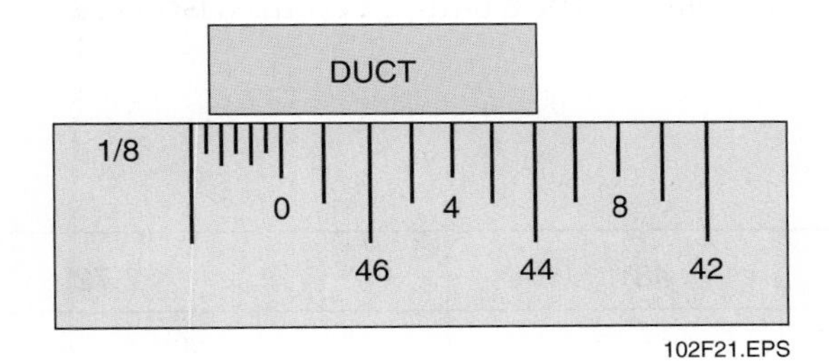

Figure 21 Duct measurement using ⅛ scale.

   Figure 22:

   b. ______

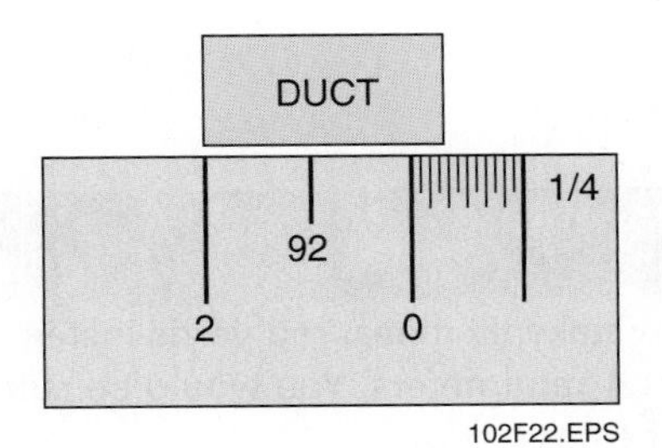

Figure 22 Duct measurement using ¼ scale.

2. Identify the marked lengths on the metric scale in *Figure 23*. Use the meter as your unit.

   a. ______

   b. ______

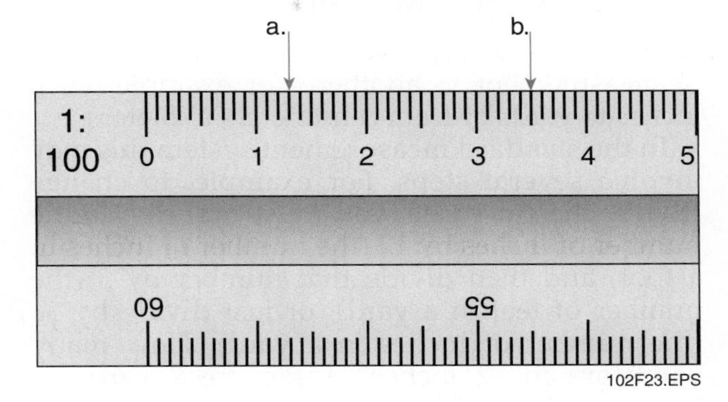

Figure 23 Meter measurements.

3. Identify the marked lengths in *Figure 24* on the engineer's scale. It has been determined that 1 inch equals 10 feet.

   a. ______

   b. ______

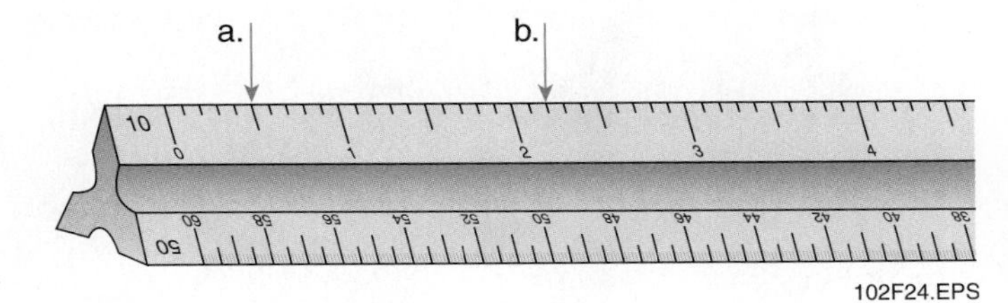

Figure 24 Engineer's scale measurements.

Did You Know?

The concave (or curve) of a measuring tape blade is designed to strengthen the blade and also to make the tape easier to read when it is placed on a surface that needs to be marked.

Once the proper measurement is established and the blade tip is secure, rotate the blade edge nearest you slightly until it lays flat on the surface, and then mark your material. Using this method makes it easier to read the measurement and to mark the material more accurately.

On-Site

Checking Scales on Drawings

Normally the scale of a drawing is marked directly on the drawing itself. If it is not, however, do not guess what the scale is. Rather, compare different scales by identifying a given length on the drawing (for example, an 8-foot, 6-inch pipe), and then find the scale that measures the pipe at 8-feet, 6-inches. Check at least two items on the drawing this way before you proceed.

3.5.0 Converting Measurements

Sometimes you may need to change from one unit of measurement to another—for example, from inches to yards or from centimeters to meters.

In the standard measurement system this may involve several steps. For example, to change from inches to yards, you must first divide the number of inches by 12 (the number of inches in a foot) and then divide that number by 3 (the number of feet in a yard), or just divide by 36 (the number of inches in a yard). How many yards are in 72 inches? There are 2 yards in 72 inches (*Figure 25*).

The metric system makes the conversion much simpler. You can simply move the decimal point, because the system is built on multiples of 10. How many meters are there in 72 centimeters?

Because 1 centimeter = 0.01 meter (move decimal two places to the left), 72 centimeters = 0.72 meters. There are 0.72 meters in 72 centimeters (*Figure 26*).

Converting measurements from the English system to the metric system, and vice versa, is more complicated. At this stage in your training, you will not be responsible for making such conversions, but it is important that you are aware of them. Many dictionaries and other reference books contain simple comparison charts that show basic equivalents between English system measurements and metric system measurements. See *Appendix C* for a sample comparison chart.

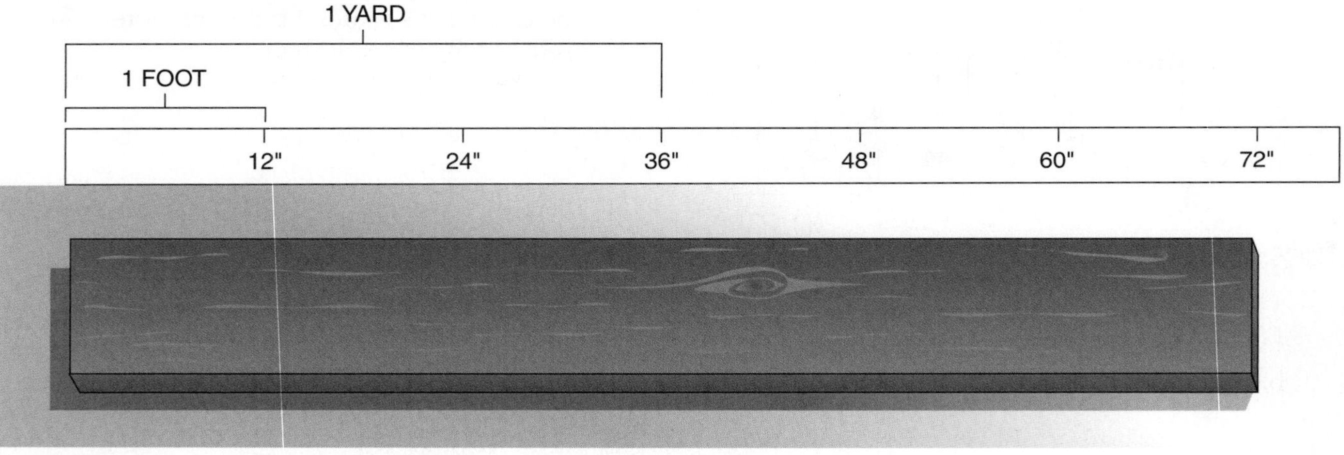

Figure 25 Converting inches to yards.

On-Site

Metrics on the Job

Imagine the problems you would have if you mistakenly measured yards instead of meters when cutting material. A yard is only 91.4 centimeters. You would be cutting everything 8.6 centimeters (or more than 3 inches) too short.

More companies are using the metric system as their unit of measure. With the increasingly international character of large firms, employees must be able to rely on a standard system of measurement.

In 1999, a $1.25 million satellite was sent to observe the planet Mars. It bounced off the Martian atmosphere and spun uselessly out into space because workers on Earth forgot to convert their measurements from feet and inches into meters before entering the data into the computer. The costly mistake set back the scientific research of Mars by years.

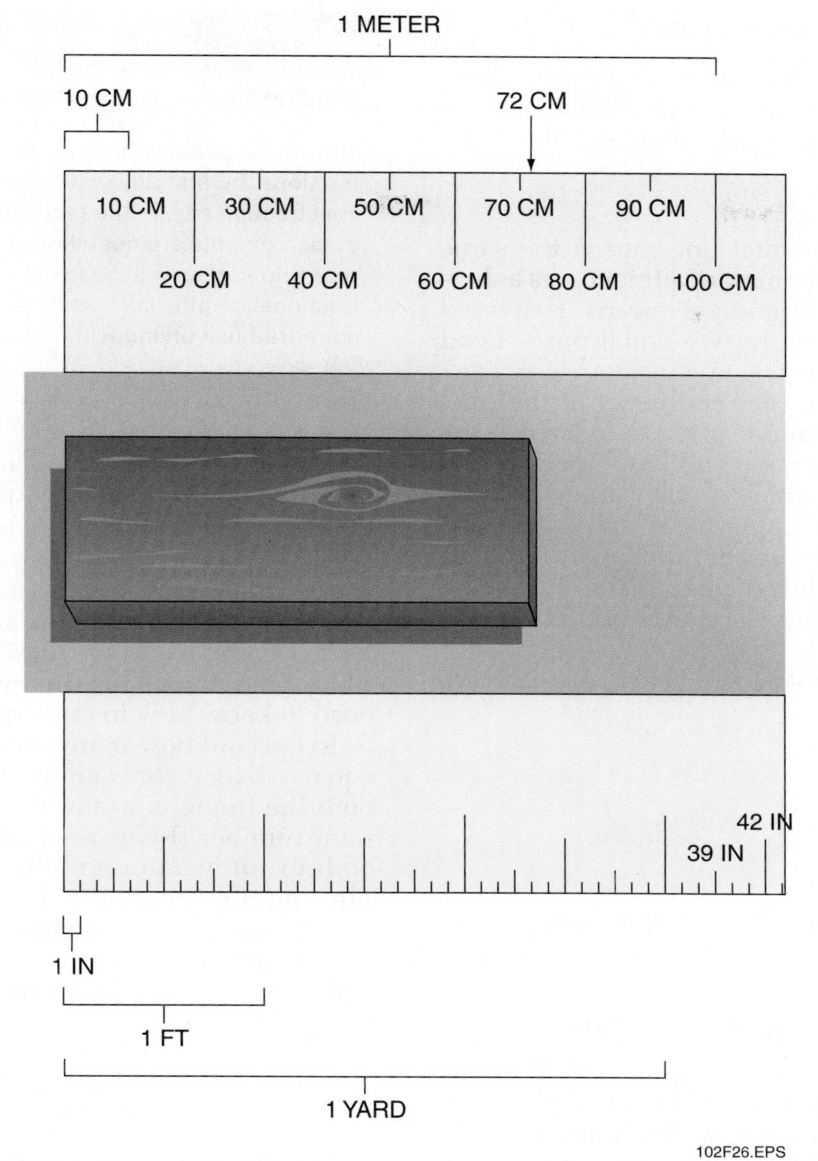

Figure 26 Converting centimeters to meters.

3.5.1 Study Problems: Converting Measurements

Find the answers to the following conversion problems without using a calculator.

1. 0.45 meter = _____ centimeter(s)
2. 3 yards = _____ inches
3. 36 inches = _____ yard(s)
4. 90 inches = _____ yards
5. 1 centimeter = _____ meters

Did You Know?

Common Terms

The most common units of measure in the metric system are the meter, kilometer, millimeter, and centimeter. Unit names for measurements larger than a kilometer are not common. For example, the term hectometer (100 meters) is not often used. In the Olympics you hear about the 200-meter and the 400-meter races, not the 2-hectometer and the 4-hectometer races.

Measurements smaller than the millimeter (0.001 meter) are usually used by scientists. The micrometer (0.000001 meter) and nanometer (0.000000001) are not used in everyday measuring.

4.0.0 What Are Fractions?

A fraction divides whole units into parts. Common fractions are written as two numbers, separated by a slash or by a horizontal line, like this:

$$1/2 \text{ or } \frac{1}{2}$$

The slash or horizontal line means the same thing as the ÷ sign. So think of a fraction as a division problem. The fraction ½ means 1 divided by 2, or one divided into two equal parts. Read this fraction as one-half.

The lower number (**denominator**) of the fraction tells you the number of parts by which the upper number (**numerator**) is being divided. The upper number is a whole number that tells you how many parts are going to be divided. In the fraction ½, the 1 is the upper number, or numerator, and the 2 is the lower number, or denominator. These numbers are also referred to as the terms of the fraction.

What measurement is the arrow in *Figure 27* pointing to?__________

a. 8⁄16
b. 2⁄4
c. 4⁄8
d. ½
e. All are correct.

The correct answer is E . . . all of the answers are correct.

4.1.0 Finding Equivalent Fractions

Equivalent fractions, such as ½, 2⁄4, and 4⁄8, have the same value. Their value can be determined by dividing the denominator by the numerator. In this case, each listed fraction's quotient is 2, and thus they are equal. If you cut off a piece of wood 8⁄16-foot long, and then cut off a piece ½-foot long, both pieces of wood would be the same length.

When you measure objects, you often need to record all measurements as the same (common) fractions—in sixteenths of an inch, for example. Doing this allows you to easily compare, add, and subtract fractional measurements. This is why you need to know how to find equivalent fractions.

To find out how many sixteenths of an inch are equal to ½ inch, for example, you need to multiply both the numerator and the denominator by the same number. (Remember that a fraction in which both the numerator and the denominator are the same number is equal to 1.)

Ask yourself what number you would multiply by 2 to get 16. The answer is 8, so you multiply both numbers by 8. (Remember that the fraction 8⁄8 is equal to 1.)

$$\frac{1}{2} \times \frac{8}{8} = \frac{8}{16}$$

The answer is 8⁄16 inch is equivalent to ½ inch.

> **On-Site**
>
> **Fractions on the Job**
>
> You will realize how important it is to understand fractions the first day you report to work on a construction site. In the real world, most measurements are not whole numbers. Typically, pipe and lumber will be measured and cut to fractional lengths such as 3⁄8, 5⁄16, or 3⁄4. Being comfortable working with fractions is an important job skill.

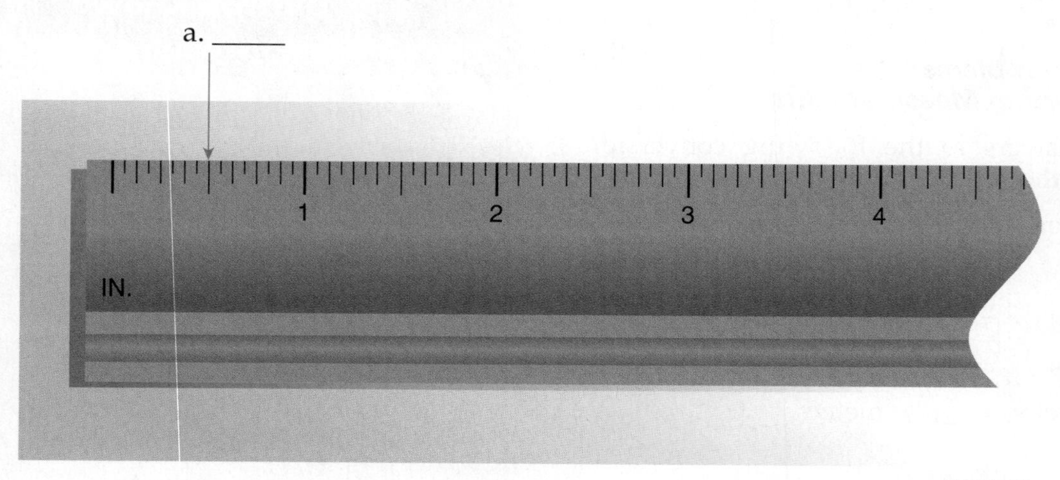

Figure 27 Fractions of an inch.

4.1.1 Study Problems: Finding Equivalent Fractions

Find the equivalents of the following measurements:

1. $\frac{1}{4}$ inch = _____$/_{16}$ inch
   a. 2
   b. 4
   c. 6
   d. 8

2. $\frac{2}{16}$ inch = _____$/_{32}$ inch
   a. 1
   b. 2
   c. 4
   d. 8

3. $\frac{3}{4}$ inch = _____$/_{8}$ inch
   a. 2
   b. 4
   c. 5
   d. 6

4. $\frac{3}{4}$ inch = _____$/_{64}$ inch
   a. 48
   b. 50
   c. 52
   d. 54

5. $\frac{3}{16}$ inch = _____$/_{32}$ inch
   a. 2
   b. 4
   c. 6
   d. 8

On-Site

Practical Application

Before moving into the processes of addition, subtraction, multiplication, and division with fractions, here is an example of how some of these processes can be used on the job.

If you have a 2-inch × 6-inch stud wall with ½-inch drywall on one side and ⅝-inch drywall on the other side, determine the width of the jamb needed for your door. Note: Use 5½ inches for the thickness of the 2 × 6 studs.

Jamb width = _____

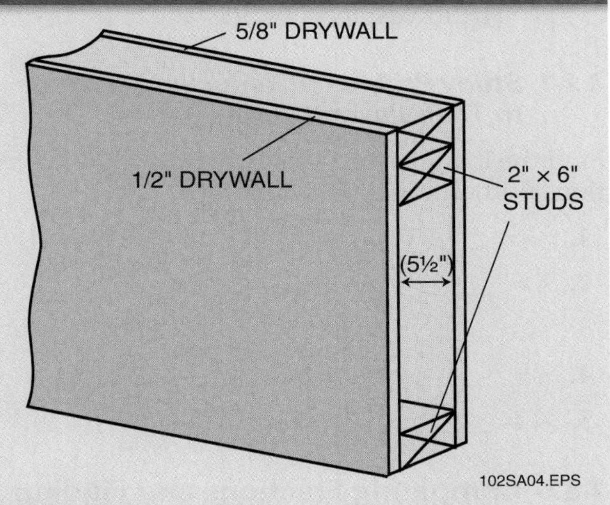

102SA04.EPS

Solution

Step 1 Calculate the combined widths of the drywall sheets using addition. Add ½-inch and ⅝-inch by finding a common denominator of 8.

$$\frac{1}{2} + \frac{5}{8}$$

$$\frac{1 \times 4 = 4}{2 \times 4 = 8} + \frac{5}{8} = \frac{9}{8} \text{ inches of drywall, or } 1\tfrac{1}{8} \text{ inches}$$

Step 2 Calculate the combined width of the drywall sheets with the width of the wall stud using addition. Add 1⅛ inches to 5½ inches by finding a common denominator of 8.

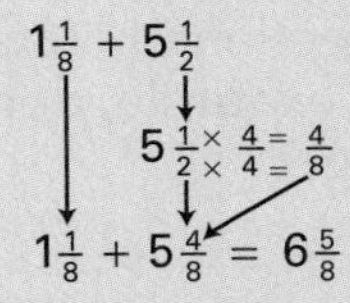

$$1\tfrac{1}{8} + 5\tfrac{1}{2}$$

$$5\,\frac{1 \times 4 = 4}{2 \times 4 = 8}$$

$$1\tfrac{1}{8} + 5\tfrac{4}{8} = 6\tfrac{5}{8}$$

Step 3 The door jamb needs to be 6⅝ inches wide.

4.2.0 Reducing Fractions to Their Lowest Terms

If you find that the measurement of something is $4/16$, you may want to reduce the measurement to its lowest terms so the number is easier to work with. To find the lowest terms of $4/16$, use division as follows:

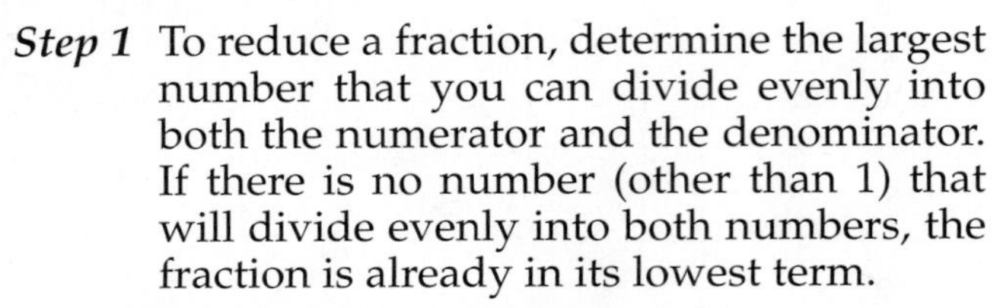

Step 1 To reduce a fraction, determine the largest number that you can divide evenly into both the numerator and the denominator. If there is no number (other than 1) that will divide evenly into both numbers, the fraction is already in its lowest term.

Step 2 Divide the numerator and the denominator by the same number. In this example, you could divide both the numerator and the denominator by 4.

$$\frac{4 \div 4}{16 \div 4} = \frac{1}{4}$$

The lowest term of $4/16$ is $1/4$.

4.2.1 Study Problems: Reducing Fractions to Their Lowest Terms

Find the lowest term of each of the following fractions without using a calculator.

1. $2/16 =$ _____
2. $2/8 =$ _____
3. $12/32 =$ _____
4. $4/8 =$ _____
5. $4/64 =$ _____

4.3.0 Comparing Fractions and Finding Lowest Common Denominators

Which measurement is larger, $3/4$ or $5/8$?

To find the answer, think about this question: Would you have a longer section of 2 × 4 if you had three sections from a board that was cut up into four equal sections (*Figure 28*) or if you had five sections of a board that was cut up into eight equal sections (*Figure 29*)?

As you can see, it's hard to compare fractions that do not have common denominators, just as it is hard to compare 2 × 4s that are sectioned off into different lengths. Using our 2 × 4 example, determine which section of 2 × 4 is longer:

$$\frac{3}{4} \text{ or } \frac{5}{8}$$

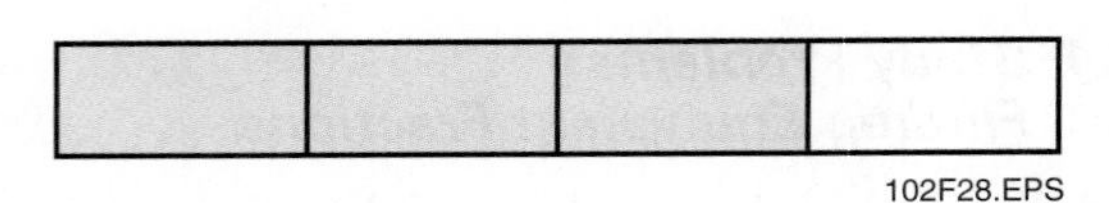

Figure 28 Four sections of a 2 × 4.

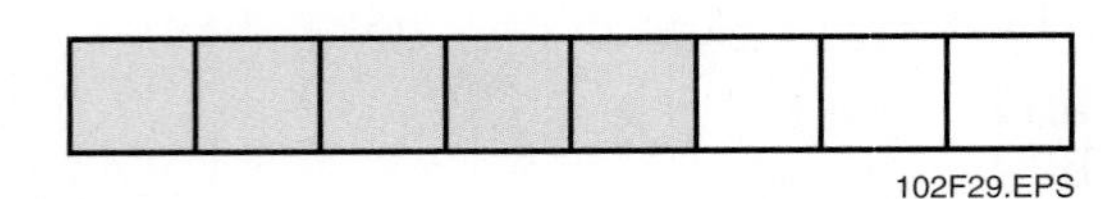

Figure 29 Eight sections of a 2 × 4.

To compare, you need to find a common denominator for the board sections. The common denominator is a number that both denominators can go into evenly.

Step 1 Multiply the two denominators together (4 × 8 = 32). This is a common denominator between the two fractions. You found a common denominator so that you can compare the pieces more easily.

Step 2 Now convert the two fractions so that they will have the same denominator of 32.

$$\frac{3}{4} \times \frac{8}{8} = \frac{24}{32}$$

$$\frac{5}{8} \times \frac{4}{4} = \frac{20}{32}$$

Now it's easy to compare the two fractions to see which is larger. You would have a longer section of 2 × 4 if you choose $3/4$ because you would have $24/32$ instead of $20/32$ of the 2 × 4.

You have found a common denominator for the 2 × 4 problem. However, working with fractions like $24/32$ or $20/32$ is difficult. To make this problem easier, you can find the lowest common denominator, which means reducing the fractions to their lowest terms.

To find the lowest common denominator, follow these steps:

Step 1 Reduce each fraction to its lowest terms.

Step 2 Find the lowest common multiple of the denominators. Sometimes this is as simple as one denominator already being a multiple of the other, meaning you can multiply by a whole number to get the larger number. If this is the case, all you have to do is find the equivalent fraction for the term with the smaller denominator.

Step 3 If neither of the denominators is a multiple of the other, you can multiply the denominators together to get a common denominator.

Let's look at the 2×4 example again where $\frac{3}{4}$ and $\frac{5}{8}$ are already in their lowest terms. When you look at the denominators, you see that 8 is a multiple of 4. So you should find the equivalent fraction for $\frac{3}{4}$ that has a denominator of 8.

$$\frac{3}{4} \times \frac{2}{2} = \frac{6}{8}$$

You can now compare $\frac{6}{8}$ to $\frac{5}{8}$ and see that $\frac{6}{8}$ is the larger fraction.

Whether you find the lowest common denominator or just multiply the denominators to find a common denominator will depend on the situation. In some applications, you may want all fractions involved to have a particular denominator. By finding the lowest common denominator, you can decrease the amount of multiplying you need to do and reduce the chances of making a mathematical error.

4.3.1 Study Problems: Finding the Lowest Common Denominator

Find the lowest common denominator for the following pairs of fractions.

1. $\frac{2}{6}$ and $\frac{3}{4}$ _____.

a. 6
b. 10
c. 12
d. 16

2. $\frac{1}{4}$ and $\frac{3}{8}$ _____.

a. 4
b. 8
c. 12
d. 18

3. $\frac{1}{8}$ and $\frac{1}{2}$ _____.

a. 3
b. 5
c. 7
d. 8

4. $\frac{1}{4}$ and $\frac{3}{16}$ _____.

a. 8
b. 16
c. 18
d. 20

5. $\frac{4}{32}$ and $\frac{5}{8}$ _____.

a. 8
b. 14
c. 21
d. 32

4.4.0 Adding Fractions

How many total inches will you have if you add $\frac{3}{4}$ inch plus $\frac{5}{8}$ inch? To answer this question, you will have to add two fractions using the following steps:

Step 1 Find the common denominator of the fractions you wish to add. A common denominator for $\frac{3}{4}$ and $\frac{5}{8}$ is 32.

Step 2 Convert the fractions to equivalent fractions with the same denominator as shown.

$$\frac{3}{4} \times \frac{8}{8} = \frac{24}{32}$$

$$\frac{5}{8} \times \frac{4}{4} = \frac{20}{32}$$

Step 3 Add the numerators of the fractions. Place this sum over the denominator.

$$\frac{24}{32} + \frac{20}{32} = \frac{44}{32}$$

Step 4 Reduce the fraction to its lowest terms. When you reduce $\frac{44}{32}$ to $\frac{11}{8}$, it becomes a fraction in which no number, other than 1, will go evenly into both the numerator and the denominator. But it is an **improper fraction**, meaning the numerator is larger than the denominator.

In this case, you need to reduce the improper fraction, $\frac{11}{8}$, to its lowest terms, $1\frac{3}{8}$. You will soon learn how to convert improper fractions to **mixed numbers**.

4.4.1 Study Problems: Adding Fractions

Find the answers to the following addition problems. Remember to reduce the sum to the lowest terms.

1. $\frac{1}{8} + \frac{4}{16}$ = _____

2. $\frac{4}{8} + \frac{6}{16}$ = _____

3. $\frac{2}{4} + \frac{1}{4}$ = _____

4. $\frac{3}{4} + \frac{2}{8}$ = _____

5. $\frac{14}{16} + \frac{3}{8}$ = _____

4.5.0 Subtracting Fractions

Subtracting fractions is very much like adding fractions. You must find a common denominator before you can subtract. For example, you have a piece of wood 7/8 of a foot long. If you use 1/4 of a foot, how much do you have left?

Step 1 Find the common denominator. In this case it is 8.

$$\frac{7}{8} \quad \frac{1}{4}$$

Step 2 Multiply each term of 1/4 by 2 to get a fraction with the denominator of 8.

$$\frac{1}{4} \times \frac{2}{2} = \frac{2}{8}$$

Step 3 Subtract the numerators; 5/8 of a foot is left.

$$\frac{7}{8} - \frac{2}{8} = \frac{5}{8}$$

You have 5/8 of a foot left.

4.5.1 Study Problems: Subtracting Fractions

Find the answers to the following subtraction problems. Reduce the differences to their lowest terms.

1. 3/8 – 5/16 = _____
2. 11/16 – 5/8 = _____
3. 3/4 – 2/6 = _____
4. 11/12 – 4/8 = _____
5. 11/16 – 1/2 = _____

4.5.2 Subtracting a Fraction from a Whole Number

Sometimes you must subtract a fraction from a whole number. For example, you need to take 1/4 of a day off from a five-day workweek. How many days will you be working that week?

Here is how to set up this type of problem:

Step 1 To subtract a fraction from a whole number, borrow 1 from the whole number to make it into a fraction.

$$\begin{array}{rcl} 5 & = & 4 + 1 \\ -\frac{1}{4} & & -\frac{1}{4} \\ \hline \end{array}$$

Step 2 Convert the 1 to a fraction having the same denominator as the number you are subtracting.

$$\begin{array}{rcl} 5 & = & 4 + \frac{4}{4} \\ -\frac{1}{4} & & -\frac{1}{4} \\ \hline \end{array}$$

Step 3 Subtract and reduce to the lowest terms.

$$\begin{array}{rcl} 5 & = & 4 + \frac{4}{4} \\ -\frac{1}{4} & & -\frac{1}{4} \\ \hline & & 4 + \frac{3}{4} \end{array}$$

You will work 4 3/4 days that week.

4.5.3 Study Problems: Subtracting a Fraction from a Whole Number

Find the answers to the following subtraction problems and reduce the fractions to their lowest terms.

1. 8 – 3/4 = _____
2. 12 – 5/8 = _____
3. Two punches are made from a bar of stock 9 7/16 inches long. If one punch is 4 1/64 inches long and the other 4 3/32 inches long, how many inches of stock are not used?
   a. 1 1/16
   b. 1 10/32
   c. 1 15/64
   d. 1 21/64
4. If you saw 12 1/16 inches off a board that is 20 3/4 inches long, you'll have _____ inches left over.
   a. 8 1/4
   b. 8 11/16
   c. 11 1/4
   d. 16 11/8
5. A rough opening for a window measures 36 3/8 inches. The window to be placed in the rough opening measures 35 15/16 inches. The total clearing that will exist between the window and the rough opening will be _____ inch(es).
   a. 1
   b. 1 7/16
   c. 7/16
   d. 1 12/16

4.6.0 Multiplying Fractions

Multiplying and dividing fractions is very different from adding and subtracting fractions. You do not have to find a common denominator when you multiply or divide fractions.

In a word problem, the words used let you know if you need to multiply. If a problem asks "What is $\frac{2}{8}$ of 9?" then think of the problem this way: $\frac{2}{8} \times \frac{9}{1}$. Note that any number (except 0) over 1 equals itself.

Using $\frac{4}{8} \times \frac{5}{6}$ as an example, follow these steps:

Step 1 Multiply the numerators together to get a new numerator. Multiply the denominators together to get a new denominator.

$$\frac{4}{8} \times \frac{5}{6} = \frac{20}{48}$$

Step 2 Reduce if possible ($\frac{20}{48}$ reduces to $\frac{5}{12}$).

Although you can multiply fractions without first reducing them to their lowest terms, keep in mind that you can reduce them before you multiply. This will sometimes make the multiplication easier, since you will be working with smaller numbers. It will also make it easier to reduce the product to the lowest terms. What may seem like an extra step can save you time in the long run.

4.6.1 Study Problems: Multiplying Fractions

Find the answers to the following multiplication problems without using a calculator. Reduce them to their lowest terms.

1. $\frac{4}{16} \times \frac{5}{8}$ = ______
2. $\frac{3}{4} \times \frac{7}{8}$ = ______
3. $\frac{2}{8}$ of 15 = ______
4. $\frac{3}{7}$ of 49 = ______
5. $\frac{8}{16}$ of $\frac{32}{64}$ = ______

4.7.0 Dividing Fractions

Dividing fractions is very much like multiplying fractions, with one difference. You must **invert**, or flip, the fraction you are dividing by. Using $\frac{1}{2} \div \frac{3}{4}$ as an example, follow these steps:

Step 1 Invert the fraction you are dividing by ($\frac{3}{4}$).

$$\frac{3}{4} \text{ becomes } \frac{4}{3}$$

Step 2 Change the division sign (÷) to a multiplication sign (×).

$$\frac{1}{2} \div \frac{3}{4} = \frac{1}{2} \times \frac{4}{3}$$

Step 3 Multiply the fraction as instructed earlier.

$$\frac{1}{2} \times \frac{4}{3} = \frac{4}{6}$$

Step 4 Reduce if possible.

$$\frac{4}{6} \text{ reduces to } \frac{2}{3}$$

If you are working with a mixed number (for example, $2\frac{1}{3}$, you must convert it to a fraction before you invert it. Do this by multiplying the denominator by the whole number (3 × 2), adding the numerator [(3 × 2) + 1], and placing the result over the denominator. It looks like this:

$$2\tfrac{1}{3} = \frac{(3 \times 2) + 1}{3} = \frac{7}{3}$$

When dividing by a whole number, place the whole number over 1 and then invert it. Remember that $\frac{4}{1}$ is the same as 4.

For example:

$$\tfrac{1}{2} \div 4 =$$
$$\tfrac{1}{2} \div \tfrac{4}{1} =$$
$$\tfrac{1}{2} \times \tfrac{1}{4} = \tfrac{1}{8}$$

4.7.1 Study Problems: Dividing Fractions

Find the answers to the following division problems without using a calculator. Reduce them to their lowest terms, including improper fractions.

1. $\frac{3}{8} \div 3$ = ______
2. $\frac{5}{8} \div \frac{1}{2}$ = ______
3. $\frac{3}{4} \div \frac{3}{8}$ = ______
4. On a scale drawing, if $\frac{1}{4}$ of an inch represents a distance of 1 foot, then a line on the drawing measuring $8\frac{1}{2}$ inches represents ______ feet.
   a. 34
   b. 36
   c. 38
   d. 40

5. You can cut _____ ⅞-inch lengths from a 7-inch strip.
   a. 5
   b. 6
   c. 7
   d. 8

4.8.0 Practical Applications with Fractions

The following examples illustrate some practical applications a tradesperson may encounter every day on the job site that require the processes of addition, subtraction, multiplication, and division of fractions to solve. Use the information provided to solve each problem. Be sure to show all of your work.

1. After you construct the girder (a beam that runs the length of the basement) of a floor frame and support it using adjustable steel posts, the next step is to install a sill plate. The sill plate (2 × 6 board) goes around the entire top perimeter of the basement walls. To install the sill plate, you need to drill holes for anchor bolts. If the anchor bolts are located in the center of the sill plate, how far in from the edge of the 2 × 6 will you need to drill the holes? Use *Figure 30* to help you find the answer. _____ inches

   Note: A 2 × 6 board is actually 5½-inches wide.

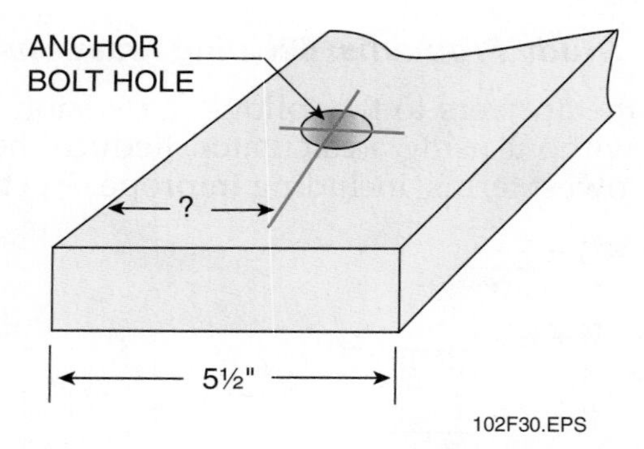

Figure 30 Drilling a sill plate for an anchor bolt.

2. Floor joists are framing members that carry the load of the floor between the sill and girder. They are usually laid out 12, 16, or 24-inch O.C. (on center). To determine how many joists you need for a 16-inch O.C. layout, divide the length of the house by ⁴⁄₃ and add 1 for the end joist. Use *Figure 31* to determine how many floor joists you will need for the whole floor. _____ joists

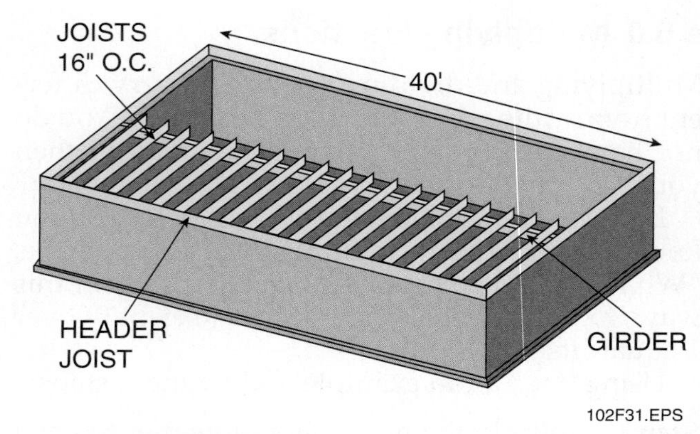

Figure 31 Floor joists.

3. In *Figure 32*, the 11-foot length of sidewalk slopes away from the front door at a rate of ⅛ inch per foot. How many inches lower will the sidewalk be at the outer edge than at the front door? _____ inch(es)

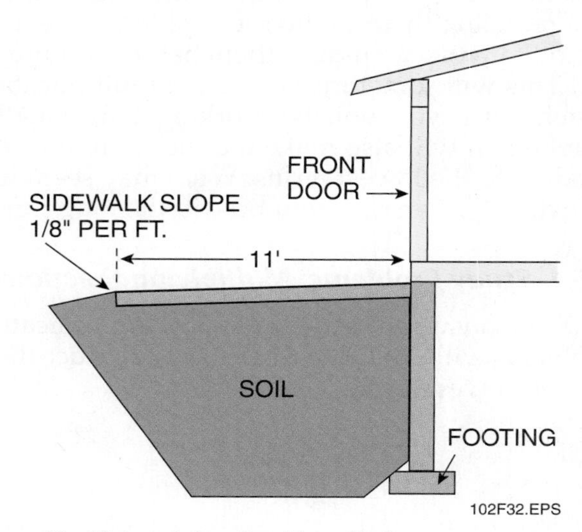

Figure 32 Determining the sidewalk slope.

5.0.0 Decimals

Earlier in this module, you learned about various types of rulers, and how decimals are used when taking and recording measurements in centimeters. Another example of a ruler that uses decimals is the **machinist's rule** (*Figure 33*). Each number shows the distance, in inches, from the squared end of the rule. The marks between the numbers divide each inch into ten equal parts. Each of these ten parts is referred to as a tenth. The nail in *Figure 33* spans one whole inch plus three-tenths of a second inch. It is one and three-tenths of an inch long. This is written as 1.3 inches.

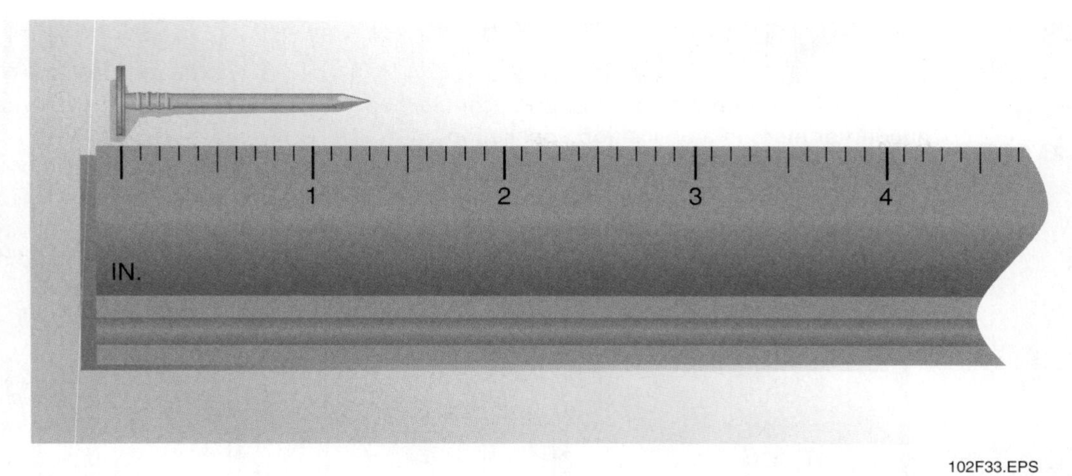

Figure 33 Using a machinist's rule to measure tenths.

5.1.0 Comparing Whole Numbers with Decimals

The following chart compares whole number place values with decimal place values:

Whole Numbers		Decimals	
1	ones	1.0	
10	tens	0.1	tenths
100	hundreds	0.01	hundredths
1000	thousands	0.001	thousandths

To read a decimal, say the number as it is written and then the name of its place value. For example, read 0.56 as fifty-six hundredths.

Mixed numbers also appear in decimals. You read 15.7 as fifteen and seven-tenths. Notice the use of the word *and* to separate the whole number from the decimal.

Did You Know?

Why Base-Ten?

When you write a number, you use a symbol for that number. This symbol is called a numeral or digit. Most of the world uses a decimal system with the numerals or digits 1, 2, 3, 4, 5, 6, 7, 8, 9, 0. The decimal system is a counting system based on the number 10 and groups of ten. The word *decimal* is from the Latin word meaning ten. That is why it is called a base-ten system. The word *digit* is from the Latin word for finger. Early people naturally used their fingers as a means of counting—a base-ten system.

5.1.1 Study Problems: Comparing Whole Numbers with Decimals

For the following problems, find the words that mean the same as the decimal or the decimal equivalent of the words.

1. 0.4 = _____

a. four
b. four-tenths
c. four-hundredths
d. four-thousandths

2. 0.05 = _____

a. five
b. five-tenths
c. five-hundredths
d. five-thousandths

3. 2.5 = _____

a. two and five-tenths
b. two and five-hundredths
c. two and five-thousandths
d. twenty-five-hundredths

4. Eighteen-hundredths = _____

a. 1.8
b. 0.18
c. 0.018
d. 0.0018

5. Five and eight-tenths = _____

a. 5.0
b. 5.8
c. 5.08
d. 5.008

On-Site

Practical Application

Before moving into the specifics on decimals, here is an example of where you may encounter them in a typical job application.

Whether you are putting up siding or drywall, constructing wall or roof frames, running electrical wiring or plumbing, you will be required to determine lengths in order to estimate the amount of material needed for the job.

What is the total length of the wall in the partial floor plan pictured, not including the openings (one door and three windows)? Note: Interior walls are 0.5 feet thick.

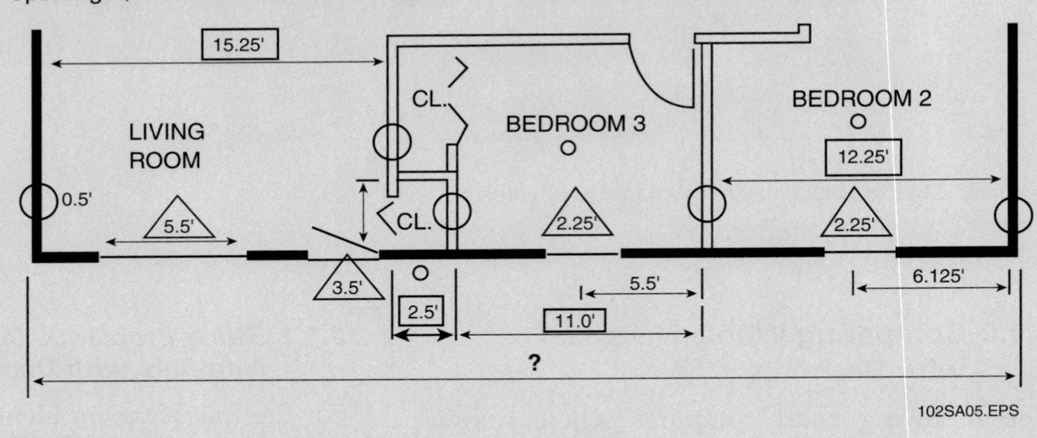

102SA05.EPS

Solution

Step 1 Identify and add the lengths of the individual rooms (identified by the rectangles).

$$\begin{array}{r} 15.25 \\ 11.00 \\ 12.25 \\ +\ 2.50 \\ \hline 41 \text{ feet} \end{array}$$

Step 2 Identify and add the widths of the interior walls (identified by the circles).

$$\begin{array}{r} .5 \\ \times\ 5 \\ \hline 2.5 \text{ feet} \end{array}$$

Step 3 Add the total widths of the interior walls to the total length of the rooms.

$$\begin{array}{r} 41.00 \\ +\ 2.50 \\ \hline 43.50 \text{ feet} \end{array}$$

Step 4 Identify and add the widths of the window and door openings (identified by the triangles).

$$\begin{array}{r} {}^{1\,1} \\ 5.50 \\ 3.50 \\ 2.25 \\ +2.25 \\ \hline 13.50 \text{ feet} \end{array}$$

Step 5 Subtract the total length of the door and window openings from the total length of the interior wall widths and room lengths.

$$\begin{array}{r} 43.50 \\ -13.50 \\ \hline 30 \text{ feet} \end{array}$$

Step 6 The length of the wall, not including door and window openings, is 30 feet.

5.2.0 Comparing Decimals with Decimals

Which decimal is the larger of the two?

0.4 or 0.42?

Here's how to compare decimals:

Step 1 Line up the decimal points of all the numbers.

0.4
0.42

Step 2 Place zeros to the right of each number until all numbers end with the same place value.

0.40
0.42

Step 3 Compare the numbers.

0.42 (42 hundredths)
is larger than
0.40 (40 hundredths)

5.2.1 Study Problems: Comparing Decimals

For the following problems, put the decimals in order from smallest to largest.

1. 0.400, 0.004, 0.044, 0.404

The answer is _____.

a. 0.400, 0.004, 0.044, 0.404
b. 0.004, 0.044, 0.404, 0.400
c. 0.004, 0.044, 0.400, 0.404
d. 0.404, 0.044, 0.400, 0.004

2. 0.567, 0.059, 0.56, 0.508

The answer is _____.

a. 0.508, 0.56, 0.567, 0.059
b. 0.059, 0.56, 0.508, 0.567
c. 0.567, 0.059, 0.56, 0.508
d. 0.059, 0.508, 0.56, 0.567

3. 0.320, 0.032, 0.302, 0.003

The answer is _____.

a. 0.003, 0.032, 0.302, 0.320
b. 0.320, 0.302, 0.032, 0.003
c. 0.302, 0.320, 0.003, 0.032
d. 0.003, 0.032, 0.320, 0.302

4. 0.867, 0.086, 0.008, 0.870

The answer is _____.

a. 0.870, 0.867, 0.086, 0.008
b. 0.008, 0.086, 0.867, 0.870
c. 0.086, 0.008, 0.867, 0.870
d. 0.008, 0.870, 0.867, 0.086

5. 0.626, 0.630, 0.616, 0.641

The answer is _____.

a. 0.616, 0.641, 0.630, 0.626
b. 0.616, 0.626, 0.630, 0.641
c. 0.641, 0.616, 0.626, 0.630
d. 0.630, 0.616, 0.626, 0.641

5.3.0 Adding and Subtracting Decimals

There is only one major rule to remember when adding and subtracting decimals: keep your decimal points lined up!

Suppose you want to add 4.76 and 0.834. Line up the problem like this, adding a 0 to help keep the numbers lined up.

$$\begin{array}{r} 4.760 \\ +0.834 \\ \hline 5.594 \end{array}$$

The same thing is true for subtraction of decimals. To subtract 2.724 from 5.6, line up the decimal points.

$$\begin{array}{r} 5.600 \\ -2.724 \\ \hline 2.876 \end{array}$$

Notice that two zeros were added to the end of the first number to make it easier to see where you need to borrow.

Did You Know?

Place-Value Systems

As mathematics advanced, counting systems became more efficient for performing calculations and solving problems. Place-value systems made it easier to represent large numbers and to simplify the process of computing. In a place-value system, the value of a particular symbol depends not only on the symbol but also on its position or place in the number.

In our decimal system, each place value is ten times greater than the place to the right. Place values in the decimal system are shown in this example:

Place –	1,000	100	10	1
5,349 =	5 ,	3	4	9

5.3.1 Study Problems: Adding and Subtracting Decimals

Find the answers to these addition and subtraction problems without using a calculator. Don't forget to line up the decimal points.

1.
```
  2.50
  4.20
 +5.00
 -----
```

2. 1.82 + 3.41 + 5.25 = _____

3. 6.43 plus 86.4 = _____

4. The combined thickness of a piece of sheet metal 0.078-inch thick and a piece of band iron 0.25-inch thick is _____.

   a. 0.308
   b. 0.328
   c. 3.08
   d. 32.8

5. Yesterday, a lumber yard contained 6.7 tons of wood. Since then, 2.3 tons were removed. The lumber yard now contains _____ tons of wood.

   a. 3.4
   b. 4.4
   c. 5.4
   d. 6.4

5.4.0 Multiplying Decimals

While unloading wood panels, you measure one panel as 4.5-feet wide. You have seven panels the same width. What is the total width if you put the panels side-by-side?

Step 1 Set up the problem just like the multiplication of whole numbers.

```
  4.5
×   7
-----
```

Step 2 Proceed to multiply.

```
  4.5
×   7
-----
  315
```

Step 3 Once you have the answer, count the number of digits to the right of the decimal point in both numbers being multiplied. In this example, there is only one number with a decimal point (4.5) and only one digit to the right of it.

Step 4 In the answer, count over the same number of digits (from right to left) and place the decimal point there.

```
  4.5
×   7
-----
 31.5
```

Did You Know?

Other Counting Systems

Ancient Babylonians developed one of the first place-value systems—a base-sixty system. It was based on the number 60, and numbers were grouped by sixties. At the beginning, there was no symbol for zero in the Babylonian system. This made it difficult to perform calculations. It was not always possible to determine if a number represented 24, 204, or 240. Although this system is no longer used today, we still use a base-sixty system for measuring time:
60 seconds in a minute and 60 minutes in an hour.

A base-twelve system requires 12 digits. This system is called a duodecimal system, from the Latin word *duodecim,* meaning 12. Since the decimal system has only ten digits, two new digits must be added to the base-twelve system. In a duodecimal system, each place value is 12 times greater than the place to the right. Although this is not a common system, a base-twelve system is used to count objects by the dozen (12) or by the gross (144 = 12 × 12).

A hexadecimal system groups numbers by sixteens. The word *hexadecimal* comes from the Greek word for six and the Latin word for ten. Just as the duodecimal system required two new place numerals, the hexadecimal system requires six additional digits. In a hexadecimal system, each place value is 16 times greater than the place to the right. Computers often use the hexadecimal system to store information. Common configurations of RAM come in multiples of 16.

128 = 8 × 16
256 = 16 × 16
512 = 32 × 16

You may have to add a zero if there are more digits to the right of the decimal points than there are in the answer, as shown in the following example.

```
    0.507
  × 0.022
  -------
     1014
    10140
  +   000
  -------
    11154  = 0.011154
```

Add the digits to the right of the decimal point in the two numbers. There are six total. Count six digits from right to left in the product. In this case, you'll need to add a zero.

5.4.1 Study Problems: Multiplying Decimals

Use the following problem to answer Questions 1 and 2.

You are machining a part. The starting thickness of the part is 6.18 inches. You take three cuts. Each cut is three-tenths of an inch.

1. You have removed _____ inches of material.
   a. 0.6
   b. 0.8
   c. 0.9
   d. 1.09

2. The remaining thickness of the part is _____ inches.
   a. 5.28
   b. 6.08
   c. 6.10
   d. 6.15

Use the following problem to answer Questions 3 and 4.

An electrician wants to know if a light circuit is overloaded. The circuit supplies two different machines.

3. The first machine has 11 bulbs lit. Each bulb uses 4.68 watts. The lights on the first machine need _____ watts.
   a. 0.5148
   b. 5.148
   c. 51.48
   d. 514.8

4. The second machine has seven bulbs lit. Each of these bulbs uses 5.14 watts. The lights on both machines need a total of _____ watts.
   a. 35.98
   b. 76.76
   c. 87.46
   d. 874.6

5. Ceramic tile weighs 4.75 pounds per square foot. Therefore, 128 square feet of ceramic tile weighs _____ pounds.
   a. 598
   b. 608
   c. 908
   d. 1108

5.5.0 Dividing with Decimals

When would you divide with decimals? Perhaps you need to cut a 44.5-inch pipe into as many 22-inch pieces as possible. How many 22-inch pieces will you be able to cut? How much will be left over?

On-Site

Measuring the Thickness of a Coating

Coating thickness is important because either too little or too much can cause problems. A coating such as paint needs a minimum thickness to prevent corrosion, withstand abrasion, and look good. A coating that is too thick may crack, flake, blister, or not cure properly.

Many jobs have requirements that specify the thickness of the coating applied to an object or surface. To ensure that the specifications are met, periodic checks of the wet-film thickness can be made using a wet-film thickness gauge. Typically these gauges use measurements in mils or microns. For example, the required wet-film thickness for a coat of paint may be 10.25 mils.

This makes understanding decimals and properly reading gauges an essential job skill.

There are three types of division problems involving decimals:

- Those that have a decimal point in the number being divided (the dividend)

```
22)44.5
```

- Those that have a decimal point in the number you are dividing by (the divisor)

```
0.22)4,450
```

- Those that have decimal points in both numbers (the dividend and the divisor)

```
0.22)44.5
```

5.5.1 Dividing with a Decimal in the Number Being Divided

For the first type of problem, let's use 44.5 ÷ 22 as an example.

Step 1 Place a decimal point directly above the decimal point in the dividend.

```
     .
22)44.5
```

Step 2 Divide as usual.

```
    2.0
22)44.5
  -44
   00.5r
```

How many 22-inch pieces of pipe will you have? The answer is 2, with 0.5 inch left over.

5.5.2 Study Problems: Dividing with Decimals, Part 1

Find the answers to the following division problems without using a calculator, and don't go any further than the hundredths (0.01) place, unless otherwise noted.

1. 45.36 ÷ 18 = _____
2. 4.536 ÷ 18 = _____
3. 0.4536 ÷ 18 = _____

Round to the nearest thousandths (0.001) place

4. 25)10.20
5. 6)31.2

5.5.3 Dividing with a Decimal in the Number You Are Dividing By

For the second type of problem, let's use 4,450 ÷ 0.22 as an example.

Step 1 Move the decimal point in the divisor to the right until you have a whole number.

0.22 becomes 22

Step 2 Move the decimal point in the dividend the same number of places (two) to the right. (You may have to add zeros.) Then divide as usual.

```
      20227.2
22)4450.00.0
  -44
   0050
  -0044
   00060
  -00044
   000160
  -000154
   0000060
  -0000044
   0000016r
```

5.5.4 Study Problems: Dividing with Decimals, Part 2

Perform the following division problems on a separate piece of paper without using a calculator. Don't go any further than the hundredths (0.01) place in your answer.

1. 282 ÷ 14.1 = _____
2. 694 ÷ 3.2 = _____
3. 99 ÷ 0.45 = _____

On-Site

Equivalents

Forty-four and one-half is the same as forty-four point five. If you measure a piece of pipe and the tape measure says it is 44½ inches, another way to say this is "forty-four point five," which is the decimal equivalent. Some other decimal equivalents are shown below:

Chart of Equivalents

½ = 0.5 ¼ = 0.25 ⅛ = 0.125 1/16 = 0.0625

4. $2.5\overline{)102}$

5. $0.6\overline{)312}$

5.5.5 *Dividing with Decimals in Both Numbers*

For the third type of problem, use 44.5 ÷ 0.22 as an example.

Step 1 Move the decimal point in the divisor to the right until you have a whole number.

$0.22\overline{)44.50}$

0.22 becomes 22

Step 2 Move the decimal point in the dividend the same number of places to the right.

$22\overline{)44.50.}$

44.5 becomes 4,450

Then divide as usual.

```
      202
  22)4450
     -44
      005
     -000
      0050
     -0044
       006r
```

The answer is 202 with a remainder of 6.

5.5.6 *Study Problems: Dividing with Decimals, Part 3*

Find the answers to the following division problems without using a calculator. Don't go any further than the hundredths (0.01) place.

1. 20.82 ÷ 4.24 = _____
2. 38.9 ÷ 3.7 = _____
3. 9.9 ÷ 0.45 = _____
4. $0.25\overline{)10.20}$
5. $0.6\overline{)31.2}$

5.6.0 Rounding Decimals

Sometimes an answer is a bit more precise than you require. As an example, if tubing costs $3.76 per foot and you spend $800, how much tubing will you buy?

The precise answer is 212.7659574 feet. You probably only need to measure it to the nearest tenth. What would you do? For this exercise, you will round 212.7659574 to the nearest tenth (0.1).

Step 1 Underline the place to which you are rounding.

212.<u>7</u>659574

Step 2 Look at the digit one place to its right.

212.<u>7</u>**6**59574

On-Site

Decimals at Work

When are you going to use decimals on the job? Here are two examples of using decimals to get your work done:

- You are installing a boiler to specifications on its concrete pad. When measuring with your level, you find that one corner of the boiler is level within 0.003-inch. Another corner is 0.005" too high, and a third corner is 0.001" too high. The concrete pad is not as even as it should be. To adjust for this, you must place shims (thin, tapered pieces of material) between the boiler base and the concrete pad. Shims come in 0.001- and 0.002-inch widths. How many shims of each size will it take to make the boiler level?
- You are working on a conveyor system. You check the lubricant by taking an oil sample and measuring the metal particles with a micrometer. (A micrometer is a precision tool that can measure to the nearest ¹⁄₁₀₀₀ of an inch.) You find metal particles that are 0.0001-inch in size. The system currently has a 20-micron filter installed to filter pieces bigger than 0.0002 inch. What size filter do you need to filter the 0.0001-inch metal particles out of the oil?

Step 3 If the digit to the right is 5 or more, you will round up by adding 1 to the underlined digit. If the digit is 4 or less, leave the underlined digit the same. In this example, the digit to the right is 6, which is more than 5, so you round up by adding 1 to the underlined digit.

212.7**6**59574

Step 4 Drop all other digits to the right.

212.8

5.6.1 Study Problems: Rounding Decimals

Solve these problems to practice rounding decimals. Round your answers to the nearest tenth.

1. You need to cut a 90.5-inch pipe into as many 3.75-inch pieces as possible. You will be able to cut _____ 3.75-inch pieces.

a. 14
b. 24
c. 34
d. 44

2. If you drove your car 622 miles on 40.1 gallons of gas, you got _____ miles per gallon.

a. 15.5
b. 15.6
c. 155.1
d. 156

3. If wire costs $4.30 per pound and you pay a total of $120.95, then you have purchased _____ pounds of wire.

a. 0.28
b. 2.8
c. 28.0
d. 28.1

4. Vent pipe is on sale at XYZ Supply Company this week for $0.37 per linear foot. If you spend $115.38, you will purchase _____ linear feet of pipe.

a. 308.11
b. 310.8
c. 311.8
d. 311.9

5. Vent pipe at XYZ Supply normally costs $0.48 per linear foot. If you spend the same amount of money ($115.38) when vent pipe is not on sale, you will purchase _____ linear feet of pipe.

a. 240
b. 240.4
c. 241
d. 241.4

5.7.0 Using the Calculator to Add, Subtract, Multiply, and Divide Decimals

Performing operations on the calculator using decimals is very much like performing the operations on whole numbers. Follow these steps using the problem 45.6 + 5.7 as an example.

Step 1 Turn the calculator on.

Step 2 Press 4, 5, . (decimal point), and 6. The number 45.6 appears in the display.

Step 3 Press the + key. The 45.6 is still displayed.

For this step, press whichever operation key the problem calls for: + to add, – to subtract, × to multiply, ÷ to divide.

Did You Know?

Scientific Notation

Scientific notation (sometimes called exponential notation) is a system that allows you to conveniently write very large or very small decimal-based numbers using an exponent. The exponent represents the number of times you multiply the multiplier by the multiplicand. Scientific notation is commonly used by scientists, mathematicians, and engineers. You may already be familiar with scientific notation if you used ft.^{2} (feet squared) in a measurement. The following are some examples of scientific notation:

Decimal Notation	Scientific Notation
1	1×10^0
50	5×10^1
7,530,000,000	7.53×10^9
–0.0000000082	-8.2×10^{-9}

Step 4 Press 5, . (decimal point), and 7. The number 5.7 is displayed.

Step 5 Press the = key. After you press the = key, whether you are adding, subtracting, multiplying, or dividing, the answer will appear on your display.

$$45.6 + 5.7 = 51.3$$
$$45.6 - 5.7 = 39.9$$
$$45.6 \times 5.7 = 259.92$$
$$45.6 \div 5.7 = 8$$

Step 6 Press the ON/C key to clear the calculator. A zero (0) appears in the display.

5.7.1 Study Problems: Using Decimals on the Calculator

Use your calculator to find the answers to the following problems, and round your answers to the nearest hundredth (0.01).

1. $\begin{array}{r} 45.89 \\ +\ 7.85 \\ \hline \end{array}$

2. $\begin{array}{r} 7.6 \\ \times 0.12 \\ \hline \end{array}$

3. $\begin{array}{r} 685.79 \\ -\ 56.266 \\ \hline \end{array}$

4. $6.45 \div 3.25 =$

5. $\begin{array}{r} 34.76 \\ +\ 3.64 \\ \hline \end{array}$

5.8.0 Practical Application

The following examples will provide you with practice in both reading a drawing and determining material costs.

1. Use *Figures 34* and *35* to complete the blanks in the materials list in *Figure 36* and to find the total cost of the plumbing fixtures. Be sure to show all your work.
2. On HVAC installations, a benchmark (BM) is established to serve as a reference point for all of the other measurements. It is usually set by the general contractor and can be marked on a column, wall, or other similar structure. Use the information in *Figure 37* to determine the distance from the benchmark to the bottom of the circular duct.

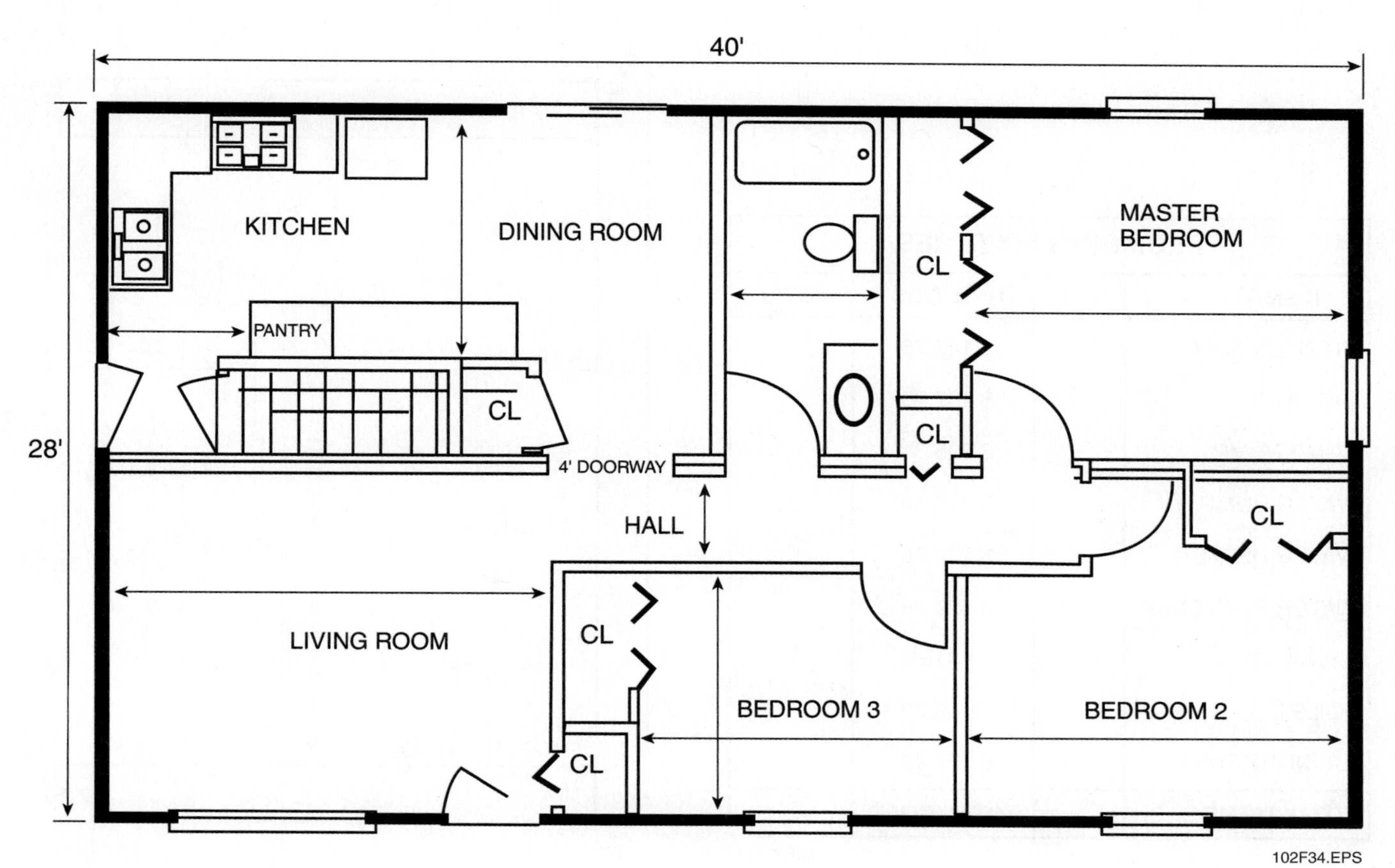

Figure 34 First floor plan.

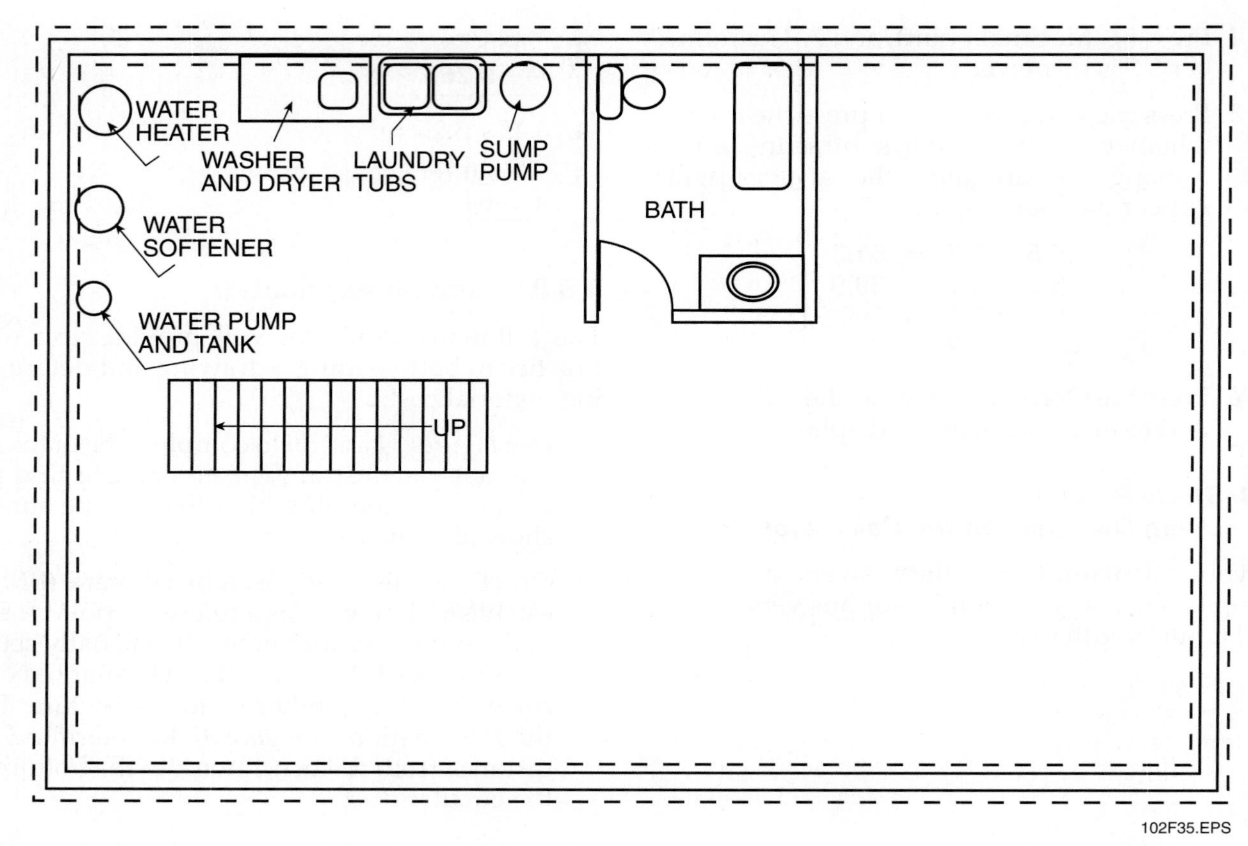

Figure 35 Foundation plan.

PLUMBING FIXTURES			
ITEMS	**QTY.**	**UNIT COST**	**TOTAL**
KITCHEN SINK		$ 142.76	
BATHTUB		$ 437.95	
SUMP PUMP		$ 205.31	
WATER PUMP		$ 375.10	
WATER HEATER		$ 232.00	
WATER SOFTENER		$ 429.99	
LAVATORY		$ 231.34	
TOILET		$ 86.59	
LAUNDRY TUBS		$ 137.43	
TOTAL ITEMS		**TOTAL COST**	

102F36.EPS

Figure 36 Materials list.

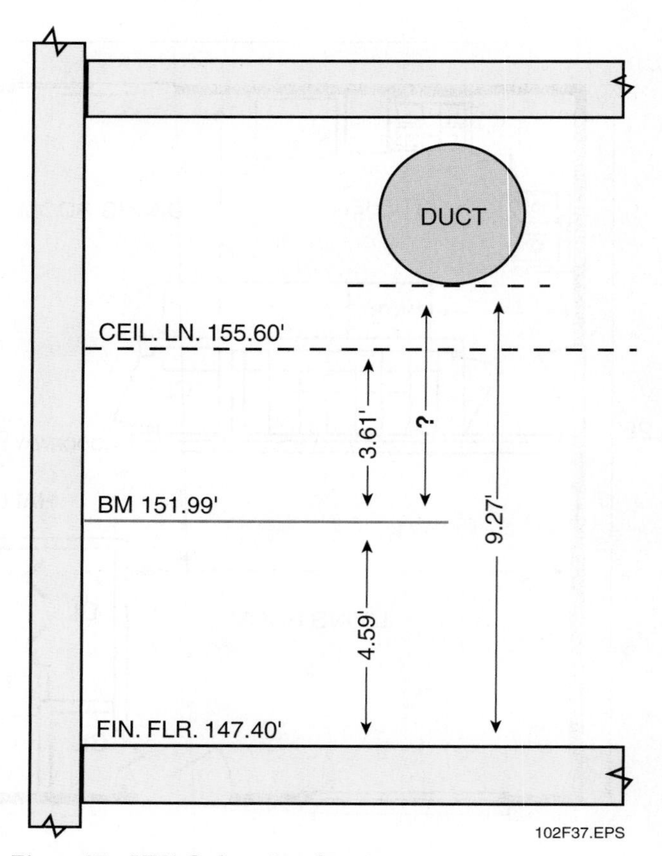

Figure 37 HVAC elevation drawing.

6.0.0 CONVERSION PROCESS

Sometimes you will need to convert some of the numbers you want to work with so that all your numbers appear in the same form. For example, some numbers may appear as decimals, some as **percentages**, and some as fractions. Decimals, percentages, and fractions are all just different ways of expressing the same thing. The decimal 0.25, the percent 25%, and the fraction ¼ all mean the same thing. In order to work with the different forms of numbers like these, you will need to know how to convert them from one form into another.

6.1.0 Converting Decimals to Percentages and Percentages to Decimals

What are percentages? Think of a whole number divided into 100 parts. You can express any part of the whole as a percentage. Refer to *Figure 38* for an example. The tank has a capacity of 100 gallons. It is now filled with 50 gallons. What percentage of the tank is filled?

If you answered 50 percent (50%), you are correct. Percentage means *out of 100*. How many gallons out of 100 does the tank contain? It contains 50 out of 100, or 50 percent. Percentages are an easy way to express parts of a whole.

Decimals and fractions also express parts of a whole. The tank in *Figure 38* is 50 percent full. If you expressed this as a fraction, you would say it is ½ full. You could also express this as a decimal and say it is 0.50 full.

Sometimes you may need to express decimals as percentages or percentages as decimals. Suppose you are preparing a gallon of cleaning solution. The mixture should contain 10 to 15 percent of the cleaning agent. The rest should be water. You have 0.12 gallon of cleaning agent. Will you have enough to prepare a gallon of the solution? To answer the question, you must convert a decimal (0.12) to a percentage by following these steps:

Step 1 Multiply the decimal by 100. (Tip: When multiplying by 100, simply move the decimal point two places to the right.)

$$0.12 \times 100 = 12$$

Step 2 Add a % sign.

12%

Recall that the mixture should be from 10 to 15 percent cleaning agent. You have 12 percent of a gallon, enough cleaning agent to make the solution.

You may also need to convert percentages to decimals. Use the following as an example. Suppose a mixture should contain 22 percent of a certain chemical by weight. You're making 1 pound of the mixture. You weigh the ingredients on a digital scale. How much of the chemical should you add? To answer this, you must convert a percentage (22%) to a decimal by following these steps:

Step 1 Drop the % sign.

22

Step 2 Divide the number by 100. (Tip: When dividing by 100 or more, move the decimal point two places to the left.)

$$22 \div 100 = 0.22$$

You would add 0.22 pound of the chemical to 0.78 pound of the other ingredient to make a 22 percent mixture.

6.1.1 Study Problems: Converting Decimals to Percentages and Percentages to Decimals

Convert these decimals to percentages.

1. 0.62 = _____
2. 0.475 = _____
3. 0.7 = _____

Convert these percentages to decimals.

4. 72% = _____
5. 12.5% = _____

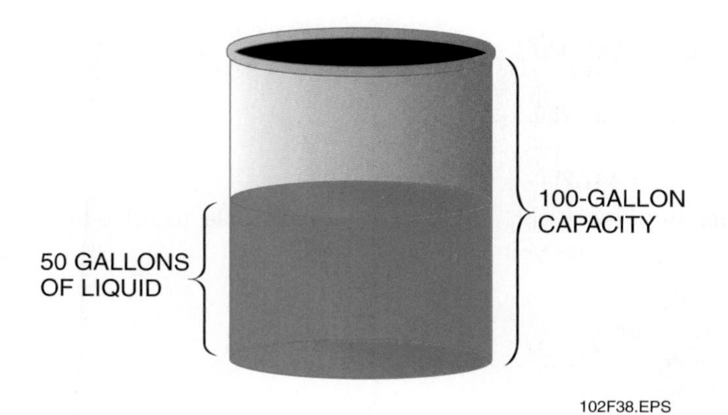

Figure 38 100-gallon-capacity tank.

6.2.0 Converting Fractions to Decimals

You will often need to change a fraction to a decimal. For example, you need ¾ of a dollar. How do you convert ¾ to its decimal equivalent?

Step 1 Divide the numerator of the fraction by the denominator.

$$4\overline{)3.0}$$

In this example, you need to put the decimal point and the zero after the number 3, because you need a number large enough to divide by 4.

Step 2 Put the decimal point directly above its location within the division symbol.

$$\begin{array}{r} . \\ 4\overline{)3.0} \end{array}$$

Step 3 Once the decimal point is in its proper place above the line, you can divide as you normally would. The decimal point holds everything in place.

$$\begin{array}{r} .75 \\ 4\overline{)3.00} \\ -2.8 \\ \hline 0.20 \\ -0.20 \\ \hline 0.00 \end{array}$$

The answer is that ¾ of a dollar is the same as $0.75.

On-Site

Practical Application

When contractors calculate the expenses for building a house, they must pay close attention to the percentages they allot to various factors. Of these factors, the main concern to the contractor is the profit to his company. For example, if a contractor is building a house that will have a selling price of $175,000, what will the profit be with the following percentage breakdown?

Profit = ?%
Overhead = 27%
Materials = 35%
Labor = 24%

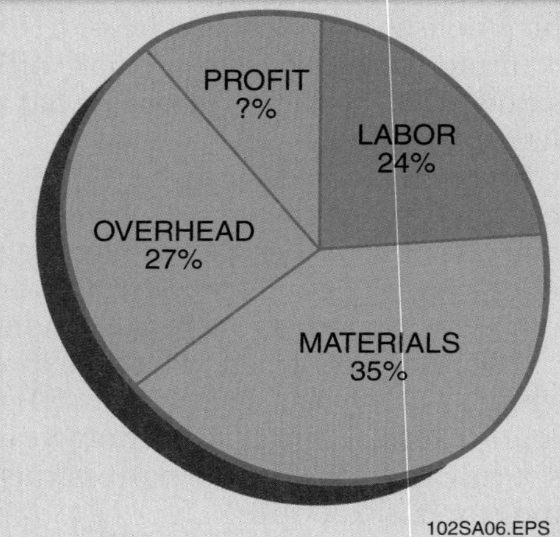

102SA06.EPS

Solution

Find the dollar amounts represented by these percentages.
Convert the percentages to decimals by dropping the percentage sign and moving the decimal two places to the left.

a. Profit = $24,500
 $175,000 × 14% = $175,000 × 0.14 = $24,500
b. Labor costs = $42,000
 $175,000 × 24% = $175,000 × 0.24 = $42,000
c. Material costs = $61,250
 $175,000 × 35% = $175,000 × 0.35 = $61,250
d. Overhead costs = $47,250
 $175,000 × 27% = $175,000 × 0.27 = $47,250

You can also use your calculator to work with percentages. For example, to solve the profit component of this problem, key it into your calculator as follows:

Step 1 Punch 175000 into the calculator.
Step 2 Press the multiplication button.
Step 3 Punch 14 into the calculator.
Step 4 Press the percent (%) button.

Profit = $175,000 × 14% = $24,500

Note that you can also use your calculator to multiply 175,000 by 0.14.

6.2.1 *Study Problems: Converting Fractions to Decimals*

Convert the following fractions to their decimal equivalents without using a calculator.

1. $\frac{1}{4}$ = _____
2. $\frac{3}{4}$ = _____
3. $\frac{1}{8}$ = _____
4. $\frac{5}{16}$ = _____
5. $\frac{20}{64}$ = _____

6.3.0 Converting Decimals to Fractions

Let's say you have 0.25 of a dollar. What fraction of a dollar is that? Follow these steps to find out:

Step 1 Say the decimal in words.

0.25 is expressed as twenty-five hundredths

Step 2 Write the decimal as a fraction.

0.25 is written as a fraction as $\frac{25}{100}$

Step 3 Reduce it to its lowest terms.

$$\frac{25}{100} = \frac{25}{100} \div \frac{25}{25} = \frac{1}{4}$$

Step 4 So 0.25 converted to a fraction is $\frac{1}{4}$. If you have 0.25 of a dollar, you have $\frac{1}{4}$ of a dollar.

6.3.1 *Study Problems: Converting Decimals to Fractions*

Convert the following decimals to fractions without using a calculator. Reduce them to their lowest terms.

1. 0.5 = _____
2. 0.12 = _____
3. 0.125 = _____
4. 0.8 = _____
5. 0.45 = _____

6.4.0 Converting Inches to Decimal Equivalents in Feet

How would you convert inches to their decimal equivalents in feet? For example, 3 inches equals what decimal equivalent in feet?

First, express the inches as a fraction that has 12 as the denominator. You use 12 because there are 12 inches in a foot. Then reduce the fraction and convert it to a decimal.

In this example, the fraction $\frac{3}{12}$ reduces to $\frac{1}{4}$. Convert the fraction $\frac{1}{4}$ to a decimal by dividing the 4 into 1.00:

$$\begin{array}{r} 0.25 \\ 4\overline{)1.00} \\ -0.8 \\ \hline 0.20 \\ -0.20 \\ \hline 0.00 \end{array}$$

The answer is that 3 inches converts to 0.25 feet.

Did You Know?

Direct and Indirect Costs

When contractors determine a percentage of profit, they must take into consideration many different costs. These costs can be grouped into one of two categories, direct costs and indirect costs.

Direct costs are expenses that can be directly attributed to completing the job. For example, all of the materials (tools, paint, insulation, lumber) and labor hours required to build a house would be considered direct costs. On larger-scale jobs, direct costs can include not only materials and labor hours, but also demolition and cleanup crews, cranes and earth-moving machinery, and even job-site security.

Indirect costs (or overhead costs), on the other hand, are expenses that cannot be directly related to building the house, but that are required to run the company on a day-to-day basis. Indirect costs can include insurance on equipment, administrative staff payroll, advertising and marketing, office supplies, and property taxes.

Can you think of any other expenses that can be considered indirect costs?

> **On-Site**
>
> **Fractions and Decimals**
>
> Remember that fractions and decimals are interchangeable. If it is hard to multiply a measurement using a fraction, convert it to decimal form. For example, it may be easier to multiply a number by 0.001 rather than by $\frac{1}{1,000}$.
>
> When using a calculator, you will have to convert fractions into decimals to perform your calculations.

6.4.1 Study Problems: Converting Inches to Decimals

Find the answers to the following conversion problems and round them to the nearest hundredth.

1. 9 inches = _____ feet
2. 10 inches = _____ feet
3. 2 inches = _____ feet
4. 4 inches = _____ feet
5. 8 inches = _____ feet

6.5.0 Practical Applications

The following examples illustrate some of the practical applications that a tradesperson encounters every day on the job site. Use the information provided to solve each conversion problem. Be sure to show all of your work.

1. Find the cost of baseboard needed for the office building shown in the floor plan in *Figure 39*. The lumber company charges $1.19 per L.F. (linear foot) of baseboard, and there is a 12% discount. All door widths are the standard 30 inches. Add tax after you reduce for the sale cost.

Costs:

a. Total L.F. needed _____
b. Initial baseboard cost $ _____
c. Discount amount $ _____
d. Baseboard cost after sale reduction $ _____
e. Amount added for 6% tax $ _____
f. Total baseboard cost $ _____

2. Use the site plan in *Figure 40* to determine the percentage of the lot that will be used for parking, the building, and for walkways. You may use a calculator.

a. Percentage for parking _____ %
b. Percentage for building _____ %
c. Percentage for walkways _____ %

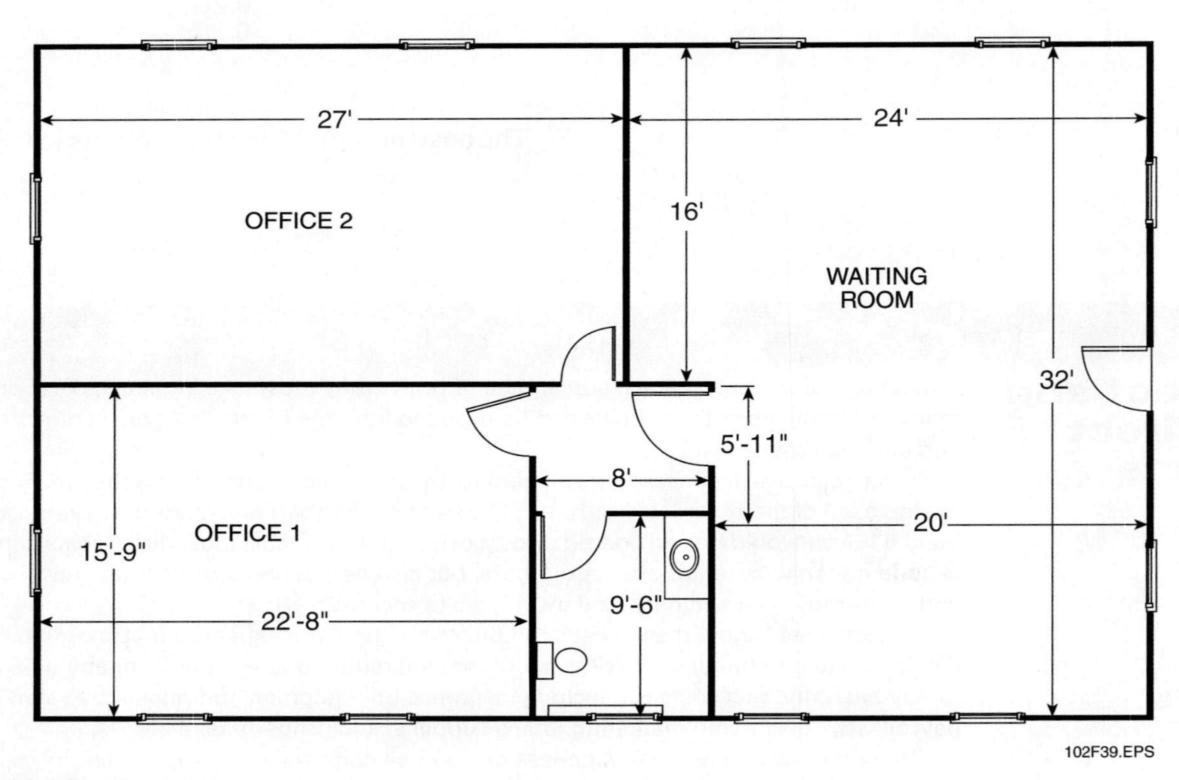

Figure 39 Office floor plan.

GOING GREEN

City Governments Thinking Green

Governing bodies in a number of cities around the world are coming up with innovative new ways to cut energy costs and lower greenhouse-gas emissions. Some of these new ideas include creating energy by harnessing the power of waves, air conditioning a building using rooftop gardens and lake water, and integrating wind turbines into a building's construction to generate energy.

Chicago

In 2001, city officials in Chicago decided to construct a 20,000-square-foot garden on the rooftop of their 11-story, 100-year-old city hall building. The building's rooftop, like many other downtown rooftops, would soar up to temperatures of 160°F (71.1°C) in the summer months. By reflecting sunlight and providing shade, the gardens lowered rooftop temperatures by 70°F (21.1°C), which reduced building cooling costs annually by 11% or around $10,000.

The gardens in Chicago have more than 100 different species of plants that can withstand the city's extreme temperatures and winds. The gardens utilize waterproof membranes to prevent leaks into the building and lightweight permeable soils to prevent rooftop overloading.

New York

In 2006, New York City Mayor Michael Bloomberg announced a plan that would cut the city's greenhouse-gas emissions by 30% by the year 2030, and would generate enough new clean energy to provide 640,000 homes with electricity. The plan outlined how bladed turbines would be submerged into New York's East River to generate power using the energy of the tidal currents. On June 11, 2007 the first five turbines were installed 30 feet (9.14 meters) into the river. These turbines alone have been generating enough electricity to meet one-third of the electricity needs of a local supermarket and parking garage.

The pilot project was such a success that city officials are planning for 30 more turbines to be installed in the near future. These additional turbines would generate enough electricity to power up to 800 homes. Planning for another 300 turbines is also in the works. Three hundred turbines would generate enough electricity to power up to 8,000 homes, which would be equal to saving 68,000 barrels of oil a year.

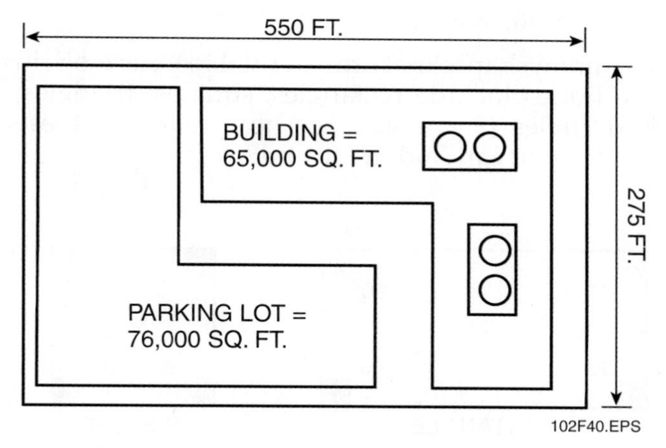

Figure 40 Site plan.

3. A general contractor is putting a bid together for the erection of a small commercial building. His bid shows the following net costs and profits added by subcontractors:

Masonry	$6,837.22 +	profit 13% (of total cost)
Plumbing	$4,881.45 +	profit 9% (of total cost)
Electrical	$2,246.89 +	profit 11% (of total cost)
Heating	$2,509.63 +	profit 8% (of total cost)

If the general contractor estimates that it will cost his company $67,238.57 plus 12% profit and he adds the costs of the subcontractors including profit, what would the total bid be? (You may use a calculator.) $ ____________

7.0.0 Introduction to Construction Geometry

Geometry might sound complicated, but it is really made up of everyday things you already know—**circles**, **triangles**, **squares**, and **rectangles**. The construction industry exists in a world of measurements. You should recognize the basic shapes and measurements that make your work possible.

7.1.0 Angles

An **angle** is an important term in the construction trades. It is used by all building trades to describe the shape made by two straight lines that meet in a point, or **vertex**. Angles are measured in **degrees**. To measure angles, you use a tool called a protractor. The following are the typical angles (*Figure 41*) you will measure in construction:

- **Acute angle** – An angle that measures between 0 and 90 degrees is an acute angle. The most common acute angles are 30, 45, and 60 degrees.
- **Right angle** – An angle that measures 90 degrees is called a right angle. The two lines that form the right angle are perpendicular to each other. Imagine the shape of a capital letter L. This is a right angle, because the sides of the L are perpendicular to one another. This is the angle used most often in the construction trade. A right angle is indicated in plans or drawings with this symbol: ⊾
- **Obtuse angle** – An angle that measures between 90 and 180 degrees is obtuse.
- **Straight angle** – A straight angle measures 180 degrees (a flat line).
- **Adjacent angles** – These angles have the same vertex and one side in common. Adjacent refers to objects that are next to each other.
- **Opposite angles** – Angles formed by two straight lines that cross are opposite. Opposite angles are always equal.

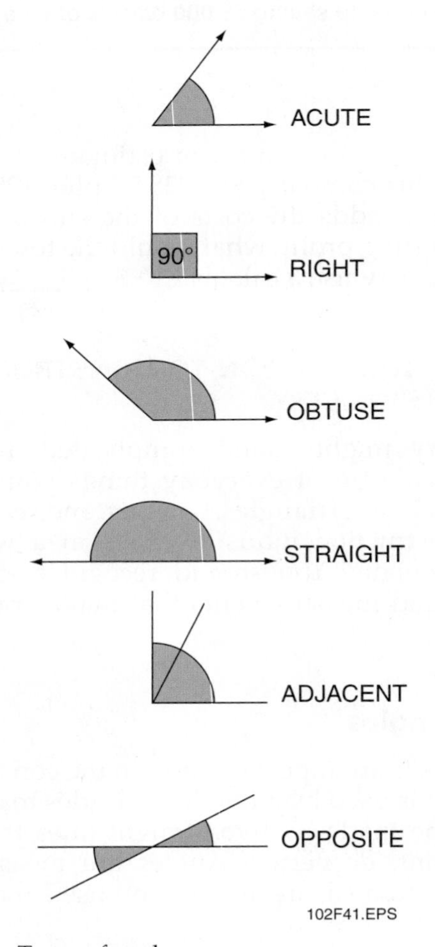

Figure 41 Types of angles.

Did You Know?

Rope Stretchers

The word *geometry* comes from two Greek words: *geos,* meaning land, and *metrein,* meaning to measure.

In ancient Egypt, most farms were located beside the Nile River. Every year during the flood season, the Nile overflowed its banks and deposited mineral-rich silt over the farmland. But the floodwaters destroyed the markers used to establish property lines between farms. When this happened, the farms had to be measured again. Men called rope stretchers re-marked the property lines.

The rope stretchers calculated distances and directions using ropes that had equally spaced knots tied along the length of the rope. Stretching the ropes to measure distances on level land was easy. In many cases, however, the rope stretchers had to measure property lines from one side of a hill to the opposite side or across a pond.

The hills and ponds made this measurement difficult. To adjust for the uneven land or ponds that stood in the way, rope stretchers determined new ways of measuring. Such discoveries became the foundation of geometry.

7.2.0 Shapes

Common shapes that are essential to your work in the trades include rectangles, squares, triangles, and circles (*Figure 42*). See how many of these shapes you can find in *Figure 43*.

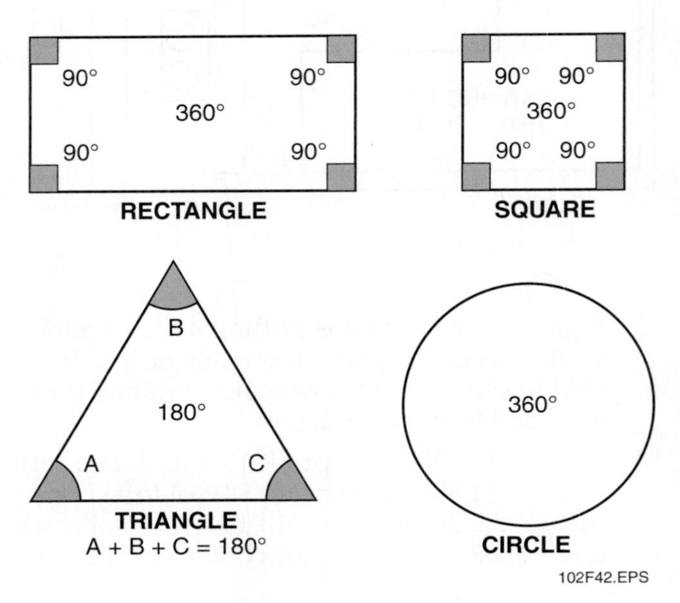

Figure 42 Common shapes.

102F43.EPS

Figure 43 What common shapes can you find at this construction site?

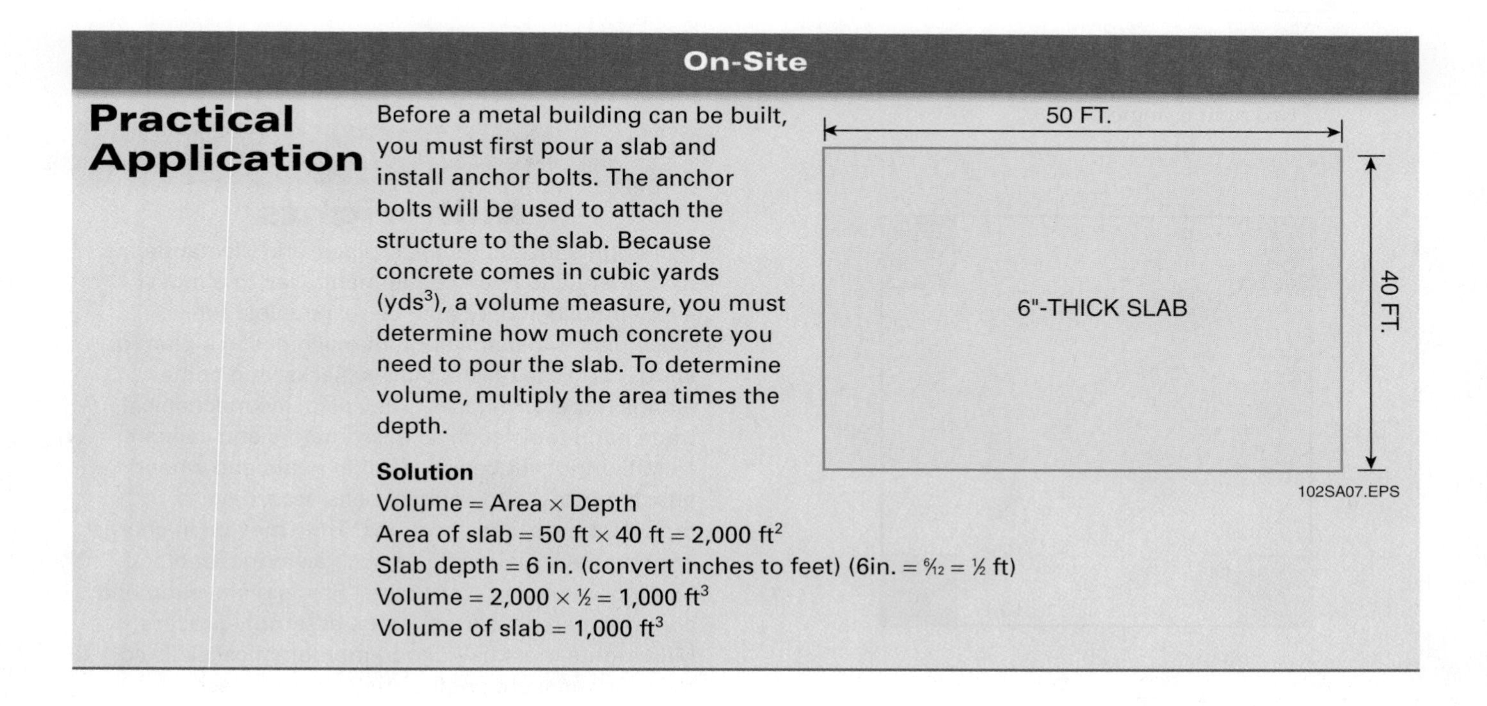

On-Site

Practical Application

Before a metal building can be built, you must first pour a slab and install anchor bolts. The anchor bolts will be used to attach the structure to the slab. Because concrete comes in cubic yards (yds^3), a volume measure, you must determine how much concrete you need to pour the slab. To determine volume, multiply the area times the depth.

Solution

Volume = Area × Depth

Area of slab = 50 ft × 40 ft = 2,000 ft^2

Slab depth = 6 in. (convert inches to feet) (6in. = $\frac{6}{12}$ = $\frac{1}{2}$ ft)

Volume = 2,000 × $\frac{1}{2}$ = 1,000 ft^3

Volume of slab = 1,000 ft^3

7.2.1 Rectangle

A rectangle is a four-sided shape with four 90-degree angles. (The sum of all four angles in any rectangle is 360 degrees.) A rectangle has two pairs of equal sides that are parallel to each other. The **diagonals** of a rectangle are always equal. Diagonals are lines connecting opposite corners. If you cut a rectangle on the diagonal, you will have two **right triangles**, as shown in *Figure 44*.

7.2.2 Square

A square is a special type of rectangle with four equal sides and four 90-degree angles. (The sum of all four angles in all squares is 360 degrees.) If you cut a square on the diagonal, you will have two right triangles. Each right triangle will have two 45-degree angles and one 90-degree angle, as shown in *Figure 45*.

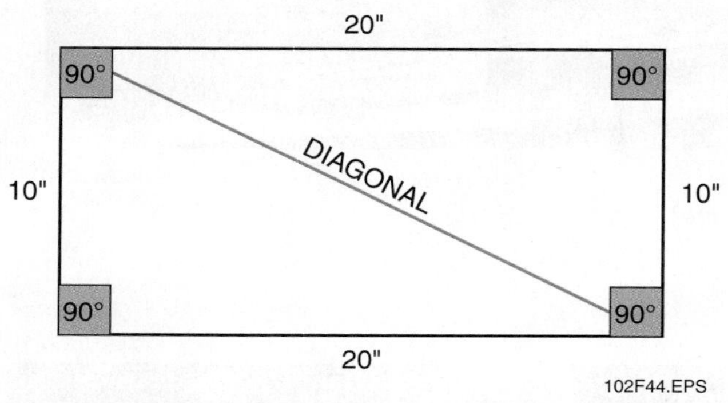

Figure 44 Cutting a rectangle on the diagonal produces two right triangles.

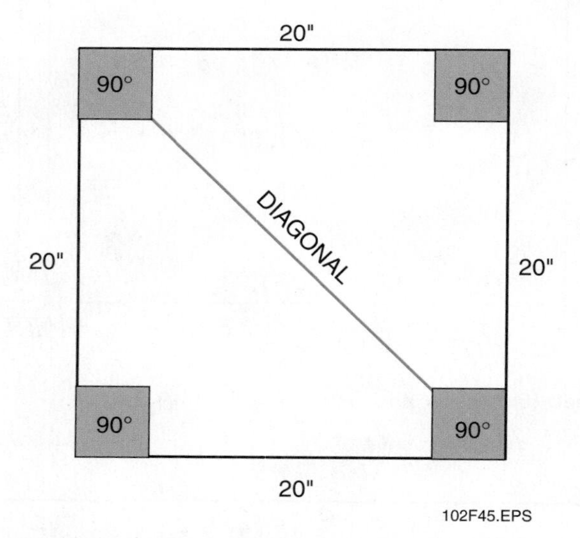

Figure 45 Cutting a square on the diagonal produces two right triangles.

When measuring the outside lines of a rectangle or a square, you are determining the perimeter. The perimeter is the sum of all four sides of a rectangle or square. You may need to calculate the perimeter of a shape so that you can measure, mark, and cut the right amount of material. For example, if you need to install shoe molding along all four walls of a room, you must know the perimeter measurement. If the room is 14 feet by 12 feet, you would calculate: 14 + 12 + 14 + 12 = 52 feet of shoe molding. Another way to calculate would be (2 × 14 feet) + (2 × 12 feet) = 52 feet.

7.2.3 Triangle

A triangle is a closed shape that has three sides and three angles. Although the angles in a triangle can vary, the sum of the three angles is always 180 degrees (*Figure 46*). The following are different types of triangles you will use in construction:

- *Right triangle* – A right triangle has one 90-degree angle.
- **Equilateral triangle** – An equilateral triangle has three equal angles and three equal sides.
- **Isosceles triangle** – An isosceles triangle has two equal angles and two equal sides. A line that **bisects** (runs from the center of the base of the triangle to the highest point) an isosceles triangle creates two adjacent right angles.
- **Scalene triangle** – A scalene triangle has three sides of unequal lengths.

On-Site

Millwrights

Millwrights install, repair, replace, and dismantle the machinery and heavy equipment used in almost every industry. They may be responsible for placement and installation of machines in a plant or shop. They use hoists, pulleys, jacks, and come-alongs to perform tasks. They also use mechanical trade hand tools such as micrometers and calipers.

Millwrights fit bearings, align gears and wheels, attach motors, and connect belts according to manufacturers' specifications. They may be in charge of preventive maintenance such as lubrication and fixing or replacing worn parts. Precision leveling and alignment are important in the assembly process. Millwrights must have good mathematical skills so they can measure angles, material thickness, and small distances.

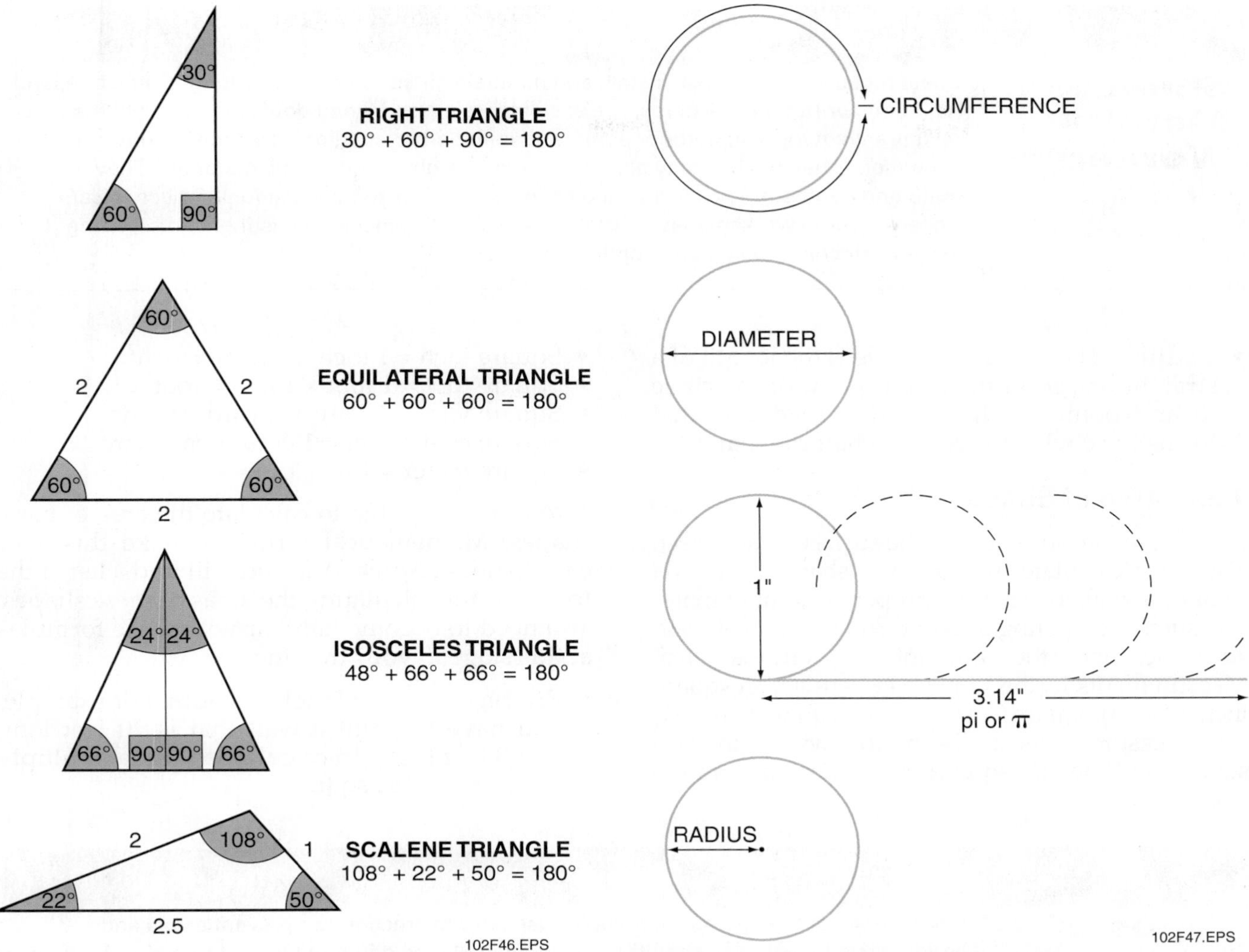

Figure 46 The sum of a triangle's three angles always equals 180 degrees.

Figure 47 Measurements that apply to circles.

7.2.4 Circle

A circle is a closed curved line around a center point. Every point on the curved line is exactly the same distance from the center point. A circle measures 360 degrees. The following measurements apply to circles (*Figure 47*):

- **Circumference** – The circumference of a circle is the length of the closed curved line that forms the circle. The **formula** for finding circumference is **pi** (3.14) × **diameter**.
- *Diameter* – The diameter of a circle is the length of a straight line that crosses from one side of the circle through the center point to a point on the opposite side. The diameter is the longest straight line you can draw inside a circle.
- *Pi* or π – pi is a mathematical constant value of approximately 3.14 (or 22⁄7) used to determine the **area** and circumference of circles.

Did You Know?

Diagonals

Diagonals have a number of uses. If you have to make sure that a surface is a true rectangle with 90-degree corners, you can measure the diagonals to find out. For example, before applying a piece of sheathing, you must make sure it is a true rectangle. Using your tape measure, find the length of the sheathing from one corner to the opposite corner. Now, find the length of the other two opposing corners. Do the diagonals match? If so, the piece of sheathing is a true rectangle. If not, the piece is not a true rectangle and will cause problems when you install it.

On-Site

Sheet Metal Workers

Sheet metal workers make, install, and maintain air conditioning, heating, ventilation, and pollution-control duct systems; roofs; siding; rain gutters and downspouts; skylights; restaurant equipment; outdoor signs; and many other building parts and products made from metal sheets. They may also work with fiberglass and plastic materials. They use math and geometry to calculate angles for fabrication and installation of mechanical systems. Some workers may specialize in testing, balancing, adjusting, and servicing existing air conditioning and ventilation systems.

- **Radius** – The radius of a circle is the length of a straight line from the center point of the circle to any point on the closed curved line that forms the circle. It is equal to half the diameter.

7.3.0 Area of Shapes

Area is the measurement of the surface of an object. You must calculate the area of a shape, such as a floor or a wall, to order the proper amount of material, such as carpeting or paint. Square units of measure describe the amount of surface area. Measurements in the English system are in square inches (sq in), square feet (sq ft), and square yards (sq yd). Measurements in the metric system include square centimeters (sq cm) and square meters (sq m).

- Square inch = 1 inch × 1 inch = $inch^2$
- Square foot = 1 foot × 1 foot = $foot^2$
- Square yard = 1 yard × 1 yard = $yard^2$
- Square centimeter = 1 cm × 1 cm = cm^2
- Square meter = 1 m × 1 m = m^2

You must be able to calculate the area of basic shapes. Mathematical formulas make this very easy to do. In *Appendix D* you will find a list of the formulas for calculating the areas of these shapes. You need to become familiar with these formulas at this stage in your training.

- *Rectangle* – Area = length × width. For example, you have to paint a wall that is 20 feet long and 8 feet high. To calculate the area, multiply 20 ft × 8 ft = 160 sq ft.

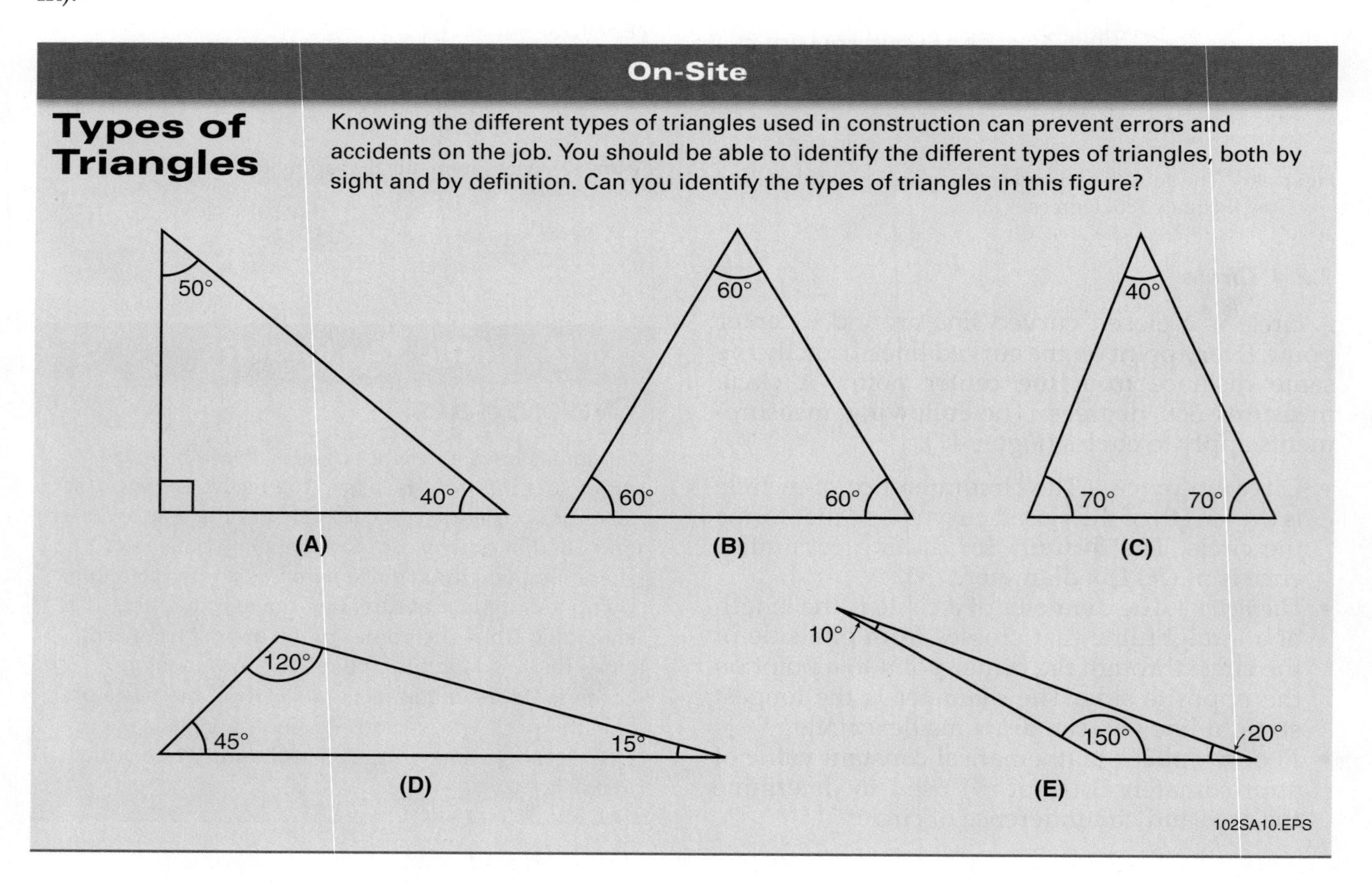

- *Square* – Area = length × width (but remember that all sides of a square are equal, so the formula can also be area = side × side, or side^2). For example, you have to tile a 12-meter square room. The area is 12 m × 12 m = 144 sq m, or $(12\text{ m})^2$ = 144 sq m.
- *Circle* – Area = pi × radius^2. In this formula, you must use the mathematical constant pi, which has an approximate value of 3.14. You multiply pi by the radius of the circle squared. For example to find the area of a circular driveway to be sealed, you must first find the radius, for example, 20 feet. The calculation is 3.14 × $(20\text{ ft})^2$ or 3.14 × 400 sq ft = 1,256 sq ft.
- *Triangle* – Area = ½ × base × height. The base is the side the triangle sits on. The height is the length of the triangle from its base to the highest point. For example, you have to install siding on a triangular section of a building. You find the triangle has a base of 2 feet and a height of 4 feet. The calculation is ½ × 2 ft × 4 ft = 4 sq ft.

7.3.1 Study Problems: Calculating the Area of a Shape

1. The area of a rectangle that is 8 feet long and 4 feet wide is _____.
   a. 12 sq ft
   b. 22 sq ft
   c. 32 sq ft
   d. 36 sq ft

2. The area of a 16-cm square is _____.
   a. 256 sq cm
   b. 265 sq cm
   c. 276 sq cm
   d. 278 sq cm

3. The area of a circle with a 14-foot diameter is _____.
   a. 15.44 sq ft
   b. 43.96 sq ft
   c. 153.86 sq ft
   d. 196 sq ft

4. The area of a triangle with a base of 4 feet and a height of 6 feet is _____.
   a. 12 sq ft
   b. 24 sq ft
   c. 32 sq ft
   d. 36 sq ft

5. The area of a rectangle that is 14 meters long and 5 meters wide is _____.
   a. 60 sq m
   b. 65 sq m
   c. 70 sq m
   d. 75 sq m

Did You Know?

3-4-5 Rule

The 3-4-5 rule is based on the Pythagorean Theorem, and it has been used in building construction for centuries. This simple method for laying out or checking 90-degree angles (right angles) requires only the use of a tape measure. The numbers 3-4-5 represent dimensions in feet that describe the sides of a right triangle. Right triangles that are multiples of the 3-4-5 triangle are commonly used, such as 9-12-15, 12-16-20, and 15-20-25. The specific multiple used is determined by the relative distances involved in the job being laid out or checked.

Refer to the figure for an example of the 3-4-5 theory using the multiples of 15-20-25. In order to square, or check, a corner, first measure and mark 15'-0" down the line in one direction, then measure and mark 20'-0" down the line in the other direction. The distance measured between the 15'-0" and 20'-0" points must be exactly 25'-0" to ensure that the angle is a perfect right (90 degree) angle.

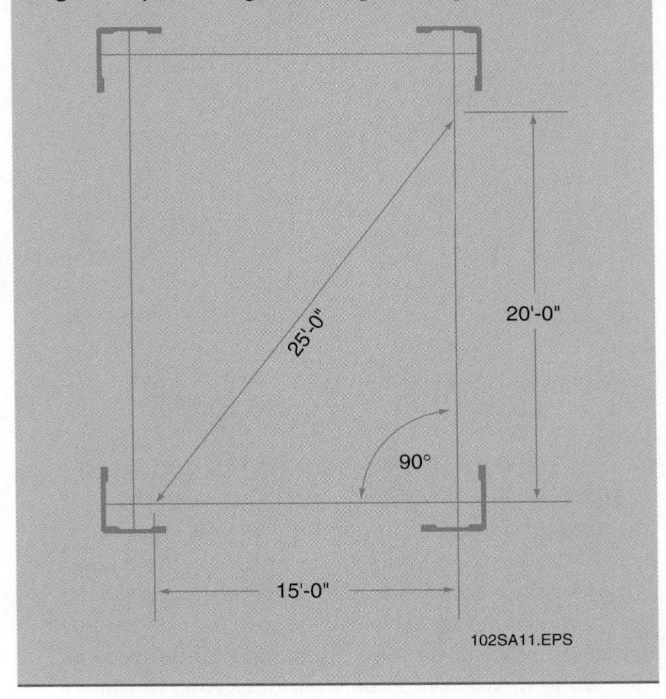

7.4.0 Volume of Shapes

Volume is the amount of space occupied in three dimensions. To measure volume, you must use three measurements: length, width, and height (depth or thickness). Cubic units of measure describe the volume of different spaces. Measurements in the English system are in cubic inches (cu in), cubic feet (cu ft), and cubic yards (cu yd). Metric measurements include cubic centimeters (cu cm) and cubic meters (cu m).

- Cubic inch = 1 inch × 1 inch × 1 inch = $inch^3$
- Cubic foot = 1 foot × 1 foot × 1 foot = $foot^3$
- Cubic yard = 1 yard × 1 yard × 1 yard = $yard^3$

> **Did You Know?**
>
> **Pythagorean Theorem**
>
> A formula (a mathematical process used to solve a problem) developed by a Greek mathematician named Pythagoras helps you find the side lengths for any right triangle. The Pythagorean Theorem states that the sum of the squares of the two shorter sides is equal to the square of the longest side, or the hypotenuse. The hypotenuse is the longest side and is always opposite the 90-degree angle of the triangle. The mathematical equation for this theorem looks like this:
>
> $$A^2 + B^2 = C^2$$
>
> 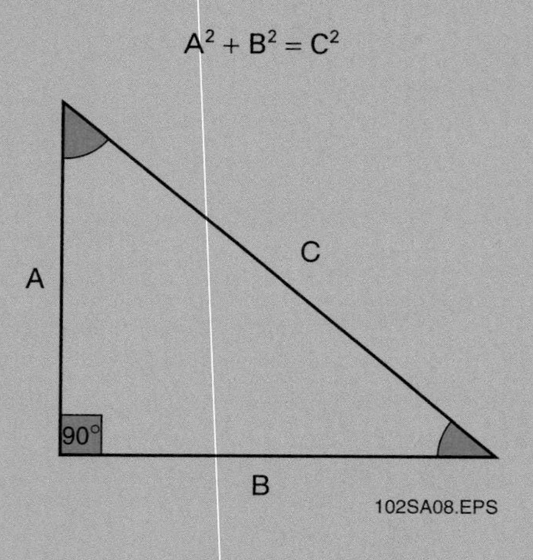
>
>
> You know that the word *square* refers to shape, but in mathematical terms, it also refers to the product of a number multiplied by itself. For example, 25 is the square of 5, and 16 is the square of 4. Another way to say this is that 25 is 5 squared, or 5 times itself, and 16 is 4 squared, or 4 times itself.
>
> In a mathematical equation this might appear as $5^2 = 25$ or $4^2 = 16$. In these examples, the numbers 5 and 4 are called the square roots, because you have to square them—or multiply them by themselves—to arrive at the squares.

- Cubic centimeter = 1 centimeter × 1 centimeter × 1 centimeter = cm^3
- Cubic meter = 1 meter × 1 meter × 1 meter = m^3

You must be able to calculate the volume of common shapes. The following are mathematical formulas that make this very easy to do. In *Appendix D* you will find a list of the formulas for calculating the volumes of these shapes. You will need to become familiar with these formulas at this stage in your training. Remember to convert dimensions before multiplying (see *On-Site: Unit Conversion*).

- *Rectangle* – Volume = length × width × depth. For example, you have to order the right amount of cubic yards of concrete for a slab that is 20 feet long and 8 feet wide and 4 inches thick (*Figure 48*). You must know the total volume of the slab. To calculate this, perform the following steps:

Step 1 Convert inches to feet.

20 ft × 8 ft × (4 in ÷ 12) =

Step 2 Multiply length × width × depth.

20 ft × 8 ft × 0.33 ft = 52.8 cu ft

Step 3 Convert cubic feet to cubic yards.

52.8 cu ft ÷ 27 (cu ft per cu yd)
= 1.96 cu yd of concrete

- *Square* – Volume = length × width × depth. For example, you have to order more concrete for a slab that is 12 feet square and 5 inches thick. The calculation for volume is as follows:

Step 1 Convert inches to feet.

12 ft × 12 ft × (5 in ÷ 12) =

Step 2 Multiply length × width × depth.

12 ft × 12 ft × 0.42 ft = 60.5 cu ft

Step 3 Convert cubic feet to cubic yards.

60.5 cu ft ÷ 27 (cu ft per cu yd)
= 2.24 cu yd of concrete

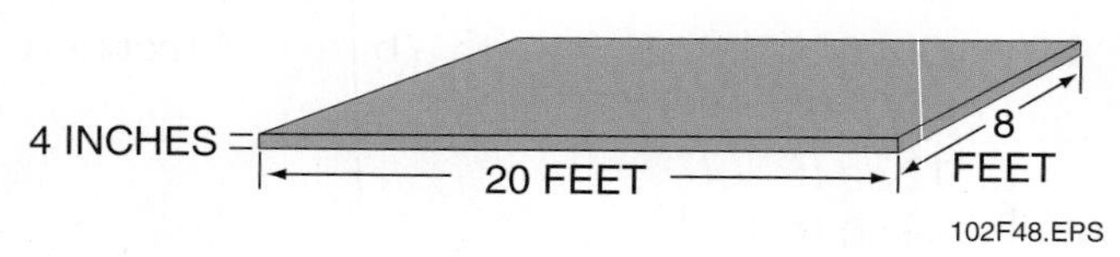

Figure 48 Volume of a rectangle.

On-Site

Pythagorean Theorem Applied

There are a number of different ways to estimate the length of common roof rafters. One way is to think of a roof section as a triangle, then use the principles behind the Pythagorean Theorem to determine your estimate.

To estimate the length of a common roof rafter, you must first become familiar with the formula for determining rafter length, as seen below. In this application you can see how the Pythagorean Theorem is being used with two knowns (run length^2 and rise length^2 or A^2 and B^2) and one unknown (rafter length^2 or C^2). The rafter length^2 step is apparent in the last step of the formula. Finding the square root of a number means you are looking for a number that, when multiplied by itself, equals a given number.

Using this picture of a house as an example, find the rafter length using the Pythagorean Theorem.

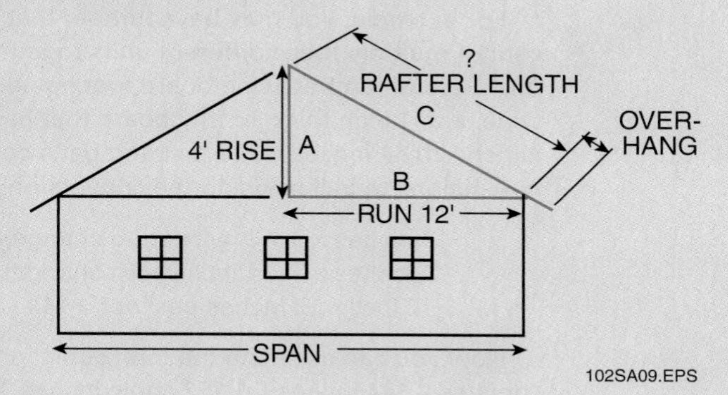

$$\text{Rafter Length} = \sqrt{\text{run}^2 + \text{rise}^2}$$

Solution

Fill in the rafter length formula using the appropriate lengths.

Run = 12 feet

Rise = 4 feet

$$\text{Rafter length} = \sqrt{12^2 + 4^2}$$

Square the rise and the run by multiplying.

Run squared = 144 $(12 \times 12 = 144)$

Rise squared = 16 $(4 \times 4 = 16)$

$$\text{Rafter length} = \sqrt{144 + 16}$$

Add the run squared and the rise squared.

$144 + 16 = 160$

$$\text{Rafter length} = \sqrt{160}$$

Rafter length = 12 feet, 8 inches

Find the square root of 160 by keying 160 into your calculator and then pressing the square root button (√). The square root of 160 = 12.64 feet or 12 feet, 8 inches if you round up to the nearest one-quarter inch (¼").

On-Site

Unit Conversion

When calculating areas or volumes of shapes, you must often work with different units of measure (inch, foot, yard) for dimensions. But you cannot use different units together in a calculation. Before doing your calculation, you must convert the measurements to the same units. Use the following conversion table to change dimensions to all inches, feet, or yards. (Also remember to convert metric measurements to the same units.)

Unit Conversion Table

Inches	Feet	Yards
1	$\frac{1}{12}$	$\frac{1}{36}$
12	1	$\frac{1}{3}$
36	3	1

For example, you may have lumber that is sized 2 inches × 4 inches × 12 feet. You cannot multiply these different units together to determine the number of board feet in this piece of lumber. (One board foot equals a board that is 12 inches long, 12 inches wide, and 1 inch thick, or any board that has the same volume.) You cannot multiply 2 inches by 4 inches by 12 feet and get a correct answer. You must convert all the dimensions to inches using the conversion table:

2 inches × 1 = 2 inches (no change)
4 inches × 1 = 4 inches (no change)
12 feet × 12 inches per foot = 144 inches (convert feet to inches)

Now you can calculate the total cubic inches in this piece of lumber: 2 inches × 4 inches × 144 inches = 1,152 cubic inches. You must now perform a final conversion to determine how many board feet this is. There are 144 cubic inches in 1 board foot. Use this number to convert the cubic inches to board feet.

1,152 cubic inches ÷ 144 cubic inches = 8 board feet

Here's another example. You must determine the area of a hallway that is 30 feet long and 36 inches wide. First, convert the dimensions to the same units. In this case, you calculate: 36 inches ÷ 12 (inches in 1 foot) = 3 feet. Now perform the calculation for area (length × width = area): 30 feet × 3 feet = 90 square feet. The area of the hallway is 90 square feet.

- *Cube (square)* – A cube is a special type of three-dimensional rectangle; its length, width, and height are equal (*Figure 49*). To find the volume of a cube, you can cube one dimension (multiply the number by itself three times). Perform the following steps to find how much concrete to order for a cube to be used as a support member of a structure:

Step 1 Determine the volume of a cube that is 8 feet cubed.

8 ft × 8 ft × 8 ft = 512 cu ft

Step 2 Convert cubic feet to cubic yards.

512 cu ft ÷ 27 = 18.96 cu yd of concrete

- *Cylinder (circle)* – Volume = pi × radius^2 × height. In this formula, you can use a shortcut: the area of a circle × height. For example, you must fill a cylinder that is 22 feet in diameter and 10 feet high (*Figure 50*):

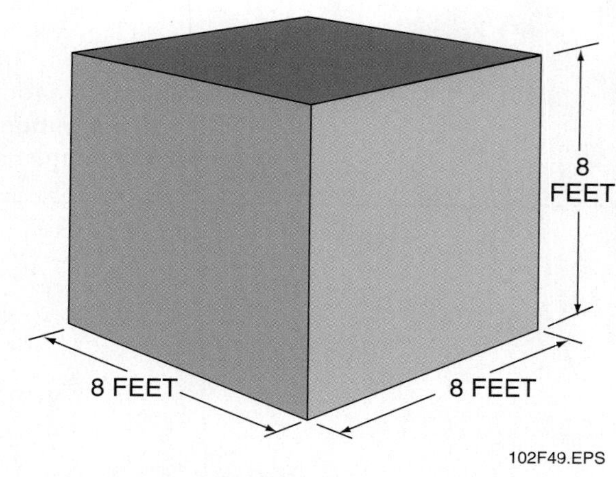

Figure 49 Volume of a cube.

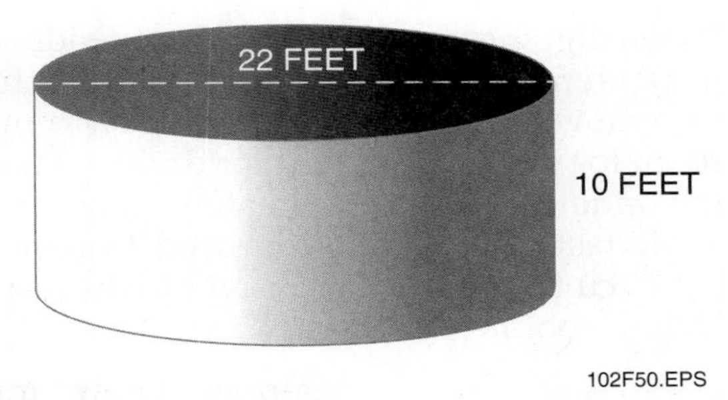

Figure 50 Volume of a cylinder.

Step 1 Calculate the area of the circle (pi)r².

$$3.14 \times 11^2 = 379.94 \text{ sq ft}$$

Step 2 Calculate the volume (area × height).

$$379.94 \text{ sq ft} \times 10 \text{ ft} = 3{,}799.4 \text{ cu ft to fill}$$

- *Triangular prism* – Volume = 0.5 × base × height × depth (thickness). For example, you must fill a triangular shape that has a base of 6 inches, a height of 12 inches, and a depth of 2 inches (*Figure 51*):

Step 1 Calculate the volume of the prism.

$$0.5 \text{ in} \times 12 \text{ in} \times 6 \text{ in} \times 2 \text{ in} = 72 \text{ cu in to fill}$$

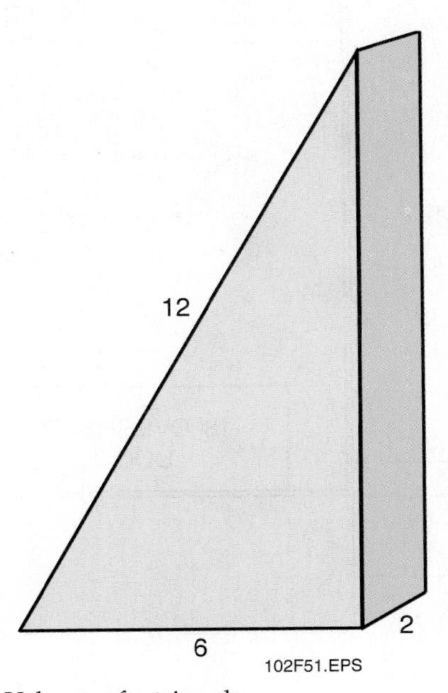

Figure 51 Volume of a triangle.

7.4.1 Study Problems: Calculating the Volume of Shapes

1. The volume of a rectangular shape 5 feet high, 6 feet thick, and 13 feet long is _____.
   a. 24 cu ft
   b. 390 cu ft
   c. 43 ft
   d. 95 ft

2. The volume of a 3-cm cube is _____.
   a. 6 cu cm
   b. 9 cu cm
   c. 12 cu cm
   d. 27 cu cm

3. The volume of a triangular prism that has a 6-inch base, a 2-inch height, and a 4-inch depth is _____.
   a. 12 sq in
   b. 24 cu in
   c. 48 sq in
   d. 36 cu in

4. The volume of a cylinder that is 6 meters in diameter and 60 centimeters high is _____.
   a. 1130.4 cu m
   b. 18 cu m
   c. 16.956 cu m
   d. 6782.4 cu m

5. If a concrete block measures 17 feet square and is 6 inches thick, its volume is _____ cubic yards.
   a. 144.5
   b. 5.35
   c. 102
   d. 3.77

7.5.0 Practical Applications

The following examples illustrate some of the practical applications a tradesperson may encounter on the job site that requires an understanding of geometry-based principals to solve. Use the information provided to solve each problem. Be sure to show all of your work.

1. Use *Figure 52* to calculate the volume of the round pipe. Volume of round pipe = ________ cubic inches.

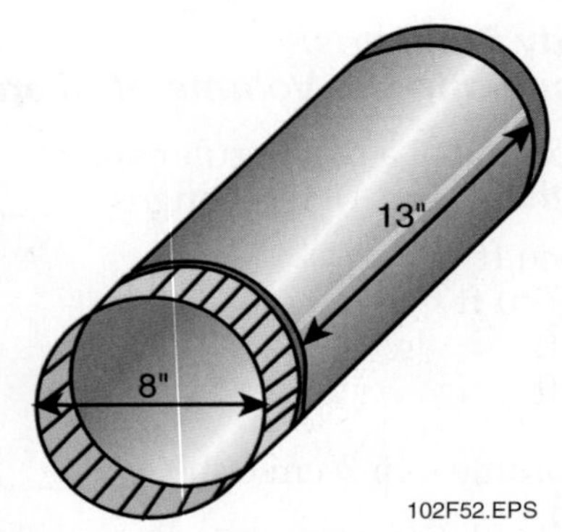

Figure 52 Finding the volume of a pipe.

2. When pouring the concrete sidewalk shown in *Figure 53*, approximately how many cubic feet of topsoil will you need to remove for the 4" thick sidewalk if the owner wants the finish surface of the sidewalk to be level with the topsoil? _____ cu ft (round your answer to the nearest ⅓ cubic foot.)

3. When digging out the topsoil for the sidewalk in *Figure 53*, you run into heavy clay. You need to remove an extra 2 inches of clay to make room for sand fill (needed for drainage under the sidewalk). Approximately how many cubic feet of clay will you need to remove? _____ cu ft (round your answer to the nearest ⅓ cubic foot.)

4. Use *Figure 54* to determine how much concrete is needed for the sidewalk, adding 5 percent extra for waste. _____ cu yd (round your answer to the nearest ½ increment).

(*Note:* You may use a calculator.)

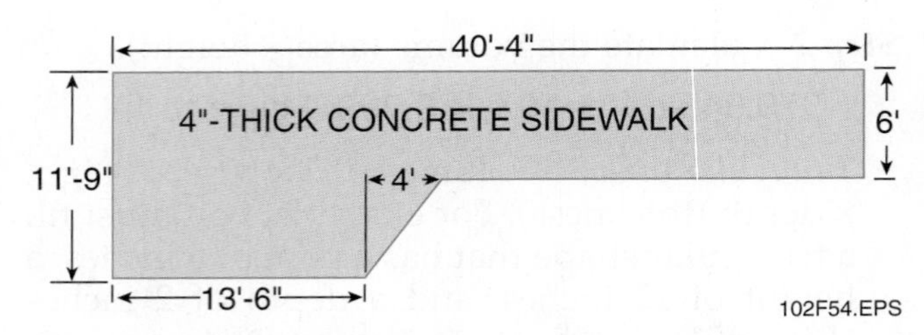

Figure 54 Estimating concrete for a sidewalk.

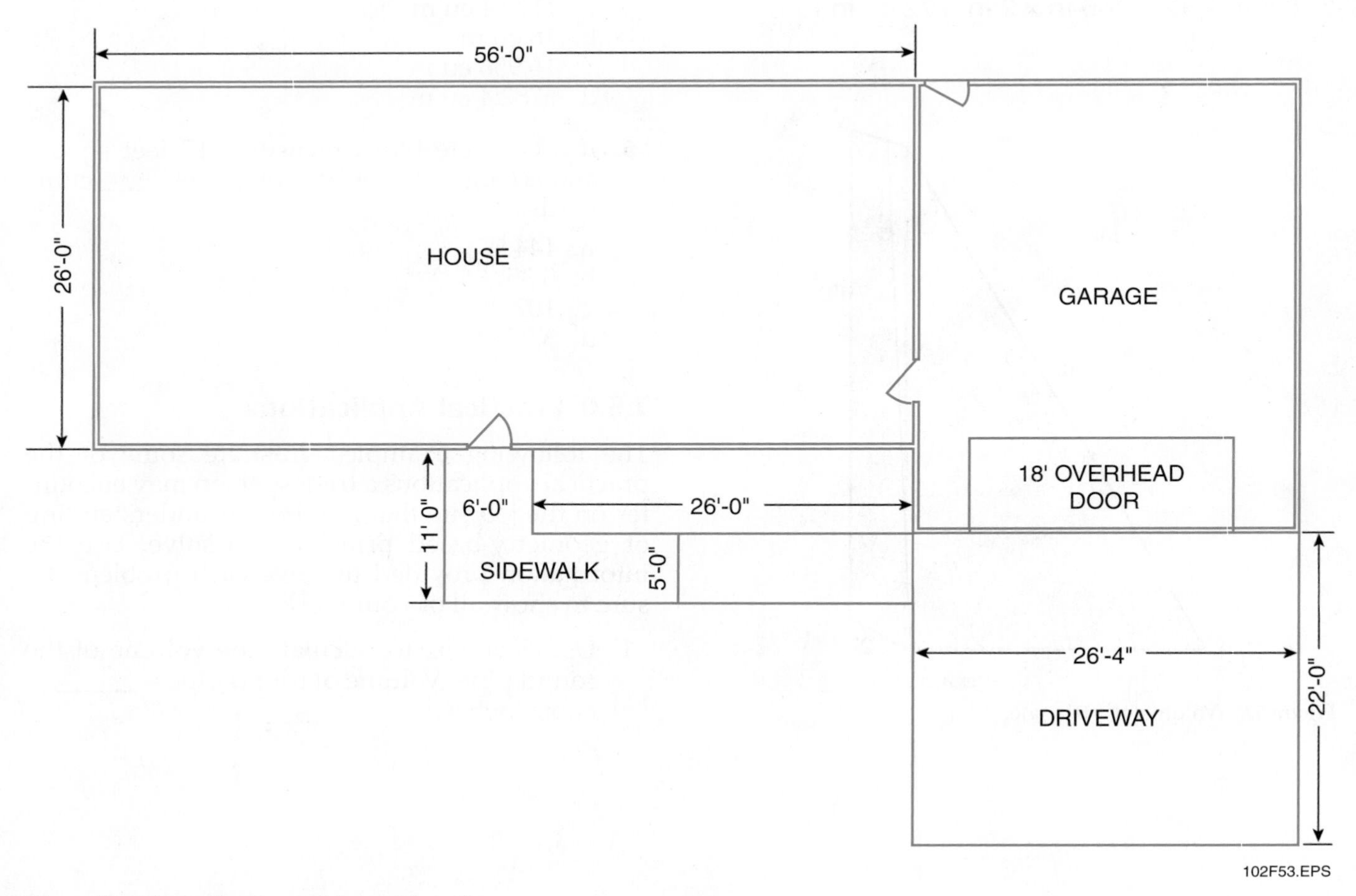

Figure 53 Removing topsoil for a sidewalk.

On-Site

Calculating the Area of a Cylinder

Let's say you have to paint the outside of a storage tank that is 10 feet in diameter and 20 feet high. How much total area will you have to paint? To know this, you have to calculate the area of a cylinder.

Step 1 The top of the storage tank is a circle, so you can use the formula to calculate the area of a circle, (pi)r^2, to find the area of the top. If the diameter is 10 feet, the radius is half of that, or 5 feet. The calculation is 3.14×5^2 or $3.14 \times 25 = 78.5$ square feet.

Step 2 What do you do about the sides? Imagine that you could unroll the tank—you would see a rectangle shape! You know that the height of the tank (which is also the width of the rectangle) is 20 feet. To calculate the length (remember, although you're visualizing a rectangle, it's still a circle), you must find the circumference of the top (pi, or 3.14, × diameter, 10 ft). Therefore, the length is 31.4 feet.

Step 3 You now know both the length and the width. To find the area (area = length × width), calculate 20 ft × 31.4 ft = 628 sq ft.

Step 4 Add the two areas together to find out how much area you must paint: 78.5 sq ft + 628 sq ft = 706.5 sq ft of tank surface.

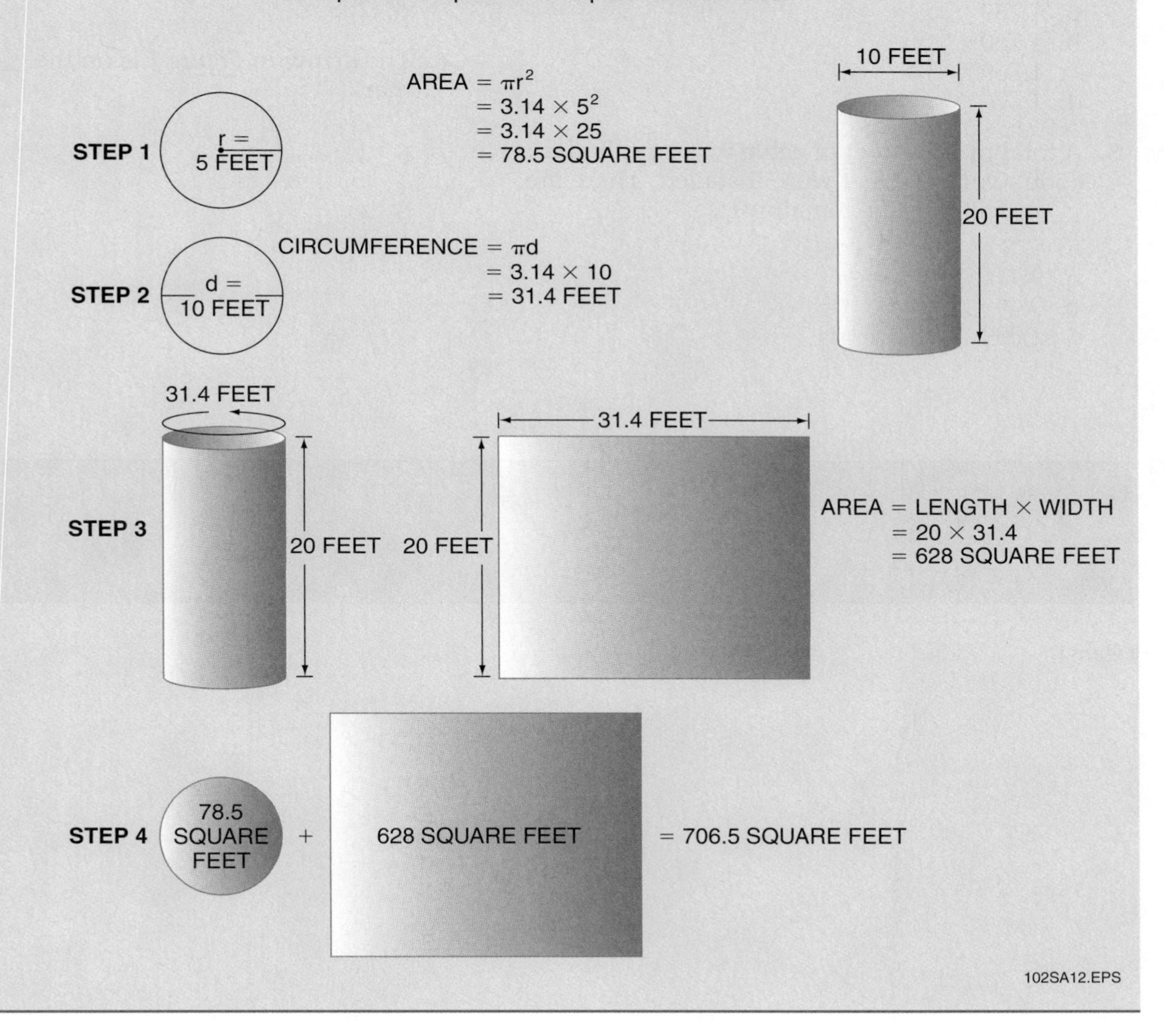

Review Questions

1. The number for the words *two thousand six hundred eighty-nine* is _____.
   a. 2,286
   b. 2,689
   c. 6,289
   d. 20,689

Solve Questions 2 through 5 without using a calculator.

2. A bricklayer lays 649 bricks the first day, 632 the second day, and 478 the third day. The bricklayer laid a total of _____ bricks in three days.
   a. 1,759
   b. 1,760
   c. 1,769
   d. 1,770

3. A total of 1,478 feet of cable was supplied for a job. Only 489 feet were installed. There are _____ feet of cable remaining.
   a. 978
   b. 980
   c. 989
   d. 1,099

4. A worker has been asked to deliver 15 scaffolds to each of 26 sites. The worker will deliver a total of _____ scaffolds.
   a. 120
   b. 240
   c. 375
   d. 390

5. Your company has 400 rolls of insulation that must go to 5 different job sites. How many rolls of insulation will go to each site?
   a. 8
   b. 20
   c. 80
   d. 208

6. The arrow in *Figure 1* is on the _____ inch(es) mark.
   a. 4⅜
   b. 4¼
   c. 4½
   d. 4⅝

Figure 1

7. The arrow in *Figure 2* is on the _____ centimeter mark.

   a. 5.2
   b. 5.3
   c. 5.4
   d. 5.5

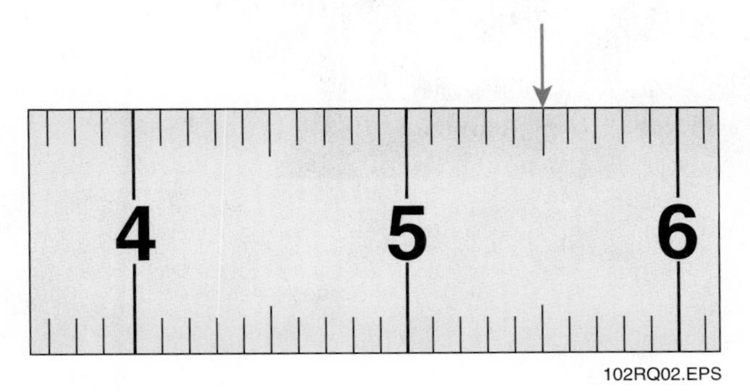

Figure 2

8. The length of the horizontal arrow in *Figure 3* represents a measurement of _____.

   a. 7 feet, 9 inches
   b. 8 feet, $\frac{3}{4}$ inches
   c. 8 feet, 9 inches
   d. 9 feet, $\frac{3}{4}$ inches

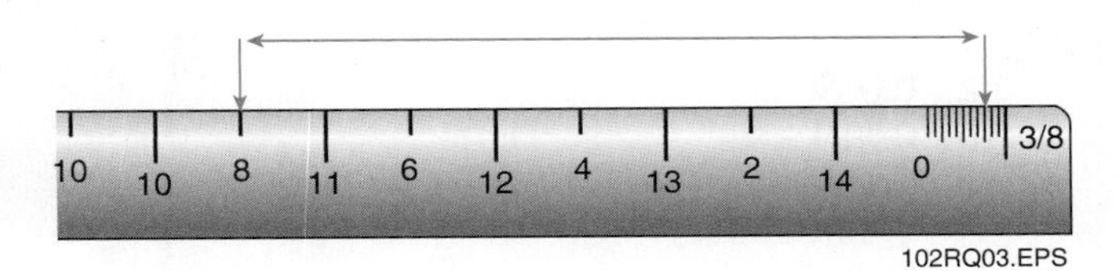

Figure 3

9. The arrow in *Figure 4* indicates a length of _____ meters.

   a. 4
   b. 8
   c. 40
   d. 80

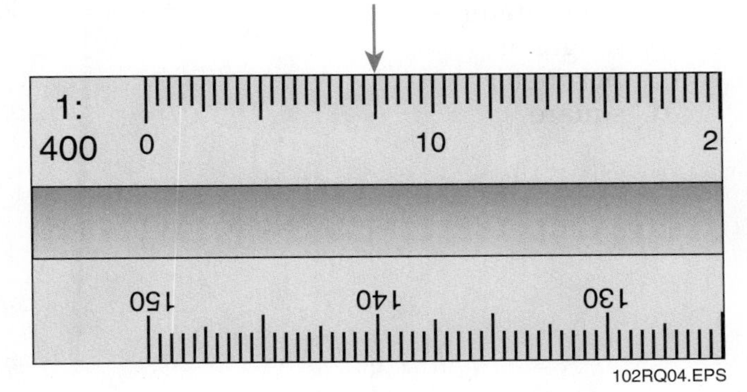

Figure 4

10. If 1 inch = 100 feet, the arrow on the ruler in *Figure 5* indicates a length of _____.

    a. 33 feet
    b. 330 feet
    c. 3,300 feet
    d. 33,000 feet

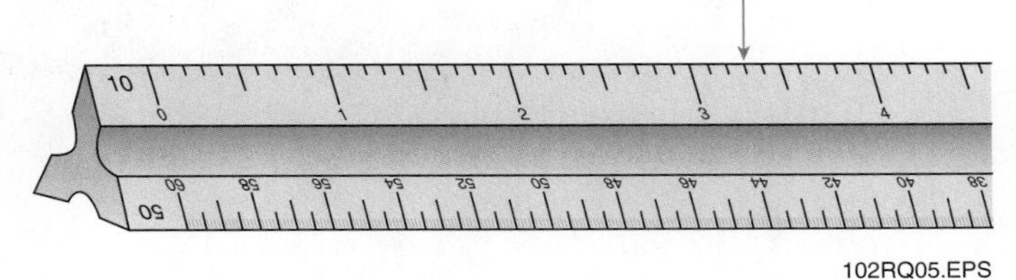

Figure 5

11. The equivalent of $\frac{3}{8}$ inch is _____ $/_{64}$ inch.

    a. 3
    b. 6
    c. 24
    d. 36

12. The lowest common denominator for the fractions $\frac{9}{64}$ and $\frac{1}{32}$ is _____.

    a. 8
    b. 16
    c. 24
    d. 32

Perform the calculations for Questions 13 and 14. Reduce the answers to their lowest terms.

13. $\frac{3}{8} + \frac{9}{16} =$ _____

    a. $\frac{7}{8}$
    b. $\frac{15}{16}$
    c. $\frac{3}{4}$
    d. $\frac{12}{16}$

14. $\frac{11}{32} - \frac{2}{8} =$ _____

    a. $\frac{9}{32}$
    b. $\frac{19}{32}$
    c. $\frac{3}{32}$
    d. $\frac{1}{8}$

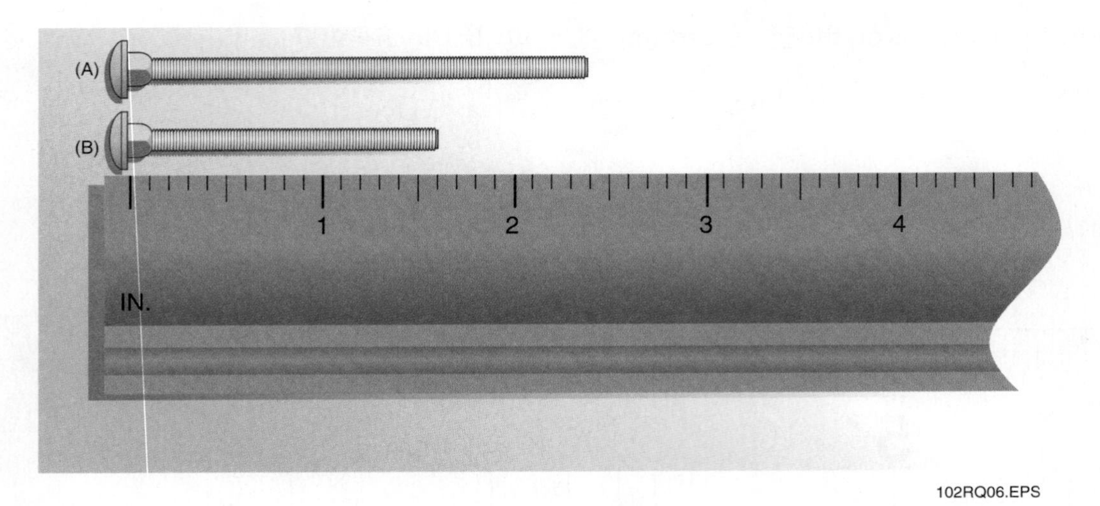

Figure 6

15. On the machinist's rule in *Figure 6*, bolt A is ____ inches long.

a. 2.4
b. 2.5
c. 2.6
d. 2.7

16. Put the following decimals in order from smallest to largest: 0.402, 0.420, 0.042, 0.442

a. 0.420, 0.042, 0.442, 0.402
b. 0.042, 0.402, 0.420, 0.442
c. 0.442, 0.420, 0.402, 0.042
d. 0.042, 0.402, 0.442, 0.420

17. Two coatings have been applied to a pipe. The first coating is 51.5 microns thick; the second coating is 89.7 microns thick. How thick is the combined coating?

a. 141.2 microns
b. 142.12 microns
c. 144.2 microns
d. 145.02 microns

18. $3.53 \times 9.75 =$ _____

a. 12.28
b. 34.4175
c. 36.1387
d. 48.13

19. It costs $2.37 to paint one square foot of wall. You need to paint a wall that measures 864.5 square feet. It will cost _____ to paint that wall. (Round your answer to the nearest hundredth.)

a. $204.88
b. $2,048.87
c. $2,848.88
d. $2,888.86

20. $89.435 \div 0.05 =$ _____

a. 1788.7
b. 17.887
c. 4.47175
d. 447.175

Solve Questions 21 and 22 without using a calculator.

21. 13.9% = _____

a. 0.009
b. 0.013
c. 0.139
d. 1.39

22. Convert 14.75 to its equivalent fraction expressed in lowest terms.

a. $14\frac{75}{100}$
b. $14\frac{3}{4}$
c. $14\frac{3}{5}$
d. $14\frac{15}{20}$

23. The _____ is the sum of all four sides of a rectangle or a square.

a. circumference
b. diagonal
c. perimeter
d. square

Use a calculator to answer Questions 24 and 25.

24. The volume of a cylindrical tank that is 23 feet high with a radius of 6.25 feet is _____. (Round your answer to the nearest hundredth.)
   a. 143.75 sq ft
   b. 898.44 cu ft
   c. 902.75 sq ft
   d. 2,821.09 cu ft

25. The volume of a triangular shape that has a 7 cm base, a 4 cm height, and a 30 mm depth is _____. (Remember to convert all measurements to the same unit.)
   a. 42 cu cm
   b. 42 cu mm
   c. 82 cu mm
   d. 82 cu cm

SUMMARY

Mathematics is not just something you need when you are in school. The construction site requires math every day to get the work done. Whether you are cutting stock, painting a wall, or installing cable, you need math skills on the job. Basic operations such as addition, subtraction, multiplication, and division are the keys to completing your tasks. Knowing how to measure, mark, and use materials and supplies increases your value to your employer, which helps ensure your job security.

Trade Terms Quiz

Fill in the blank with the correct trade term that you learned from your study of this module.

1. The ____________ is equal to half the diameter of a circle.
2. When you solve the problem 46 – 9, you must ____________ from the tens column.
3. A(n) ____________ measures between 0 and 90 degrees.
4. When you ____________ an angle, you divide it into equal parts.
5. When you add larger, 2-digit numbers in a column you may need to ____________ one amount to the next column.
6. ____________ can be measured in sq ft, sq in, and sq cm.
7. When multiplying 3 inches by 7 feet, you must ____________ the measurements to the same units.
8. A(n) ____________ is the shape made by two straight lines coming together at a point.
9. Volume can be measured in ____________ feet or meters.
10. In the number 3.45, .45 is the ____________ part of the number.
11. A(n) ____________ is a closed curved line around a central point.
12. ____________ is a unit of measurement for angles.
13. In the fraction ¾, 3 is the ____________ and 4 is the ____________.
14. A line drawn from one corner of a rectangle to the opposite corner is called a(n) ____________.
15. The ____________ is the longest straight line you can draw inside a circle.
16. In the problem 17 – 8 = 9, the number 9 is the ____________.
17. The numerical symbols from 0 to 9 are called ____________.
18. A(n) ____________, also called a(n) ____________, is an instrument that measures English measurements.
19. In the number 7,890,342, the ____________ of the 3 is three hundred.
20. Fractions having different numerators and denominators but equal values, such as ¼ and 2/8, are called ____________.
21. The ____________ for finding the area of a triangle is ½ b × h.
22. The formula for finding ____________ is pi × diameter.
23. ¼ and ⅛ are examples of ____________.
24. The ____________ is the base unit of length in the metric system.
25. 6/5 and 7/4 are examples of ____________.
26. To solve the problem ¾ ÷ ½, you must ____________ the second fraction and multiply.
27. A(n) ____________ is used to translate large measurements of real structures into smaller measurements for drawings.
28. In the problem 25 + 25 = 50, the numeral 50 is the ____________.
29. A(n) ____________ divides each inch into ten equal parts.
30. In the problem 86 ÷ 4, the answer is 21 with a(n) ____________ of 2.
31. A(n) ____________ is an instrument that measures metric lengths.
32. A(n) ____________ measures between 90 and 180 degrees.
33. A(n) ____________ is a combination of a whole number with a fraction or decimal.
34. ____________ are formed by two straight lines that cross, while ____________ are angles that have the same vertex and one side in common.
35. A(n) ____________ is a part of one hundred.
36. To determine the ____________, measure the distance around the outside of any closed shape.
37. Complete units without fractions or decimals are called ____________.
38. Numbers less than zero are ____________, and numbers greater than zero are ____________.

39. A(n) ____________ and a(n) ____________ have two pairs of equal sides that are parallel to each other.

40. A(n) ____________ has sides of unequal lengths, a(n) ____________ has one 90-degree angle, a(n) ____________ has three equal sides and three equal angles, and a(n) ____________ has two equal sides and two equal angles.

41. A(n)____________ measures 90 degrees.

42. A(n) ____________ measures 180 degrees.

43. A(n) ____________ has three sides and three angles.

44. Two or more lines or curves come together at the ____________.

45. ____________ equals approximately 3.14 or 22/7.

46. To calculate the ____________ of a cube, multiply the length by the width by the height.

47. A(n) ____________ is used because it can accommodate very large measurements.

48. The answer to a multiplication problem is called the ____________. Similarly, in a division problem the answer is the ____________.

Trade Terms

Acute angle
Adjacent angles
Angle
Architect's scale
Area
Bisect
Borrow
Carry
Circle
Circumference
Convert
Cubic
Decimal
Degree
Denominator
Diagonal
Diameter
Difference
Digit
Engineer's scale
English ruler
Equilateral triangle
Equivalent fractions
Formula
Fraction
Improper fraction
Invert
Isosceles triangle
Machinist's rule
Meter
Metric ruler
Metric scale
Mixed number
Negative numbers
Numerator
Obtuse angle
Opposite angles
Percent
Perimeter
Pi
Place value
Positive numbers
Product
Quotient
Radius
Rectangle
Remainder
Right angle
Right triangle
Scalene triangle
Square
Standard ruler
Straight angle
Sum
Triangle
Vertex
Volume
Whole numbers

Cornerstone of Craftsmanship

Joe Beyer

Monroe #1 BOCES
Building Trades and Carpentry Instructor
Penfield, New York

Joe was born in Rochester, New York, and developed his love of carpentry and teaching from his parents at an early age. Joe earned his BA from the School for American Craftsmen of the Rochester Institute of Technology and went on to earn a master's degree in educational administration from the University of New Hampshire. He also holds a certificate of advanced study from Oswego State University, where he completed five certifications. Joe then taught math, science, and career and technical education classes at area high schools. He is also an OSHA-certified instructor.

The BOCES program (Board of Cooperative Educational Services) allows teenage and adult students to explore alternatives to a traditional academic education. Students may work toward NCCER certification, union apprenticeship, or college credits in their chosen area.

How did you become interested in the construction industry?
I've been interested in furniture and teaching since I was very young. My mother had an antique furniture collection, and my father's family owned sawmills and also manufactured furniture. I started toward a career in carpentry, but decided to focus more on teaching when I became involved in a project to disassemble and move a special barn built around 1815 or 1820 and reassemble it on a museum site. I got very involved with running the crew and doing the work. I loved working with a timber-frame building, including pulling the pins, labeling all the parts, putting in the foundation, and building the barn to last for another 200 years. I saw the tremendous workmanship that had gone into the building, and I decided that this type of skill should continue to be taught.

How did you decide on your current position?
I taught in high schools for 18 years. I began to feel that the construction education programs there were being watered down, so I went into career education. I believed that I could integrate math, science, business, work ethic, and other important skills into the 2½-hour construction class. And I have been able to do that.

What are some of the things you do in your job?
The program provides students with training in hands-on and social skills as they work their way through projects. They perform a variety of smaller projects and, during their junior and senior years, they build a house. In the past, the completed homes were auctioned off, but recently I have worked to find a buyer up front. It's less complicated than an auction, and it adds to the students' knowledge base. They not only have to learn to work together in teams, but they also have to interact with an owner, which is a learning experience they will find useful in the real world. Students learn how to work through the same problems they are going to face on the job.

The modular format of the NCCER courses works really well for me. The students are in 2½-hour classes five days a week, and I break up the material into what I call profiles. When I teach Contren®'s *Tools for Success,* I combine soft skills with hands-on projects. For example, an electrical skills instructor and I build and wire a 4' × 3' × 6' cube to demonstrate team skills and the importance of learning more than one craft. The students then work in teams to build and wire their own cubes. Everything in the soft skills, from communication skills to getting along with others to solving problems, gets pulled into this training.

The courses demand a high level of energy from students and from me as well. The students have to do quite a bit of remedial work in the 3 Rs (reading, (w)riting, and (a)rithmetic), and that makes the teaching more challenging. But when you see a kid who has been struggling to find success, when you see that light go on and you know you had a role in that experience, that's all the reward you need. Students seem to learn faster when you can relate training to a practical application. Usually they'll catch on at some point that I've actually been teaching them math or English right along with their craft skills; but by then they have learned something.

What do you like most about your job?
I really enjoy passing on information and skills to the next generation, and I also enjoy working with the variety of students that come through our program, whether they are adults or high school students. I love carpentry, but I had wanted to be a teacher since I was in the ninth grade. Now I have the best of both worlds: making furniture over the summer and teaching carpentry and cabinetmaking the rest of the year.

Approximately 20 to 30 percent of my students go on to a two-year vocational college, and 30 to 40 percent go straight to work. That latter number has been higher in recent years because the demand for skilled construction workers is high and the pay and benefits are excellent. Many former students return for a visit, and some talk about how my teaching has affected their lives. And some past students even enroll their own children in the course. It does make you feel a little bit old, but it's a nice feeling to see the results of my work!

What do you think it takes to be a success in the construction trades?
The first thing you must have is knowledge of the subject matter, and that includes not only book knowledge but also working knowledge. You need to know how to hold a hammer so that it doesn't hurt your body when you use it. If you have only the working knowledge, then as codes change and you decide to keep doing things the way you want, you will cause major problems. And that's why I like the Contren® *Core Curriculum, Tools for Success*, and the multi-level arrangement of the different curricula. If you do the activities and follow the book, you get both types of knowledge.

What would you say to someone entering the trades today?
Be honest and hard working. Meet the customer's needs, and even exceed them when you can, because word of mouth is what's going to keep you employed. You may have the smarts and the ability, but you need good people skills to meet the customer's needs. That way, they get the good job they deserve and they will refer you to the next person. Construction is tougher to do than a lot of jobs, but if you're a bright person, it is not that difficult. It's very concrete, you can visualize it, and you get to see what you're doing. You get instant gratification.

Safety is very important. I invite other OSHA-certified instructors to talk to students about how their clothing and attitudes can affect workplace safety. In my 29 years as a trainer, I have had only two minor accidents on student projects. One student, an adult, shot a finishing nail through his fingernail. The other accident involved a visitor to a project worksite. This fellow was wearing baggy pants and started to climb a ladder before anyone could stop him. Both he and the students got a quick lesson in wearing appropriate clothing when he fell off the second rung.

I believe that you must demonstrate the proper work ethic. If you don't, then you will not have a chance. And that goes for me, too, as an instructor. Students will use you as either an excuse or an example. It reminds me of songs dating back to when I was a kid: "If you're not part of the solution, you're part of the problem," and "Fix it or get out of the way." It takes a while to pick those words out, though, because a lot of times they're hard to understand!

Cornerstone of Craftsmanship

Frank Branham

Tri-Counties Multi Trade Centers
Executive Director

How did you choose a career in the field?
I was very fortunate at the age of 16 to receive my introduction to the construction trades through a thoughtful neighborhood contractor I knew as Mr. Whitaker. Mr. Whitaker took me under his wing and taught me skills I would end up using the rest of my life. These included skills like rough and finish carpentry, and kitchen, basement, bathroom, and addition remodeling. I became a custom cabinet maker and am also skilled in electrical and plumbing applications.

What types of training have you been through?
I started construction training through Turner Construction Company in Cleveland, Ohio, and then received additional training through the National Center for Construction Education and Research (NCCER). I eventually became a certified instructor in NCCER's Core and Residential Carpentry courses, and have since become certified as an NCCER Master Trainer. In addition, I have taught a Builder Licensing Course at local community colleges throughout southwest Michigan.

What kinds of work have you done in your career?
I have over forty years of experience in the construction industry. Most of my work has been in residential and new construction—everything from foundation to roofs to floor framing, including vinyl, ceramic, and tile installations. In 1979 I started my own company, Branham Construction Company. We are a full-service contractor licensed through the state of Michigan.

Tell us about your present job and what you like about it.
Today, in addition to owning Branham Construction, I am also the owner of Tri-Counties Multi Trade Centers, which is a trade school licensed by the state of Michigan and accredited by NCCER. Our carpentry, electrical, and plumbing programs are registered through the United States Department of Labor Office of Apprenticeship Training (see *Construction Education Newsletter*, Spring 2006 edition).

What factors have contributed most to your success?
I had no idea as a sixteen-year-old young man that my career would become what it is today. I have told many students at career fairs how the construction industry offers great opportunities. To be able to train and re-train individuals throughout my community and to give back what Mr. Whitaker taught me is what has made me successful.

What advice would you give to those new to the field?
I can sum up my advice in three words: patience, practice, and discipline.

Tell us some interesting career-related facts or accomplishments:
I am a certified Teaching Instructor Master Trainer, Owner of Branham Construction Company, owner of Tri-Counties Multi Trade Centers, and Co-Owner of Tri-Counties Construction Associations (Training Sponsored).

Trade Terms Introduced in this Module

Acute angle: Any angle between 0 degrees and 90 degrees.

Adjacent angles: Angles that have the same vertex and one side in common.

Angle: The shape made by two straight lines coming together at a point. The space between those two lines is measured in degrees.

Architect's scale: A specialized ruler used in making or measuring reduced scale drawings. It is marked with a range of calibrated ratios used for laying out distances, with scales indicating feet, inches, and fractions of inches. Used on drawings other than site plans.

Area: The surface or amount of space occupied by a two-dimensional object such as a rectangle, circle, or square. To calculate the area for rectangles and squares, multiply the length and width. To calculate the area for circles, multiply the radius squared and pi.

Bisect: To divide into equal parts.

Borrow: To move numbers from one value column (such as the tens column) to another value column (such as units) to perform subtraction problems.

Carry: To transfer an amount from one column to another column.

Circle: A closed curved line around a central point. A circle measures 360 degrees.

Circumference: The distance around the curved line that forms a circle.

Convert: To change from one unit of expression to another. For example, convert a decimal to a percentage: 0.25 to 25%; or convert a fraction to an equivalent: $\frac{3}{4}$ to $\frac{6}{8}$.

Cubic: Measurement found by multiplying a number by itself three times; it describes volume measurement.

Decimal: Part of a number represented by digits to the right of a point, called a decimal point. For example, in the number 1.25, .25 is the decimal part of the number.

Degree: A unit of measurement for angles. For example, a right angle is 90 degrees, an acute angle is between 0 and 90 degrees, and an obtuse angle is between 90 and 180 degrees.

Denominator: The part of a fraction below the dividing line. For example, the 2 in $\frac{1}{2}$ is the denominator.

Diagonal: Line drawn from one corner of a rectangle or square to the farthest opposite corner.

Diameter: The length of a straight line that crosses from one side of a circle, through the center point, to a point on the opposite side. The diameter is the longest straight line you can draw inside a circle.

Difference: The result you get when you subtract one number from another. For example, in the problem $8 - 3 = 5$, the number 5 is the difference.

Digit: Any of the numerical symbols 0 to 9.

Engineer's scale: A tool for measuring distances and transferring measurements at a fixed ration of length. It is used to make drawings to scale. Usually used for land measurements on site plans. A straightedge measuring device divided uniformly into multiples of 10 divisions per inch so drawings can be made with decimal values.

English ruler: Instrument that measures English measurements; also called the standard ruler. Units of English measure include inches, feet, and yards.

Equilateral triangle: A triangle that has three equal sides and three equal angles.

Equivalent fractions: Fractions having different numerators and denominators, but equal values, such as $\frac{1}{2}$ and $\frac{2}{4}$.

Formula: A mathematical process used to solve a problem. For example, the formula for finding the area of a rectangle is side A times side B = Area, or $A \times B$ = Area.

Fraction: A number represented by a numerator and a denominator, such as $\frac{1}{2}$.

Improper fraction: A fraction whose numerator is larger than its denominator. For example, $\frac{8}{4}$ and $\frac{6}{3}$ are improper fractions.

Invert: To reverse the order or position of numbers. In fractions, to turn upside down, such as $\frac{3}{4}$ to $\frac{4}{3}$. When you are dividing by fractions, one fraction is inverted.

Isosceles triangle: A triangle that has two equal sides and two equal angles.

Machinist's rule: A ruler that is marked so that the inches are divided into 10 equal parts, or tenths.

Meter: The base unit of length in the metric system; approximately 39.37 inches.

Metric ruler: Instrument that measures metric lengths. Units of measure include millimeters, centimeters, and meters.

Metric scale: A straightedge measuring device divided into centimeters, with each centimeter divided into 10 millimeters. Usually used for architectural drawings.

Mixed number: A combination of a whole number with a fraction or decimal. Examples of mixed numbers are 3⁷⁄₁₆, 5.75, and 1¼.

Negative numbers: Numbers less than zero. For example, –1, –2, and –3 are negative numbers.

Numerator: The part of a fraction above the dividing line. For example, the 1 in ½ is the numerator.

Obtuse angle: Any angle between 90 degrees and 180 degrees.

Opposite angles: Two angles that are formed by two straight lines crossing. They are always equal.

Percent: Of or out of one hundred. For example, 8 is 8 percent (%) of 100.

Perimeter: The distance around the outside of any closed shape, such as a rectangle, circle, or square.

Pi: A mathematical value of approximately 3.14 (or ²²⁄₇) used to determine the area and circumference of circles. It is sometimes symbolized by π.

Place value: The exact quantity of a digit, determined by its place within the whole number or by its relationship to the decimal point.

Positive numbers: Numbers greater than zero. For example, 1, 2, and 3 are positive numbers.

Product: The answer to a multiplication problem. For example, the product of 6×6 is 36.

Quotient: The result of a division. For example, when dividing 6 by 2, the quotient is 3.

Radius: The distance from a center point of a circle to any point on the curved line, or half the width (diameter) of a circle.

Rectangle: A four-sided shape with four 90-degree angles. Opposite sides of a rectangle are always parallel and the same length. Adjacent sides are perpendicular and are not equal in length.

Remainder: The leftover amount in a division problem. For example, in the problem $34 \div 8$, 8 goes into 34 four times ($8 \times 4 = 32$) and 2 is left over; in other words, it is the remainder.

Right angle: An angle that measures 90 degrees. The two lines that form a right angle are perpendicular to each other. This is the angle used most in the trades.

Right triangle: A triangle that includes one 90-degree angle.

Scalene triangle: A triangle with sides of unequal lengths.

Square: (1) A special type of rectangle with four equal sides and four 90-degree angles. (2) The product of a number multiplied by itself. For example, 25 is the square of 5; 16 is the square of 4.

Standard ruler: An instrument that measures English lengths (inches, feet, and yards). See *English ruler.*

Straight angle: A 180-degree angle or flat line.

Sum: The total in an addition problem. For example, in the problem $7 + 8 = 15$, 15 is the sum.

Triangle: A closed shape that has three sides and three angles.

Vertex: A point at which two or more lines or curves come together.

Volume: The amount of space occupied in three dimensions (length, width, and height/depth/thickness).

Whole numbers: Complete units without fractions or decimals.

Appendix A

Multiplication Table

Trace across and down from the numbers that you want to multiply, and find the answer. In the example, $7 \times 7 = 49$.

	2	**3**	**4**	**5**	**6**	**7**	**8**	**9**	**10**	**11**	**12**
2	4	6	8	10	12	14	16	18	20	22	24
3	6	9	12	15	18	21	24	27	30	33	36
4	8	12	16	20	24	28	32	36	40	44	48
5	10	15	20	25	30	35	40	45	50	55	60
6	12	18	24	30	36	42	48	54	60	66	72
7	14	21	28	35	42	49	56	63	70	77	84
8	16	24	32	40	48	56	64	72	80	88	96
9	18	27	36	45	54	63	72	81	90	99	108
10	20	30	40	50	60	70	80	90	100	110	120
11	22	33	44	55	66	77	88	99	110	121	132
12	24	36	48	60	72	84	96	108	120	132	144

Appendix B

INCHES CONVERTED TO DECIMALS OF A FOOT

Inches	Decimals of a Foot
1/16	0.005
1/8	0.010
3/16	0.016
1/4	0.021
5/16	0.026
3/8	0.031
7/16	0.036
1/2	0.042
9/16	0.047
5/8	0.052
11/16	0.057
3/4	0.063
13/16	0.068
7/8	0.073
15/16	0.078
1	0.083
1 1/16	0.089
1 1/8	0.094
1 3/16	0.099
1 1/4	0.104
1 5/16	0.109
1 3/8	0.115
1 7/16	0.120
1 1/2	0.125
1 9/16	0.130
1 5/8	0.135
1 11/16	0.141
1 3/4	0.146
1 13/16	0.151
1 7/8	0.156
1 15/16	0.161
2	0.167
2 1/16	0.172
2 1/8	0.177
2 3/16	0.182
2 1/4	0.188
2 5/16	0.193
2 3/8	0.198
2 7/16	0.203
2 1/2	0.208
2 9/16	0.214
2 5/8	0.219
2 11/16	0.224
2 3/4	0.229
2 13/16	0.234
2 7/8	0.240
2 15/16	0.245
3	0.250
3 1/16	0.255
3 1/8	0.260
3 3/16	0.266
3 1/4	0.271
3 5/16	0.276
3 3/8	0.281
3 7/16	0.286
3 1/2	0.292
3 9/16	0.297
3 5/8	0.302
3 11/16	0.307
3 3/4	0.313
3 13/16	0.318
3 7/8	0.323
3 15/16	0.328
4	0.333
4 1/16	0.339
4 1/8	0.344
4 1/16	0.349
4 1/4	0.354
4 5/16	0.359
4 3/8	0.365
4 7/16	0.370
4 1/2	0.374
4 9/16	0.380
4 5/8	0.385
4 11/16	0.391
4 3/4	0.396
4 13/16	0.401
4 7/8	0.406
4 15/16	0.411
5	0.417
5 1/16	0.422
5 1/8	0.427
5 3/16	0.432
5 1/4	0.438
5 5/16	0.443
5 3/8	0.448
5 7/16	0.453
5 1/2	0.458
5 9/16	0.464
5 5/8	0.469
5 11/16	0.474
5 3/4	0.479
5 13/16	0.484
5 7/8	0.490
5 15/16	0.495
6	0.500

INCHES CONVERTED TO DECIMALS OF A FOOT (Continued)

Inches	Decimals of a Foot	Inches	Decimals of a Foot	Inches	Decimals of a Foot
$6\frac{1}{16}$	0.505	$8\frac{1}{16}$	0.672	$10\frac{1}{16}$	0.839
$6\frac{1}{8}$	0.510	$8\frac{1}{8}$	0.677	$10\frac{1}{8}$	0.844
$6\frac{3}{16}$	0.516	$8\frac{3}{16}$	0.682	$10\frac{3}{16}$	0.849
$6\frac{1}{4}$	0.521	$8\frac{1}{4}$	0.688	$10\frac{1}{4}$	0.854
$6\frac{5}{16}$	0.526	$8\frac{5}{16}$	0.693	$10\frac{5}{16}$	0.859
$6\frac{3}{8}$	0.531	$8\frac{3}{8}$	0.698	$10\frac{3}{8}$	0.865
$6\frac{7}{16}$	0.536	$8\frac{7}{16}$	0.703	$10\frac{7}{16}$	0.870
$6\frac{1}{2}$	0.542	$8\frac{1}{2}$	0.708	$10\frac{1}{2}$	0.875
$6\frac{9}{16}$	0.547	$8\frac{9}{16}$	0.714	$10\frac{9}{16}$	0.880
$6\frac{5}{8}$	0.552	$8\frac{5}{8}$	0.719	$10\frac{5}{8}$	0.885
$6\frac{11}{16}$	0.557	$8\frac{11}{16}$	0.724	$10\frac{11}{16}$	0.891
$6\frac{3}{4}$	0.563	$8\frac{3}{4}$	0.729	$10\frac{3}{4}$	0.896
$6\frac{13}{16}$	0.568	$8\frac{13}{16}$	0.734	$10\frac{13}{16}$	0.901
$6\frac{7}{16}$	0.573	$8\frac{7}{8}$	0.740	$10\frac{7}{8}$	0.906
$6\frac{15}{16}$	0.578	$8\frac{15}{16}$	0.745	$10\frac{15}{16}$	0.911
7	0.583	9	0.750	11	0.917
$7\frac{1}{16}$	0.589	$9\frac{1}{16}$	0.755	$11\frac{1}{16}$	0.922
$7\frac{1}{8}$	0.594	$9\frac{1}{8}$	0.760	$11\frac{1}{8}$	0.927
$7\frac{3}{16}$	0.599	$9\frac{3}{16}$	0.766	$11\frac{3}{16}$	0.932
$7\frac{1}{4}$	0.604	$9\frac{1}{4}$	0.771	$11\frac{1}{4}$	0.938
$7\frac{5}{16}$	0.609	$9\frac{5}{16}$	0.776	$11\frac{5}{16}$	0.943
$7\frac{3}{8}$	0.615	$9\frac{3}{8}$	0.781	$11\frac{3}{8}$	0.948
$7\frac{7}{16}$	0.620	$9\frac{7}{16}$	0.786	$11\frac{7}{16}$	0.953
$7\frac{1}{2}$	0.625	$9\frac{1}{2}$	0.792	$11\frac{1}{2}$	0.958
$7\frac{9}{16}$	0.630	$9\frac{9}{16}$	0.797	$11\frac{9}{16}$	0.964
$7\frac{5}{8}$	0.635	$9\frac{5}{8}$	0.802	$11\frac{5}{8}$	0.969
$7\frac{11}{16}$	0.641	$9\frac{11}{16}$	0.807	$11\frac{11}{16}$	0.974
$7\frac{3}{4}$	0.646	$9\frac{3}{4}$	0.813	$11\frac{3}{4}$	0.979
$7\frac{13}{16}$	0.651	$9\frac{13}{16}$	0.818	$11\frac{13}{16}$	0.984
$7\frac{7}{8}$	0.656	$9\frac{7}{8}$	0.823	$11\frac{7}{8}$	0.990
$7\frac{15}{16}$	0.661	$9\frac{15}{16}$	0.828	$11\frac{15}{16}$	0.995
8	0.667	10	0.833	12	1.000

Appendix C

Conversion Tables

How to Convert Units of Volume

Metric to English

From	Multiply By	To Obtain
Liters	1.0567	Quarts
Liters	2.1134	Pints
Liters	0.2642	Gallons

English to Metric

From	Multiply By	To Obtain
Quarts	0.946	Liters
Pints	0.473	Liters
Gallons	3.785	Liters

How to Convert Units of Weight

Metric to English

From	Multiply By	To Obtain
Grams	0.0353	Ounces
Grams	15.4321	Grains
Kilograms	2.2046	Pounds
Kilograms	0.0011	Tons (short)
Tons (metric)	1.1023	Tons (short)

English to Metric

From	Multiply By	To Obtain
Pounds	0.4536	Kilograms
Pounds	453.6	Grams
Ounces	28.35	Grams
Grains	0.0648	Grams
Tons (short)	0.9072	Tons (metric)

How to Convert Units of Length

Metric to English

From	Multiply By	To Obtain
Meters	39.37	Inches
Meters	3.2808	Feet
Meters	1.0936	Yards
Centimeters	0.3937	Inches
Millimeters	0.03937	Inches
Kilometers	0.6214	Miles

English to Metric

From	Multiply By	To Obtain
Inches	2.54	Centimeters
Inches	0.0254	Meters
Inches	25.4	Millimeters
Miles	1,609,344	Millimeters
Feet	0.3048	Meters
Feet	30.48	Centimeters
Yards	0.9144	Meters
Yards	91.44	Centimeters
Miles	1.6093	Kilometers

Appendix D

AREA AND VOLUME FORMULAS

Area Formulas

Rectangle:	area = length × width
Square:	area = length × width or area = side^2
Circle:	area = pi × radius^2
Triangle:	area = 0.5 × base × height

Volume Formulas

Rectangle:	volume = length × width × depth
Square:	volume = length × width × depth
Cube (square):	volume = side^3
Cylinder (circle):	volume = pi × radius^2 × height = area of a circle × height
Triangle:	volume = 0.5 × base × height × depth (thickness) = area of a triangle × depth

Appendix E

Using a Calculator

It is important to be able to perform calculations (such as addition and subtraction) in your head even if you have a calculator. It allows you to estimate the answer before you use the calculator. Why is this important? So you can double-check the answer against your estimate. If you press a wrong key on the calculator, you might be out hundreds of dollars or off by several inches.

The calculator is a marvelous tool for saving time. Let's look at the most frequently used operations of the calculator: adding, subtracting, multiplying, and dividing whole numbers. *Figure A-1* shows the parts of a common calculator.

Using the Calculator to Add Whole Numbers

Adding numbers is easy with a calculator. Just follow these steps, using 5 + 4 to practice.

Step 1 Press the ON/C key to turn the calculator on.

Step 2 Press 5. A 5 appears in the display.

Step 3 Press the + key. The 5 is still displayed.

Step 4 Press the 4 key. A 4 is displayed.

Step 5 Press the = key. The sum, 9, appears in the display.

Step 6 Press the ON/C key to clear the calculator. A zero (0) appears in the display.

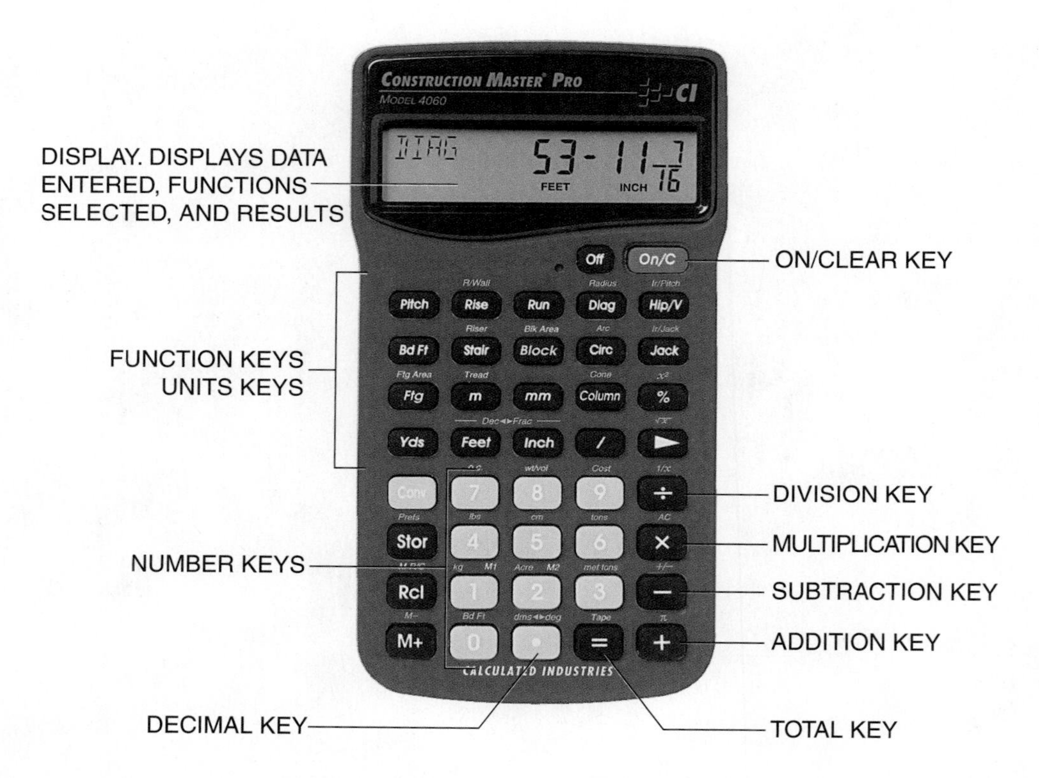

Figure A-1 Parts of a calculator.

USING A CALCULATOR (Continued)

Using the Calculator to Subtract Whole Numbers

Subtracting with a calculator is as easy as adding. Here are the steps, using the problem 25 – 5 to practice.

Step 1 Press the ON/C key to turn the calculator on.

Step 2 Press the 2 key and then the 5 key. A 25 appears in the display.

Step 3 Press the – key. The 25 is still displayed.

Step 4 Press the 5 key. A 5 is displayed.

Step 5 Press the = key. The difference, 20, appears in the display.

Step 6 Press the ON/C key to clear the calculator. A zero (0) appears in the display.

Using the Calculator to Multiply Whole Numbers

Multiplying with a calculator is as easy as adding and subtracting. Here are the steps, using the problem 6 × 5 to practice.

Step 1 Press the ON/C key to turn the calculator on.

Step 2 Press 6. A 6 appears in the display.

Step 3 Press the × key. The 6 is still displayed.

Step 4 Press the 5 key. A 5 is displayed.

Step 5 Press the = key. The answer, 30, appears in the display.

Step 6 Press the ON/C key to clear the calculator. A zero (0) appears in the display.

Using the Calculator to Divide Whole Numbers

Dividing with a calculator is as easy as the other operations. Here are the steps, using 12 ÷ 4 to practice:

Step 1 Press the ON/C key to turn the calculator on.

Step 2 Press the 1 key and then the 2 key. A 12 appears in the display.

Step 3 Press the ÷ key. The 12 is still displayed.

Step 4 Press the 4 key. A 4 is displayed.

Step 5 Press the = key. The answer, 3, appears in the display.

Step 6 Press the ON/C key to clear the calculator. A zero (0) appears in the display.

Expressing a Remainder as a Whole Number

When one number does not go into another number evenly, you are left with a remainder. For example, use your calculator to figure this problem: There is a piece of wood 6 feet long. How many 4-foot pieces can a worker cut from it? How many feet will be left over?

Step 1 Press the ON/C key to turn the calculator on.

Step 2 Press 6. A 6 appears in the display.

Step 3 Press the ÷ key. The 6 is still displayed.

Step 4 Press the 4 key. A 4 is displayed.

Step 5 Press the = key.

Step 6 The total, 1.5, appears in the display (0.5 is a decimal, a part of a number, represented by digits to the right of a point called the decimal point).

Step 7 Press the ON/C key to clear the calculator. A zero (0) appears in the display.

Step 8 To express the 0.5 as a whole number rather than a decimal, multiply it by the number you divided by (4). The remainder expressed as a whole number is 2 feet.

The answer to the problem, "How many 4-foot pieces can a worker cut from a 6-foot piece of wood and how many feet will be left over?" is one piece, with 2 feet left over.

Additional Resources

This module is intended to present thorough resources for task training. The following reference works are suggested for further study. These are optional materials for continued education rather than for task training.

All the Math You'll Ever Need, 1999. Stephen Slavin. New York: John Wiley & Sons.

Applied Construction Math: A Novel Approach, 2006. National Center for Construction Education and Research. Upper Saddle River, NJ: Prentice Hall.

Basic Construction Math Review: A Manual of Basic Construction Mathematics for Contractor and Tradesman License Exams. Printcorp Business Printing. Construction Book Express.

Math for the Building Trades. Homewood, IL: American Technical Publishers (ATP).

Math to Build On: A Book for Those Who Build, 1997. Johnny and Margaret Hamilton. Clinton, NC: Construction Trades Press.

CONTREN® LEARNING SERIES – USER UPDATE

NCCER makes every effort to keep these textbooks up-to-date and free of technical errors. We appreciate your help in this process. If you have an idea for improving this textbook, or if you find an error, a typographical mistake, or an inaccuracy in NCCER's Contren® textbooks, please write us, using this form or a photocopy. Be sure to include the exact module number, page number, a detailed description, and the correction, if applicable. Your input will be brought to the attention of the Technical Review Committee. Thank you for your assistance.

Instructors – If you found that additional materials were necessary in order to teach this module effectively, please let us know so that we may include them in the Equipment/Materials list in the Annotated Instructor's Guide.

Write: Product Development and Revision
National Center for Construction Education and Research
3600 NW 43rd St., Bldg. G, Gainesville, FL 32606

Fax: 352-334-0932

E-mail: curriculum@nccer.org

Craft ______ Module Name ______

Copyright Date ______ Module Number ______ Page Number(s) ______

Description ______

______

______

______

(Optional) Correction ______

______

______

(Optional) Your Name and Address ______

______

______

Introduction to Hand Tools

00103-09

CORE CURRICULUM

00109-09 Introduction to Materials Handling

00108-09 Basic Employability Skills

00107-09 Basic Communication Skills

00106-09 Basic Rigging

00105-09 Introduction to Construction Drawings

00104-09 Introduction to Power Tools

00103-09 Introduction to Hand Tools

00102-09 Introduction to Construction Math

00101-09 Basic Safety

This course map shows all of the modules in the *Core Curriculum: Introductory Craft Skills*. The suggested training order begins at the bottom and proceeds up. Skill levels increase as you advance on the course map. The local Training Program Sponsor may adjust the training order.

Note that Module 00106-09, *Basic Rigging*, is an elective. It is not a requirement for level completion, but it may be included as part of your training program.

Objectives

When you have completed this module, you will be able to do the following:

1. Recognize and identify some of the basic hand tools and their proper uses in the construction trade.
2. Visually inspect hand tools to determine if they are safe to use.
3. Safely use hand tools.

Trade Terms

Adjustable end wrench
Ball peen hammer
Bell-faced hammer
Bevel
Box-end wrench
Carpenter's square
Cat's paw
Chisel
Chisel bar
Claw hammer
Combination square
Combination wrench
Dowel
Fastener
Flat bar
Flats
Foot-pounds
Hex key wrench
Inch-pounds
Joint
Kerf
Level
Miter joint
Nail puller
Open-end wrench
Peening
Pipe wrench
Planed
Pliers
Plumb
Points
Punch
Rafter angle square
Ripping bar
Round off
Square
Striking (or slugging) wrench
Strip
Tang
Tempered
Tenon
Torque
Try square
Weld

Prerequisites

Before you begin this module, it is recommended that you successfully complete the following: *Core Curriculum,* Modules 00101-09 through 00102-09. Module 00106-09 is an elective and is not a requirement for completion of this course.

Contents

Topics to be presented in this module include:

1.0.0 Introduction . . . 3.1
1.1.0 Safety . . . 3.1
2.0.0 Hammers . . . 3.1
2.1.0 The Claw Hammer . . . 3.1
2.1.1 How to Use a Claw Hammer to Drive a Nail . . . 3.1
2.1.2 How to Use a Claw Hammer to Pull a Nail . . . 3.2
2.2.0 The Ball Peen Hammer . . . 3.2
2.3.0 Safety and Maintenance . . . 3.2
2.4.0 Sledgehammers . . . 3.3
2.4.1 How to Use a Sledgehammer . . . 3.3
2.4.2 Safety and Maintenance . . . 3.4
3.0.0 Ripping Bars and Nail Pullers . . . 3.5
3.1.0 Ripping Bars . . . 3.5
3.2.0 Nail Pullers . . . 3.6
3.2.1 How to Use a Nail Puller (Cat's Paw) . . . 3.6
3.3.0 Safety and Maintenance . . . 3.6
4.0.0 Chisels and Punches . . . 3.6
4.1.0 Chisels . . . 3.6
4.1.1 How to Use a Wood Chisel . . . 3.6
4.1.2 How to Use a Cold Chisel . . . 3.8
4.1.3 Safety and Maintenance . . . 3.8
4.2.0 Punches . . . 3.8
5.0.0 Screwdrivers . . . 3.9
5.1.0 How to Use a Screwdriver . . . 3.10
5.2.0 Safety and Maintenance . . . 3.10
6.0.0 Pliers and Wire Cutters . . . 3.11
6.1.0 Slip-Joint (Combination) Pliers . . . 3.11
6.1.1 How to Use Slip-Joint Pliers . . . 3.11
6.2.0 Long-Nose (Needle-Nose) Pliers . . . 3.12
6.2.1 How to Use Long-Nose Pliers . . . 3.12
6.3.0 Lineman Pliers (Side Cutters) . . . 3.12
6.3.1 How to Use Lineman Pliers . . . 3.12
6.4.0 Tongue-and-Groove Pliers . . . 3.13
6.4.1 How to Use Tongue-and-Groove Pliers . . . 3.13
6.5.0 Locking Pliers . . . 3.13
6.5.1 How to Use Locking Pliers . . . 3.13
6.6.0 Safety and Maintenance . . . 3.14
7.0.0 Wrenches . . . 3.14
7.1.0 Nonadjustable Wrenches . . . 3.14
7.1.1 How to Use a Nonadjustable Wrench . . . 3.15
7.2.0 Adjustable Wrenches . . . 3.15
7.2.1 How to Use an Adjustable Wrench . . . 3.16
7.3.0 Safety and Maintenance . . . 3.16
8.0.0 Sockets and Ratchets . . . 3.17
8.1.0 How to Use Sockets and Ratchets . . . 3.17
8.2.0 Safety and Maintenance . . . 3.17
9.0.0 Torque Wrenches . . . 3.17
9.1.0 How to Use a Torque Wrench . . . 3.18
9.2.0 Safety and Maintenance . . . 3.18

10.0.0 Rules and Other Measuring Tools 3.18
10.1.0 Steel Rule 3.18
10.2.0 Measuring Tape. 3.19
10.3.0 Wooden Folding Rule 3.19
10.4.0 Laser Measuring Tools 3.19
10.5.0 Safety and Maintenance 3.20
11.0.0 Levels. 3.20
11.1.0 Spirit Levels 3.20
11.1.1 How to Use a Spirit Level. 3.21
11.1.2 Safety and Maintenance 3.21
11.2.0 Digital (Electronic) Levels 3.21
11.3.0 Laser Levels 3.22
12.0.0 Squares 3.23
12.1.0 The Carpenter's Square. 3.23
12.2.0 The Combination Square 3.23
12.2.1 How to Use a Combination Square 3.24
12.3.0 Safety and Maintenance 3.25
13.0.0 Plumb Bob 3.25
13.1.0 How to Use a Plumb Bob 3.26
14.0.0 Chalk Lines. 3.26
14.1.0 How to Use a Chalk Line 3.26
15.0.0 Utility Knives 3.27
15.1.0 How to Use a Utility Knife. 3.27
15.2.0 Safety and Maintenance 3.27
16.0.0 Saws 3.27
16.1.0 Handsaws 3.28
16.1.1 How to Use a Crosscut Saw 3.28
16.1.2 How to Use a Ripsaw 3.29
16.2.0 Safety and Maintenance 3.29
17.0.0 Files and Rasps 3.30
17.1.0 How to Use a File 3.31
17.2.0 Safety and Maintenance 3.32
18.0.0 Clamps 3.32
18.1.0 How to Use a Clamp 3.34
18.2.0 Safety and Maintenance 3.35
19.0.0 Chain Falls and Come-Alongs 3.35
19.1.0 Chain Falls 3.35
19.2.0 Come-Alongs 3.35
19.3.0 Safety and Maintenance 3.36
20.0.0 Shovels 3.36
20.1.0 How to Use a Shovel. 3.37
20.2.0 Safety and Maintenance 3.37
21.0.0 Pick 3.38
21.1.0 How to Use a Pick 3.38
21.2.0 Safety and Maintenance 3.38

1.0.0 Introduction

Every profession has its tools. A surgeon uses a scalpel, a teacher uses a chalkboard, and an accountant uses a calculator. The construction trade has a whole collection of hand tools, such as hammers, screwdrivers, and **pliers**, that everyone uses. Even if you are already familiar with some of these tools, you need to learn to maintain them and use them safely. The better you use and maintain your tools, the better you will be in your craft.

This module shows you how to safely use and maintain some of the most common hand tools of the construction trade. It also highlights some specialized crafts and hand tools.

1.1.0 Safety

To work safely, you must think about safety. Before you use any tool, you should know how it works and some of the possible dangers of using it the wrong way. Always read and understand the procedures and safety tips in the manufacturer's guide for every tool you use. Make sure every tool you use is in good condition. Never use worn or damaged tools.

WARNING!

Always protect yourself when you are using tools by wearing appropriate personal protective equipment (PPE), such as safety gloves and eye protection.

2.0.0 Hammers

Hammers are made in different sizes and weights for specific types of work. The safest hammers are those with heads made of alloy and drop-forged steel. Two of the most common hammers are the **claw hammer** (*Figure 1*) and the **ball peen hammer** (*Figure 2*).

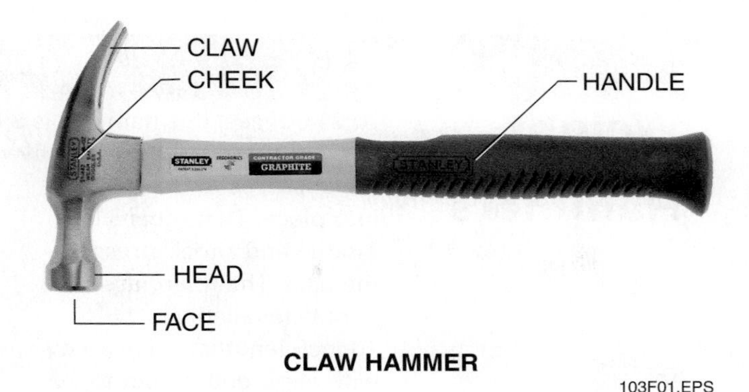

Figure 1 Claw hammer.

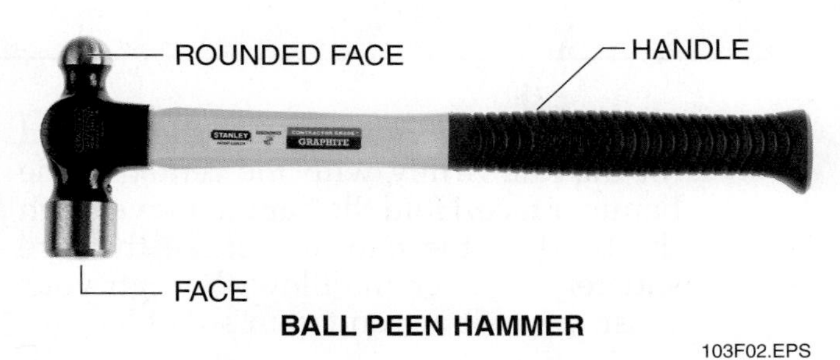

Figure 2 Ball peen hammer.

Did You Know?

Hammers

The quality of a hammer is important. The strongest (and safest) hammers have heads made from tough alloy (a mixture of two or more metals) and drop-forged steel (a strong steel formed by pounding and heating). Hammers with cast heads—heads formed by being poured or pressed into a mold—are more brittle. They are not suited for construction work because they tend to chip and break. Hammers with heads made of tough alloy and drop-forged steel tend to be more expensive than hammers with cast heads. When it comes to tools, it pays to invest in quality equipment.

2.1.0 Claw Hammer

The claw hammer has a steel head and a handle made of wood, steel, or fiberglass. You use the head to drive nails, wedges, and **dowels**. You use the claw to pull nails out of wood. The face of the hammer may be flat or rounded. It's easier to drive nails with the flat face (plain) claw hammer, but the flat face may leave hammer marks when you drive the head of the nail flush (even) with the surface of the work.

A claw hammer with a slightly rounded (or convex) face is called a **bell-faced hammer**. A skilled worker can use it to drive the nail head flush without damaging the surface of the work.

2.1.1 How to Use a Claw Hammer to Drive a Nail

Remember these simple steps to use a claw hammer properly when driving a nail:

Step 1 Rest the face of the hammer on the nail.

Step 2 Draw the hammer back and give the nail a few light taps to start it.

On-Site

Weight-Forward Hammers

At 21 ounces, this hammer is a little heavier than most standard hammers. However, its curved, extended handle delivers greater striking force to the square head, so it takes fewer strikes to drive nails into place. That means less fatigue and shock stress for the user. The fiberglass handle (available in 14- or 16-inch lengths) is covered with neoprene rubber to ensure a good grip.

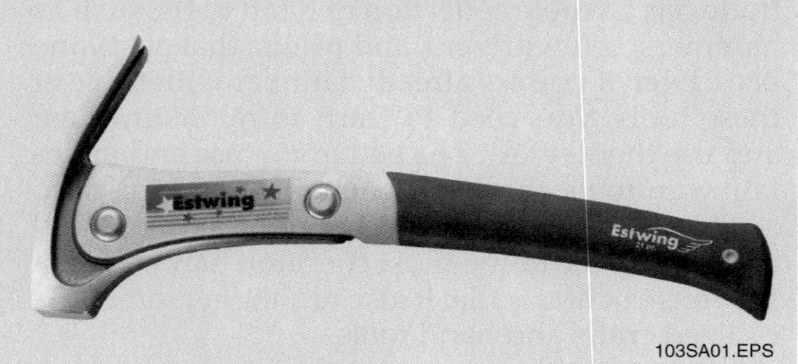

103SA01.EPS

Step 3 Move your fingers away from the nail and hit the nail firmly with the center of the hammer face. Hold the hammer **level** with the head of the nail and strike the face squarely. Deliver the blow through your wrist, your elbow, and your shoulder.

2.1.2 How to Use a Claw Hammer to Pull a Nail

Pulling a nail with a claw hammer is as easy as driving one. Follow these steps:

Step 1 Slip the claw of the hammer under the nail head and pull until the handle is nearly straight up (vertical) and the nail is partly drawn out of the wood.

Step 2 Pull the nail straight up from the wood.

2.2.0 Ball Peen Hammer

A ball peen hammer is a type of hammer used in metalworking. It has a flat face for striking and a spherical or hemispherical head for **peening** (rounding off) metal or rivets. You use this hammer with **chisels** and **punches** (discussed later in this module). In welding operations, the ball peen hammer is used to reduce stress in the **weld** by peening or striking the **joint** as it cools.

Ball peen hammers are also known as engineer's hammers or machinist's hammers. They are classified by weight, and can weigh 6 ounces to 2½ pounds.

When you use a ball peen hammer, it is important that you not strike the material as though you were striking a nail with a claw hammer. A ball peen hammer should strike the material in a controlled manner so that you do not damage the material. Never use a ball peen hammer to drive in nails, because the head of the hammer is made of tough, but relatively soft steel. Continual pounding on nails can deform or damage the hammer's head. A claw hammer is not interchangeable with a ball peen hammer.

2.3.0 Safety and Maintenance

To keep from hurting yourself or a co-worker, you must focus on your work. Make sure you are aware of these guidelines for safety and maintenance when using all types of hammers:

- Make sure there are no splinters in the handle of the hammer.
- Make sure the handle is set securely in the head of the hammer.
- Replace cracked or broken handles.
- Make sure the face of the hammer is clean.

On-Site

Physics and the Hammer

The hammer is designed to produce a certain amount of force on the object it strikes. If you hold the hammer incorrectly, you cancel out the benefits of its design. Always remember to hold the end of the handle even with the lower edge of your palm. The distance between your hand and the head of the hammer affects the force you use to drive a nail. The closer you hold the hammer to the head, the harder you will need to swing to achieve the desired force. Make it easier on yourself by holding the hammer properly; it takes less effort to drive the nail.

- Hold the hammer properly. Grasp the handle firmly near the end and hit the nail squarely.
- Don't hit with the cheek or side of the hammer head.
- Don't use hammers with chipped, mushroomed (overly flattened by use), or otherwise damaged heads.
- Don't use a hammer with a cast head.

CAUTION

A chip could easily break off a cast head and injure you or a co-worker.

- Don't strike hammer heads together.

CAUTION

Never use a hammer to strike the head of another hammer. Flying fragments from drop-forged alloy steel are dangerous.

2.4.0 Sledgehammers

A sledgehammer is a heavy-duty tool used to drive posts or other large stakes. You can also use it to break up cast iron or concrete. The head of the sledgehammer is made of high-carbon steel and weighs 2 to 20 pounds. The shape of the head depends on the job the sledgehammer is designed to do. Sledgehammers can be either long-handled or short-handled, depending on the jobs for which they are designed. *Figure 3* shows two types of sledgehammers: the double-face and the cross-peen.

2.4.1 How to Use a Sledgehammer

A sledgehammer can cause injury to you or to anyone working near you. You must use a sledgehammer properly, and you must focus on what you are doing the entire time you use one. Refer to *Figure 4* and follow these steps to use a sledgehammer safely:

Step 1 Wear appropriate personal protective equipment, including safety gloves and eye protection.

Step 2 Inspect the sledgehammer to ensure that there are no defects.

Step 3 Be sure there is no one in the area nearby.

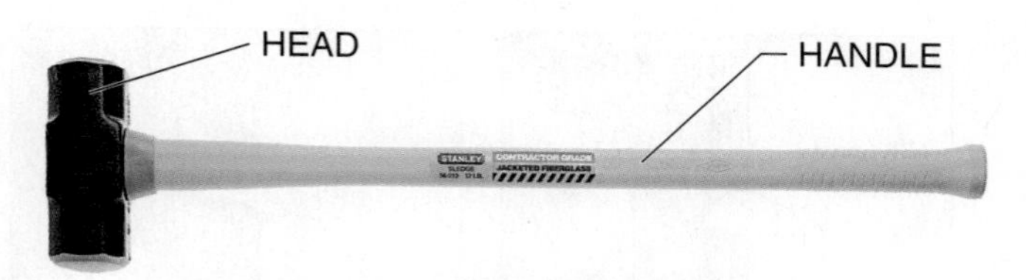

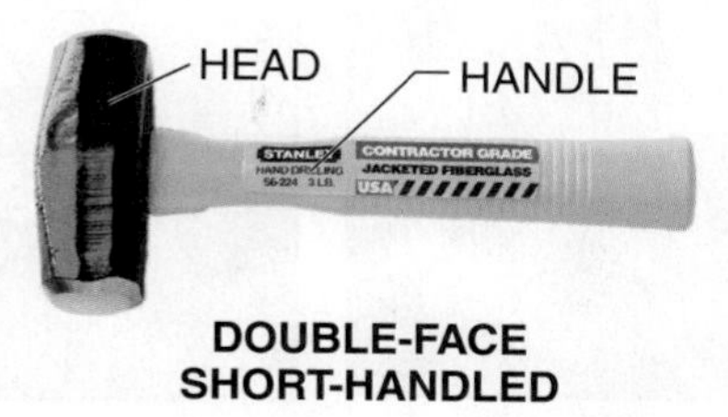

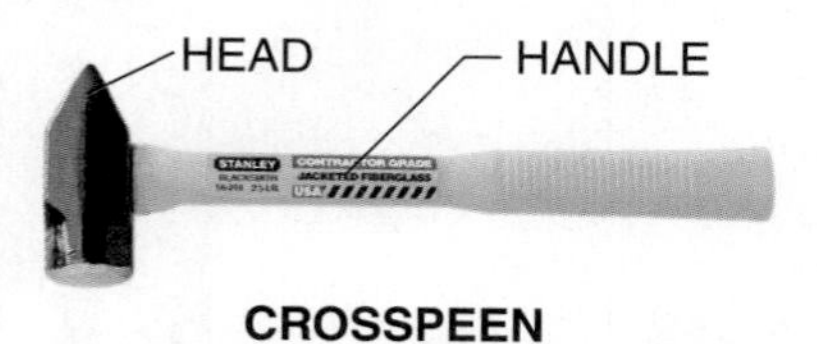

Figure 3 Sledgehammers.

Step 4 Hold the sledgehammer with both hands apart (hand over hand).

Step 5 Stand directly in front of the object you want to drive.

Step 6 Lift the sledgehammer straight up above the target.

Step 7 Set the head of the sledgehammer on the target.

Step 8 Begin delivering short blows to the target and gradually increase the length and force of the stroke.

CAUTION

Hold the sledgehammer with both hands. Never use your hands to hold an object while someone else drives with a sledgehammer. Doing so could result in serious injury, such as crushed or broken bones.

CAUTION

Do not swing a sledgehammer with your hands beyond your head. Doing so may cause injury to your back and could limit the control you have directing the blow to the target.

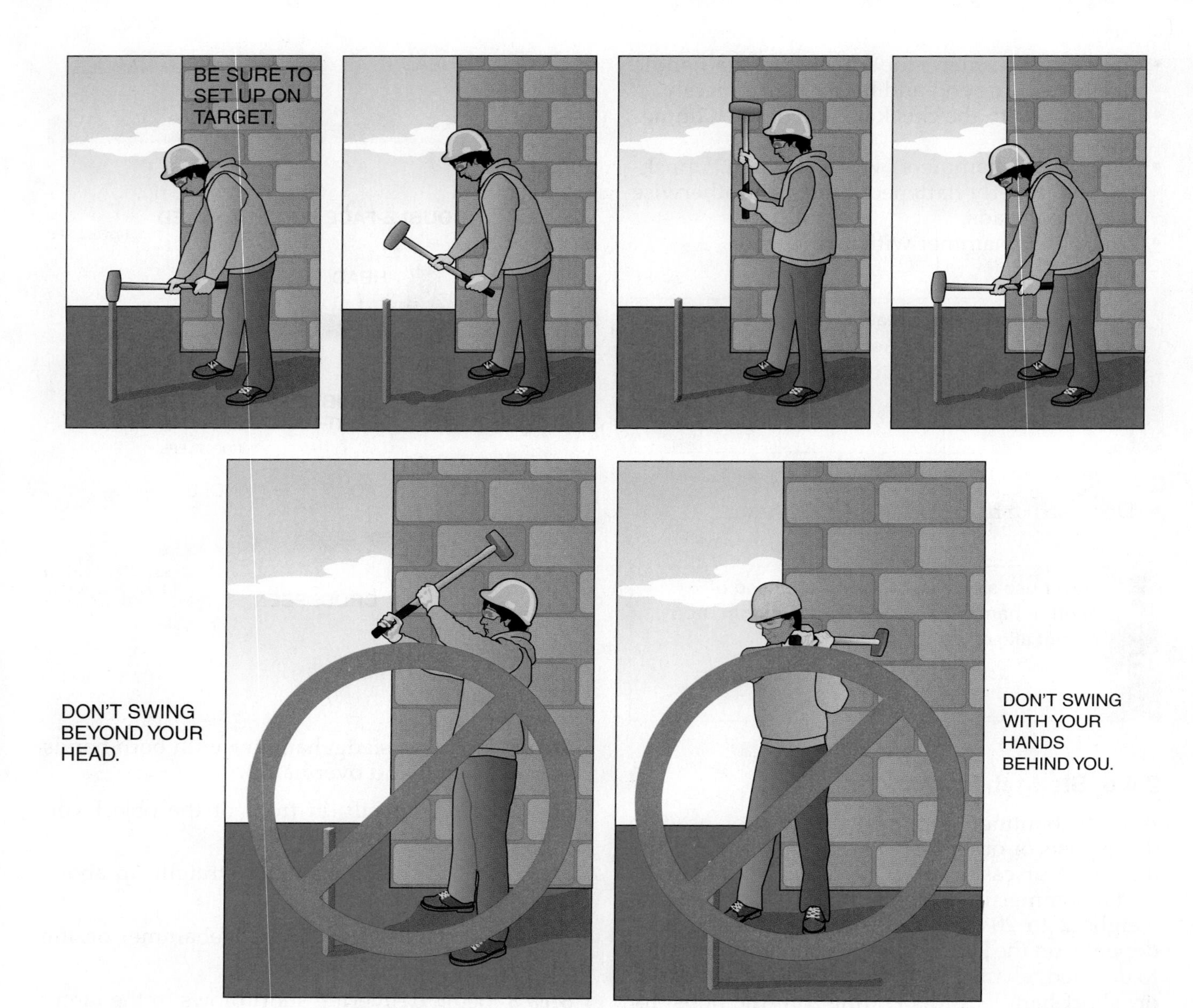

Figure 4 Proper use of a long-handled sledgehammer.

2.4.2 Safety and Maintenance

Remember that using a sledgehammer the correct way not only gets the job done right, but also keeps everyone in the area safe, including you. Here are other considerations for safety and maintenance when working with this tool:

- Wear eye protection when you are using a sledgehammer. It's also a good idea to wear safety gloves.
- Replace cracked or broken handles before you use the sledgehammer.
- Make sure the handle is secured firmly at the head.
- Use the right amount of force for the job.
- Keep your hands away from the object you are driving.
- Don't swing until you have checked behind you to make sure you have enough room and no one is behind you.

On-Site

Stone-masons

Stone used to be one of our primary building materials. Because of its strength, stone was often used for dams, bridges, fortresses, foundations, and monumental buildings. Today, steel and concrete have replaced stone as a basic construction material. Stone is used primarily as sheathing for buildings, for flooring in high-traffic areas, and for decorative uses.

A stonemason's job requires precision. Stones have uneven, rough edges that must be trimmed and finished before each stone can be set. The process of trimming projections and jagged edges is called dressing the stone. This requires skill and experience using specialized hand tools, such as chisels and sledgehammers. Many craftworkers consider stonework an art.

Stonemasons build stone walls as well as set stone exteriors and floors, working with natural cut and artificial stones. These include marble, granite, limestone, cast concrete, marble chips, or other masonry materials. Stonemasons usually work on structures such as houses, churches, hotels, and office buildings. Special projects include zoos, theme parks, and movie sets.

3.0.0 Ripping Bars and Nail Pullers

A number of tools are made to rip and pry apart woodwork as well as to pull nails. In this section, you will learn about **ripping bars** and **nail pullers** (*Figure 5*). These tools are necessary in the construction trade because often the job involves building where something else already exists. The existing structure needs to be torn apart before the building can begin. That is where the ripping bar and the nail puller come in.

3.1.0 Ripping Bars

The ripping bar—also called a pinch, pry, or wrecking bar—can be 12 to 36 inches long. This bar is used for heavy-duty dismantling of woodwork, such as tearing apart building frames or concrete forms. The ripping bar has an octagonal (eight-sided) shaft and two specialized ends. A deeply curved nail claw at one end is used as a nail puller. An angled, wedge-shaped face at the other end is used as a prying tool to pull apart materials that are nailed together.

When using a ripping bar, make sure that you always wear the appropriate personal protective equipment. Eye protection is extremely important due to the risk of flying debris. Use the angled prying end to force apart pieces of wood or use the heavy claw to pull large nails and spikes.

CAUTION

When you are using a ripping bar or a nail puller, a piece of material can break off and fly through the air. Wear a hard hat, safety glasses, and gloves to protect yourself from flying debris. Make sure others around you are similarly protected.

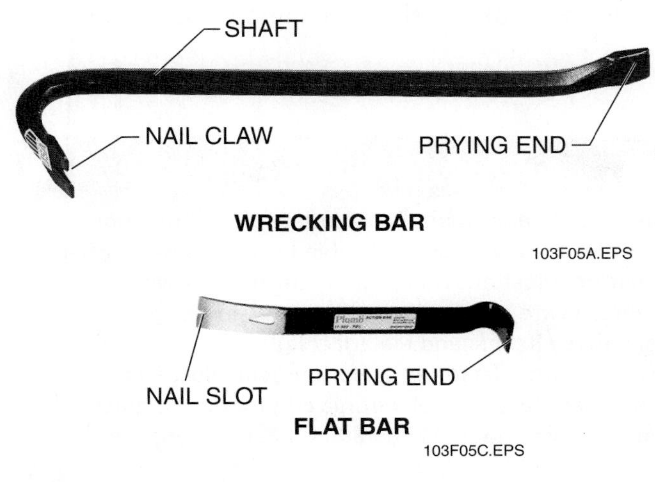

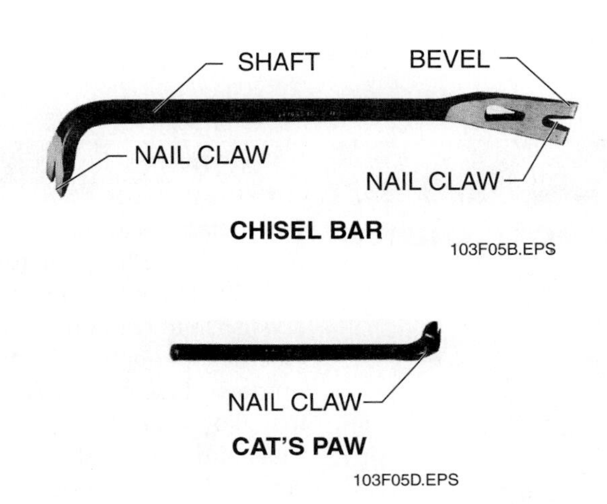

Figure 5 Ripping bars and nail pullers.

3.2.0 Nail Pullers

There are three main types of nail-pulling tools: **cat's paw** (also called nail claws and carpenter's pincers), **chisel bars**, and **flat bars**.

The cat's paw is a straight steel rod with a curved claw at one end. It is used to pull nails that have been driven flush with the surface of the wood or slightly below it. You use the cat's paw to pull nails to just above the surface of the wood so they can be pulled completely out with the claw of a hammer or a pry bar.

The chisel bar has a claw at each end and is ground to a chisel-like **bevel** (slant) on both ends. You can use it like a claw hammer to pull nails. You can also drive it into wood to split and rip apart the pieces.

The flat bar (ripping chisel, wonder bar, action bar) has a nail slot at the end to pull nails out from tightly enclosed areas. It can also be used as a small pry bar. The flat bar is usually 2 inches wide and 15 inches long.

3.2.1 How to Use a Nail Puller (Cat's Paw)

Take the following steps when using a nail puller:

Step 1 Wear appropriate personal protective equipment.

Step 2 Drive the claw into the wood, grabbing the nail head.

Step 3 Pull the handle of the bar to lift the nail out of the wood.

CAUTION

When manipulating the bar, make sure that you never pull the bar toward your face; always pull the bar toward your shoulder. Also, never push in a direction away from your body, because this may cause you to lose your balance.

3.3.0 Safety and Maintenance

The following are safety guidelines for ripping and nail pulling:

- Wear appropriate personal protective equipment, including eye protection, a hard hat, and safety gloves.
- Use two hands when ripping; this helps to ensure that you keep even pressure on your back as you pull.
- To prevent injury when pulling a nail, be sure the material holding the nail is braced securely before you pull.

Most accidents with prying tools occur when a pry bar slips and the worker falls to the ground. Be sure to keep balanced footing and a firm grip on the tool. This technique also helps reduce damage to materials that must be reused, such as concrete forms.

4.0.0 CHISELS AND PUNCHES

Chisels are used to cut and shape wood, stone, or metal. Punches are used to indent metal, drive pins, and align holes.

4.1.0 Chisels

A chisel is a metal tool with a sharpened, beveled (sloped) edge. It is used to cut and shape wood, stone, or metal. You will learn about two kinds of chisels in this section: the wood chisel and the cold chisel (*Figure 6*). Both chisels are made from steel that is heat-treated to make it harder. A chisel can cut any material that is softer than the steel of the chisel.

4.1.1 How to Use a Wood Chisel

Use the wood chisel to make openings or notches in wooden material. For instance, you can use it to

On-Site

Electricians

Electricians read blueprints to install electrical systems in factories, office buildings, homes, and other structures. They may also install coaxial cable for television or fiber-optic cable for computers and telecommunications equipment. Electricians who specialize in residential work may install wire and hardware in a new home, such as electrical panel boxes, receptacles, light switches, and electrical light fixtures, or replace outdated fuse boxes. Those who work in large factories as commercial electricians may install or repair motors, transformers, generators, or electronic controllers on machine tools and industrial robots. They use many hand tools, including pliers, wrenches, screwdrivers, hammers, and saws.

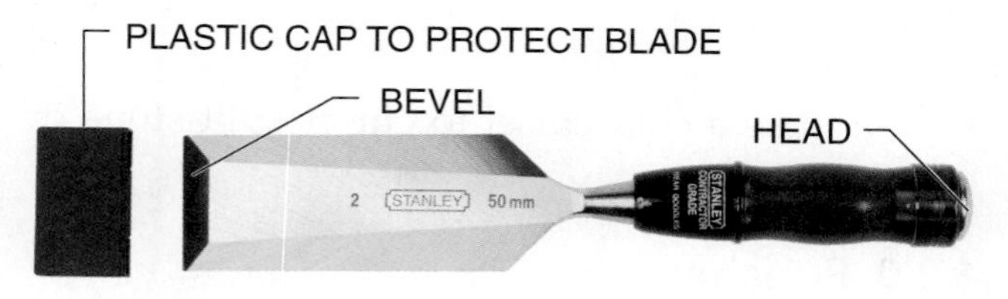

Figure 6 Cold and wood chisels.

make a recess for butt-type hinges, such as the hinges in a door. Follow these steps to use a wood chisel properly:

Step 1 Wear appropriate personal protective equipment. It is especially important to wear eye protection when using a cold chisel.

Step 2 Outline the opening (recess) to be chiseled.

Step 3 Set the chisel at one end of the outline, with its edge on the cross-grain line and the bevel facing the recess to be made.

Step 4 Strike the chisel head lightly with a mallet.

Step 5 Repeat this process at the other end of the outline, again with the bevel of the chisel blade toward the recess. Then make a series of cuts about ¼-inch apart from one end of the recess to the other.

Step 6 To pare (trim) away the notched wood, hold the chisel bevel-side down to slice inward from the end of the recess (*Figure 7*).

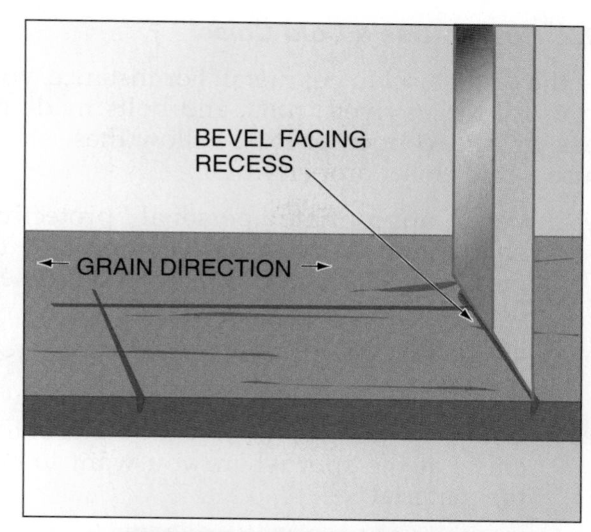

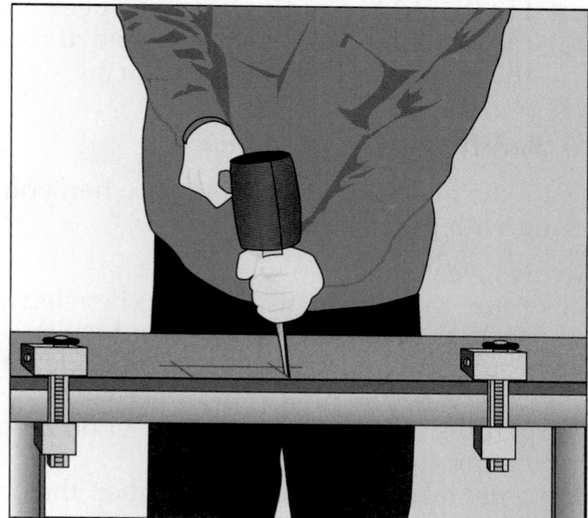

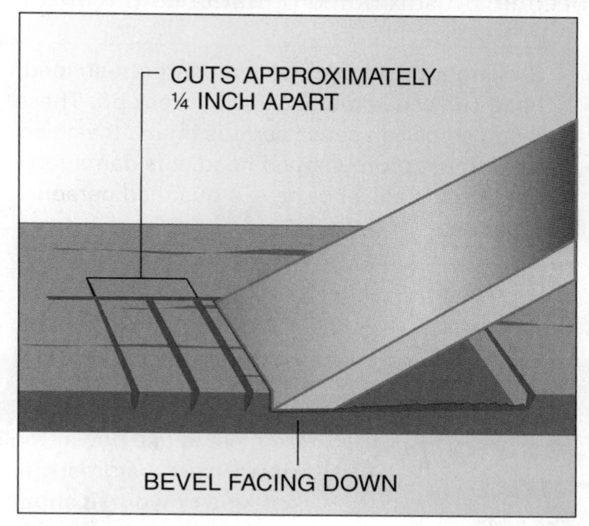

Figure 7 Proper use of a wood chisel.

4.1.2 How to Use a Cold Chisel

Use the cold chisel to cut metal. For instance, you can use it to cut rivets, nuts, and bolts made of brass, bronze, copper, or iron. Follow these steps to use a cold chisel properly:

Step 1 Wear appropriate personal protective equipment. It is especially important to wear appropriate eye protection when using a cold chisel.

Step 2 Secure the object you want to cut in a vise, if possible.

Step 3 Using a holding tool, place the blade of the chisel at the spot where you want to cut the material.

Step 4 Hit the chisel handle with a ball peen hammer to force the chisel into and through the material. Repeat if necessary.

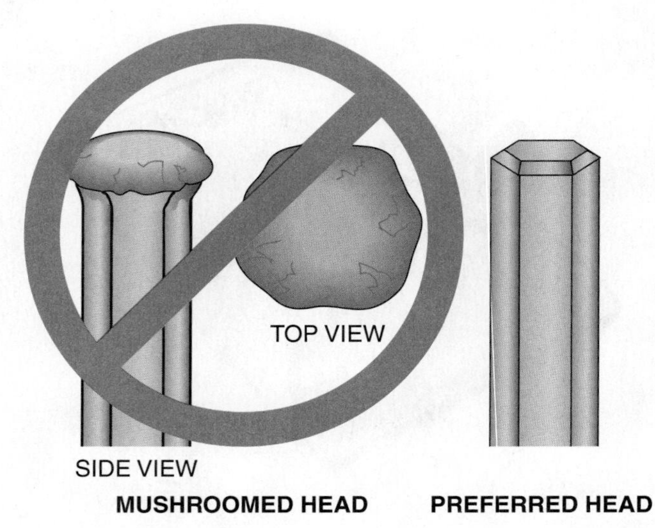

Figure 8 Chisel damage.

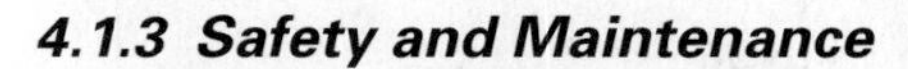

4.1.3 Safety and Maintenance

Here are the guidelines to remember when you're working with punches and chisels:

- Always wear safety goggles.
- Make sure the wood chisel blade is beveled at a precise 25-degree angle so it will cut well.
- Make sure the cold chisel blade is beveled at a 60-degree angle so it will cut well.
- Sharpen the cutting edge of a chisel on an oilstone to produce a keen edge.
- Don't use a chisel head or hammer that has become mushroomed or flattened (*Figure 8*).

> **CAUTION**
>
> Striking a chisel that has a mushroom-shaped head can cause metal chips to break off. These flying chips can cause serious injury. If a chisel has a mushroom-shaped head, it is damaged. Replace the chisel or have a qualified person repair it.

- Don't use a cold chisel to cut or split stone or concrete.

4.2.0 Punches

A punch (*Figure 9*) uses the impact of a hammer to indent metal before you drill a hole, to drive pins, and to align holes in two parts that are mates. Punches are made of hardened and **tempered** steel. They come in various sizes.

Three common types of punches are the center punch, the prick punch, and the straight punch. The center and prick punches are used to make small locating **points** for drilling holes. The straight punch is used to punch holes in thin sheets of metal.

> **On-Site**
>
> **Caring for Hand Tools**
>
> You have to use power tools to care for some types of hand tools, such as chisels, screwdrivers, hammers, and punches. If the edge or striking surface of a hand tool is damaged or worn, it should be ground back to its desired shape using a grinder. Grinding a hammer face or a punch point will remove unwanted burrs or mushrooming. For a chisel to cut well, the blade needs to be beveled (sloped) at a precise angle. A grinder can be used to remove nicks. The cutting edge must then be sharpened on an oilstone to produce a keen, precise edge. Screwdriver blades can also be cleaned up using a grinder.

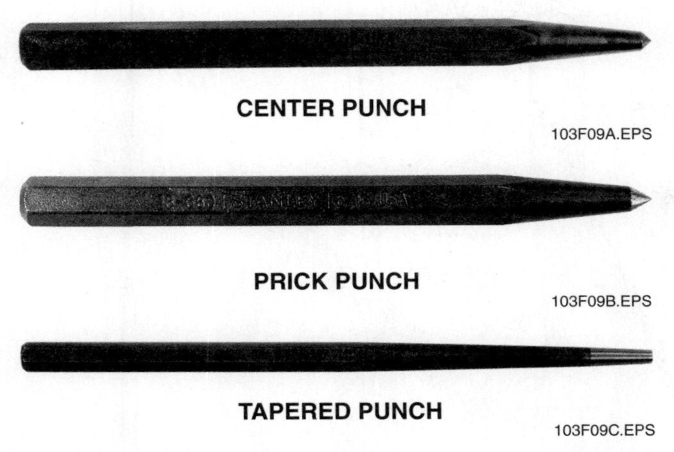

Figure 9 Punches.

5.0.0 Screwdrivers

A screwdriver is used to tighten or remove screws. It is identified by the type of screw it fits. The most common screwdrivers are slotted (also known as straight-blade, flat, or standard tip) and Phillips head screwdrivers. You may also use more specialized screwdrivers such as a clutch-drive, Torx®, Robertson®, and Allen head (hex). *Figure 10* shows six common types of screw heads.

- *Slotted* – This is the most common type of standard screwdriver. It fits slotted screws.
- *Phillips* – This is the most common type of crosshead screwdriver. It fits Phillips head screws.
- *Clutch-drive* – This screwdriver has an hourglass-shaped tip that is especially useful when you need extra holding power, as when working on cars or appliances.
- *Torx®* – This screwdriver has a star-shaped tip that is useful for replacing such parts as tailgate lenses. It is widely used in automobile repair work. Torx® screws are also used in household appliances, as well as lawn and garden equipment.
- *Robertson® (square)* – This screwdriver has a square drive that provides high **torque** power. Usually color coded according to size, it can reach screws that are sunk below the surface.
- *Allen (hex)* – This screwdriver works with screws that can also be operated with hex keys. It is suitable for socket-head screws that are recessed.

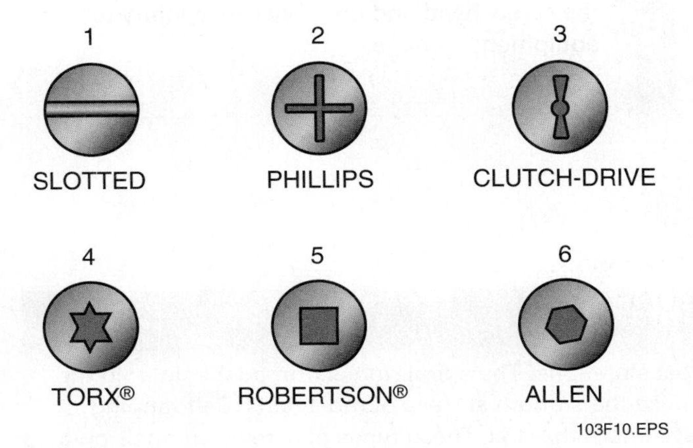

Figure 10 Common screw heads.

To choose the right screwdriver and use it correctly, you have to know a little bit about the sections of a screwdriver. Each section has a name, as shown in *Figure 11*. The handle is designed to give you a firm grip. The shank is the hardened metal portion between the handle and the blade. The shank can withstand a lot of twisting force. The blade is the formed end that fits into the head of a screw. Industrial screwdriver blades are made of tempered steel to resist wear and to prevent bending and breaking.

It is important to choose the right screwdriver for the screw. The blade should fit snugly into the screw head and not be too long, too short, loose, or tight (see *Figure 12*). If you use the wrong size blade, you might damage the screwdriver or the screw head.

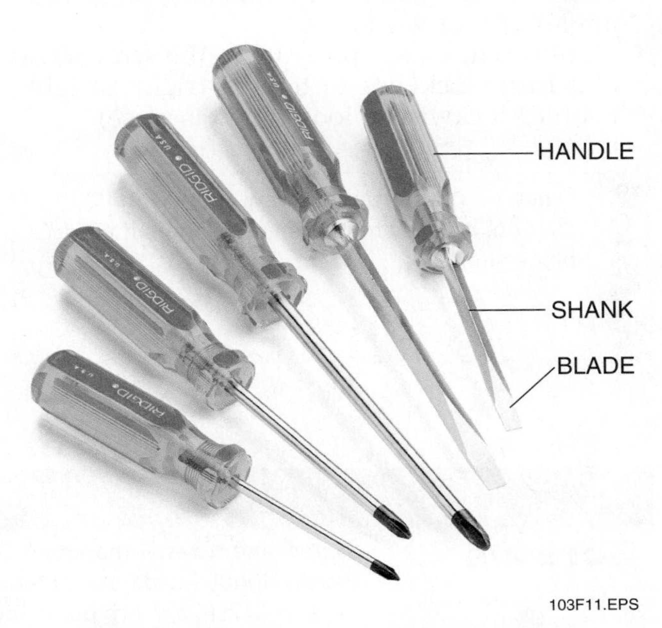

Figure 11 Slotted and Phillips head screwdrivers.

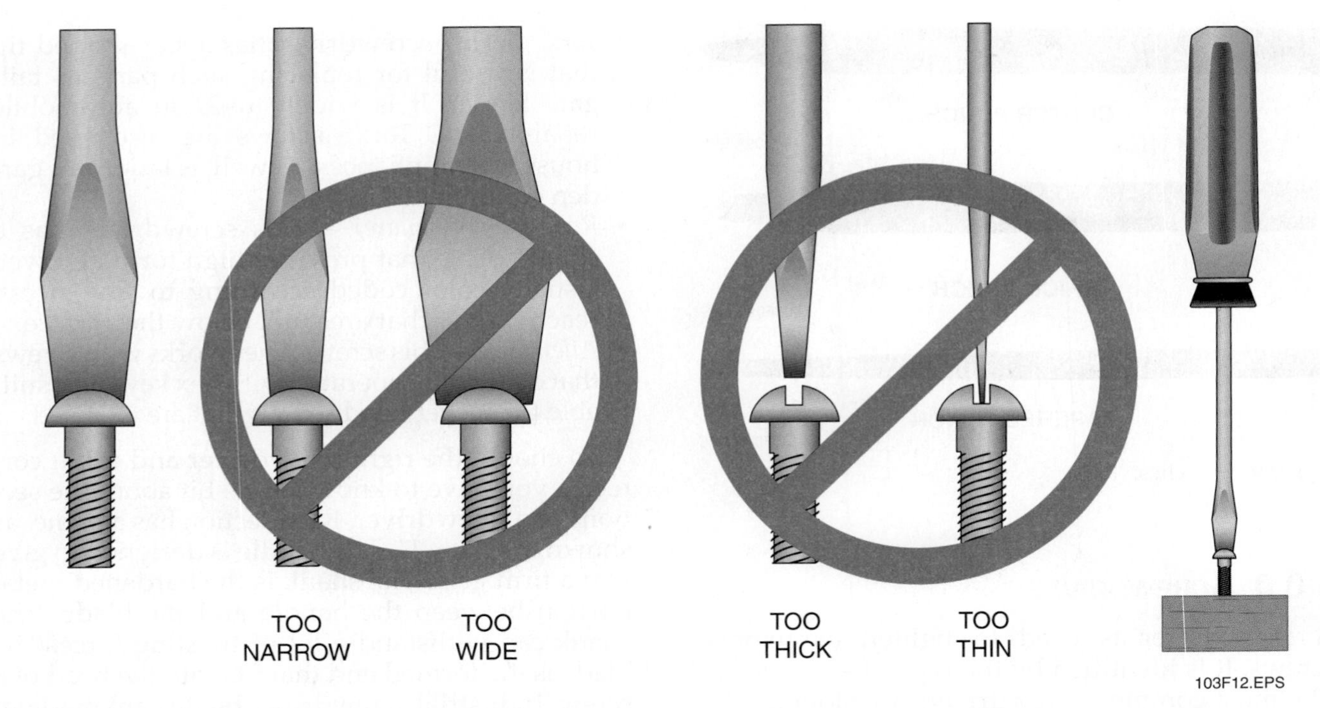

Figure 12 Proper use of a screwdriver.

5.1.0 How to Use a Screwdriver

It is very important to use a screwdriver correctly. Using one the wrong way can damage the screwdriver or **strip** the screw head. Follow these steps:

- Choose the right size and type of blade for the screw head.
- Position the shank perpendicular (at a right angle) to your work.
- Apply firm, steady pressure to the screw head and turn clockwise to tighten (right is tight), counterclockwise to loosen (left is loose).

CAUTION

When you're starting a screw, work with caution. It's easy to injure your fingers if the blade slips.

5.2.0 Safety and Maintenance

You will be effective with a screwdriver if you follow the steps in this section. You also need to know how to work safely. The following are guidelines for your own safety and the safety of others, as well as for maintaining your screwdriver:

- Keep the screwdriver free of dirt, grease, and grit so the blade will not slip out of the screwhead slot.

CAUTION

Keep the screwdriver clean. A dirty or greasy screwdriver can slip out of your hand or out of the screw head and possibly cause injury or equipment damage.

On-Site

Screws

Screws hold better than nails in most situations. The spiral ridges (threads) help hold the screw tightly inside the material, unlike the smooth surface of most nails. Self-tapping screws end in a sharp point and have sharp threads. These types of screws cut their own threads in the material, and you do not need to drill a starter hole. In woodworking, however, making a small starter hole with a drill helps keep the wood from splitting.

Did You Know?

Types of Screws

The demand for different types of screw heads came about due to a number of specific needs in the workplace. For example, Torx® head screws were developed to be compatible with robotic assembly line equipment used in a number of production applications. Phillips heads were developed with four-point contact so that a higher torque could be applied and the head countersunk into the material. Tamper-resistant (snake eye or pig nose) screws were developed to prevent vandalism to public facilities such as school and park restrooms.

- Don't ever use the screwdriver as a punch, chisel, or pry bar.
- Don't ever use a screwdriver near live wires or as an electrical tester.
- Don't expose a screwdriver to excessive heat.
- Don't use a screwdriver that has a worn or broken handle.

CAUTION

Visually inspect your screwdriver before using it. If the handle is worn or damaged, or the tip is not straight and smooth, the screwdriver should be repaired or replaced.

- Don't point the screwdriver blade toward yourself or anyone else.

6.0.0 Pliers and Wire Cutters

Pliers are a special type of adjustable wrench. They are scissor-shaped tools with jaws. The jaws usually have teeth to help grip objects. The jaws are adjustable because the two legs (or handles) move on a pivot. You will generally use pliers to hold, cut, and bend wire and soft metals. Do not use pliers on nuts or bolt heads. They will **round off** the edges of the hex (six-sided) head, and wrenches will no longer fit properly.

High-quality pliers are made of hardened steel. Pliers come in many different head styles, depending on their use (see *Figure 13*). The following types of pliers are the most commonly used:

- Slip-joint (combination) pliers
- Long-nose (needle-nose) pliers
- Lineman pliers (side cutters)
- Tongue-and-groove (CHANNELLOCK®) pliers
- Locking pliers (Vise-Grip®)

This section will explain how to use the correct set of pliers for the appropriate application.

6.1.0 Slip-Joint (Combination) Pliers

You use slip-joint (or combination) pliers to hold and bend wire and to grip and hold objects during assembly operations. They have adjustable jaws. There are two jaw settings: one for small materials and one for larger materials.

6.1.1 How to Use Slip-Joint Pliers

When using slip-joint pliers, be sure to wear appropriate personal protective equipment. To use slip-joint pliers properly, place the jaws on the object to be held. Then, squeeze the handles until the pliers grip the object.

GOING GREEN

Recycling Equipment from Western Countries

In Nigeria, merchants and blacksmiths are working together to recycle old machinery and automobile parts into simple hand tools.

The recycled parts for the tools come from automobiles and other machinery that are manufactured in Western countries (including the United States) and then sent to Nigeria to help with their economic development. When it is no longer economical to service or repair the equipment, it is stripped down and many of the parts are used to repair similar pieces of machinery. Once all the reusable pieces are salvaged, the blacksmiths and merchants browse through the remaining scraps. They consider the objects' natural shape and strength, and then convert them into hand tools that are as effective as their imported counterparts, and lower in cost. The following are some of the tools that they make, and the recycled parts used to create them:

- *Axes* – Blades made from leaf springs, handles made from the shock rods of old struts.
- *Chisels and punches* – Made from engine valves and bolts.
- *Hammers* – Made from rocker arms of large diesel engines.
- *Crow bar* – Made from rack and pinion steering.
- *Pick* – Rod made from aluminum wire, handle made by dipping and twirling the rod in melted plastic.

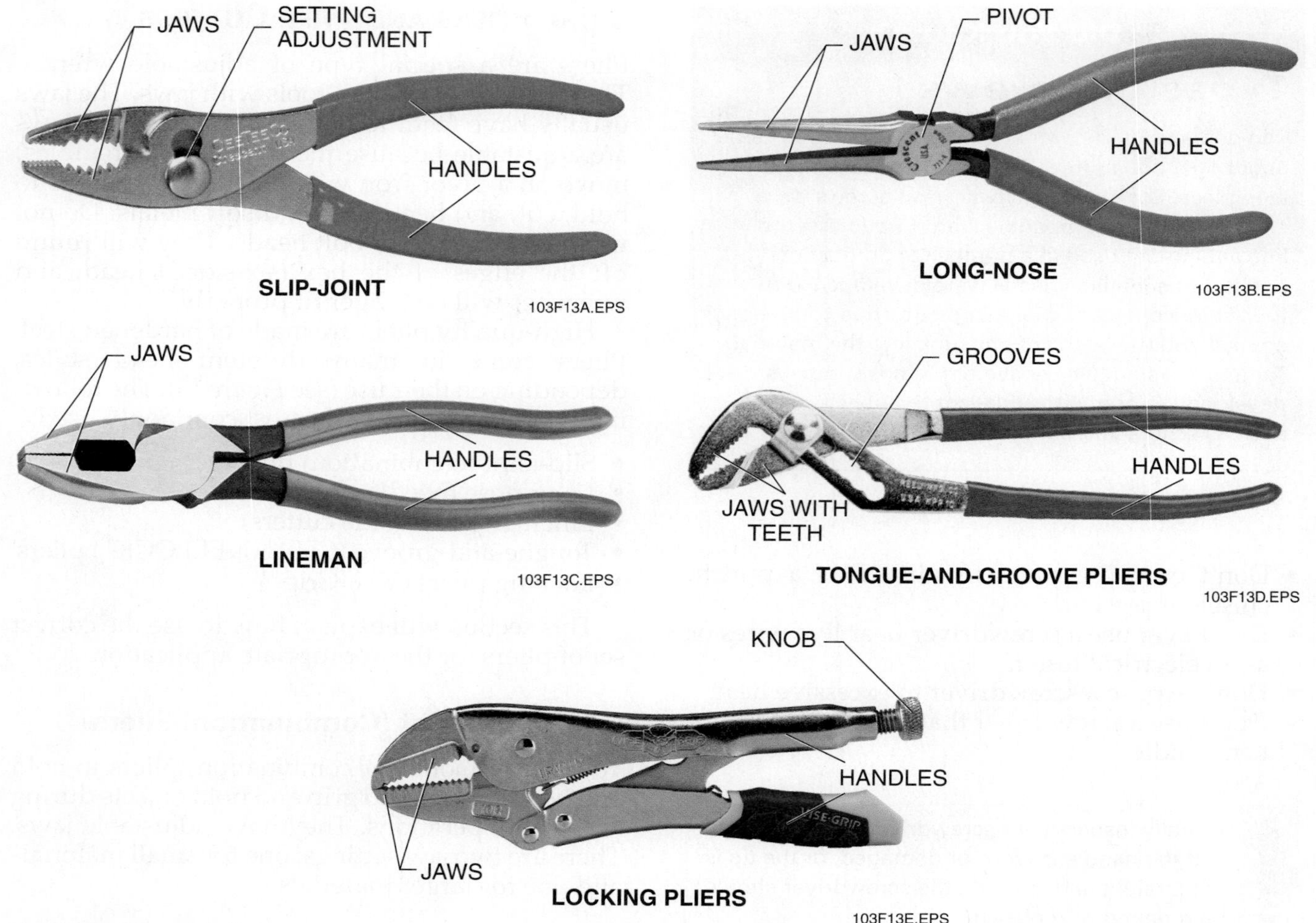

Figure 13 Types of pliers.

6.2.0 Long-Nose (Needle-Nose) Pliers

Long-nose (or needle-nose) pliers are used to get into tight places where other pliers won't reach, or to grip parts that are too small to hold with your fingers. These pliers are useful for bending angles in wire or narrow metal strips. They have a sharp wire cutter near the pivot. Long-nose pliers, like many other types of pliers, are available with spring openers. This is a spring-like device between the handles that keep the handles apart—and the jaws open—unless you purposely close them. This device can make long-nose pliers easier to use.

6.2.1 How to Use Long-Nose Pliers

Be sure to wear appropriate personal protective equipment when working with long-nose pliers. The following are guidelines for using long-nose pliers properly:

- If the pliers do not have a spring between the handles to keep them open, place your third or little finger inside the handles to keep them open.
- To cut a wire, squeeze the handles to cut at a right angle to the wire.

6.3.0 Lineman Pliers (Side Cutters)

Lineman pliers (or side cutters) have wider jaws than slip-joint pliers do. You use them to cut heavy or large-gauge wire, and to hold work. The wedged jaws reduce the chance that wires will slip, and the hook bend in both handles gives you a better grip.

6.3.1 How to Use Lineman Pliers

Be sure to wear appropriate personal protective equipment when working with lineman pliers. To properly cut wire with lineman pliers, always point the loose end of the wire down. Squeeze the handles to cut at a right angle to the wire.

6.4.0 Tongue-and-Groove Pliers

Tongue-and-groove pliers have serrated teeth that grip flat, square, round, or hexagonal objects. You can set the jaws in any of five positions by slipping the curved ridge into the desired groove (see *Figure 14*). Large tongue-and-groove pliers are often used to hold pipes, because the longer handles give more leverage. The jaws stay parallel and give a better grip than slip-joint pliers.

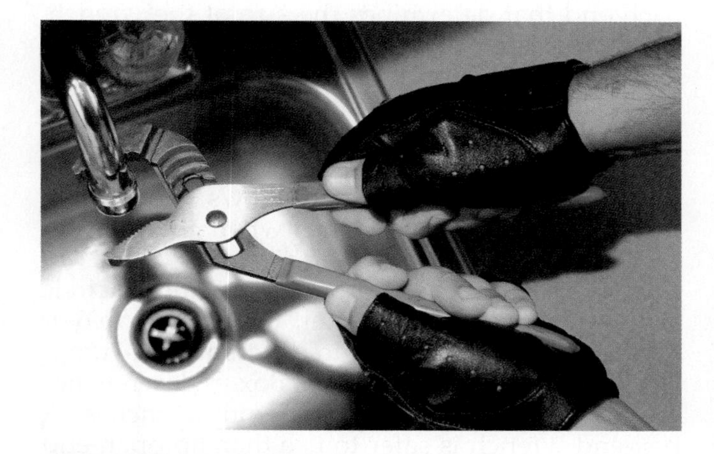

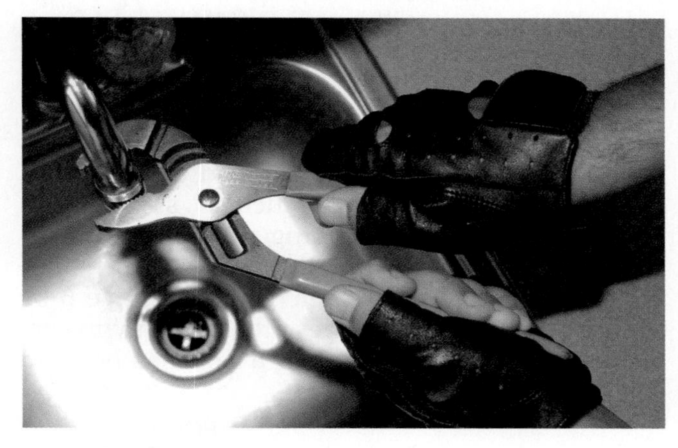

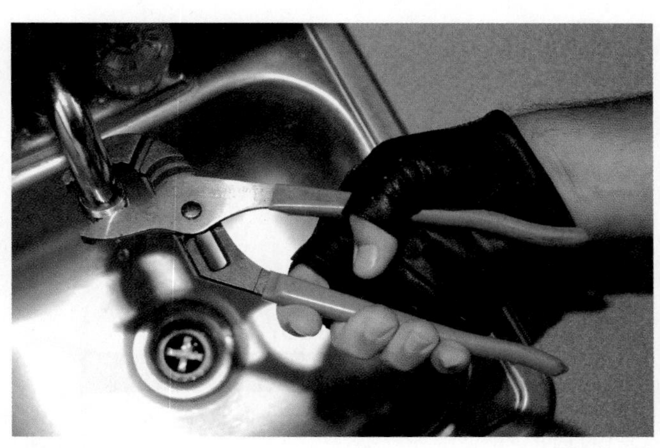

103F14.EPS

Figure 14 Proper use of tongue-and-groove pliers.

6.4.1 How to Use Tongue-and-Groove Pliers

Be sure to wear appropriate personal protective equipment when using tongue-and-groove pliers. Take the following steps to use tongue-and-groove pliers properly:

Step 1 With pliers open to the widest position, place the upper jaw on the object to be held.

Step 2 Determine which groove provides the proper position.

Step 3 Squeeze the handles until the pliers grip the object.

6.5.0 Locking Pliers

Locking pliers clamp firmly onto objects the way a vise does. A knob in the handle controls the width and tension of the jaws (*Figure 15*). You close the handles to lock the pliers, and release the pliers by pressing the lever to open the jaws.

6.5.1 How to Use Locking Pliers

As with all the types of pliers, be sure to use appropriate personal protective equipment. Take the following steps to use locking pliers properly:

Step 1 Place the jaws on the object to be held.

Step 2 Turn the adjusting screw in the handle until the pliers grip the object.

Step 3 Squeeze the handles together to lock the pliers.

Step 4 Squeeze the release lever when you want to remove the pliers.

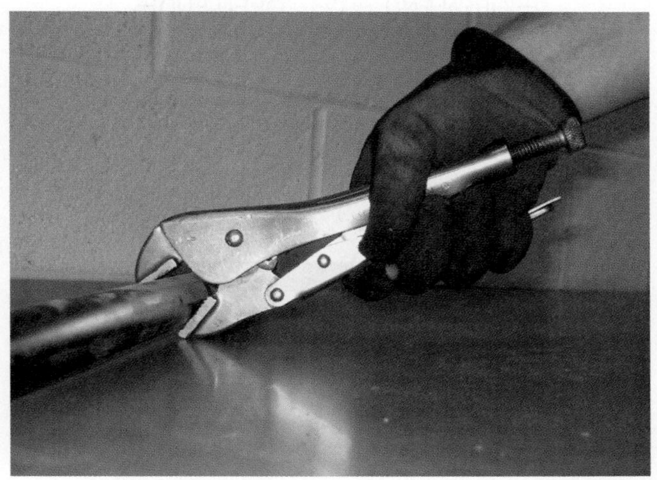

103F15.EPS

Figure 15 Proper use of locking pliers.

6.6.0 Safety and Maintenance

You might not think that misusing pliers could cause injury, but it can. Proper safety precautions when using pliers, and proper maintenance of your pliers are very important. Here are some guidelines to remember when using pliers:

- Hold pliers close to the end of the handles to avoid pinching your fingers in the hinge.
- Don't extend the length of the handles for greater leverage. Use a larger pair of pliers instead.
- Wear appropriate personal protective equipment, especially if you cut wire.
- Hold the short ends of wires to avoid flying metal bits when you cut.
- Always cut at right angles. Don't rock the pliers from side to side or bend the wire back and forth against the cutting blades. Loose wire can fly up and injure you or someone else.
- Oil pliers regularly to prevent rust and to keep them working smoothly.
- Don't use pliers around energized electrical wires. Although the handles may be plastic-coated, they are not insulated against electrical shock.
- Don't expose pliers to direct heat.
- Don't use pliers to turn nuts or bolts; they are not wrenches.
- Don't use pliers as hammers.

7.0.0 Wrenches

Wrenches are used to hold and turn screws, nuts, bolts, and pipes. There are many types of wrenches, but they fall into two main categories: nonadjustable and adjustable. Nonadjustable wrenches fit only one size nut or bolt. They come in both standard (English) and metric sizes. Adjustable wrenches can be expanded to fit different-sized nuts and bolts.

7.1.0 Nonadjustable Wrenches

Nonadjustable wrenches (*Figure 16*) include the **open-end wrench**, the **box-end wrench**, the **hex key** (Allen®) **wrench**, the **striking** (or **slugging**) **wrench**, and the **combination wrench**.

The open-end wrench is one of the easiest nonadjustable wrenches to use. It has an opening at each end that determines the size of the wrench. Often, one wrench has two different-sized openings, such as 7/16-inch and 1/2-inch, one on each end. These sizes measure the distance between the **flats** (straight sides or jaws of wrench opening) of the wrench and the distance across the head of the **fastener** used. The open end allows you to slide the tool around the fastener when there is not enough room to fit a box-end wrench.

Box-end wrenches form a continuous circle around the head of a fastener. The ends have 6 or 12 points. The ends come in different sizes, ranging from 3/8-inch to 15/16-inch. Box-end wrenches offer a firmer grip than open-end wrenches. A box-end wrench is safer to use than an open-end wrench because it will not slip off the sides of certain kinds of bolts. The handles of box-end wrenches are available in a range of lengths.

Striking or slugging wrenches (*Figure 17*) are similar to box-end wrenches in that they have an enclosed circular opening designed to lock on to the fastener when the wrench is struck. The wrenches have a large striking surface so you can hit them more accurately, usually with a mallet or handheld sledgehammer. The ends have 6 or 12 points. Striking wrenches are used only in certain situations, such as when a bolt has become

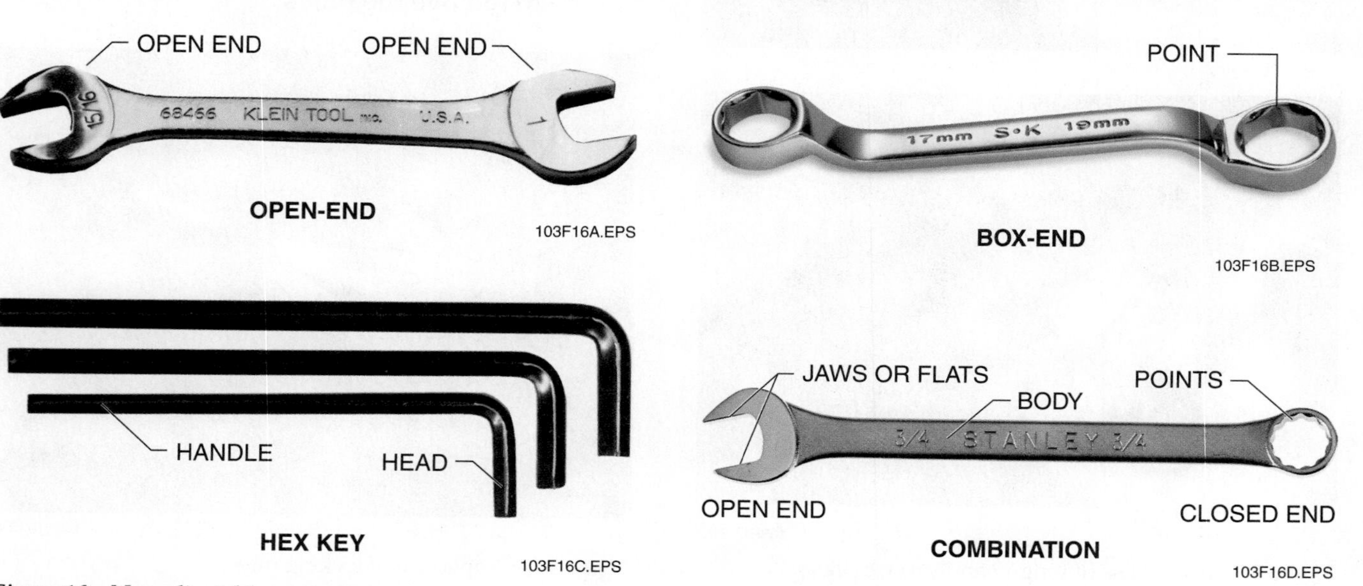

Figure 16 Nonadjustable wrenches.

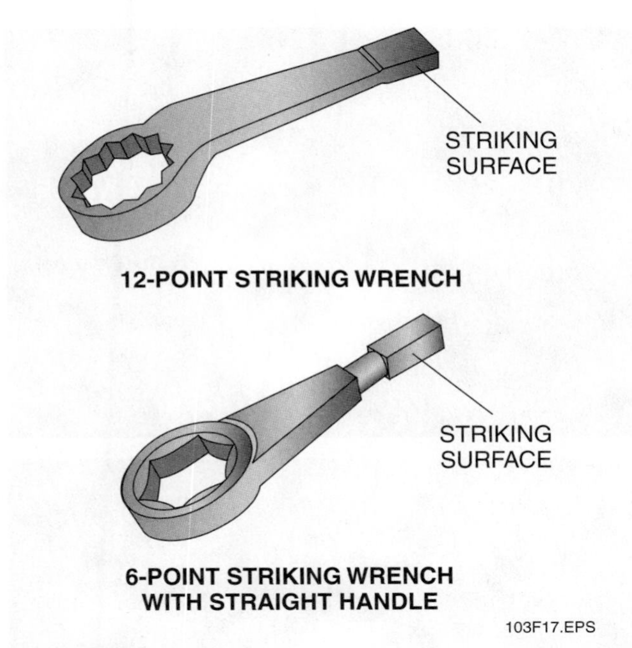

Figure 17 Striking wrenches.

stuck to another material through rust or corrosion. Striking wrenches can damage screw threads and bolt heads. If you are ever in doubt about whether or not to use a striking wrench, ask your instructor or immediate supervisor.

Hex key wrenches are L-shaped, hexagonal (six-sided) steel bars. Both ends fit the socket of a screw or bolt. The shorter length of the L-shape is called the head, and the longer length is the handle. These wrenches generally have a 1/16-inch to 3/4-inch diameter. You might use them with setscrews. Setscrews are used in tools and machinery to set two parts tightly together so they don't move from the set position.

Combination wrenches are, as the name implies, a combination of two types of wrenches. One end of the combination wrench is open and the other is closed, or box-end. Combination wrenches can speed up your work because you don't have to keep changing wrenches.

7.1.1 How to Use a Nonadjustable Wrench

When using a nonadjustable wrench, use the correct size wrench for the nut or bolt. Always pull the wrench toward you. Pushing the wrench can cause injury.

> **CAUTION**
>
> Be sure that the fit of the wrench is snug and square (exactly adjusted) around the nut, bolt, or other fastener. If the fit of the wrench is too loose, it will slip and round off or strip the points of the nut or bolt head. Stripped points may make it impossible to remove the fastener.

7.2.0 Adjustable Wrenches

Adjustable wrenches are used to tighten or remove nuts and bolts and all types and sizes of pipes. They have one fixed jaw and one movable jaw. The adjusting nut on the wrench joins the teeth in the body of the wrench and moves the adjustable jaw. These wrenches come in lengths from 4 to 24 inches, and open as wide as 2 7/16 inches. Common types include **pipe wrenches**, spud wrenches, and **adjustable end wrenches** (*Figure 18*). Using an adjustable wrench may save time when you're working with different sizes of nuts and bolts.

Pipe wrenches (often called monkey wrenches) are used to tighten and loosen all types and sizes of threaded pipe. You adjust the upper jaw of the wrench by turning the adjusting nut. Both jaws have serrated teeth for gripping power. The jaw is spring-loaded and slightly angled so you can release the grip and reposition the wrench without having to readjust the jaw.

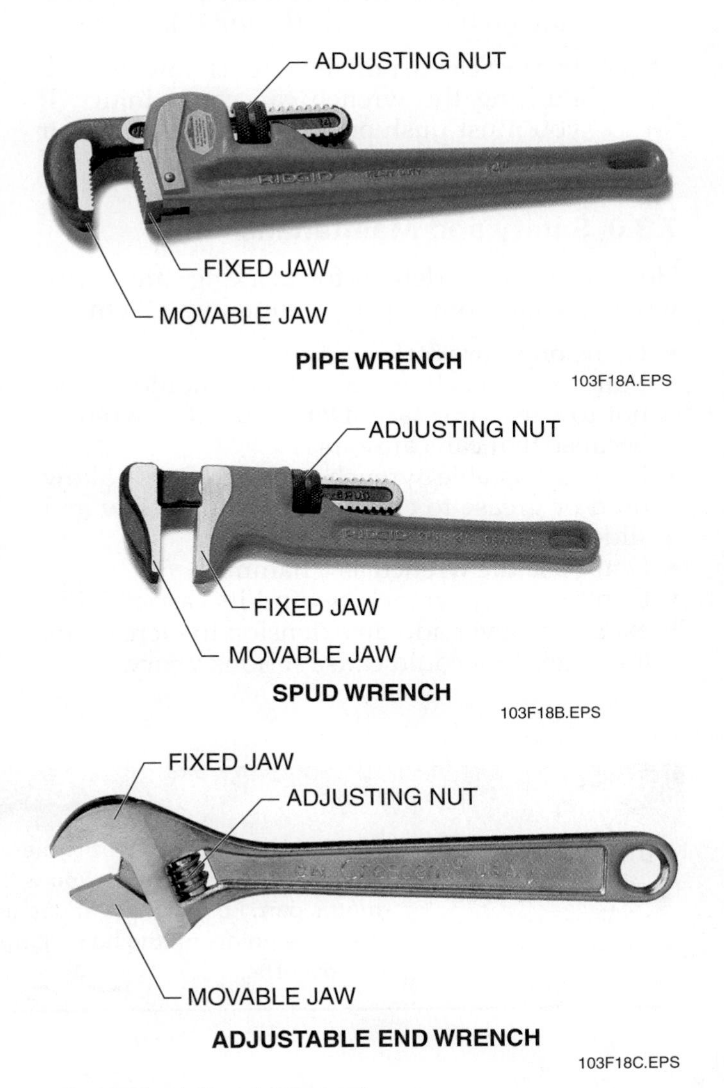

Figure 18 Adjustable wrenches.

Adjustable wrenches are smooth-jawed for turning nuts, bolts, small pipe fittings, and chrome-plated pipe fittings.

7.2.1 How to Use an Adjustable Wrench

To use an adjustable wrench properly, follow these steps (see *Figure 19*):

Step 1 Set the jaws to the correct size for the nut, bolt, or pipe.

> **CAUTION**
>
> If the jaws are improperly adjusted, you could be injured. The wrench could slip, causing you to hurt your hand or lose your balance.

Step 2 Be sure the jaws are fully tightened on the work.

Step 3 Turn the wrench so you are putting pressure on the fixed jaw (*Figure 19*).

Step 4 In most cases, pull the wrench toward you. Pushing the wrench can cause injury. If you must push on the wrench, keep your hand open to avoid getting pinched.

7.3.0 Safety and Maintenance

Here are some guidelines for working safely with wrenches, and for properly maintaining them:

- Focus on your work.
- Pull the wrench toward your shoulder and not toward your face. Don't push the wrench, because that can cause injury.
- Keep adjustable wrenches clean. Don't allow mud or grease to clog the adjusting screw and slide; oil these parts frequently.
- Don't use the wrench as a hammer.
- Don't use any wrench beyond its capacity. For example, never add an extension to increase its leverage. This could cause serious injury.

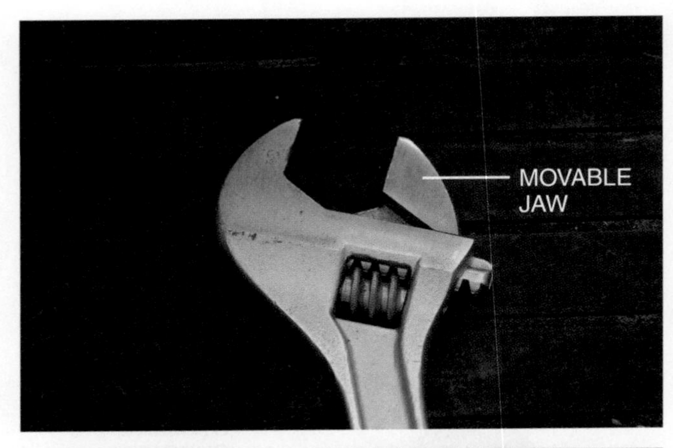

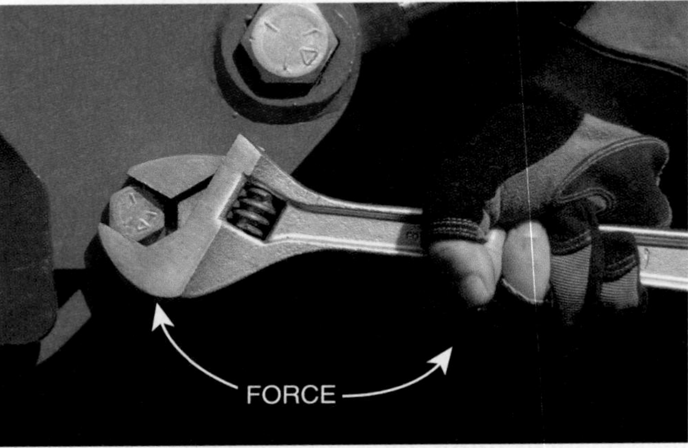

103F19.EPS

Figure 19 Proper use of an adjustable wrench.

On-Site

Metrics and Tools

You must know whether the materials you are working with are made using metric or standard measurements. You will not get a proper fit if you try to use a standard tool on a metric part. For example, if you use a standard-measure socket on a metric bolt, you may wear the points off the head. It may then be necessary to drill the bolt out in order to remove it.

8.0.0 Sockets and Ratchets

Socket wrench sets include different combinations of sockets (the part that grips the nut or bolt) and ratchets (handles) that are used to turn the sockets.

Most sockets (*Figure 20*) have 6 or 12 gripping points. The end of the socket that fits into the handle is square. Sockets also come in different lengths. The longer socket is called a deep socket. It is used when normal sockets will not reach down over the end of the bolt to grip the nut.

Socket sets contain different types of handles for different uses. The ratchet handle (*Figure 21*) has a small lever that you can use to change the turning direction.

8.1.0 How to Use Sockets and Ratchets

Follow these steps to use sockets and ratchets properly:

Step 1 Select a socket that fits the fastener (nut or bolt) that you want to tighten or loosen.

Step 2 Place the square end of the socket over the spring-loaded button on the ratchet shaft.

Step 3 Place the socket over the nut or bolt.

Step 4 Pull on the handle in the appropriate direction to turn the nut. To reverse the direction of the socket, use the adjustable lock mechanism.

Figure 20 Sockets.

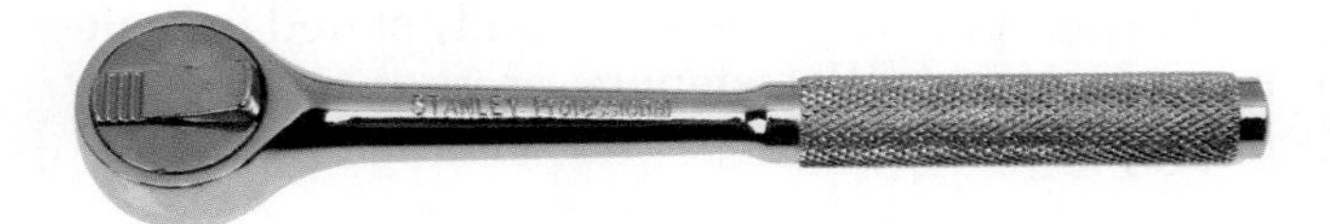

Figure 21 Ratchet handle.

8.2.0 Safety and Maintenance

Follow these guidelines to use sockets and ratchets safely, and to keep them in good working order:

- Never force the ratchet handle beyond hand-tight. This could break the head off the fastener.
- Don't use a cheater pipe (a longer piece of pipe slipped over the ratchet handle to provide more leverage). This could snap the tool or break the head off the bolt or nut.

9.0.0 Torque Wrenches

Torque wrenches (*Figure 22*) measure resistance to turning. You need them when you are installing fasteners that must be tightened in sequence without distorting the workpiece. You will use a torque wrench only when a torque setting is specified for a particular bolt.

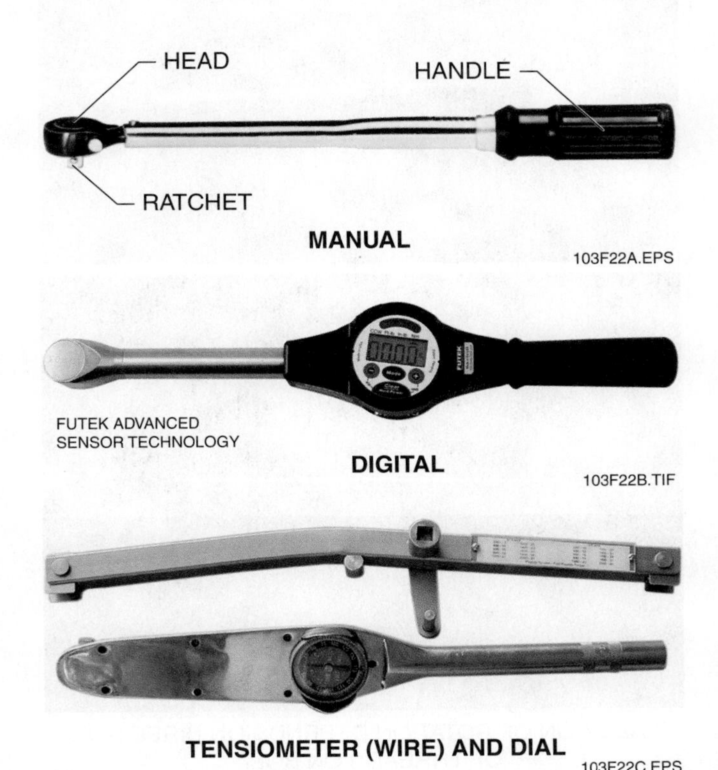

Figure 22 Torque wrenches.

Torque specifications are usually stated in **inch-pounds** for small fasteners or **foot-pounds** for large fasteners.

9.1.0 How to Use a Torque Wrench

Take these steps to use a torque wrench properly:

Step 1 Determine how many inch-pounds or foot-pounds you need to torque to.

Step 2 Set the controls on the wrench to the desired torque level (wrench models vary).

Step 3 Place the torque wrench on the object to be fastened, such as a bolt. Hold the head of the wrench with one hand to support the bolt and to make sure it is properly aligned (*Figure 23*).

Step 4 Watch the torque indicator or listen for the click (depending on the model of the wrench) as you tighten the bolt.

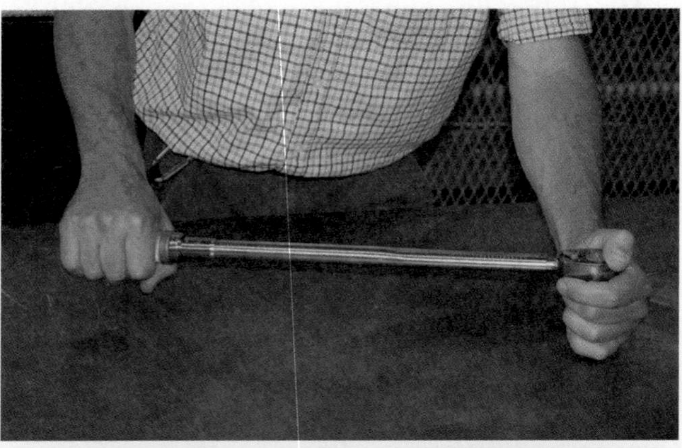

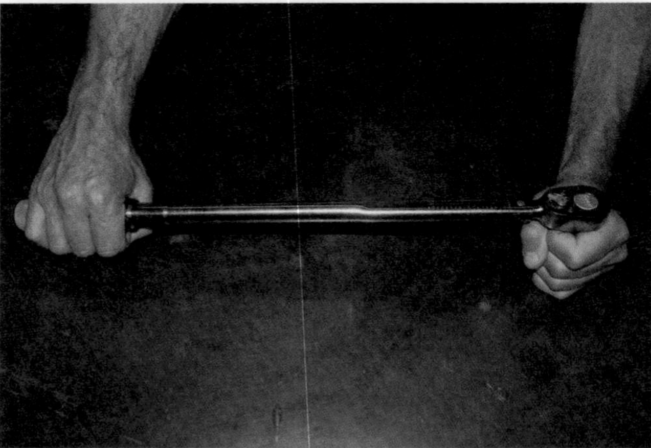

DIRECTION OF ROTATION DEPENDS ON DIRECTION OF THREADS ON BOLT

103F23.EPS

Figure 23 Proper use of a torque wrench.

> **NOTE**
> To get accurate settings, make sure all threaded fasteners are clean and undamaged.

9.2.0 Safety and Maintenance

A torque wrench can damage materials and result in injury if used incorrectly. Follow these guidelines when using a torque wrench:

- Always follow the manufacturer's recommendations for safety, maintenance, and calibration.
- Always store the wrench in its case.
- Never use the wrench as a ratchet or for anything other than its intended purpose.

10.0.0 Rules and Other Measuring Tools

In *Introduction to Construction Math*, you learned how to read various types of measuring tools. This section presents information on these and other types of measuring tools, and includes guidelines for their use.

Craftworkers use four basic types of measuring tools:

- Flat steel rule
- Tape measure
- Wooden folding rule
- Laser measuring tool

When choosing a measuring tool, keep the following in mind:

- It must be accurate.
- It should be easy to use.
- It should be durable.
- The numbers should be easy to read (preferably black on yellow or off-white).

10.1.0 Steel Rule

The flat steel rule (*Figure 24*) is the simplest and most common measuring tool. Flat steel rules are usually 6 or 12 inches long, but longer sizes are available. Steel rules can be flexible or nonflexible, thin or wide. The thinner the rule, the more accurately it measures, because the division marks are closer to the work. Generally, a steel rule has four sets of standard (English) marks, two on each side of the rule. Depending on the increments, they are read like the standard rulers discussed in the *Introduction to Construction Math* module.

103F24.EPS

Figure 24 Steel rule.

10.2.0 Measuring Tape

The standard tape measure blade (*Figure 25*) is marked similarly to a 1/16-inch standard ruler. Along with the ½-inch, ¼-inch, ⅛-inch, and 1/16-inch markings, the standard tape measure may also include metric increments. Tape measures are also available with markings used for framing out walls.

The concave (or curve) of a tape measure blade is designed to strengthen the blade and makes the tape easier to read when it is placed on a surface that needs to be marked. Once the proper measurement is established and the blade tip is secure, rotate the blade edge nearest you slightly until it lies flat on the surface, then mark the material. Using this method makes it easier to read the measurement and mark the material more accurately.

10.3.0 Wooden Folding Rule

A wooden folding rule (*Figure 26*) is usually marked in sixteenths of an inch on both edges of each side. Folding rules come in 6- and 8-foot lengths. Because of its stiffness, a folding rule is better than a cloth or steel tape for measuring vertical distance. This is because, unlike tape, it can be held straight up. This makes it easier to measure some distances, such as those where you might need a ladder to reach one end. Like a tape measure, the folding rule can also have special marks at the 16-, 24-, and 19.2-inch increments to make wall framing easier.

10.4.0 Laser Measuring Tools

A laser measuring tool (*Figure 27*) can be considered a battery-powered, electronic version of a tape measure. This hand-held tool works by pointing it at a specific object and then pressing a measurement button on the control panel. When the button is pressed, a laser shoots out at the object and a reading is sent back to the instrument and displayed on the display screen. Laser measuring tools can be designed to register and record in both standard and metric measurements.

The following are some of the advantages a laser measuring tool has over a traditional tape measure:

- Measurements required at higher elevations can be taken from ground level.
- Longer measurements can be taken. Most construction laser measuring tools measure from 0.05 meters to 100 meters (0.1 yards to 109.4 yards). A target plate may be required for longer distances with higher power lasers.
- Some laser measuring tools have special buttons on the control panel that make length-related calculations, such as addition, subtraction, and area, easier to perform.
- A number of measurements can be electronically stored on the tool, so that measurements can be written down all at once after the task is completed.
- Some laser measuring tools can have built-in electronic levels and spirit levels.

103F25.EPS

Figure 25 Standard tape measure.

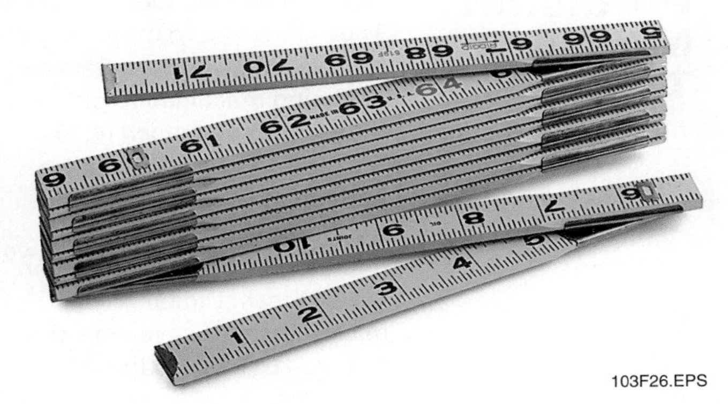

103F26.EPS

Figure 26 Wooden folding rule.

103F27.EPS

Figure 27 Laser measuring tool.

Some disadvantages of a laser measuring tool are that it can have an accuracy range of ±3 to 10 millimeters, it can take time to learn to use, and it is much more costly than a tape measure.

Laser instruments are precision instruments, so make sure that you always handle them with care and store them appropriately. Never toss a laser instrument and never store it in a gang box. If you do store it in your toolbox, make sure it will not bounce around or be crushed by other tools.

10.5.0 Safety and Maintenance

There are some safety concerns, as well as some needed maintenance, for rules and measuring tools. Here are guidelines to remember:

- Occasionally apply a few drops of light oil on the spring joints of a wooden folding rule and steel tape.
- Wipe moisture off steel tape to keep it from rusting.
- Don't kink or twist steel tape, because this could cause it to break.
- Don't use steel tape near exposed electrical parts.
- Don't let laser measuring tools get wet.

11.0.0 Levels

A level is a tool used to determine both how level a horizontal surface is and how **plumb** a vertical surface is. If a surface is described as level, it means it is exactly horizontal. If a surface is described as plumb, it means it is exactly vertical. Levels are used to determine how near to exactly horizontal or exactly vertical a surface is.

Types of levels range from simple spirit levels to electronic and laser instruments. The spirit level (*Figure 28*) is the most commonly used level in the construction trade.

11.1.0 Spirit Levels

Most levels are made of tough, lightweight metals, such as magnesium or aluminum. The spirit level has three vials filled with alcohol. The center

On-Site

Precision Measuring Tools

Laser measuring tools are becoming increasingly common in the construction industry. They allow you to make very precise and accurate measurements. Precision measuring tools, such as micrometers and calipers, make it possible to accurately measure parts that are being machined to one thousandth of an inch (0.001"). Micrometers can be used for both outside and inside measurements. A standard micrometer's smallest division is 0.001 inch. Digital micrometers are available that can read to 0.00005 inch.

103SA02.EPS

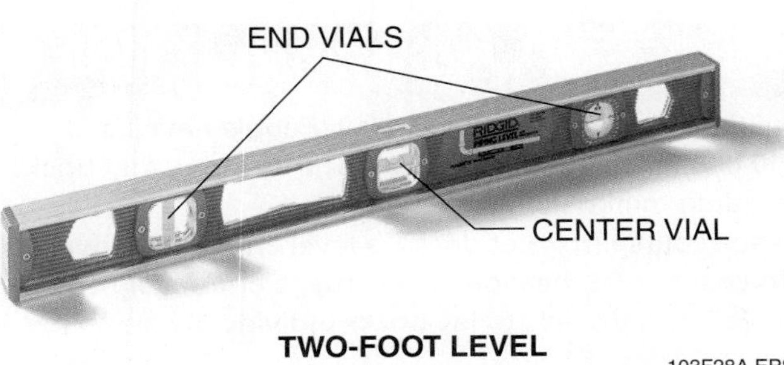

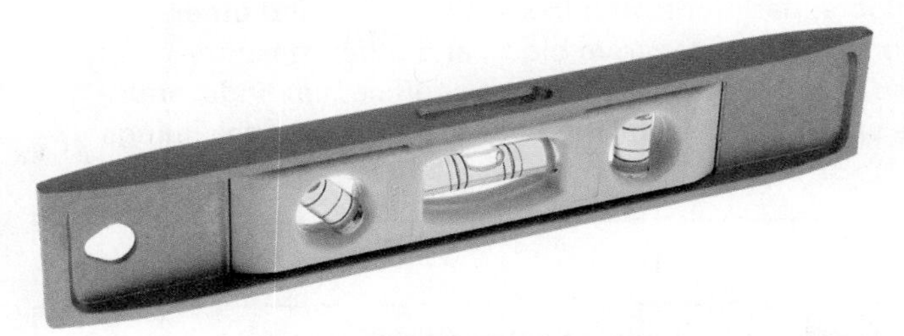

Figure 28 Spirit levels.

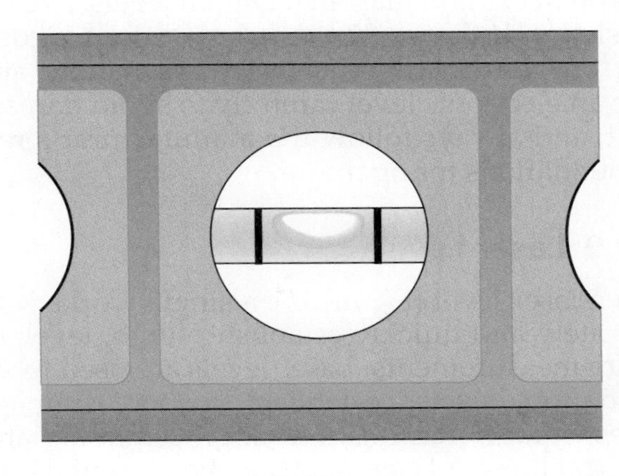

Figure 29 An air bubble centered between the lines shows level or plumb.

vial is used to check for level, and the two end vials are used to check for plumb.

The amount of liquid in each vial is not enough to fill it, so there is always a bubble in the vial. When the bubble is centered precisely between the lines on the vial, the surface is either level or plumb, depending on the vial (*Figure 29*).

Spirit levels come in a variety of sizes. The longer the level, the greater its accuracy.

11.1.1 How to Use a Spirit Level

Using a spirit level is very simple. It just requires a careful eye to be sure you are reading it correctly. Follow these steps to use a spirit level properly:

Step 1 Ensure the surface is free of debris.

Step 2 Put the spirit level on the object you are checking.

Step 3 Look at the air bubble. If the bubble is centered between the lines, the object is level (or plumb).

Did You Know?

The Spirit Level

The spirit level got its name because the vials in it are filled with alcohol, which is sometimes called spirits. Alcohol is used because it does not freeze.

11.1.2 Safety and Maintenance

Levels are precision instruments that must be handled with care. Although there is little risk of personal injury when working with levels, there is a chance of damaging or breaking the level. Remember these guidelines when working with levels:

- Replace a level if a crack or break appears in any of the vials.
- Keep levels clean and dry.
- Don't bend or apply too much pressure on a level.
- Don't drop or bump a level.

11.2.0 Digital (Electronic) Levels

Digital (electronic) levels feature a simulated bubble display plus a digital readout of degrees of slope, inches per foot for rise and run of stairs and

On-Site

Bricklayers

Working with your hands to create buildings is a time-honored craft. People have used bricks in construction for more than 10,000 years. Archaeological records prove that bricks were one of the earliest man-made building materials.

In bricklaying, it is important that each course (row) of bricks is level and that the wall is straight. An uneven wall is weak. To ensure that the work stays true, a bricklayer uses a straight level and a plumb line. Bricklayers use trowels to lay bricks individually along the length of the structure.

Today, bricklayers build walls, floors, partitions, fireplaces, chimneys, and other structures with brick, precast masonry panels, concrete block, and other masonry materials. They lay brick for houses, schools, baseball stadiums, office buildings, and other structures. Some bricklayers specialize in installing heat-resistant firebrick linings inside huge industrial furnaces.

roofs, and percentage of slope for drainage problems on decks and masonry. Digital levels, like the one shown in *Figure 30*, are becoming more common on construction sites. Always handle and store an electronic level carefully to avoid damaging it, and always follow the manufacturer's recommendations for operation.

11.3.0 Laser Levels

With a laser level (*Figure 31*), a single worker can accurately and quickly establish plumb, level, or square measurements. Laser levels are used to set foundation levels, establish proper drainage slopes, square framing, and align plumbing and electrical lines. A laser level may be mounted on a tripod, fastened onto pipes or framing studs, or suspended from ceiling framing. Levels for professional construction jobs are housed in sturdy casings designed to withstand jobsite conditions. These tools come in a variety of sizes and weights, depending on the application. Always handle and store laser levels carefully, and follow the manufacturer's recommendations for use.

WARNING!

Never look at the laser that is generated by a laser tool because it can impair your vision and damage your eyes.

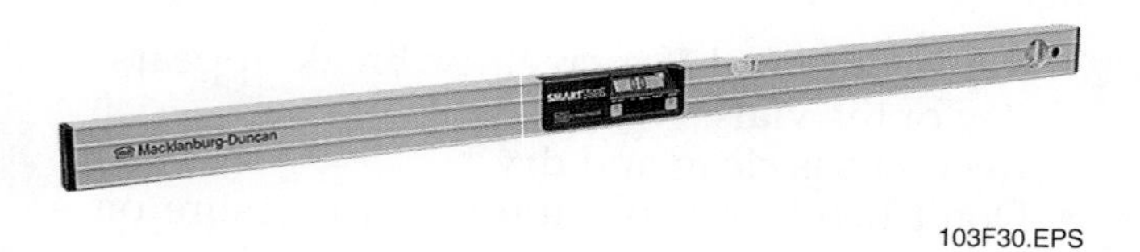

103F30.EPS

Figure 30 Digital (electronic) level.

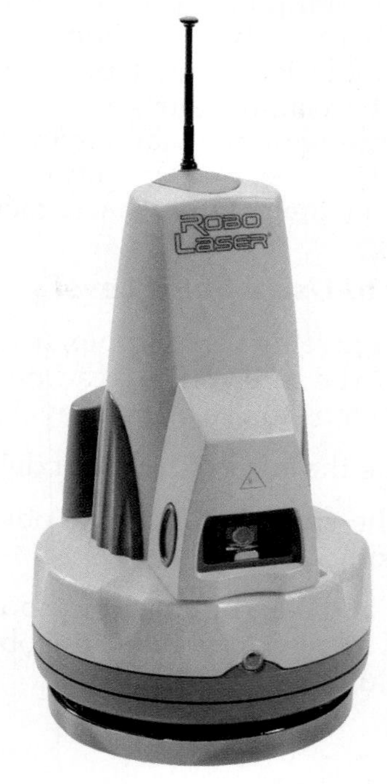

103F31.EPS

Figure 31 Laser level.

12.0.0 Squares

Squares (*Figure 32*) are used for marking, checking, and measuring. The type of square you use depends on the type of job and your preference. Common squares are the **carpenter's square**, **rafter angle square** (also called the speed square or magic square), **try square**, and **combination square**.

12.1.0 The Carpenter's Square

The carpenter's square (framing square) is shaped like an L and is used mainly for squaring up sections of work such as wall studs and sole plates; that is, to ensure that they are at right angles to each other. It can also be used for laying out common rafters, hip rafters, and stairs. The carpenter's square has a 24-inch blade and a 16-inch tongue, forming a right (90-degree) angle. The blade and tongue are marked with inches and fractions of an inch. You can use the blade and the tongue as a rule or a straightedge. Tables and formulas are printed on the blade for making quick calculations such as determining area and volume.

The rafter angle square is another type of carpenter's square, frequently made of cast aluminum. It is a combination protractor, try square, and framing square. It is marked with degree gradations for fast, easy layout. The square is small, so it's easy to store and carry. By clamping the square on a piece of lumber, you can use it as a guide when cutting with a portable circular saw.

The try square is a fixed, 90-degree angle and is used mainly for woodworking. You can use it to lay out cutting lines at 90-degree angles (*Figure 33*), to check (or try) the squareness of adjoining surfaces, to check a joint to make sure it is square, and to check if a **planed** piece of lumber is warped or cupped (bowed).

12.2.0 The Combination Square

The combination square has a ruled blade that moves through a head. The head is marked with 45-degree and 90-degree angle measures. Some squares also contain a small spirit level and a carbide scriber, which is a sharp, pointed tool for marking metal. The combination square is one of

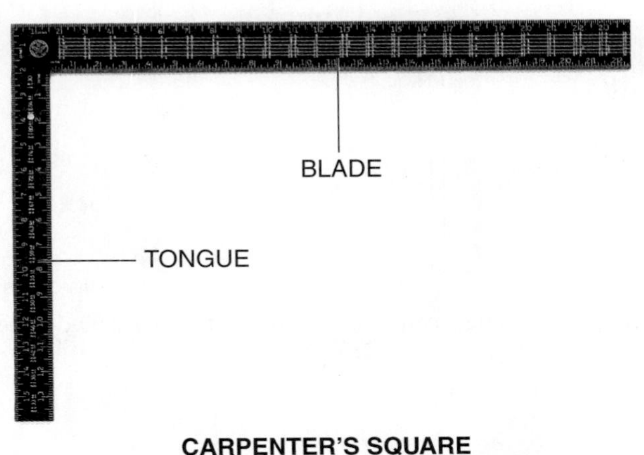

CARPENTER'S SQUARE

103F32A.EPS

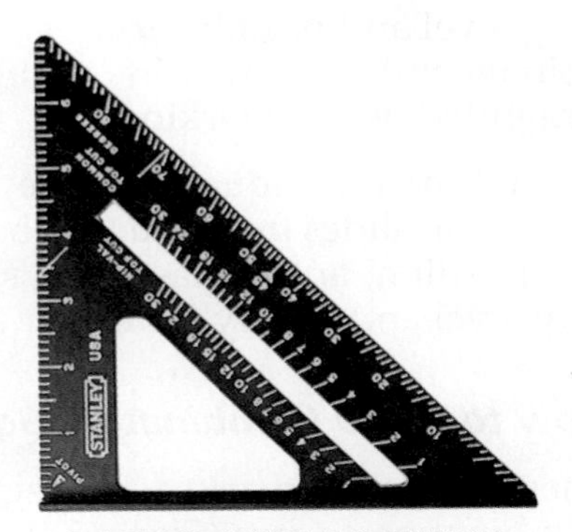

RAFTER ANGLE SQUARE (SPEED SQUARE)

103F32B.EPS

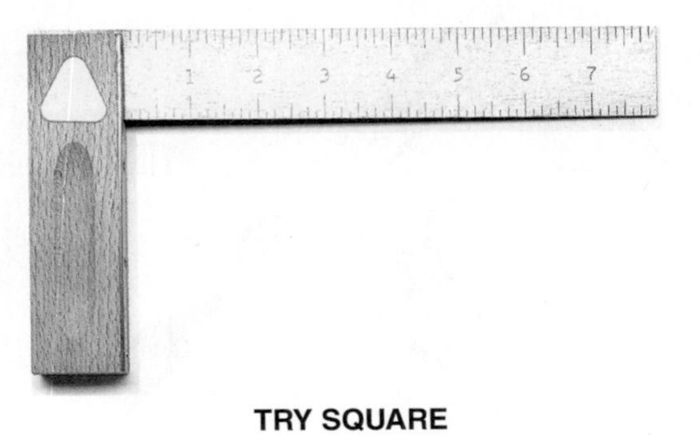

TRY SQUARE

103F32C.EPS

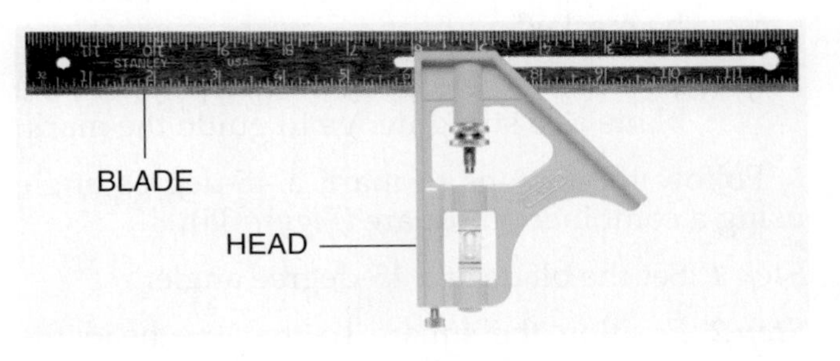

COMBINATION SQUARE

103F32D.EPS

Figure 32 Types of squares.

103F33.EPS

Figure 33 Marking a line for cutting.

the most useful tools for layout work. You can use it for any of the following tasks:

- Testing work for squareness
- Marking 90-degree and 45-degree angles

> **NOTE**
> To mark angles other than 45 and 90 degrees, slide the protractor part of the square onto the blade and dial in the desired angle.

- Checking level and plumb surfaces
- Measuring lengths and widths
- As a straightedge and marking tool

Good combination squares have all-metal parts, a blade that slides freely but can be clamped securely in position, and a glass tube spirit level that is truly level and tightly fastened.

12.2.1 How to Use a Combination Square

Follow these steps to mark a 90-degree angle using a combination square (*Figure 34*):

Step 1 Set the blade at a right angle (90 degrees).

Step 2 Position the square so that the head fits snugly against the edge of the material to be marked.

Step 3 Starting at the edge of the material, use the blade as a straightedge to guide the mark.

Follow these steps to mark a 45-degree angle using a combination square (*Figure 35*):

Step 1 Set the blade at a 45-degree angle.

Step 2 Position the square so that the head fits snugly against the edge of the material to be marked.

Step 3 Starting at the edge of the material, use the blade as a straightedge to guide the mark.

> **Did You Know?**
> ## Geometry and the Combination Square
> The combination square set is used to measure and mark 30-, 45-, 60-, and 90-degree angles. When you use the combination square to measure and mark materials, you are applying basic geometric principles to your work.

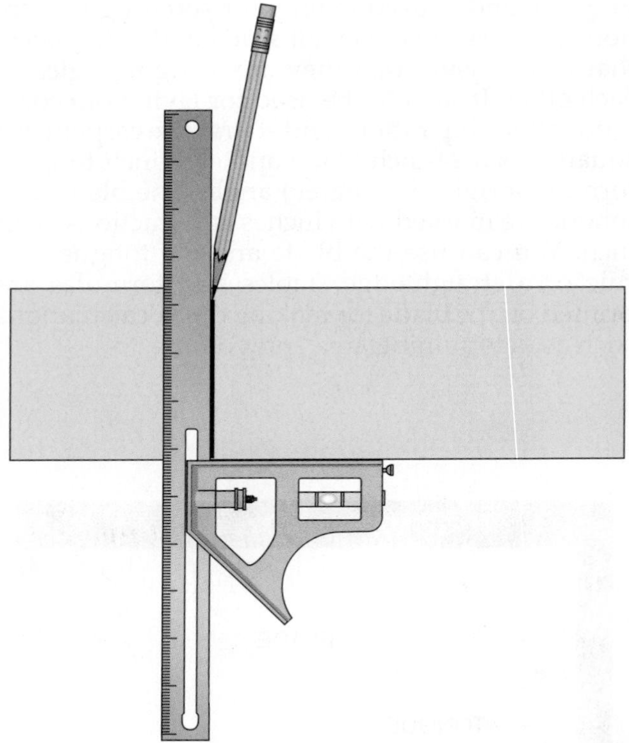

103F34.EPS

Figure 34 Using a combination square to mark a 90-degree angle.

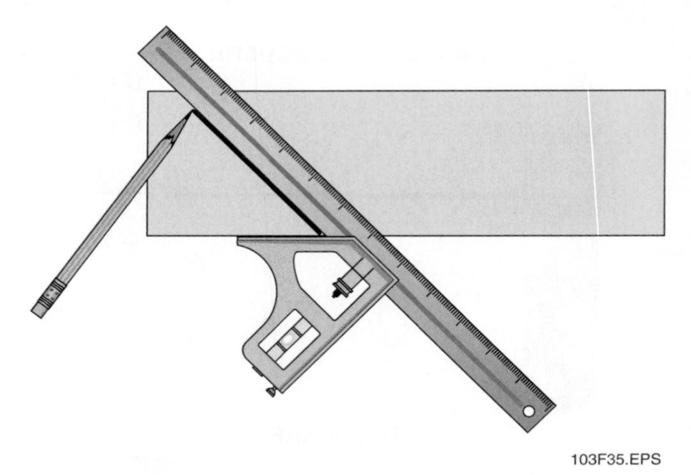

103F35.EPS

Figure 35 Using a combination square to mark a 45-degree angle.

12.3.0 Safety and Maintenance

Here are the safety and maintenance guidelines to remember when using squares:

- Keep the square dry to prevent it from rusting.
- Use a light coat of oil on the blade, and occasionally clean the blade's grooves and the setscrew.
- Don't use a square for something it wasn't designed for, especially prying or hammering.
- Don't bend a square or use one for any kind of horseplay.
- Don't drop or strike the square hard enough to change the angle between the blade and the head.

13.0.0 Plumb Bob

The plumb bob, which is a pointed weight attached to a string (*Figure 36*), uses the force of gravity to make the line hang vertical, or plumb. Plumb bobs come in different weights, with 12 ounces, 8 ounces, and 6 ounces being the most common.

When the weight is allowed to hang freely, the string is plumb (*Figure 37*). You can use a plumb bob to make sure a wall or a doorjamb is vertical. Or, suppose you want to install a post under a beam. A plumb bob can show what point on the floor is directly under the section of the beam you need to support.

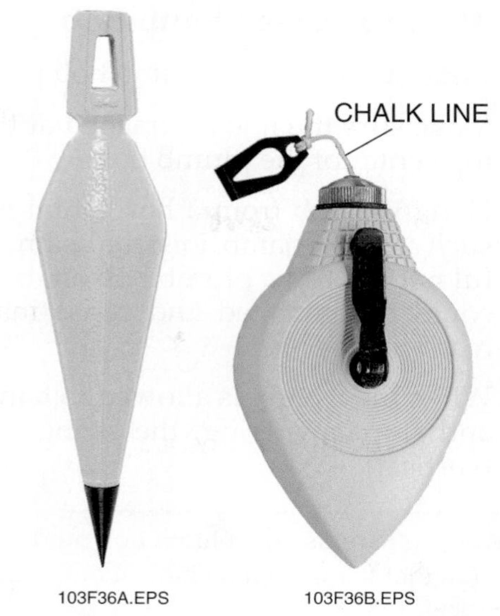

Figure 36 Plumb bobs.

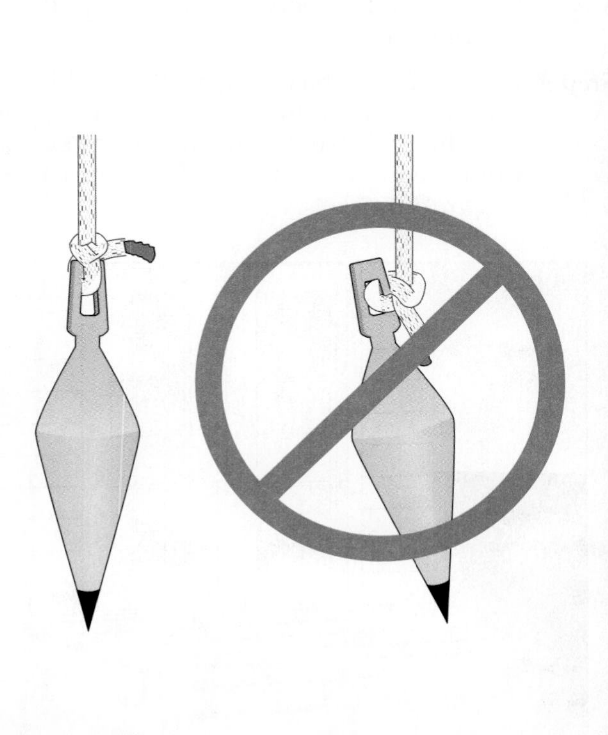

Figure 37 Proper use of a plumb bob.

13.1.0 How to Use a Plumb Bob

Follow these steps to use a plumb bob properly:

Step 1 Make sure the line is attached at the exact top center of the plumb bob.

Step 2 Hang the bob from a horizontal member, such as a doorjamb, joist, or beam. Be careful not to drop a plumb bob on its point; it could be damaged and cause inaccurate readings.

Step 3 When the weight is allowed to hang freely and stops swinging, the string is plumb (vertical).

> **NOTE**
> When you are using a plumb bob outdoors, be aware that the wind may blow it out of true vertical.

Step 4 Mark the point directly below the tip of the plumb bob. This point is precisely below the point where you attached the bob.

14.0.0 Chalk Lines

A chalk line is a piece of string or cord that is coated with chalk. You stretch the line tightly between two points and then snap it to release a chalky line to the surface.

For frequent use, a mechanical self-chalking line is more convenient. A mechanical self-chalking line is a box containing a line on a reel (*Figure 38*). The box is filled with colored chalk powder. The line is automatically chalked each time you pull it out of the box. Some models have a point on the end of the box so it can also be used as a plumb bob.

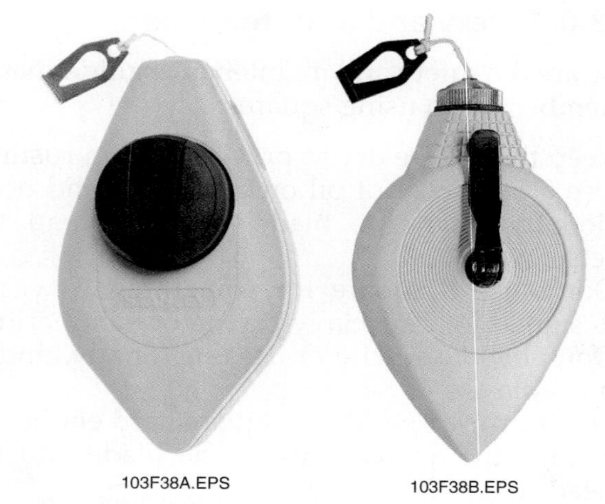

Figure 38 Mechanical self-chalkers.

> **CAUTION**
> Store the chalk line in a dry place. Damp or wet chalk is unusable.

14.1.0 How to Use a Chalk Line

Follow these steps to use a chalk line properly:

Step 1 Pull the line from the case and secure one end.

Step 2 Stretch the line between the two points to be connected.

Step 3 After the line has been pulled tight, pull the string straight away from the work and then release it. This marks the surface underneath with a straight line of chalk (*Figure 39*).

Figure 39 Proper use of a chalk line.

15.0.0 UTILITY KNIVES

A utility knife is used for a variety of purposes including cutting roofing felt, fiberglass or asphalt shingles, vinyl or linoleum floor tiles, fiberboard, and gypsum board. You can also use it for trimming insulation.

The utility knife has a replaceable razor-like blade. It has a handle about six inches long, made of cast-iron or plastic, to hold the blade. The handle is made in two halves, held together with a screw (*Figure 40*).

With many utility knives, you can lock the blade in the handle in one, two, or three positions, depending on the type of knife. Models that have a retractable blade (the blade pulls into the handle when not in use) are the safest. Some models have different blades for cutting different materials.

15.1.0 How to Use a Utility Knife

Follow these steps to use a utility knife safely and properly:

Step 1 Unlock the knife blade. Push the blade out.

Step 2 Lock the blade in the open position.

Step 3 Place some scrap, such as a piece of wood, under the object you are cutting. This will protect the surface under the object.

Step 4 As soon as you have finished cutting, unlock the blade, pull it back to the closed position, and lock the blade.

15.2.0 Safety and Maintenance

Here are some guidelines for safely using and maintaining a utility knife:

- Replace dull blades.
- Always keep the blade closed and locked when you are not using the knife.
- Be sure to position yourself properly and to make the cut in the appropriate direction.

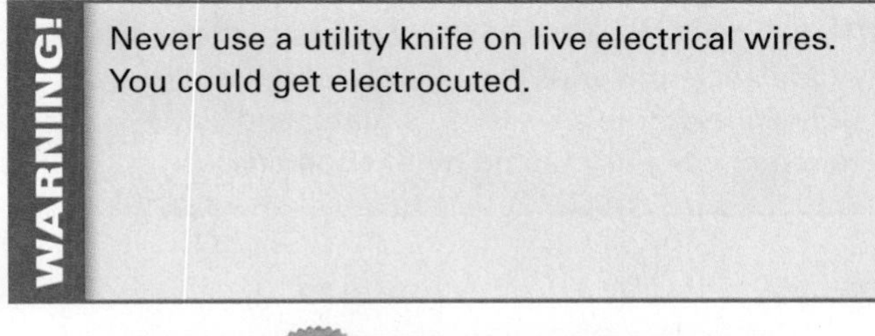

WARNING!

Never use a utility knife on live electrical wires. You could get electrocuted.

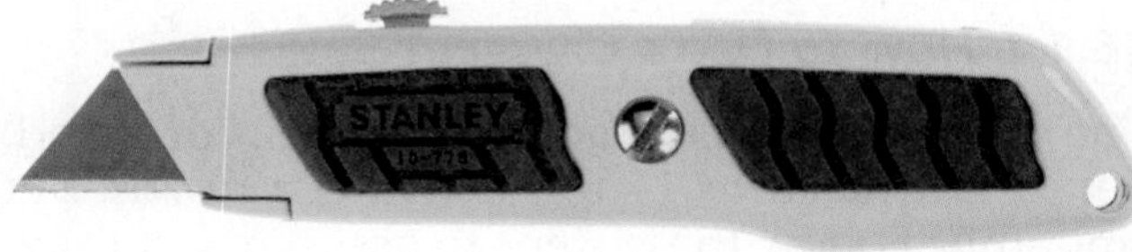

103F40.EPS

Figure 40 Utility knife.

16.0.0 SAWS

Using the right saw for the job makes cutting easy. The main differences between types of saws are the shape, number, and pitch of their teeth. These differences make it possible to cut across or with the grain of wood, along curved lines, or through metal, plastic, or wallboard. Generally, the fewer **points**, or teeth per inch (tpi), the coarser and faster the cut. The more tpi, the slower and smoother the cut.

Figure 41 shows several types of saws and their parts. The following are descriptions of common types of saws:

- *Backsaw* – The standard blade of this saw is 8 to 14 inches long with 11 to 14 tpi. A backsaw has a broad, flat blade and a reinforced back edge. It is used for cutting joints, especially **miter joints** and **tenons**.
- *Compass (keyhole) saw* – The standard blade of this saw is 12 to 14 inches long with 7 or 8 tpi. This saw cuts curves quickly in wood, plywood, or wallboard. It is also used to cut holes for large-diameter pipes, vents, and plugs or switch boxes.
- *Coping saw* – This saw has a narrow, flexible 6¾-inch blade attached to a U-shaped frame. Holders at each end of the frame can be rotated so you can cut at angles. Standard blades range from 10 to 20 tpi. The coping saw is used for making irregular-shaped moldings fit together cleanly.
- *Drywall saw* – A drywall saw is a long, narrow saw used to cut softer building materials, such as drywall. Drywall saws can have a fixed blade or a retractable blade held to either a wood or plastic handle with thumb screws. The blade on a drywall saw has a very sharp point, so that you can easily poke a hole to start your cut without drilling a starter hole. The jigsaw and spiral saw are power tools that are sometimes used in the same applications as a drywall saw.
- *Hacksaw* – The standard blade of this saw is 8 to 16 inches long with 14 to 32 tpi. It has a sturdy frame and a pistol-grip handle. The blade is tightened using a wing nut and bolt. The hacksaw is used to cut through metal, such as nails, bolts, or pipe. When installing a hacksaw blade, be sure that the teeth face away from, not toward, the saw handle. Hacksaws are designed to cut on the push stroke, not on the pull stroke.
- *Handsaw (crosscut saw or ripsaw)* – The standard blade of this saw is 26 inches long with 8 to 14 tpi for a crosscut saw and 5 to 9 tpi for a ripsaw. You will learn how to use handsaws in this module.

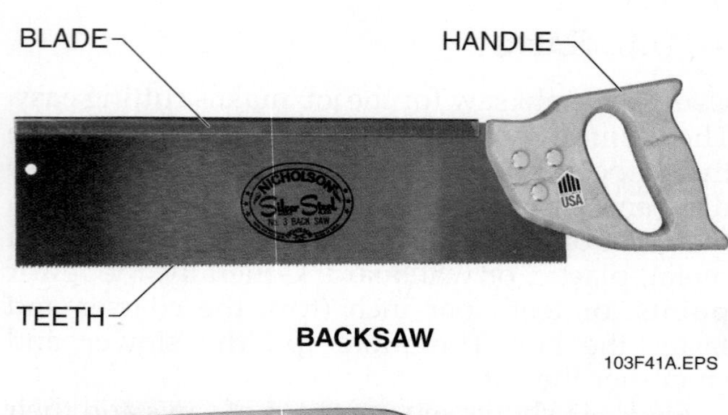

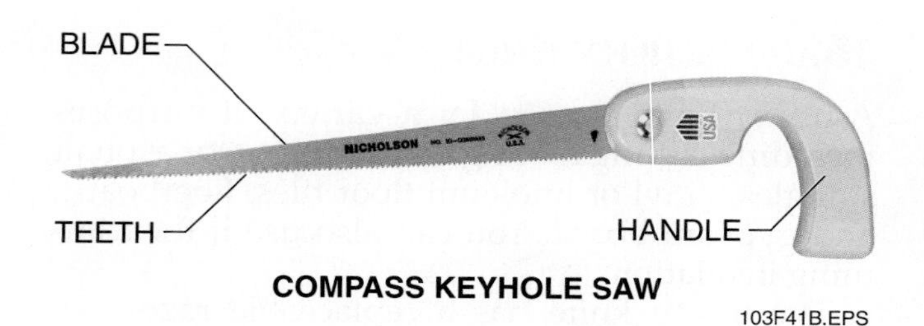

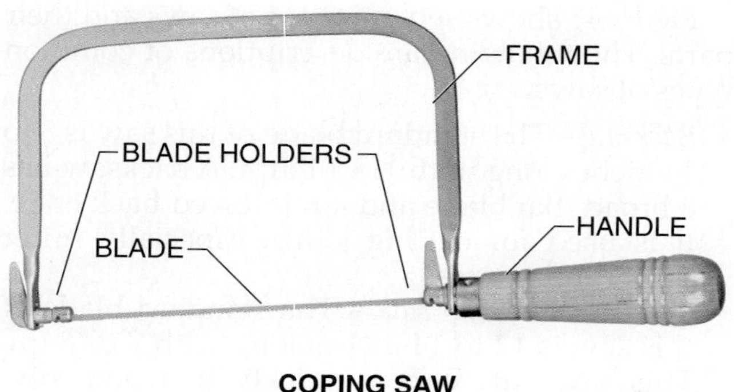

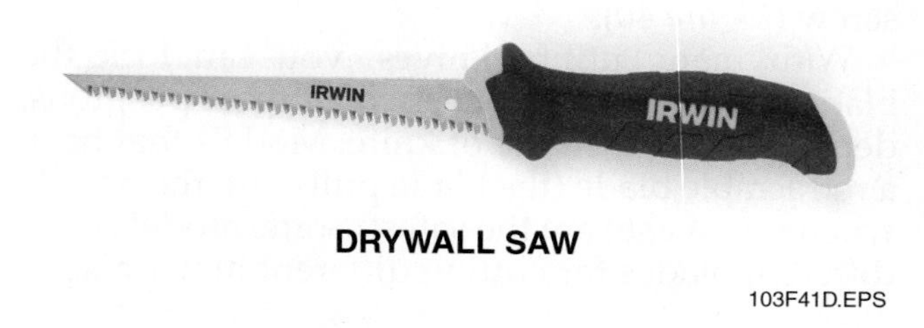

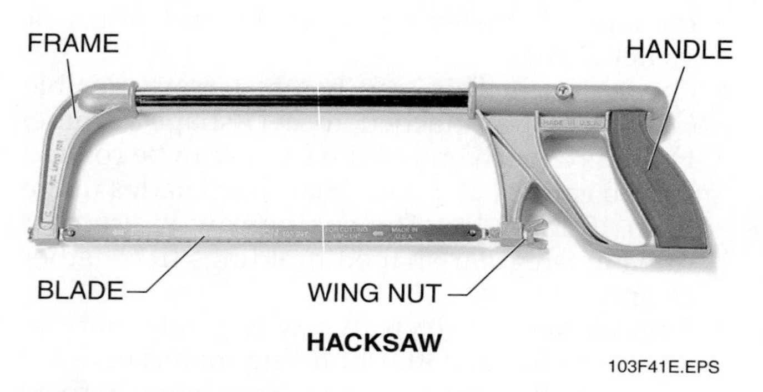

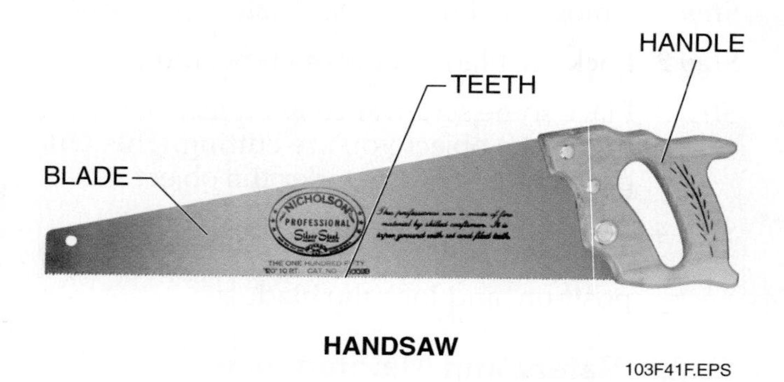

Figure 41 Types of saws.

16.1.0 Handsaws

The handsaw's blade is made of tempered steel so it will stay sharp and will not bend or buckle. Handsaws are classified mainly by the number, shape, size, slant, and direction of the teeth. Saw teeth are set or angled alternately in opposite directions to make a cut (or **kerf**) slightly wider than the thickness of the saw blade itself. Two common types of handsaws are the crosscut saw and the ripsaw.

The crosscut saw, which has 8 to 14 tpi, is designed to cut across the grain (the direction of the fibers) of wood. Blade lengths range from 20 inches to 28 inches. For most general uses, 24 inches or 26 inches is a good length.

The ripsaw has 5 to 9 tpi. The ripsaw, designed to cut with the grain (parallel to the wood fibers), meets less resistance than a saw cutting across the grain.

Did You Know?

The Romans and Their Tools

Ancient Egyptians and Greeks used a variety of tools. The Romans, however, are known as the toolmakers of the ancient world. The plane, metal-cutting saw, drawknife, frame saw, level, square, and claw hammer were all created by the Romans.

16.1.1 How to Use a Crosscut Saw

The crosscut saw cuts across the grain of wood. Because it has 8 to 14 tpi, it will cut slowly but smoothly. Follow these steps to use a crosscut saw properly:

Step 1 Mark the cut to be made with a square or other measuring tool.

Step 2 Make sure the piece to be cut is well-supported (on a sawhorse, jack, or other support). Support the scrap end as well as the main part of the wood to keep it from splitting as the kerf nears the edge. With short pieces of wood, you can support the scrap end of the piece with your free hand. With longer pieces, you will need additional support.

Step 3 Place the saw teeth on the edge of the wood farthest from you, just at the outside edge of the mark.

Step 4 Start the cut with the part of the blade closest to the handle end of the saw, because you will pull the first stroke toward your body.

Step 5 Use the thumb of the hand that is not sawing to guide the saw so it stays vertical to the work.

Step 6 Place the saw at about a 45-degree angle to the wood, then pull the saw to make a small groove.

Step 7 Start sawing slowly, increasing the length of the stroke as the kerf deepens.

NOTE

Don't push or ride the saw into the wood. Let the weight of the saw set the cutting rate. It's easier to control the saw and less tiring that way.

Step 8 Continue to saw with the blade at a 45-degree angle to the wood.

NOTE

If the saw starts to wander from the line, angle the blade toward the line. If the saw blade sticks in the kerf, wedge a thin piece of wood into the cut to hold it open.

16.1.2 How to Use a Ripsaw

The ripsaw cuts along the grain of wood. Because it has fewer points (5 to 9 tpi) than the crosscut saw, it will make a coarser, but faster, cut. To use a ripsaw properly, mark and start a ripping cut the same way you would start cutting with a crosscut saw (*Figure 42*). Once you've started the kerf, saw with the blade at a steeper angle to the wood—about 60 degrees.

Did You Know?

Emery Cloth

An emery cloth is a maintenance and cleaning tool often used in the construction industry. It is normally used for cleaning tools made of metal, like handsaws. It may be used wet or dry in a manner very similar to sandpaper. Emery cloth is coated with a substance called powdered emery, which is a granular form of pure carborundum. After using an emery cloth, always be sure to properly lubricate the tool to prevent rusting.

16.2.0 Safety and Maintenance

You must maintain your saws for them to work properly. Also, it is very important to focus on your work when you are sawing—saws can be dangerous if used incorrectly or if you are not paying attention. Here are the guidelines for working with handsaws:

- Clean your saw blade with a fine emery cloth and apply a coat of light machine oil if it starts to rust—rust ruins the saw blade.
- Always lay a saw down gently.
- Have your saw sharpened by an experienced sharpener.
- Brace yourself when sawing so you are not thrown off balance on the last stroke.
- Don't let saw teeth come in contact with stone, concrete, or metal.

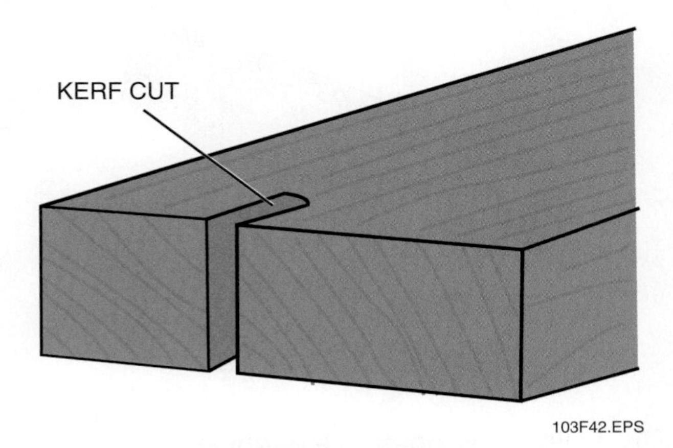

Figure 42 Kerf cut with the wood grain.

17.0.0 FILES AND RASPS

Files and rasps are used to cut, smooth, or shape metal parts. You can also use them to finish and shape all metals except hardened steel, and to sharpen many tools. A variety of files and rasps are shown in *Figure 43*.

Files have slanting rows of teeth. Rasps have individual teeth. Files and rasps are usually made from a hardened piece of high-grade steel. Both

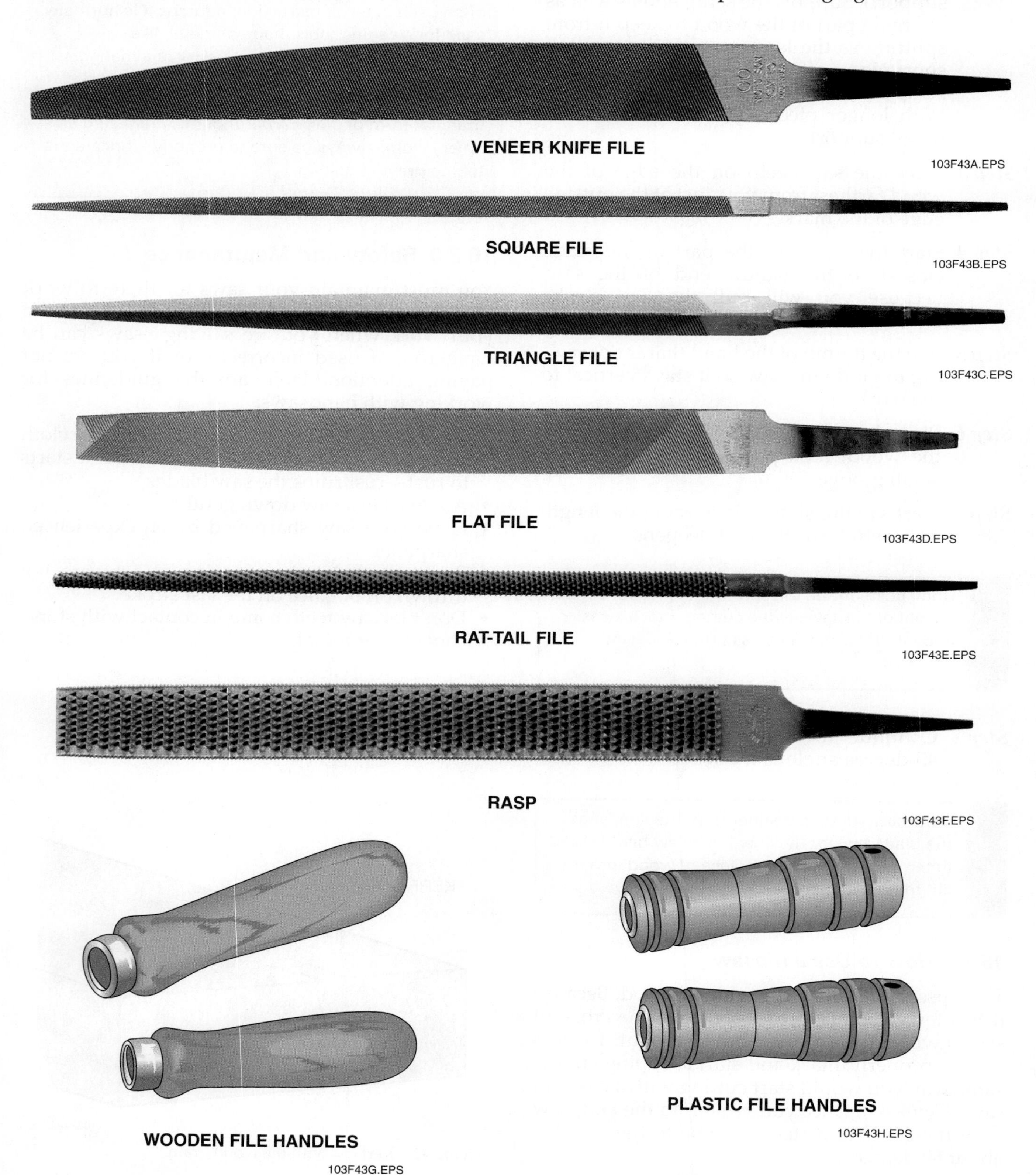

Figure 43 Types of files, rasps, and handles.

are sized by the length of the body (see *Figure 44*). The size does not include the handle because the handle is generally separate from the file or rasp. For most sharpening jobs, files and rasps range from 4 inches to 14 inches in length.

Choose a file or rasp with a shape that fits the area you are filing. Files and rasps are available in round, square, flat, half-round, and triangular shapes. For filing large concave (curved inward) or flat surfaces, you might use a half-round shape. For filing small curves or for enlarging and smoothing holes, you might use a round shape with a tapered end, called a rat-tail file. For filing angles, you might use a triangular file.

There is a specific type of file for each of the common soft metals, hard metals, plastics, and wood. In general, the teeth of files for soft materials are very sharp and widely spaced. Those for hard materials are less sharp, and closer together. The shape of the teeth also varies depending on the material to be worked.

If you use a file designed for soft material on hard material, the teeth will quickly chip and dull. If you use a file designed for hard material on soft material, the teeth will clog.

Most files are sold without a handle. You can use a single handle for different files. The sharp metal point at the end of the file, the **tang**, fits into the handle. You can tighten the handle to prevent the tang from coming loose.

Files are classified by the cut of their teeth. File classifications include the following:

- Single-cut and double-cut
- Rasp-cut
- Curved-tooth

Table 1 lists types of files and some uses for each. Rasps are also classified by the size of their teeth: coarse, medium, and fine.

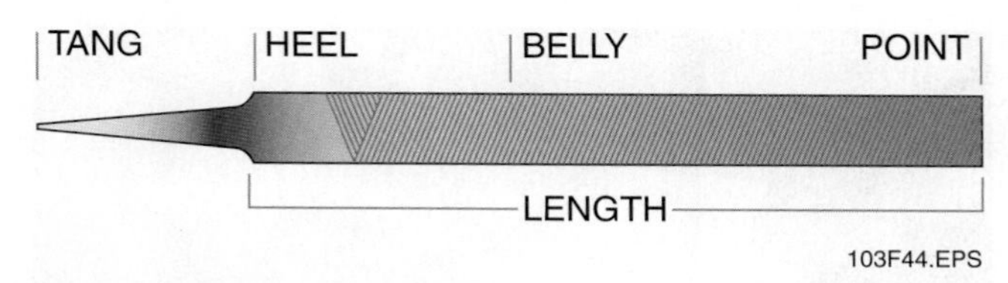

Figure 44 Parts of a file.

17.1.0 How to Use a File

Trying to use a file the wrong way is inefficient and can be frustrating. Refer to *Figure 45* and follow these steps to use a file properly:

Step 1 Mount the work you are filing in a vise at about elbow height.

Step 2 Do not lean directly over your work. Stand back from the vise a little with your feet about 24 inches apart, the right foot ahead of the left. (If you are left-handed, put your left foot ahead of the right.)

Step 3 Hold the file handle with one hand and the tip of the blade with your other hand.

Step 4 For average work, hold the tip with your thumb on top of the blade and your first two fingers under it. For heavy work, use a full-hand grip on the tip.

Step 5 Apply pressure only on the forward stroke.

Step 6 Raise the file from the work on the return stroke to keep from damaging the file.

Step 7 Keep the file flat on the work. Clean it by tapping lightly at the end of each stroke.

Table 1 Types and Uses of Files

Type	Description	Uses
Rasp-cut file	The teeth are individually cut; they are not connected to each other.	Gives a very rough surface. Used mostly on aluminum, lead, and other soft metals to remove waste materials. Also used on wood.
Single-cut file	Has a single set of straight-edged teeth running across the file at an angle.	Used to sharpen edges, such as rotary mower blades.
Double-cut file	Two sets of teeth crisscross each other. Types are bastard (roughest cut), second cut, and smooth.	Used for fast cutting.

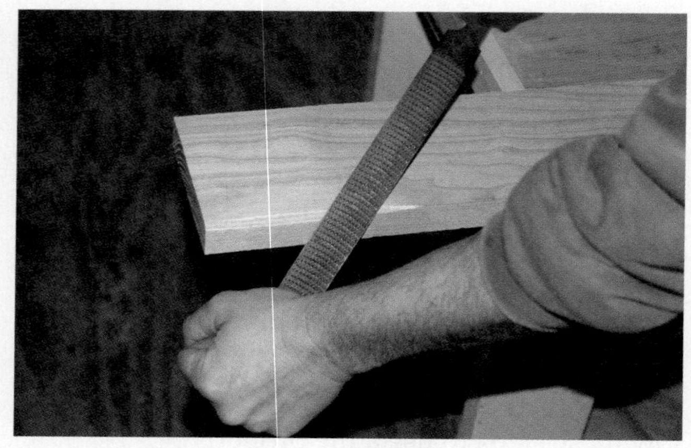

103F45.EPS

Figure 45 Proper use of a file.

17.2.0 Safety and Maintenance

Files will become worthless without proper maintenance. Here are some guidelines for the use and maintenance of files:

- Use the correct file for the material being worked.
- Always put a handle on a file before using it—most files have handle attachments.
- After you have used the file, brush the filings from between the teeth with a file card (*Figure 46*), pushing in the same direction as the line of the teeth.
- Store files in a dry place and keep them separated so that they won't chip or damage each other.
- Don't let the material vibrate in the vise, because it dulls the file teeth.

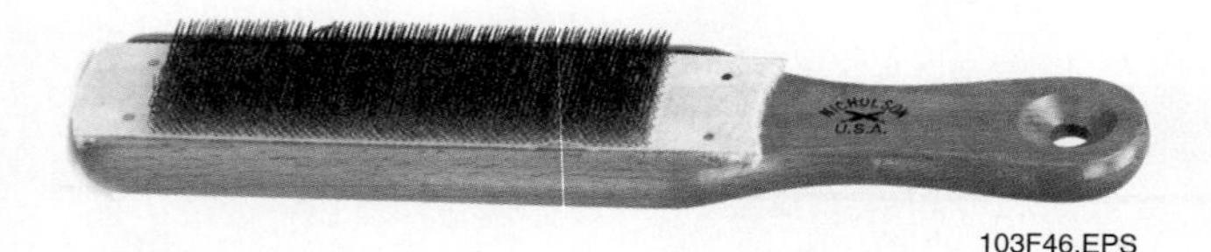

103F46.EPS

Figure 46 File card.

18.0.0 Clamps

There are many types and sizes of clamps, each designed to satisfy a different holding requirement (*Figure 47*). Clamps are sized by the maximum opening of the jaw. The depth (or throat) of the clamp determines how far from the edge of the work the clamp can be placed. The following are common types of clamps:

- *C-clamp* – This multipurpose clamp has a C-shaped frame. The clamp has a metal shoe at the end of a screw. Using a T-bar, you tighten the clamp so that it holds material between the metal jaw of the frame and the shoe.
- *Locking C-clamp* – This clamp works like locking pliers. A knob in the handle controls the width and tension of the jaws. You close the handles to lock the clamp and release the clamp by pressing the lever to open the jaws.
- *Spring clamp* – You use your hand to open the spring-operated clamp. When you release the handles, the spring holds the clamp tightly shut, applying even pressure to the material. The jaws are usually made of steel, some with plastic coating to protect the material's surface against scarring.
- *Bar clamp* – A rectangular piece of steel or aluminum is the spine of the bar clamp. It has a fixed jaw at one end and a sliding jaw (tail slide) with a spring-locking device that moves along the bar. Position the fixed jaw against the object you want to hold and then move the sliding jaw into place. The screw set is tightened as with a C-clamp.

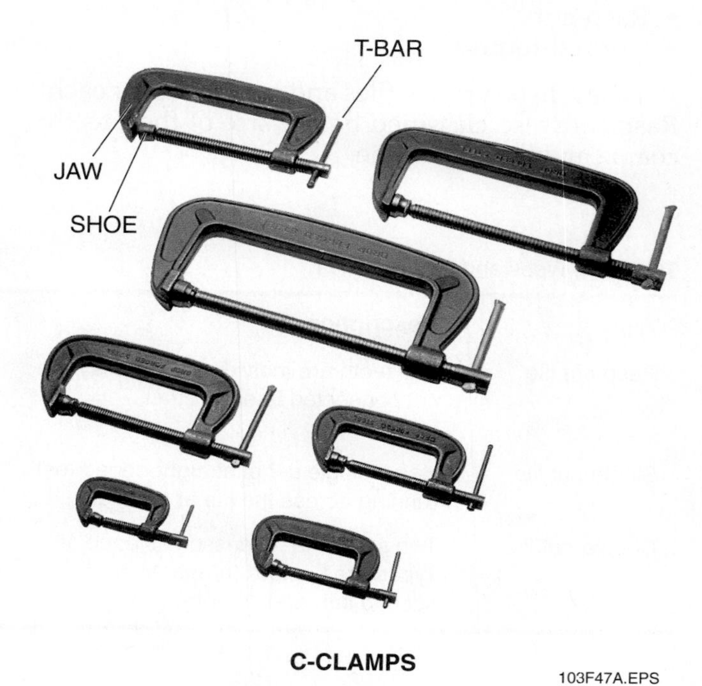

C-CLAMPS

103F47A.EPS

Figure 47 Types of clamps. (1 of 2)

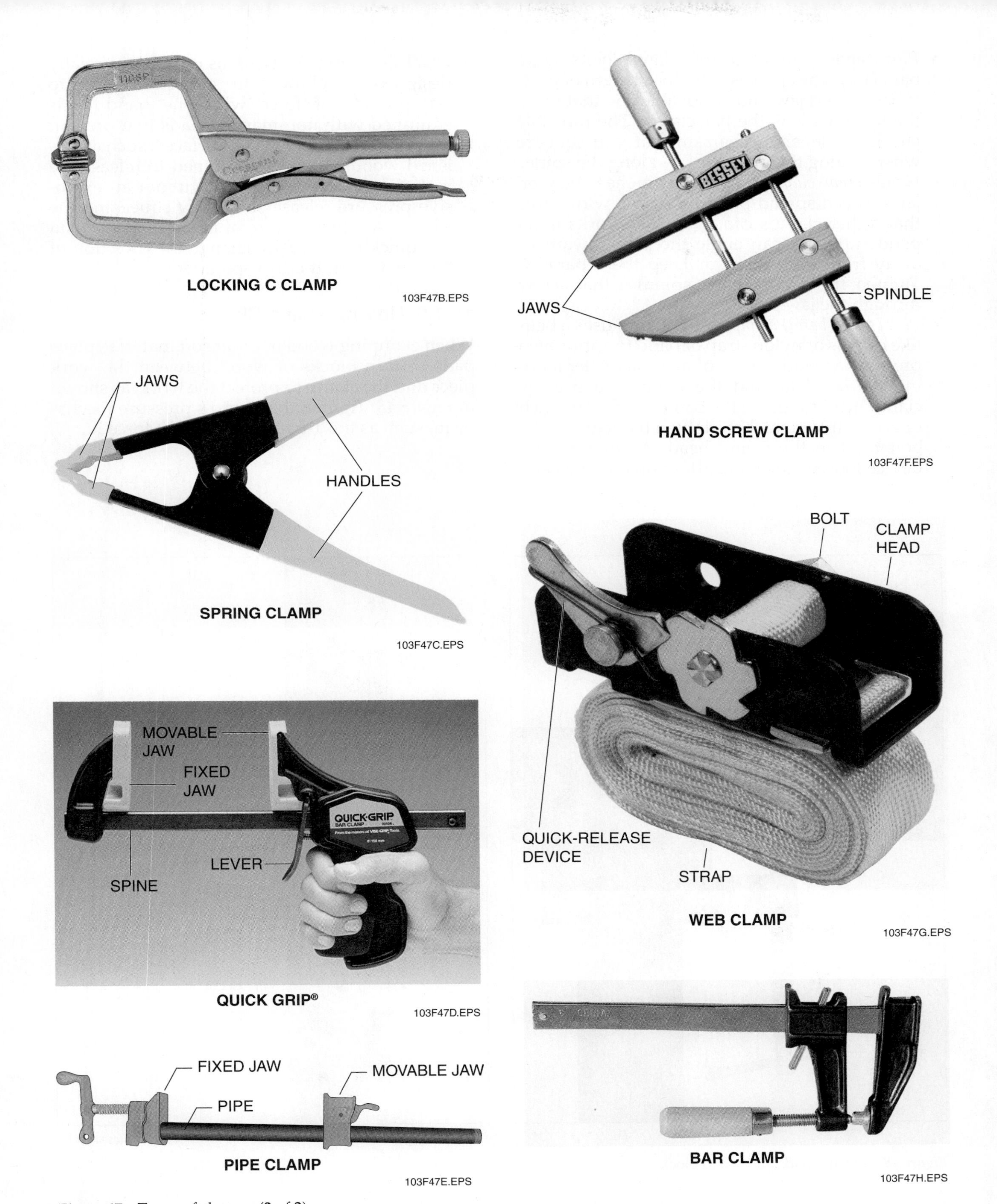

Figure 47 Types of clamps. (2 of 2)

- *Pipe clamp* – Although this clamp looks like a bar clamp, the spine is actually a length of pipe. It has a fixed jaw and a movable jaw that work the same way as the bar clamp. The movable jaw has a lever mechanism that you squeeze when sliding the movable jaw along the spine.
- *Hand-screw clamp* – This clamp has wooden jaws. It can spread pressure over a wider area than other clamps can. Each jaw works independently. You can angle the jaws toward or away from each other or keep them parallel. Tighten the clamp using spindles that screw through the jaws.
- *Web (strap, band) clamp* – This clamp uses a belt-like canvas or nylon strap or band to apply even pressure around a piece of material. After looping the band around the work, you use the clamp head to secure the band. Using a wrench or screwdriver, ratchet (tighten by degrees) the bolt tight in the clamp head. A quick-release device loosens the band after you are finished.
- *Quick Grip® bar clamp* – This clamp is specially designed to allow you to squeeze up to 600 pounds of force with one hand. It is equipped with non-marring pads that prevent it from damaging delicate surfaces, such as finished woods. It is also designed to release the material quickly and easily without an explosive pressure release simply by squeezing the trigger. Another feature of this tool is that you can quickly and easily change the direction of the jaw to turn it into a spreader bar.

18.1.0 How to Use a Clamp

When clamping wood or other soft material, place pads or thin blocks of wood between the work piece and the clamp to protect the work, as shown in *Figure 48*. Tighten the clamp's pressure mechanism, such as the T-bar handle. Don't force it.

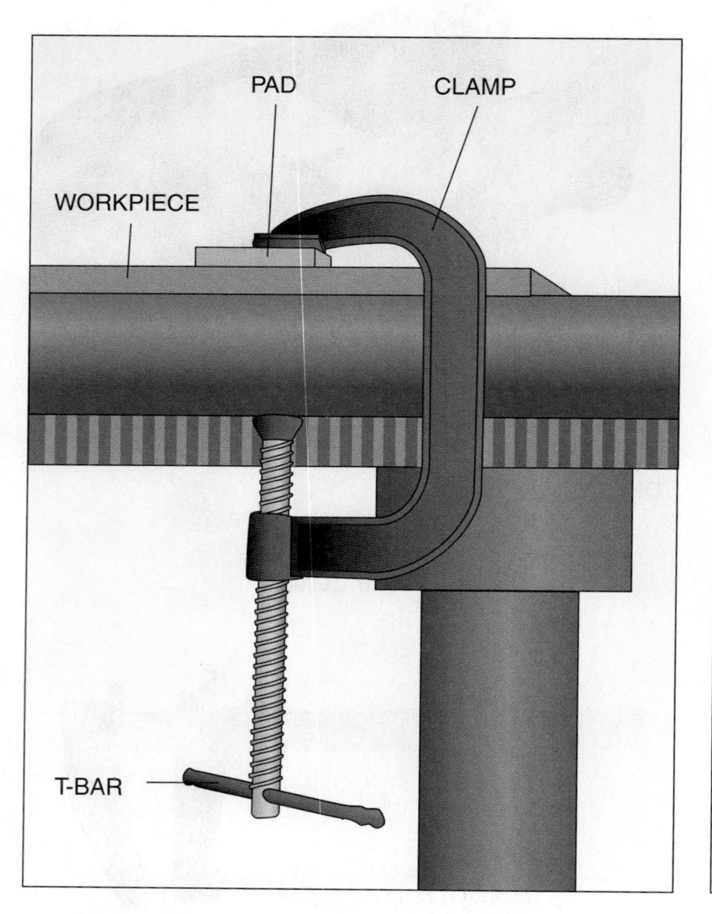

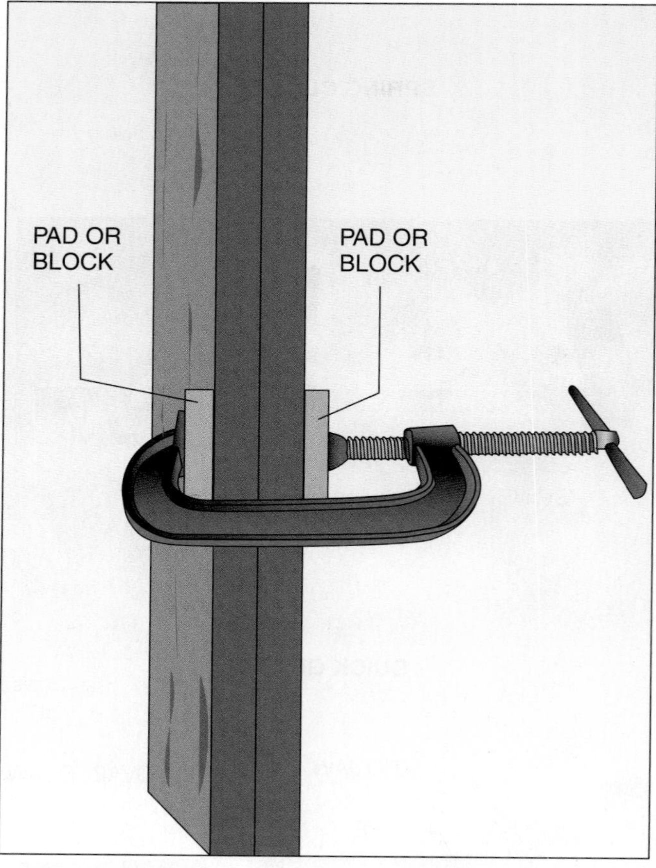

103F48.EPS

Figure 48 Placing pads and wood blocks.

18.2.0 Safety and Maintenance

Here are some guidelines to remember when using clamps:

- Store clamps by clamping them to a rack.
- Use pads or thin wood blocks when clamping wood or other soft materials.
- Discard clamps with bent frames.
- Clean and oil threads.
- Check the shoe at the end of the screw to make sure it turns freely.
- Don't use a clamp for hoisting (pulling up) work.
- To maintain proper control, don't use pliers or pipe on the handle of a clamp.

> **CAUTION**
>
> When tightening a clamp, do not use pliers or a section of pipe on the handle to extend your grip or gain more leverage. Doing so means you will have less control over the clamp's tightening mechanism.

- Don't overtighten clamps.

> **NOTE**
>
> If you are clamping work that has been glued, do not tighten the clamps so much that all the glue is squeezed out of the joint.

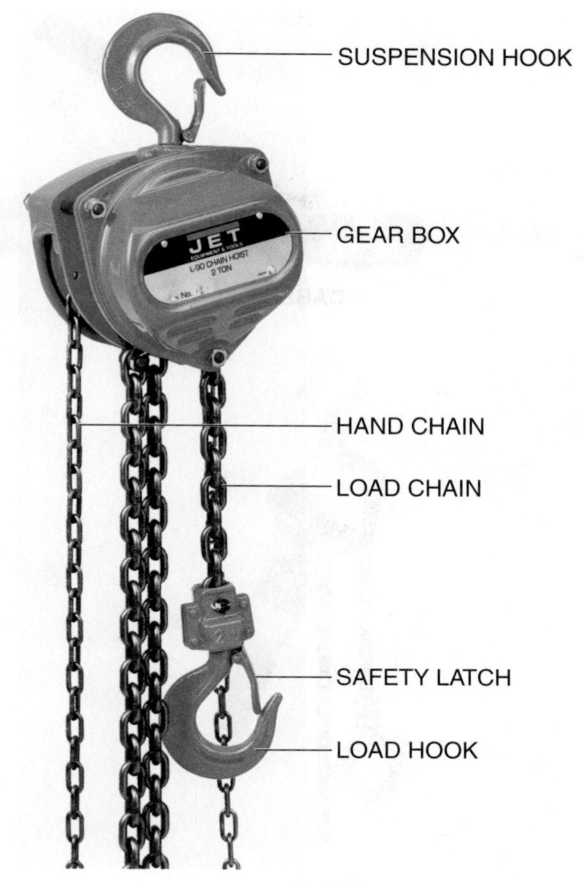

Figure 49 Parts of a manual chain fall.

19.0.0 Chain Falls and Come-Alongs

Chain falls and come-alongs are used to move heavy loads safely. A chain fall, also called a chain block or chain hoist, is a tackle device fitted with an endless chain used for hoisting heavy loads by hand. It is usually suspended from an overhead track. A come-along is used to move loads horizontally over the ground for short distances.

19.1.0 Chain Falls

The chain fall (*Figure 49*) has an automatic brake that holds the load after it is lifted. As the load is lifted, a screw forces fiber discs together to keep the load from slipping. The brake pressure increases as the loads get heavier. The brake holds the load until the lowering chain is pulled. Manual chain falls are operated by hand. Electrical chain falls are operated from an electrical control box.

The suspension hook is a steel hook used to hang the chain fall. It is one size larger than the load hook. The gear box contains the gears that provide lifting power. The hand chain is a continuous chain used to operate the gearbox. The load chain is attached to the load hook and used to lift loads. A safety latch prevents the load from slipping off the load hook, which is attached to the load.

19.2.0 Come-Alongs

Come-alongs, also called cable pullers, use a ratchet handle to position loads or to move heavy loads horizontally over short distances (*Figure 50*). They can support loads from 1 to 6 tons. Some come-alongs use a chain for moving their loads; others use wire ropes.

When using a chain come-along, use the ratchet handle to take up the chain and the ratchet release to allow the chain to be pulled out. You can also use the fast-wind handle to take up or let out slack in the chain without using the ratchet handle.

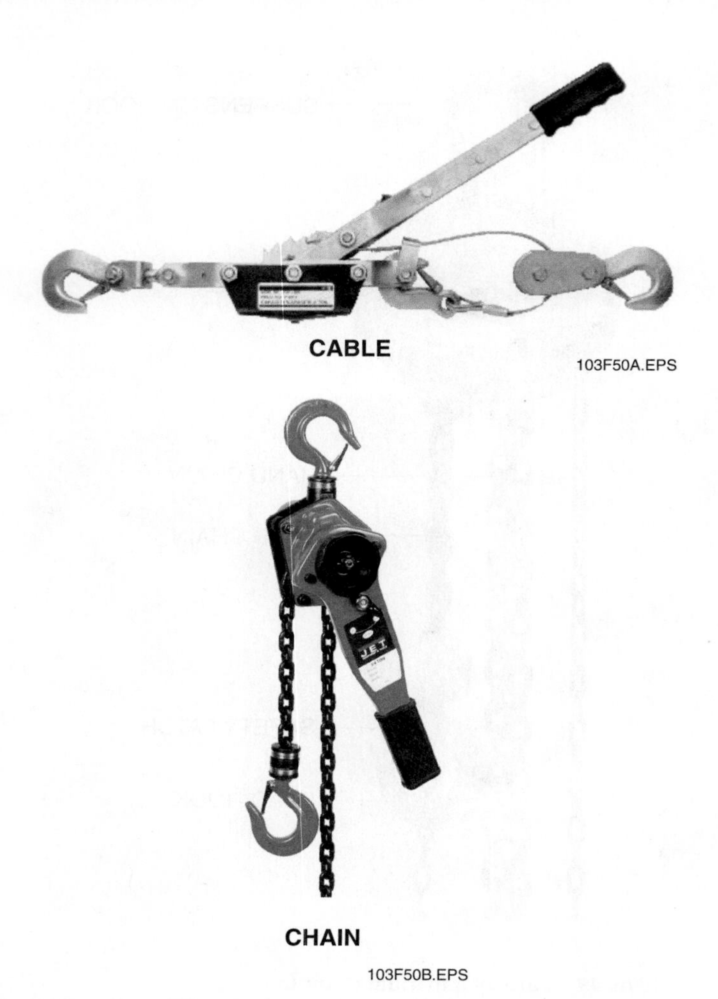

Figure 50 Come-alongs.

WARNING!

Never use a cable come-along for vertical overhead lifting. Use a cable come-along only to move loads horizontally over the ground for short distances. Cable come-alongs are not equipped with the safety features, such as a pawl to check the ratchet's motion, that would ensure the safety of anyone underneath a load.

19.3.0 Safety and Maintenance

Here are the guidelines for maintaining and safely using chain falls and come-alongs:

- Follow the manufacturer's recommendations for lubricating the chain fall or come-along.
- Inspect a chain fall and come-along for wear before each use.
- Try out a chain fall or come-along on a small load first.
- Have a qualified person ensure that the support rigging is strong enough to handle the load.
- Don't get lubricant on the clutches.
- Don't ever stand under a load.
- Don't put your hands near pinch-points on the chain.

20.0.0 SHOVELS

Shovels are used by many different construction trades. An electrician running underground wiring may dig a trench. A concrete mason may dig footers for a foundation. A carpenter may clear dirt from an area for concrete form-building. A plumber may dig a ditch to lay pipe. A welder may use a shovel to clean up scrap metal and slag after the job is finished.

There are three basic shapes of shovel blades: round, square, and spade (*Figure 51*). Use a round-bladed shovel to dig holes or remove large amounts of soil. Use a square-bladed shovel to move gravel or clean up construction debris. Use a spade to move large amounts of soil or dig trenches that need smooth, straight sides.

Shovels can have wooden or fiberglass handles. They generally come in two lengths. A long handle is usually 47 to 48 inches long. A short handle is usually 27 inches long.

Figure 51 Shapes of shovel blades.

20.1.0 How to Use a Shovel

Refer to *Figure 52*, and follow these steps to shovel properly and safely:

Begin by selecting the type of shovel that is best for the job. For a round shovel or spade:

Step 1 Place the tip of the shovel blade or spade at the point where you will begin digging or removing soil.

Step 2 With your foot balanced on the turned step (ridge), press down and cut into the soil with the blade.

For a square shovel:

Step 1 Place the leading edge of the shovel blade against the gravel or construction debris and push until the shovel is loaded.

20.2.0 Safety and Maintenance

Here are some guidelines for working safely with shovels:

- Always check the handle before using a shovel. There should be no cracks or splits.
- Use appropriate personal protective equipment when digging, trenching, or clearing debris. Wear steel-toed boots to protect your feet from dropped materials.
- Don't let dirt or debris build up on the blade. Always rinse off the shovel blade after using it.

ROUND SHOVEL OR SPADE

SQUARE SHOVEL

103F52.EPS

Figure 52 Proper use of a shovel.

21.0.0 PICK

A pick is a swinging tool similar to an ax (*Figure 53*). A pick consists of a wooden handle that is 36- to 45-inches in length, and a forged steel head that can weigh between 2 and 3 pounds. Depending on the size and strength of the pick, it can be used to break hardened or rocky soil, to level out stones and pavers, to chop tree roots and loosen soil, and to break up stones and concrete. Long-handled picks (45 inches) are used for tasks that require a normal amount of swing force, such as that used for digging a hole. Short-handled picks are used when a maximum amount of swinging force is required, such as that used for breaking concrete.

103F53.EPS

Figure 53 Pick.

21.1.0 How to Use a Pick

Follow these steps to use a pick safely and correctly:

Step 1 Select the correct pick for your height and strength.

Step 2 Place one hand at the end of the handle and your dominant hand about two-thirds of the way up the handle.

Step 3 Depending on the task, use one of the two following swinging motions.

For short-handled, hard strikes: Raise the pick up and over your head, like an axe, then rapidly bend your back to plunge the tool into the ground.

For long-handled, normal strikes: Raise the pick up to chest height and then swing it back toward the ground, using the weight of the tool head and the leverage of the long handle to produce the strike.

21.2.0 Safety and Maintenance

Here are some guidelines for working safely with picks:

- Always check to ensure that the head is fixed firmly to the handle.
- Make sure that there are no other workers in your swing path before beginning the work.
- Always use a pick that is of the appropriate length and weight for your size.
- Only use maximum force swings (over your head) when necessary, because they put more strain on your back and shoulders than do normal swings (chest height).
- Be sure to wear appropriate eye protection.

Review Questions

1. The safest hammers are those with heads that are _____.
   a. welded and alloyed
   b. cast steel and chiseled
   c. chiseled and drop forged
   d. alloy and drop-forged steel

2. A chisel bar can be used to _____.
   a. pry apart steel beams
   b. split and rip apart pieces of wood
   c. break apart concrete
   d. make ridges in wood beams

3. For safety's sake, industrial screwdriver blades are made of _____.
   a. tempered steel
   b. Torx®
   c. clutch-driven steel
   d. fiberglass

4. Pliers should not be used on a nut or bolt because _____.
   a. they will round off the edges of the hex head
   b. they are not strong enough
   c. they are designed only for tightening
   d. their jaws will not open wide enough

5. Using an adjustable wrench can save time when you are working with _____.
   a. nuts and bolts that are all the same size
   b. stripped heads
   c. different sizes of nuts and bolts
   d. nails and plywood

6. Torque wrenches are used when you need to install _____ that must be tightened in sequence.
   a. torques
   b. clean threads
   c. ratchets
   d. fasteners

7. An advantage of a laser measuring tool is that _____.
   a. it can be stored in your toolbox without breaking
   b. it can take longer measurements
   c. it has a large accuracy window
   d. it is cheap to purchase

8. When determining whether a surface is level, you check the _____.
   a. vertical surface
   b. spirit
   c. horizontal surface
   d. amount of bubbles

9. The try square is a fixed _____ angle.
   a. 45-degree
   b. 180-degree
   c. 90-degree
   d. 360-degree

10. When something is plumb, it is _____.
    a. vertically straight
    b. horizontally level
    c. at a 30-degree angle
    d. bobbed

11. Once you have pushed the utility knife blade out, always _____.
    a. lock the blade in the closed position
    b. lock the blade in the open position
    c. scuff the blade
    d. sharpen the blade

12. Files have slanting rows of teeth and rasps have _____ teeth.
    a. smooth
    b. individual
    c. coarse
    d. wire

13. A hand-screw clamp has _____ jaws.
    a. metal
    b. nylon
    c. wooden
    d. fiberglass

14. Chain falls are used to _____.
    a. transport light loads safely
    b. supplement come-along pulls
    c. rig light loads safely
    d. safely move heavy loads vertically

15. A spade is used to _____.
    a. dig holes or remove large amounts of soil
    b. move gravel or clean up construction debris
    c. tamp down soil along a building's foundation
    d. move large amounts of soil or dig trenches with straight sides

SUMMARY

As a craft professional, your tools are essential to your success. In this module, you learned to identify and work with many of the basic hand tools commonly used in construction. Learning to properly use and maintain your tools is an essential skill for every craftworker. Although you may not work with all of the tools introduced in this module, you will use many of them as you progress in your career, regardless of what craft area you choose to work in.

When you use tools properly, you are working safely and efficiently. You are not only preventing accidents that can cause injuries and equipment damage, you are showing your employer that you are a responsible, safe worker.

The same pride you take in using your tools to do a job well is important when it comes to maintaining your tools. When you maintain your tools properly, they last longer, work better, and function more safely. The simple act of maintaining your tools will help you prevent accidents, make your tools last longer, and help you perform your job better. Taking the time to learn to use and maintain these tools properly now will help keep you safe and save you time and money down the road.

Trade Terms Quiz

Fill in the blank with the correct key term that you learned from your study of this module.

1. Used mainly for woodworking, the ____________ is a fixed, 90-degree angle.
2. A(n) ____________ is an L-shaped, hexagonal steel bar.
3. The ____________ has a flat face for striking and a rounded face for peening (rounding off) metal and rivets.
4. Usually 2 inches wide and 15 inches long, the ____________ has a nail slot at the end to pull nails out from tightly enclosed areas.
5. Shaped like an L, the ____________ is used to make sure wall studs and sole plates are at right angles to each other.
6. A(n) ____________ is a metal tool with a sharpened, beveled edge that is used to cut and shape wood, stone, or metal.
7. The ____________ is used to drive nails and to pull nails out of wood.
8. To ____________ is to cut on a slant at an angle that is not a right angle.
9. The ____________ has a 12-inch blade that moves through a head that is marked with 45-degree and 90-degree angle measures.
10. If you use a screwdriver incorrectly, you can damage the screwdriver or ____________ the screw head.
11. Use a(n) ____________ to turn nuts, bolts, small pipe fittings, and chrome-plated pipe fittings.
12. To fasten or align two pieces or material, you can use a(n) ____________, which is a pin that fits into a corresponding hole.
13. A(n) ____________ is a device such as a nut or bolt used to attach one material to another.
14. Use a(n) ____________ for heavy-duty dismantling of woodwork.
15. The straight sides or jaws of a wrench opening are called the ____________.
16. A(n) ____________ is a claw hammer with a slightly rounded face.
17. ____________ is a unit of measure used to describe the torque needed to tighten a large object.
18. ____________ is a unit of measure used to describe the torque needed to tighten a small object.
19. The point at which members or the edges of members are joined is called the ____________.
20. The ____________ is the cut or channel made by a saw.
21. Using a(n) ____________ can speed up your work because it has an open wrench at one end and a box-end at the other.
22. Use a(n) ____________ to determine if a surface is exactly horizontal.
23. You make a(n) ____________ by fastening together usually perpendicular parts with the ends cut at an angle.
24. A(n) ____________ is a tool used to remove nails.
25. A(n) ____________ has an opening at each end that determines its size.
26. To reduce stress in a weld, use a special type of hammer for ____________ the joint as it cools.
27. Used for marking, checking, and measuring, a(n) ____________ comes in several types: carpenter's, rafter angle, try, and combination.
28. A(n) ____________ has serrated teeth on both jaws for gripping power.
29. ____________, which is the turning force applied to an object, is measured in inch-pounds or foot-pounds.
30. The ____________ is used to pull nails that have been driven flush with the surface of the wood or slightly below it.
31. A box-end wrench has 6 or 12 ____________.
32. The ____________, a nonadjustable wrench, forms a continuous circle around the head of a fastener.

33. To indent metal before you drill a hole, to drive pins, or to align holes in two parts that are mates, use a(n) ____________.

34. Also called a speed square or magic square, the ____________ is a combination protractor, try square, and framing square.

35. Using pliers on nuts or bolt heads may ____________ the edges of the hex head and cause wrenches to no longer fit properly.

36. The ____________ is a tool with a claw at each end, commonly used to pull nails.

37. A special type of adjustable wrench, ____________ are scissor-shaped tools with jaws.

38. The ____________ fits into a wooden file handle.

39. Some tools are made of ____________ steel so that they resist wear and do not bend or break.

40. A(n) ____________ piece of lumber is one that has had its surface made smooth.

41. If a surface is ____________, it is exactly vertical.

42. A(n) ____________ is a piece that projects out of wood so it can be placed into a hole or groove to form a joint.

43. A(n) ____________ is a joint that has been created by heating pieces of metal.

44. A(n) ____________ is a nonadjustable wrench with an enclosed, circular opening designed to lock onto the fastener when the wrench is struck.

Trade Terms

Adjustable end wrench
Ball peen hammer
Bell-faced hammer
Bevel
Box-end wrench
Carpenter's square
Cat's paw
Chisel
Chisel bar
Claw hammer
Combination square
Combination wrench
Dowel
Fastener
Flat bar
Flats
Foot-pounds
Hex key wrench
Inch-pounds
Joint
Kerf
Level
Miter joint
Nail puller
Open-end wrench
Peening
Pipe wrench
Planed
Pliers
Plumb
Points
Punch
Rafter angle square
Ripping bar
Round off
Square
Striking (or slugging) wrench
Strip
Tang
Tempered
Tenon
Torque
Try square
Weld

Cornerstone of Craftsmanship

Larry Thurston

Gulf States, Inc. Trainer
Midland, Michigan

How did you choose a career in the field?
I was very fortunate to have had the influence of my father. He was a boilermaker/welder at the Dow Chemical Company. There were constant projects going on in his shop at home that we were always involved in. In 1966, I was hired by Dow Chemical as an operator. In 1969, I was awarded an apprenticeship as a pipefitter, working and training in the Capital Construction Projects division. From there it became a career in construction and maintenance. Having always been interested in mechanical projects, it was a great start for me.

What types of training have you been through?
I am a journeyman insulator, pipefitter, and welder and have been through apprenticeships in all three trades—not to mention the training I received from some great journeymen.

What kinds of work have you done in your career?
I started out as an operator learning how processes worked, then I started my trades. Over the years, I have worked in several aspects of the construction and maintenance field, including design, take-off, fabrication, and installation. I retired from Dow Chemical in 1998. I decided to return to work and was employed at Gulf States Inc., in Midland, Michigan as a pipefitter/welder. There I was asked if I would start a training program for tradesmen. Training in several trades has been both a challenge and learning experience. Passing on to younger journeymen what I was taught over the years has been more than fun and is very rewarding. I'm also an instructor at Associated Builders and Contractors. There I am able to teach apprentices all levels of pipefitting.

Tell us about your present job and what you like about it.
Having spent many years as a tradesman and learning my crafts, it is rewarding to give others the knowledge that was given to me. Training, being a never-ending opportunity as new technology is presented, gives me a chance to coach other craftworkers. Working for a great company, which has a vision to train and add to the skill levels of craft workers, gives me the chance to not only provide better-trained people for the company, but also, in turn, provide tradesmen more value to themselves.

What factors have contributed most to your success?
There have been many times I have thought about this and I always have the same answer. I worked for a company that saw value in training its tradesmen from within the workforce it employed. I was rewarded with three apprenticeships. The power of learning cannot be replaced.

What advice would you give to those new to the field?
First would be attitude. A good attitude makes people want to be around you and that makes your learning and working experience much easier. Second is work ethic. Be at work every day on time. A good attendance record is very important to the employer. Knowing you can be counted on every day makes a more valuable employee. Learn your craft well. Knowledge cannot be replaced or taken away and, as a result, helps you gain respect from your peers as a tradesman. Always work safely so you and your fellow workers go home with a feeling of accomplishment, not a feeling of pain. A can-do attitude will take a person a long way. The construction field is wide open for advancement to those who learn their trade well and have a willingness to lead through example.

Tell us some interesting career-related facts or accomplishments:
I have made a very good living for many years for myself and my family. I believe the tradesman of the future will do even better because of the fewer amount of people entering the trades.

Trade Terms Introduced in This Module

Adjustable end wrench: A smooth-jawed adjustable wrench used for turning nuts, bolts, and pipe fittings. Often referred to as a Crescent® wrench.

Ball peen hammer: A hammer with a flat face that is used to strike cold chisels and punches. The rounded end—the peen—is used to bend and shape soft metal.

Bell-faced hammer: A claw hammer with a slightly rounded, or convex, face.

Bevel: To cut on a slant at an angle that is not a right angle (90 degrees). The angle or inclination of a line or surface that meets another at any angle but 90 degrees.

Box-end wrench: A wrench, usually double-ended, that has a closed socket that fits over the head of a bolt.

Carpenter's square: A flat, steel square commonly used in carpentry.

Cat's paw: A straight steel rod with a curved claw at one end that is used to pull nails that have been driven flush with the surface of the wood or slightly below it.

Chisel: A metal tool with a sharpened, beveled edge used to cut and shape wood, stone, or metal.

Chisel bar: A tool with a claw at each end, commonly used to pull nails.

Claw hammer: A hammer with a flat striking face. The other end of the head is curved and divided into two claws to remove nails.

Combination square: An adjustable carpenter's tool consisting of a steel rule that slides through an adjustable head.

Combination wrench: A wrench with an open end and a closed end.

Dowel: A pin, usually round, that fits into a corresponding hole to fasten or align two pieces.

Fastener: A device such as a bolt, clasp, hook, or lock used to attach or secure one material to another.

Flat bar: A prying tool with a nail slot at the end to pull nails out in tightly enclosed areas. It can also be used as a small pry bar.

Flats: The straight sides or jaws of a wrench opening. Also, the sides on a nut or bolt head.

Foot-pounds: Unit of measure used to describe the amount of pressure exerted (torque) to tighten a large object.

Hex key wrench: A hexagonal steel bar that is bent to form a right angle. Often referred to as an Allen® wrench.

Inch-pounds: Unit of measure used to describe the amount of pressure exerted (torque) to tighten a small object.

Joint: The point where members or the edges of members are joined. The types of welding joints are butt joint, corner joint, and T-joint.

Kerf: A cut or channel made by a saw.

Level: Perfectly horizontal; completely flat; also, a tool used to determine if an object is level.

Miter joint: A joint made by fastening together usually perpendicular parts with the ends cut at an angle.

Nail puller: A tool used to remove nails.

Open-end wrench: A nonadjustable wrench with an opening at each end that determines the size of the wrench.

Peening: The process of bending, shaping, or cutting material by striking it with a tool.

Pipe wrench: A wrench for gripping and turning a pipe or pipe-shaped object; it tightens when turned in one direction.

Planed: Describing a surface made smooth by using a tool called a plane.

Pliers: A scissor-shaped type of adjustable wrench equipped with jaws and teeth to grip objects.

Plumb: Perfectly vertical; the surface is at a right angle (90 degrees) to the horizon or floor and does not bow out at the top or bottom.

Points: Teeth on the gripping part of a wrench. Also refers to the number of teeth per inch on a handsaw.

Punch: A steel tool used to indent metal.

Rafter angle square: A type of carpenter's square made of cast aluminum that combines a protractor, try square, and framing square.

Ripping bar: A tool used for heavy-duty dismantling of woodwork, such as tearing apart building frames or concrete forms.

Round off: To smooth out threads or edges on a screw or nut.

Square: Exactly adjusted; any piece of material sawed or cut to be rectangular with equal dimensions on all sides; a tool used to check angles.

Striking (or slugging) wrench: A non-adjustable wrench with an enclosed, circular opening designed to lock on to the fastener when the wrench is struck.

Strip: To damage the threads on a nut or bolt.

Tang: Metal handle-end of a file. The tang fits into a wooden or plastic file handle.

Tempered: Treated with heat to create or restore hardness in steel.

Tenon: A piece that projects out of wood or another material for the purpose of being placed into a hole or groove to form a joint.

Torque: The turning or twisting force applied to an object, such as a nut, bolt, or screw, using a socket wrench or screwdriver to tighten it. Torque is measured in inch-pounds or foot-pounds.

Try square: A square whose legs are fixed at a right angle.

Weld: To heat or fuse two or more pieces of metal so that the finished piece is as strong as the original; a welded joint.

Additional Resources

This module is intended to present thorough resources for task training. The following reference works are suggested for further study. These are optional materials for continued education rather than for task training.

Field Safety, 2003. NCCER. Upper Saddle River, NJ: Prentice Hall.

Hand Tolls & Techniques, 1999. Minneapolis, MN: Handyman Club of America.

The Long and Short of It: How to Take Measurements. Video. Charleston, WV: Cambridge Vocational & Technical, 800-468-4227.

National Institute for Occupational Safety and Health (NIOSH), DHHS Publication No. 2004-164, "Easy Ergonomics: A Guide to Selecting Non-Powered Hand Tools." http://www.cdc.gov/niosh/docs/2004-164/pdfs/2004-164.pdf

Reader's Digest Book of Skills and Tools, 1996. Pleasantville, NY: Reader's Digest.

CONTREN® LEARNING SERIES – USER UPDATE

NCCER makes every effort to keep these textbooks up-to-date and free of technical errors. We appreciate your help in this process. If you have an idea for improving this textbook, or if you find an error, a typographical mistake, or an inaccuracy in NCCER's Contren® textbooks, please write us, using this form or a photocopy. Be sure to include the exact module number, page number, a detailed description, and the correction, if applicable. Your input will be brought to the attention of the Technical Review Committee. Thank you for your assistance.

Instructors – If you found that additional materials were necessary in order to teach this module effectively, please let us know so that we may include them in the Equipment/Materials list in the Annotated Instructor's Guide.

Write: Product Development and Revision
National Center for Construction Education and Research
3600 NW 43rd St., Bldg. G, Gainesville, FL 32606

Fax: 352-334-0932

E-mail: curriculum@nccer.org

Craft ____________ Module Name ____________

Copyright Date ____________ Module Number ____________ Page Number(s) ____________

Description ____________

(Optional) Correction ____________

(Optional) Your Name and Address ____________

Introduction to Power Tools

00104-09

CORE CURRICULUM

00109-09 Introduction to Materials Handling

00108-09 Basic Employability Skills

00107-09 Basic Communication Skills

00106-09 Basic Rigging

00105-09 Introduction to Construction Drawings

00104-09 Introduction to Power Tools

00103-09 Introduction to Hand Tools

00102-09 Introduction to Construction Math

00101-09 Basic Safety

This course map shows all of the modules in the *Core Curriculum: Introductory Craft Skills*. The suggested training order begins at the bottom and proceeds up. Skill levels increase as you advance on the course map. The local Training Program Sponsor may adjust the training order.

Note that Module 00106-09, *Basic Rigging,* is an elective. It is not a requirement for level completion, but it may be included as part of your training program.

Objectives

When you have completed this module, you will be able to do the following:

1. Identify power tools commonly used in the construction trades.
2. Use power tools safely.
3. Explain how to maintain power tools properly.

Trade Terms

Abrasive
AC (alternating current)
Auger
Booster
Carbide
Chuck
Chuck key
Countersink
DC (direct current)
Electric tools
Ferromagnetic
Grit
Ground fault circuit interrupter (GFCI)
Ground fault protection
Hazardous materials
Hydraulic tools
Masonry
Pneumatic tools
Reciprocating
Revolutions per minute (rpm)
Ring test
Shank
Trigger lock

Required Trainee Materials

Appropriate personal protective equipment.

Prerequisites

Before you begin this module, it is recommended that you successfully complete the following: *Core Curriculum*, Modules 00101-09 through 00103-09. Module 00106-09 is an elective and is not a requirement for completion of this course.

Contents

Topics to be presented in this module include:

1.0.0 Introduction . . . 4.1
2.0.0 Electric, Pneumatic, and Hydraulic Tools . . . 4.1
2.1.0 Safety . . . 4.1
3.0.0 Power Drills . . . 4.1
3.1.0 Types of Power Drills . . . 4.2
3.1.1 How to Use a Power Drill . . . 4.2
3.1.2 Safety and Maintenance . . . 4.4
3.2.0 Cordless Drills . . . 4.5
3.2.1 How to Use a Cordless Drill with a Keyless Chuck . . . 4.6
3.2.2 Safety and Maintenance . . . 4.6
3.3.0 Hammer Drills . . . 4.6
3.3.1 How to Use a Hammer Drill . . . 4.7
3.3.2 Safety and Maintenance . . . 4.7
3.4.0 Electromagnetic Drills . . . 4.7
3.4.1 How to Set Up an Electromagnetic Drill . . . 4.8
3.4.2 Safety and Maintenance . . . 4.8
3.5.0 Pneumatic Drills . . . 4.10
3.5.1 How to Use a Pneumatic Drill . . . 4.10
3.5.2 Safety and Maintenance . . . 4.11
4.0.0 Saws . . . 4.11
4.1.0 Circular Saws . . . 4.11
4.1.1 How to Use a Circular Saw . . . 4.12
4.1.2 Safety and Maintenance . . . 4.13
4.2.0 Saber Saws (Jig Saws) . . . 4.14
4.2.1 How to Use a Saber Saw . . . 4.14
4.2.2 Safety and Maintenance . . . 4.15
4.3.0 Reciprocating Saws . . . 4.15
4.3.1 How to Use a Reciprocating Saw . . . 4.16
4.3.2 Safety and Maintenance . . . 4.16
4.4.0 Portable Handheld Bandsaw . . . 4.16
4.4.1 How to Use a Portable Bandsaw . . . 4.17
4.4.2 Safety and Maintenance . . . 4.17
4.5.0 Power Miter Saw . . . 4.18
4.5.1 How to Use a Power Miter Saw . . . 4.18
4.5.2 Safety and Maintenance . . . 4.18
4.6.0 Abrasive Cutoff Saw . . . 4.19
4.6.1 How to Use an Abrasive Cutoff Saw . . . 4.19
4.6.2 Safety and Maintenance . . . 4.19
5.0.0 Grinders and Sanders . . . 4.19
5.1.0 Angle Grinders, End Grinders, and Detail Grinders . . . 4.20
5.1.1 How to Use an Angle Grinder, End Grinder, or Detail Grinder . . . 4.20
5.1.2 Safety and Maintenance . . . 4.20
5.2.0 Bench Grinders . . . 4.21
5.2.1 How to Use a Bench Grinder . . . 4.22
5.2.2 Safety and Maintenance . . . 4.22
6.0.0 Miscellaneous Power Tools . . . 4.23
6.1.0 Pneumatically Powered Nailers (Nail Guns) . . . 4.23
6.1.1 How to Use a Power Nailer . . . 4.25
6.1.2 Safety and Maintenance . . . 4.25

6.2.0 Powder-Actuated Fastening Systems 4.26
6.2.1 How to Set Up and Use a Powder-Actuated Fastening Tool . . . 4.27
6.2.2 Safety and Maintenance 4.27
6.3.0 Air Impact Wrench 4.27
6.3.1 How to Set Up and Use an Air Impact Wrench 4.28
6.3.2 Safety and Maintenance 4.28
6.4.0 Pavement Breakers 4.28
6.4.1 How to Set Up and Use a Pavement Breaker 4.28
6.4.2 Safety and Maintenance 4.28
6.5.0 Hydraulic Jack 4.30
6.5.1 How to Use a Hydraulic Jack 4.30
6.5.2 Safety and Maintenance 4.30

1.0.0 INTRODUCTION

Power tools are used in almost every construction trade to make holes; cut, smooth, and shape materials; and even demolish pavement. As a construction worker, you will probably use power tools on the job. Knowing how to identify and use power tools safely and correctly is very important. This module provides an overview of the various types of power tools and how they work. You will also learn the proper safety techniques required to operate these tools.

2.0.0 ELECTRIC, PNEUMATIC, AND HYDRAULIC TOOLS

This module introduces three kinds of power tools: electric, pneumatic, and hydraulic.

- *Electric tools* – These tools are powered by electricity. They are operated from either an **alternating current (AC)** source (such as a wall receptacle) or a **direct current (DC)** source (such as a battery). Belt sanders and circular saws are examples of electric tools.
- *Pneumatic tools* – These tools are powered by air. Electric or gasoline-powered compressors produce the air pressure. Air hammers and pneumatic nailers are examples of pneumatic tools.
- *Hydraulic tools* – These tools are powered by fluid pressure. Hand pumps or electric pumps are used to produce the fluid pressure. Pipe benders, jackhammers, and Porta-Powers® are examples of hydraulic tools.

> GOING GREEN
>
> **Green Plug®**
>
> Every day around the world, billions of devices such as power tools, cell phones, laptops, and printers use unique power converters to convert between 90V (volts) and 254V of wall power to device-specific power. This nonregulated distribution of power is inconvenient and inefficient, resulting in wasted resources and pollution. However, it also presents a very real and significant opportunity to reduce such waste and pollution.
>
> That's where Green Plug® comes in. Green Plug® is the developer of Greentalk®—a secure, digital protocol for real-time collaboration between electronic devices and their power sources. Green Plug® is developing highly efficient power adapter hubs that are able to simultaneously power multiple devices, each with its own energy demand. With Green Plug® technology, intelligent power supplies communicate with electronic devices and agree upon device power requirements for all of the devices connected to them. When these devices work in collaboration with their power supply, an extraordinary amount of monitoring, control, and optimization becomes possible.
>
> Green Plug® technology maximizes resources, minimizes solid waste, and reduces wasted energy, all of which impact the environment and the bottom line of any business or project in a positive way.

2.1.0 Safety

You must complete the *Basic Safety* module before you take this course. It is easy to hurt yourself or others if you use a power tool incorrectly or unsafely. Safety issues for each tool are covered in this module, but general safety issues—such as safety in the work area, safety equipment, and working with electricity—are covered in the *Basic Safety* module in this book. This information is vital for working with power tools.

> **WARNING!**
>
> If you have not completed the *Basic Safety* module, stop here! You must complete the *Basic Safety* module first. Also, you must wear appropriate personal protective equipment when you operate any power tool or when you are near someone else who is operating a power tool.

One of the most important rules about working with power tools is to always disconnect the power source for any tool before you replace parts such as bits, blades, or discs. Always disconnect the power source before you perform maintenance on any power tool. Never activate the **trigger lock** on any power tool.

> **WARNING!**
>
> Always be sure to read and follow the manufacturer's recommendations when using power hand tools.

3.0.0 POWER DRILLS

The power drill is used often in the construction industry. It is most commonly used to make holes by spinning drill bits into wood, metal, plastic, and other materials. However, with different attachments and accessories, the power drill can be used as a sander, polisher, screwdriver, grinder, or **countersink**— it can even be used as a saw.

3.1.0 Types of Power Drills

In this section, you will learn about various types of power drills, including the following:

- Electric drills
- Cordless drills
- Hammer drills
- Electromagnetic drills
- Pneumatic drills (air hammers)
- Electric screwdrivers

Most of these drills are similar, so you will first learn about what they have in common. Most power drills have a pistol grip with a trigger switch for controlling power (*Figure 1*).

The harder you pull on the trigger of a variable speed drill, the faster the speed. Drills also have reversing switches that allow you to back the drill bit out if it gets stuck in the material while drilling. Most drills have replaceable bits for use on different kinds of jobs (*Figure 2*). On most power drills, you can insert a screwdriver bit in place of a drill bit and use the drill as a screwdriver. Be sure to use screwdriver bits that are designed for use in a power drill.

Twist drill bits are used to drill wood and plastics at high speeds or to drill metal at a lower speed. A forstner bit is used on wood and is particularly good for boring a flat bottom hole. A paddle bit or spade bit is also used in wood. The bit size is measured by the paddle's diameter, which generally ranges from ½-inch to 1½-inch. A **masonry** bit, which has a **carbide** tip, is used in concrete, stone, slate, and ceramic. The **auger** drill bit is used for drilling wood and other soft materials, but not for drilling metal. As a rule, the point of a bit should be sharper for softer materials than for harder ones. All bits are held in the drill by the drill **chuck**. Chucks can be either keyed or keyless (*Figure 3*).

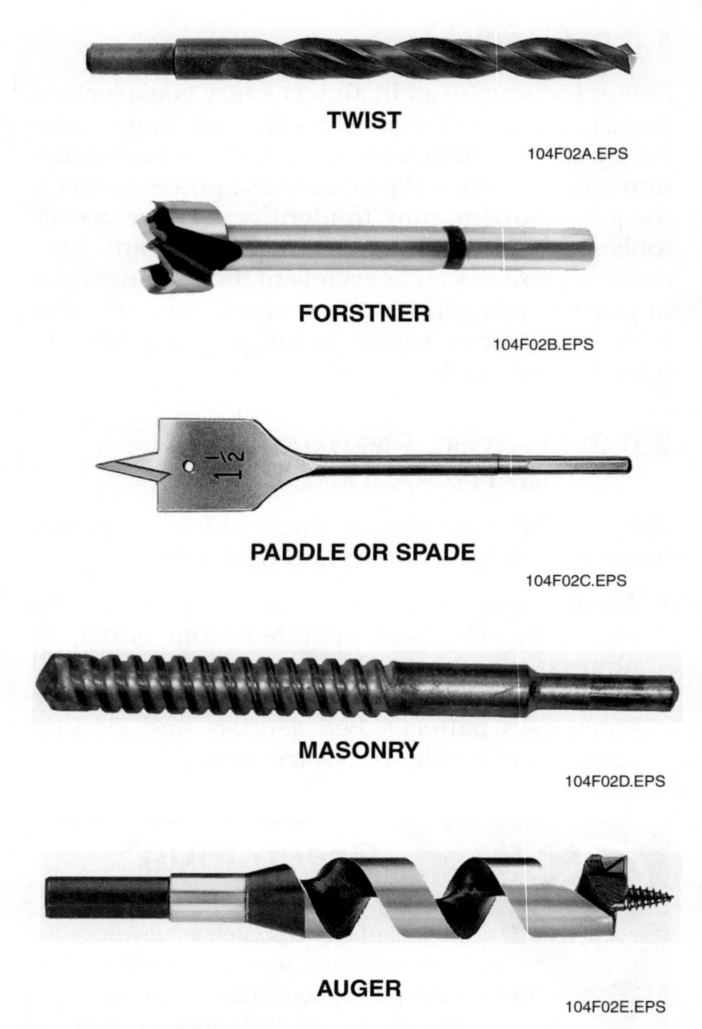

Figure 2 Drill bits.

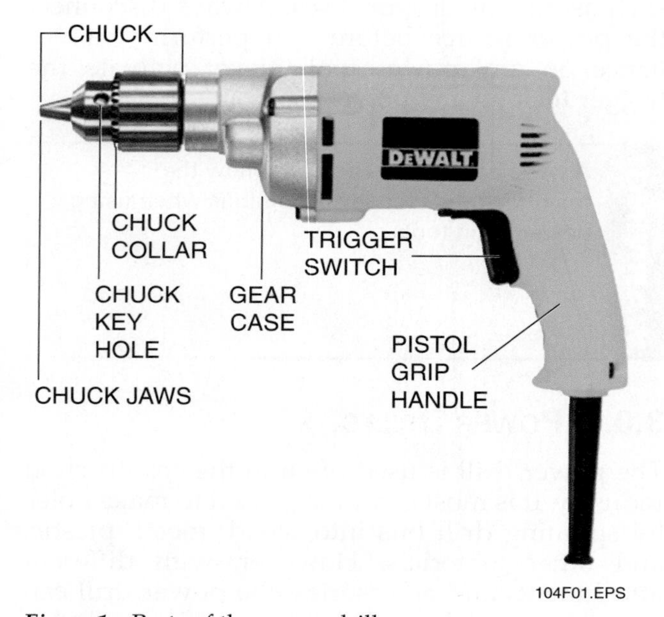

Figure 1 Parts of the power drill.

3.1.1 How to Use a Power Drill

Power drills can be dangerous if you do not use them properly. Always wear personal protective equipment, including appropriate eye, head, and hand protection.

Follow these steps to load a bit in an electric power drill:

Step 1 Disconnect the power. Open the chuck and turn it counterclockwise (to the left) until the chuck opening is large enough for you to insert the bit **shank** (*Figure 4A*). The shank is the smooth part of the bit.

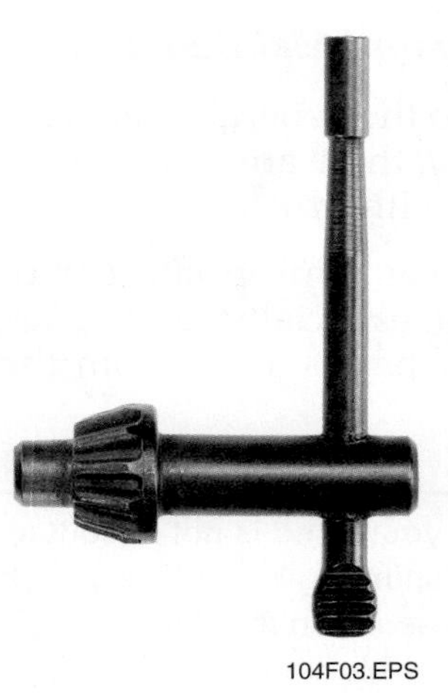

104F03.EPS

Figure 3 Chuck key.

Step 2 Insert the bit shank.

Step 3 Tighten the chuck by hand until the jaws grip the bit shank. Keep the bit centered as you tighten it. It should not be leaning to one side but should be straight in the chuck.

> **NOTE**
> Stop here if you are using a keyless chuck.

Step 4 Insert the **chuck key** (*Figure 4B*) in one of the holes on the side of the chuck. You will notice that the chuck key has a grooved ring called a gear. Make sure that the chuck key's gear meshes with the matching gears on the geared end of the chuck. In larger drills, tighten the bit by inserting the chuck key into each of the holes in the three-jawed chuck. This ensures that all the jaws close uniformly tight around the bit.

Step 5 Turn the chuck key clockwise (right) to tighten the grip on the bit.

Step 6 Remove the key from the chuck.

> **WARNING!**
> Always remember to remove the key from the chuck. Otherwise, when you start the drill, the key could fly out and injure you or a co-worker.

104F04A.EPS

(A) INSERT THE BIT SHANK INTO THE CHUCK OPENING.

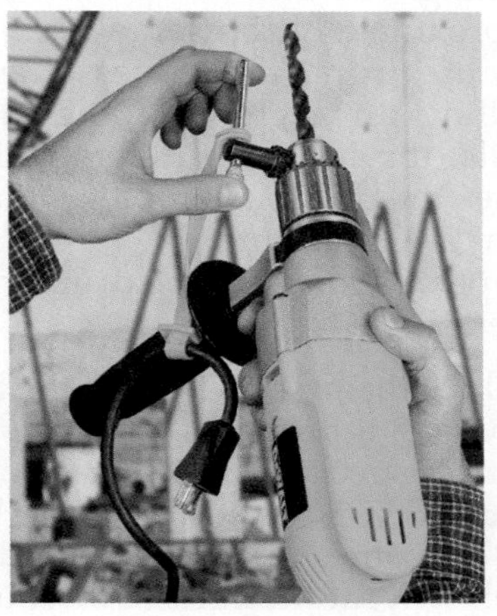

104F04B.EPS

(B) TIGHTEN WITH THE CHUCK KEY.

104F04C.EPS

(C) HOLD THE DRILL PERPENDICULAR TO THE MATERIAL AND START THE DRILL.

Figure 4 Proper drill use.

Follow these steps to use a power drill to drill a hole.

Step 1 Make a small indent exactly where you want the hole drilled.
- In wood, use a small punch to make an indent.
- In metal, first use a center punch.

Step 2 Firmly clamp or support the work that is being drilled.

Step 3 Hold the drill perpendicular (at a right angle) to the material surface and start the drill motor. Be sure the drill is rotating in the right direction (with the bit facing away from you it should be turning clockwise). Hold the drill with both hands and apply only moderate pressure when drilling. The drill motor should operate at approximately the same **revolutions per minute (rpm)** as it does when it is not drilling through anything. For power drills, the term *rpm* refers to how many times the drill bit completes one full rotation every minute. *Figure 4C* shows the proper way to hold the drill when you are operating it.

Step 4 Reduce the pressure when the bit is about to come through the other side of the work, especially when you are drilling metal. If you are still pressing hard when the bit comes out the other side, the drill itself will hit the surface of your material. This could damage or dent the metal surface. If the drill bit gets stuck in the material while you are drilling, release the trigger, use the reversing switch to change the direction of the drill, and back it gently out of the material. When you are finished backing it out, switch back to your original drilling position.

3.1.2 Safety and Maintenance

In addition to the general safety rules you learned in *Basic Safety*, there are some specific safety rules for working with drills:

- Always wear appropriate personal protective equipment, especially safety glasses.
- Keep your hands away from the drill bit and chuck.

WARNING!

Be sure your hand is not in contact with the drill bit. The spinning bit will cut your hand. Keep an even pressure on the drill to keep the drill from twisting or binding.

- To prevent an electrical shock, operate only those tools that are double-insulated electric power tools with proper **ground fault protection**. Using a **ground fault circuit interrupter (GFCI)** device protects the equipment from continued electrical current in case of a circuit fault. The GFCI monitors the current flow and opens the circuit (which stops the flow of electricity) if it detects a difference between positive and negative flow. The interruption typically takes place in less than one-tenth of a second.
- Before you connect to the power source, make sure the trigger is not turned on. It should be off. Always disconnect the power source before you change bits or work on the drill.
- Find out what is inside the wall or on the other side of the work material before you cut through a wall or partition. Avoid hitting water lines or electrical wiring.

On-Site

Drilling Metal

When you are drilling metal, lubricate the bit to help cool the cutting edges and produce a smoother finished hole. A very small amount of cutting oil that is not combustible (capable of catching fire and burning) makes a good lubricant for drilling softer metals. No lubrication is needed for wood drilling. When you are drilling deep holes, pull the drill bit partly out of the hole every so often. This helps to clear the hole of shavings.

WARNING!

Before you start drilling into or through a wall, find out what is on the other side, and take steps to avoid hitting anything that would endanger your safety or cause damage. Spaces between studs (upright pieces in the walls of a building) often contain electrical wiring, plumbing, or insulation, for instance. If you are not careful, you will drill directly into the wiring, pipes, or insulation.

- Ensure that electric tools with two-prong plugs are double insulated. If a tool is not double insulated, its plug must have a third prong to provide grounding.
- Use the right bit for the job.
- Always use a sharp bit.
- Make sure the drill bit is tightened in the chuck before you start the drill.
- Make sure the chuck key is removed from the chuck before you start the drill.
- Hold the drill with both hands, when appropriate, and apply steady pressure. Let the drill do the work.
- Never ram the drill while you are drilling. This chips the cutting edge and damages the bearings.
- Never use the trigger lock. The trigger lock is a small lever, switch, or part that you push or pull to lock the trigger in operating mode.
- Drills do not need much maintenance, but they should be kept clean. Many drills have gears and bearings that are lubricated for life. Some drills have a small hole in the case for lubricating the motor bearings. Apply about three drops of oil occasionally, but don't overdo it. Extra lubricant can leak onto electrical contacts and burn the copper surfaces.
- Keep the drill's air vent clean with a small brush or small stick. Airflow is crucial to the maintenance and safety of a drill.
- Attach the chuck key to the power cord when you are not using the key, so it does not get lost.
- Do not overreach when using a power drill.

WARNING!

Do not use electric drills around combustible materials. Motors and bits can create sparks, which can cause an explosion.

3.2.0 Cordless Drills

Cordless power drills (*Figure 5*) are useful for working in awkward spaces or in areas where a power source is hard to find.

Cordless drills contain a rechargeable battery pack that runs the motor. The pack can be detached and plugged into a battery charger any time you are not using the drill. Some chargers can recharge the battery pack in an hour, while others require more time. Workers who use cordless drills often carry an extra battery pack with them. Some cordless drills have adjustable clutches so that the drill motor can also serve as a power screwdriver. Many cordless drills are now available with keyless chucks.

GOING GREEN

Recycling Rechargeable Batteries

Recycling your rechargeable batteries once their power supply has become exhausted is no longer a luxury; it has become a matter of necessity. In fact, there are now federal and state laws in place regulating the disposal of some types of rechargeable batteries.

The U.S. EPA estimates that more than 350 million rechargeable batteries are purchased in the United States while hundreds of millions of rechargeable batteries and cell phones are retired each year. Rechargeable batteries are made using heavy metals such as nickel, cadmium, mercury, and lead that can be toxic to our health and our environment if not disposed of properly. Most dumps are not designed to handle the toxic metals that will eventually leak out of all batteries.

Rechargeable batteries are commonly found in:

- Cordless power tools
- Cellular and cordless phones
- Two-way radios
- Laptop computers
- Digital cameras
- Camcorders
- Remote control toys

If your battery is rechargeable, then it's recyclable! Not only are rechargeable batteries better for the environment if properly discarded, they are more cost effective, too. Using rechargeable batteries can help save you money and protect the environment at the same time.

Recently, most cities have added hazardous waste collection centers that collect both rechargeable and regular batteries, along with paint, oil, refrigerant, and other hazardous wastes.

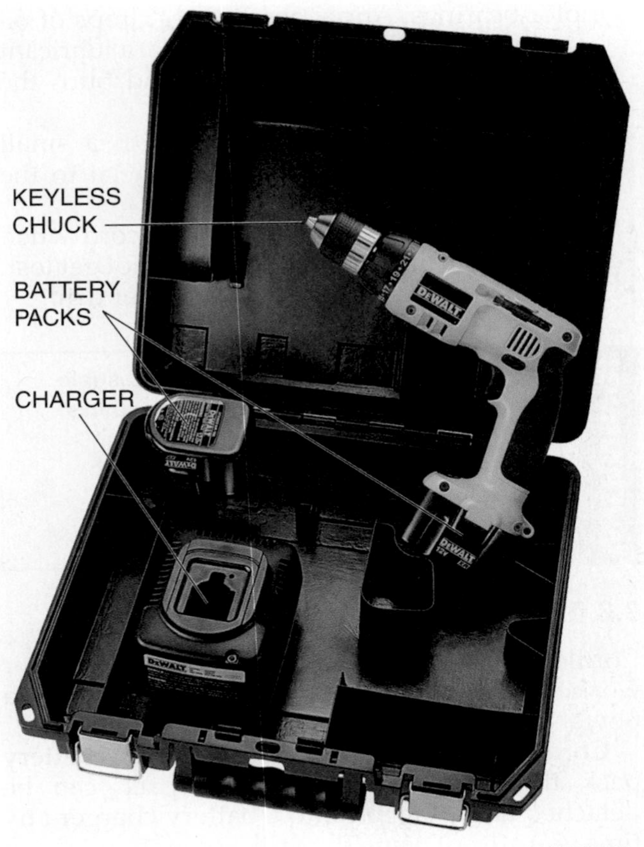

104F05B.EPS

Figure 5 Cordless drill.

3.2.1 How to Use a Cordless Drill with a Keyless Chuck

When using a cordless drill with a keyless chuck, always wear personal protective equipment, including appropriate eye, head, and hand protection.

Follow these steps to load a bit on a cordless drill with a keyless chuck (*Figure 6*):

Step 1 Remove the power pack/battery. Open the chuck by turning it counterclockwise until the jaws are wide enough for you to insert the bit shank.

Step 2 Tighten the chuck by hand until the jaws grip the bit shank. Be sure to keep the bit centered as you tighten it. It should not be leaning to one side, but should be straight in the chuck.

To operate the drill, follow the procedures previously outlined for the power drill.

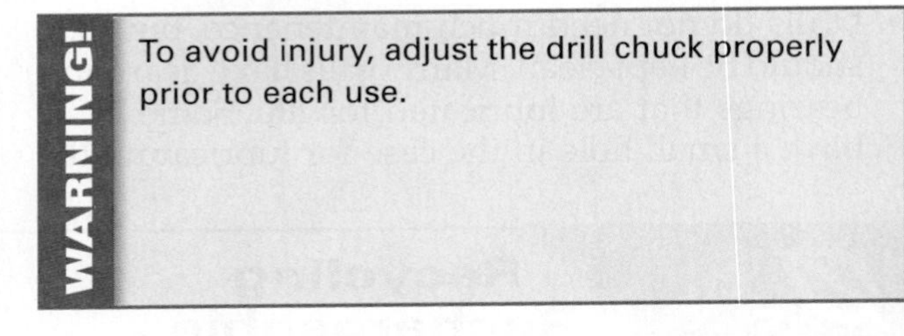

To avoid injury, adjust the drill chuck properly prior to each use.

3.2.2 Safety and Maintenance

When using a cordless drill, observe the safety practices that were presented for electric power drills.

3.3.0 Hammer Drills

The hammer drill (*Figure 7*) has a pounding action that lets you drill into concrete, brick, or tile. The bit rotates and hammers at the same time, allowing you to drill much faster than you could with a regular drill. The depth gauge on a hammer drill can be set to the depth of the hole you want to drill.

(A) INSERT THE BIT SHANK.

104F06A.EPS

(B) TIGHTEN THE CHUCK.

104F06B.EPS

Figure 6 Loading the bit on a cordless drill.

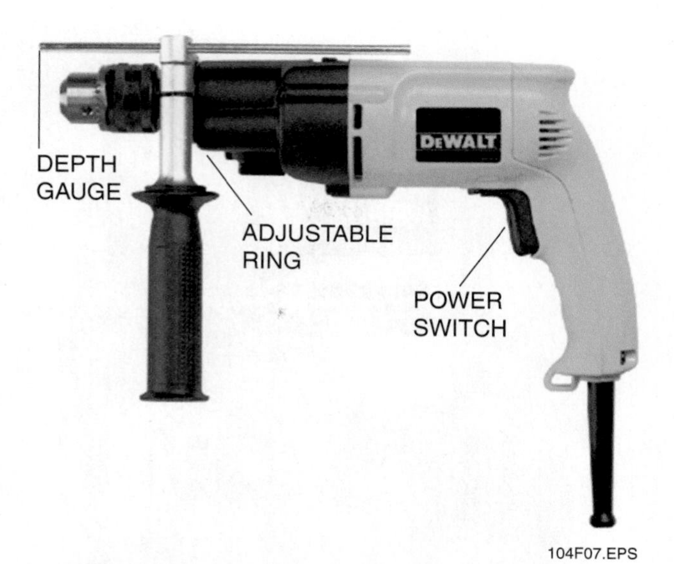

104F07.EPS

Figure 7 Hammer drill.

You need special hammer drill bits that can take the pounding. Some hammer drills use percussion and masonry bits (*Figure 8*).

3.3.1 How to Use a Hammer Drill

When using a hammer drill, always wear personal protective equipment, including appropriate eye, head, and hand protection. Follow the procedures that were presented for using a power drill.

Most hammer drills will not hammer until you put pressure on the drill bit (*Figure 9*). You can adjust the drill's blows per minute by turning the adjustable ring (refer to *Figure 7*). The hammer action stops when you stop applying pressure to the drill.

3.3.2 Safety and Maintenance

When using a hammer drill, observe the safety practices that were presented for electric power drills.

3.4.0 Electromagnetic Drills

The electromagnetic drill (*Figure 10*) is a portable drill mounted on an electromagnetic base. It is used for drilling thick metal. When the drill is placed on metal and the power is turned on, the magnetic base will hold the drill in place for drilling. Some drills can also be rotated on the base.

A switch on the junction box controls the electromagnetic base. When the switch is turned on, the magnet holds the drill in place on a **ferromagnetic** metal surface. (Ferromagnetic refers to

104F08.EPS

Figure 8 Hammer drill bits.

substances, especially metals, which have magnetic properties.) The switch on the top of the drill turns the drill on and off. You can also set the depth gauge to the depth of the hole you are drilling.

3.4.1 How to Set Up an Electromagnetic Drill

The use of this tool is explained in detail in the specific craft areas that use it. For now, you will learn only the setup procedures for using the electromagnetic drill.

Step 1 Wear appropriate personal protective equipment.

Step 2 Place the drill face down into the metal holder.

Step 3 Put the electromagnetic switch (not the drill) in the ON position. Doing this holds the drill in place by magnetizing the base of the drill directly onto the metal to be drilled.

104F09.EPS

Figure 9 Proper use of a hammer drill.

Step 4 Lock the drill in place.

Step 5 Set the depth gauge to the depth of the hole you are going to drill.

Step 6 Fasten the work securely on the drilling surface with clamps.

Step 7 Proceed to drill.

WARNING!

Expect the unexpected. Use a safety chain to secure the electromagnetic drill in case the power is shut off. If there is no power, you lose the electromagnetic field that holds the base to the metal being drilled.

3.4.2 Safety and Maintenance

In addition to the general safety rules you learned in *Basic Safety*, there are some specific safety rules for working with electromagnetic drills, as follows:

- Clamp the material securely. Unsecured materials can become deadly flying objects.
- Make sure the electrical power is not interrupted. Put a DO NOT UNPLUG tag on the cord (*Figure 10*).
- Safety attachments, such as shields to block flying objects and safety lines to keep the drill from falling if the power is cut off, are available. In some states, they are required. Ask your instructor or supervisor about requirements for safety attachments in your area.

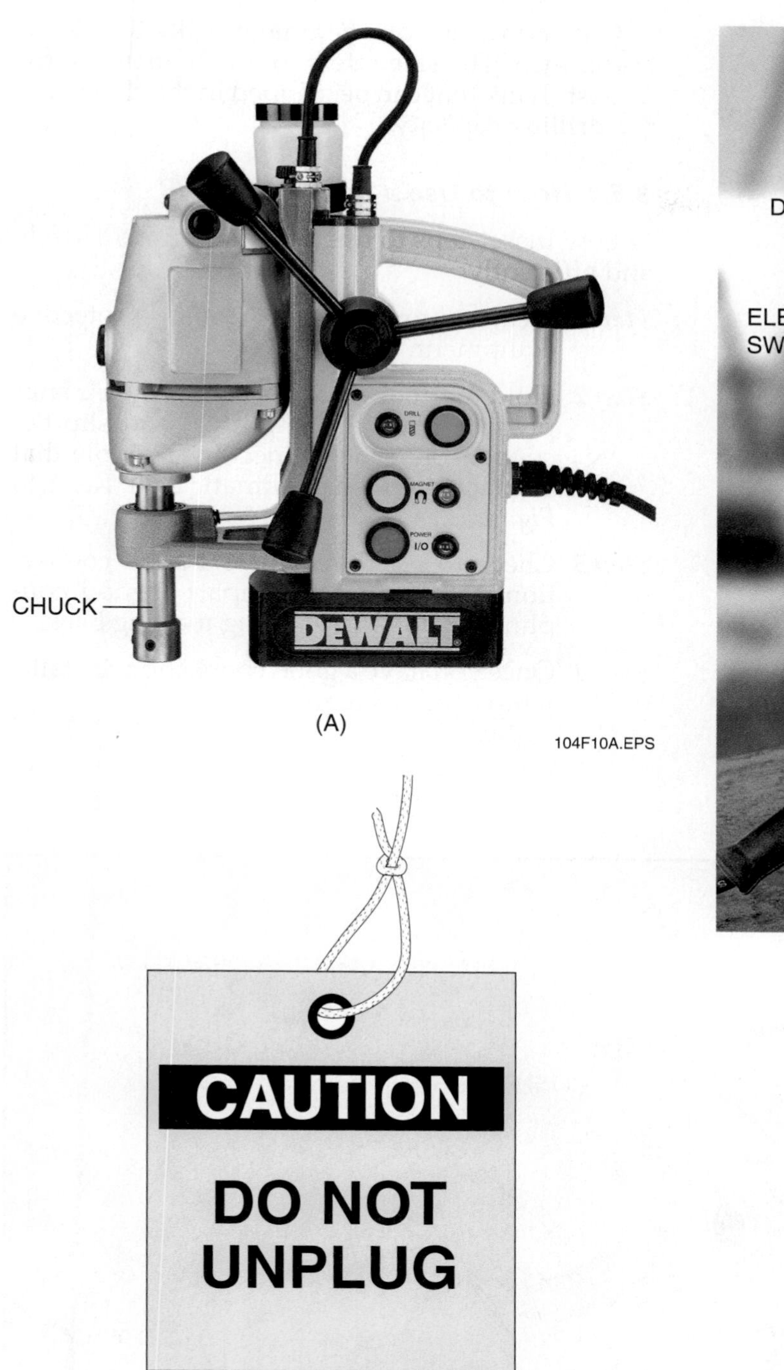

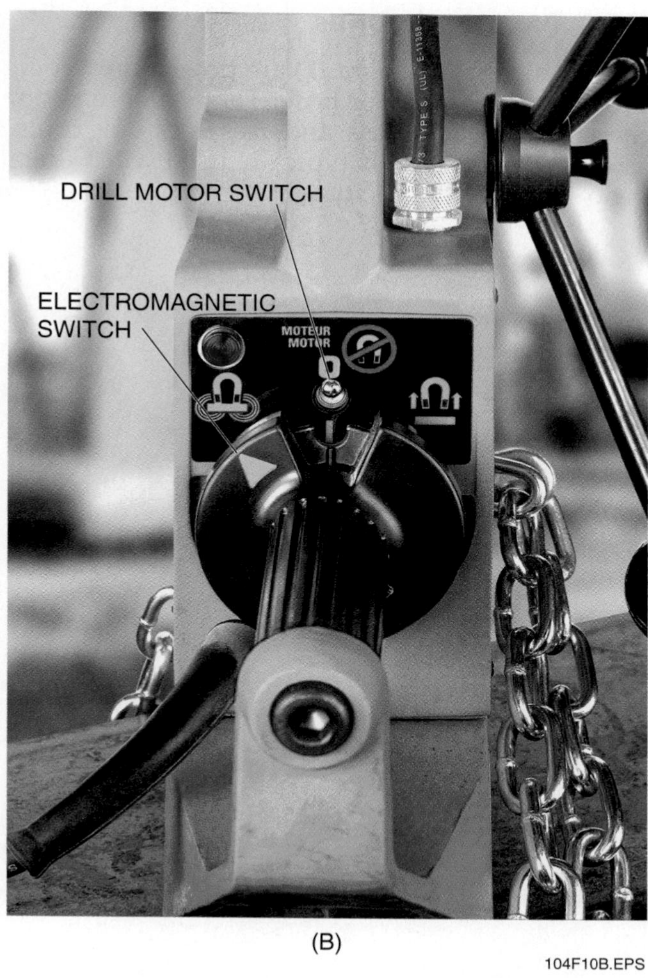

Figure 10 Electromagnetic drill.

- Support the drill before you turn it off. It will fall over if you do not hold it when you turn off the power.
- Use a safety chain to secure the electromagnetic drill in case power is shut off or lost.

WARNING!

When you are working near combustible materials, be sure to use a nonsparking drill. A drill that gives off sparks could start a fire.

3.5.0 Pneumatic Drills

Pneumatic drills (*Figure 11*) are powered by compressed air from an air hose. They have many of the same parts, controls, and uses as electric drills. The pneumatic drill is typically used when there is no source of electricity.

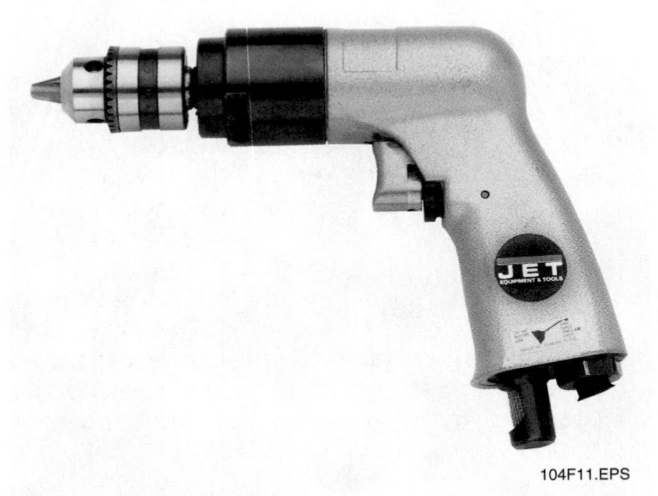

Figure 11 Pneumatic drill.

Common sizes of pneumatic drills are ¼-, ⅜-, and ½-inch. The size refers to the diameter of the largest shank that can be gripped in the chuck, not the drilling capacity.

3.5.1 How to Use a Pneumatic Drill

Follow these steps to use a pneumatic drill safely and efficiently:

Step 1 Wear appropriate personal protective equipment.

Step 2 Hold the coupler at the end of the air supply line, slide the ring back, and slip the coupler onto the connector or nipple that is attached to the pneumatic drill. Refer to *Figure 12*.

Step 3 Check to see if you have a good connection. You cannot take apart a good coupling without first sliding the ring back.

Step 4 Once you have a good connection, install a whip check as required.

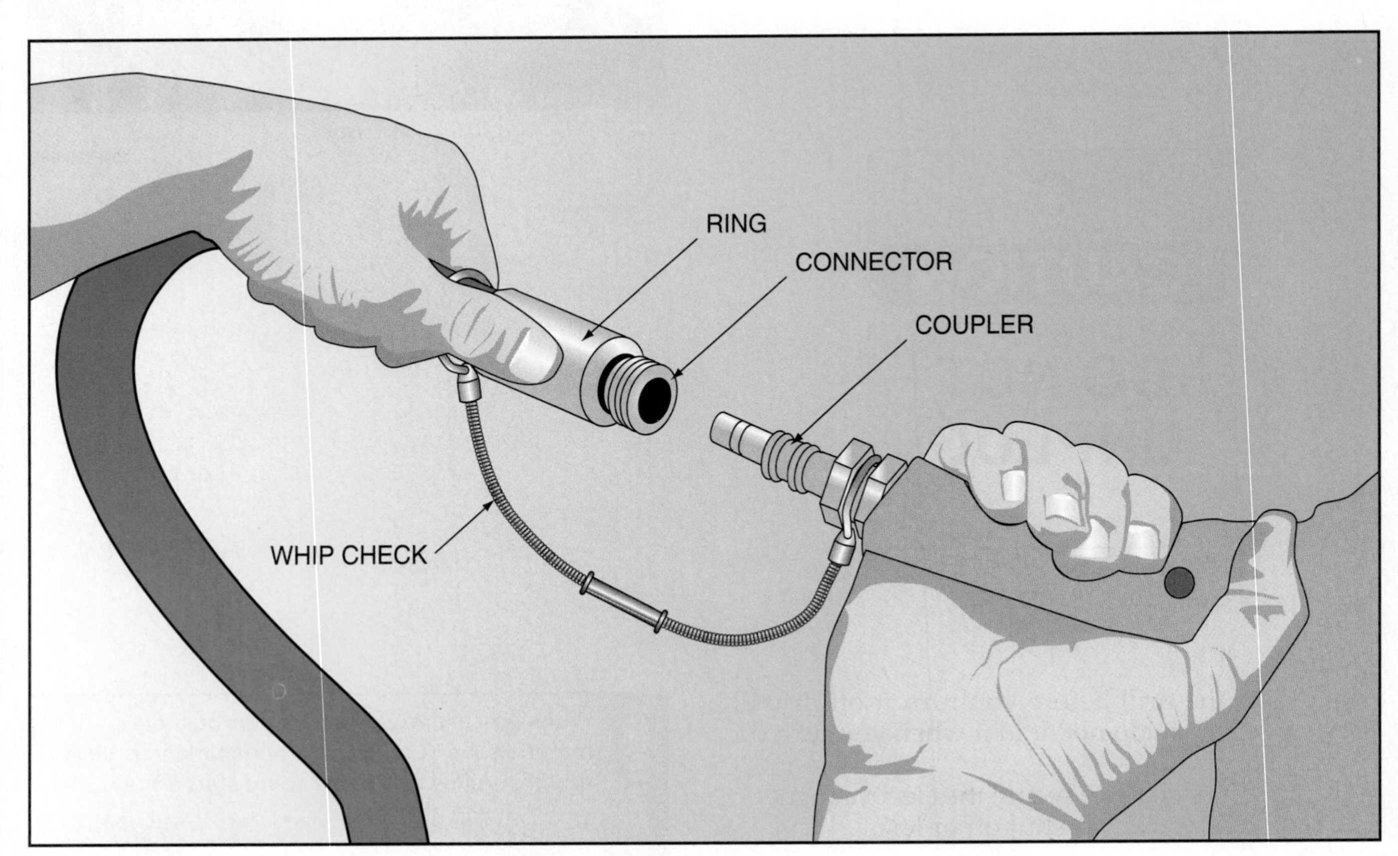

Figure 12 Proper use of a pneumatic drill.

> **NOTE**
> A whip check is a safety attachment that is used to prevent whiplashing in hoses that are inadvertently uncoupled.

Step 5 Proceed to drill as needed.

Step 6 When your work is completed, disconnect the drill from the hose.

3.5.2 Safety and Maintenance

Follow the safety practices that were presented for power drills in this module.

> **WARNING!**
> The amount of air pressure used to power a pneumatic drill is higher than the OSHA recommendation for cleaning a work area or oneself. Improper use of pressurized air can lead to injuries as well as damage to the tool or materials.

4.0.0 SAWS

Using the right saw for the job will make your work much easier. Always make sure that the blade is right for the material being cut. In this section, you will learn about the following types of power saws:

- Circular saws
- Saber saws
- Reciprocating saws
- Portable handheld bandsaws
- Power miter box saws

4.1.0 Circular Saws

Many years ago, a company named Skil® made power-tool history by introducing the portable circular saw. Today many different companies make dozens of models, but a lot of people still call any portable circular saw a Skilsaw. Other names you might hear are utility saw, electric handsaw, and builder's saw. The portable circular saw (*Figure 13*) is designed to cut lumber and boards to size for a project.

Saw size is measured by the diameter of the circular blade. Saw blade diameters range from 3⅜ to 16¼ inches. The 7¼-inch size is the most popular. A typical circular saw weighs between 9 and 12 pounds. The handle of the circular saw has a trigger switch that starts the saw. The motor is protected by a rigid plastic housing. Blade speed when the blade is not engaged in cutting is given in rpm. The teeth of the blade point in the direction of the rotation. The blade is protected by two guards. On top, a rigid plastic guard protects you from flying debris and from touching the spinning blade if you lean forward accidentally. The lower guard is spring-loaded—as you push the saw forward, it retracts up and under the top guard to allow the saw to cut.

> **WARNING!**
> Never use the saw unless the lower blade guard is properly attached. The guard protects you from the blade and from flying particles.

On-Site

Using Saw Blades

Having a variety of blades will allow you to adapt your saw to different projects. Blades fall into two categories: standard steel, which must be sharpened regularly, and carbide-tipped. You must use the appropriate type of saw blade for the job. Some common types of saw blades include the following:

- *Rip* – These blades are designed to cut with the grain of the wood. The square chisel teeth cut parallel with the grain and are generally larger than other types of blade teeth.
- *Crosscut* – These blades are designed to cut across the grain of the wood (at a 90-degree angle). Crosscut teeth cut at an angle and are finer than rip blade teeth.
- *Combination* – These blades are designed to cut hard or soft wood, either with or across the grain. The combination blade features both rip and crosscut teeth with deep troughs (gullets) between the teeth.
- *Nail cutter* – This blade has large carbide-tipped teeth that can make rough cuts through nails that may be embedded in the work.
- *Nonferrous metal cutter* – This blade has carbide-tipped teeth for cutting aluminum, copper, lead, and brass. It should be lubricated with oil or wax before each use.

Always follow the manufacturer's instructions when using saw blades.

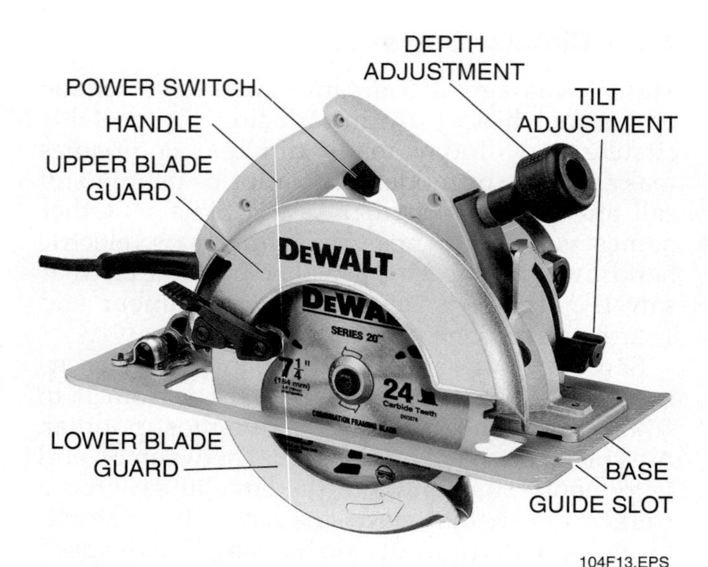

Figure 13 Circular saw.

4.1.1 How to Use a Circular Saw

Follow these steps to use a circular saw safely and efficiently:

Step 1 Wear appropriate personal protective equipment.

Step 2 Secure the material to be cut and ensure it is properly supported. If the work isn't heavy enough to stay in position without moving, weight or clamp it down.

Step 3 Make your cut mark with a pencil or other marking tool.

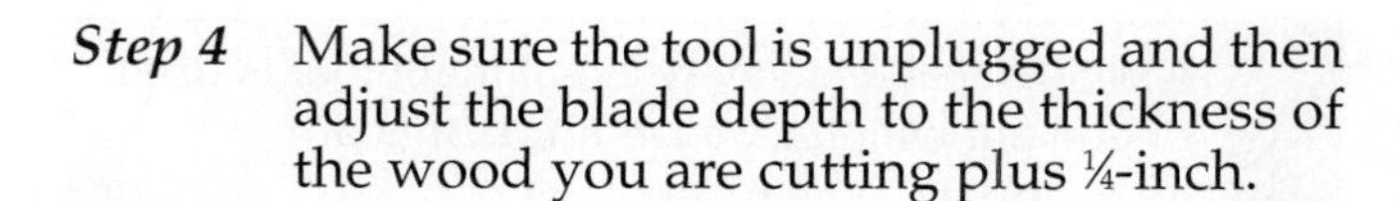

Step 4 Make sure the tool is unplugged and then adjust the blade depth to the thickness of the wood you are cutting plus ¼-inch.

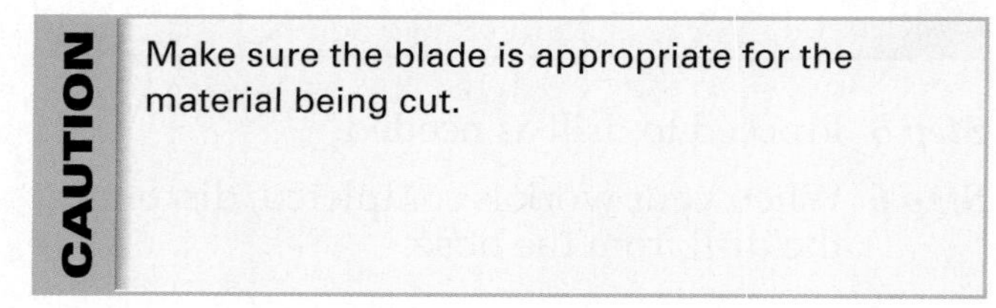

CAUTION

Make sure the blade is appropriate for the material being cut.

Step 5 Place the front edge of the baseplate on the work so the guide notch and the blade are in line with the cut mark.

Step 6 Start the saw. After the blade has revved up to full speed, move the saw forward to start cutting. The lower blade guard will automatically rotate up and under the top guard when you push the saw forward.

Step 7 While cutting with the saw, grip the saw handles firmly with two hands, as shown in *Figure 14*.

Step 8 If the saw cuts off the line, stop, back out, and restart the cut. Do not force the saw.

Step 9 As you get to the end of the cut, the guide notch on the baseplate will move off the end of the work. Use the blade as your guide.

Step 10 Release the trigger switch. The blade will stop rotating.

Step 11 Make sure the blade has stopped before setting the saw down.

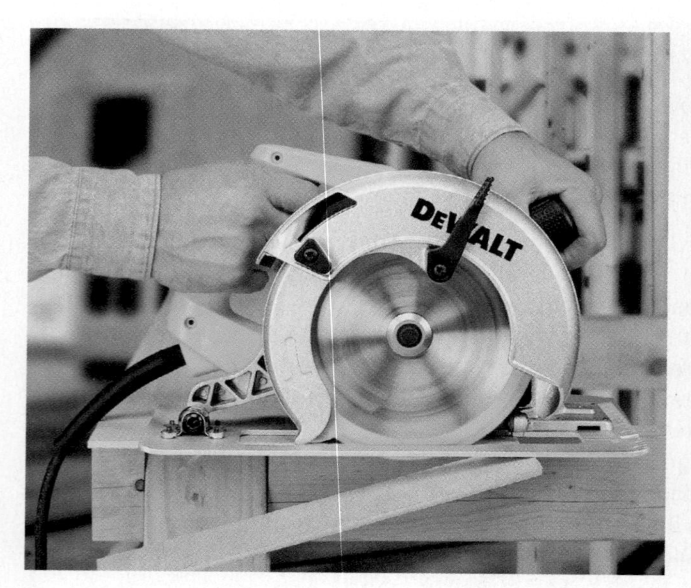

Figure 14 Proper use of a circular saw.

On-Site

The Worm-Drive Saw

The worm-drive saw is a heavy-duty type of circular saw. Most circular saws have a direct drive. That is, the blade is mounted on a shaft that is part of the motor. With a worm-drive saw, the motor drives the blade from the rear through two gears. One gear (the worm gear) is cylindrical and threaded like a screw. The worm gear drives a wheel-shaped gear (the worm wheel) that is directly attached to the shaft to which the blade is fastened. This setup delivers much more rotational force (torque), making it easier to cut a double thickness of lumber. The worm-drive saw is almost twice as heavy as a conventional circular saw. This saw should be used only by an experienced craftworker.

104SA01.EPS

4.1.2 Safety and Maintenance

To use a circular saw safely and ensure its long life, follow these guidelines:

- Wear appropriate personal protective equipment.
- Ensure that the blade is tight.
- Check that the blade guard is working correctly before you connect the saw to the power source.
- Before you cut through a wall or partition, find out what is inside the wall or on the other side of the partition. Avoid hitting water lines or electrical wiring.
- Whenever possible, keep both hands on the saw grips while you are operating the saw.
- Never force the saw through the work. This causes binding and overheating and may cause injury.
- Never reach underneath the work while you are operating the saw.
- Never stand directly behind the work. Always stand to one side of it.
- Do not use your hands to try to secure small pieces of material to be cut. Use a clamp instead.
- Know where the power cord is located. You don't want to cut through the power cord by accident and electrocute yourself!
- The most important maintenance on a circular saw is at the lower blade guard. Sawdust builds up and causes the guard to stick. If the guard sticks and does not move quickly over the blade after it makes a cut, the bare blade may still be turning when you set the saw down and may cause damage. Remove sawdust from the blade guard area. Remember to always disconnect the power source before you do maintenance.
- To avoid injury to you, your materials, and your co-workers, check often to make sure the guard snaps shut quickly and smoothly. To ensure smooth operation of the guard, disconnect the saw from its power source, allow it to cool, and clean foreign material from the track. Be aware of fire hazards when using cleaning liquids such as isopropyl alcohol.

On-Site

Cutting

Most circular saw blades have a kerf (a cut or channel) ⅛-inch thick. Be sure to cut on the waste (unused) side of the material, or your finished piece will be ⅛-inch short. Mark an X on the waste side after you make the cut mark. This will help you remember the side of the mark on which to cut.

- Do not lubricate the guard with oil or grease. This could cause sawdust to stick in the mechanism.
- Always keep blades clean and sharp to reduce friction and kickback. Blades can be cleaned with hot water or mineral spirits. Be careful with mineral spirits; they are very flammable.
- Do not hold material to be ripped with your hands.

WARNING!
When using a circular saw, never hold material to be cut with your hands; always use a clamp instead.

4.2.0 Saber Saws (Jig Saws)

Saber saws, sometimes referred to as jig saws, have very fine blades, which makes the saw an effective tool for doing delicate and intricate work, such as cutting out patterns or irregular shapes from wood or thin, soft metals. They are also some of the best tools for cutting circles.

The saber saw (*Figure 15*) is a very useful portable power tool. It can make straight or curved cuts in wood, metal, plastic, wallboard, and other materials. It can also make its own starting hole if a cut must begin in the middle of a board. The saber saw cuts with a blade that moves up and down, unlike the spinning circular saw blade. This means that each cutting stroke (upward) is followed by a return stroke (downward), so the saw is cutting only half the time it is in operation. This is called up-cutting or cleancutting.

Many models are available with tilting baseplates for cutting beveled edges. Models come with a top handle or a barrel handle. Some cordless models are available.

The saber saw has changeable blades that enable it to cut many different materials, from wood and metal to wallboard and ceramic tile. Most saber saws can be operated at various blade speeds. Types of saber saws include single-speed, two-speed, and variable-speed. The variable-speed saber saw can cut at low and high speeds. The low-speed setting is for cutting hard materials, and the high-speed setting is for soft materials. An important part of the saber saw is the baseplate (shoeplate or footplate). Its broad surface helps to keep the blade lined up. It keeps the work from vibrating and allows the blade teeth to bite into the material.

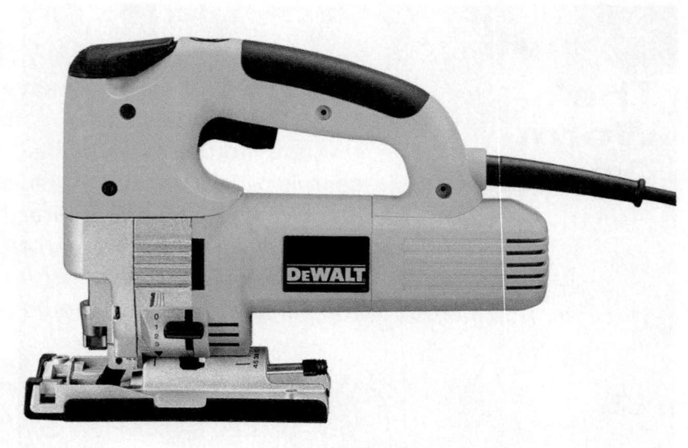

TOP HANDLE

104F15A.EPS

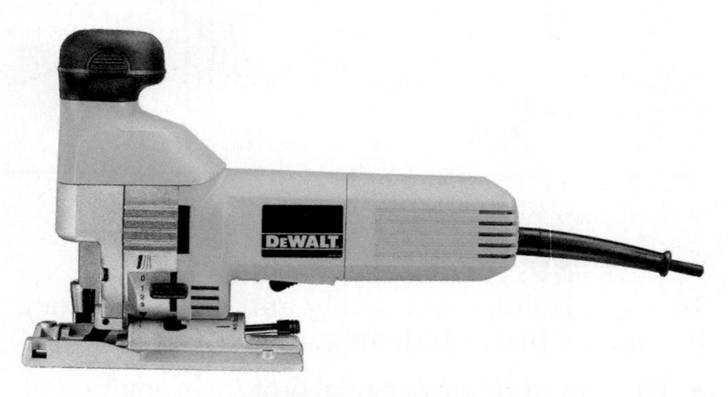

BARREL HANDLE

104F15B.EPS

Figure 15 Saber saws.

CAUTION
Do not lift the blade out of the work while the saw is still running. If you do, the tip of the blade may hit the wood surface, marring the work and possibly breaking the blade.

4.2.1 How to Use a Saber Saw

Follow these steps to ensure that you use a saber saw safely and efficiently:

Step 1 Wear appropriate personal protective equipment.

Step 2 To avoid vibration, clamp the work to a pair of sawhorses or hold the work in a vise.

Step 3 Check the blade to see that it is the right blade for the job and that it is sharp and undamaged.

Step 4 Measure and mark the work.

Step 5 When you cut from the edge of a board or panel, be sure the front of the baseplate is resting firmly on the surface of the work before you start the saw. The blade should not be touching the work at this stage.

Step 6 Start the saw (pull the trigger) and move the blade gently but firmly into the work. Continue feeding the saw into the work as fast as possible without forcing it. Do not push the blade into the work. *Figure 16* shows the proper way to use a saber saw.

Step 7 When the cut is finished, release the trigger and let the blade come to a stop before you remove it from the work.

4.2.2 Safety and Maintenance

When using a saber saw, follow these guidelines:

- Always wear appropriate personal protective equipment.
- Secure the material you are working with to reduce vibration and ensure safety.
- Before you plug the saw into a power source, make sure the switch is in the OFF position.
- Before you cut through a wall or partition, find out what is inside the wall or on the other side of the partition. Avoid hitting water lines or electrical wiring.
- Always use a sharp blade and never force the blade through the work.
- Do not force or lean into the blade. You could lose your balance and fall forward, or your hands could slip onto the work surface and you could cut yourself.
- When cutting metal pieces, use a metal-cutting blade. Lubricate the blade with an agent, such as beeswax, to help make tight turns and to reduce the chance of breaking the blade.
- When you are replacing a broken blade, look for any pieces of the blade that may be stuck inside the collar.
- When you install a blade in the saw, make sure it is in as far as it will go, and tighten the setscrew securely. Always disconnect the power source before you change blades or perform maintenance.

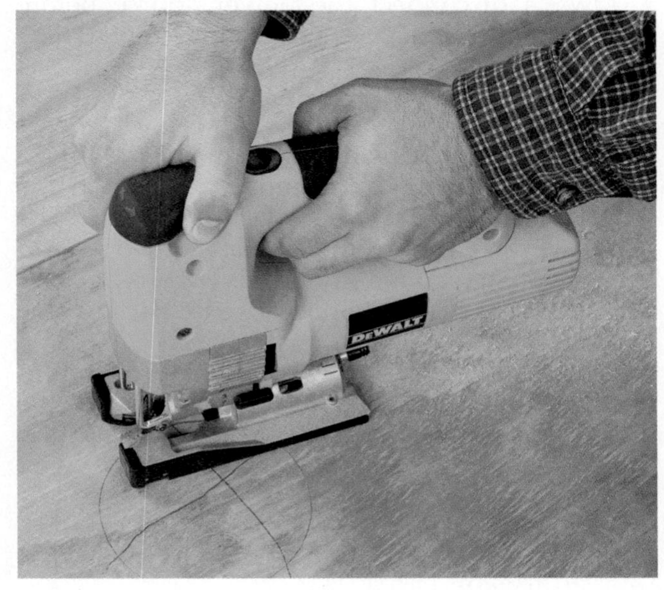

104F16.EPS

Figure 16 Proper use of a saber/jig saw.

4.3.0 Reciprocating Saws

A reciprocating saw, regardless of the manufacturer, is often referred to as a SawZall® because it is one of the most popular brands on the market. Both the saber saw and the reciprocating saw can make straight and curved cuts. They are used to cut irregular shapes and holes in plaster, plasterboard, plywood, studs, metal, and most other materials that can be cut with a saw.

Both saws have straight blades that move as you guide them in the direction of the cut. But here's the difference: The saber saw's blade moves up and down, whereas the reciprocating saw's blade moves back and forth. The reciprocating saw is designed for more heavy-duty jobs than the saber saw is. It can use longer and tougher blades than a saber saw. Also, because of its design, you can get into more places with it. The reciprocating saw (*Figure 17*) is used for jobs that require brute strength. It can saw through walls or ceilings and create openings for windows, plumbing lines, and more. It is a basic tool in any demolition work.

Like the saber saw, reciprocating saws come in single-speed, two-speed, and variable-speed models. The two-speed reciprocating saw can cut at low and high speeds. The low-speed setting is best for metal work. The high-speed setting is for sawing wood and other soft materials.

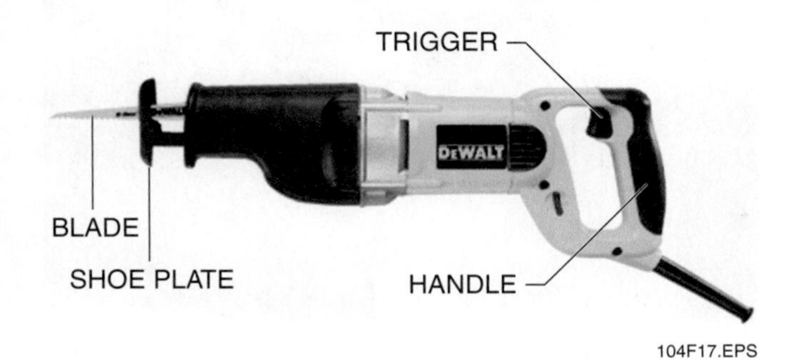

104F17.EPS

Figure 17 Reciprocating saw.

On-Site

Blade Safety

If you are cutting into the middle of a piece with a saber saw, first make a starter hole using a power drill. Once you have a drilled hole, tip the saber saw forward on the front of its baseplate, positioning the blade over your drilled hole. Press the trigger and slowly tip the baseplate and the blade down toward the surface. When it strikes the surface, the blade may jump. Keep a steady hand, and the gentle pressure will eventually push the blade through the workpiece. Plunging the blade into the work with sudden force is one of the most common causes of broken blades. The other cause of broken blades is pushing a saber saw too fast. The common result of too much pressure too fast is a snapped blade.

The baseplate (shoeplate or footplate) may have a swiveling action, or it may be fixed. Whatever the design, the baseplate is there to provide a brace or support point for the sawing operation.

4.3.1 *How to Use a Reciprocating Saw*

Follow these steps to use a reciprocating saw safely and efficiently:

Step 1 Wear appropriate personal protective equipment.

Step 2 To avoid vibration, clamp the work to a pair of sawhorses or secure it in a vise.

Step 3 Set the saw to the desired speed. Remember these guidelines:
- Use lower speeds for sawing metal.
- Use higher speeds for sawing wood and other soft materials.

Step 4 Grip the saw with both hands. Place the baseplate firmly against the workpiece (*Figure 18*).

104F18.EPS

Figure 18 Proper use of a reciprocating saw.

Step 5 Squeeze the trigger ON switch. The blade moves back and forth, cutting on the backstroke.

CAUTION

Use both hands to grip the saw firmly. Otherwise, the pull created by the blade's grip might jerk the saw out of your grasp.

4.3.2 *Safety and Maintenance*

Follow these guidelines to ensure safety for yourself and your co-workers, and a long life for the saw:

- Always wear appropriate personal protective equipment.
- Before you cut through a wall or partition, find out what is inside the wall or on the other side of the partition. Avoid hitting water lines or electrical wiring.
- Always disconnect the power source before you change blades or perform maintenance.

4.4.0 Portable Handheld Bandsaw

The portable handheld bandsaw (*Figure 19*) is used when it is better to move the saw to the work than to move the work to the saw. The bandsaw can cut pipe, metal, plastics, wood, and irregularly shaped materials. It is especially good for cutting heavy metal, but it will also do fine cutting work.

The bandsaw has a one-piece blade that runs in one direction around guides at either end of the saw. The blade is a thin, flat piece of steel. It is sized according to the diameter of the revolving pulleys that drive and support the blade. The saw often works at various speeds.

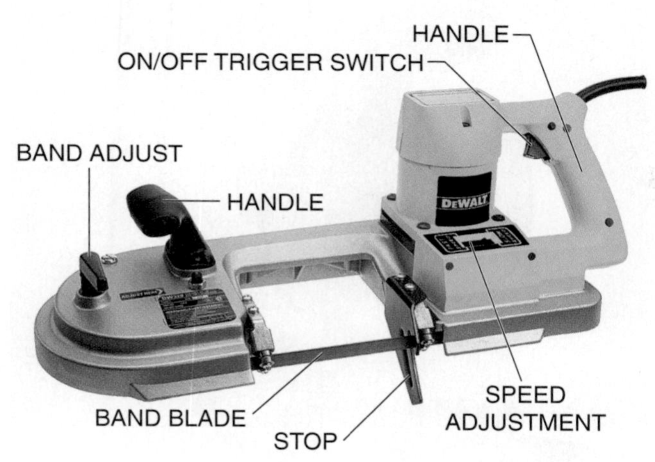

104F19.EPS

Figure 19 Portable handheld bandsaw.

CAUTION

The portable bandsaw cuts on the pull, not the push. You must be especially careful because, in some situations, the saw blade might be moving directly toward your body. Always wear appropriate personal protective equipment and keep your mind focused on the work in front of you.

4.4.1 How to Use a Portable Bandsaw

Follow these steps to use a bandsaw safely and efficiently:

Step 1 Wear appropriate personal protective equipment.

Step 2 Place the stop firmly against the object to be cut (*Figure 20A*). This will keep the saw from bouncing against the object and breaking the band.

Step 3 Gently pull the trigger. Only a little pressure is needed to make a good clean cut because the weight of the saw gives you more leverage for cutting. *Figure 20B* shows the proper way to use a portable bandsaw.

4.4.2 Safety and Maintenance

Follow these guidelines to ensure safety for yourself and your co-workers, and a long life for the saw:

- Always wear appropriate personal protective equipment.
- Use only a bandsaw that has a stop.
- Before you cut through a wall or partition, find out what is inside the wall or on the other side of the partition. Avoid hitting water lines or electrical wiring.

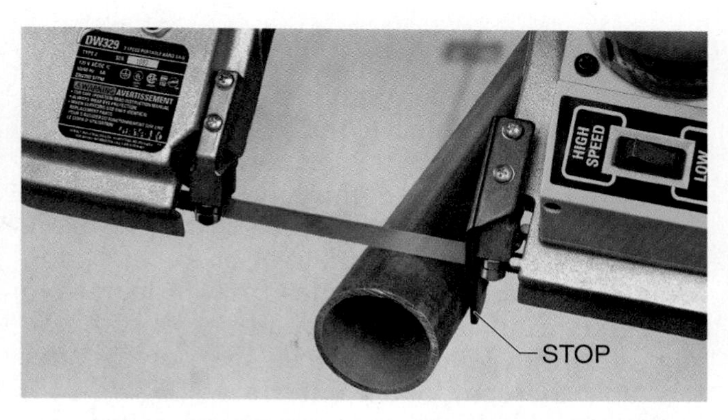

(A) PLACE THE STOP FIRMLY AGAINST THE OBJECT.

104F20A.EPS

(B) APPLY ONLY A LITTLE PRESSURE TO MAKE A CUT.

104F20B.EPS

Figure 20 Proper use of a portable bandsaw.

- The blade of a portable bandsaw gets stuck very easily. Never force a portable bandsaw. Let the saw do the cutting.
- The blades should be waxed with an appropriate lubricant, such as the one recommended by the blade's manufacturer. Always disconnect the power source before you do maintenance.

Did You Know?

For Portable Bandsaws, Low Speed Works Best

The portable bandsaw cuts best at a low speed. Using a high speed will cause the blade's teeth to rub rather than cut. This can create heat through friction, which will cause the blade to wear out quickly.

4.5.0 Power Miter Saw

The power miter saw combines a miter box with a circular saw, allowing it to make straight and miter cuts. There are two types of power miter boxes: power miter saws and compound miter saws.

In a power miter saw (*Figure 21A*), the saw blade pivots horizontally from the rear of the table and locks in position to cut angles from 0 degrees to 45 degrees right and left. Stops are set for common angles. The difference between the power miter saw and the compound miter saw (*Figure 21B*) is that the blade on the compound miter saw can be tilted vertically, allowing the saw to be used to make a compound cut (combined bevel and miter cut).

Similar to a power miter saw and compound miter saw is the compound slide miter saw (*Figure 21C*).

(A) POWER MITER BOX

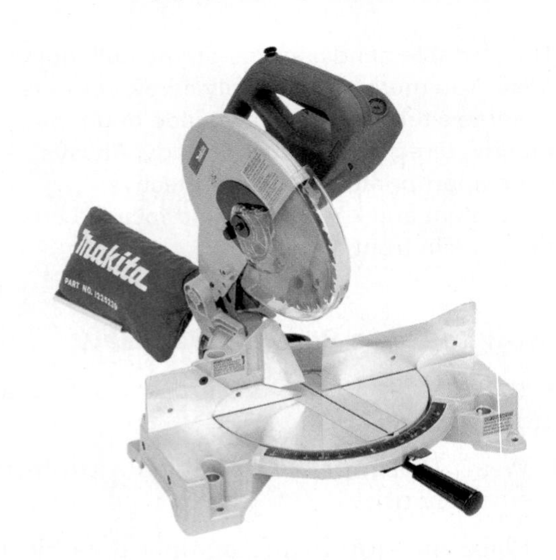

(B) COMPOUND MITER SAW

(C) COMPOUND SLIDE MITER SAW

104F21.EPS

Figure 21 Miter saws.

4.5.1 How to Use a Power Miter Saw

Follow these steps to use a power miter saw safely and efficiently:

Step 1 Wear appropriate personal protective equipment.

Step 2 Be sure the saw blade has reached its maximum speed before starting the cut.

Step 3 Hold the workpiece firmly against the fence when making the cut.

Step 4 Turn off the saw immediately after making the cut.

4.5.2 Safety and Maintenance

Follow these guidelines to ensure safety for yourself and your co-workers, and a long life for the miter saw:

- Always check the condition of the blade and be sure the blade is secure before starting the saw.
- Keep your fingers clear of the blade.
- Be sure the blade guards are in place and working properly.
- Never make adjustments while the saw is running.
- Never leave a saw until the blade stops.
- Be sure the saw is sitting on a firm base and is properly fastened to the base.
- Be sure the saw is securely locked at the correct angle.
- If working on long stock, have a helper support the end of the stock.

4.6.0 Abrasive Cutoff Saw

An **abrasive** cutoff saw (also referred to as a chop saw or cutoff saw) (*Figure 22*) can be used to make straight cuts or angular cuts through thicker materials such as angle iron, flat bar, and channel. Cutoff saws can be either stationary or portable. Portable cutoff saws are convenient because they can be quickly transported between the shop and the field.

The circular, abrasive blade on a cutoff saw can be between 10 and 18 inches in diameter and is ¼-inch thick. When the saw is in operation, the blade spins at such a high speed that the resulting friction is hot enough to burn through the material.

4.6.1 How to Use an Abrasive Cutoff Saw

Follow these steps to use an abrasive cutoff saw safely and efficiently:

Step 1 Wear appropriate personal protective equipment.

Step 2 Make sure the material to be cut is clamped firmly in place, especially if it is a short piece.

Step 3 Be sure the saw blade has reached its maximum speed before starting the cut.

Step 4 Turn off the saw immediately after making the cut.

104F22.EPS

Figure 22 Abrasive cutoff saw.

4.6.2 Safety and Maintenance

Follow these guidelines to ensure safety for yourself and your co-workers, and a long life for the abrasive cutoff saw:

- Make sure the chop saw is completely shielded by a guard.
- Always wear safety goggles and a face mask, because the blade produces sparks when it is making the cut.
- Wear long sleeves and gloves to protect yourself from sparks.
- Make sure the work area is clear of flammable materials such as chemicals and rags.
- Always unplug the saw before changing the blade.
- Never allow anyone to stand nearby while you operate the saw.
- Never wear a watch or jewelry while operating the saw because they can get caught in the machinery.
- Never retract the safety guard on the saw to see the piece you are cutting.

5.0.0 Grinders and Sanders

Grinding tools can power all kinds of abrasive wheels, brushes, buffs, drums, bits, saws, and discs. These wheels come in a variety of materials and **grits**. They can drill, cut, smooth, and polish; shape or sand wood or metal; mark steel and glass; and sharpen or engrave. They can even be used on plastics.

Sanders can shape workpieces, remove imperfections in wood and metal, and create the smooth surfaces needed before finishing work can begin. Sanding is an essential part of all finish carpentry. Sanding gives a smooth, professional look to the completed work regardless of whether or not it will be painted.

WARNING!

Always wear safety goggles and a face shield when working with grinders and sanders. Make sure that the work area is free of combustible materials such as rags or flammable liquids and that a fire extinguisher is easily accessible. Be sure that your clothes are snug and comfortable and free of cuffs at the wrists and ankles. Wearing excessively loose clothing on the worksite can be extremely dangerous.

5.1.0 Angle Grinders, End Grinders, and Detail Grinders

These types of grinders are grouped together because they are all handheld.

- *Angle grinders* (also called side grinders) – Used to grind away hard, heavy materials and to grind surfaces such as pipes, plates, or welds (*Figure 23*).
- *End grinders* – Also called horizontal grinders or pencil grinders. These smaller grinders are used to smooth the inside of materials, such as pipe (*Figure 24*).
- *Detail grinders* – Use small attachments, also called points, to smooth and polish intricate metallic work (*Figure 25A*). These attachments, some of which are shown in *Figure 25B*, are commonly made in sizes ranging from 1⁄16- to 1⁄4-inch.

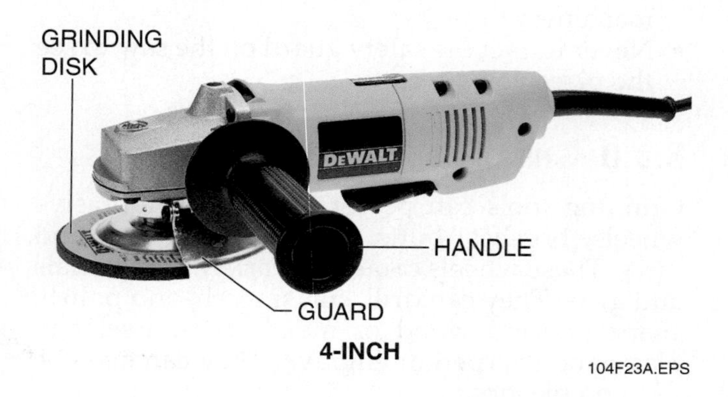

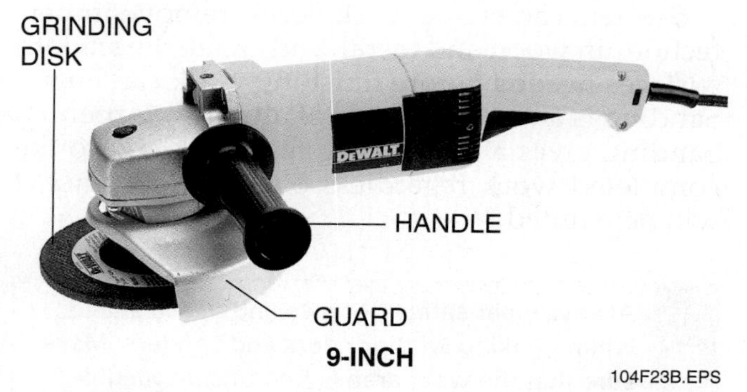

Figure 23 Angle grinders.

The angle grinder has a rotating grinding disc set at a right angle to the motor shaft. The grinding disc on the end grinder rotates in line with the motor shaft. Grinding is also done with the outside of the grinding disc. The detail grinder has a shank that extends from the motor shaft; points of different sizes and shapes can be mounted on the shank.

5.1.1 How to Use an Angle Grinder, End Grinder, or Detail Grinder

Follow these steps to use an angle, end, or detail grinder safely and efficiently:

Step 1 Wear appropriate personal protective equipment.

Step 2 If it is not already secured, secure the material in a vise or clamp it to the bench.

Step 3 To use an angle grinder, place one hand on the handle of the grinder and one on the trigger. To use an end grinder or detail grinder, grip the grinder at the shaft end with one hand and cradle the opposite end of the tool in your other hand.

Step 4 Finish the work by removing any loose material with a wire brush.

5.1.2 Safety and Maintenance

Follow these guidelines to ensure safety for yourself and your co-workers, and a long life for the grinder:

- Always wear appropriate personal protective equipment.
- Never use an angle grinder, end grinder, or detail grinder unless it is equipped with the guard that surrounds the grinding wheel.
- Choose a grinding disc that is appropriate for the type of work you are doing.
- Make sure that you are using a disc that is properly sized for the grinder.
- Before you start the grinder, make sure the grinding disc is secured and is in good condition.
- Make sure all guards are in place.

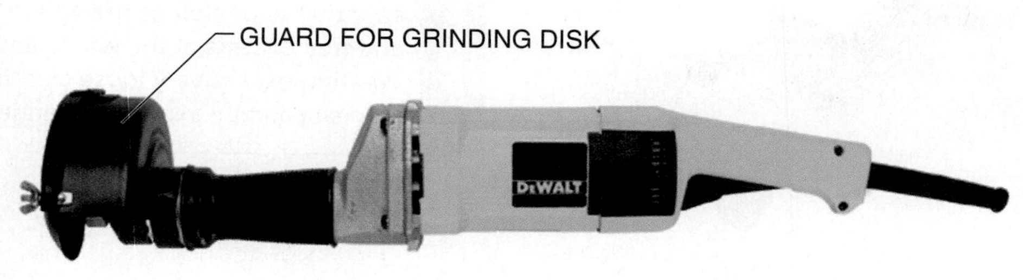

Figure 24 End grinder.

DETAIL GRINDER

104F25A.EPS

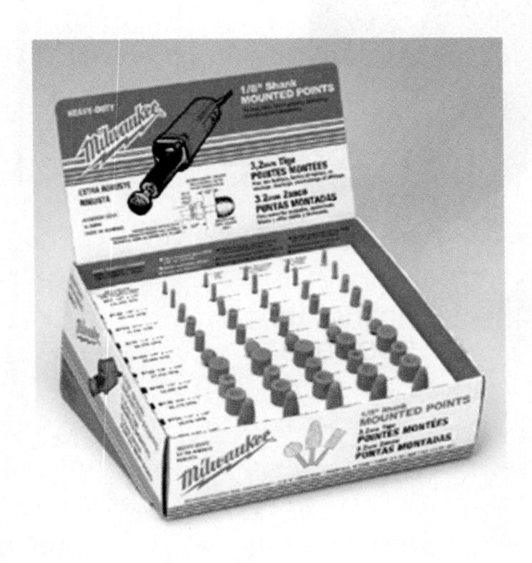

1/8-INCH SHANK-MOUNTED POINTS

104F25B.EPS

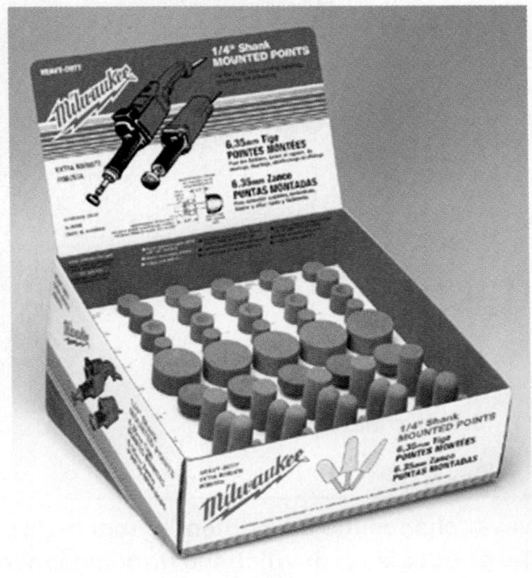

1/4-INCH SHANK-MOUNTED POINTS

104F25C.EPS

Figure 25 Detail grinder and points.

- Be sure to have firm footing and a firm grip before you use a grinder. Grinders have a tendency to pull you off balance.
- Always hold the grinder with both hands.
- Always use a spark deflector (shield) as well as proper eye protection.
- Direct sparks and debris away from people and any **hazardous materials**.
- When you are grinding on a platform, use a flame-retardant blanket to catch falling sparks.
- When you shut off the power, do not leave the tool until the grinding disc has come to a complete stop.
- Always disconnect the power source before you do maintenance.

WARNING!

Grinding discs can explode if used when they are cracked. Inspect the disc for cracks before using the grinder.

5.2.0 Bench Grinders

Bench grinders (*Figure 26*) are electrically powered stationary grinding machines. They usually have two grinding wheels that are used for grinding, rust removal, and metal buffing. They are also great for renewing worn edges and maintaining the sharp edges of cutting tools. Remember learning about the danger of mushroomed cold chisel heads in *Introduction to Hand Tools*? The bench grinder can smooth these heads.

Heavy-duty grinder wheels range from 6¾ to 10 inches in diameter. Each wheel's maximum speed is given in rpm. Never use a grinding wheel above its rated maximum speed. Bench grinders come with an adjustable tool rest. This is the surface on which you position the material you are grinding, such as cold chisel heads. There should be a distance of only ⅛-inch between the tool rest and the wheel. Attachments for the bench grinder include knot-wire brushes for removing rust, scale, and file marks from metal surfaces, and cloth buffing wheels for polishing and buffing metal surfaces.

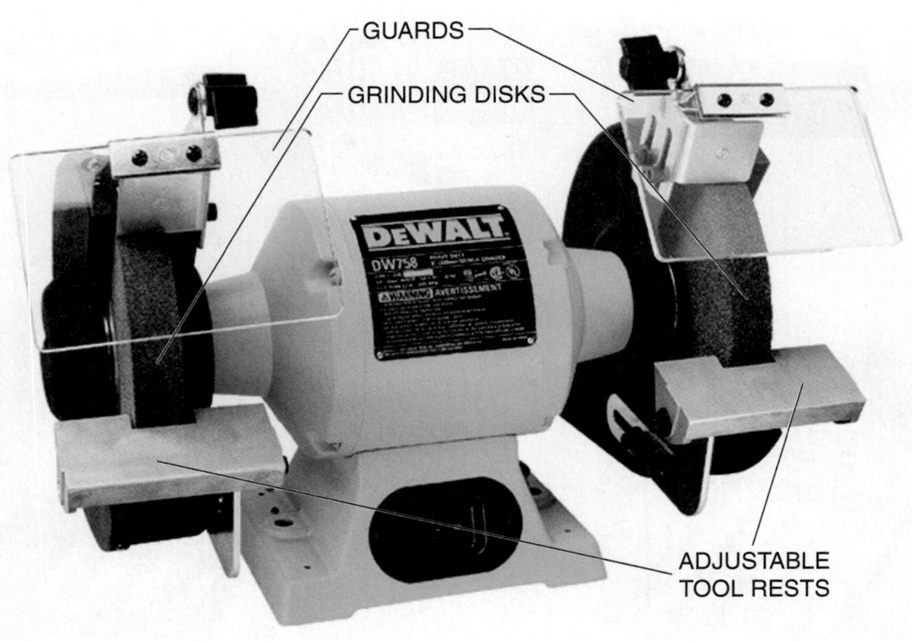

Figure 26 Bench grinder.

WARNING!

Never change the adjustment of tool rests when the grinder is on or when the grinding wheels are spinning. Doing so may damage the work or cause injury to you or another worker.

5.2.1 How to Use a Bench Grinder

Follow these steps to use a bench grinder safely and efficiently:

Step 1 Wear appropriate personal protective equipment. A face shield is essential.

Step 2 Always use the adjustable tool rest as a support when you are grinding or beveling metal pieces. There should be a maximum gap of ⅛-inch between the tool rest and the wheel and ¼-inch between the top guard and wheel. Make sure the bench grinder is placed on a secure surface.

Step 3 Let the wheel come up to full speed before you touch the work.

Step 4 Keep the metal you are grinding cool. If the metal gets too hot, it can destroy the temper (hardness) of the material you are grinding.

Step 5 Whenever possible, work on the face of the wheel. For many jobs, you must work on the side of the wheel, but inspect the wheel frequently to be sure you do not reduce the thickness so much that it can break. *Figure 27* shows the proper way to use a bench grinder.

5.2.2 Safety and Maintenance

Follow these guidelines to ensure safety for yourself and your co-workers, and a long life for the grinder:

- Always wear appropriate personal protective equipment.
- Never wear loose clothing or jewelry when you are grinding. It can get caught in the wheels.
- Grinding metal creates sparks, so keep the area around the grinder clean.
- Always adjust the tool rests so they are within ⅛-inch of the wheel. This reduces the chance of getting the work wedged between the rest and the wheel.
- Keep your hands away from the grinding wheels.
- Let the wheel come up to full speed before you touch the work.
- Never use a grinding wheel above its rated maximum speed.

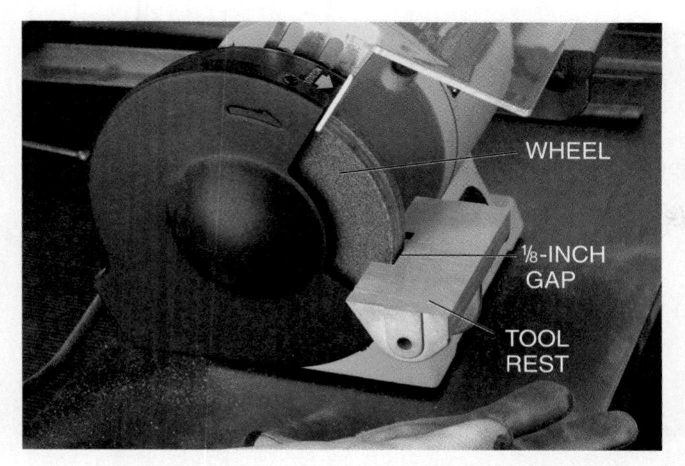

THERE SHOULD BE A ⅛-INCH GAP BETWEEN THE TOOL REST AND THE WHEEL AND A ¼-INCH GAP BETWEEN THE TOP GUARD AND THE WHEEL.

104F27A.EPS

104F27B.EPS

WHENEVER POSSIBLE, WORK ON THE FACE OF THE WHEEL.

Figure 27 Proper use of a bench grinder.

- When you are finished using the bench grinder, shut it off.
- Always make sure the bench grinder is disconnected before you change grinding wheels.
- Perform a **ring test** before you mount a wheel. When performing a ring test, look for chipped edges and cracks. Then, mount the wheel on a rod that you pass through the wheel hole. Tap the wheel gently on the side with a piece of wood. The wheel will ring clearly if it is in good condition. A dull thud may mean that there is a crack that you can't see. Get rid of the wheel if this happens.

WARNING!

Never grind soft metals or wood on a grinding wheel. The material will wedge itself between the grit, creating stresses that may cause the wheel to shatter.

6.0.0 Miscellaneous Power Tools

It is common to see several different types of power tools on a construction site. In this section, you will learn about other power tools frequently used on the job site, including the following:

- Pneumatically powered nailers (nail guns)
- Powder-actuated fastening systems
- Air impact wrenches
- Pavement breakers
- Hydraulic jacks

6.1.0 Pneumatically Powered Nailers (Nail Guns)

Pneumatically powered nailers (*Figure 28*), or nail guns, are common on construction jobs. They greatly speed up the installation of materials such as wallboard, molding, framing members, and shingles.

Nail guns are driven by compressed air traveling through air lines connected to an air compressor. Nailers are designed for specific purposes,

104F28.EPS

Figure 28 Pneumatic nailer.

On-Site

Power Nailer Safety

Pneumatic nailers are designed to fire when the trigger is pressed and the tool is pressed against the material being fastened. An important safety feature of all pneumatic nailers is that they will not fire unless pressed against the material.

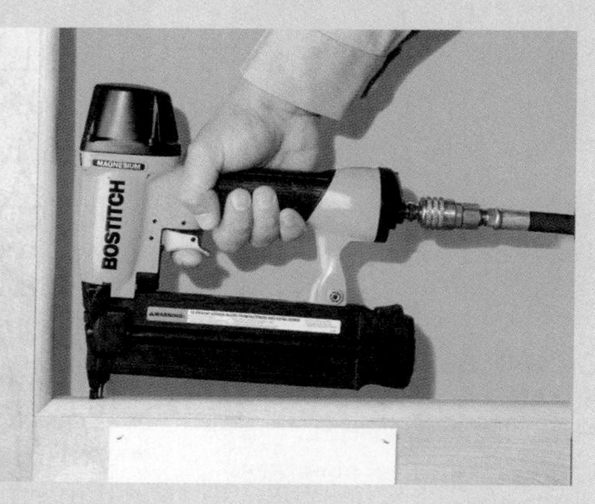

104SA02.EPS

On-Site

Power Screw-drivers

This tool also uses a power source (this model uses a battery) to speed production in a variety of applications, such as drywalling, floor sheathing and underlayment, decking, fencing, and cement board installation. A chain of screws feeds automatically into the firing chamber. Most models incorporate a back-out feature to drive out screws as well as a guide that keeps the screw feed aligned and tangle free. This tool can accept Phillips or square slot screws and weighs an average of six pounds.

104SA03.EPS

such as roofing, framing, siding, flooring, sheathing, trim, and finishing. Nailers use specific types of nails depending on the material to be fastened. The nails come in coils and in strips and are loaded into the nail gun.

WARNING!

Never exceed the maximum specified operating pressure of a pneumatic nailer. Doing so will damage the pneumatic nailer and cause injury.

6.1.1 How to Use a Power Nailer

Follow these steps to use a power nailer safely and efficiently:

Step 1 Read the manufacturer's instructions before using a pneumatically powered nailer. Wear appropriate personal protective equipment.

Step 2 Inspect the nailer for damage and loose connections.

Step 3 Load the nails into the nailer. Be sure to use the correct type of nail for the job.

Step 4 Ensure that hoses are connected properly.

Step 5 Check the air compressor and adjust the pressure.

Step 6 Try a test nail in scrap material. Most nailers operate at pressures of 70 to 120 pounds per square inch (psi).
If nail penetration is not correct, follow the manufacturer's instructions for adjusting the particular gun.

Step 7 When nailing wall materials, locate and mark wall studs before nailing. Otherwise, you won't be able to feel a missed nail that penetrates the wallboard but misses the stud.

Step 8 Hold the nailer firmly against the material to be fastened, then press the trigger (*Figure 29*).

Step 9 Disconnect the air hose as soon as you finish and never leave the nailer connected and unattended.

WARNING!

A nail gun is not a toy. Playing with a nail gun can cause serious injury. Nails can easily pierce a hand, leg, or eye. Never point a nail gun at anyone or carry one with your finger on the trigger. Use the nail gun only as directed.

104F29.EPS

Figure 29 Proper use of a nailer.

6.1.2 Safety and Maintenance

Follow these guidelines to ensure safety for yourself and your co-workers, and a long life for the nail gun:

- Always wear appropriate personal protective equipment.
- Never aim the nail gun toward your body.
- Review the operating manual before using any nailer.
- Keep the nailer oiled according to the manufacturer's instructions. Add a few drops to the air inlet before each use, according to the manufacturer's recommendations.
- Use the correct nailer for the job. Use the correct size and type of nail for the job.
- Never load the nailer with the compressor hose attached.
- Never leave the nailer connected when not in use.
- If the nailer is not firing, disconnect the air hose before you attempt repairs.
- Keep all body parts and co-workers away from the nail path to avoid serious injury. Nails can go through paneling and strike someone on the other side.
- Check for pipes, electrical wiring, vents, and other materials behind wallboard before nailing.

On-Site

Power Nailer Safety

Pneumatic nailers are designed to fire when the trigger is pressed and the tool is pressed against the material being fastened. An important safety feature of all pneumatic nailers is that they will not fire unless pressed against the material.

6.2.0 Powder-Actuated Fastening Systems

The use of powder-actuated anchor or fastening systems has been increasing rapidly in recent years. They are used for anchoring static loads to steel and concrete beams, walls, and so forth.

A powder-actuated tool is a low-velocity fastening system powered by gunpowder cartridges called **boosters**. The tools are used to drive steel pins or threaded steel studs directly into masonry and steel (*Figure 30*).

> **WARNING!**
> OSHA requires that all operators of powder-actuated tools be qualified and certified by the manufacturer of the tool. You must carry a certification card when using the tool.

> **WARNING!**
> Avoid firing a powder-actuated tool into easily penetrated materials. The fastener may pass through the material and become a flying missile on the other side.

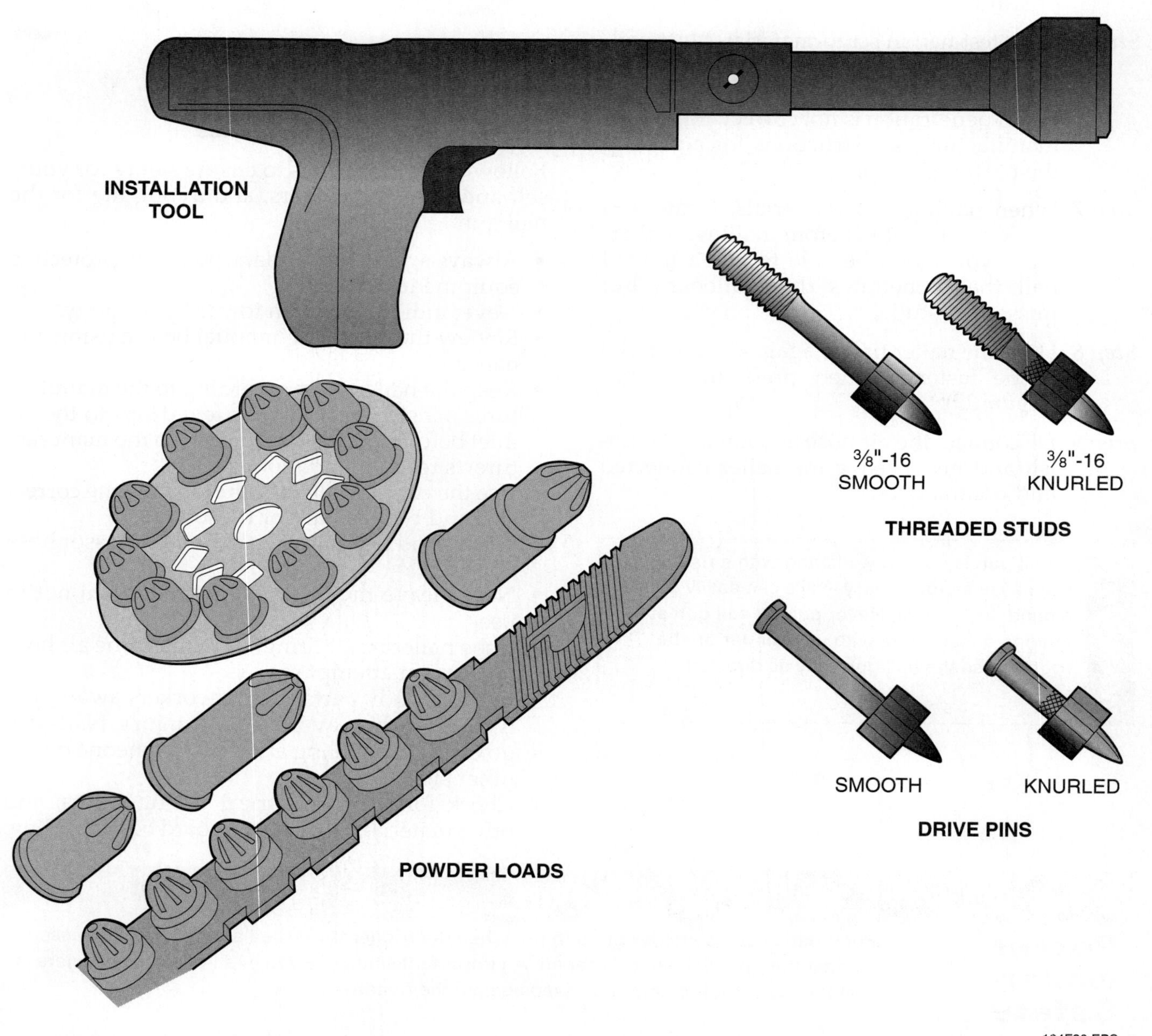

Figure 30 Powder-actuated fastening system.

WARNING!

If a powder load fails to fire, wait ten seconds, remove it from the installation tool, and then dispose of it in a bucket of water or oil.

6.2.1 How to Set Up and Use a Powder-Actuated Fastening Tool

Follow these steps to use a powder-actuated tool safely and efficiently:

Step 1 Wear the appropriate personal protective equipment, including safety goggles, ear protection, and a hard hat.

Step 2 Feed the pin or stud into the piston.

Step 3 Feed the gunpowder cartridge (booster or charge) into position.

Step 4 Position the tool in front of the item to be fastened and press it against the mounting surface. This pressure releases the safety lock.

Step 5 Pull the trigger handle to fire the booster charge (*Figure 30*).

6.2.2 Safety and Maintenance

Follow these guidelines to ensure safety for yourself and your co-workers, and a long life for the powder-actuated tool:

- Always wear appropriate personal protective equipment, including ear protection, safety goggles, and a hard hat.
- Do not use a powder-actuated tool until you are certified on the model you will be using.
- Follow all safety precautions in the manufacturer's instruction manual.
- Do not load the tool until you are prepared to complete the firing.
- Use the proper size pin for the job you are doing.
- When loading the driver, put the pin in before you load the charge.
- Use the correct booster charge according to the manufacturer's instructions for the tool being used.
- Never hold the end of the barrel against any part of your body or cock the tool against your hand.
- Never hold your hand behind the material you are fastening.
- Do not fire the tool close to the edge of concrete. Pieces of concrete may chip off and strike someone, or the projectile could continue past the concrete and strike a co-worker.
- Never try to pry the booster out of the magazine with a sharp instrument.

6.3.0 Air Impact Wrench

Air impact wrenches (*Figure 31*) are power tools that are used to fasten, tighten, and loosen nuts and bolts. The speed and strength (torque) of these wrenches can easily be adjusted depending on the type of job. Air impact wrenches are powered pneumatically (with compressed air). In order to operate an air wrench, it must be attached with a hose to an air compressor.

104F31A.EPS

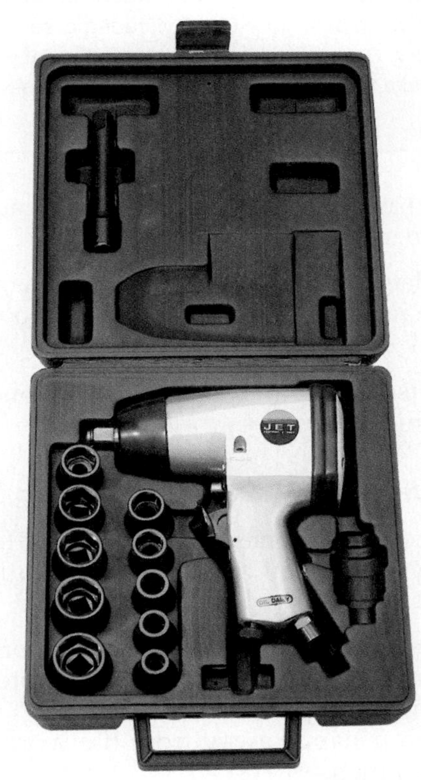

104F31B.EPS

Figure 31 Air impact wrench.

6.3.1 How to Set Up and Use an Air Impact Wrench

Follow these steps to use an air impact wrench safely and efficiently:

Step 1 Read the manufacturer's instructions before using an air impact wrench. Wear appropriate personal protective equipment, including eye and ear protection.

Step 2 Inspect the wrench for damage.

Step 3 Select the appropriate impact socket.

> **WARNING!**
> Using handheld sockets can damage property and cause injury. Use only impact sockets made for air impact wrenches.

Step 4 Connect the wrench to the appropriate air hose.

> **WARNING!**
> The air hose must be connected properly and securely. An unsecured air hose can come loose and whip around violently, causing serious injury. Some fittings require the use of whip checks to keep them from coming loose.

Step 5 Turn on the air compressor and adjust the pressure.

Step 6 Place the impact socket firmly against the material to be fastened, removed, or loosened, and press the trigger.

Step 7 Disconnect the air hose as soon as you finish.

6.3.2 Safety and Maintenance

Follow these guidelines to ensure safety for yourself and your nearby co-workers, and a long life for the air impact wrench:

- Always wear appropriate personal protective equipment, including eye and ear protection.
- Keep your body stance balanced.
- Keep your hands away from the working end of the wrench.
- Ensure that the workpiece is secure.
- Always use clean, dry air at the proper pressure.
- Always turn off the air supply and disconnect the air supply hose before performing any maintenance on the wrench.

6.4.0 Pavement Breakers

Several large-scale demolition tools are frequently used in construction. They include pavement breakers, clay spades, rock drills, and core borers (*Figure 32*). These tools do not rotate like hammer drills. They reciprocate (move back and forth). The name jackhammer comes from a trade name, but has come to refer to almost any of the handheld impact tools. There are differences in the tools and their uses, however. In this section, we will look at the pavement breaker.

The pavement breaker is used for large-scale demolition work, such as tearing down brick and concrete walls and breaking up concrete or pavement.

A pavement breaker weighs from 50 to 90 pounds. On most pavement breakers, a throttle is located on the T-handle. When you push the throttle, compressed air operates a piston inside the tool. The piston drives the steel-cutting shank into the material you want to break up. You can use attachments, such as spades or chisels, for different tasks.

6.4.1 How to Set Up and Use a Pavement Breaker

Follow these steps to use a pavement breaker safely and efficiently:

Step 1 Wear appropriate personal protective equipment.

Step 2 Make sure that the air pressure is shut off at the main air outlet.

Step 3 Hold the coupler at the end of the air supply line, slide the ring back, and slip the coupler on the connector, or nipple, that is attached to the air drill.

Step 4 Check to see if you have a good connection. (A good coupling cannot be taken apart without first sliding the ring back.)

Step 5 Add a whip check.

Step 6 Once you have a good connection, turn on the air supply valve. The pavement breaker is now ready to use.

6.4.2 Safety and Maintenance

Follow these guidelines to ensure safety for yourself and your co-workers, and a long life for the demolition tool:

- Always wear appropriate personal protective equipment. Because some of these tools make a lot of noise, you must wear hearing protection (earplugs).

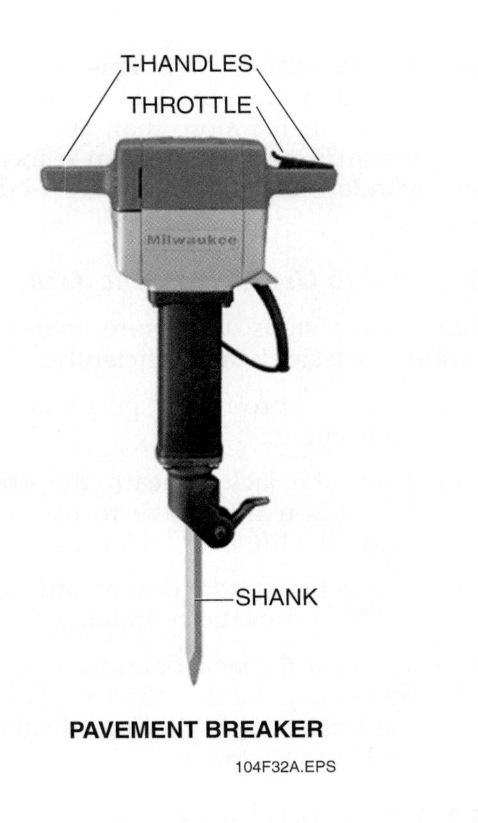

PAVEMENT BREAKER

104F32A.EPS

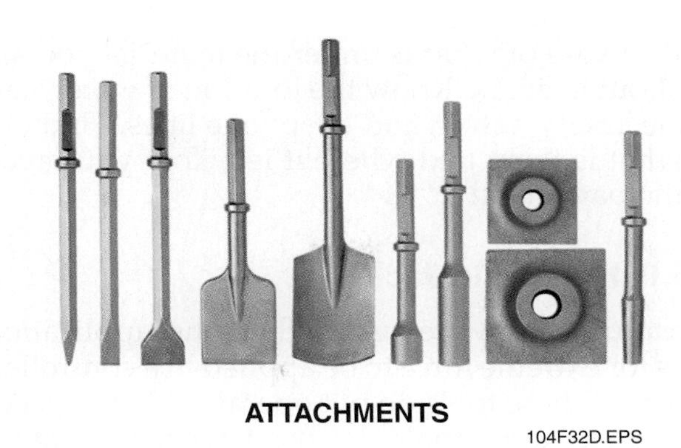

ATTACHMENTS

104F32D.EPS

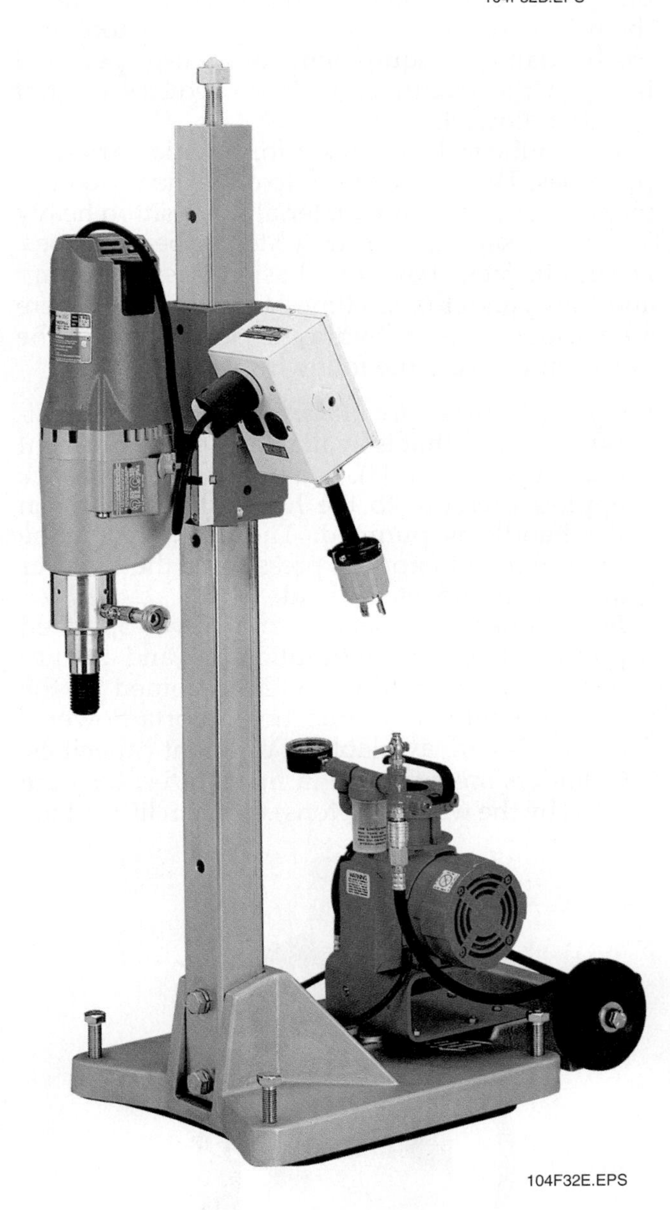

104F32E.EPS

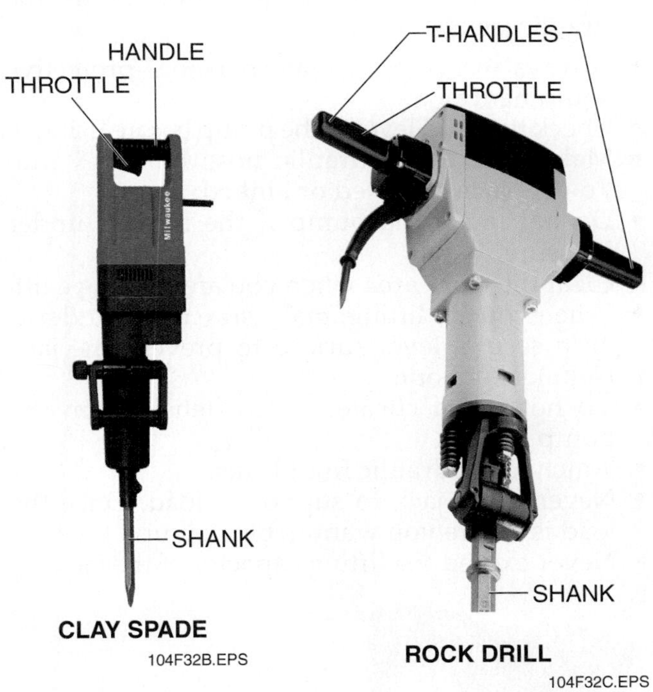

CLAY SPADE

104F32B.EPS

ROCK DRILL

104F32C.EPS

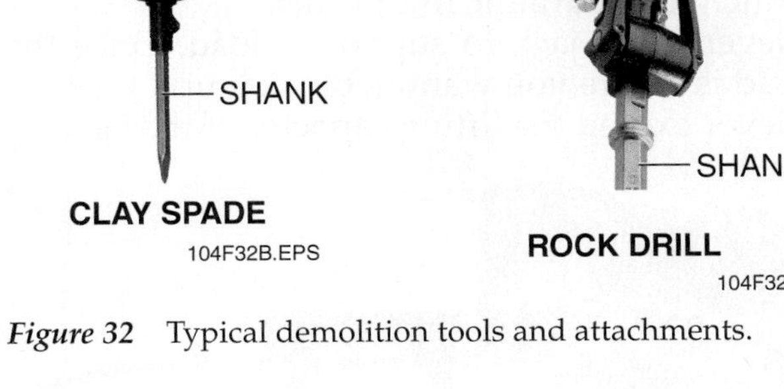

Figure 32 Typical demolition tools and attachments.

- Be aware of what is under the material you are about to break. Know the location of water, gas, electricity, sewer, and telephone lines. Find out what is there and where it is before you break the pavement!

6.5.0 Hydraulic Jack

Hydraulic tools are used when the application calls for extreme force to be applied in a controlled manner. These tools do not operate at high speed, but great care should be used when operating them. The forces generated by hydraulic tools can easily damage equipment or cause personal injury if the manufacturer's procedures are not strictly followed.

Hydraulic jacks are used for a wide variety of purposes. They can be used to move heavy equipment and other heavy material, to position heavy loads precisely, and to straighten or bend frames. Hydraulic jacks have two basic parts: the pump and the cylinder (sometimes called a ram). There are various types of hydraulic jacks. Some of the most common are the following:

- *Hydraulic jacks with internal pumps* – A general-purpose jack that is available in many different capacities (*Figure 33*). The pump inside the jack applies pressure to the hydraulic fluid when the handle is pumped. The pressure on the hydraulic fluid applies pressure to the cylinder and lifts or moves the load.
- *Porta-Power®* – Consists of a lever-operated pump, a length of hydraulic hose, and a cylinder. The pump and cylinder are joined by the high-pressure hydraulic hose. Porta-Powers® (*Figure 34*) are available in different capacities. Cylinders are available in many sizes; they are rated by the weight (in tons) they can lift and the distance they can move it. This distance is called stroke and is measured in inches. Hydraulic cylinders can lift more than 500 tons. Strokes range from ¼-inch to more than 48 inches. Different cylinder sizes and ratings are used for different jobs.

104F33.EPS

Figure 33 Portable hydraulic jack.

6.5.1 How to Use a Hydraulic Jack

Follow these steps to ensure that you use a hydraulic jack safely and efficiently:

Step 1 Wear appropriate personal protective equipment.

Step 2 Place the jack beneath the object to be lifted. You may have to use a wedge to begin the lift.

Step 3 Pump the handle down, and then release it. This raises the cylinder.

Step 4 To lower the jack, open the return passage by turning the thumbscrew. The weight of the load pushes the fluid in the cylinder back into the pump.

6.5.2 Safety and Maintenance

Follow these guidelines to ensure safety for yourself and your co-workers, and a long life for the hydraulic jack:

- Always wear appropriate personal protective equipment.
- Check the fluid level in the pump before using it.
- Make sure the hydraulic hose on the Porta-Power® is not twisted or kinked.
- Do not move the pump if the hose is under pressure.
- Clear the work area when you are making a lift.
- When you are lifting, make sure the cylinder is on a secure, level surface to prevent the jack from kicking out.
- Do not use a cheater bar (extension) on the pump handle.
- Watch for hydraulic fluid leaks.
- Never use a jack to support a load. Once the load is where you want it, block it up.
- Never exceed the lifting capacity of the jack.

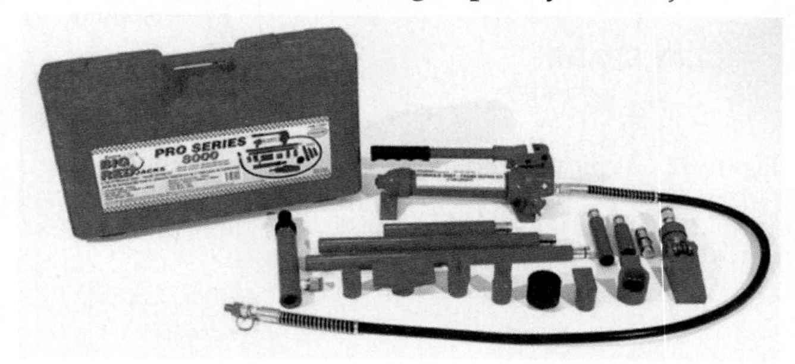

104F34.EPS

Figure 34 Porta-Power®.

Review Questions

1. Pneumatic tools get their power from _____.
   a. air pressure
   b. fluid pressure
   c. hand pumps
   d. AC power sources

2. The most common use of the power drill is to _____.
   a. cut wood, metal, and plastic
   b. drive nails into wood, metal, and plastic
   c. make holes in wood, metal, and plastic
   d. carve letters in wood, metal, and plastic

3. Hammer drills are designed to drill into _____.
   a. wood, metal, and plastic
   b. concrete, brick, and tile
   c. drywall
   d. roofing shingles

4. The electromagnetic drill is a _____.
   a. handheld drill used on wood
   b. cordless drill used on masonry and tile
   c. portable drill used on thick metal
   d. pneumatic drill that has a pounding action

5. When cutting with a circular saw, grip the saw handles _____.
   a. and pull the saw toward you
   b. firmly with one hand
   c. loosely with one hand
   d. firmly with two hands

6. When using a saber saw, avoid vibration by _____.
   a. holding the workpiece down with your free hand
   b. setting a heavy object on the workpiece
   c. using a low-speed setting
   d. using a clamp or vise to hold the work

7. Before you cut through a wall or partition, always _____.
   a. remove the lower blade guard
   b. find out what is on the other side
   c. increase the revolutions per minute
   d. lubricate the guard with oil or grease

8. The high speed setting on a reciprocating saw is used for _____.
   a. sawing wood and other soft materials
   b. metal work
   c. cutting through drywall
   d. grinding surfaces

9. Use only a bandsaw that has a _____.
   a. breastplate with a broad surface
   b. battery pack
   c. thick, three-piece blade
   d. stop

10. The end grinder is used to _____.
    a. polish intricate work
    b. grind surfaces
    c. smooth the work before painting
    d. smooth the inside of materials, such as pipe

11. A detail grinder uses _____ to smooth and polish intricate metallic work.
    a. points
    b. abrasive paper
    c. wire brushes
    d. grinding wheels

12. Powder-actuated fastening systems are used to _____.
    a. penetrate drywall
    b. hammer nails
    c. anchor static loads to steel beams
    d. remove nails

13. Before you begin setting up a pavement breaker for use, make sure that the air pressure is _____.
    a. shut off at the coupler
    b. turned on only halfway
    c. turned on full
    d. shut off at the main air outlet

14. Porta-Power® cylinders are rated by how much weight they can lift and by _____.
    a. their torque
    b. the amount of electromagnetic material they have
    c. how much they weigh
    d. the distance they can move the weight

15. Hydraulic jacks are used when the application calls for _____.
    a. operation at high speed
    b. extreme force to be applied
    c. quiet operation
    d. manually-assisted lifting

SUMMARY

Power tools are a necessity in the construction industry. You might not use all of the tools covered in this module during your career, but it is still important for you to understand how they work and what they do. In fact, it's likely that you'll find yourself working around other craftworkers who use them. You and your co-workers will be safer if everyone is familiar with the tools being used on the job site.

You must also learn how to maintain your power tools properly, whether they belong to you or your employer. The better care you take of your tools, the better and more safely they will function, and the longer they'll last. Proper maintenance of power tools saves you and your employer time and money.

As you progress in your chosen field within the construction industry, you will learn to use the power tools for your specialized area. Although some of these specific tools might not be covered in this module, the basic safety and usage concepts are always applicable. Remember to always read the manufacturer's manual for any new power tool you use and never to use a tool on which you have not been properly trained. Following the basic use and safety guidelines explained in this module, maintaining your tools well, and educating yourself before using any new equipment will help you progress in your career, work efficiently, and stay safe.

Trade Terms Quiz

Fill in the blank with the correct key term that you learned from your study of this module.

1. Activate the ____________ to make the trigger stay in operating mode even without your finger on the trigger.
2. ____________ reverses its direction at regularly recurring intervals; this type of current is delivered through wall plugs.
3. A(n) ____________ saw's straight blades move back and forth.
4. A(n) ____________ powers a powder-actuated tool.
5. ____________ must be accompanied by material safety data sheets.
6. Masonry bits and nail cutter saw blades have a(n) ____________ tip.
7. A(n) ____________ is a substance, such as sandpaper, that is used to wear away material.
8. A(n) ____________ is used to open and close the chuck on a power drill.
9. ____________ is the number of times a drill bit completes one full rotation in a minute.
10. A(n) ____________ is used to set the head of a screw at or below the surface of the material.
11. ____________ flows in one direction, from the negative to the positive terminal of the source.
12. Belt sanders and circular saws are examples of ____________.
13. Use a(n) ____________ to bore holes in wood and other materials.
14. An electromagnet holds an electromagnetic drill in place on a(n) ____________ metal surface.
15. ____________ is applied to the surface of a grinding wheel to give it a nonslip finish.
16. The ____________ of the drill holds the drill bit.
17. To prevent an electrical shock, do not operate electric power tools without proper ____________.
18. A Porta-Power® is an example of a(n) ____________.
19. ____________ refers to building material such as stone, brick, or concrete block.
20. Air hammers and pneumatic nailers are examples of ____________.
21. Perform a(n) ____________ to check the condition of a grinding wheel.
22. The ____________ is the smooth part of a drill bit that fits into the chuck.
23. A(n) ____________ protects people from electric shock and protects equipment from damage by interrupting the flow of electricity if a circuit fault occurs.

Trade Terms

Abrasive
AC (alternating current)
Auger
Booster
Carbide
Chuck
Chuck key
Countersink
DC (direct current)
Electric tools
Ferromagnetic
Grit
Ground fault circuit interrupter (GFCI)
Ground fault protection
Hazardous materials
Hydraulic tools
Masonry
Pneumatic tools
Reciprocating
Revolutions per minute (rpm)
Ring test
Shank
Trigger lock

Cornerstone of Craftsmanship

Peter J. Klapperich

Associated Training Services Superintendent,
National Training Director/Affiliate School Development &
Safety Director NCCER Compliance Officer

How did you choose a career in the field?
I am a fourth-generation carpenter. I tried other fields, but came back to construction after a year and a half of college.

What types of training have you been through?
I have been through a number of NCCER curricula, including Core craft instructor for Carpentry, Master Trainer, and Construction Site Safety Supervisor. I have been honored to be a subject matter expert reviewer for Carpentry and Core. I am OSHA-30 certified and will soon complete my OSHA-500 trainer course. I have complemented my craft training skills by taking many seminars on training, coaching, selling, and motivation. I also hold a first aid/CPR Red Cross instructor's card.

What kinds of work have you done in your career?
I have been able to have variety in my career, including carpentry, general contracting, and heavy equipment operation. I have been a business owner with a construction and remodeling company. And I am a professional trainer of craftworkers and salespeople.

Tell us about your present job and what you like about it.
The best part of my position is that I can train both students and instructors. I also get to write training programs, policies, and procedures. Our training changes people's lives, and to be part of that process is exciting, extremely gratifying, and special. I also teach two nights a week at a vocational technical school when I am not on-site at one of our 10 schools. The interaction with students and instructors is why I do what I do.

What factors have contributed most to your success?
Passion: for not only the trades, but also a passion to pass on that knowledge and help shape people's lives for the better. Also, my own appetite for additional information or study, related to construction and/or training, has been a key success factor. And I practice my good listening skills.

What advice would you give to those new to the field?
Keep your eyes, ears, and mind open to all areas of your trade. Pursue any and all forms of training.

Tell us some interesting career-related facts or accomplishments:
My entire career has been very exciting and I count my accomplishments by the people I reach, even in any small way.

Cornerstone of Craftsmanship

Rick Klepin

Corbins Electric, LLC Career Path Training Director

How did you choose a career in the field?
I started as a duct insulator in my uncle's sheet metal shop in Houston, Texas, and from there kept soaking up knowledge wherever I could find it. I kept taking on new challenges so I could become a better worker, and the pay wasn't bad either.

What types of training have you been through?
Some of my training was on-the-job training, but I did complete training with the Local 54 in Houston, where I became a journeyman. I worked on many of the high-rise buildings you see today in the downtown area. I then decided I wanted to go back to a shop environment and worked as a layout bench-man in various sheet metal shops. Through the years, I have been trained in everything from sheet metal, OSHA 500, various management courses, CAD, Safety, and various training through the National Center for Construction Education and Research (NCCER).

What kinds of work have you done in your career?
I have been a sheet metal journeyman, shop foreman, safety director, craft training director, and an NCCER sponsor representative. I am currently an OSHA 500 certified trainer, NCCER instructor, master trainer, assessment administrator, and the career path training director for Corbins Service Electric in Phoenix, Arizona.

Tell us about your present job and what you like about it.
I am responsible for training all 300 employees as the Career Path Training Director. One of the best feelings is watching the face of an apprentice when the "light bulb goes on" and knowing you have reached that individual through your training. It is one of the greatest rewards as a trainer.

What factors have contributed most to your success?
I attribute most of my success to my family and a sense of morals that was bestowed upon me by my parents, Richard and Marie Klepin. Their belief in life values and morals helped me through the years, in both easy times and the tough ones; we all have them. I would like to thank my grandfathers, Joe and Nick, both steel workers from Youngstown, Ohio, for the tradeworker in me, and my grandmothers, Theresa and Mary, for my common sense abilities.

Most of all thanks to my wife, Sherri, who has always told me to "go for it," especially when I was afraid to. I have always tried to be successful, and one way to do that is to try things you really didn't think you would like to do. I have always tried to live my life to the fullest and find something to smile about each day. Try something different; you might just like it!

What advice would you give to those new to the field?
Try to learn something new each day. My Uncle Tom would tell me that he learned something new the first day he worked and learned something the day that he retired from the sheet metal trade. Love your family. They will be there for you when you need them. Work hard, be on time and get along with your co-workers. You spend the majority of your life working with people; you don't need to like them, but you have to be able to work with them. It just makes life a little easier, and chances are it will help you in your career successes.

Tell us some interesting career-related facts or accomplishments:
I have been a Craft Training Director for two different companies in Phoenix, Arizona: Mechanical Corporation and Corbins Service Electrical. I have had the privilege of learning many things by meeting and working with many people. Twenty-eight years ago, when I first moved to Houston to become a sheet metal instructor, I never thought I would be a training director for a leading electrical contractor in the Southwest. Life has many surprises.

Trade Terms Introduced in this Module

Abrasive: A substance, such as sandpaper, that is used to wear away material.

AC (alternating current): An electrical current that reverses its direction at regularly recurring intervals; the current delivered through wall plugs.

Auger: A tool with a spiral cutting edge for boring holes in wood and other materials.

Booster: Gunpowder cartridge used to power powder-actuated fastening tools.

Carbide: A very hard material made of carbon and one or more heavy metals. Commonly used in one type of saw blade.

Chuck: A clamping device that holds an attachment; for example, the chuck of the drill holds the drill bit.

Chuck key: A small, T-shaped steel piece used to open and close the chuck on power drills.

Countersink: A bit or drill used to set the head of a screw at or below the surface of the material.

DC (direct current): Electrical current that flows in one direction, from the negative (2) to the positive (1) terminal of the source, such as a battery.

Electric tools: Tools powered by electricity. The electricity is supplied by either an AC source (wall plug) or a DC source (battery).

Ferromagnetic: Having magnetic properties. Substances such as iron, nickel, cobalt, and various alloys are ferromagnetic.

Grit: A granular abrasive used to make sandpaper or applied to the surface of a grinding wheel to give it a nonslip finish. Grit is graded according to its texture. The grit number indicates the number of abrasive granules in a standard size (per inch or per cm). The higher the grit number, the finer the abrasive material.

Ground fault circuit interrupter (GFCI): A circuit breaker designed to protect people from electric shock and to protect equipment from damage by interrupting the flow of electricity if a circuit fault occurs.

Ground fault protection: Protection against short circuits; a safety device cuts power off as soon as it senses any imbalance between incoming and outgoing current.

Hazardous materials: Materials (such as chemicals) that must be transported, stored, applied, handled, and identified according to federal, state, or local regulations. Hazardous materials must be accompanied by material safety data sheets (MSDSs).

Hydraulic tools: Tools powered by fluid pressure. The pressure is produced by hand pumps or electric pumps.

Masonry: Building material such as stone, brick, or concrete block.

Pneumatic tools: Air-powered tools. The power is produced by electric or fuel-powered compressors.

Reciprocating: Moving back and forth.

Revolutions per minute (rpm): The number of times (or rate) a motor component or accessory (drill bit) completes one full rotation every minute.

Ring test: A method of testing the condition of a grinding wheel. The wheel is mounted on a rod and tapped. A clear ring means the wheel is in good condition; a dull thud means the wheel is in poor condition and should be disposed of.

Shank: The smooth part of a drill bit that fits into the chuck.

Trigger lock: A small lever, switch, or part that you push or pull to activate a locking catch or spring. Activating the trigger lock causes the trigger to stay in the operating mode even without your finger on the trigger.

Additional Resources

This module is intended to present thorough resources for task training. The following reference works are suggested for further study. These are optional materials for continued education rather than for task training.

29 CFR 1926, OSHA Construction Industry Regulations, latest edition. Washington, DC: Occupational Safety and Health Administration, U.S. Department of Labor, U.S. Government Printing Office.

All About Power Tools. 2002. Des Moines, IA: Meredith Books.

Hand & Power Tool Training. Video. All About OSHA. Surprise, AZ.

Power Tools. 1997. Minnetonka, MN: Handyman Club of America.

Powered Hand Tool Safety: Handle with Care. Video. 20 minutes. Coastal Training Technologies Corp. Virginia Beach, VA.

Reader's Digest Book of Skills and Tools, 1993 edition. Pleasantville, NY: Reader's Digest.

CONTREN® LEARNING SERIES – USER UPDATE

NCCER makes every effort to keep these textbooks up-to-date and free of technical errors. We appreciate your help in this process. If you have an idea for improving this textbook, or if you find an error, a typographical mistake, or an inaccuracy in NCCER's Contren® textbooks, please write us, using this form or a photocopy. Be sure to include the exact module number, page number, a detailed description, and the correction, if applicable. Your input will be brought to the attention of the Technical Review Committee. Thank you for your assistance.

Instructors – If you found that additional materials were necessary in order to teach this module effectively, please let us know so that we may include them in the Equipment/Materials list in the Annotated Instructor's Guide.

Write: Product Development and Revision
National Center for Construction Education and Research
3600 NW 43rd St., Bldg. G, Gainesville, FL 32606

Fax: 352-334-0932

E-mail: curriculum@nccer.org

Craft ______________________ Module Name ______________________

Copyright Date ____________ Module Number ____________ Page Number(s) ____________

Description

____

____

____

____

(Optional) Correction

____

____

____

(Optional) Your Name and Address

____

____

____

Introduction to Construction Drawings

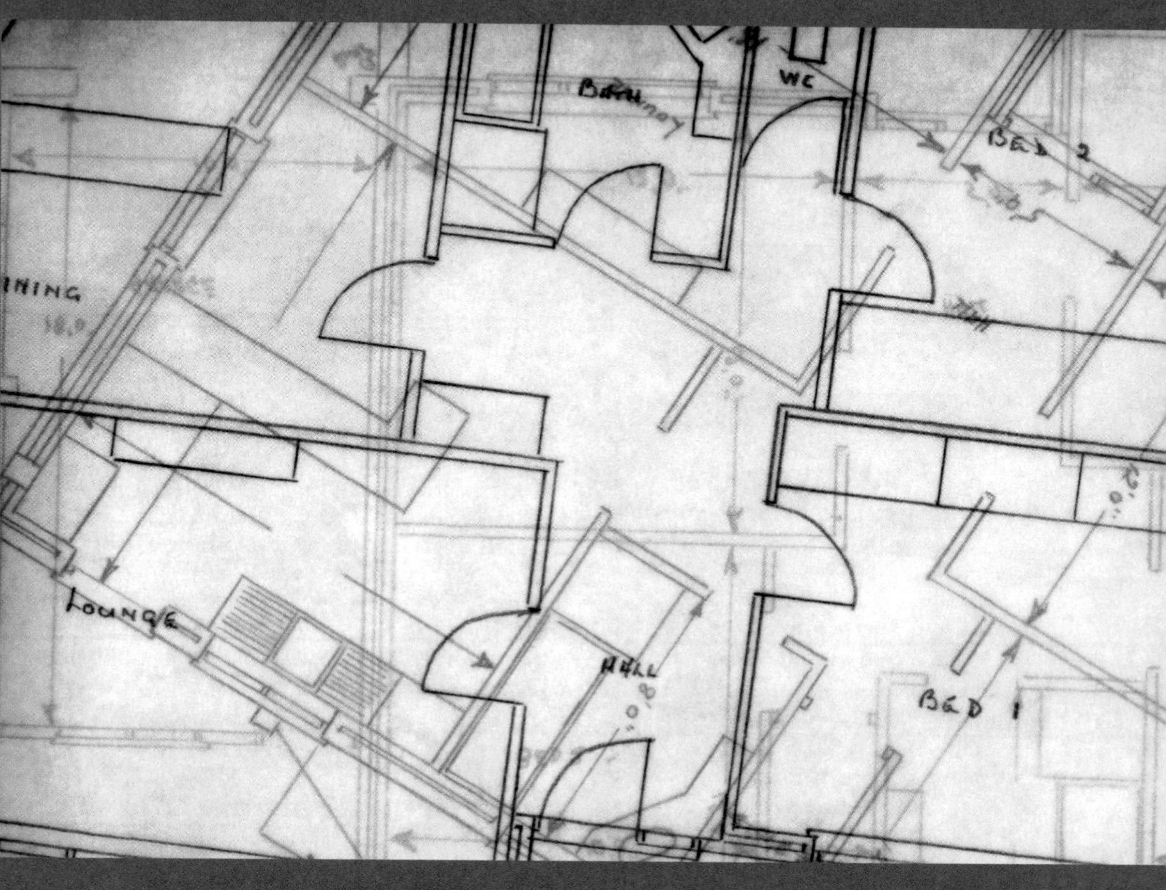

00105-09

CORE CURRICULUM

00109-09
Introduction to Materials Handling

00108-09
Basic Employability Skills

00107-09
Basic Communication Skills

00106-09
Basic Rigging

00105-09
Introduction to Construction Drawings

00104-09
Introduction to Power Tools

00103-09
Introduction to Hand Tools

00102-09
Introduction to Construction Math

00101-09
Basic Safety

This course map shows all of the modules in the *Core Curriculum: Introductory Craft Skills*. The suggested training order begins at the bottom and proceeds up. Skill levels increase as you advance on the course map. The local Training Program Sponsor may adjust the training order.

Note that Module 00106-09, *Basic Rigging*, is an elective. It is not a requirement for level completion, but it may be included as part of your training program.

Objectives

When you have completed this module, you will be able to do the following:

1. Recognize and identify basic construction drawing terms, components, and symbols.
2. Relate information on construction drawings to actual locations on the print.
3. Recognize different classifications of construction drawings.
4. Interpret and use drawing dimensions.

Trade Terms

Architect
Architectural plans
Beam
Blueprints
Civil plans
Computer-aided drafting (CAD)
Construction drawing
Contour lines
Detail drawings
Dimension line
Dimensions
Electrical plans
Elevation (EL)
Elevation drawing
Engineer
Fire protection plan
Floor plan
Foundation plan
Heating, ventilating, and air conditioning (HVAC)
Hidden line
Isometric drawing
Leader
Legend
Mechanical plans
Not to scale (NTS)
Piping and instrumentation drawings (P&IDs)
Plumbing
Plumbing plans
Request for information (RFI)
Roof plan
Scale
Schematic
Section drawing
Specifications
Structural plans
Symbol
Title block

Required Trainee Materials

Construction Drawing Plans included with this module

Prerequisites

Before you begin this module, it is recommended that you successfully complete the following: *Core Curriculum: Introductory Craft Skills*, Modules 00101-09 through 00104-09. Module 00106-09 is an elective and is not required for successful level completion.

DRAWINGS

Contents

Topics to be presented in this module include:

1.0.0 Introduction . . . 5.1
2.0.0 The Drawing Set . . . 5.1
2.1.0 Basic Components of Construction Drawings . . . 5.1
2.1.1 Title Block . . . 5.2
2.1.2 Border . . . 5.2
2.1.3 Drawing Area . . . 5.2
2.1.4 Revision Block . . . 5.3
2.1.5 Legend . . . 5.3
3.0.0 Six Types of Construction Drawings . . . 5.4
3.1.0 Civil Plans . . . 5.4
3.2.0 Architectural Plans . . . 5.4
3.3.0 Structural Plans . . . 5.8
3.4.0 Mechanical Plans . . . 5.11
3.5.0 Plumbing/Piping Plans . . . 5.11
3.6.0 Electrical Plans . . . 5.16
3.7.0 Fire Protection Plans . . . 5.16
3.8.0 Specifications . . . 5.16
3.9.0 Request for Information . . . 5.16
4.0.0 Scale . . . 5.16
5.0.0 Lines of Construction . . . 5.34
6.0.0 Abbreviations, Symbols, and Keynotes . . . 5.34
7.0.0 Using Gridlines to Identify Plan Locations . . . 5.39
8.0.0 Dimensions . . . 5.39

NOTE

Many of the sample construction drawings shown in this module are borrowed from Rinker Hall at the University of Florida (Gainesville). Rinker Hall is an example of green construction, achieving a LEED Gold rating from the U.S. Green Building Council in 2004. For more information on Rinker Hall and green construction, see the Going Green article in this module.

1.0.0 Introduction

Construction drawings are architectural or working drawings used to represent a structure or system. These were traditionally referred to as **blueprints**, because years ago the lines on a blueprint were white and the background was blue. Construction drawings are also called prints. Today, most prints are created by **computer-aided drafting (CAD)**, and they have blue or black lines on a white background.

Various kinds of construction drawings, including residential drawings, commercial drawings, landscaping plans, shop drawings, and industrial drawings, are used in construction. In this module, you will learn about some basic types of drawings.

Construction drawings, together with the set of **specifications** (specs), detail what is to be built and what materials are to be used.

2.0.0 The Drawing Set

The set of construction drawings forms the basis of agreement and understanding that a building will be built as detailed in the drawings. Therefore, everyone involved in planning, supplying, and building any structure should be able to read construction drawings. For any building project, also consult the civil engineering plans for that location, including sewer, highway, and water installation plans.

2.1.0 Basic Components of Construction Drawings

A set of construction drawing plans almost always includes six major types of drawings (*Figure 1*). These types of drawings will be discussed in greater detail later in this module. They include the following:

- Civil
- Architectural
- Structural
- Mechanical
- **Plumbing**
- Electrical
- Fire Protection (may be included)

Most construction drawings are laid out in a fairly standardized format. In this section, you will learn about the five parts of a construction drawing, which include the following:

- **Title block**
- Border
- Drawing area
- Revision block
- **Legend**

On-Site

What Is Computer-Aided Drafting?

The use of computers is a cost-effective way to increase drafting productivity because the computer program automates much of the repetitive work. A CAD system generates drawings from computer programs. Using CAD has the following advantages over hand drawing construction drawings:

- It is automated.
- The computer performs calculations quickly and easily.
- Changes can be made quickly and easily.
- Commonly used symbols can be easily retrieved.
- CAD can include three-dimensional modeling of the structure.

105SA01.EPS

2.1.1 Title Block

When you look at any drawing, the first thing to look at is the title block. The title block is normally in the lower right-hand corner of the drawing or across the right edge of the paper (*Figure* 2). The title block has two purposes. First, it gives information about the structure or assembly. Second, it is numbered so the print can be filed easily.

Different companies put different information in the title block. Generally, it contains the following:

- *Company logo* – Usually preprinted on the drawing.
- *Sheet title* – Identifies the project.
- *Date* – Date the drawing was checked and readied for seal, or issued for construction.
- *Drawn by* – Initials of the person who drafted the drawing.
- *Drawing number* – Code numbers assigned to a project.
- *Scale* – The ratio of the size of the object as drawn to the object's actual size.
- *Revision blocks* – Information on revisions, including (at a minimum) the date and the initials of the person making the revision. Other information may include descriptions of the revision and a revision number.

Every company has its own system for such things as project numbers and departments. Every company also has its own placement locations for the title and revision blocks. Your supervisor should explain your company's system to you.

> **Did You Know?**
>
> **How Blueprints Started**
>
> The process for making blueprints was developed in 1842 by an English astronomer named Sir John F. Herschel. The method involved coating a paper with a special chemical. After the coating dried, an original hand drawing was placed on top of the paper. Both papers were then covered with a piece of glass and set in the sunlight for about an hour. The coated paper was developed much like a photograph. After a cold-water wash, the coated paper turned blue, and the lines of the drawing remained white.

2.1.2 Border

The border is a clear area of approximately half an inch around the edge of the drawing area. It is there so that everything in the drawing area can be printed or reproduced on printing machines with no loss of information.

2.1.3 Drawing Area

The drawing area presents the information for constructing the project: the **floor plan**, elevations of the building, sections, and details.

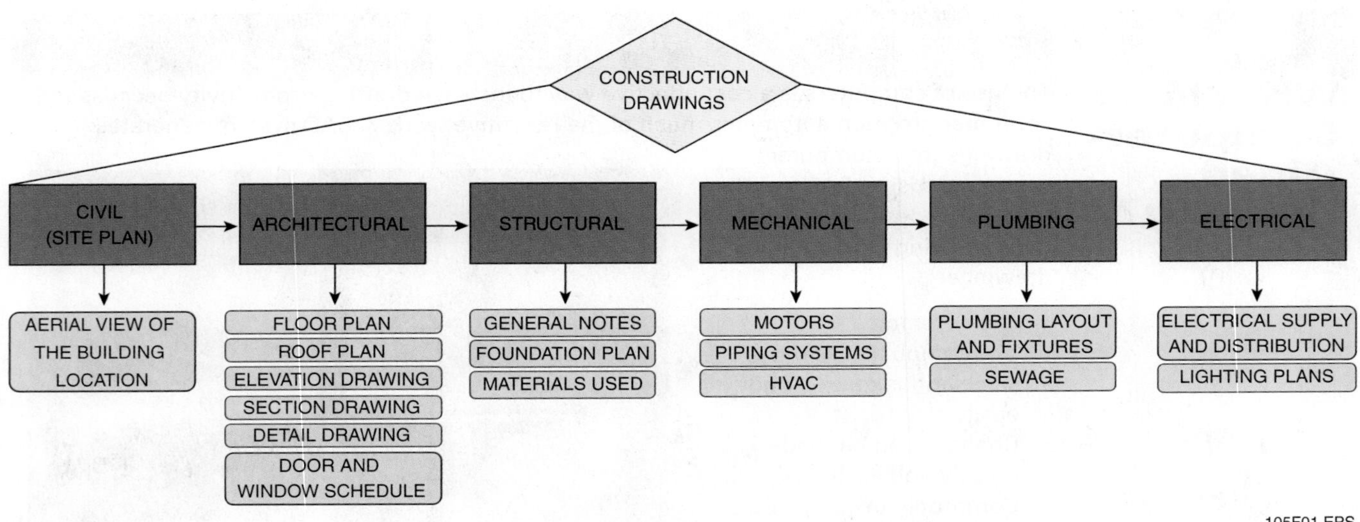

Figure 1 Types of construction drawings.

Figure 2 The title block of a construction drawing.

2.1.4 Revision Block

A revision block is located in the drawing area, usually in the lower right corner inside the title block or near it. Different companies put the revision block in different places. This block is used to record any changes (revisions) to the drawing. It typically contains the revision number, a brief description, the date, and the initials of the person who made the revisions (*Figure 3*). All revisions must be noted in this block and dated and identified by a letter or number.

CAUTION

It is essential to note the revision designation on a construction drawing and to use only the latest version. Otherwise, costly mistakes may result.

2.1.5 Legend

Each line on a construction drawing has a specific design and thickness that identifies it. (Some of the lines may be used to identify off-site utilities. There may be an identification on the cover sheet or on the **civil plans** where they are used.) The identification of these lines and other **symbols** is called the legend. Although a legend doesn't automatically appear on every construction drawing, when it does, it explains or defines symbols or special marks used in the drawing (*Figure 4*). Be aware that legends are specific only to the set of drawings in which they are contained.

Did You Know?

Legality of Construction Drawings

Construction drawings are incorporated into building contracts by reference, making them part of the legal documents associated with a project. When describing the project to be completed, the legal contract between the builder and the owner refers to the accompanying construction drawings for details that would be too lengthy to write out. That makes tracking changes or revisions to construction drawings over the course of a project vitally important. If an error is made along the way, either the owner or the builder must be able to find the discrepancy by reading the drawings. Taking care of construction drawings over the course of the project makes good business sense.

PROJ	NO	REVISION	RVSD	CHKD	APPD	DATE
3483	01	RELEASED FOR CONSTRUCTION		APD	NWS	JULY 92
3483	02	DELETED PART OF LINE 12037		APD	NWS	AUG 92
3483	03	A ADDED WELDING SYMBOL		APD	NWS	AUG 92

105F03.EPS

Figure 3 The revision block of a construction drawing.

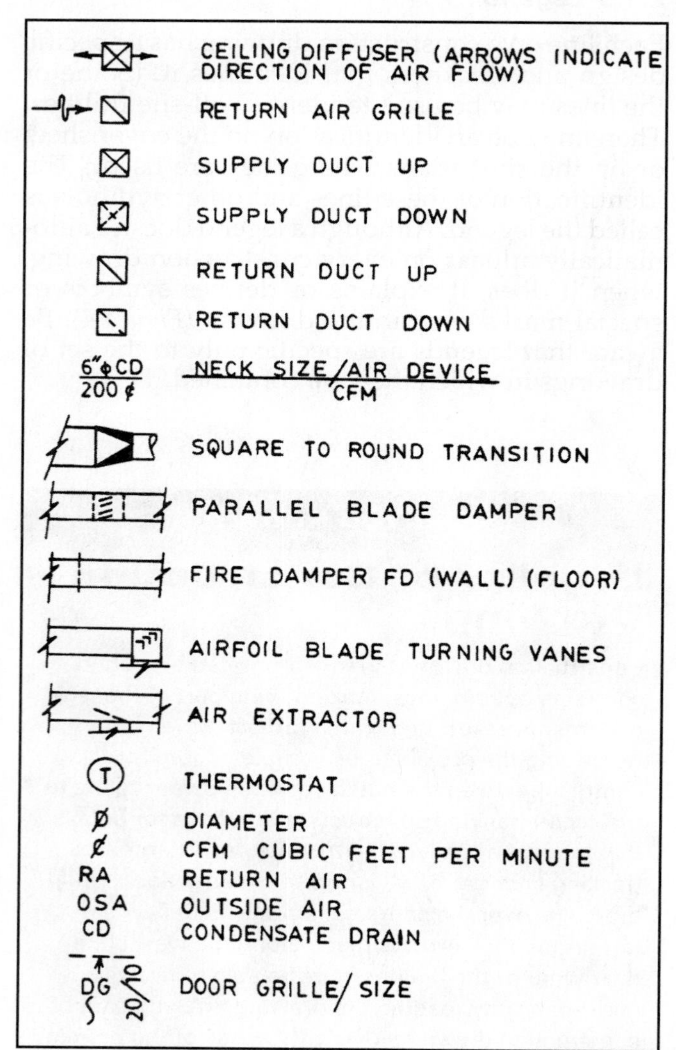

105F04.EPS

Figure 4 Sample legend.

3.0.0 Six Types of Construction Drawings

As stated earlier in this module, a complete set of construction drawing plans includes civil, architectural, structural, mechanical, plumbing, electrical, and sometimes fire protection drawings. This section will examine in greater detail the various characteristics of each type of drawing.

3.1.0 Civil Plans

Civil plans are used for work that has to do with construction in or on the earth. Civil plans are also called site plans, survey plans, or plot plans. They show the location of the building on the site from an aerial view (*Figure 5*). A civil plan also shows the natural contours of the earth, represented on the plan by **contour lines**. The civil plans can also include a landscape plan (*Figure 6*) that shows any trees on the property; construction features such as walks, driveways, or utilities; the **dimensions** of the property; and possibly a legal description of the property.

This is where it all starts. If the site is not acceptable, there is no reason to continue building!

3.2.0 Architectural Plans

Architectural plans (also called architectural drawings) show the design of the project. One part of an architectural plan is a floor plan, also known as a plan view (refer to Drawing 1, *First Floor Plan*, in the drawing package). Any drawing made looking down on an object is commonly

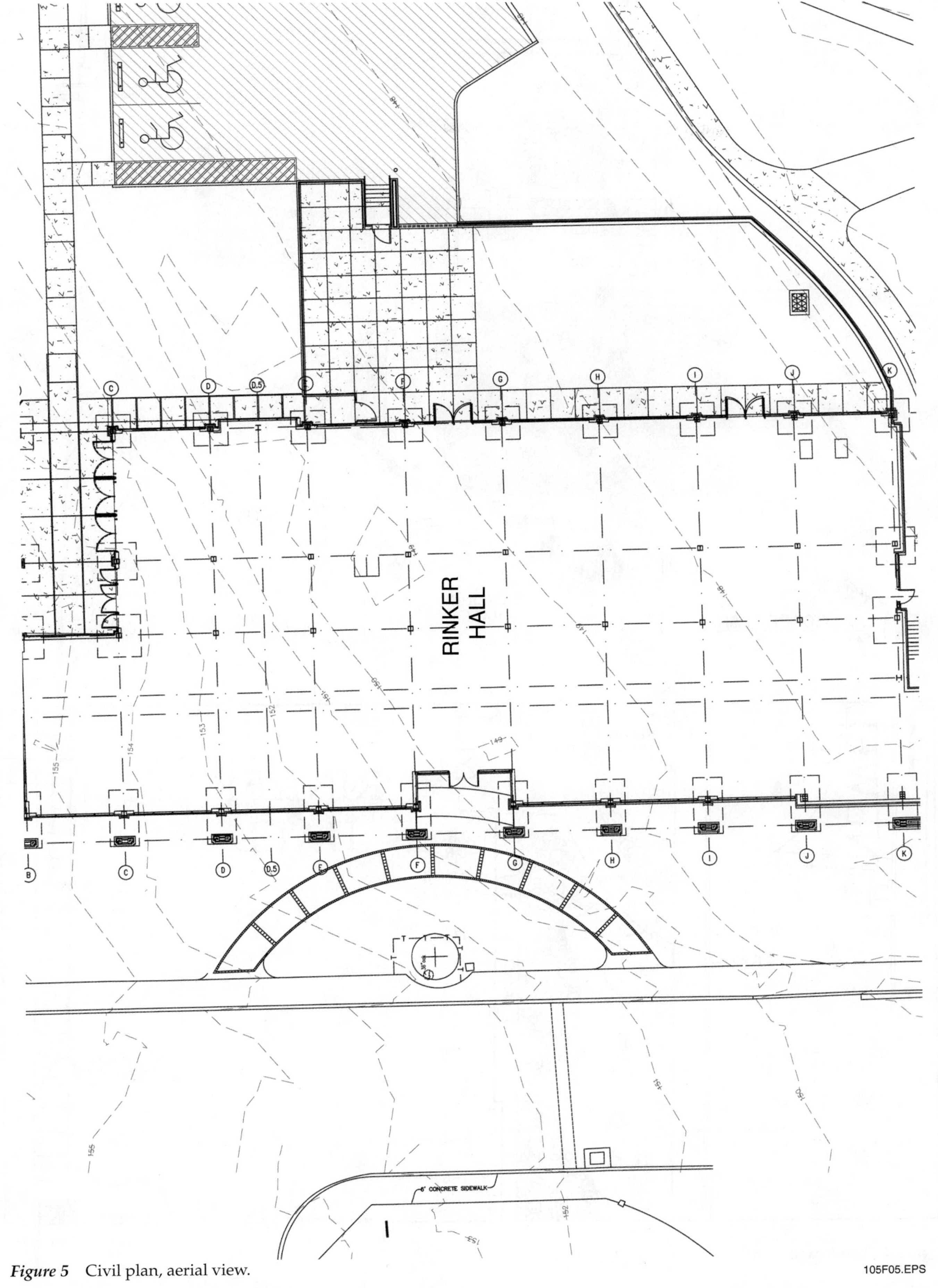

Figure 5 Civil plan, aerial view.

105F05.EPS

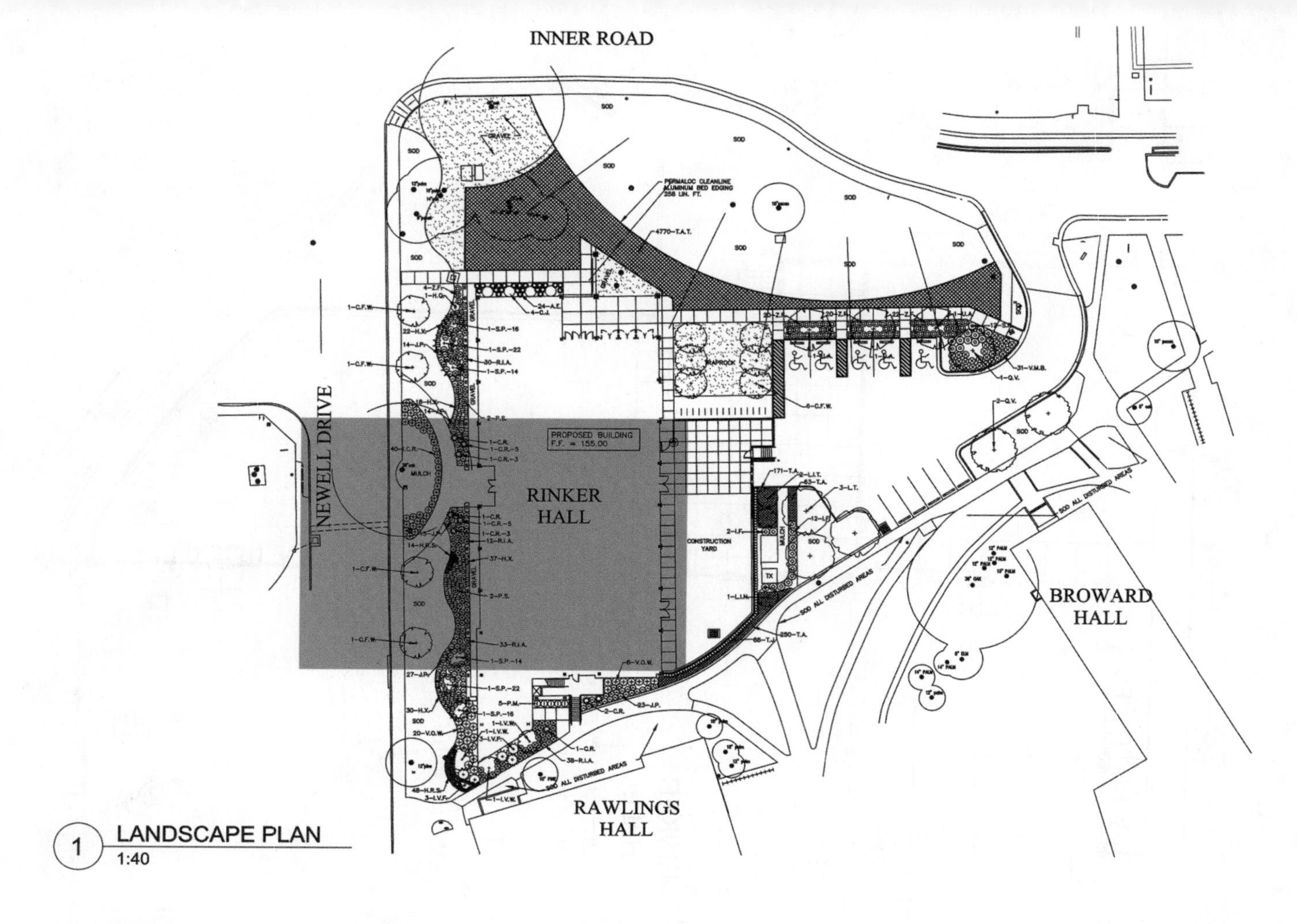

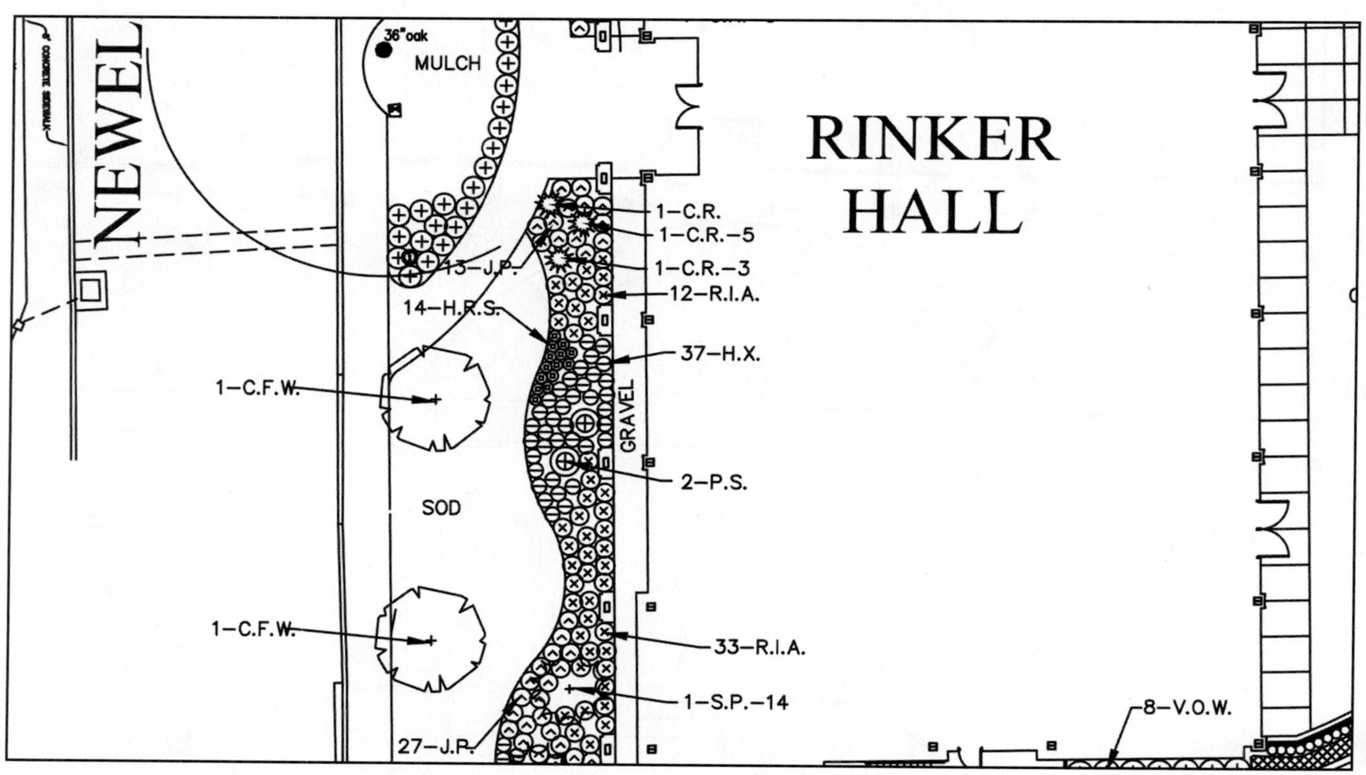

105F06.EPS

Figure 6 Landscape plan.

called a plan view. The floor plan is an aerial view of the layout of each room. It provides the most information about the project. It shows exterior and interior walls, doors, stairways, and mechanical equipment. The floor plan shows the floor as you would see it from above if the upper part of the building were removed.

An architectural plan also includes a **roof plan** (*Figure 7*), which is a view of the roof from above the building. It shows the shape of the roof and the materials that will be used to finish it.

Elevation (EL) is another element of architectural drawings. **Elevation drawings** are side views. They are called elevations because they show height. On a building drawing, there are standard names for different elevations. For example, the side of a building that faces south is called the south elevation. Exterior elevations (*Figure 8*) show the size of the building; the style of the building; and the placement of doors, windows, chimneys, and decorative trim.

Another element of the architectural plan is **section drawings**, which show how the structure is to be built. Section drawings (*Figure 9*) are cross-sectional views that show the inside of an object or building. They show what construction materials to use and how the parts of the object or building fit together. They normally show more detail than plan views. Compare the information on the drawing in *Figure 9* with the information on the drawing in *Figure 8*.

Even more detail is shown in **detail drawings** (*Figure 10*), which are enlarged views of some special features of a building, such as floors and walls. They are enlarged to make the details clearer.

On-Site

Importance of Architectural Symbols

When you look at a section drawing, pay close attention to the way different parts are drawn. Each part of the drawing represents a method of construction or a type of material. These symbols are covered in more detail later in this module.

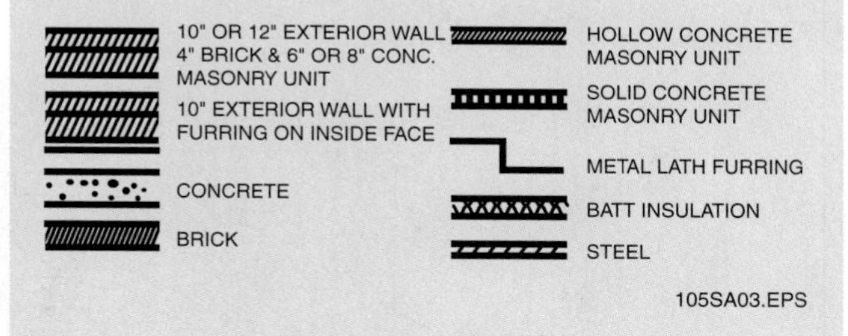

Often the detail drawings are placed on the same sheet where the feature appears in the plan, but sometimes they are placed on separate sheets and referred to by a number on the plan view.

The architectural plan also shows the finish schedules to be used for the doors and windows of the building. Door and window schedules, for example, are tables that list the sizes and other information about the various types of doors and windows used in the project. *Figure 11* shows an example of a window schedule and an elevation drawing for the Type D1 window. Be aware that in some sets of drawings the window schedule and elevation drawings for each type of window may

On-Site

Building Information Modeling (BIM)

Traditionally, Building Information Modeling (BIM) was a system of paper-based documents and construction drawings designed to monitor and track specific information about a building.

In the last decade, BIM has increasingly moved toward computer technology, which allows the various trades and the building owner to track the lifecycle of the building from the start of planning, through construction, to completion. BIM allows building owners and trades to easily track and identify the various components and assemblies of individual systems (electrical, HVAC, plumbing) and building materials within a specific location or locations of a building.

BIM also has the capabilities of simulating the lifecycles of various systems within the building in order to provide information on determining the wear on and lifespan of individual components. Additionally, BIM software allows the trades and the building owner to input and save information on the building as repairs, updates, or renovations are made to the building. Information such as part sizes, manufacturers, part numbers, and related drawings, or any other information ever researched in the past can be stored in a BIM system.

Another advantage of BIM lies in its capabilities of allowing documentation and information on a building to be easily transferred and communicated between all the members of a project.

On-Site

Topographic Maps

Topographic (topo) maps provide a representation of vertical dimension that allows you to get a feel for the shape or contour of a piece of land. Topo maps identify physical features such as mountains, lakes, and streams. These maps indicate where highways and railroads run. They often include information on drainage and land use such as orchards and woodland.

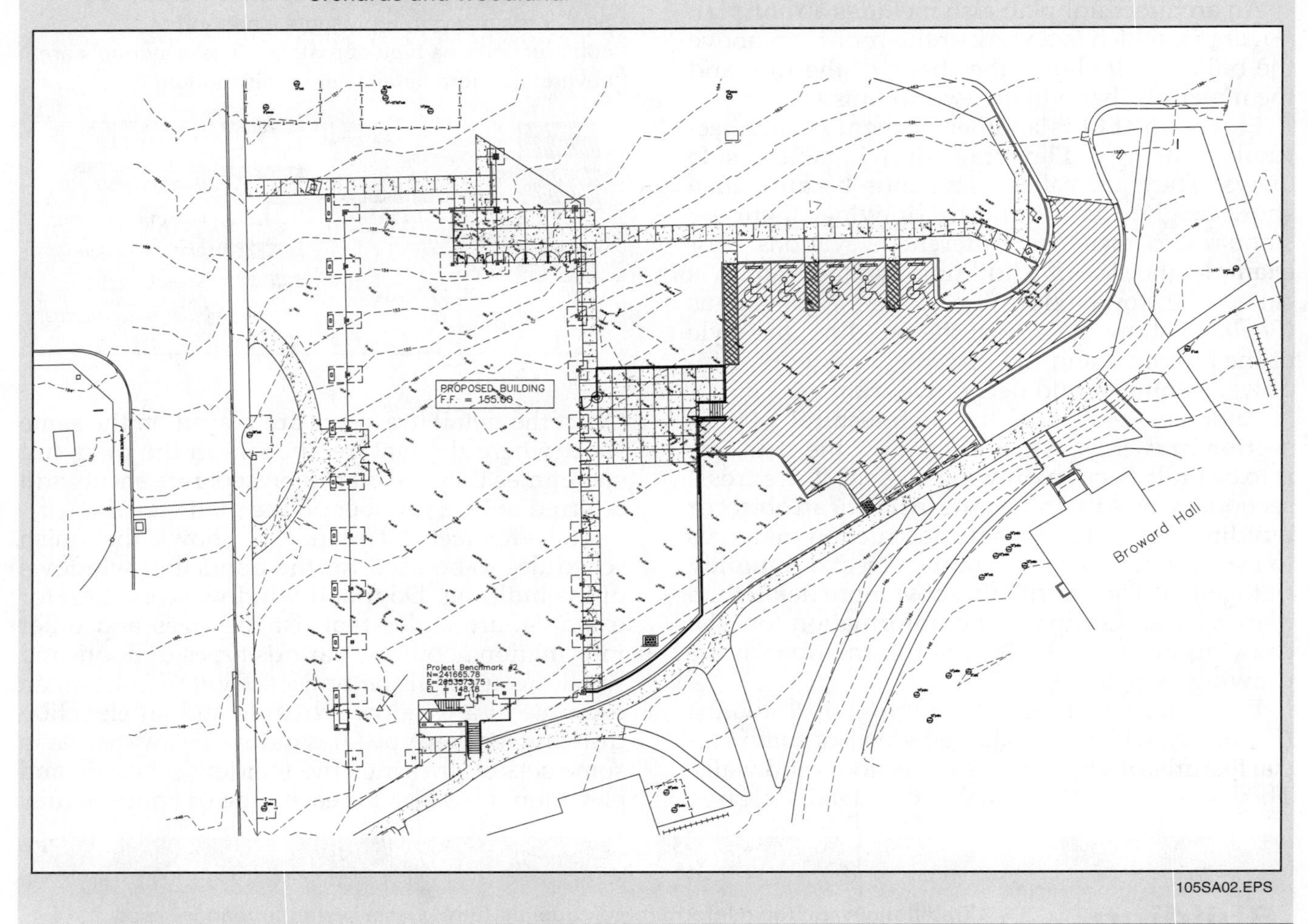

105SA02.EPS

On-Site

Importance of Architectural Drawings

Look at architectural plans first, because all other drawings follow from them. Architectural plans are the most general; they show how all the parts of the project fit together.

appear on the same sheet. Schedules may also be included for finish hardware and fixtures. These schedules are not drawings, but they are usually included in a set of working drawings.

3.3.0 Structural Plans

The **structural plans** are a set of engineered drawings used to support the architectural design. The first part of the structural plans is the general notes (*Figure 12*). These notes give details of the materials to be used and the requirements to be followed in order to build the structure that the architectural plan depicts. The notes, for instance, might specify the type and strength of concrete required for the foundation, the loads that the roof and stairs must be built to accommodate, and codes that contractors must follow. General notes may be on a separate general notes sheet or may be part of individual plan sheets.

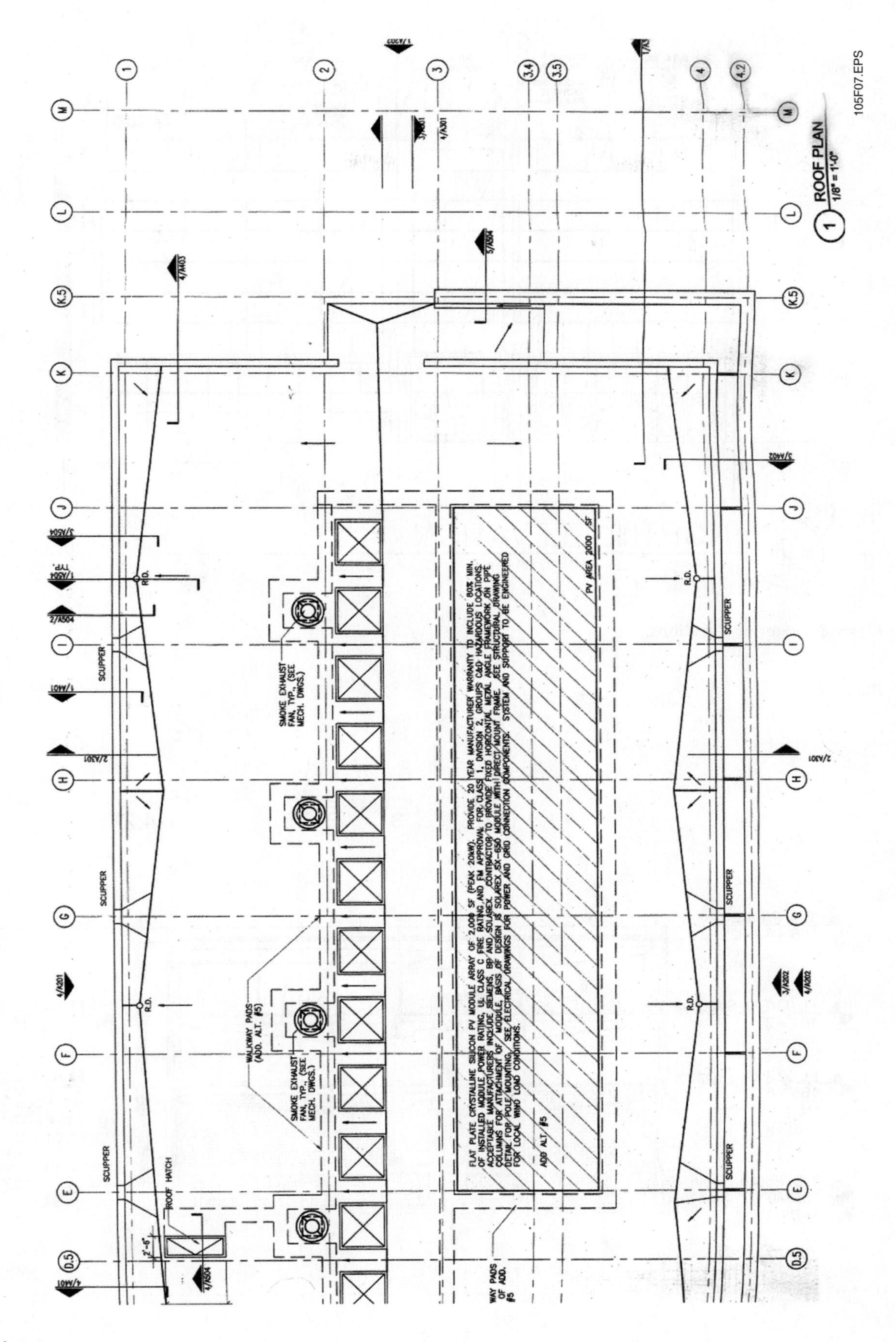

Figure 7 Roof plan.

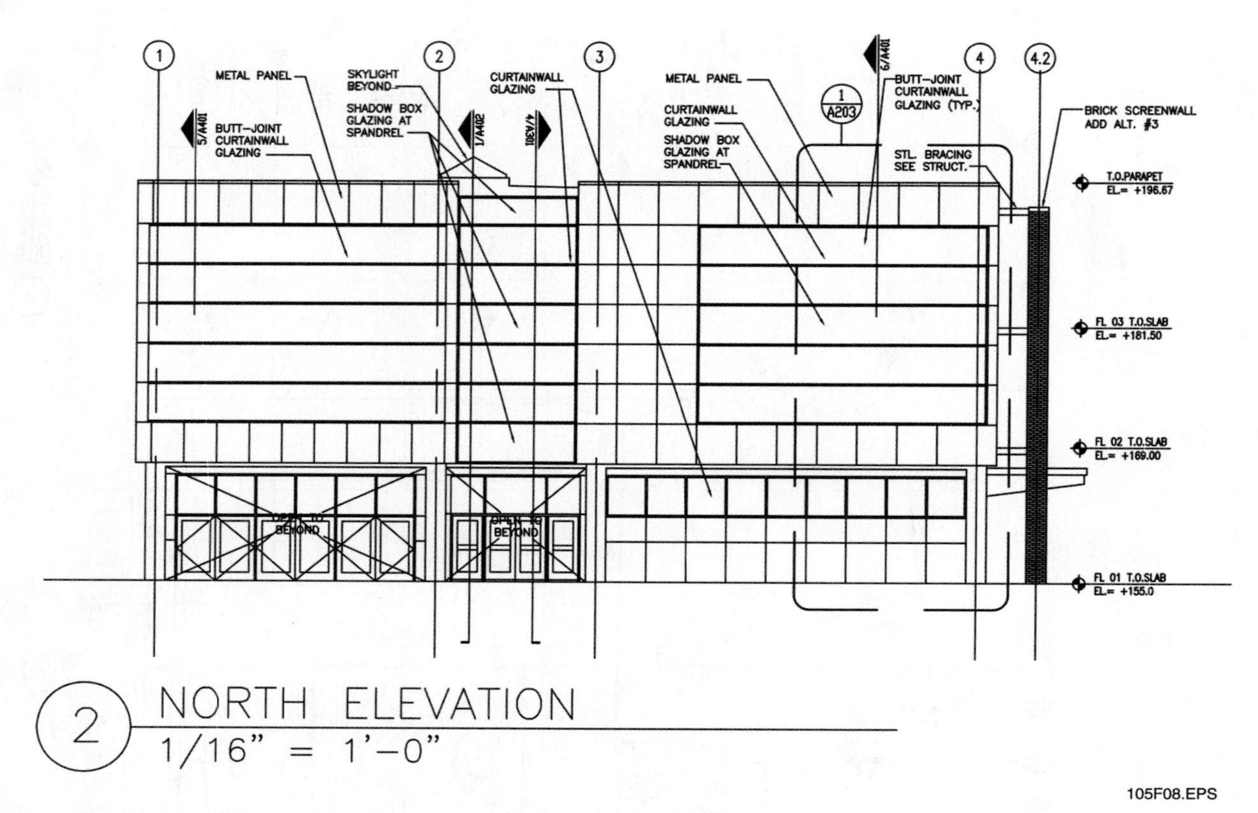

Figure 8 Exterior elevations.

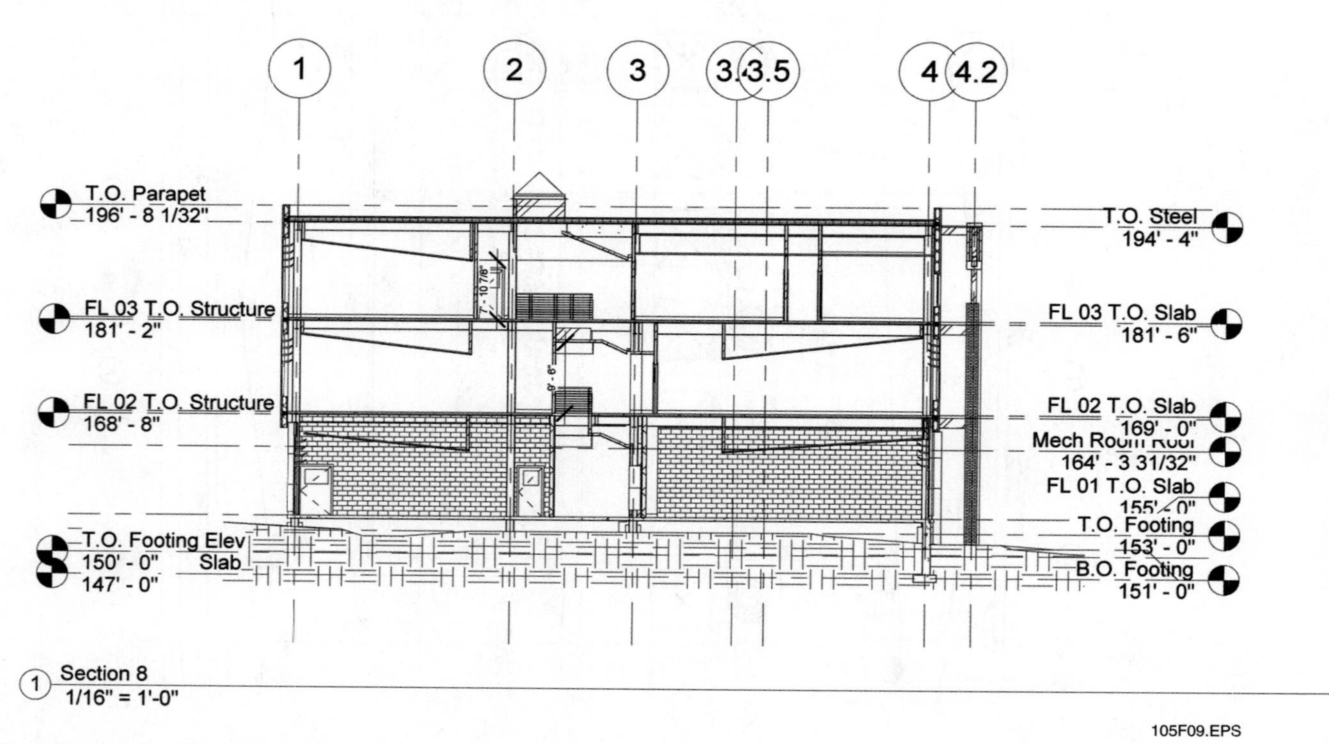

Figure 9 Section drawing (wall section).

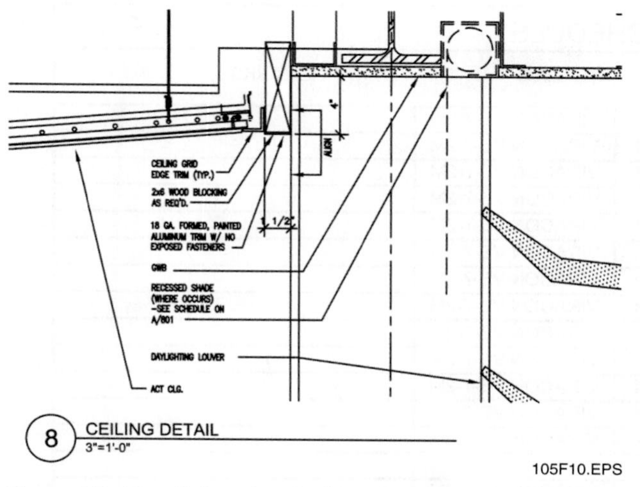

Figure 10 Detail drawing (ceiling detail).

The structural plans also include a **foundation plan** (*Figure 13*), which shows the lowest level of the building, including concrete footings, slabs, and foundation walls. They also may show steel girders, columns, or **beams**, as well as detail drawings to show where and how the foundation must be reinforced. Column and spread footing schedules, and foundation notes may be included on the foundation plan. A related element is the structural floor plan, which depicts the framing, made of either wood or metal joists, and the underlayment of each floor of the structure.

The structural plans show the materials to be used for the walls, whether concrete or masonry, and whether the framing is wood or steel. Structural plans also include a roof-framing plan, showing what kinds of ceiling joists and roof rafters are to be used and where trusses are to be placed (refer to Drawing 2, *Roof Framing Plan*, in the drawing package). Notes for the framing plan are usually found on the same sheet as the drawing.

The structural plans include structural section drawings (*Figure 14*), which are similar to the architectural section drawings but show only the structural requirements. Miscellaneous structural details may also be shown in these sections to provide a better understanding of such things as connections and attachments of accessories.

3.4.0 Mechanical Plans

Mechanical plans are engineered plans for motors, pumps, piping systems, and piping equipment. These plans incorporate general notes (*Figure 15*) containing specifications ranging from what the contractor is to provide to how the contractor determines the location of grilles and registers. A mechanical legend (*Figure 16*) defines the symbols used on the mechanical plans. A list of abbreviations (*Figure 17*) spells out abbreviations found on the plans.

Piping and instrumentation drawings, or **P&IDs** (*Figure 18*), are **schematic** diagrams of a complete piping system that show the process flow. They also show all the equipment, pipelines, valves, instruments, and controls needed to operate the system. P&IDs are not drawn to scale because they are meant only to give a representation, or a general idea, of the work to be done. Additionally, P&IDs do not indicate north, south, east, and west directions.

For more complex jobs, a separate **heating, ventilating, and air conditioning (HVAC)** plan is added to the set of plans. Piping system plans for gas, oil, or steam heat may be included in the HVAC plan. The mechanical plans include the layout of the HVAC system, showing specific requirements and elements for that system, including a floor, a reflected ceiling, or a roof. HVAC drawings (*Figure 19*) include an electrical schematic that shows the electrical circuitry for the HVAC system. HVAC plans are both mechanical and electrical drawings in one plan. Be aware that a page with a series of mechanical detail drawings may be included in the mechanical plans. These drawings show specific details of certain components within the mechanical system. *Figure 20* is an example of a mechanical detail drawing.

3.5.0 Plumbing/Piping Plans

Plumbing plans (*Figure 21*) are engineered plans showing the layout for the plumbing system that supplies the hot and cold water, for the sewage disposal system, and for the location of plumbing fixtures. For commercial projects, each system may be on a separate plan.

A plumbing isometric is part of the plumbing plan. A plumbing isometric is an **isometric drawing** that depicts the plumbing system. *Figure 22* shows a plumbing isometric drawing for the sanitary riser system.

> **Did You Know?**
>
> **The North Arrow**
>
> Most construction drawings have an arrow indicating north. It may be located near the title block or in some other conspicuous place. The north arrow allows you to describe the locations of objects, walls, and building parts shown on the construction drawing using a common reference point.

WINDOW SCHEDULE				
TYPE	FRAME SIZE	FRAME FINISH	GLAZING	REMARKS
A	4' 2" x 4' 2"	CLEAR ANODIZED ALUMINUM	VIRACON VE-7-2M	
B	24' 2-1/2" x 11' 0"	CLEAR ANODIZED ALUMINUM	VIRACON VE-7-2M	
B-1	12' 10" x 11' 0"	CLEAR ANODIZED ALUMINUM	VIRACON VE-7-2M	
C	15'-11-1/2" x 8' 3-1/2"	CLEAR ANODIZED ALUMINUM	VIRACON VE-7-2M	
C-1	15'-11-1/2" x 4' 1-1/2"	CLEAR ANODIZED ALUMINUM	VIRACON VE-7-2M	
D	15 11-1/2'-0" x 11' 1-1/2"	CLEAR ANODIZED ALUMINUM	VIRACON VE-7-2M	
D-1	15 11-1/2'-0" x 11' 1-1/2"	CLEAR ANODIZED ALUMINUM	VIRACON VE-7-2M	
D-2	31' 11-1/2" x 4' 4"	CLEAR ANODIZED ALUMINUM	VIRACON VE-7-2M	
E	14' 4" x 7' 8"	CLEAR ANODIZED ALUMINUM	VIRACON VE-7-2M	
F	15'-11-1/2" x 4' 4 "	CLEAR ANODIZED ALUMINUM	VIRACON VE-7-2M	
H	11' 11-1/2" x 8' 1/4"	CLEAR ANODIZED ALUMINUM	VIRACON VE-7-2M	
H-1	11' 11-1/2" x 10' 1-3/4"	CLEAR ANODIZED ALUMINUM	VIRACON VE-7-2M	
I	15'-11-1/2" x 4' 1-1/2 "	CLEAR ANODIZED ALUMINUM	VIRACON VE-7-2M	
CW - 1	12' 0" x 27' 11-1/4"	CLEAR ANODIZED ALUMINUM	VIRACON VE-7-2M	
CW - 3	15' 11 1/2" x 37' 3 1/4"	CLEAR ANODIZED ALUMINUM	VIRACON VE-7-2M	
CW - 4	36' 10" x 20' 9 1/2"	CLEAR ANODIZED ALUMINUM	VIRACON VE-7-2M	
CW - 5	28' 11-7/8" x 20' 9-1/2"	CLEAR ANODIZED ALUMINUM	VIRACON VE-7-2M	

NOTES

1. FRAME SIZES HAVE BEEN INDICATED ASSUMING 1/2" JOINT ADJACENT TO METAL PANEL SYSTEM. ALIGNMENT OF JOINTS W/ METAL PANEL MODULE IS REQUIRED. INTERMEDIATE MULLIONS TO BE CENTERED ON ADJACENT METAL PANEL JOINTS

2. "SG" INDICATES SAFETY GLAZING (TEMPERED GLASS REQUIRED)

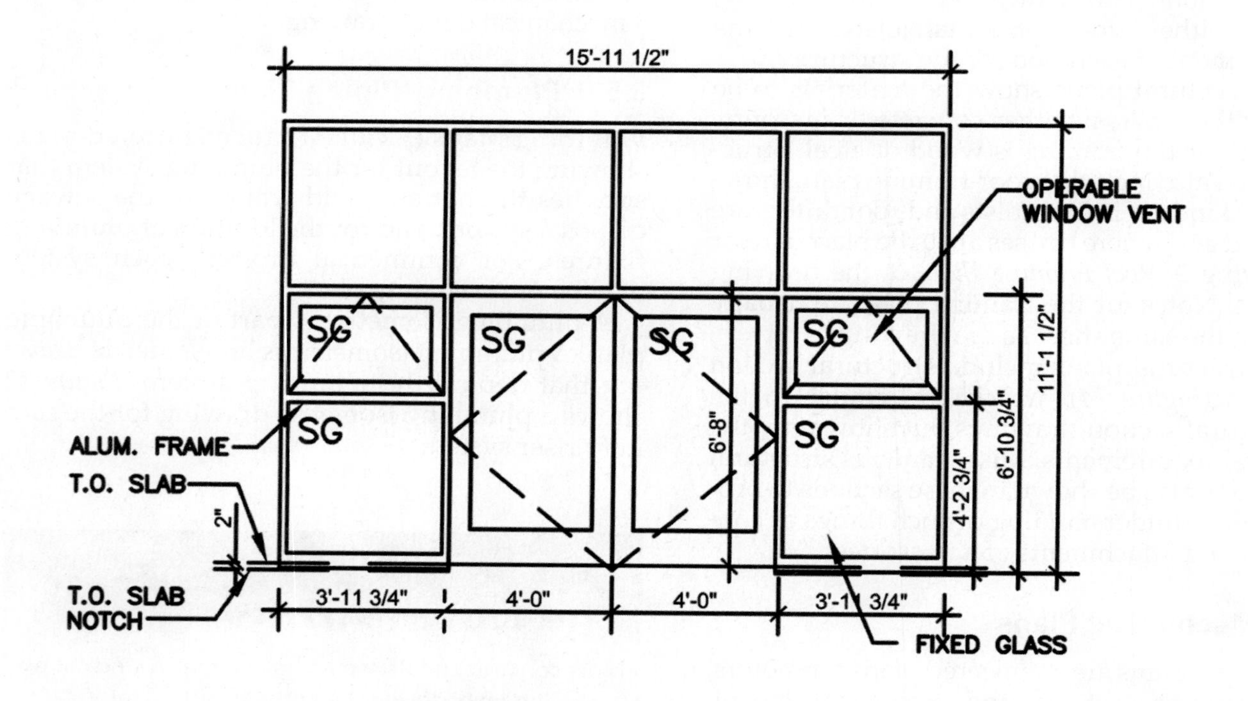

105F11.EPS

Figure 11 Window schedule and window detail.

GENERAL NOTES

I. DESIGN CRITERIA

A. GENERAL BUILDING CODE
The Contract Documents are based on the requirements of the:
1. Standard Building Code, 1997 edition.

B. DEAD LOADS

1. Partitions. An allowance of 20 PSF has been made for partitions as a uniformly distributed dead load.

2. Hanging Ceiling and Mechanical Loads. An allowance of 10 PSF has been made for hanging ceiling and mechanical equipment loads such as duct work and sprinkler pipes.

C. LIVE LOADS
1. Design live loads are based on the more restrictive of the uniform load listed below or the concentrated load listed acting over an area 2.5 feet square.

CATEGORY	UNIFORM LOAD (PSF)	CONCENTRATED LOAD (LB)
1. Roof	20	N/A
2. Elevated Floors	50	0
3. Terraces, Lobbies	100	0
4. Stairways, Exit Facilities	100	0
5. Elevator Machine Rooms	100	Assumed Eqp. Wt.
6. Mechanical Rooms, typical	150	Assumed Eqp. Wt.

NOTES:
1. Live Load Reduction. Live loads have been reduced on any member supporting more than 150 square feet, including flat slabs, except for floors in places of public assembly and for live loads greater than 100 pounds per square foot in accordance with the following formula:

$$R = r(A-150)$$

The reduction,, R, shall not exceed 40 percent for members supporting one level only, 60 percent for other members, or R as calculated in the following formula:

$$R = 23.1\ (1 + \frac{D}{L})$$

R = Reduction in percent.
r = Rate of reduction equal to .08 percent for floors.
A = Area of floor supported by the member.
D = Total dead load supported by the member.
L = Total, unreduced, live load supported by the member.

2. For storage loads exceeding 100 pounds per square foot, no reduction has been made, except that design live loads on columns have been reduced 20 percent.

D. ELEVATOR LOADS
Machine Beam, Car Buffer, Counterweight Buffer, and Guide Rail Loads. Assumed elevator loads to the supporting structure are shown on the drawings, including machine beam reactions, car buffer reactions, counterweight buffer reactions, and horizontal and vertical guide rail loads. The General Contractor shall submit to the Structural Engineer final elevator shop drawings showing all loads to the structure prior to the installation of the elevators for verification of load carrying capacity.

E. MECHANICAL EQUIPMENT LOADS
The General Contractor shall submit actual weights of equipment to be used in the project to the Structural Engineer for verification of loads used in the design at least three weeks prior to fabrication and construction of the supporting structure.

F. WIND LOADS
1. Wind pressures are based on the American Society of Civil Engineers, Minimum Design Loads for Buildings and Other Structures, ASCE 7–98 with a Wind Speed = 110 MPH (3 sec. gust), Exposure C, Importance Factor 1.15.
2. Wind pressures used in the design of the cladding are shown on these Drawings.

II. FOUNDATION

A. GEOTECHNICAL REPORT
Foundation design is based on the geotechnical investigation report as follows:
1. Reports of Geotechnical exploration, M.E. Rinker Sr. Hall (Revised Location) Near the southeast Corner of Newell Drive and Inner Road, Gainesville, Alachua County, Florida. Law Engineering and Environmental Services, Inc. January 2, 2001.

The geotechnical report is available to the General Contractor upon request to the Owner. The information included therein may be used by the General Contractor for his general information only. The Architect and Engineer will not be responsible for the accuracy or applicability of such data therein.

B. FOUNDATION TYPE

1. Spread Footing.

a. Design Pressures:
1. All footings have been designed assuming an allowable bearing pressure of 4000 PSF.
Allowable pressures are increased 33% for combined gravity and wind loads.

C. SLAB–ON–GRADE
Radon resistant construction guidlines are being followed on this project. The details and specifications for slab–on–grade construction must be adhered to without deviation.
Slab–on–Grade shall be immediately underlain by a 8 mil. vapor barrier. Seams shall be lapped 12 inches and sealed with 2" wide pressure sensitive vinyl tape. All penetrations shall be sealed with tape.

D. CONSTRUCTION DEWATERING
The Contractor shall determine the extent of construction dewatering required for the excavation. The Contractor shall submit to the Geotechnical Engineer for review the proposed plan for construction dewatering, prior to beginning the excavation.

III. REINFORCED CONCRETE

A. CLASSES OF CONCRETE
All concrete shall conform to the requirements as specified in the table below unless noted otherwise on the drawings:

Usage	28 Day Comp. Strength (PSI)	Conc Type	Max Size Agg.	W/C Ratio
1. Elevated Floors	4000	NWT	3/4"	0.48
2. Spread Footings	3000	NWT	1"	0.55
3. Slab–On–Grade	4000	NWT	1"	0.48
4. Fnd. Walls & Plinths	4000	NWT	1"	0.48

All concrete shall be proportioned for a maximum allowable unit shrinkage of 0.03% measured at 28 days after curing in lime water as determined by ASTM C 157 (using air storage).

B. HORIZONTAL CONSTRUCTION JOINTS IN CONCRETE POURS
There shall be no horizontal construction joints in any concrete pours unless shown on the drawings. The Architect/Engineer shall approve all deviations or additional joints in writing.

C. REINFORCING STEEL SPECIFICATION

1. All Reinforcing Steel shall be ASTM A615 Grade 60 unless noted otherwise on the drawings or in these notes.
2. Welded Reinforcing Steel. Provide reinforcing steel conforming to ASTM A706 for all reinforcing steel required to be welded and where noted on the drawings.
3. Galvanized Reinforcing Steel. Provide reinforcing steel galvanized according to ASTM A767 Class II (2.0 oz. zinc PSF where noted on the drawings.
4. Deformed Bar Anchors. ASTM A496 minimum yield strength 70,000 PSI as noted on the drawings. Reinforcing bars shall not be substituted for deformed bar anchors.
5. Welded Wire Fabric. Welded smooth wire fabric, ASTM A 185, yield strength 65,000 PSI where noted on the drawings. Welded deformed wire fabric for, ASTM A 497, yield strength 70,000 PSI where noted on the drawings.

D. PLACEMENT OF WELDED WIRE FABRIC
Wherever welded wire fabric is specified as reinforcement, it shall be continuous across the entire concrete surface and not interrupted by beams or girders and properly lapped one cross wire spacing plus 2".

E. REINFORCEMENT IN TOPPING SLABS
Provide welded smooth wire fabric minimum 6 x 6 W2.9 x W2.9 in all topping slabs unless specified otherwise on the drawings.

F. REINFORCEMENT IN HOUSEKEEPING PADS
Provide welded smooth wire fabric 6 x 6 W2.9 x W2.9 minimum in all housekeeping pads supporting mechanical equipment whether shown on the drawings or not unless heavier reinforcement is called for on the drawings.

G. REINFORCING STEEL COVERAGE
Concrete Cover for reinforcement layer nearest to the surface unless specified otherwise on the drawings.
1. Concrete surfaces cast against and permanently exposed to earth. 3 inches
2. Concrete surfaces exposed to earth or weather or where noted on the drawings 2 inches.
3. Concrete surfaces not exposed to weather or in contact with the ground.
a. #3 to #11 bars 1 inch

H. SPLICES IN REINFORCING STEEL
1. All unscheduled splices shall be Class A tension splice.

IV. STRUCTURAL STEEL

V. STEEL DECKS

VI. CURTAIN WALL

VII. CONCRETE MASONRY

VIII. MISCELLANEOUS

IX. SUBMITTALS

X. DRAWING INTERPRETATION

105F12.EPS

Figure 12 General notes for structural plans.

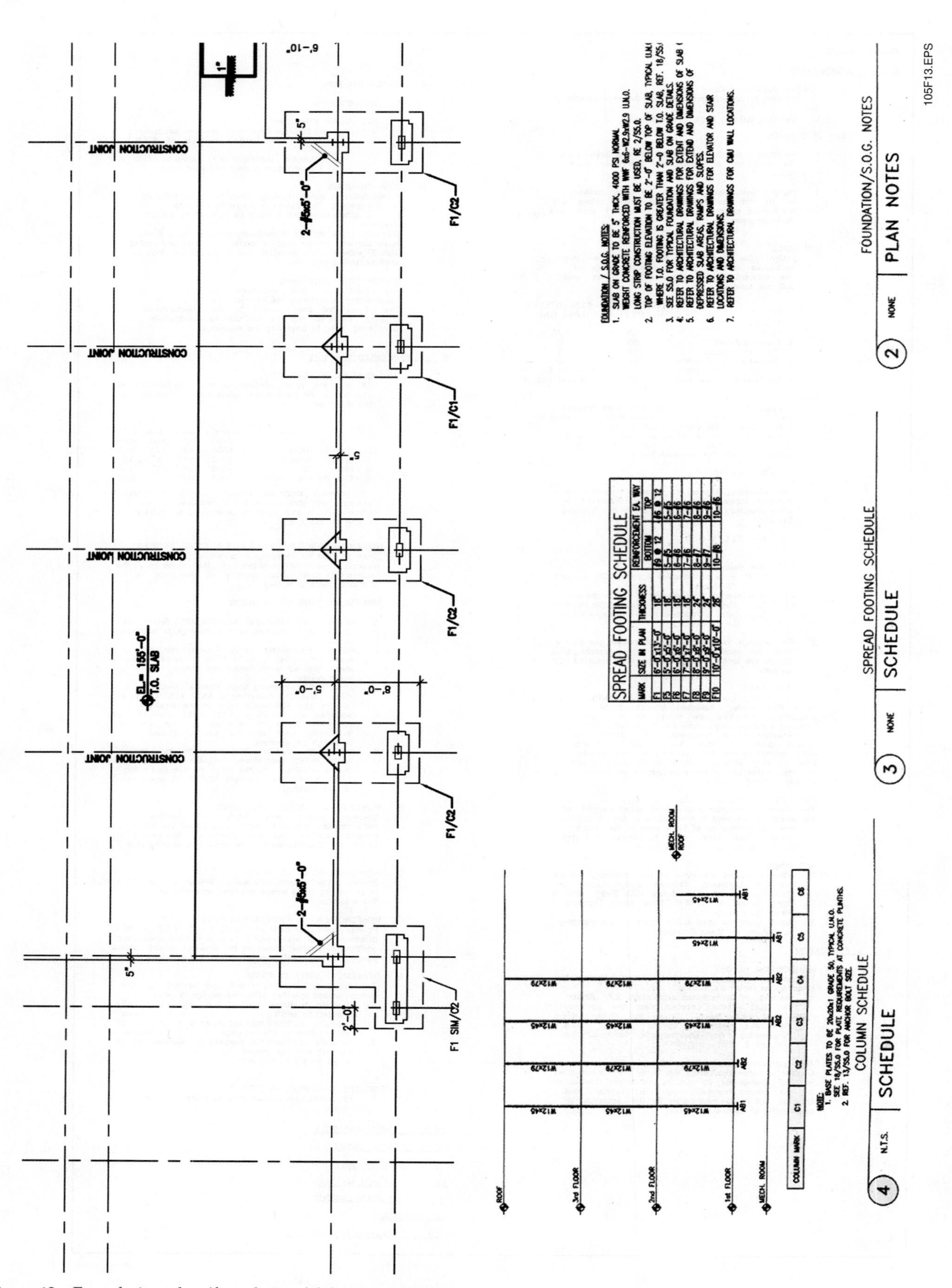

Figure 13 Foundation plan (foundation/slab-on-grade plan).

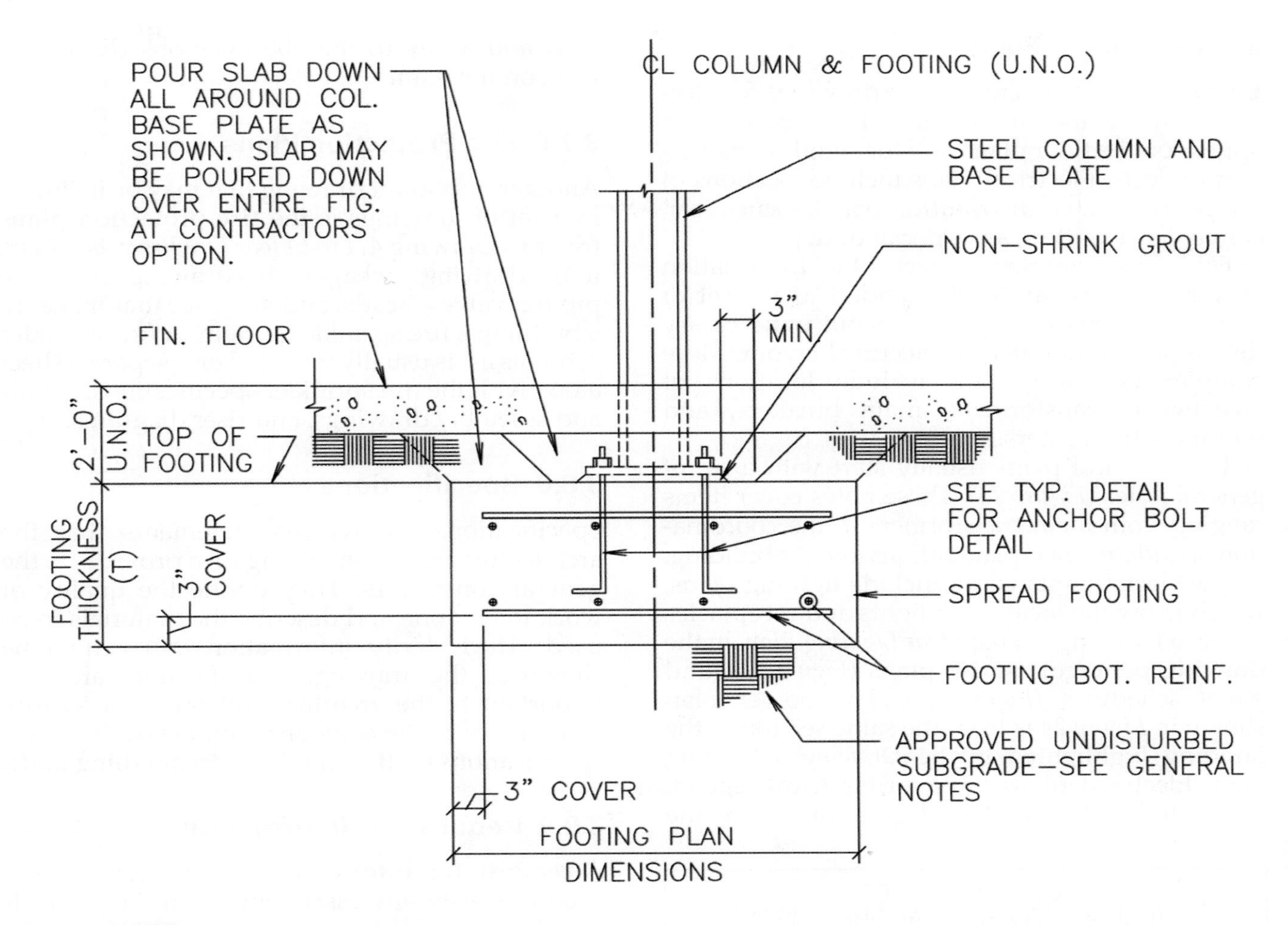

Figure 14 Structural section drawing (foundation details).

On-Site

Visualize Before Building

It is important for a builder to be able to visualize a finished product before starting it. Detail drawings help you see in your head, or visualize, different parts of the structure before you measure a board or hammer a nail. By visualizing, you can plan ahead and can anticipate potential problems. This saves time and money on the job.

3.6.0 Electrical Plans

Electrical plans are engineered drawings for electrical supply and distribution. These plans may appear on the floor plan itself for simple construction projects. Electrical plans include locations of the electric meter, distribution panel, switchgear, convenience outlets, and special outlets.

For more complex projects, the information may be on a separate plan added to the set of plans. This separate plan leaves out unnecessary details and shows just the electrical layout. More complex electrical plans include locations of switchgear, transformers, main breakers, and motor control centers.

The electrical plans usually start with a set of general notes (*Figure 23*). These notes cover items ranging from main transformers to the coordination of underground penetrations into the building.

The electrical plans can include lighting plans, which show the location of lights and receptacles (refer to Drawing 3, *First Floor Lighting Plan*, in the drawing package), power plans (*Figure 24*), and panel schedules (*Figure 25*). The power plan shown in *Figure 24* is from the same section of the building highlighted in the *Drawing 3* lighting plan. Electrical plans have an electrical legend, which defines the symbols (*Figure 26*) used on the plan and a key to the abbreviations (*Figure 27*) used on the plan.

GENERAL NOTES
(FOR ALL MECHANICAL DRAWINGS)

1. CONTRACTOR IS TO PROVIDE COMPLETE CONNECTIONS TO ALL NEW AND RELOCATED OWNER FURNISHED EQUIPMENT.
2. CONTRACTOR TO COORDINATE THE LOCATION OF ALL DUCTWORK AND DIFFUSERS WITH REFLECTED CEILING PLAN AND STRUCTURE PRIOR TO BEGINNING WORK.
3. DIMENSIONS FOR INSULATED OR NON-INSULATED DUCT ARE OUTSIDE SHEET METAL DIMENSIONS.
4. DRAWINGS ARE NOT TO BE SCALED FOR DIMENSIONS. TAKE ALL DIMENSIONS FROM ARCHITECTURAL DRAWINGS, CERTIFIED EQUIPMENT DRAWINGS AND FROM THE STRUCTURE ITSELF BEFORE FABRICATING ANY WORK. VERIFY ALL SPACE REQUIREMENTS COORDINATING WITH OTHER TRADES, AND INSTALL THE SYSTEMS IN THE SPACE PROVIDED WITHOUT EXTRA CHARGES TO THE OWNER.
5. LOCATION OF ALL GRILLES, REGISTERS, DIFFUSERS AND CEILING DEVICES SHALL BE DETERMINED FROM THE ARCHITECTURAL REFLECTED CEILING PLANS.
6. THE OWNER AND DESIGN ENGINEER ARE NOT RESPONSIBLE FOR THE CONTRACTOR'S SAFETY PRECAUTIONS OR TO MEANS, METHODS, TECHNIQUES, CONSTRUCTION SEQUENCES, OR PROCEDURES REQUIRED TO PERFORM HIS WORK.
7. ALL WORK SHALL BE INSTALLED IN ACCORDANCE WITH PLRC'S SAFETY PLAN AND ALL APPLICABLE STATE AND LOCAL CODES.
8. ALL EXTERIOR WALL AND ROOF PENETRATIONS SHALL BE SEALED WEATHERPROOF. REFERENCE SPECIFICATION SECTION 15050.
9. ALL MECHANICAL WORK UNDER THIS CONTRACT IS TO FIVE (5) FEET OUTSIDE THE BUILDING.

105F15.EPS

Figure 15 Mechanical plan general notes.

3.7.0 Fire Protection Plans

Another important drawing that may be included in a set of drawings is the fire protection plans (refer to Drawing 4, *First Floor Fire Protection Plan*, in the drawing package). This drawing shows the piping, valves, heads, and switches that make up a building's fire sprinkler system. A fire sprinkler symbols list is usually included on a separate sheet along with the fire sprinkler specifications, details and assembly drawings, and riser diagrams.

3.8.0 Specifications

Specifications are written statements that the architectural and engineering firm provides to the general contractors. They define the quality of work to be done and describe the materials to be used. They clarify information that cannot be shown on the drawings. Specifications are very important to the architect and owner to ensure compliance to the standards set. *Figure 28* shows specifications for the building's air handling units.

3.9.0 Request for Information

A request for information (RFI) (*Figure 29*) is used to clarify any discrepancies in the plans. If you notice a discrepancy, you should notify the foreman. The foreman will write up an RFI, explaining the problem as specifically as possible and putting the date and time on it. The RFI is submitted to the superintendent, who passes it to the general contractor, who passes it to the architect or engineer, who then resolves the discrepancy.

Always refer to specifications and the RFI when deciding how to interpret drawings.

4.0.0 Scale

The scale of a drawing tells the size of the object drawn compared with the actual size of the object represented. The scale is shown in one of the spaces in the title block, beneath the drawing itself, or in both places. The type of scale used on a drawing depends on the size of the objects being shown, the space available on the paper, and the type of plan.

On a site plan, the scale may read SCALE: 1" = 20'-0". This means that every 1 inch on the drawing represents 20 feet, 0 inches. The scale used to develop site plans is an engineer's scale.

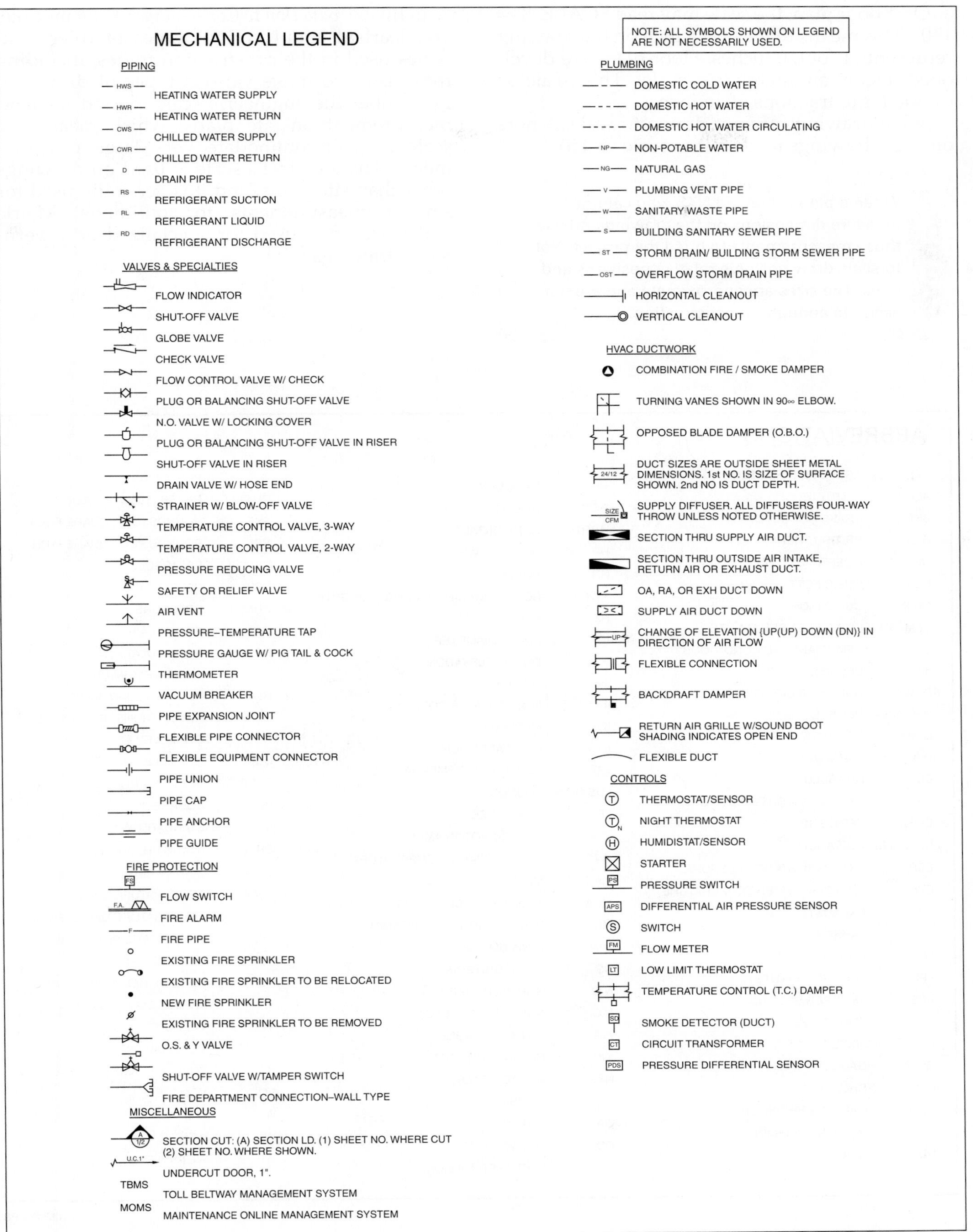

Figure 16 Mechanical plan legend.

On a floor plan, the scale may read SCALE: ¼" = 1'-0". This means that every ¼ inch on the drawing represents 1 foot, 0 inches. Floor plans are developed using an architect's scale. This scale is divided into fractions of an inch.

Some drawings are not drawn to scale. A note on such drawings reads **not to scale (NTS)**.

> **CAUTION**
>
> When a plan is marked NTS, you cannot measure dimensions on the drawing and use those measurements to build the project. Not-to-scale drawings give relative positions and sizes. The sizes are approximate and are not accurate enough for construction.

In the *Introduction to Construction Math* module, you learned about various types of rulers and scales used in the construction trades, including those used to make various types of drawings. Remember the engineer's scale is used for land measurement on site plans, which means the scale must accommodate very large measurements. The architect's scale is used on drawings other than site plans, and it is generally used for smaller measurements for buildings. Metric scales are often used for specialized or government drawings.

ABBREVIATIONS

AFF	ABOVE FINISHED FLOOR
ALT	ALTITUDE
BHP	BRAKE HORSEPOWER
BTU	BRITISH THERMAL UNIT
Cv	COEFFICIENT, VALVE FLOW
CU FT	CUBIC FEET
CU IN	CUBIC INCH
CFM	CUBIC FEET PER MINUTE
SCFM	CFM, STANDARD CONDITIONS
dB	DECIBEL
DCW	DOMESTIC COLD WATER
DEG OR∞	DEGREE
DHW	DOMESTIC HOT WATER
DIA	DIAMETER
DB	DRY-BULB
EAT	ENTERING AIR TEMPERATURE
EFF	EFFICIENCY
ELEV or EL	ELEVATION
ESP	EXTERNAL STATIC PRESSURE
EWT	ENTERING WATER TEMPERATURE
EXH	EXHAUST
F	FAHRENHEIT
FLA	FULL LOAD AMPS
FPM	FEET PER MINUTE
FPS	FEET PER SECOND
FT	FOOT OR FEET
FU	FIXTURE UNITS
GA	GAUGE
GAL	GALLONS
GPH	GALLONS PER HOUR
GPM	GALLONS PER MINUTE
HD	HEAD
HG	MERCURY
HGT	HEIGHT
HORZ	HORIZONTAL
HP	HORSEPOWER
HR	HOUR(S)
HWC	HOT WATER CIRCULATING (DOMESTIC)
HZ	HERTZ
ID	INSIDE DIAMETER
IE	INVERT ELEVATION
IN	INCHES
IN W.C.	INCHES WATER COLUMN
KW	KILOWATT
KWH	KILOWATT HOUR
LAT	LEAVING AIR TEMPERATURE
LBS OR #	POUNDS
LF	LINEAR FEET
LRA	LOCKED ROTOR AMPS
LWT	LEAVING WATER TEMPERATURE
MAX	MAXIMUM
MCA	MINIMUM CIRCUIT AMPS
MBH	BTU PER HOUR (THOUSAND)
MIN	MINIMUM
NC	NOISE CRITERIA
N.O.	NORMALLY OPEN
N.C.	NORMALLY CLOSED
N/A	NOT APPLICABLE
NIC	NOT IN CONTRACT
NTS	NOT TO SCALE
NO	NUMBER
OA	OUTSIDE AIR
OD	OUTSIDE DIAMETER
PPM	PARTS PER MILLION
%	PERCENT
PH OR f	PHASE (ELECTRICAL)
PSF	POUNDS PER SQUARE FOOT
PSI	POUNDS PER SQUARE INCH
PSIA	PSI ABSOLUTE
PSIG	PSI GAUGE
PRESS	PRESSURE
RA	RETURN AIR
RECIRC	RECIRCULATE
RH	RELATIVE HUMIDITY
RLA	RUNNING LOAD AMPS
RPM	REVOLUTIONS PER MINUTE
SL	SEA LEVEL
SENS	SENSIBLE
SPEC	SPECIFICATION
SQ	SQUARE
STD	STANDARD
SP	STATIC PRESSURE
SA	SUPPLY AIR
TEMP	TEMPERATURE
TD	TEMPERATURE DIFFERENCE
TSP	TOTAL STATIC PRESSURE
TSTAT	THERMOSTAT
TONS	TONS OF REFRIGERATION
VAV	VARIABLE AIR VOLUME
VEL	VELOCITY
VERT	VERTICAL
V	VOLT
VOL	VOLUME
W	WATT
WT	WEIGHT
WB	WET-BULB

105F17.EPS

Figure 17 Mechanical plan list of abbreviations.

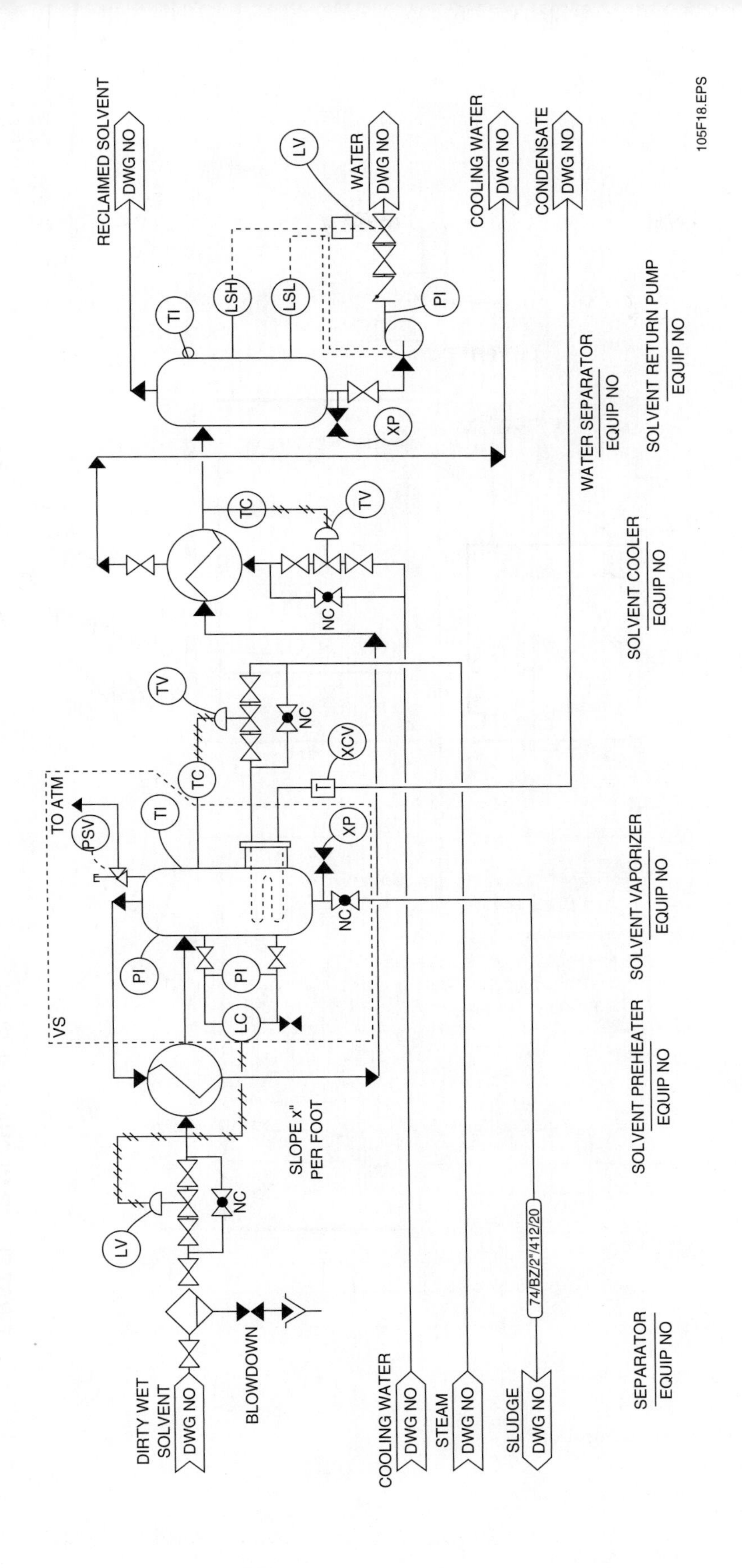

Figure 18 P&ID.

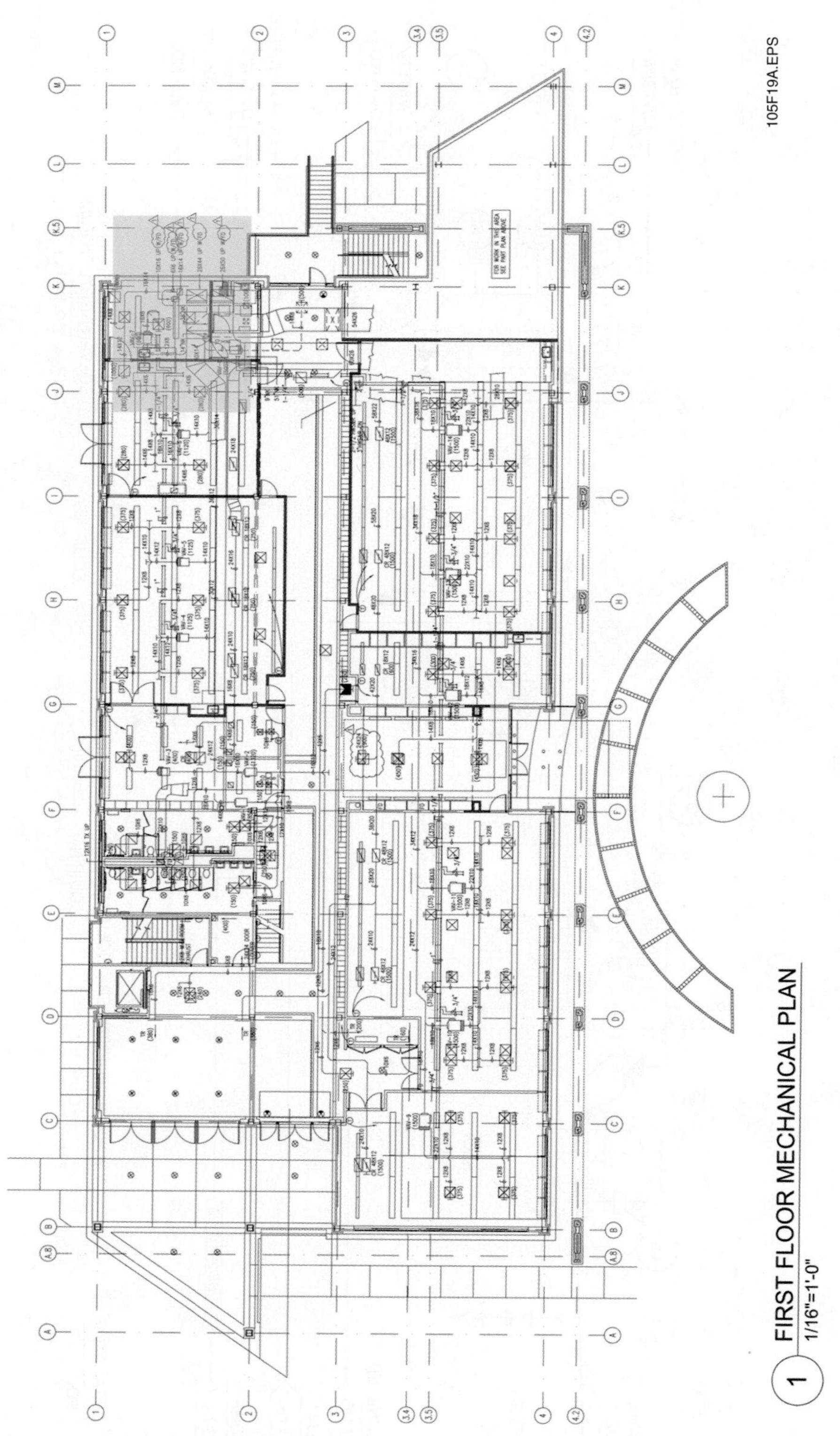

Figure 19 HVAC drawing (1 of 2).

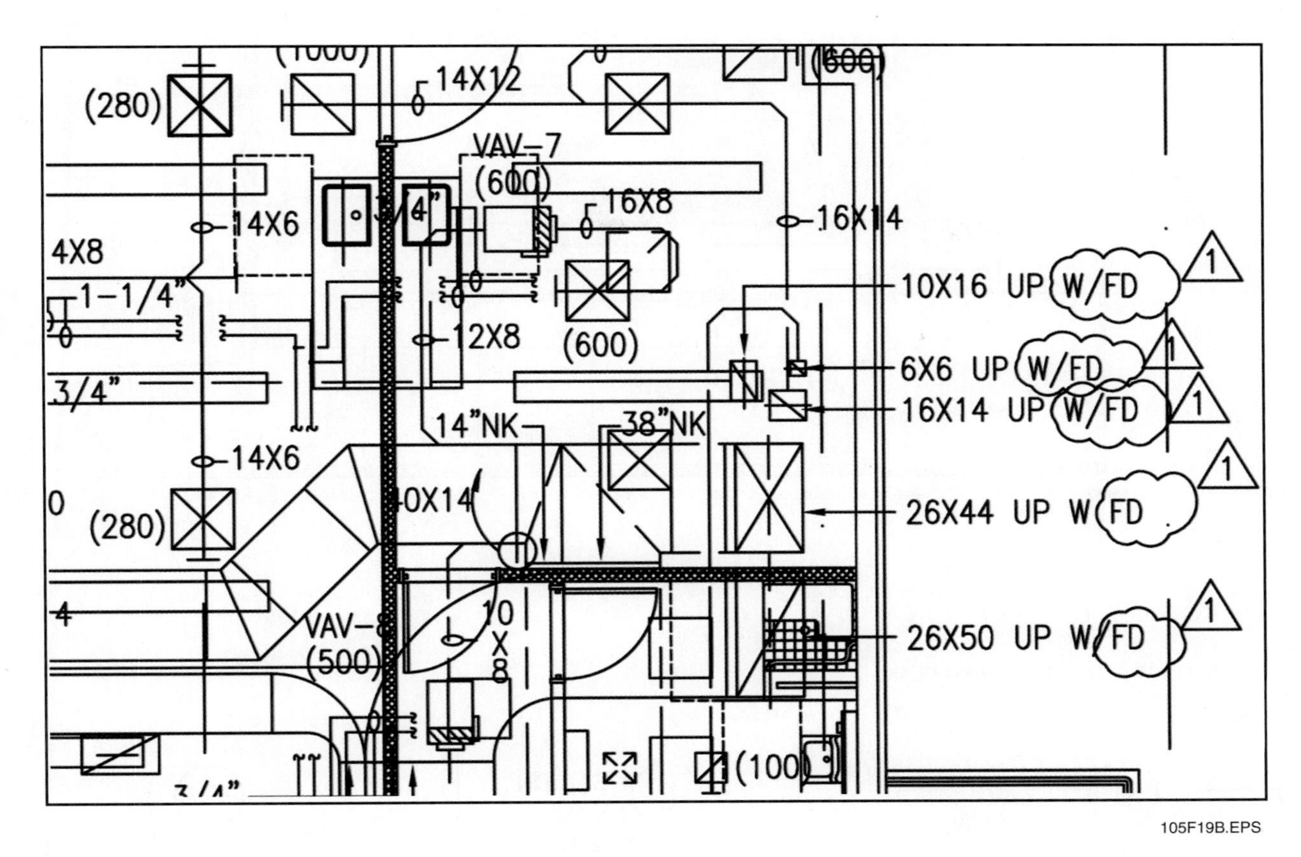

105F19B.EPS

Figure 19 HVAC drawing detail (2 of 2).

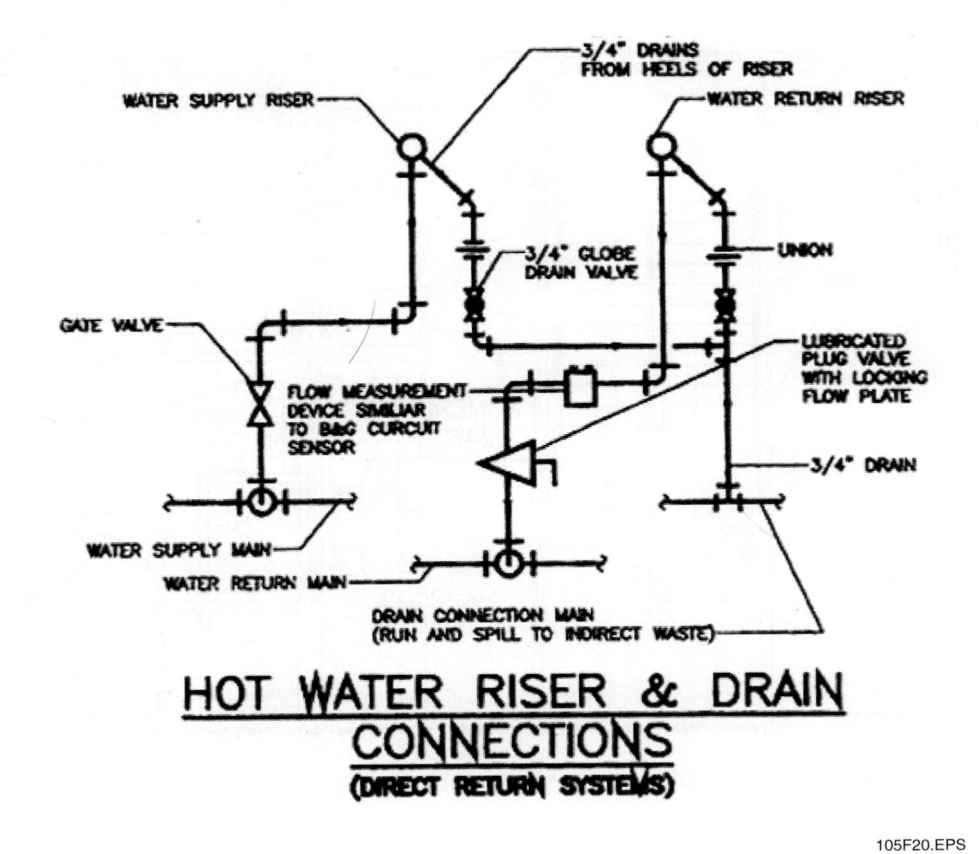

105F20.EPS

Figure 20 Mechanical detail drawing for hot water riser and drain connections.

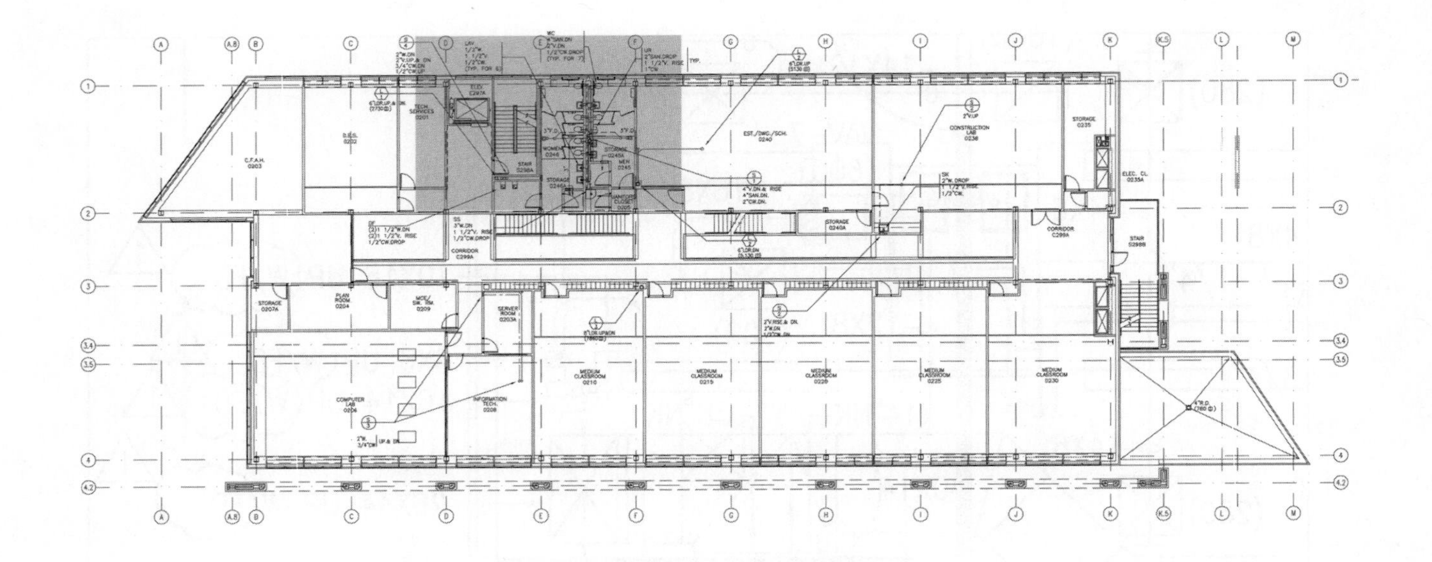

1 FIRST FLOOR PLUMBING PLAN
1/16"=1'-0"

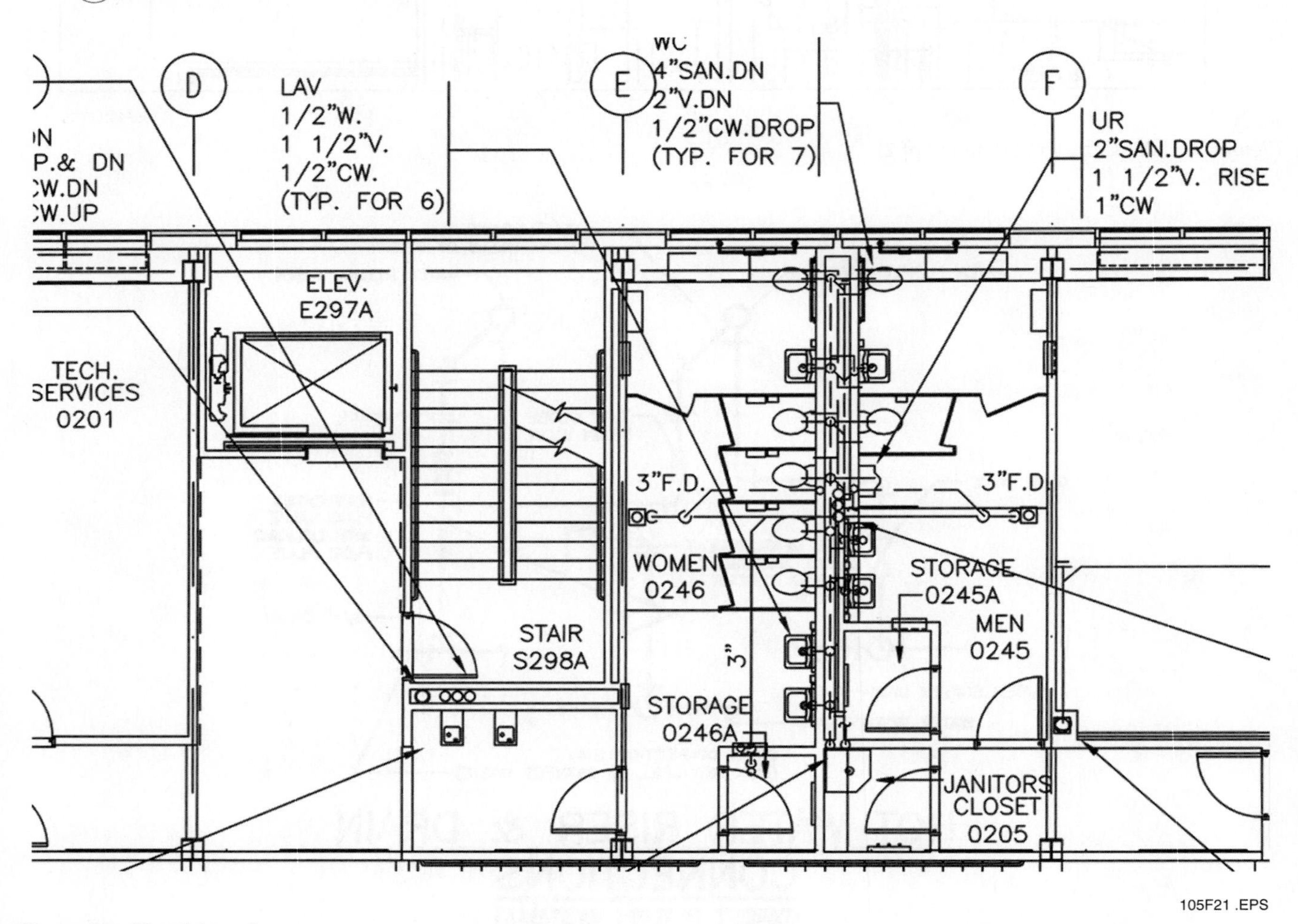

Figure 21 Plumbing plan.

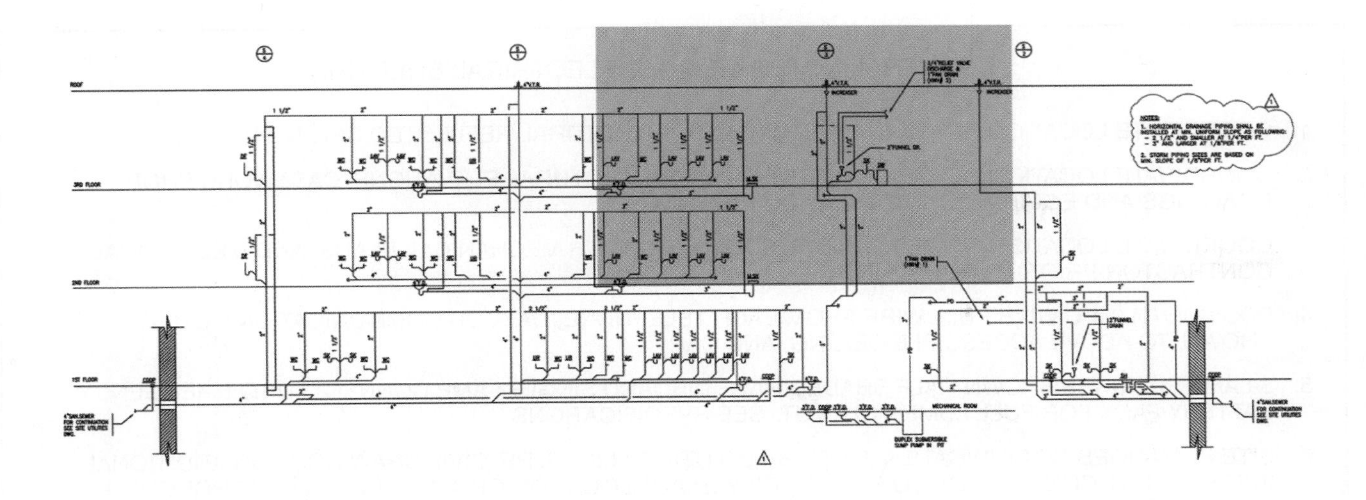

1 SANITARY RISER DIAGRAM
N.T.S.

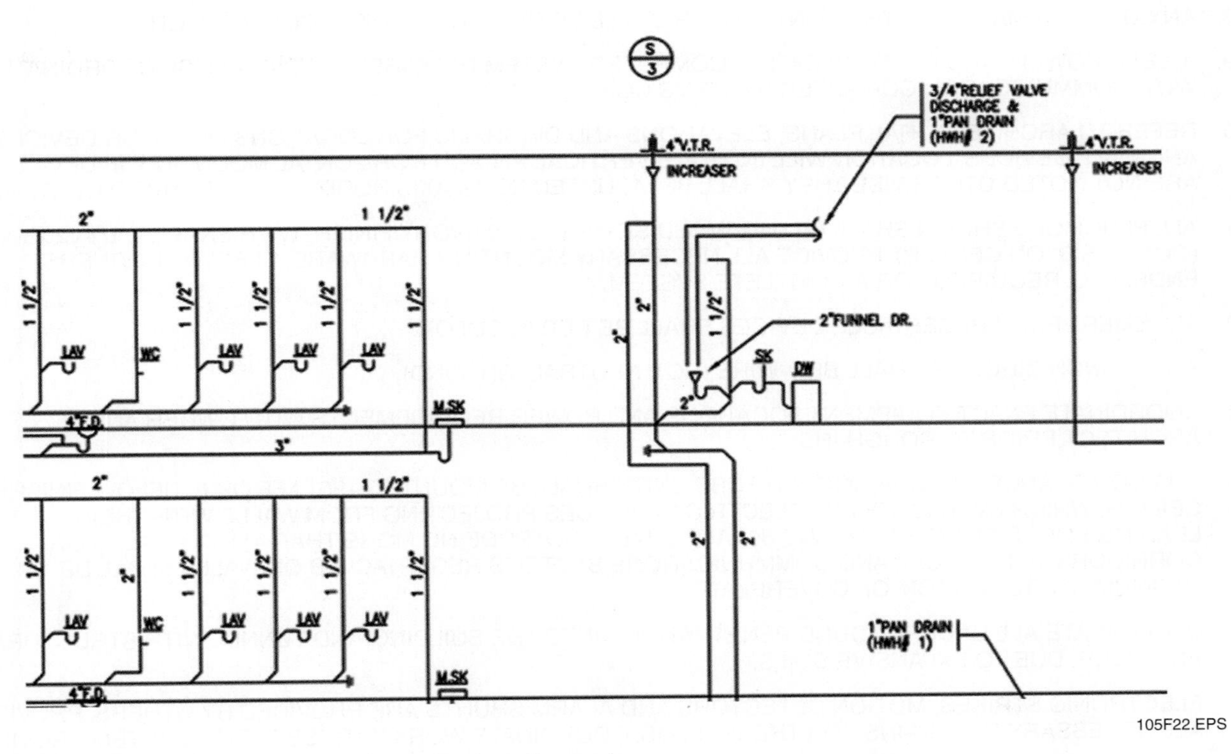

Figure 22 Plumbing isometric drawing (sanitary riser diagram).

GENERAL NOTES (FOR ALL ELECTRICAL SHEETS)

1. COORDINATE LOCATION OF LUMINARIES WITH ARCHITECTURAL REFLECTED CEILING PLANS.
2. COORDINATE LOCATION OF ALL OUTLETS WITH ARCHITECTURAL ELEVATIONS, CASEWORK SHOP DRAWINGS AND EQUIPMENT INSTALLATION DRAWINGS.
3. COORDINATE LOCATION OF MECHANICAL EQUIPMENT WITH MECHANICAL PLANS AND MECHANICAL CONTRACTOR PRIOR TO ROUGH-IN.
4. PROVIDE (1) 3/4"C WITH PULL WIRE FROM EACH TELEPHONE, DATA OR COMMUNICATION OUTLET SHOWN, TO ABOVE ACCESSIBLE CEILING, AND CAP.
5. 3-LAMP FIXTURES SHOWN HALF SHADED HAVE INBOARD SINGLE LAMP CONNECTED TO EMERGENCY BATTERY PACK FOR FULL LUMEN OUTPUT. SEE SPECIFICATIONS.
6. SITE PLAN DOES NOT INDICATE ALL OF THE UG UTILITY LINES, RE: CIVIL DRAWINGS FOR ADDITIONAL INFORMATION. CONTRACTOR TO FIELD VERIFY EXACT LOCATION OF ALL EXISTING UNDERGROUND UTILITY LINES OF ALL TRADES PRIOR TO ANY SITE WORK.
7. THE LOCATIONS OF ALL SMOKE DETECTORS SHOWN ARE CONSIDERED TO BE SCHEMATIC ONLY. THE ACTUAL LOCATIONS (SPACING TO ADJACENT DETECTORS, WALLS, ETC.) ARE REQUIRED TO MEET NFPA 72.
8. ANY ITEMS DAMAGED BY THE CONTRACTOR SHALL BE REPLACED BY THE CONTRACTOR.
9. "CLEAN POWER" AND COMMUNICATION/COMPUTER SYSTEM REQUIREMENTS SHALL BE COORDINATED WITH COMMUNICATION/COMPUTER SYSTEMS CONTRACTOR.
10. REFER TO ARCHITECTURAL PLANS, ELEVATIONS AND DIAGRAMS FOR LOCATIONS OF FLOOR DEVICES AND WALL DEVICES. LOCATION WILL INDICATE VERTICAL AND/OR HORIZONTAL MOUNTING. IF DEVICES ARE NOT NOTED OTHERWISE THEY SHALL BE MOUNTED LONG AXIS HORIZONTAL AT +16" TO CENTER.
11. ALL PLUGMOLD SHOWN SHALL BE WIREMOLD SERIES V2000 (IVORY FINISH) WITH SNAPICOIL #V20GB06 (OUTLETS 6" ON CENTER). PROVIDE ALL NECESSARY MOUNTING HARDWARE, ELBOWS, CORNERS, ENDS, ETC. REQUIRED FOR A COMPLETE SYSTEM.
12. ALL EMERGENCY RECEPTACLE DEVICES SHALL BE RED IN COLOR.
13. ALL BRANCH CIRCUITS SHALL BE 3-WIRE (HOT, NEUTRAL, GROUND).
14. COORDINATE EXACT EQUIPMENT LOCATIONS AND POWER REQUIREMENTS WITH OWNER AND ARCHITECT PRIOR TO ROUGH-INS.
15. ADA COMPLIANCE: ALL ADA HORN/STROBE UNITS SHALL BE MOUNTED +90" AFF OR 6" BELOW FINISHED CEILING, WHICH EVER IS LOWER. ELECTRICAL DEVICES PROJECTING FROM WALLS WITH THEIR LEADING EDGES BETWEEN 27" AND 80" AFF SHALL PROTRUDE NO MORE THAN 4" INTO WALKS OR CORRIDORS. ELECTRICAL AND COMMUNICATIONS SYSTEMS RECEPTACLES ON WALLS SHALL BE 15" MINIMUM AFF TO BOTTOM OF COVERPLATE.
16. COORDINATE ALL UNDERGROUND PENETRATIONS INTO THE BUILDING AND TUNNEL WITH STRUCTURAL ENGINEER, DUE TO EXPANSIVE SOILS.
17. ELECTRONIC STRIKES, MOTION DETECTORS AND ALARM SHUNTS ARE PROVIDED BY OTHERS. PROVIDE ALL NECESSARY ROUGH-INS FOR THESE ITEMS. COORDINATE WORK WITH SECURITY SYSTEM PROVIDER.

105F23.EPS

Figure 23 Electrical plan general notes.

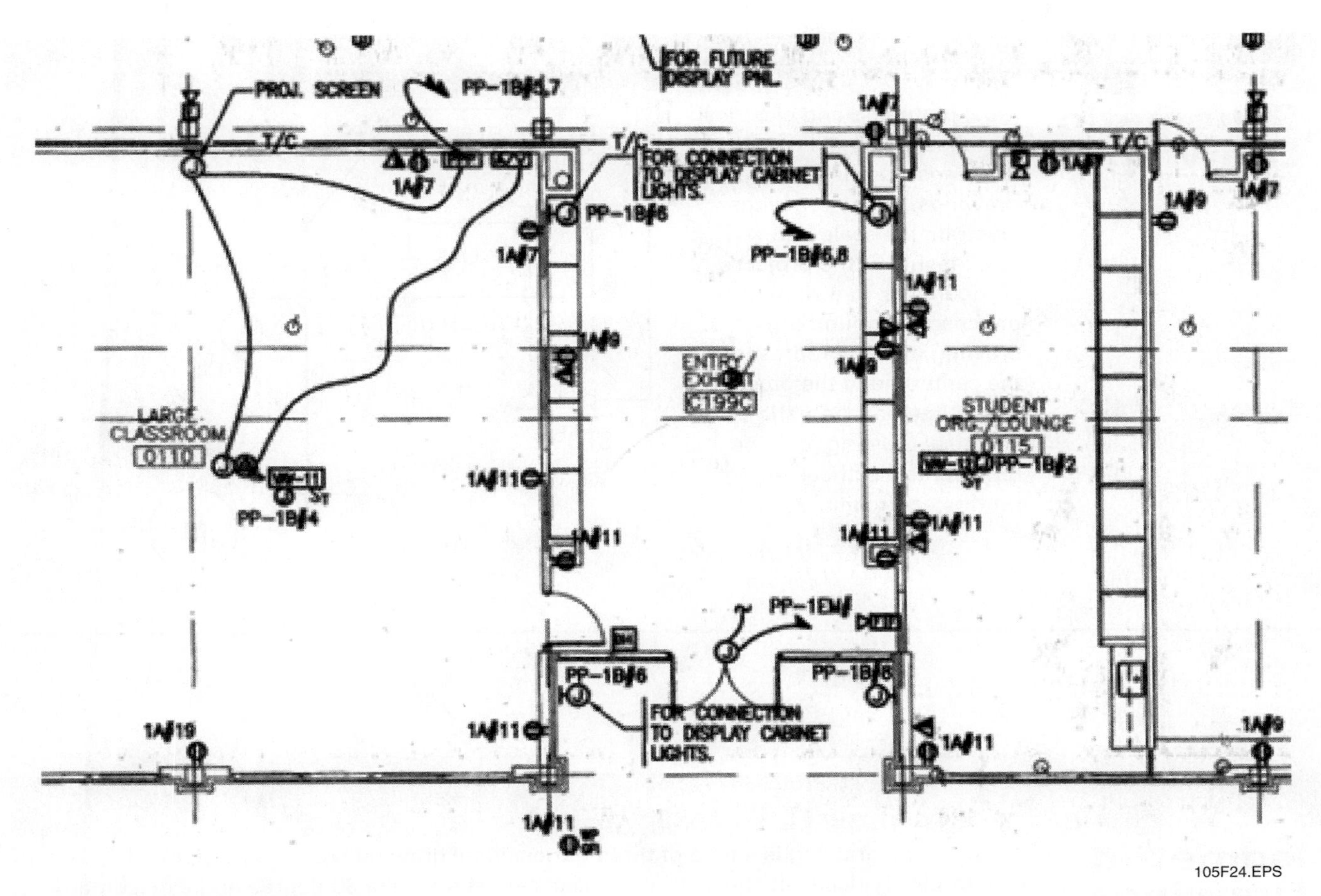

Figure 24 Power plan (first floor).

GOING GREEN

Green Construction

Green construction refers to a method of designing and building structures using materials and techniques that help minimize the stress on our natural resources and the environment.

With only 5% of the world's population, the United States manages to consume about 26% of the world's energy (buildings account for 36% of this consumption) and produces 136 million tons of construction and demolition debris a year.

In response to these numbers, and a general increase in environmental awareness, organizations such as the U.S. Green Building Council (USGBC) created environmental assessment systems, such as the LEED (Leadership in Energy and Environmental Design) Green Building Rating System. Systems like LEED provide green standards for the construction industry to follow, and they also officially certify structures (nonresidential) that meet USGBC's strict criteria.

The LEED program is a voluntary national standard that awards points for incorporating green strategies into areas such as site planning, safeguarding water quality and efficiency, efficiency in energy use and recycling, conservation of resources and materials, and the design, quality, and efficiency of the indoor environment.

In some cases, there may be more costs upfront to build a green building than there would be with a traditional building. However, the benefits over time more than make up for those costs, both monetarily and environmentally. Some of these benefits include a decrease in environmental impact, decreased operating costs, an increase in productivity due to increased occupant comfort and health, a potential reduction in insurance costs, and less strain on the local infrastructure.

The LEED program was launched in 1988 and has already certified hundreds of projects nationally, with hundreds more registered and awaiting certification.

On-Site

Schematic Drawings

Most plumbing and electrical sketches are single-line drawings or schematic drawings. These drawings illustrate the scale and relationship of the project's components. In a single-line or schematic plumbing drawing, the line represents the centerline of the pipe. In a single-line or schematic electrical drawing, the line represents electrical wiring routing or circuit.

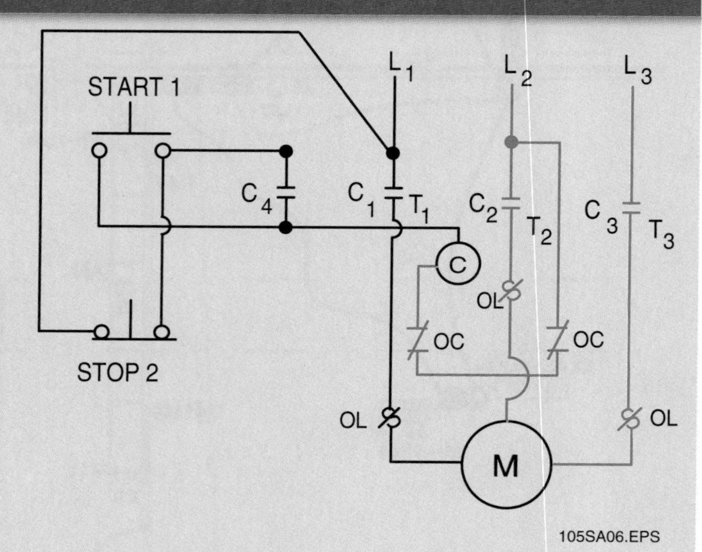

105SA06.EPS

On-Site

Isometric Drawings

An isometric drawing is a type of three-dimensional drawing known as a pictorial illustration. Typically in construction, objects are shown at a 30-degree angle in isometric drawings to provide a three-dimensional perspective, rather than as a flat, two-dimensional view.

BILL OF MATERIAL

P.M.	REQ'D	SIZE	DESCRIPTION
1		1-1/2"	PIPE SCH/40 ASTM-A-120 GR.B
2		3/4"	PIPE SCH/40 ASTM-A-120 GR.B
3	5	1-1/2"	90° ELL ASTM-A-197 BW
4	1	1-1/2"	TEE
5	2	3/4"	45° ELL
6	1	1-1/2" × 3/4"	BELL RED. CONC.
7	2	1-1/2"	GATE VA. BW ASTM-B62
8	1	1-1/2"	CHECK VA. SWING BW

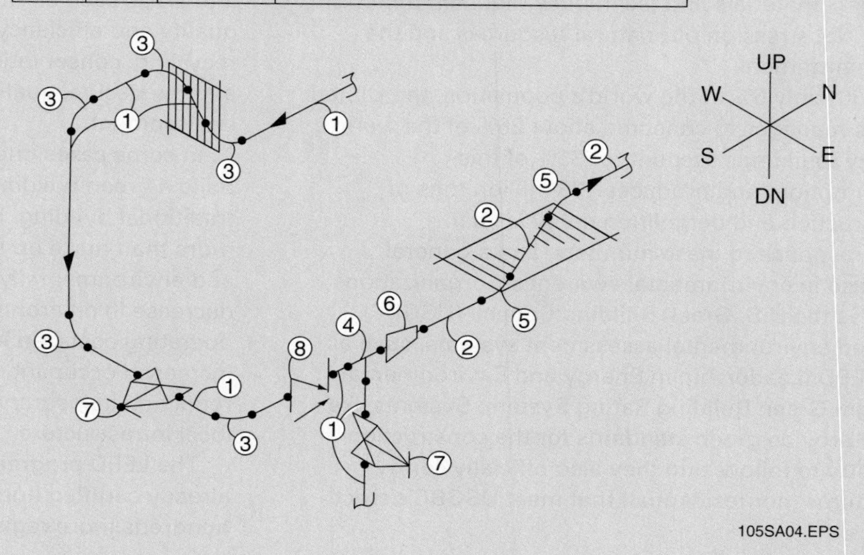

105SA04.EPS

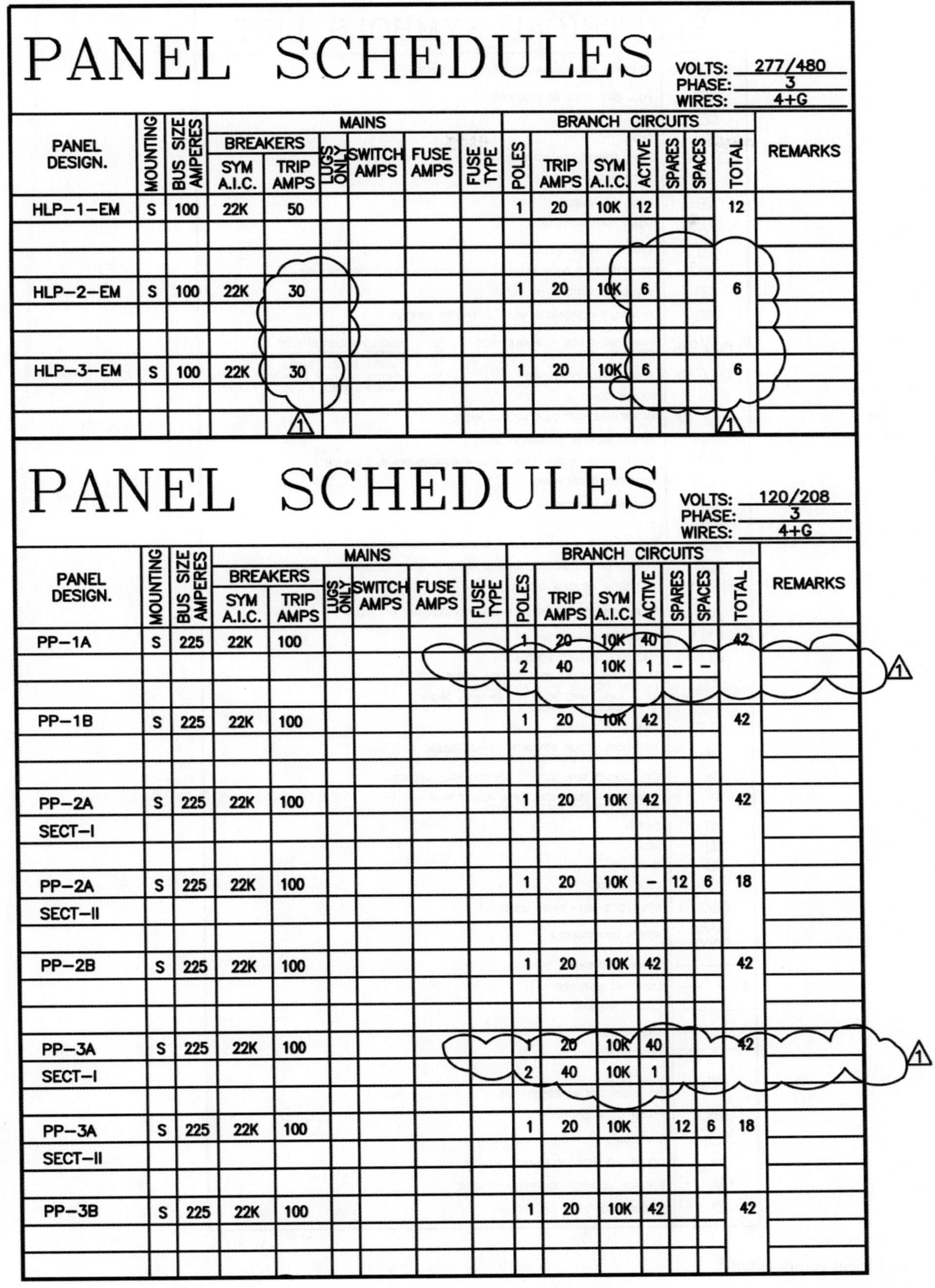

PANEL SCHEDULES

VOLTS: 277/480
PHASE: 3
WIRES: 4+G

PANEL DESIGN.	MOUNTING	BUS SIZE AMPERES	MAINS						BRANCH CIRCUITS							REMARKS
			BREAKERS SYM A.I.C.	BREAKERS TRIP AMPS	LUGS ONLY	SWITCH AMPS	FUSE AMPS	FUSE TYPE	POLES	TRIP AMPS	SYM A.I.C.	ACTIVE	SPARES	SPACES	TOTAL	
HLP-1-EM	S	100	22K	50					1	20	10K	12			12	
HLP-2-EM	S	100	22K	30					1	20	10K	6			6	
HLP-3-EM	S	100	22K	30					1	20	10K	6			6	

PANEL SCHEDULES

VOLTS: 120/208
PHASE: 3
WIRES: 4+G

PANEL DESIGN.	MOUNTING	BUS SIZE AMPERES	MAINS						BRANCH CIRCUITS							REMARKS
			BREAKERS SYM A.I.C.	BREAKERS TRIP AMPS	LUGS ONLY	SWITCH AMPS	FUSE AMPS	FUSE TYPE	POLES	TRIP AMPS	SYM A.I.C.	ACTIVE	SPARES	SPACES	TOTAL	
PP-1A	S	225	22K	100					1	20	10K	40			42	
									2	40	10K	1	–	–		
PP-1B	S	225	22K	100					1	20	10K	42			42	
PP-2A SECT-I	S	225	22K	100					1	20	10K	42			42	
PP-2A SECT-II	S	225	22K	100					1	20	10K	–	12	6	18	
PP-2B	S	225	22K	100					1	20	10K	42			42	
PP-3A SECT-I	S	225	22K	100					1	20	10K	40			42	
									2	40	10K	1				
PP-3A SECT-II	S	225	22K	100					1	20	10K		12	6	18	
PP-3B	S	225	22K	100					1	20	10K	42			42	

105F25.EPS

Figure 25 Panel schedules.

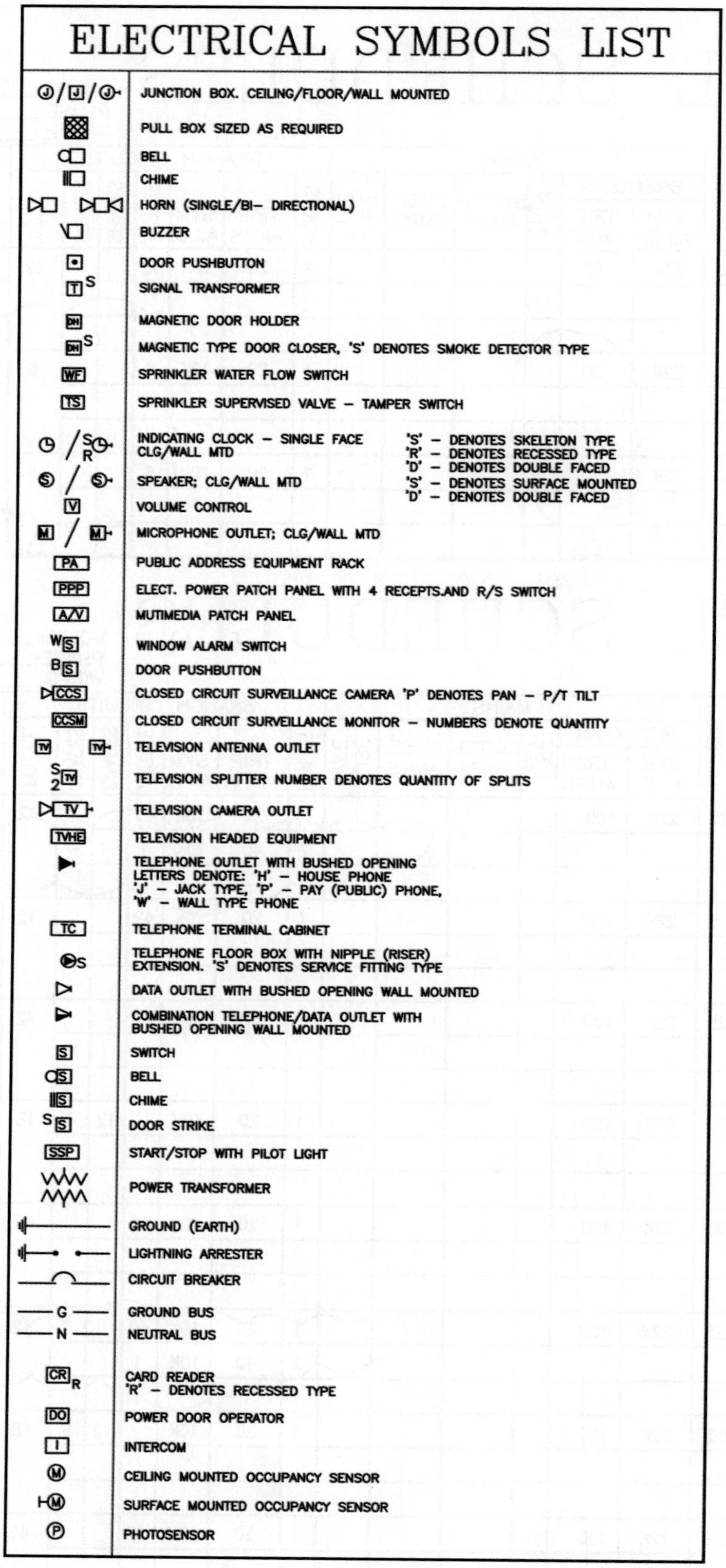

ELECTRICAL SYMBOLS LIST

J / J / J	JUNCTION BOX. CEILING/FLOOR/WALL MOUNTED
	PULL BOX SIZED AS REQUIRED
	BELL
	CHIME
	HORN (SINGLE/BI– DIRECTIONAL)
	BUZZER
	DOOR PUSHBUTTON
T S	SIGNAL TRANSFORMER
DH	MAGNETIC DOOR HOLDER
DH S	MAGNETIC TYPE DOOR CLOSER, 'S' DENOTES SMOKE DETECTOR TYPE
WF	SPRINKLER WATER FLOW SWITCH
TS	SPRINKLER SUPERVISED VALVE – TAMPER SWITCH
S / R	INDICATING CLOCK – SINGLE FACE CLG/WALL MTD — 'S' – DENOTES SKELETON TYPE; 'R' – DENOTES RECESSED TYPE; 'D' – DENOTES DOUBLE FACED
S / S	SPEAKER; CLG/WALL MTD — 'S' – DENOTES SURFACE MOUNTED; 'D' – DENOTES DOUBLE FACED
V	VOLUME CONTROL
M / M	MICROPHONE OUTLET; CLG/WALL MTD
PA	PUBLIC ADDRESS EQUIPMENT RACK
PPP	ELECT. POWER PATCH PANEL WITH 4 RECEPTS.AND R/S SWITCH
A/V	MUTIMEDIA PATCH PANEL
W S	WINDOW ALARM SWITCH
B S	DOOR PUSHBUTTON
CCS	CLOSED CIRCUIT SURVEILLANCE CAMERA 'P' DENOTES PAN – P/T TILT
CCSM	CLOSED CIRCUIT SURVEILLANCE MONITOR – NUMBERS DENOTE QUANTITY
TV TV	TELEVISION ANTENNA OUTLET
S 2 TV	TELEVISION SPLITTER NUMBER DENOTES QUANTITY OF SPLITS
TV	TELEVISION CAMERA OUTLET
TVHE	TELEVISION HEADED EQUIPMENT
	TELEPHONE OUTLET WITH BUSHED OPENING LETTERS DENOTE: 'H' – HOUSE PHONE 'J' – JACK TYPE, 'P' – PAY (PUBLIC) PHONE, 'W' – WALL TYPE PHONE
TC	TELEPHONE TERMINAL CABINET
S	TELEPHONE FLOOR BOX WITH NIPPLE (RISER) EXTENSION. 'S' DENOTES SERVICE FITTING TYPE
	DATA OUTLET WITH BUSHED OPENING WALL MOUNTED
	COMBINATION TELEPHONE/DATA OUTLET WITH BUSHED OPENING WALL MOUNTED
S	SWITCH
S	BELL
S	CHIME
S S	DOOR STRIKE
SSP	START/STOP WITH PILOT LIGHT
	POWER TRANSFORMER
	GROUND (EARTH)
	LIGHTNING ARRESTER
	CIRCUIT BREAKER
G	GROUND BUS
N	NEUTRAL BUS
CR R	CARD READER 'R' – DENOTES RECESSED TYPE
DO	POWER DOOR OPERATOR
I	INTERCOM
M	CEILING MOUNTED OCCUPANCY SENSOR
M	SURFACE MOUNTED OCCUPANCY SENSOR
P	PHOTOSENSOR

105F26.EPS

Figure 26 Electrical symbols list.

ELECTRICAL ABBREVIATIONS

A	AMPERE(S)
AC	ALTERNATING CURRENT
ACB	AIR CIRCUIT BREAKER
AFF	ABOVE FINISHED FLOOR
AFG	ABOVE FINISHED GRADE
AL	ALUMINUM
ALT	ALTERNATE
ASYM	ASYMMETRICAL
ATS	AUTOMATIC TRANSFER SWITCH
AWG	AMERICAN WIRE GAUGE
BC	BOTTOM CONDUIT
BD	BUS DUCT
BFG	BELOW FINISHED GRADE
BIL	BASIC IMPULSE LEVEL
BLDG	BUILDING
BX	ARMORED CABLE
C	CONDUIT
CATV	CABLE ANTENNA TELEVISION SYSTEM
CCAB	CONTROL CABINET
CCTV	CLOSED CIRCUIT TELEVISION
CH	CABINET HEATER
CKT	CIRCUIT
CKT BKR/CB	CIRCUIT BREAKER
CL	CLOSET
CLG	CEILING
COND	CONDUCTOR
CO	CONDUIT ONLY
CT	CURRENT TRANSFORMER
CU	COPPER
DB	DUCT BANK
DMB	DIMMER BOARD
DC	DIRECT CURRENT
DH	DUCT HEATER
DIM	DIMMER CONTROL
DISC	DISCONNECT
DM	DAMPER MOTOR
DN	DOWN
DP	DISTRIBUTION POWER PANEL(BOARD)
DT	DOUBLE THROW
DWG	DRAWING
EA	EACH
E.C.	ELECTRICAL CONTRACTOR
E.HTR	ELECTRIC HEATER
ELEV	ELEVATOR
EL	ELECTRIC
EM	EMERGENCY
EMT	ELECTRICAL METALLIC TUBING
ENT	ELECTRICAL NON-METALLIC TUBING
EWC	ELECTRIC WATER COOLER
EX	EXISTING
F	FUSE
FA	FIRE ALARM
FACP	FIRE ALARM CONTROL PANEL
FBO	FURNISHED BY OTHERS
FCC	FLAT CONDUCTOR CABLE
FCU	FAN COIL UNIT
FDR	FEEDER
FL	FLOOR
FLUOR	FLUORESCENT
F.P.C.	FIRE PROTECTION CONTRACTOR
FS	FUSIBLE SWITCH
F.S.C.	FOOD SERVICE CONTRACTOR
FT	FEET OR FOOT
G.C.	GENERAL CONTRACTOR
GEN	GENERATOR
GF	GROUND FAULT
GG	GROUND GRID
GRD	GROUND
HC	HUNG CEILING
H.I.D.	HIGH INTENSITY DISCHARGE
HP	HORSEPOWER
H.P.S.	HIGH PRESSURE SODIUM
HPU	HEAT PUMP UNIT
HT	HEIGHT
HV	HIGH VOLTAGE
HW	HEAVY WALL RIGID CONDUIT
HZ	FREQUENCY IN CYCLES PER SECOND
IC	INTERRUPTING CAPACITY
IG	ISOLATED GROUND
IMC	INTERMEDIATE METALLIC CONDUIT
INC	INCANDESCENT
JB	JUNCTION BOX
K	KEY OPERATED
kVA	KILOVOLT AMPERE(S)
kVAR	KILOVAR(S)
kW	KILOWATT(S)
kWhr	KILOWATT(S) HOUR(S)
LP	LIGHTING PANEL
L.P.S.	LOW PRESSURE SODIUM
LTG	LIGHTING
LV	LOW VOLTAGE
MATV	MASTER ANTENNA TELEVISION SYSTEM
MC	METAL CLAD CABLE
MCC	MOTOR CONTROL CENTER
MCM	THOUSAND CIRCULAR MIL(S)
MCP	MOTOR CONTROL PANEL
M.C.	MECHANICAL CONTRACTOR
MH	MANHOLE
MIC	MICROPHONE
MIN	MINIMUM
MS	MAGNETIC STARTER
MTD	MOUNTED
MTG	MOUNTING
MTR	MOTOR
MTS	MANUAL TRANSFER SWITCH
N	NEUTRAL
NA	NON-AUTOMATIC
NC	NORMALLY CLOSED
NF	NON-FUSE
N.I.C.	NOT IN CONTRACT
NL	NIGHT LIGHT
NO	NORMALLY OPEN
NP	NETWORK PROTECTOR
NTS	NOT TO SCALE
OC	ON CENTER
OL	OVERLOAD ELEMENT
P	POLE
PA	PUBLIC ADDRESS
PB	PULL BOX
P.C.	PLUMBING CONTRACTOR
PF	POWER FACTOR
Ø	PHASE
PL	PILOT (INDICATOR) LIGHT
PNL	PANEL (PANELBOARD)
PP	POWER PANEL
PRI	PRIMARY
PT	POTENTIAL TRANSFORMER
PVC	POLYVINYL CHLORIDE
PWR	POWER
R	RECESSED
RC	REMOTE CONTROL
REC	RECEPTACLE
S	SURFACE
SC	SEPARATE CIRCUIT
SDB	SUB-DISTRIBUTION BOARD
SEC	SECONDARY
SMR	SURFACE METAL RACEWAY
SP	SINGLE POLE
SPK	SPEAKER
ST	SINGLE THROW
SW	SWITCH
SWBD	SWITCHBOARD
SYM	SYMMETRICAL
T	THERMOSTAT
TEL	TELEPHONE
TB	TERMINAL BOX
TC	TOP CONDUIT
TCAB	TELEPHONE CABINET
T.C.C.	TEMPERATURE CONTROL CONTRACTOR
TP	TAMPER PROOF
TV	TELEVISION
TYP	TYPICAL
UG	UNDERGROUND
UH	UNIT HEATER
UNG	UNGROUNDED
UON	UNLESS OTHERWISE NOTED
UPS	UNINTERRUPTED POWER SYSTEM
V	VOLT(S)
VA	VOLTAMP(S)
VAR	VOLT AMPERES REACTIVE
VP	VAPORPROOF
W	WATT(S)
WP	WEATHERPROOF
WT	WATERTIGHT
XFR	TRANSFORMER
XP	EXPLOSION PROOF

105F27.EPS

Figure 27 Electrical abbreviations.

M.E. RINKER SR. HALL
SCHOOL OF BUILDING CONSTRUCTION
BR-191

AIR HANDLING UNITS
SECTION 15760

SECTION 15760 - AIR HANDLING UNITS

PART 1 - GENERAL

The requirements of the Contract Documents, including the General and Supplementary General Condition, and Division 1 - General Requirements shall apply to the work of this section.

1.1 DESCRIPTION

A. Work Included: Furnish and install air handling units in accordance with the contract drawings.

B. Related Work Specified Elsewhere

1. Vibration Control: Section 15628

2. Insulation: Section 15800

3. Electrical Energy. Division 16

1.2 QUALITY ASSURANCE

A. Standards and Codes

1. AMCA

2. NEMA

3. NEC

1.3 SUBMITTALS

A. Shop drawings indicating size, location, details and installation requirements.

B. Product Data: Manufacturer's printed data, catalog cuts, test data, manufacturer's recommendations.

C. Operational and Maintenance Manuals: Manufacturer's instructions for operation and maintenance.

PART 2 - PRODUCTS

GEA 0300-0600

15760 - 1
Revision No. 1
Dec. 19, 2001

105F28A.EPS

Figure 28 Specification (air handling units) (1 of 2).

M.E. RINKER SR. HALL
SCHOOL OF BUILDING CONSTRUCTION
BR-191

AIR HANDLING UNITS
SECTION 15760

2.1 AIR HANDLING UNITS

A. Units shall be of the type, size and capacity as set forth in the schedule. The fan outlet velocities and coil and filter face velocities shall be within 5% of the values specified in the schedule. Units shall be double wall McQuay, or approved equal.

B. The units, as assembled, shall be complete with fans, coils, insulated casing, filters, drives and accessories. Each unit, including the fan enclosure, shall have essentially constant cross-sectional dimensions as to width and height. Internal baffles shall be provided as required to prevent bypassing of coils and filters.

C. The casing shall consist of an independent structural steel frame, properly reinforced and braced for maximum rigidity, having individually removable, flush mounted, insulated panels. The casing shall be of sectionalized construction, consisting basically of individual fan section, coil section, access sections, filter section, and drain pan. Sections shall be joined with continuous gasketing to form an air tight closure. Sections shall be so designed that the method of joining can be performed with relative ease and without damage to the insulation and vapor barrier.

The framework shall be constructed of AISC structural rolled shapes having minimum thickness of 1/8" (3 mm) or die formed sheet steel having the minimum gauges set forth in the following schedules:

Maximum Individual Casing Cross-Section	Minimum Framework Gauge
Up to 30 sq.ft. (2.8 sq.m.)	14
30.5 sq.ft. to 47 sq.ft. (2.81 to 4.4 sq.m.)	12
48 sq.ft. (4.45 sq.m.) and up	10

D. Framework shall be designed with recesses suitable to receive enclosure panels, providing neat appearance, airtight enclosure, and ease of panel removal.

E. Enclosure panels 12 sq. ft. in area and larger shall be constructed of not less than 18 gauge die formed sheet steel. Should the sides or top of a casing section exceed 20 sq. ft. in area, the panels shall be fabricated of more than one piece, with the individual panels recessed into intermediate structural members.

Protection for the insulation edges shall be provided around the perimeter of each

GEA 0300-0600

15760 - 2
Revision No. 1
Dec. 19, 2001

105F28B.EPS

Figure 28 Specification (air handling units) (2 of 2).

On-Site

Orthographic Drawings

An orthographic drawing is a construction drawing showing straight-on views of the different sides of an object. Orthographic drawings show dimensions that are proportional to the actual physical dimensions. In orthographic drawings, the designer draws lines that are scaled-down representations of real dimensions. Every 12 inches, for example, may be represented by ¼ inch on the drawing. This type of drawing is used for elevation drawings.

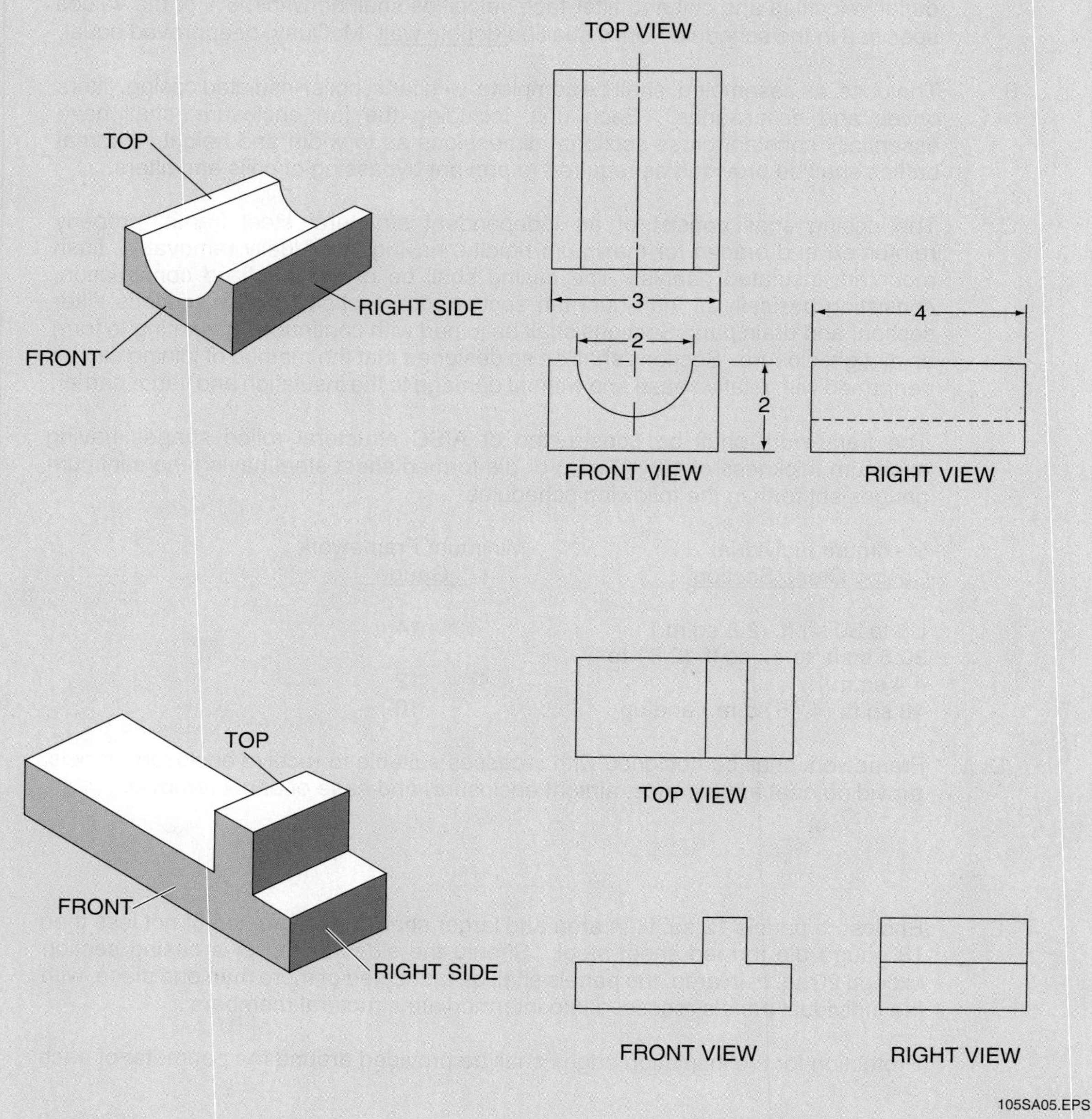

DATE 12/07/02 RFI NO. 1

PROJECT NAME GERMANS FROM RUSSIA PROJECT NO. 15-1593

REQUEST: REF D.W.G.NO. M2 REV. DETAIL 1/M2 OTHER

WILL THE 16 X 10 INTAKE AIR DUCTWORK RUNNING THROUGH RM 116 REQUIRE WALL MOUNTED FIRE DAMPERS ON ALL 4 EXIT CORRIDOR WALL PENETRATIONS AND WILL THE 14 X 10 TRANSFER DUCTWORK REQUIRE THE INSTALLATION OF A FIRE DAMPER AS WELL?

BY: LARRY MAYRE REPLY BY (DATE): 12/20/02

REPLY:

ANSWER: ALL DUCTWORK IS ABOVE CEILING, SO OK AS IS.

DATE: 12/19/02

105F29.EPS

Figure 29 Sample RFI.

5.0.0 Lines of Construction

It is very important to understand the meanings of lines on a drawing. The lines commonly used on a drawing are sometimes called the Alphabet of Lines. Here are some of the more common types of lines (*Figure 30*):

- **Dimension lines** – Establish the dimensions (sizes) of parts of a structure. These lines end with arrows (open or closed), dots, or slashes at a termination line drawn perpendicular to the dimension line.
- **Leaders** *and arrowheads* – Identify the location of a specific part of the drawing. They are used with words, abbreviations, symbols, or keynotes.
- *Property lines* – Indicate land boundaries.
- *Cut lines* – Lines around part of a drawing that is to be shown in a separate cross-sectional view.
- *Section cuts* – Show areas not included in the cutting line view.
- *Break lines* – Show where an object has been broken off to save space on the drawing.
- **Hidden lines** – Identify part of a structure that is not visible on the drawing. You may have to look at another drawing to see the part referred to by the lines.
- *Centerlines* – Show the measured center of an object, such as a column or fixture.
- *Object lines* – Identify the object of primary interest or the closest object.

6.0.0 Abbreviations, Symbols, and Keynotes

Architects and engineers use systems of abbreviations, symbols, and keynotes to keep plans uncluttered, making them easier to read and understand. Throughout this module you'll find examples of some of these items. Following is additional information on these items and some variants on examples of each that you may come across in other sets of drawings.

Each trade has its own symbols, and you should learn to recognize the symbols used by other trades. For example, if you are an electrician,

On-Site

Regional and Company Differences

Although most symbols are standard, you may find slight variations in different regions of the country. Always check the title sheet or other introductory drawing to verify the symbols you find on the project drawings.

Your company may also use some special symbols or terms. Your instructor or supervisor will tell you about conventions that are unique to your company.

On-Site

Check All the Plans

Always double-check all the plans for a project. Be familiar with other trade work that may affect your job. Determine whether an electrical conduit is close to your plumbing pipes or whether your framing member is placed too close to an HVAC duct. By knowing all the work planned for a particular area, you can prevent errors and save time and money.

If a particular dimension is missing on the plan you will be using, check the other plans to see if it is included there. The part of the building you are working on will also be shown on other plans.

When taking measurements from plans, make sure that the dimensions of each part equal the total measurement. For instance, if you are looking at a drawing of a wall with one window, add up the measurement from the left end of the wall to the left side of the window, the width of the window, and the measurement from the right side of the window to the right end of the wall. Make sure this total matches the measurement of the entire wall.

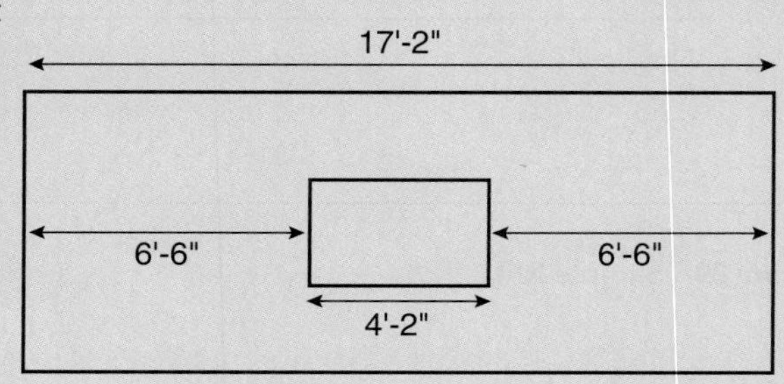

you should understand a carpenter's symbols. If you are a carpenter, you should understand a plumber's symbols, and so on. Then, no matter what symbols you see when you are working on a project, you will understand what they mean.

Abbreviations used in construction drawings are short forms of common construction terms. For example, the term On Center is abbreviated O.C. Some common abbreviations are listed in *Figure 31*.

Abbreviations should always be written in capital letters. Abbreviations for each project should be noted on the title sheet or other introductory drawing page such as the legend page. Books that list construction abbreviations and their meanings are available. You do not need to memorize these abbreviations. You will start to remember them as you use them.

Symbols are used on a drawing to tell what material is required for that part of the project. A combination of these symbols, expanded and drawn to the same size, makes up the pictorial view of the plan. There are architectural symbols (*Figure 32*), civil and structural engineering symbols (*Figure 33*), mechanical symbols (*Figure 34*),

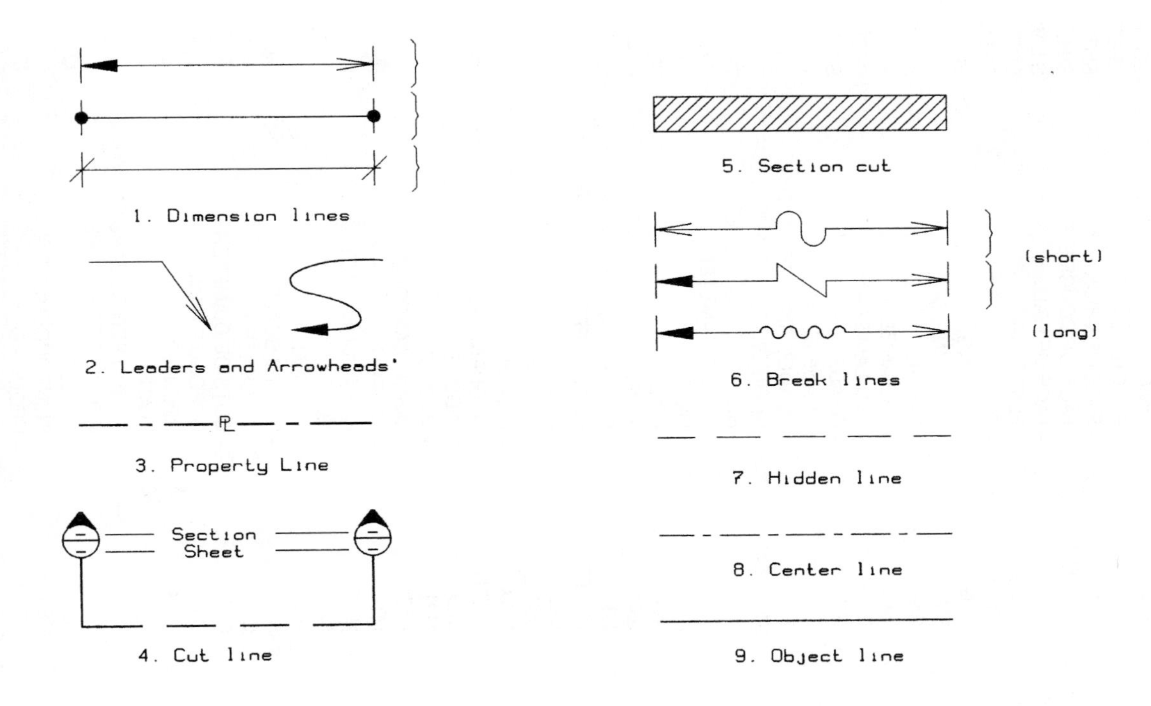

Figure 30 Lines of construction (Alphabet of Lines).

On-Site

Care of Construction Drawings

Construction drawings are valuable records and must be cared for. Follow these rules when you handle construction drawings:

- Never write on a construction drawing without authorization.
- Keep drawings clean. Dirty drawings are hard to read and can cause errors.
- Fold drawings so that the title block is visible.
- Fold and unfold drawings carefully to avoid tearing.
- Do not lay sharp tools or pointed objects on construction drawings.
- Keep drawings away from moisture.
- Make copies for field use; don't use originals.

ABBREVIATIONS

A.B. – ANCHOR BOLT
ADD'L – ADDITIONAL
ADJ. – ADJACENT
A.I.S.C. – AMERICAN INSTITUTE OF STEEL CONSTRUCTION
ALT. – ALTERNATE
ARCH. – ARCHITECTURAL
A.S.T.M. – AMERICAN SOCIETY FOR TESTING & MATERIALS
BLDG. – BUILDING
BM. – BEAM
B.O. – BOTTOM OF
BOT. – BOTTOM
BSMT. – BASEMENT
BTWN. – BETWEEN
CANT. – CANTILEVER
CB. – CARDBOARD
CH. – CHAMFER
C.J. – CONTROL/CONSTRUCTION JOINT
CLR. – CLEAR, CLEARANCE
C.M.U. – CONCRETE MASONRY UNIT
COL. – COLUMN
CONC. – CONCRETE
CONN. – CONNECTION
CONST. – CONSTRUCTION
CONT. – CONTINUOUS
CONTR. – CONTRACTOR
CTRD. – CENTERED
DET. – DETAIL
DIAG. – DIAGONAL
DIAM. – DIAMETER
DIM. – DIMENSION
DISCONT. – DISCONTINUOUS
DWG. – DRAWING
EA. – EACH
E.F. – EACH FACE
EL. – ELEVATION
ELECT. – ELECTRICAL
ELEV. – ELEVATOR
EQ. – EQUAL
E.W.B. – END WALL BARS
E.W. – EACH WAY
EXIST. – EXISTING
EXP. JNT. – EXPANSION JOINT
EXT. – EXTERIOR
F.D. – FLOOR DRAIN

FDN. – FOUNDATION
FIN. – FINISH
FLR. – FLOOR
F.O.B. – FACE OF BRICK
F.O.CONC. – FACE OF CONCRETE
F.O.W. – FACE OF WALL
FS. – FLAT SLAB
FT. – FOOT
FTG. – FOOTING
F.W. – FILLET WELD
GA. – GAUGE
GAL. – GALVANIZED
G.L. – GLU-LAM BEAM
GR. – GRADE
GR. BM. – GRADE BEAM
H.A.S. – HEADED ANCHOR STUD
HORIZ. – HORIZONTAL
H.S.B. – HIGH STRENGTH BOLT
I.D. – INSIDE DIAMETER
IN. – INCH
INT. – INTERIOR
JNT. – JOINT
LB. – POUND
LIN. FT. – LINEAL FEET
L.L.V. – LONG LEG VERTICAL
MAT'L. – MATERIAL
MAX. – MAXIMUM
MECH. – MECHANICAL
MID. – MIDDLE
MIN. – MINIMUM
MISC. – MISCELLANEOUS
MTL. – METAL
N.I.C. – NOT IN CONTRACT
NO. – NUMBER
NOM – NOMINAL
N.T.S. – NOT TO SCALE
O.C. – ON CENTER
O.D. – OUTSIDE DIAMETER
O.H. – OPPOSITE HAND
OPNG. – OPENING
PL – PLATE
P.S.F. – POUND PER SQUARE FOOT
P.S.I. – POUND PER SQUARE INCH
R. – RADIUS
REINF. – REINFORCEMENT
REQ'D. – REQUIRED

RM. – ROOM
SCHED. – SCHEDULE
SECT. – SECTION
SHT. – SHEET
SIM. – SIMILAR
S.L.V. – SHORT LEG VERTICAL
SPC. – SPACE
SPEC. – SPECIFICATION
SQ. – SQUARE
STD. – STANDARD
STIFF. – STIFFENER
STL. – STEEL
STOR. – STORAGE
SYM. – SYMMETRICAL
T.&B. – TOP AND BOTTOM
THK. – THICKNESS
T.O. – TOP OF
TYP. – TYPICAL
U.N.O. – UNLESS NOTED OTHERWISE
VAR. – VARIES
VERT. – VERTICAL
V.I.F. – VERIFY IN FIELD
WT. – WEIGHT

SYMBOLS

Symbol	Meaning
℄	CENTER LINE
⌀	DIAMETER
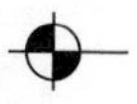 ⯐	ELEVATION
&	AND
W/	WITH
PL	PLATE
X	BY
#	NUMBER
◎	AT
⌻	SQUARE
L	ANGLE

105F31.EPS

Figure 31 Abbreviations.

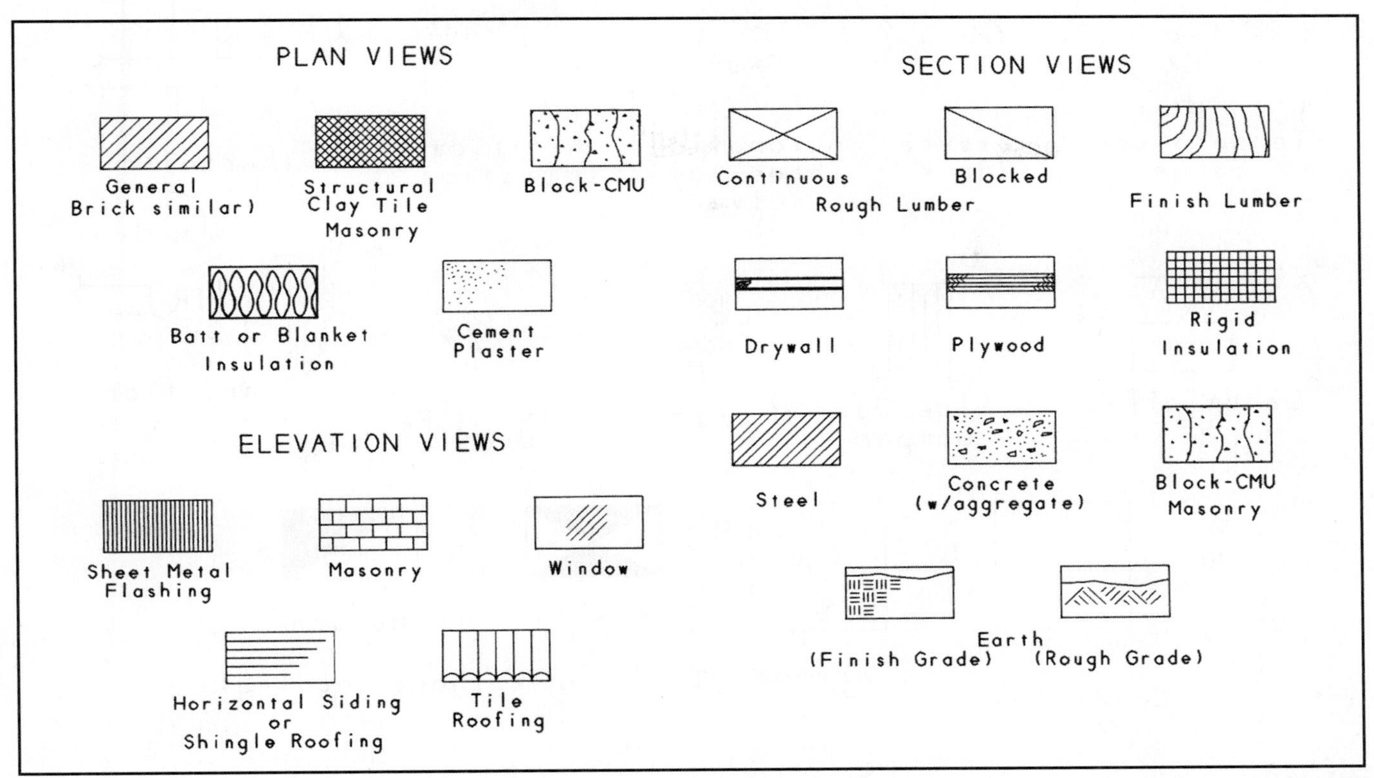

105F32.EPS

Figure 32 Architectural symbols.

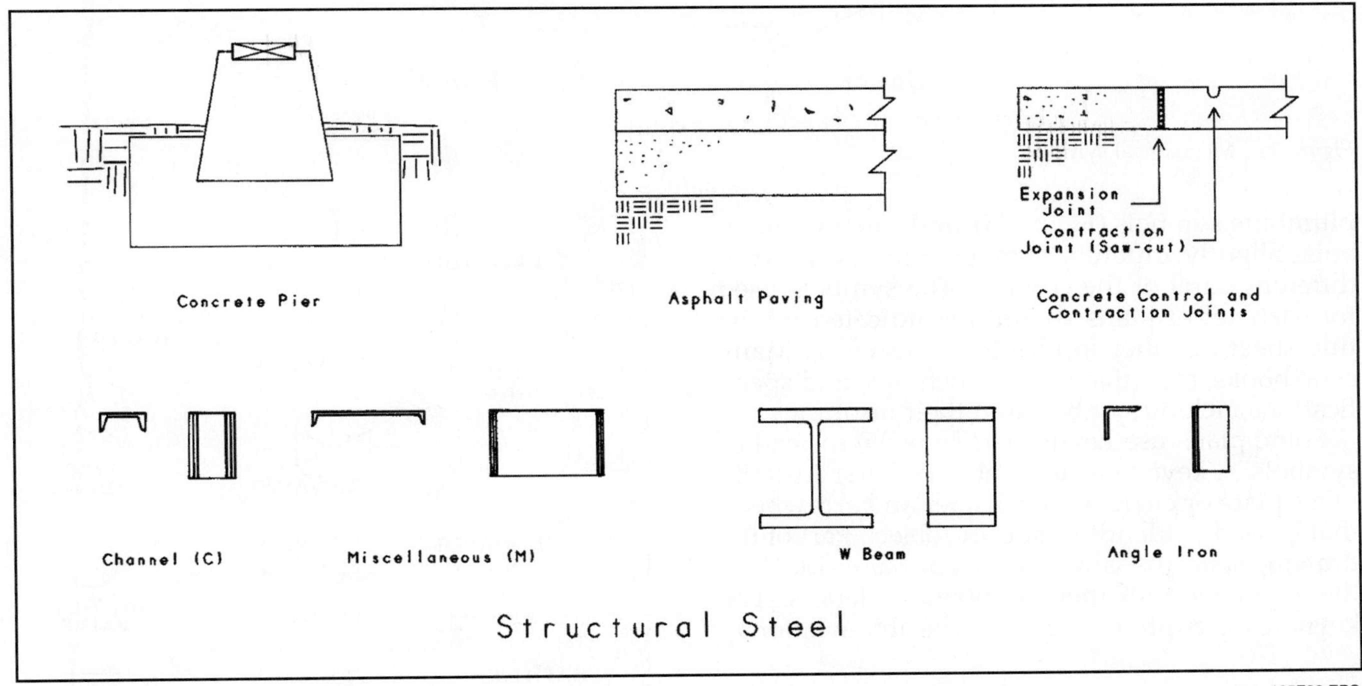

105F33.EPS

Figure 33 Civil and structural engineering symbols.

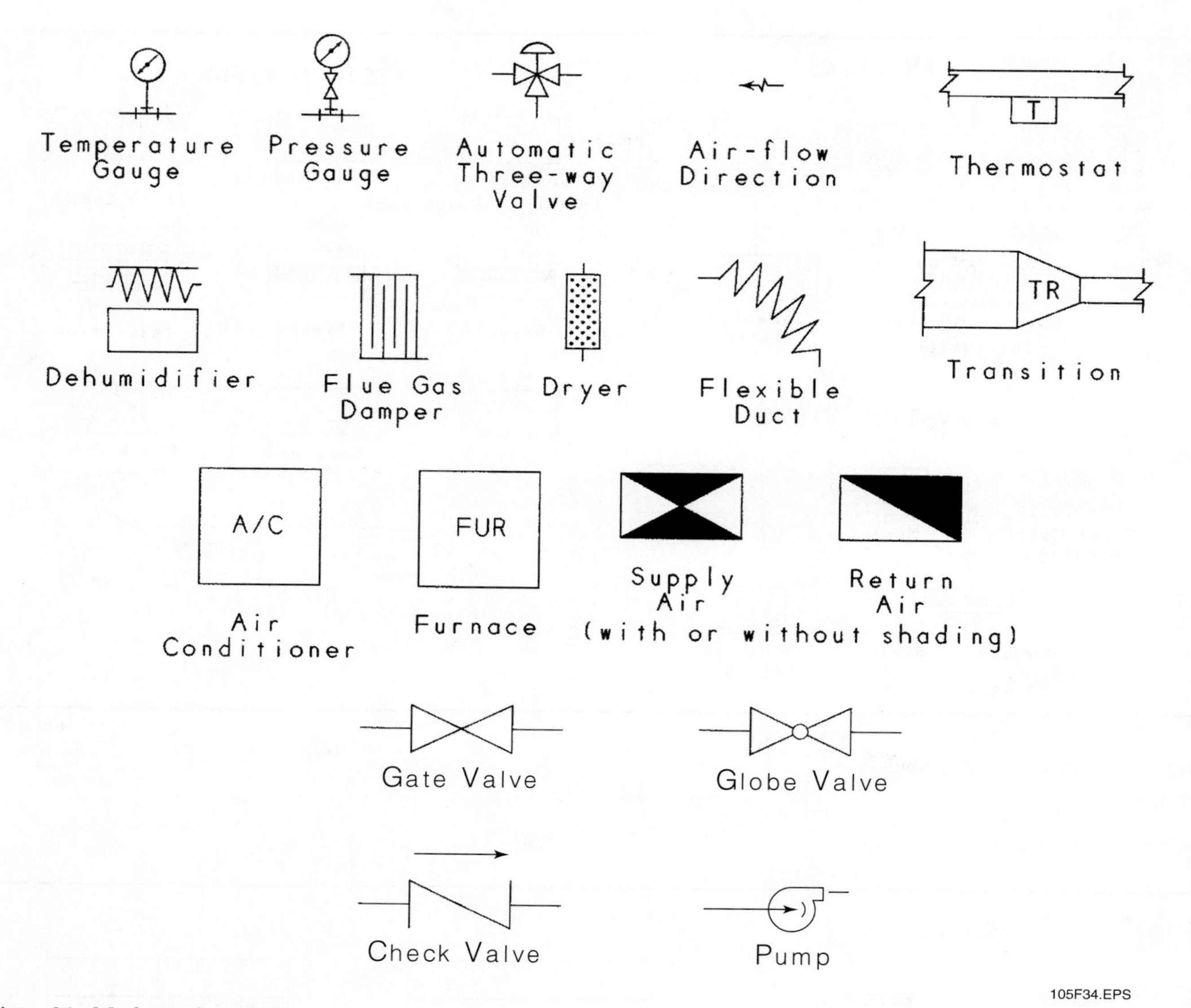

Figure 34 Mechanical symbols.

plumbing symbols (*Figure 35*), and electrical symbols. Slightly different symbols may be used in different parts of the country. The symbols used for each set of plans should be indicated on the title sheet or other introductory drawing. Many code books, manufacturers' brochures, and specifications include symbols and their meanings.

Some plans use keynotes (*Figure 36*) instead of symbols. A keynote is a number or letter (usually in a square or circle) with a leader and arrowhead that is used to identify a specific object. Part of the drawing sheet (usually on the right-hand side) lists the keynotes with their numbers or letters. The keynote descriptions normally use abbreviations.

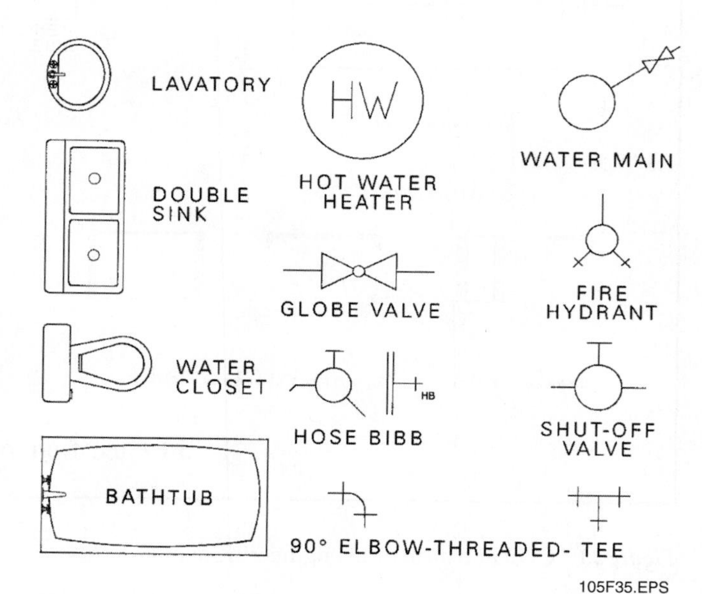

Figure 35 Plumbing symbols.

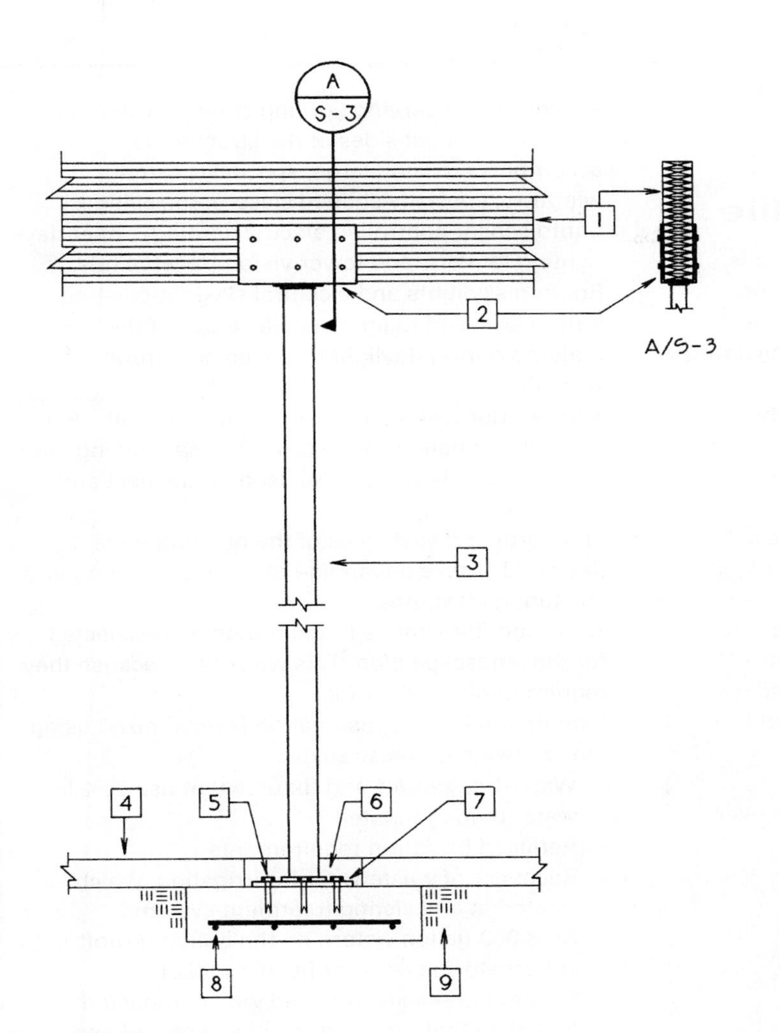

KEYNOTES

1. 6⅛" x18" glulam beam
2. 6⅛" end cap
3. 6"DIA pipe column
4. 6" concrete slab w/6x6-W2.9 x W2.9 WWF
5. ⅝" DIA expansion bolt (typical of 6)
6. Seismic opening (to be filled with concrete)
7. ½" steel plate column base
8. 1'-0" x 4'-0" concrete footing w/4-#5 rebar each way
9. Natural or compacted grade

105F36.EPS

Figure 36 Keynotes.

7.0.0 Using Gridlines to Identify Plan Locations

Have you ever used a map to find a street? The map probably used a grid to make locating a detailed area easier. The index might have referred you to section B-3. You located B along the side of the map and 3 along the top. Then you located the intersection of the two and found your street.

The gridline system shown on a plan (*Figure 37*) is used like the grid on a map. On a drawing such as a floor plan, a grid divides the area into small parts called bays.

The numbering and lettering system begins in the upper left-hand corner of the floor plan. The numbers are normally across the top and the letters are along the side. To avoid confusion, certain letters and the symbol for zero are not used. Omitted from the gridline system are letters I, O, and Q; and numbers 1 and 0.

A gridline system makes it easy to refer to specific locations on a plan. Suppose you want to refer to one outlet, but there are a dozen on a plan. Simply refer to the outlet in bay C-8.

8.0.0 Dimensions

Dimensions are the parts of the drawings that show the size and the placement of the objects that will be built or installed. Dimension lines can have arrowheads or slashes at both ends, with the dimension itself written near the middle of the line. The dimension is a measurement written as a number, and it may be written in inches with fractions (6½"), in feet with inches (1'-2"), in inches with decimals (3.2"), or in millimeters (9mm) if the metric system is used. Feet are always in whole numbers on construction drawings.

To do accurate work, you need to know how to read dimensions on construction drawings. This means you need to know whether the dimensions measure to the exterior or the interior of an object. To understand the difference, look at *Figure 38*, which shows a piece of pipe. There are

GOING GREEN

Recycling Rinker Hall, University of Florida in Gainesville

An example of a LEED Gold rated building is Rinker Hall on the campus of the University of Florida in Gainesville. The structure is part of the University of Florida's School of Building Construction and occupies 47,000 square feet. The building accommodates up to 450 students and 100 faculty members in a variety of classrooms, teaching labs, construction labs, faculty and administrative offices, and student facilities.

During the planning phase of the building, special consideration was given to land use and the use of recycled and recyclable materials. Additionally, special attention was also given to planning the incorporation of the most recent green technologies in the areas of water consumption and conservation, heating and cooling energies, and other power needs required to run the various systems of a building.

The following is a list of some of the green strategies incorporated into the construction of Rinker Hall:

- A majority of the materials used in the structure came from recycled materials and were selected based on the longevity of their service life, low maintenance requirements, low toxicity, and the material's ability to be recycled or reused.
- The structure was built on already-disturbed land and in an orientation to nearby buildings that would preserve the maximum amount of green space.
- The structure is oriented on a pure north-south axis allowing it to utilize low-angle light for day-lighting. Computer modeling software was used to simulate daylight in order to establish the most energy-efficient position for the structure in terms of lighting, heating, and cooling needs.
- An energy conserving lighting control system on the east and west sides of the structure is accomplished using large, specially glazed windows, specially shaped classroom ceilings, photo sensor-controlled electric lighting, upper day-lighting louvers, and lower vision-panel blinds.
- Rooftop skylights and a central skylight-covered atrium are used to light specific areas of the building during daylight hours using natural sunlight.
- The exterior wall system of the building was designed to balance moisture, thermal loading, and daylight. Shade walls were used on the east and west sides.
- The north and east faces of the building were designed to take advantage of thermal shading and sheltering attributes.
- Trees and flora native to the region were selected for the landscape plan. This was done because they require minimal watering.
- Interior water use conservation is maximized using the following green strategies:
  - Water-free urinals and fixtures that use 20% less water than mandated
  - Reduced irrigation requirements
  - Recovery of wastewater for irrigation, which is treated at an existing treatment system
  - An 8,000-gallon cistern on the building rooftop to collect storm water for flushing toilets
  - Outdoor areas are surfaced with compacted gravel so that rainwater can be captured and directed to the storm water system. From there, the water is used for irrigation.

With the various green strategies, materials selection, and building techniques, Rinker Hall and all of its systems use up to 57% less energy than a similar structure designed in minimal compliance with the American Society of Heating, Refrigerating, and Air-Conditioning Engineers (ASHRAE).

105SA08.EPS

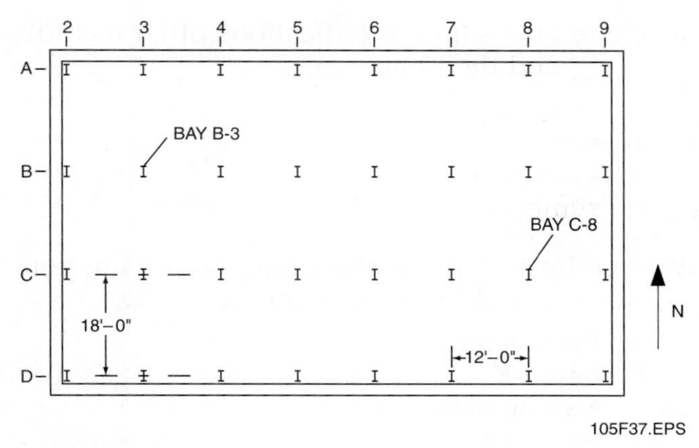

Figure 37 Grid.

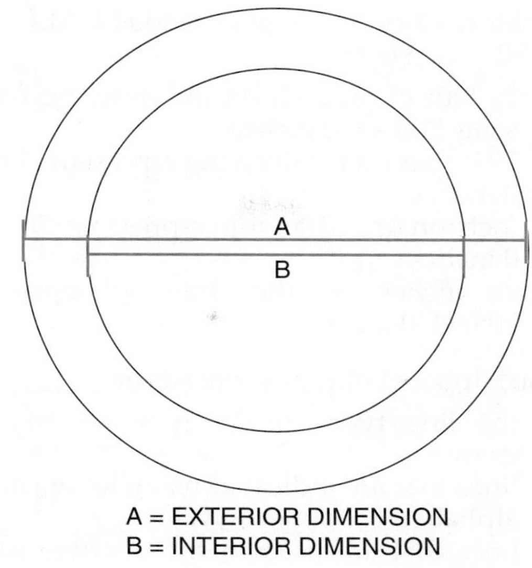

Figure 38 Exterior and interior dimensions on pipe.

two measurements you could take to get the pipe's dimensions.

The first measurement is from the pipe's exterior edge on one side directly across to its exterior edge on the other side. The second measurement is from the pipe's inside (interior) edge on one side directly across to its interior edge on the other side. Even though the difference between these two dimensions may be only a fraction of an inch (the thickness of the pipe), they are still two completely different dimensions. This is important to remember because any dimensioning inaccuracy or miscalculation in one place will affect the accuracy of calculations in other places.

Review Questions

1. The title block generally contains _____.

a. revision blocks
b. special marks used in the drawing
c. the mechanical plans
d. the legend

2. The latest revision date on a set of construction drawings can be found _____.

a. inside the detail
b. in the schedule
c. inside the title block
d. in the legend

3. Revisions to the drawing are entered in the revision block and must include _____.

a. the date the drawing was approved
b. the project description, engineering approvals, and intended date of completion
c. the date and the initials of the person who made the revision
d. customer approval

4. A site plan _____.

a. does not show features such as trees and driveways
b. shows the location of the building from an aerial view
c. is drawn after the floor plan is drawn, before the HVAC is designed
d. includes information about plumbing fixtures and equipment

5. _____ plans show the layout of the HVAC system.

a. Plumbing
b. Structural
c. Mechanical
d. Foundation

6. If the scale on a site plan reads SCALE: 1" = 20'-0", then every _____.
   a. 1/20th of an inch on the drawing represents 20 feet, 0 inches
   b. 20 inches on the drawing represents 1 foot, 0 inches
   c. inch on the drawing represents 20 feet, 0 inches
   d. 20 inches on the drawing represents 20 feet, 0 inches

7. The Alphabet of Lines consists of _____.
   a. the line types used on a construction drawing
   b. lines that are indicated using letters of the alphabet
   c. lines that are used to show where pages match up when a large page has been broken down to several smaller pages
   d. lines that indicate land boundaries on the site plan

8. Code books and specifications often include _____ and their meanings.
   a. drawings
   b. references
   c. glossaries
   d. symbols

9. To refer to a particular outlet on a plan, you can refer to the outlet in _____.
   a. the mechanical plan
   b. bay C-8
   c. the site plan
   d. Segment A-1

10. On construction drawings, dimensions in feet are always shown in _____.
    a. decimals
    b. a different color
    c. whole numbers
    d. bold type

SUMMARY

As a result of studying this module, you are now familiar with basic construction drawing terms used to refer to the types of information shown on construction drawings. You can distinguish among types of construction drawings commonly found on a job site—civil, architectural, structural, mechanical, plumbing, electrical plans, and fire protection plans—and can describe why each type of drawing is important for the completion of a project. You can identify standardized information that is included on all construction drawings, such as the drawing and revision dates, title of the project, legend, and scale used on the drawings. You understand the importance of consulting the specifications and understanding the RFI properties when deciding how to interpret the drawings.

All construction drawings use symbols to convey information. Each craft has symbols particular to the type of work performed. You should recognize and understand the symbols used by your own trade as well as others. This helps you visualize the entire project and may alert you to inconsistencies or errors on the plans.

Construction drawings represent actual components of a building project. By carefully studying a construction drawing, you can locate the part of the building referenced. You may do this by identifying symbols for parts of the structure and by measuring from a given point to locate something like the center line of a window or the edge of a stairway. The information on a construction drawing helps you visualize where the window will go, what type of window is used, and the materials used to make the window.

Dimensions allow you to transfer the outlines of the building from the construction drawing to the actual building site. This allows you to lay out the work properly and avoid mistakes.

The information you learned in this module enables you to use construction drawings on a job site. Your skill in reading and interpreting construction plans will grow the more you practice. Construction drawing reading and drawing are both valuable skills. Mastering the creation and use of construction drawings opens more career possibilities for you to explore.

Trade Terms Quiz

Fill in the blank with the correct trade term that you learned from your study of this module.

1. A(n) ____________ is a side view that shows height.
2. A(n) ____________ usually has an arrowhead at both ends, with the measurement written near the middle of the line.
3. A(n) ____________ is a qualified, licensed person who creates and designs drawings for a construction project.
4. ____________, which show the design of the project, include many parts, such as the floor plan, roof plan, elevation drawings, and section drawings.
5. ____________, together with the set of specifications, detail what is to be built and what materials are to be used.
6. Also called site plans or survey plans, ____________ show the location of the building from an aerial view, as well as the natural contours of the earth.
7. Almost all construction drawings today are made by ____________.
8. ____________ are solid or dashed lines showing the elevation of the earth on a civil drawing.
9. ____________ are enlarged views of some special features of a building, such as floors and walls.
10. A(n) ____________ applies scientific principles in design and construction.
11. To do accurate work, you need to know whether the ____________ measure to the exterior or the interior of the object.
12. ____________, or engineered drawings for electrical supply and distribution, include locations of the electric meter, switchgear, and convenience outlets.
13. Structural plans may show this large, horizontal support made of concrete, steel, stone, or wood ____________.
14. Schematic drawings called ____________ show all the equipment, pipelines, valves, instruments, and controls needed to operate a piping system.
15. An element of architectural drawings, ____________ refers to the height above sea level or other defined surface.
16. Architectural, civil and structural engineering, mechanical, and plumbing ____________ may be used on a drawing to tell what material is required for that part of the project.
17. Also called a plan view, a(n) ____________ is an aerial view of the layout of each room.
18. A(n) ____________ is a one-line drawing showing the flow path for electrical circuitry.
19. Part of the structural plans, the ____________ shows the lowest level of the building.
20. Piping system plans for gas, oil, or steam heat may be included in the ____________ plan.
21. When a plan is marked ____________, it means that the drawing gives approximate positions and sizes.
22. A(n) ____________ is a dashed line on a plan showing an object obstructed from view by another object.
23. Known also as a pictorial illustration, a(n) ____________ lets you see an object as it really is, rather than as a flat, two-dimensional view.
24. In drafting, an arrowhead is placed on a(n) ____________ in order to identify a component.
25. ____________ are written statements provided by the architectural and engineering firm to define the quality of work to be done and to describe the materials to be used.
26. The ____________ defines the symbols used in architectural plans.
27. ____________ are engineered plans for motors, pumps, piping systems, and piping equipment.
28. A(n) ____________ is a cross-sectional view that shows the inside of an object or building.

29. The term ____________ refers to both water supply and all liquid waste disposal.

30. If you notice a discrepancy in the plans, notify the foreman, who will write up a(n) ____________, explaining the problem and noting the date and time.

31. A(n) ____________ shows the shape of the roof and the materials that will be used to finish it.

32. The ____________ of a drawing tells the size of the object drawn compared with the actual size of the object represented.

33. ____________ show the layout for the plumbing system that supplies hot and cold water, for the sewage disposal system, and for the location of plumbing fixtures.

34. The ____________, which are used to support the architectural design, include the general notes, a foundation plan, a roof framing plan, and structural section drawings.

35. Part of the construction drawing, the ____________ gives information about the structure and is numbered for easy filing.

36. ____________ are the traditional name for construction drawings.

37. The drawing that makes up a building's piping, valves, and switches is a(n) ____________.

Trade Terms

Architect
Architectural plans
Beam
Blueprints
Civil plans
Computer-aided drafting (CAD)
Construction drawing
Contour lines
Detail drawings
Dimension line
Dimensions
Electrical plans
Elevation (EL)
Elevation drawing
Engineer
Fire protection plan
Floor plan
Foundation plan
Heating, ventilating, and air conditioning (HVAC)
Hidden line
Isometric drawing
Leader
Legend
Mechanical plans
Not to scale (NTS)
Piping and instrumentation drawings (P&IDs)
Plumbing
Plumbing plans
Request for information (RFI)
Roof plan
Scale
Schematic
Section drawing
Specifications
Structural plans
Symbol
Title block

Cornerstone of Craftsmanship

Jason C. Roberts

Associated Builders and Contractors
Director of Education
Metro Washington, DC

How did you choose a career in the field?
My mother told me, "If you do what you love you will never work a day in your life." I have stood by that to choose the direction I have taken.

My family has been in the construction industry, primarily as carpenters and general contractors, since the late 1800s. I was asked right out of high school to work for an uncle as a drywall finisher, and then moved on to carpentry.

What types of training have you been through?
I have received a degree in secondary education, as well as several industry-recognized national certifications: NCCER master trainer and instructor, CEDIA certified installer, and MD State Licensed Master Electrician Restricted Low Voltage.

What kinds of work have you done in your career?
I have had a varied career. I have been a drywall finisher and a carpenter. I am an industry trainer and, as noted, a Director of Education. And I've had the opportunity to own my own business, as a construction company owner and a multimedia installation owner.

Tell us about your present job and what you like about it.
My title is Director of Education with Associated Builders and Contractors, Metro Washington Chapter. I direct and support members in training programs for the construction industry. I have enjoyed training adults for years, and this opportunity to direct a training facility and apply my skills is very rewarding.

What factors have contributed most to your success?
Family, friends, persistence, and laughter have created my success.

What advice would you give to those new to the field?
Never be afraid of asking questions, think outside the box, follow your gut, and always be professional.

Tell us some interesting career-related facts or accomplishments:
I founded two start-up companies: Jason C. Roberts Construction and Enhanced Cinema Installations, and have been successfully operating them for six and seven years, respectively. I created the number one Electronic Systems Technician division in Lincoln Educational Services EST history. I've participated in the Electronic System Professional Alliance job task analysis team and the National Systems Contractors Association as a member of their Education Analysis Board. I've also participated multiple times as a subject matter expert to review NCCER's curricula.

Cornerstone of Craftsmanship

Thomas (Tom) G. Murphy

Alfred State SUNY College of Technology,
School of Applied Technology
Building Trades Instructor and Assistant to the Dean

How did you choose a career in the field?
After 20 years in the construction industry, I had an opportunity to teach apprenticeship and craft training. While doing this, my employer asked if I would be interested in teaching inner city job training programs to minorities. I taught carpentry for the inner city adult training programs for a number of years. In 1997, I worked with Alfred State on various construction curricula. In the fall of 1998, Alfred State offered me a teaching position. I am a tenured faculty member at the college in the Building Trades Department. I still teach classes in various minority communities. This type of teaching is my passion.

What types of training have you been through?
I have had four years of carpentry apprenticeship training, two years' training in heavy equipment operations, two years' training in supervision and estimating, and two years' training in historic rehabilitation/renovation. I am also an OSHA Construction Outreach trainer, master trainer for the NCCER, carpentry instructor for the NCCER, and Alfred State College trainer for the Anti-Defamation League's® A World of Difference Institute.

What kinds of work have you done in your career?
I started my construction career in 1975. I worked in commercial construction, high-end residential construction, high-rise construction, and historic renovation.

Tell us about your present job and what you like about it.
I work with high schools and technical centers. I travel and discuss opportunities in applied technology. I assist high school guidance counselors with articulation agreements to our programs. I like a number of things about my job. I still teach minority construction trade programs, meeting lots of students, and am able to excite them about careers in applied technology. I am blessed to have the best support staff and colleagues to work with anywhere. I work for a college that is the premier technology college in New York, having supportive administrators and helping students obtain careers in the construction industry.

What factors have contributed most to your success?
I have wonderful parents who never gave up and always discussed the value of education and hard work and a wife who has always been supportive. While growing up, I was always told you can achieve anything you put your mind to.

What advice would you give to those new to the field?
Work hard, take training and ongoing education classes, be patient, and you never know where the journey will take you.

Trade Terms Introduced in This Module

Architect: A qualified, licensed person who creates and designs drawings for a construction project.

Architectural plans: Drawings that show the design of the project. Also called architectural drawings.

Beam: A large, horizontal structural member made of concrete, steel, stone, wood, or other structural material to provide support above a large opening.

Blueprints: The traditional name used to describe construction drawings.

Civil plans: Drawings that show the location of the building on the site from an aerial view, including contours, trees, construction features, and dimensions.

Computer-aided drafting (CAD): The making of a set of construction drawings with the aid of a computer.

Construction drawings: Architectural or working drawings used to represent a structure or system.

Contour lines: Solid or dashed lines showing the elevation of the earth on a civil drawing.

Detail drawings: Enlarged views of part of a drawing used to show an area more clearly.

Dimension line: A line on a drawing with a measurement indicating length.

Dimensions: Measurements such as length, width, and height shown on a drawing.

Electrical plans: Engineered drawings that show all electrical supply and distribution.

Elevation (EL): Height above sea level, or other defined surface, usually expressed in feet.

Elevation drawing: Side view of a building or object, showing height and width.

Engineer: A person who applies scientific principles in design and construction.

Fire protection plan: A drawing that shows the details of the building's sprinkler system.

Floor plan: A drawing that provides an aerial view of the layout of each room.

Foundation plan: A drawing that shows the layout and elevation of the building foundation.

Heating, ventilating, and air conditioning (HVAC): Heating, ventilating, and air conditioning.

Hidden line: A dashed line showing an object obstructed from view by another object.

Isometric drawing: A type of three-dimensional drawing of an object.

Leader: In drafting, the line on which an arrowhead is placed and used to identify a component.

Legend: A description of the symbols and abbreviations used in a set of drawings.

Mechanical plans: Engineered drawings that show the mechanical systems, such as motors and piping.

Not to scale (NTS): Describes drawings that show relative positions and sizes.

Piping and instrumentation drawings (P&IDs): Schematic diagrams of a complete piping system.

Plumbing: A general term used for both water supply and all liquid waste disposal.

Plumbing plans: Engineered drawings that show the layout for the plumbing system.

Request for information (RFI): A means of clarifying a discrepancy in the construction drawings.

Roof plan: A drawing of the view of the roof from above the building.

Scale: The ratio between the size of a drawing of an object and the size of the actual object.

Schematic: A one-line drawing showing the flow path for electrical circuitry.

Section drawing: A cross-sectional view of a specific location, showing the inside of an object or building.

Specifications: Precise written presentation of the details of a plan.

Structural plans: A set of engineered drawings used to support the architectural design.

Symbol: A drawing that represents a material or component on a plan.

Title block: A part of a drawing sheet that includes some general information about the project.

Additional Resources

This module is intended to present thorough resources for task training. The following reference works are suggested for further study. These are optional materials for continued education rather than for task training.

Blueprint Reading for Construction, Second Edition. 2003. James Fatzinger. Upper Saddle River, NJ: Prentice Hall.

Blueprint Reading for the Construction Trades, Second Edition. 2005. Peter A. Mann. Micro-press.com.

Reading Architectural Plans for Residential and Commercial Construction, Fifth Edition. 2001. Ernest R. Weidhaas. Englewood Cliffs, NJ: Prentice Hall Career & Technology.

CONTREN® LEARNING SERIES – USER UPDATE

NCCER makes every effort to keep these textbooks up-to-date and free of technical errors. We appreciate your help in this process. If you have an idea for improving this textbook, or if you find an error, a typographical mistake, or an inaccuracy in NCCER's Contren® textbooks, please write us, using this form or a photocopy. Be sure to include the exact module number, page number, a detailed description, and the correction, if applicable. Your input will be brought to the attention of the Technical Review Committee. Thank you for your assistance.

Instructors – If you found that additional materials were necessary in order to teach this module effectively, please let us know so that we may include them in the Equipment/Materials list in the Annotated Instructor's Guide.

Write: Product Development and Revision
National Center for Construction Education and Research
3600 NW 43rd St., Bldg. G, Gainesville, FL 32606

Fax: 352-334-0932

E-mail: curriculum@nccer.org

Craft ________ Module Name ________

Copyright Date ________ Module Number ________ Page Number(s) ________

Description ________

____

____

____

(Optional) Correction ________

____

____

(Optional) Your Name and Address ________

____

____

Basic Rigging

00106-09

CORE CURRICULUM

00109-09
Introduction to Materials Handling

00108-09
Basic Employability Skills

00107-09
Basic Communication Skills

00106-09
Basic Rigging

00105-09
Introduction to Construction Drawings

00104-09
Introduction to Power Tools

00103-09
Introduction to Hand Tools

00102-09
Introduction to Construction Math

00101-09
Basic Safety

This course map shows all of the modules in the *Core Curriculum: Introductory Craft Skills*. The suggested training order begins at the bottom and proceeds up. Skill levels increase as you advance on the course map. The local Training Program Sponsor may adjust the training order.

Note that Module 00106-09, *Basic Rigging*, is an elective. It is not a requirement for level completion, but it may be included as part of your training program.

Objectives

When you have completed this module, you will be able to do the following:

1. Identify and describe the use of slings and common rigging hardware.
2. Describe basic inspection techniques and rejection criteria used for slings and hardware.
3. Describe basic hitch configurations and their proper connections.
4. Describe basic load-handling safety practices.
5. Demonstrate proper use of American Society of Mechanical Engineers (ASME) hand signals.

Trade Terms

ASME hand signals
Block and tackle
Bridle
Bull ring
Core
Cribbing
Eyebolt
Grommet sling
Hitch
Hoist
Lifting clamp
Load
Load control
Load stress
Master link
One rope lay
Pad eye
Plane
Rated capacity
Rejection criteria
Rigging hook
Risk management
Shackle
Sheave
Side pull
Sling
Sling angle
Sling legs
Sling reach
Sling stress
Splice
Strand
Stress
Tag line
Tattle-tail
Threaded shank
Warning yarn
Weight capacity
Wire rope

Prerequisites

Before you begin this module, it is recommended that you successfully complete the following: *Core Curriculum*, Modules 00101-09 through 00105-09. This *Basic Rigging* module is an elective and is not required for successful completion of this course. Trainees can obtain further training and a rigging completion certificate from the Contren® Learning Series *Rigging* curriculum.

Contents

Topics to be presented in this module include:

1.0.0 Introduction 6.1
2.0.0 Slings 6.3
2.1.0 Tagging Requirements 6.3
2.2.0 Synthetic Slings 6.4
2.2.1 Synthetic Web Sling Design and Characteristics 6.4
2.2.2 Types of Synthetic Web Slings 6.5
2.2.3 Round Sling Design and Characteristics 6.5
2.2.4 Synthetic Sling Inspection 6.7
2.2.5 Synthetic Sling Rejection Criteria 6.7
2.3.0 Alloy Steel Chain Slings 6.10
2.3.1 Alloy Steel Chain Sling Design and Characteristics 6.10
2.3.2 Alloy Steel Chain Inspection 6.11
2.3.3 Alloy Steel Chain Rejection Criteria 6.11
2.4.0 Wire Rope Slings 6.11
2.4.1 Wire Rope Sling Design and Characteristics 6.11
2.4.2 Wire Rope Sling Inspection 6.13
2.4.3 Wire Rope Sling Rejection Criteria 6.13
3.0.0 Hitches 6.15
3.1.0 Vertical Hitch 6.15
3.2.0 Choker Hitch 6.16
3.3.0 Basket Hitch 6.18
4.0.0 Rigging Hardware 6.19
4.1.0 Shackles 6.19
4.1.1 Shackle Design and Characteristics 6.20
4.1.2 Types of Shackles 6.20
4.1.3 Specialty Shackles 6.20
4.1.4 Shackle Inspection and Rejection Criteria 6.21
4.2.0 Eyebolts 6.21
4.2.1 Eyebolt Design and Characteristics 6.21
4.2.2 Unshouldered Eyebolts 6.21
4.2.3 Shouldered Eyebolts 6.22
4.2.4 Swivel Eyebolts 6.22
4.2.5 Eyebolt Inspection and Rejection Criteria 6.23
4.3.0 Lifting Clamps 6.23
4.3.1 Lifting Clamp Design and Characteristics 6.23
4.3.2 Types of Lifting Clamps 6.25
4.3.3 Lifting Clamp Inspection and Rejection Criteria 6.25
4.4.0 Rigging Hooks 6.25
4.4.1 Rigging Hook Design and Characteristics 6.26
4.4.2 Rigging Hook Inspection and Rejection Criteria 6.27
5.0.0 Sling Stress 6.28
6.0.0 Hoists 6.29
6.1.0 Operation of Chain Hoists 6.29
6.2.0 Hoist Safety and Maintenance 6.29
7.0.0 Rigging Operations and Practices 6.32
7.1.0 Rated Capacity 6.32
7.2.0 Sling Attachment 6.33
7.3.0 Hardware Attachment 6.33
7.3.1 Shackles 6.33
7.3.2 Eyebolts 6.33
7.3.3 Hooks 6.34

7.4.0 Load Control 6.35
7.4.1 American Society of Mechanical Engineers (ASME) Hand Signals 6.35
7.4.2 Load Path, Load Control, and Tag Lines 6.36
7.4.3 Rigging Safety 6.36
7.4.4 Risk Management 6.39
7.4.5 Safely Handling a Load 6.41

1.0.0 Introduction

Rigging is the planned movement of material and equipment from one location to another, using **slings**, **hoists**, or other types of equipment. Some rigging operations use a loader to move materials around a job site. Other operations require cranes to lift such **loads**. There are two basic types of cranes: overhead (*Figure 1*) and mobile (*Figure 2*).

This module provides basic information on the following:

- Slings
- **Hitches**
- Rigging hardware
- The principles and practices behind safe and efficient rigging operations

At this early stage in your career, you will not be asked to perform rigging tasks without supervision. You must not attempt any rigging operations on your own. All rigging operations must be done under the supervision of a competent person. Even so, it is important that you understand the fundamentals of the following aspects of rigging:

- The types of slings and equipment you may eventually be using.
- How to determine whether your equipment is fit for use.
- The proper use of the most common types of slings and hardware.
- How to select appropriate rigging equipment and how factors, such as **load stress**, affect that selection.

Core-level rigging is designed to introduce you to the basic principles of rigging. These basic principles are true whether you are rigging a single piece of pipe, a bundle of lumber, or an 850-ton steam generator.

Rigging operations can be extremely complicated and dangerous. Do not experiment with rigging operations, and never attempt a lift on your own, without the supervision of an officially recognized qualified person. A lift may appear simple while it is in progress. That is because the people performing the lift know exactly what they are doing. There is no room for guesswork!

On-Site

World's Largest Mobile Cranes

Some of the world's largest cranes are built by Mammoet in the Netherlands. There are three of them in the fleet. One crane is called the Platform Twinring. The other two are referred to as PTCs. PTC stands for Platform Twinring Containerised. These two can be disassembled into 20- and 40-ft. shipping containers, and every piece of the crane can be lifted with a normal container crane on any dock anywhere in the world. The fittings are already molded into the ends, so the PTCs can go from any container crane onto any truck trailer that normally handles shipping containers. These giant cranes have a maximum load capacity of 1,600 tons.

106SA01.EPS

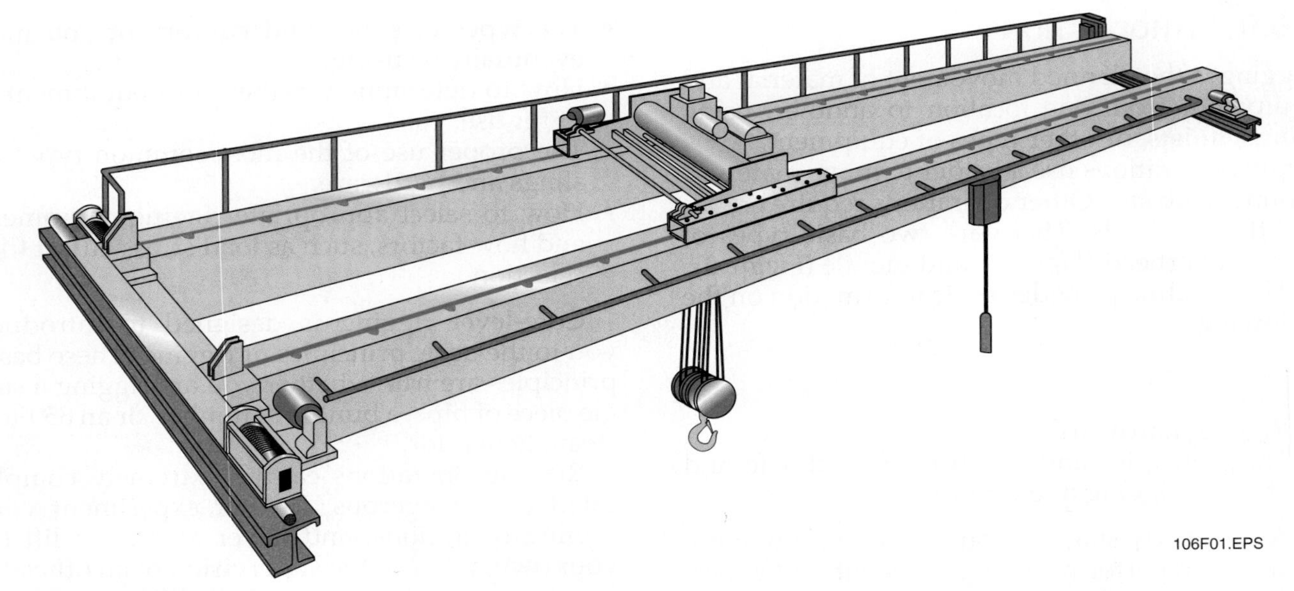

106F01.EPS

Figure 1 Overhead crane.

No matter whether rigging operations involve a simple vertical lift, a powered hoist, or a highly complicated apparatus, only qualified persons may perform them without supervision.

WARNING!

Although this module introduces you to the criteria that are used to make rigging decisions, you will not be making the decisions yourself at this stage in your training. If you ever have any questions about whether a synthetic sling is defective or unsafe, ask your instructor or your immediate supervisor. Always refer to the manufacturer's instructions for the sling rejection criteria for that particular sling.

106F02A.EPS

106F02B.EPS

Figure 2 Mobile cranes.

On-Site

Regulations and Site Procedures

The information in this module is intended as a general guide. The techniques shown here are not the only methods you can use to perform a lift. Many techniques can be safely used to rig and lift different loads (the total amount being lifted).

Some of the techniques for certain kinds of rigging and lifting are spelled out in requirements issued by federal government agencies. Some will be provided at the job site, where you will see written site procedures that address any special conditions that affect lifting procedures on that site. If you have questions about any of these procedures, ask the supervisor at the site.

2.0.0 Slings

During a rigging operation, the load being moved must be connected to the device, such as a crane, that is doing the moving. The connector—the link between the load and the lifting device—is often a sling made of synthetic, chain, or **wire rope** materials.

In this section, you will learn about three types of slings:

- Synthetic slings
- Alloy steel chain slings
- Wire rope slings

2.1.0 Tagging Requirements

All slings are required to have identification tags (*Figure 3*). An identification tag must be securely attached to each sling and clearly marked with the information required for that type of sling. For all three types of slings, that information will include the manufacturer's name or trademark and the **rated capacity** of the type of hitch used with that sling. The rated capacity is the maximum load weight that the sling was designed to carry. Rated capacity charts for various types of slings appear in the *Appendix*. Synthetic, alloy steel chain, and wire rope slings are covered in this module.

The following are the tagging requirements for synthetic slings:

- Manufacturer's name or trademark
- Manufacturer's code or stock number (unique for each sling)
- Rated capacities for the types of hitches used
- Type of synthetic material used in the manufacture of the sling

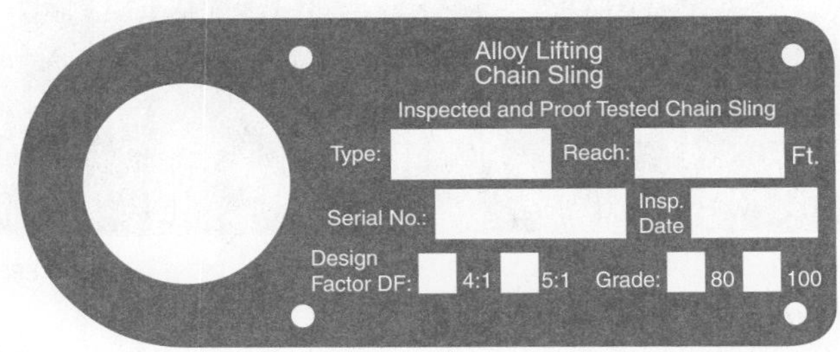

Figure 3 Identification tag.

The following are the tagging requirements for alloy steel chain slings:

- Manufacturer's name or trademark
- Manufactured grade of steel
- Link size (diameter)
- Rated load and the angle on which the rating is based
- 2003 ASME Standard requires clarification as to the angle upon which the capacity is based
- **Sling reach**
- Number of **sling legs**

The following are the tagging recommendations for wire rope slings:

- Manufacturer's name or trademark
- Rated capacity in a vertical hitch (other hitches optional)
- Size of wire rope (diameter)
- Manufacturer's code or stock number (unique for each sling)

WARNING!

All slings and hardware have a rated capacity, also called lifting capacity, working capacity, working load limit (WLL), or safe working load (SWL). Under no circumstances should you ever exceed the rated capacity! Overloading may result in catastrophic failure. Rated capacity is defined as the maximum load weight a sling or piece of hardware or equipment can hold or lift.

Did You Know?

Sling Identification Tags

The identification tags that are required on all slings must be securely attached to the slings and clearly marked with the information required for each type of sling.

2.2.0 Synthetic Slings

Synthetic slings are widely used to lift loads, especially easily damaged ones. In this section, you will learn about two types of synthetic slings: synthetic web slings and round slings.

2.2.1 Synthetic Web Sling Design and Characteristics

Synthetic web slings provide several advantages over other types of slings:

- They are soft and wider than wire rope or chain slings. Therefore, they do not scratch or damage machined or delicate surfaces (*Figure 4*).
- They do not rust or corrode and therefore will not stain the loads they are lifting.
- They are lightweight, making them easier to handle than wire rope or chain slings. Most synthetic slings weigh less than half as much as a wire rope that has the same rated capacity. Some new synthetic fiber slings weigh one-tenth as much as wire rope.
- They are flexible. They mold themselves to the shape of the load (*Figure 5*).
- They are very elastic, and they stretch under a load much more than wire rope does. This stretching allows synthetic slings to absorb shocks and to cushion the load.
- Loads suspended in synthetic web slings are less likely to twist than those in wire rope or chain slings.

106F04.EPS

Figure 4 Surface protection.

Synthetic web slings should not be exposed to temperatures above 180°F. They are also susceptible to cuts, abrasions, and other wear-and-tear damage. To prevent damage to synthetic web slings, riggers use protective pads (*Figure 6*).

CAUTION

If the sling does not come with protective pads, use other kinds of softeners of sufficient strength or thickness to protect the sling. Pieces of old sling, fire hose, canvas, or rubber can be used.

Most synthetic web slings are manufactured with red core **warning yarns**. These are used to let the rigger know whether the sling has suffered too much damage to be used. When the yarns are exposed, the synthetic web sling should not be used (*Figure 7*). Red core yarns should not be used exclusively.

106F05.EPS

Figure 5 Synthetic web sling shaping.

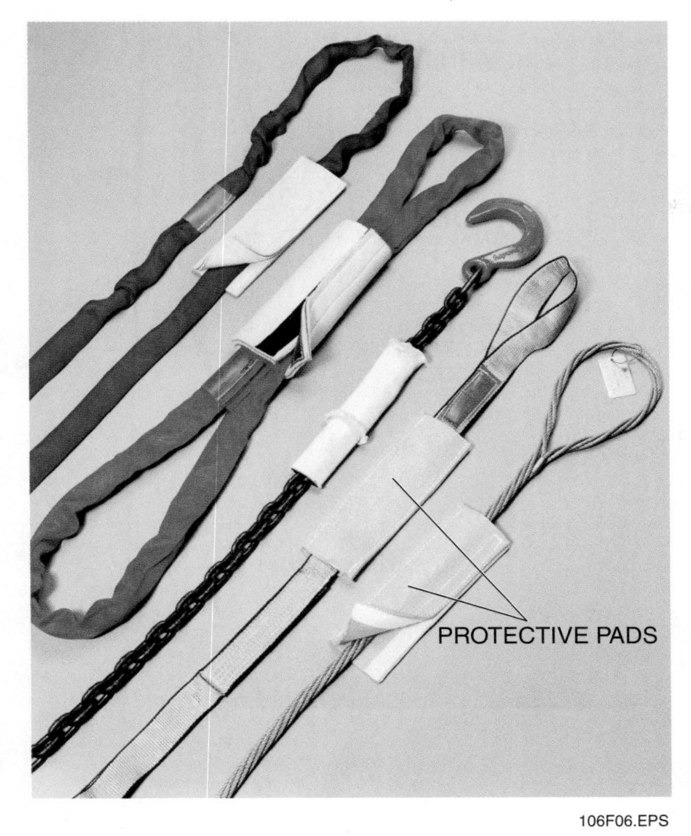

Figure 6 Protective pads.

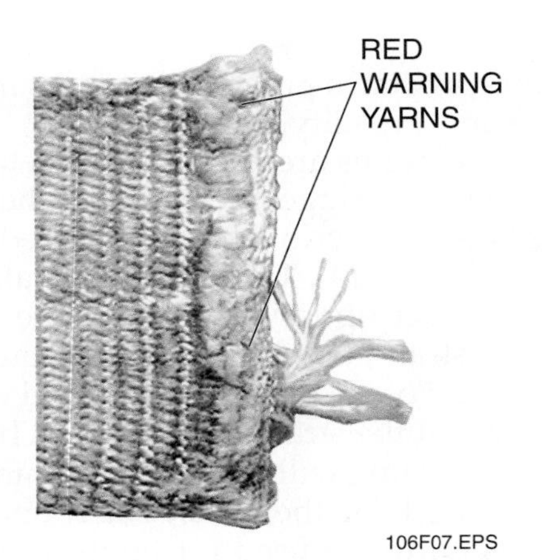

Figure 7 Synthetic web sling warning yarns.

2.2.2 Types of Synthetic Web Slings

Synthetic web slings are available in several designs. The most common are the following (*Figure 8*):

- Endless web slings, which are also called **grommet slings**.
- Round slings, which are endless and made in a continuous circle out of polyester filament yarn. The yarn is then covered by a woven sleeve.

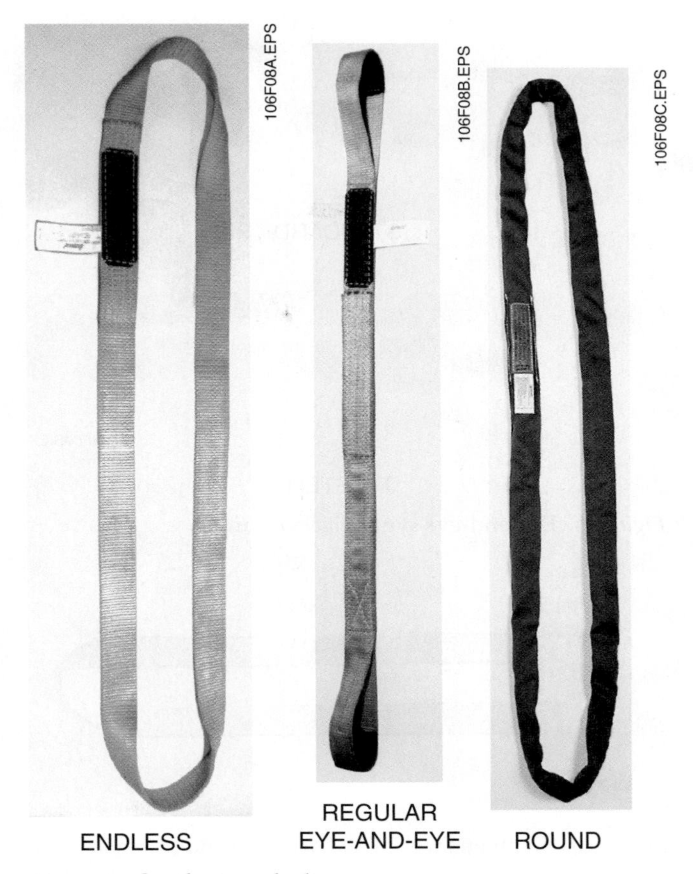

Figure 8 Synthetic web slings.

- Synthetic web eye-and-eye slings, which are made by sewing an end of the sling directly to the sling body. Standard eye-and-eye slings have eyes on the same **plane**; twisted eye-and-eye slings have eyes at right angles to each other (*Figure 9*) and are primarily used for choker hitches (which you will learn about later).

Synthetic web eye-and-eye slings are also available with hardware end fittings instead of fabric eyes. The standard end fittings are made of either aluminum or steel. They come in male and female configurations (*Figure 10*).

2.2.3 Round Sling Design and Characteristics

Twin-Path® slings are made by wrapping a synthetic yarn around a set of spindles to form a loop. A protective jacket encases the core yarn (*Figure 11*).

Twin-Path® slings are made of a synthetic fiber, such as polyester. The material used to make up the **strands** and the number of wraps in a loop determines the rated capacity of the sling, or how much weight it can handle. Aramid round slings, for example, have a greater rated capacity for their size than the web-type slings do.

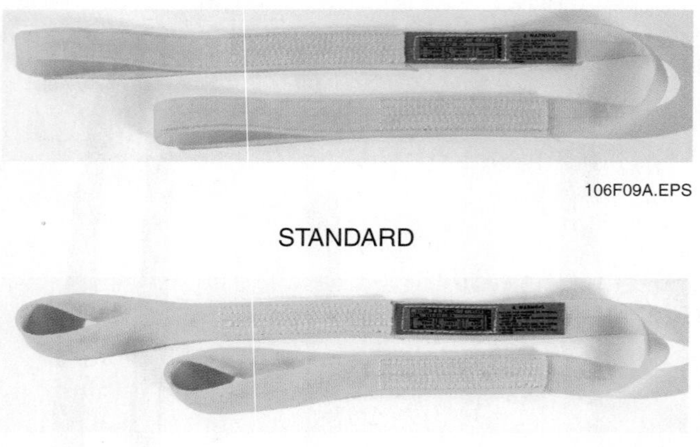

Figure 9 Eye-and-eye synthetic web slings.

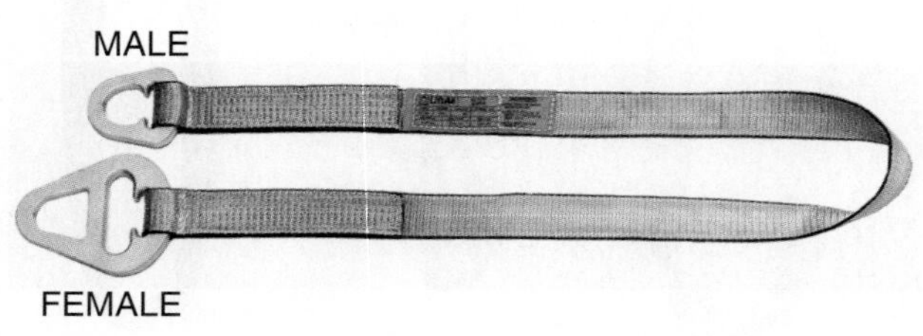

Figure 10 Synthetic web sling hardware end fittings.

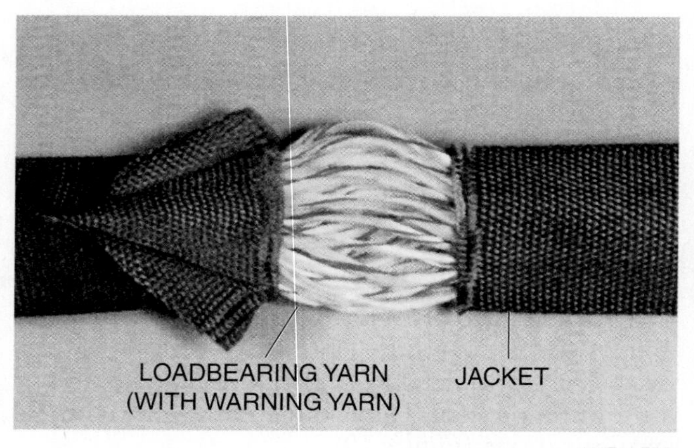

Figure 11 Synthetic endless-strand jacketed sling.

Twin-Path® slings are also available in a design with two separate wound loops of strand jacketed together side-by-side (*Figure 12*). This design greatly increases the lifting capacity of the sling.

The jackets of these slings are available in several materials for various purposes, including heat-resistant Nomek®, polyester, and bulked nylon (Covermax™) (*Figure 13*). Twin-Path® slings featuring K-Spec® yarn weigh at least

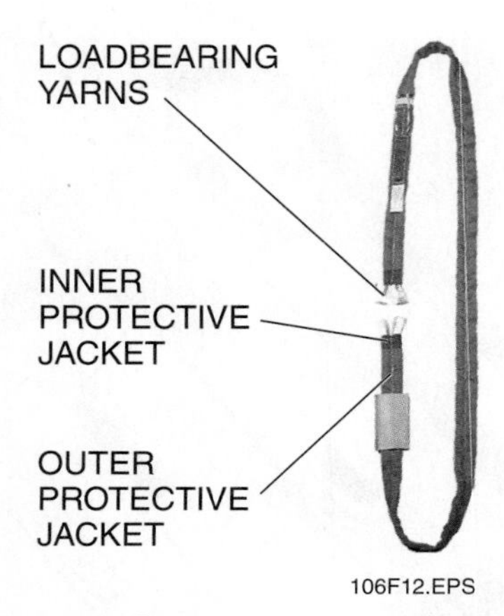

Figure 12 Twin-Path® sling.

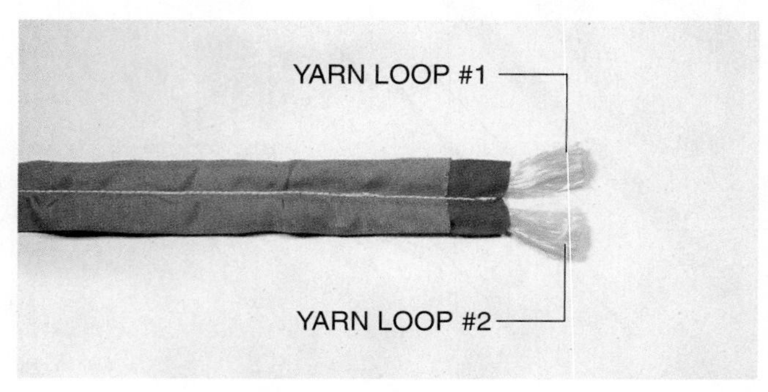

Figure 13 Twin-Path® sling makeup.

50 percent less than a polyester round sling of the same size and capacity.

Twin-Path® slings are equipped with **tattle-tail** yarns to help the rigger determine whether the sling has become overloaded or stretched beyond a safe limit (*Figure 14*). These slings are also available with a fiber-optic inspection cable running through the strand (*Figure 15*). When you direct a light at one end of the fiber, the other end will light up to show that the strand has not been broken.

The way the sling will be used determines what material is used for the jackets of round slings. Polyester is normally used for light- to medium-duty sling jackets. Covermax™, a high-strength material with much greater resistance to cutting and abrasion, is used to make a sturdier jacket for heavy-duty uses.

Did You Know?

Any sling without an identification tag must be removed from service immediately. It may be turned over to a qualified rigging inspector.

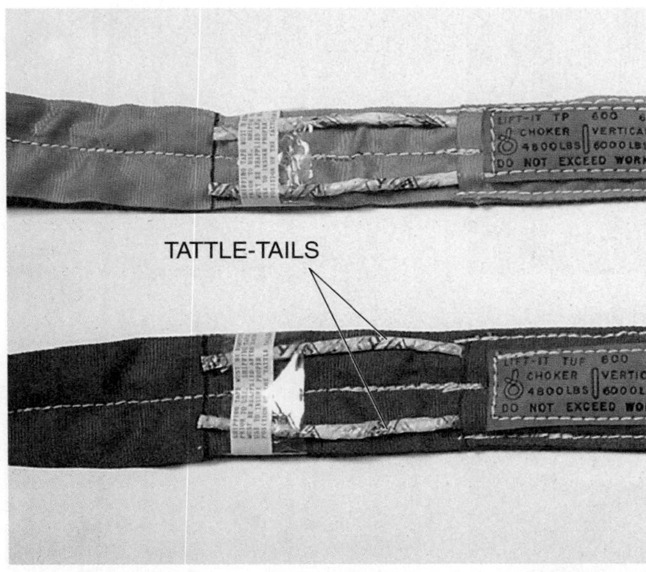

Figure 14 Tattle-tails.

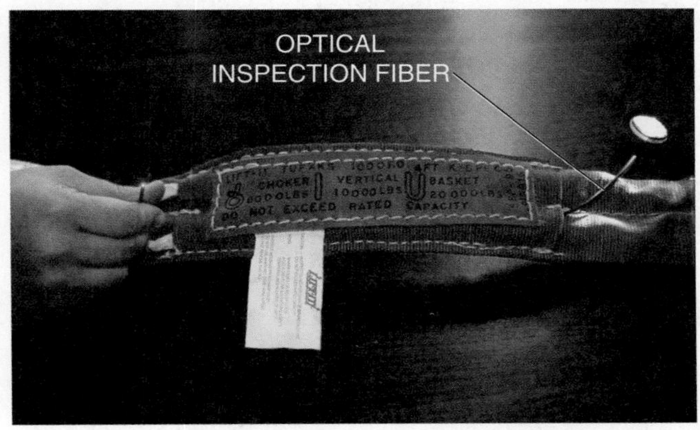

Figure 15 Fiber-optic inspection cable.

2.2.4 Synthetic Sling Inspection

Like all slings, synthetic slings must be inspected before each use to determine whether they are in good condition and can be used. They must be inspected along the entire length of the sling, both visually (looking at them) and manually (feeling them with your hand). If any **rejection criteria** are met, the sling must be removed from service.

> **WARNING!** The rigger must always inspect synthetic slings before using them, every time. The rigger must inspect them by looking at them as well as by feeling them, because sometimes you can feel damage that you cannot see.

2.2.5 Synthetic Sling Rejection Criteria

If any synthetic sling meets any of the rejection criteria presented in this section, it must be removed from service immediately. In addition, the rigger has to exercise sound judgment. Along with looking for any single major problem, the rigger needs to watch for combinations of relatively minor defects in the synthetic sling. Combinations of minor damage may make the sling unsafe to use, even though such defects may not be listed in the rejection criteria.

If you are helping to inspect synthetic slings, alert your supervisor if you suspect any defects at all, especially if you find any of the following synthetic sling damage rejection criteria (*Figure 16*):

- A missing identification tag or an identification tag that cannot be read. Any synthetic sling without an identification tag must be removed from service immediately.
- Abrasion that has worn through the outer jacket or that has exposed the loadbearing yarn of the sling, or that has exposed the warning yarn of a web sling.
- A cut that has severed the outer jacket or exposed the loadbearing yarn (single-layer jacket) of a round sling or that has exposed the warning yarn of a web sling.
- A tear that has exposed the inner jacket or the loadbearing yarns (single-layer jacket) of a round sling or that has exposed the warning yarn of a web sling.
- A puncture.
- Broken or worn stitching in the **splice** or stitching of a web sling.
- A knot that cannot be removed by hand in either a web or round sling.
- A snag in the sling that reveals the warning yarn of a web sling or a snag that tears through the outer jacket or exposes the loadbearing yarns of a round sling.
- Crushing of either a web sling or a round sling. Crushing in a web sling feels like a hollow pocket or depression in the sling. Crushing in a round sling feels like a hard, flat spot underneath the jacket.
- Damage from overload (called tensile damage or overstretching) in a web or round sling. Tensile damage in a web sling is evident when the weave pattern of the fabric begins to pull apart. Twin-Path® slings have tattle-tails. When the tail has been pulled into the jacket, it indicates that the sling may have been overloaded.
- Chemical damage, including discoloration, burns, and melting of the fabric or jacket.

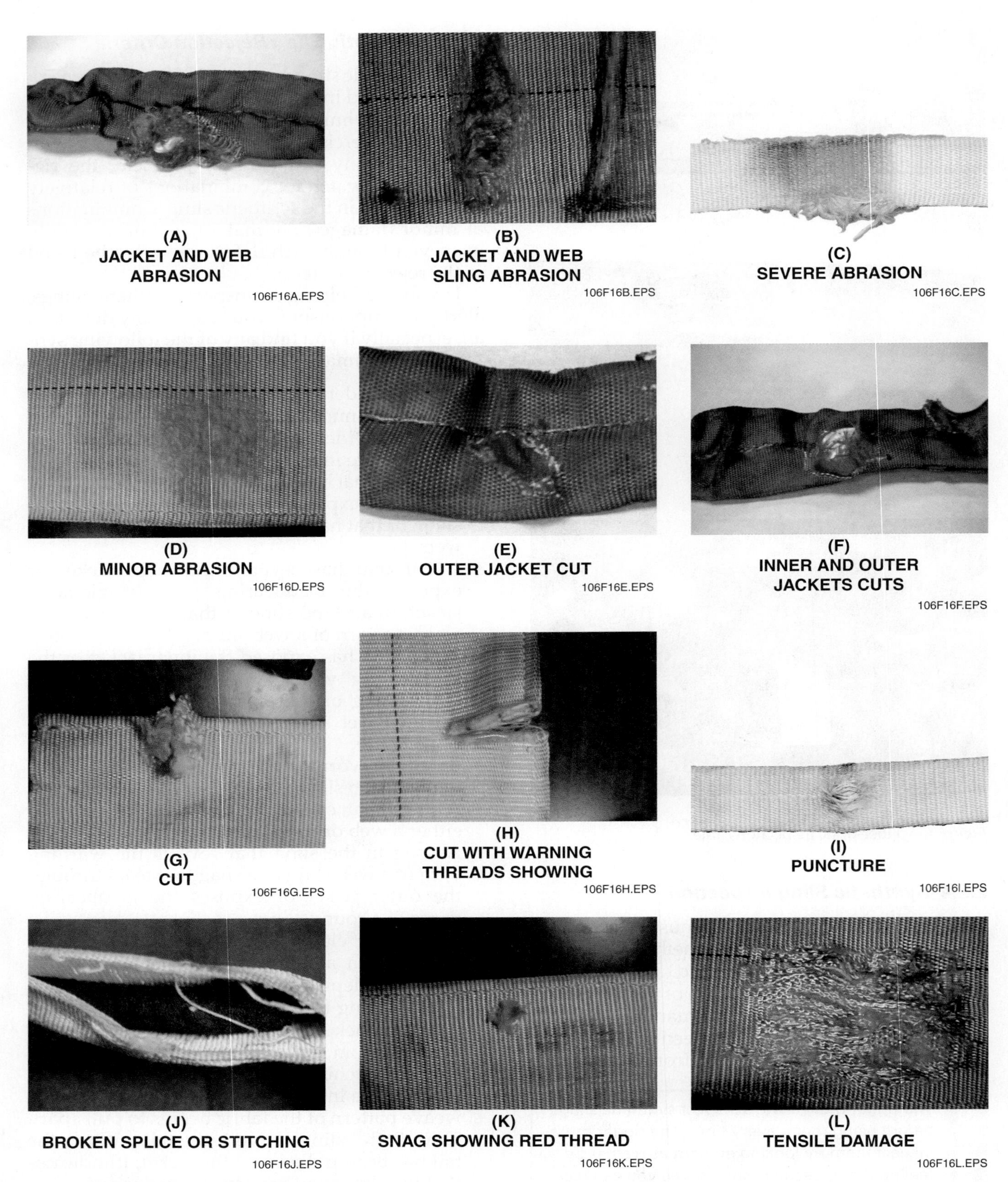

Figure 16 Sling damage rejection criteria. (1 of 2)

(M)
TENSILE BREAK

106F16M.EPS

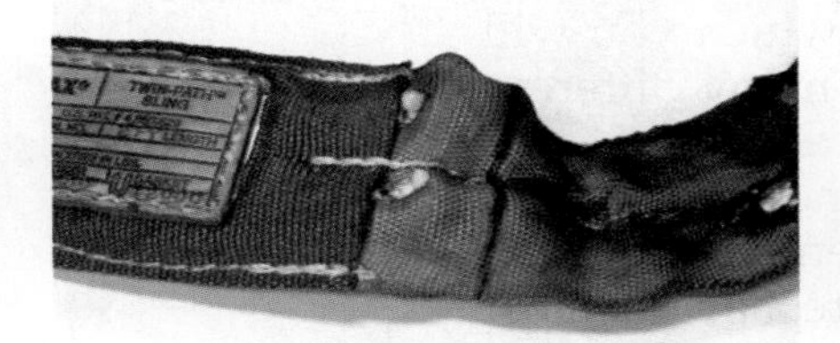

(N)
OVERLOAD DAMAGE
(TATTLE-TAILS PULLED IN)

106F16N.EPS

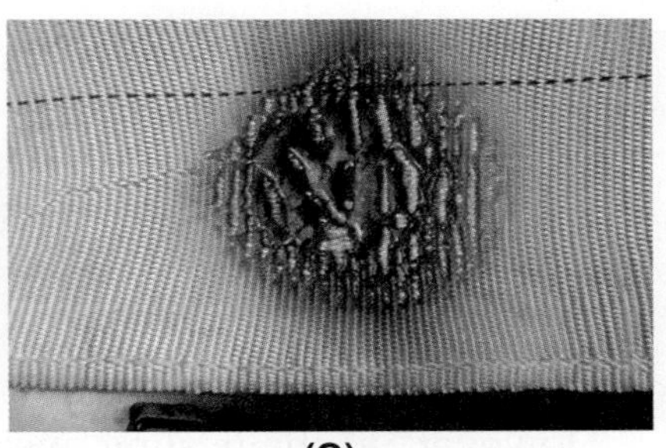

(O)
FRICTION BURN FROM
ABRASION AND HEAT DAMAGE

106F16O.EPS

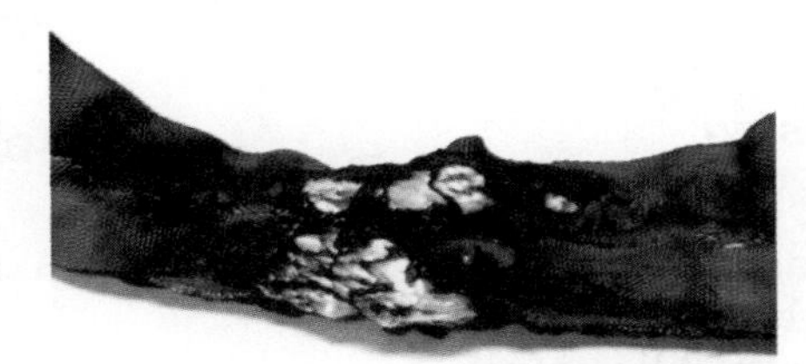

(P)
SEVERE HEAT DAMAGE

106F16P.EPS

Figure 16 Sling damage rejection criteria. (2 of 2)

- Heat damage, ranging from friction burns to melting of the sling material, the loadbearing strands, or the jacket. Friction burns give the webbing material a crusty or slick texture. Heat damage to the jacket of a round sling looks like glazing or charring. In round slings with heat-resistant jackets, you may not be able to see heat damage to the outside of the sling; however, the internal yarns may have been damaged. You can detect that damage by carefully handling the sling and feeling for brittle or fused fibers inside the sling or by flexing and folding the sling and listening for the sound of fused yarn fibers cracking or breaking.
- Ultraviolet (UV) damage. The evidence of UV damage is a bleaching-out of the sling material, which breaks down the synthetic fibers. Web slings with UV damage will give off a powder-like substance when they are flexed and folded. Round slings, especially those made of Cordura nylon and other specially treated synthetic fibers, have a much greater resistance to UV damage. In round slings, UV damage shows up as a roughening of the fabric texture where no other sign of damage, such as abrasion, can be found.
- Loss of flexibility caused by the presence of dirt or other abrasives. You can tell that a sling has lost flexibility when the sling becomes stiff. Abrasive particles embedded in the sling material act like tiny blades that cut apart the internal fibers of the sling every time the sling is stretched, flexed, or wrapped around a load.

On-Site

Sling Jackets

If only the outer jacket of a sling is damaged, under the rejection criteria, that sling can be sent back to the manufacturer for a new jacket. It must be tested and certified before it is used again. It is much less costly to do this than to replace the sling. However, slings that are removed because of heat- or tension-related jacket damage cannot be returned for repair and must be disposed of by a qualified person.

2.3.0 Alloy Steel Chain Slings

Alloy steel chain, like wire rope (which you will learn about later), can be used in many different rigging operations. Chain slings are often used for lifts in high heat or rugged conditions. Chain slings can be adjusted over your center of gravity, making it a versatile sling. They are also the most durable.

However, a chain sling weighs much more than a wire rope sling. It also may be harder to inspect. You will encounter both types of slings in the field, so you must decide which type of sling is best for each situation.

2.3.1 Alloy Steel Chain Sling Design and Characteristics

Steel chain slings used for overhead lifting must be made of alloy steel. The higher the alloy grade, the safer and more durable the chain is. Alloy steel chains commonly used for most overhead lifting are marked with the number 8, the number 80, the number 800, or the letter A (*Figure 17*).

Steel chain slings have two basic designs with many variations:

- Single- and double-basket slings (*Figure 18*) do not require end-fitting hardware. The chain is attached to the **master link** in a permanent basket hitch or hitches.
- Chain **bridle** slings are available with two to four legs (*Figure 19*). Chain bridle slings require some type of end-fitting hardware—eye hooks, grab hooks, plate hooks, sorting or pipe hooks (*Figure 20*), or links.

Alloy steel chain slings are used for lifts when temperatures are high or where the slings will be subjected to steady and severe abuse. Most alloy steel chain slings can be used in temperatures up to 500°F with little loss in rated capacity.

Even though alloy steel chain slings can withstand extreme temperature ranges and abusive working conditions, these slings can be damaged if loads are dropped on them, if they are wrapped around loads with sharp corners (unless you use softeners), or if they are exposed to intense temperatures.

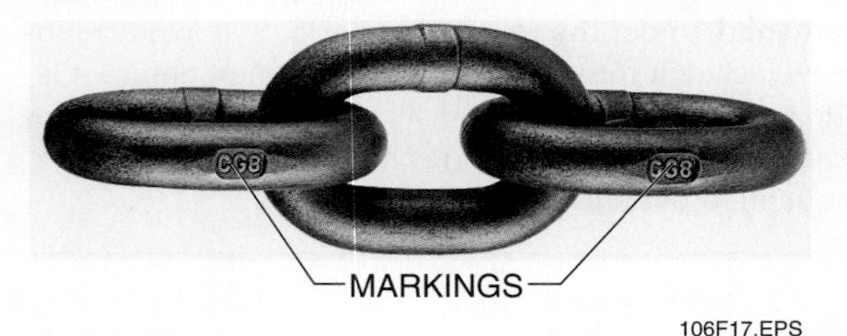

Figure 17 Markings on alloy steel chain slings.

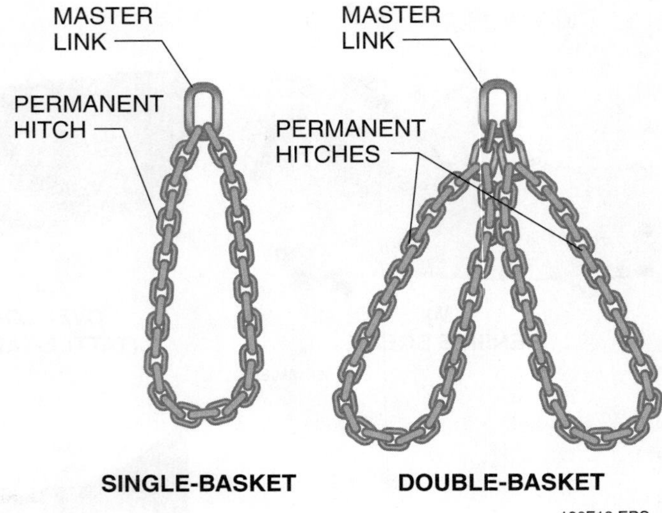

Figure 18 Chain slings.

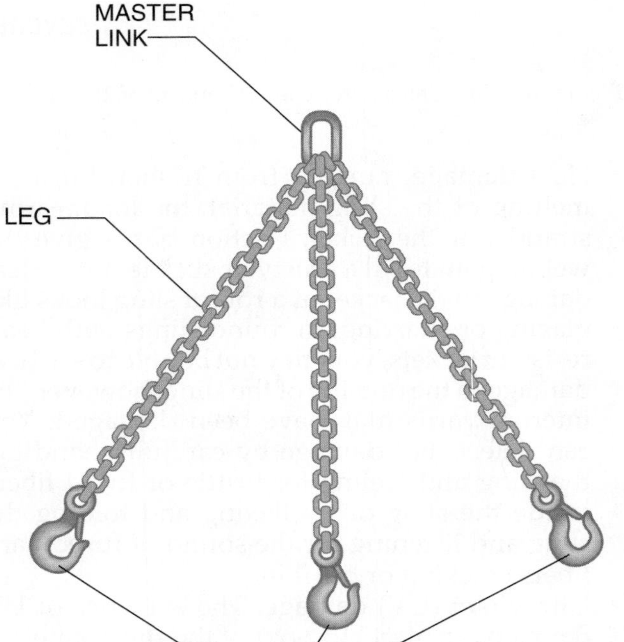

Figure 19 Three-leg chain bridle.

CAUTION

Never drag alloy steel chain slings across hard surfaces, especially concrete. Friction with abrasive surfaces wears out the chain and weakens the sling.

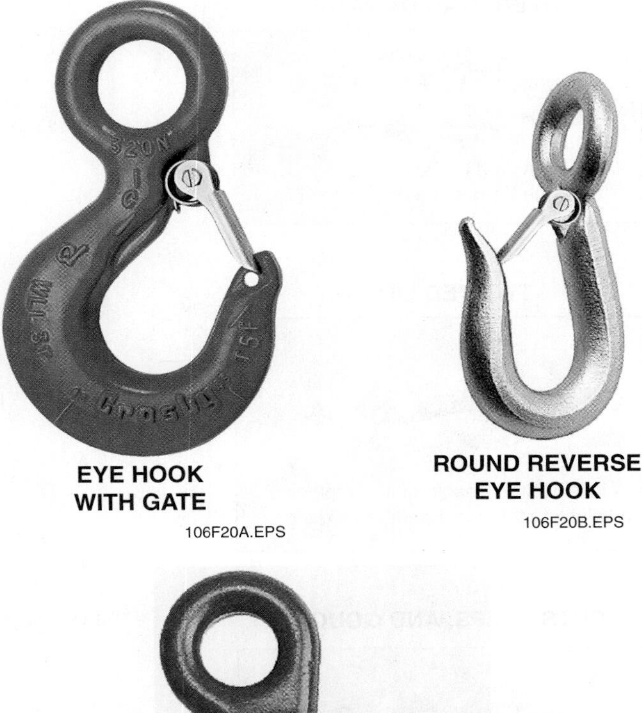

Figure 20 Eye and sorting hooks.

2.3.2 Alloy Steel Chain Inspection

Before each use, the rigger must inspect the alloy steel chain sling to determine whether it is safe to use. This module introduces you to the criteria involved in making this decision, but at this point in your training you will not make the decision yourself. If you have any questions about whether a sling is defective or unsafe and should not be used, ask your instructor or your immediate supervisor.

Steel chain slings must be visually inspected before each lift. A chain must be removed from service if a deficiency in the chain matches any of the rejection criteria. A chain may also need to be removed from service if something shows up that does not exactly match the rejection criteria.

2.3.3 Alloy Steel Chain Rejection Criteria

An alloy steel chain sling must be removed from service for any of the following defects (*Figure 21*):

- Missing or illegible identification tag
- Cracks
- Heat damage
- Stretched links. Damage is evident when the link grows long and when the barrels—the long sides of the links—start to close up
- Bent links
- Twisted links
- Excessive rust or corrosion, meaning rust or corrosion that cannot be easily removed with a wire brush
- Cuts, chips, or gouges resulting from impact on the chain
- Damaged end fittings, such as hooks, clamps, and other hardware
- Excessive wear at the link-bearing surfaces
- Scraping or abrasion

2.4.0 Wire Rope Slings

Wire rope slings (*Figure 22*) are made of high-strength steel wires formed into strands wrapped around a supporting core. They are lighter than chain, can withstand substantial abuse, and are easier to handle than chain slings. They can also withstand relatively high temperatures. However, because wire rope slings can slip, the use of synthetic slings is preferred. Wire rope is still being used, though, so you need to learn about the design and characteristics, applications, inspection, and maintenance of wire rope slings.

2.4.1 Wire Rope Sling Design and Characteristics

There are many types of wire ropes. They all consist of a core that supports the rope, and center strand wires, each with many high-grade steel wires wound around them to form the strands.

Wire rope is designed to operate like a self-adjusting machine. This means a wire rope has moving parts, as a machine does. A wire rope's moving parts are the wires that make up the strands and the core of the rope itself. These moving parts interact with one another by sliding and adjusting. This sliding and adjusting compensates for the ever-changing stresses placed upon a working rope. Because of this, a wire rope's rated capacity depends on it being in good condition. This means its wires must be able to move the way they were designed to.

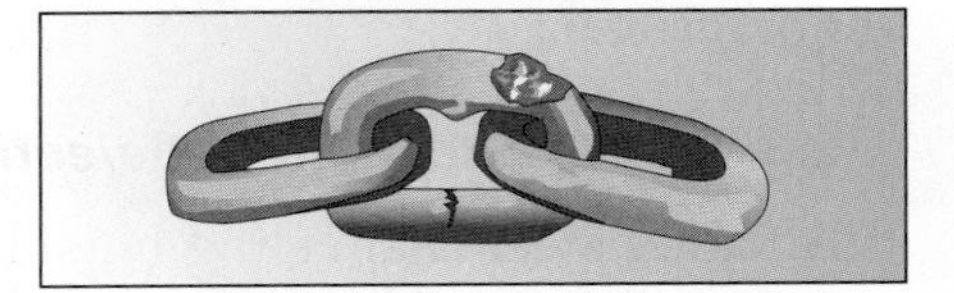

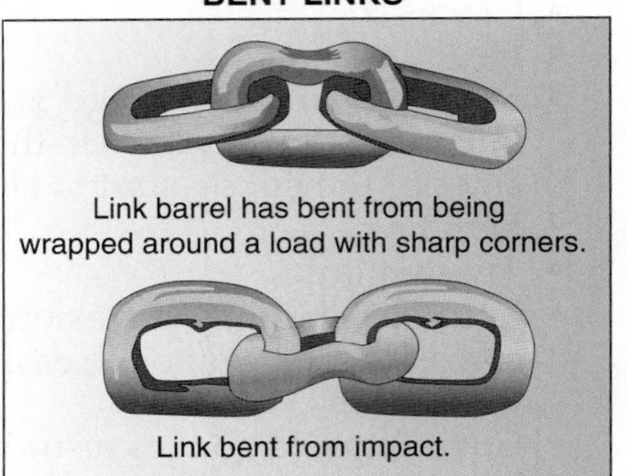

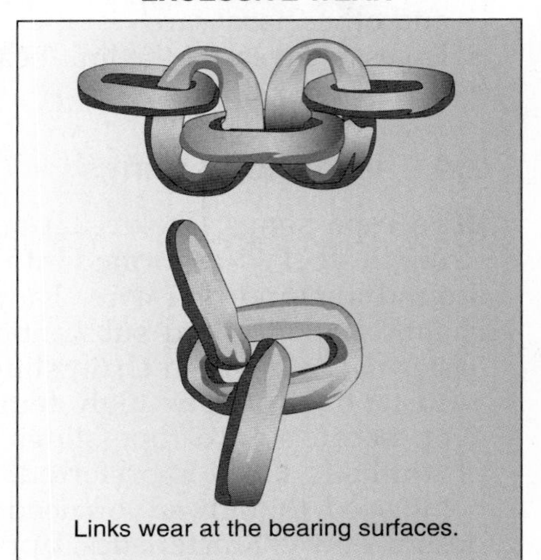

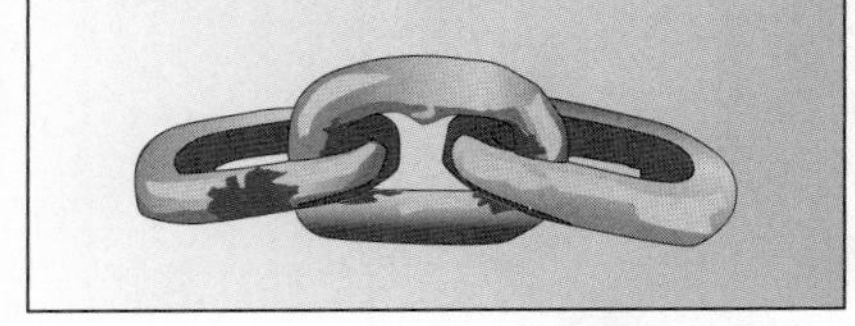

Figure 21 Damage to chains.

Figure 22 Wire rope sling.

The three basic components of a wire rope are as follows (*Figure 23*):

- A supporting core
- High-grade steel wires
- Multiple center wires

There are three basic types of supporting cores for wire rope (*Figure 24*): fiber cores, strand cores, and independent wire rope cores. Fiber cores are usually made of synthetic fibers, but they also can be made of natural vegetable fibers, such as sisal. Strand cores are made by using one strand of the same size and type as the rest of the strands of rope. Independent wire rope cores are made of a separate wire rope with its own core and strands. The core rope wires are much smaller and more delicate than the strand wires in the outer rope.

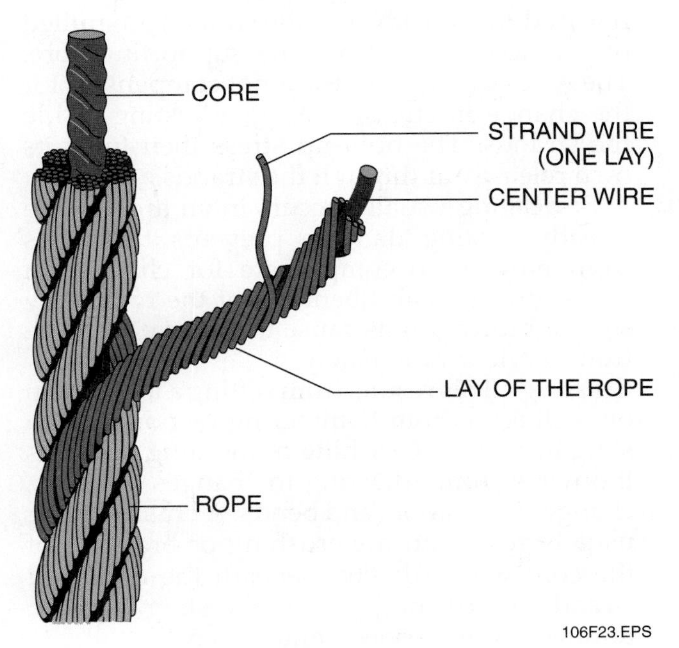

Figure 23 Wire rope components.

The various materials used to form the supporting core of a wire rope have both desirable characteristics and drawbacks, depending on how they are to be used. Fiber core ropes, for example, may be damaged by heat at relatively low temperatures (180°F to 200°F) as well as by exposure to caustic chemicals.

The function of the core in a wire rope is to support the strands so that the strands keep their original shapes. When the core supports the strands, their high-grade steel wires can slide and adjust against each other. The ability to adjust makes them less likely to be damaged by stress when the rope is bent around **sheaves** or loads, or when it is placed at an angle during rigging.

2.4.2 Wire Rope Sling Inspection

Like other slings, wire rope slings must be inspected before each use. If the wire rope is damaged, it must be removed from service. Always ask your instructor or immediate supervisor if you have any questions about whether a wire rope sling is suitable for use.

2.4.3 Wire Rope Sling Rejection Criteria

Following an inspection, a wire rope sling may be rejected based on several common types of damage (*Figure 25*), including broken wires, kinks, birdcaging, crushing, corrosion and rust, and heat damage. At this stage in your training, you need to be able to demonstrate that you understand the rejection criteria for wire ropes. Only a qualified person can actually make the decision to use a wire rope in a rigging operation or to discard it if it is damaged. If you think a wire rope may be damaged, bring it to the attention of your instructor

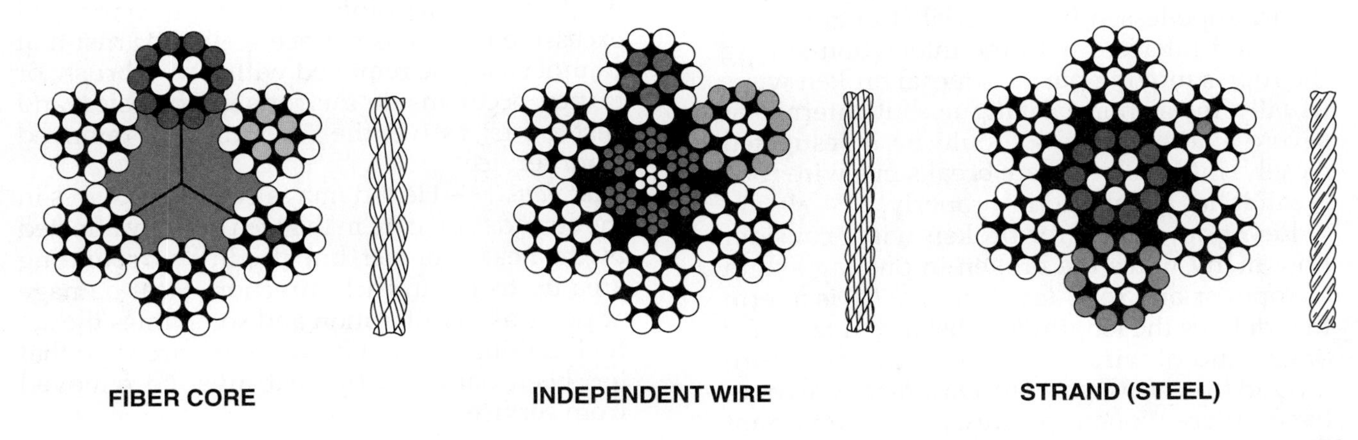

Figure 24 Wire rope supporting cores.

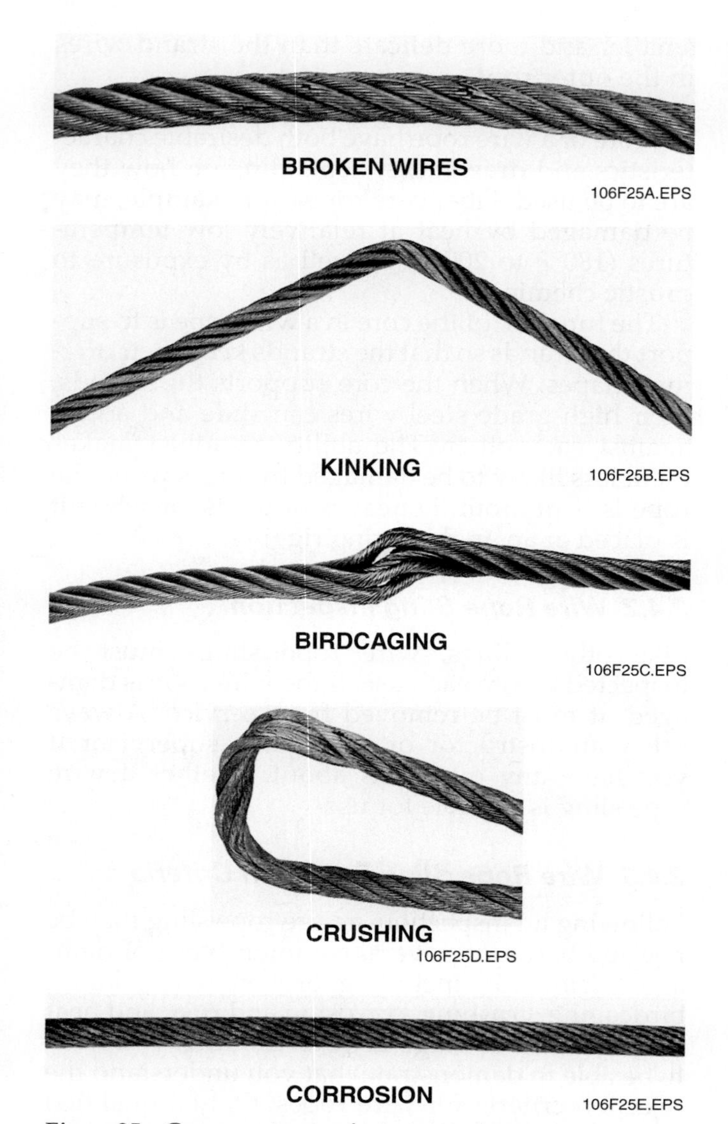

Figure 25 Common types of wire rope damage.

or immediate supervisor. Additional information about possible types of damage is as follows.

- *Broken wires* – Broken wires in the strands of a wire rope lessen the material strength of the rope and interfere with the interaction among the rope's moving parts. External broken wires usually mean normal fatigue, but internal or severe external breaks should be investigated closely. Internal or severe breaks in a wire rope mean it has been used improperly.

Rejection criteria for broken wires consider how many wires are broken in one lay length of rope, or **one rope lay**. *One rope lay* is a term that defines the lengthwise distance it takes for one strand of wire to make one complete turn around the core (*Figure 26*). Different wire ropes have different one rope lays, so it is important to inspect each wire rope closely when looking for broken wires.

Figure 26 One rope lay.

- *Kinks* – Kinking, or distortion of the rope, is a very common type of damage. Kinking can result in serious accidents. Sharp kinks restrict or prevent the movement of wires in the strands at the area of the kink. This means the rope is damaged and must not be used. Ropes with kinks in the form of large, gradual loops in a corkscrew configuration must be removed from service.
- *Birdcaging* – This damage occurs when a load is released too quickly and the strands are pulled or bounced away from the supporting core. The wires in the strands cannot compensate for the change in stress level by adjusting inside the strands. The built-up stress then finds its own release out through the strands.

  Birdcaging usually occurs in an area where already-existing damage prevents the wires from moving to compensate for changes in stress, position, and bending of the rope. Any sign of birdcaging is cause to remove the rope from service immediately.
- *Crushing* – This results from setting a load down on a sling or from hammering or pounding a sling into place. Crushing of the sling prevents the wires from adjusting to changes in stress, changes in position, and bends. A crushed sling usually results in the crushing or breaking of the core wires directly beneath the damaged strands. If crushing occurs, the sling must be removed from service immediately.
- *Corrosion and rust* – Corrosion and rust of wire rope are the result of improper or insufficient lubrication. Corrosion and rust are considered excessive if there is surface scaling or rust that cannot easily be removed with a wire brush, or if they occur inside the rope. If corrosion and rust are excessive, the rope must be removed from service.
- *Heat damage* – Heat damage embrittles wires in the strands and core in the area directly affected by the heat's contact, but also in a surrounding area up to 120 in each direction. Heat damage appears as discoloration and sometimes the actual melting of the wire rope. A wire rope that has been damaged by heat must be removed from service.

3.0.0 HITCHES

As you have learned, the link between the load and the lifting device is often a sling made of synthetic, alloy steel chain, or wire rope material. The way the sling is arranged to hold the load is called the rigging configuration or hitch. Hitches can be made using just the sling or by using connecting hardware, as well. There are three basic types of hitches:

- Vertical
- Choker
- Basket

One of the most important parts of the rigger's job is making sure that the load is held securely. The type of hitch the rigger uses depends on the type of load to be lifted. Different hitches are used to secure, for example, a load of pipes, a load of concrete slabs, or a load of heavy machinery.

Controlling the movement of the load once the lift is in progress is another extremely important part of the rigger's job. Therefore, the rigger must also consider the intended movement of the load when choosing a hitch. For example, some loads are lifted straight up and then straight down. Other loads are lifted up, turned 180 degrees in midair, and then set down in a completely different place. In this section, you will learn about how each of the three basic types of hitches—vertical, choker, and basket—is used to both secure the load and control its movement.

WARNING! All rigging operations are dangerous, and extreme care must be used at all times. A straight-up-and-down vertical lift is every bit as dangerous as a lift that involves rotating a load in midair and moving it to a different place. Only a qualified person may select the hitch to be used in any rigging operation.

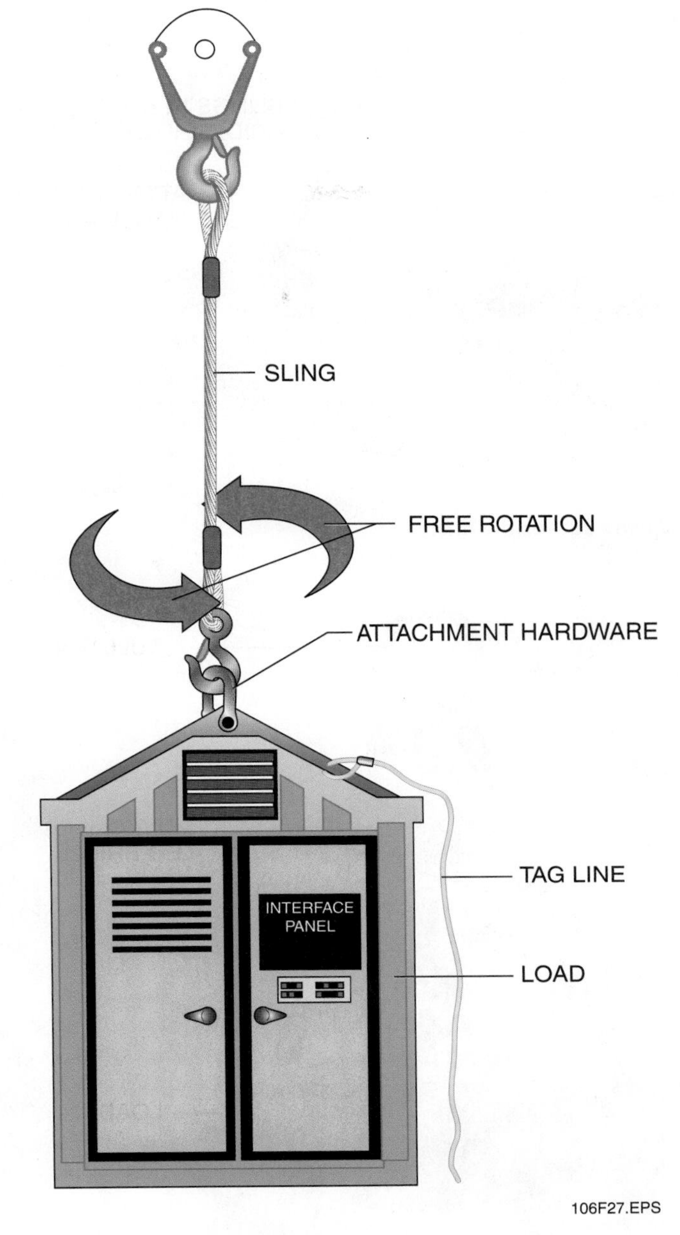

Figure 27 Single vertical hitch.

3.1.0 Vertical Hitch

The single vertical hitch is used to lift a load straight up. It forms a 90-degree angle between the hitch and the load. With this hitch, some type of attachment hardware is needed to connect the sling to the load (*Figure 27*). The single vertical hitch allows the load to rotate freely. If you do not want the load to rotate freely, some method of **load control** must be used, such as a **tag line**.

Another classification of vertical hitch is the bridle hitch (*Figure 28*). The bridle hitch consists of two or more vertical hitches attached to the same hook, master link, or **bull ring**. The bridle hitch allows the slings to be connected to the same load without the use of such devices as a spreader beam, which is a stiff bar used when lifting large objects with a crane hook.

The multiple-leg bridle hitch (*Figure 29*) consists of three or four single hitches attached to the same hook, master link, or bull ring. Multiple-leg bridle hitches provide increased stability for the load being lifted. A multiple-leg bridle hitch is always considered to have only two of the legs supporting the majority of the load and the rest of the legs balancing it.

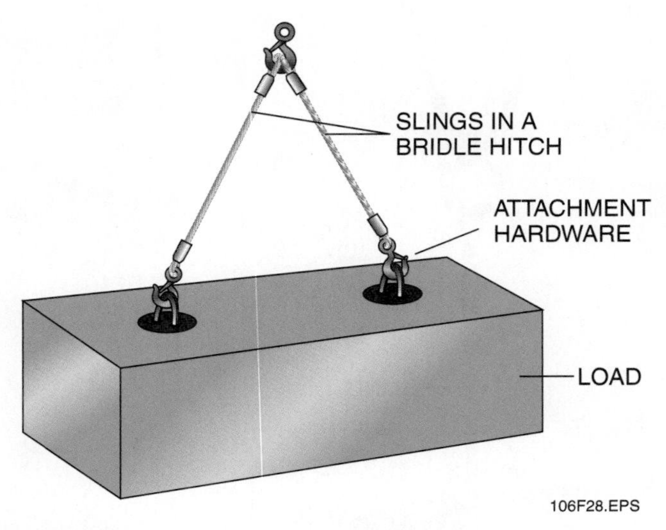

Figure 28 Bridle hitch.

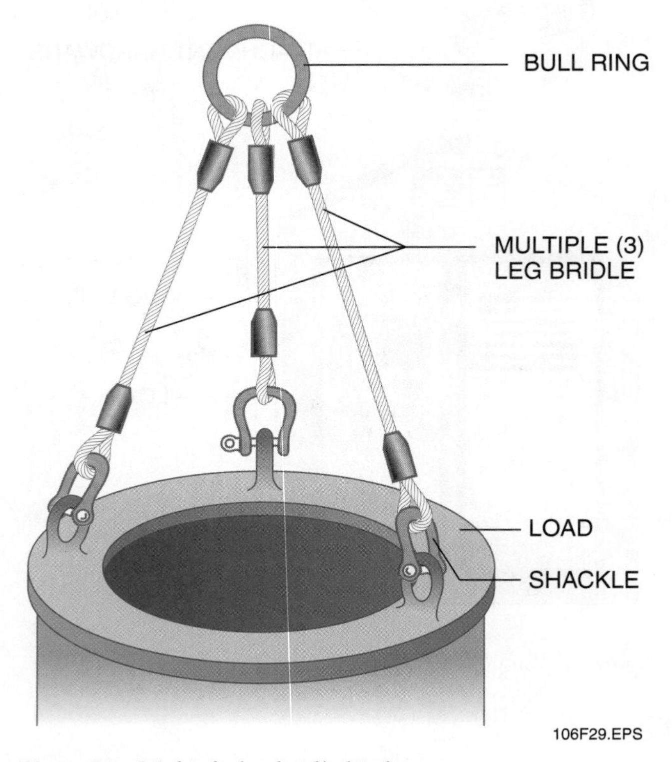

Figure 29 Multiple-leg bridle hitch.

3.2.0 Choker Hitch

The choker hitch is used when a load has no attachment points or when the attachment points are not practical for lifting (*Figure 30*). The choker hitch is made by wrapping the sling around the load and passing one eye of the sling through a **shackle** to form a constricting loop around the load. It is important that the shackle used in a choker hitch be oriented properly, as shown in *Figure 30*. The choker hitch affects the capacity of the sling, reducing it by a minimum of 25 percent. This reduction must be considered when choosing the proper sling.

The choker hitch does not grip the load securely. It is not recommended for loose bundles of materials because it tends to push loose items up and out of the choker. Many riggers use the choker hitch for bundles, mistakenly believing that forcing the choke down provides a tight grip. In fact, it serves only to drastically increase the stress on the choke leg (*Figure 31*).

When an item more than 12 feet long is being rigged, the general rule is to use two choker hitches spaced far enough apart to provide the stability needed to transport the load (*Figure 32*).

To lift a bundle of loose items, or to maintain the load in a certain position during transport, a double-wrap choker hitch (*Figure 33*) may be useful. The double-wrap choker hitch is made by wrapping the sling completely around the load, and then wrapping the choke end around again and passing it through the eye like a conventional

GOING GREEN

Biodiesel

Look around your job site and you'll probably see vehicles such as trucks and backhoes, or machinery such as generators or rigging equipment, running on diesel fuel. These vehicles and machines could go green and run on biodiesel instead. Biodiesel is a plant oil-based fuel made from oils such as soybean, canola, and other waste vegetable oils. It is even possible to make biodiesel from recycled frying oil from your local restaurant. Biodiesel is considered a green fuel since it is made using renewable resources and waste products. Biodiesel can be combined with regular diesel at any ratio or be run completely on its own. This means you can use any combination of biodiesel and regular petroleum diesel or switch back and forth as needed.

Why Biodiesel?

- It's environmentally friendly. Biodiesel is sustainable and a much more efficient use of our resources than diesel.
- It's non-toxic. Biodiesel reduces health risks such as asthma and water pollution linked with petroleum diesel.
- It produces lower greenhouse gas emissions. Biodiesel is almost carbon-neutral, contributing very little to global warming.
- It can improve engine life. Biodiesel provides excellent lubricity and can significantly reduce wear and tear on your engine.

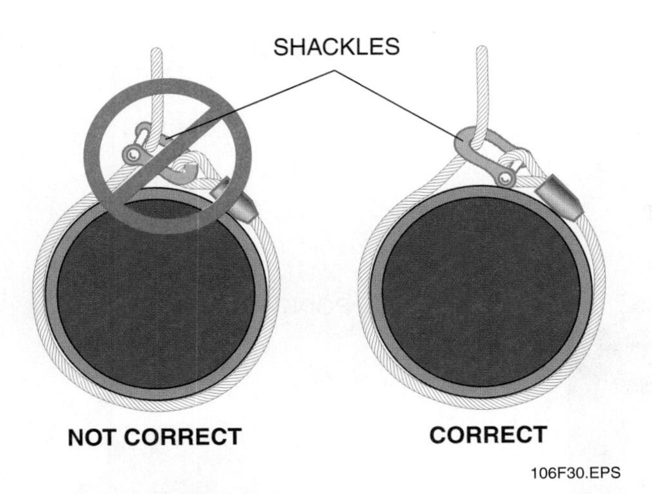

Figure 30 Choker hitch.

choker hitch. This enables the load weight to produce a constricting action that binds the load into the middle of the hitch, holding it firmly in place throughout the lift.

Forcing the choke down will drastically increase the stress placed on the sling at the choke point. To gain gripping power, use a double-wrap choker hitch. The double-wrap choker uses the load weight to provide the constricting force, so there is no need to force the sling down into a tighter choke.

A double-wrap choker hitch is ideal for lifting bundles of items, such as pipes and structural steel. It will also keep the load in a certain position, which makes it ideal for equipment installation lifts. Lifting a load longer than 12 feet requires two of these hitches.

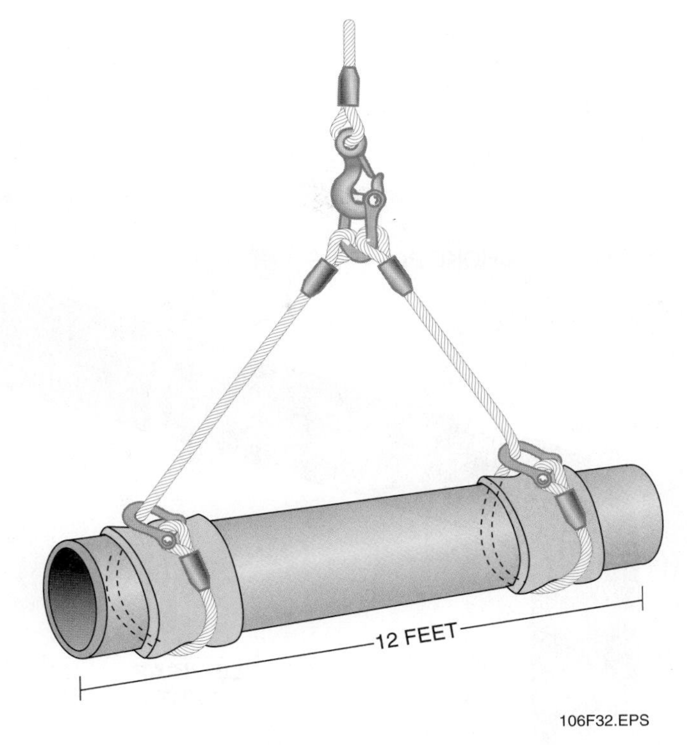

Figure 32 Double choker hitch.

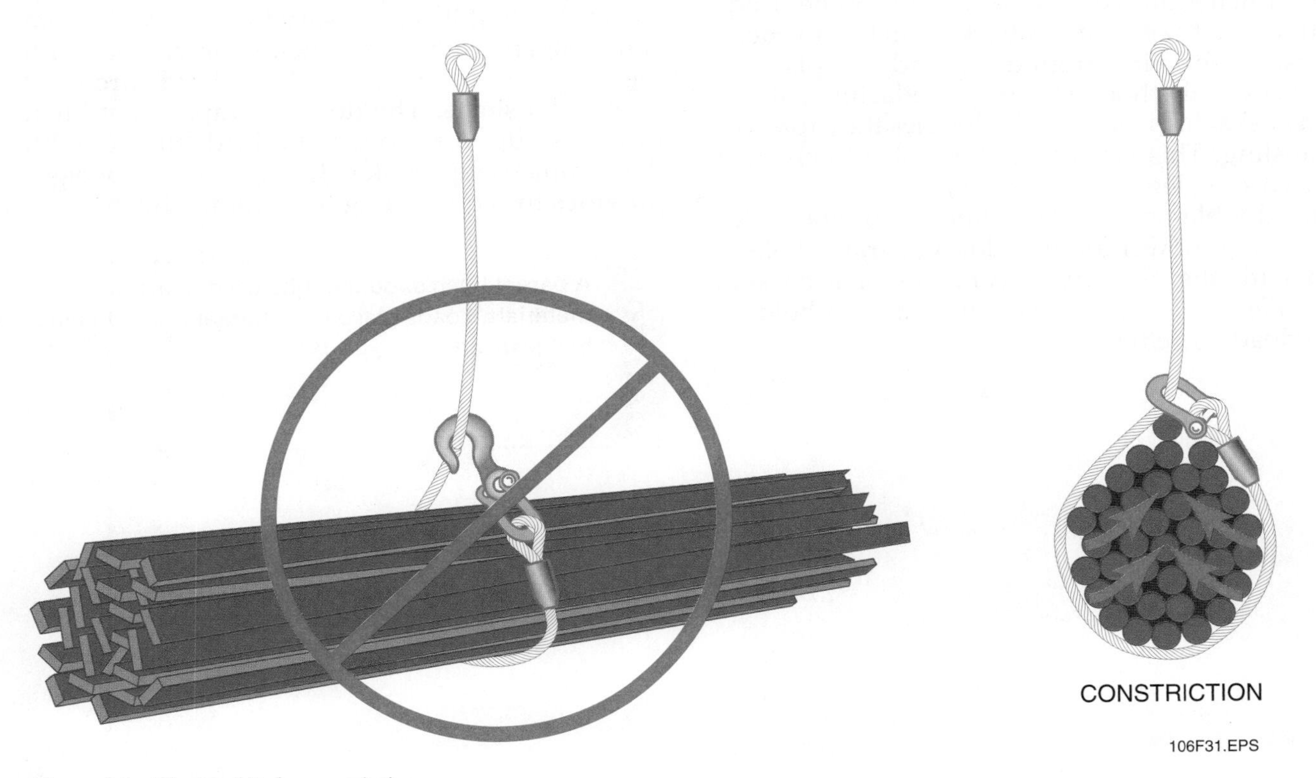

Figure 31 Choker hitch constriction.

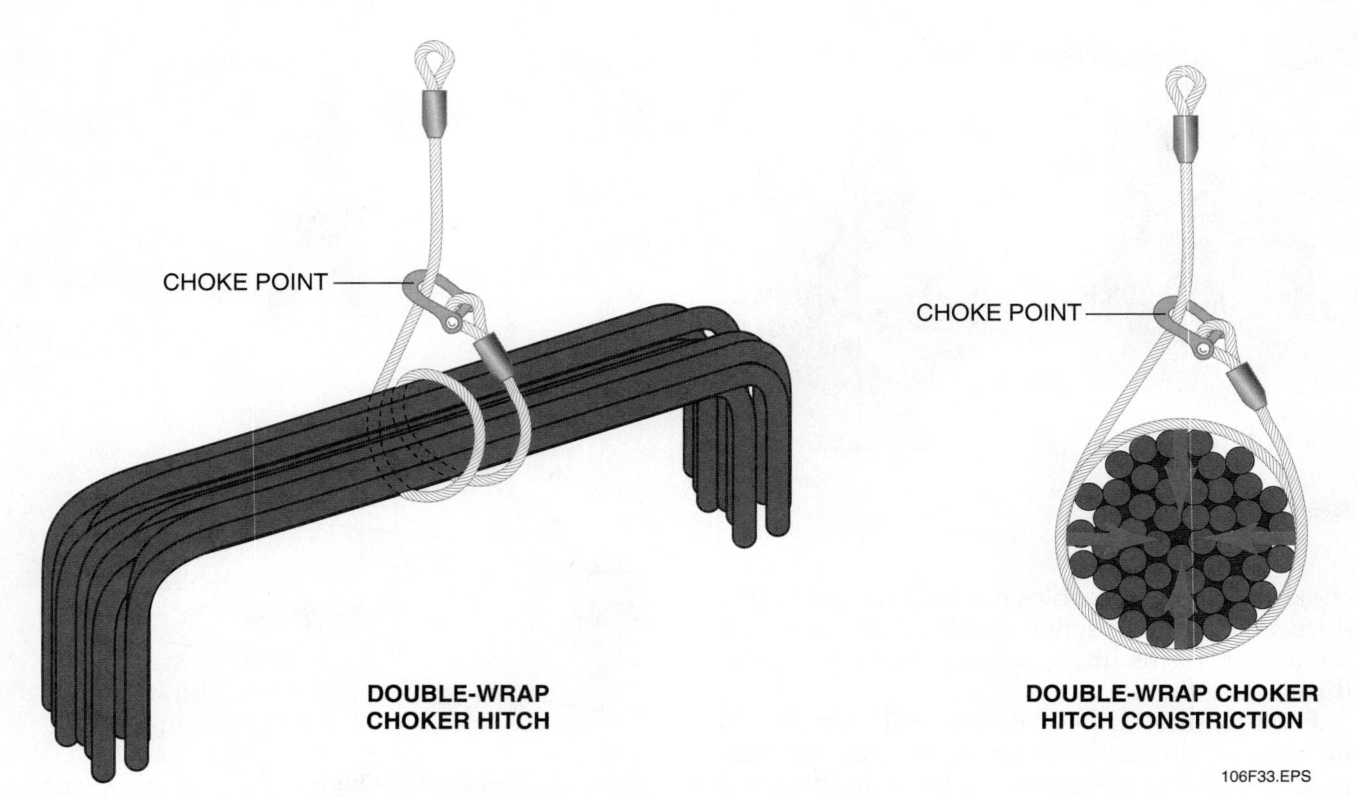

Figure 33 Double-wrap choker hitch and double-wrap choker hitch constriction.

3.3.0 Basket Hitch

Basket hitches are very versatile and can be used to lift a variety of loads. A basket hitch is formed by passing the sling around the load and placing both eyes in the hook (*Figure 34*). Placing a sling into a basket hitch effectively doubles the capacity of the sling. This is because the basket hitch creates two sling legs from one sling.

The double-wrap basket hitch combines the constricting power of the double-wrap choker hitch with the capacity advantages of a basket hitch (*Figure 35*). This means it is able to hold a larger load more tightly.

The double-wrap basket hitch requires a considerably longer sling length than a double-wrap choker hitch. If it is necessary to join two or more slings together, the load must be in contact with the sling body only, not with the hardware used to join the slings. The double-wrap basket hitch provides support around the load. Just as with the double-wrap choker hitch, the load weight provides the constricting force for the hitch.

CAUTION

A basket hitch should not be used to lift loose materials. Loads placed in a basket hitch should be balanced.

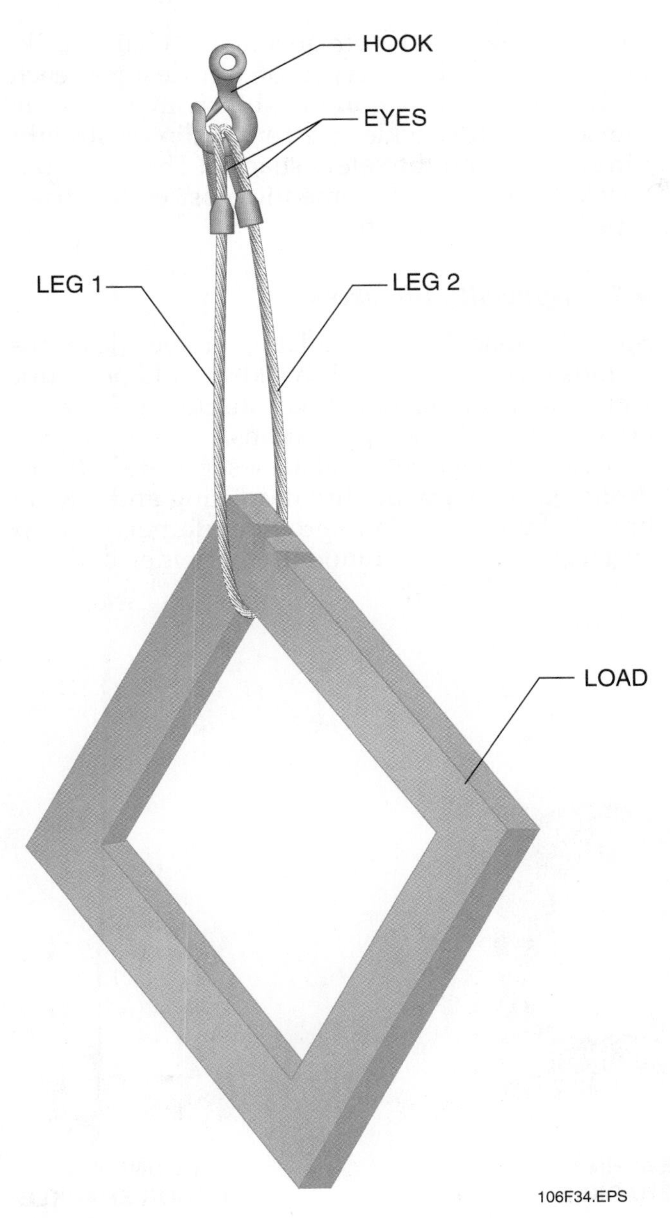

Figure 34 Basket hitch.

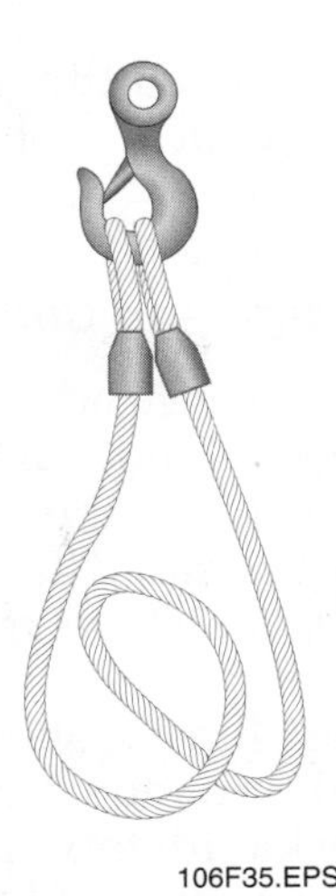

Figure 35 Double-wrap basket hitch.

4.0.0 Rigging Hardware

Rigging hardware is as crucial as the crane, the slings, or any specially designed lifting frame or hoisting device. If the hardware that connects the slings to either the load or the master link were to fail, the load would drop just as it would if the crane, hoist, or slings were to fail. Hardware failure related to improper attachment, selection, or inspection contributes to a great number of the deaths, serious injuries, and property damages in rigging accidents. The importance of hardware selection, maintenance, inspection, and proper use cannot be stressed enough. The regulations and requirements for rigging hardware are as stringent as those governing cranes and slings.

4.1.0 Shackles

A shackle is an item of rigging hardware used to attach an item to a load or to couple slings together. A shackle can be used to couple the

On-Site

Load Oscillation

Because only three or four small points (the points where the hitches connect to the load) transfer the entire weight of the load, a load will oscillate (swing back and forth like a pendulum) as it is moved. Because of this load oscillation, at any given moment it is impossible to tell which slings are supporting the weight and which ones are providing stability.

To understand load oscillation, imagine the following example. A glass of water filled to the rim is placed on a table. Four people each place one finger under each corner of the table. They try to lift the table and walk across the room without spilling any of the water. The table under the glass oscillates as the people move the table, and a large amount of the water is probably spilled on the way across the room. This is because only a small amount of surface area of their fingers is in contact with the table, which causes the weight of the table to shift back and forth, or oscillate.

Now imagine that each person is allowed to use both hands on a corner of the table, thereby spreading the contact stress over a larger surface area. Now they could probably make it all the way across the room without spilling much, if any, of the water.

end of a wire rope to eye fittings, hooks, or other connectors. It consists of a U-shaped body and a removable pin.

4.1.1 Shackle Design and Characteristics

Shackles used for overhead lifting should be made from forged steel, not cast steel. Quenched and tempered steel is the preferred material because of its increased toughness, but at a minimum, shackles must be made of drop- or hammer-forged steel.

All shackles must have a stamp that is clearly visible, showing the manufacturer's trademark, the size of the shackle (determined by the diameter of the shackle's body, not by the diameter of the pin), and the rated capacity of the shackle.

4.1.2 Types of Shackles

Shackles are available in two basic classes, identified by their shapes: anchor shackles and chain shackles (*Figure 36*). Both anchor and chain shackles have three basic types of pin designs, each with a unique function, as shown in *Figure 36*: the screw pin shackle, the round pin or straight pin shackle, and the safety shackle. The screw pin shackle design has become the most widely used type in general industry.

4.1.3 Specialty Shackles

Specialty shackles are available for specific applications where a standard shackle would not work well. For example, wide-body shackles (*Figure 37*) are for heavy lifting applications.

Synthetic web sling shackles (*Figure 38*) are designed with a wide throat opening and a wide bow that is contoured to provide a larger, nonslip surface area to accommodate the wider body of synthetic web slings.

SCREW PIN ANCHOR SHACKLE

106F36A.EPS

SCREW PIN CHAIN SHACKLE

106F36B.EPS

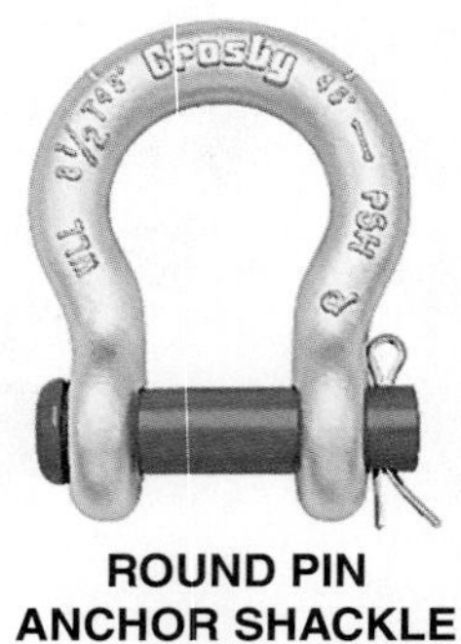

ROUND PIN ANCHOR SHACKLE

106F36C.EPS

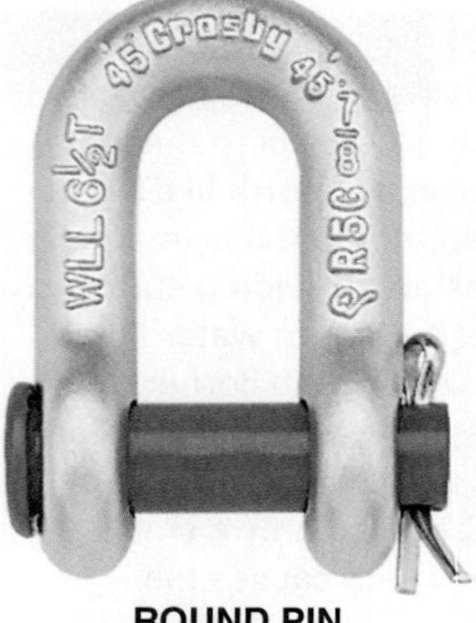

ROUND PIN CHAIN SHACKLE

106F36D.EPS

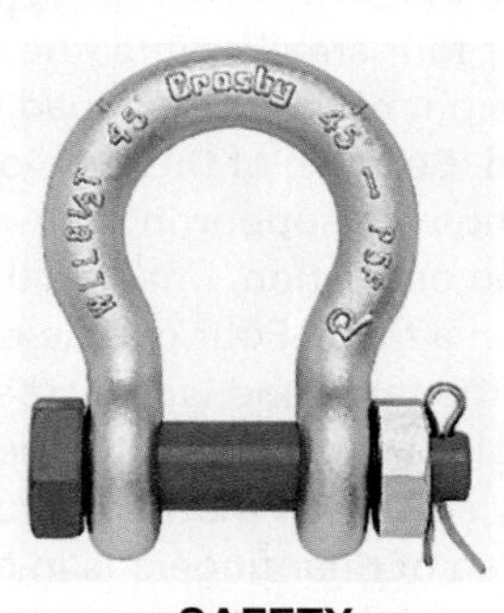

SAFETY ANCHOR SHACKLE

106F36E.EPS

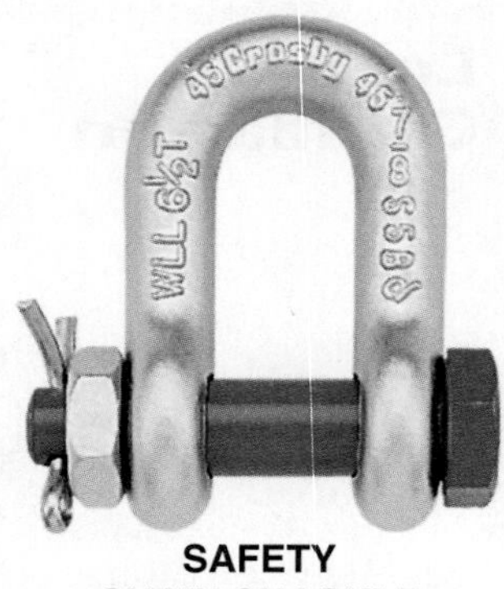

SAFETY CHAIN SHACKLE

106F36F.EPS

Figure 36 Shackles.

On-Site

Using Screw Pin Shackles

One of the nation's leading manufacturers of rigging hardware has judged the former practice of backing the pin off a quarter turn to be unsafe, because it allows the shackle to stretch under the load. The new practice is to engage the pin but not to tighten it fully. If you tighten or torque the pin fully, the load stress from the sling may torque it even more. This will stretch and jam the pin's threads.

Pins should never be swapped in shackles. The threadings of different brands of shackles and of different brands of pins may not engage properly with one another. In addition, there is no easy way to tell how much reserve strength is left in a pin or a shackle. Also, the rated capacities of different shackles and pins may not be compatible. The shackle pin should be placed and secured into the shackle between uses.

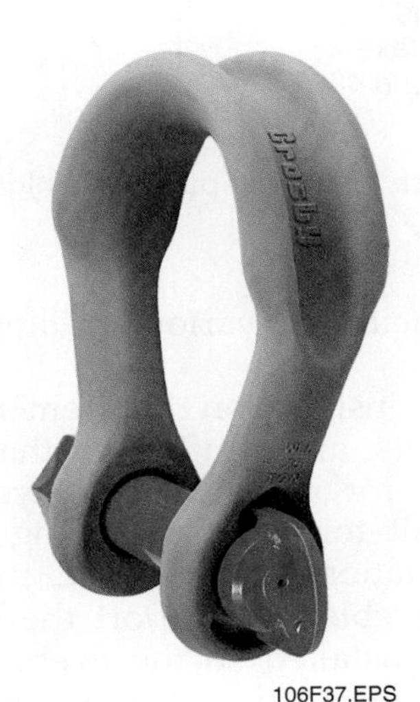

106F37.EPS

Figure 37 Wide-body shackle.

106F38.EPS

Figure 38 Synthetic web sling shackle.

4.1.4 Shackle Inspection and Rejection Criteria

Shackles, like any other type of hardware, must be inspected by the rigger before each use to make sure there are no defects that would make the shackle unsafe. Each lift may cause some degree of damage or may further reveal existing damage.

If any of the following conditions exists, a shackle must be removed from use:

- Bends, cracks, or other damage to the shackle body
- Incorrect shackle pin or improperly substituted pin
- Bent, broken, or loose shackle pin
- Damaged threads on threaded shackle pin
- Missing or illegible capacity and size markings

4.2.0 Eyebolts

An **eyebolt** is an item of rigging hardware with a **threaded shank**. The eyebolt's shank end is attached directly to the load, and the eyebolt's eye end is used to attach a sling to the load.

4.2.1 Eyebolt Design and Characteristics

Eyebolts for overhead lifting should be made of drop- or hammer-forged steel. Eyebolts are available in three basic designs with several variations (*Figure 39*):

- Unshouldered eyebolts, designed for straight vertical pulls only
- Shouldered eyebolts, with a shoulder that is used to help support the eyebolt during pulls that are slightly angular
- Swivel eyebolts, also called hoist rings, designed for angular pulls from 0 to 90 degrees from the horizontal plane of the load

4.2.2 Unshouldered Eyebolts

Unshouldered eyebolts are designed for vertical pulls only, not angular pulls. In the installation of an unshouldered eyebolt, the sling must not pull perpendicular to the eyebolt (*Figure 40*). This will cause the eyebolt to bend and probably break off.

Figure 39 Eyebolt variations.

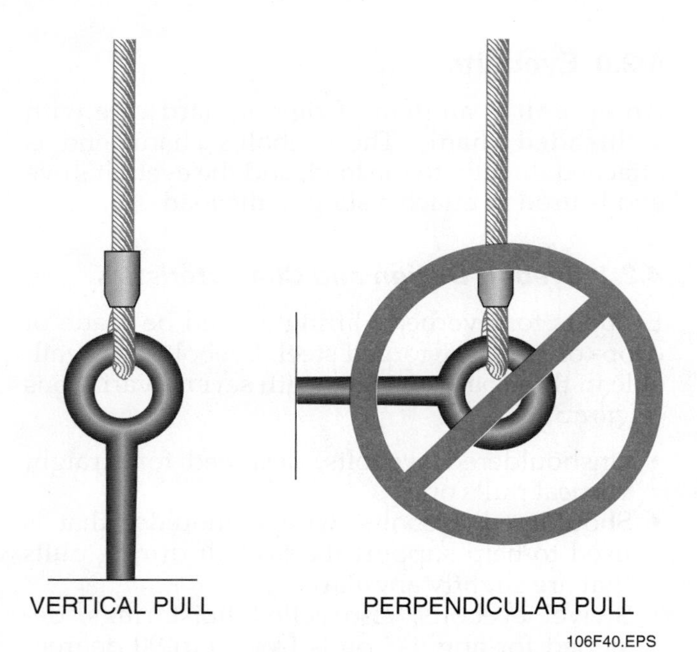

Figure 40 Installing an unshouldered eyebolt.

4.2.3 Shouldered Eyebolts

Shouldered eyebolts can be used for some angular pulls. Some manufacturers set the limit for angular pulling at 45 degrees from the horizontal plane of the load. Shouldered eyebolts lose capacity as the angle of the pull deviates from the vertical (*Figure 41*). Always check manufacturer ratings for load capacities at various pulling angles and eyebolt sizes.

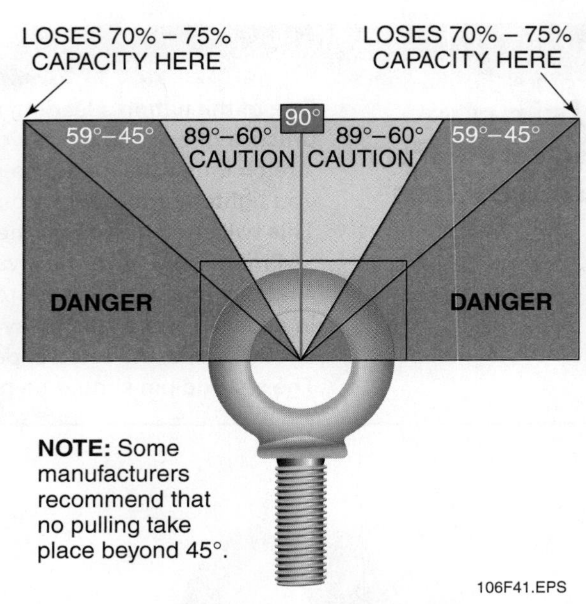

Figure 41 Effects of angular pull on shouldered eyebolts.

Most of the installation requirements for shouldered eyebolts are similar to the installation requirements for unshouldered eyebolts. However, the limit for angular pulling depends on proper installation. For example, in order for a shouldered eyebolt to support the load weight, it has to be installed so the eyebolt's shoulder is flush with the load surface (*Figure 42*). If the shoulder is not flush with the load surface, the eyebolt acts the same as an unshouldered eyebolt. Shouldered eyebolts must be positioned so the sling is in the same plane as the eyebolt.

4.2.4 Swivel Eyebolts

Swivel eyebolts are specially designed for angular pulls. They are available in through-bolt and machine-bolt styles. A through-bolt swivel eyebolt is designed so that the shank passes completely through the members it connects. A machine-bolt swivel eyebolt has a straight shank with a conventional head, such as a square or hexagonal head.

A swivel eyebolt's seating surface must be flush with the surface of the load or it will not function any better than an unshouldered eyebolt. Because the seating base of the swivel eyebolt is considerably larger than that of the unshouldered or shouldered eyebolt, it often will not install into the same area that a shouldered or unshouldered eyebolt would. Swivel eyebolts are self-aligning; their bases rotate 360 degrees, and the bail swivels 180 degrees. The rated capacity of a swivel eyebolt is based upon its 0- to 90-degree pull in

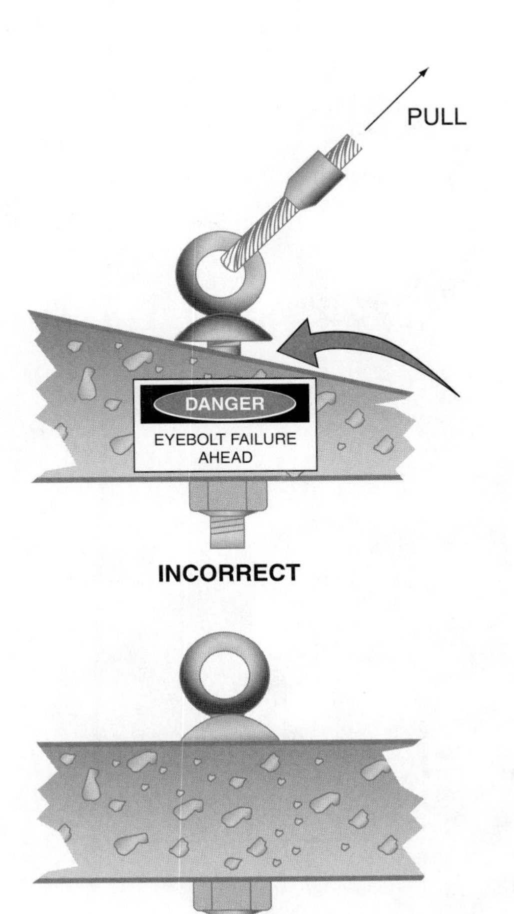

Figure 42 Proper installation of shouldered eyebolts.

THREAD SIZE (inches)	SAMPLE RATED CAPACITY FROM 0°–180° (pounds)
½"	2,500
⅝"	4,000
¾"	5,000
⅞"	8,000
1"	10,000
1½"	24,000
2"	30,000

BOLT SPECIFICATION IS GRADE 8 ALLOY SOCKET HEAD CAP SCREW TO ASTM A 574.

106F43.EPS

Figure 43 Effects of angular pull on swivel eyebolts.

any direction. There is no reduction factor for angular pulls (*Figure 43*).

4.2.5 Eyebolt Inspection and Rejection Criteria

Eyebolts must be visually inspected before each use, and a qualified rigging inspector must inspect eyebolts once a year. They should be in like-new condition, although minor surface rust, superficial scraping, or minor nicks are permissible. Any degree of defect beyond these may be reason to remove eyebolts from service, even if the defect is not described in the rejection criteria below. Rejection criteria for eyebolts include the following (*Figure 44*):

- Scraping or abrasion that results in a noticeable loss of material
- A bent shank or distorted threads
- Stress cracks
- Rust or corrosion that cannot be easily removed with a wire brush
- Elongation from overload of the eye or bottle-necking of the shank
- Damaged or worn threads
- Deformation or twist from side loading
- Wear

4.3.0 Lifting Clamps

Lifting clamps are used to move loads such as steel plates or concrete panels without the use of slings. Loads are moved one item at a time only.

4.3.1 Lifting Clamp Design and Characteristics

Lifting clamps are designed to bite down on a load and use the jaw tension to secure the load during transport (*Figure 45*). They are available in a wide variety of designs. Lifting clamps must be made of forged steel, and they must be stamped with their rated capacity.

Some clamps use the weight of the load to produce and sustain the clamping pressure. Others

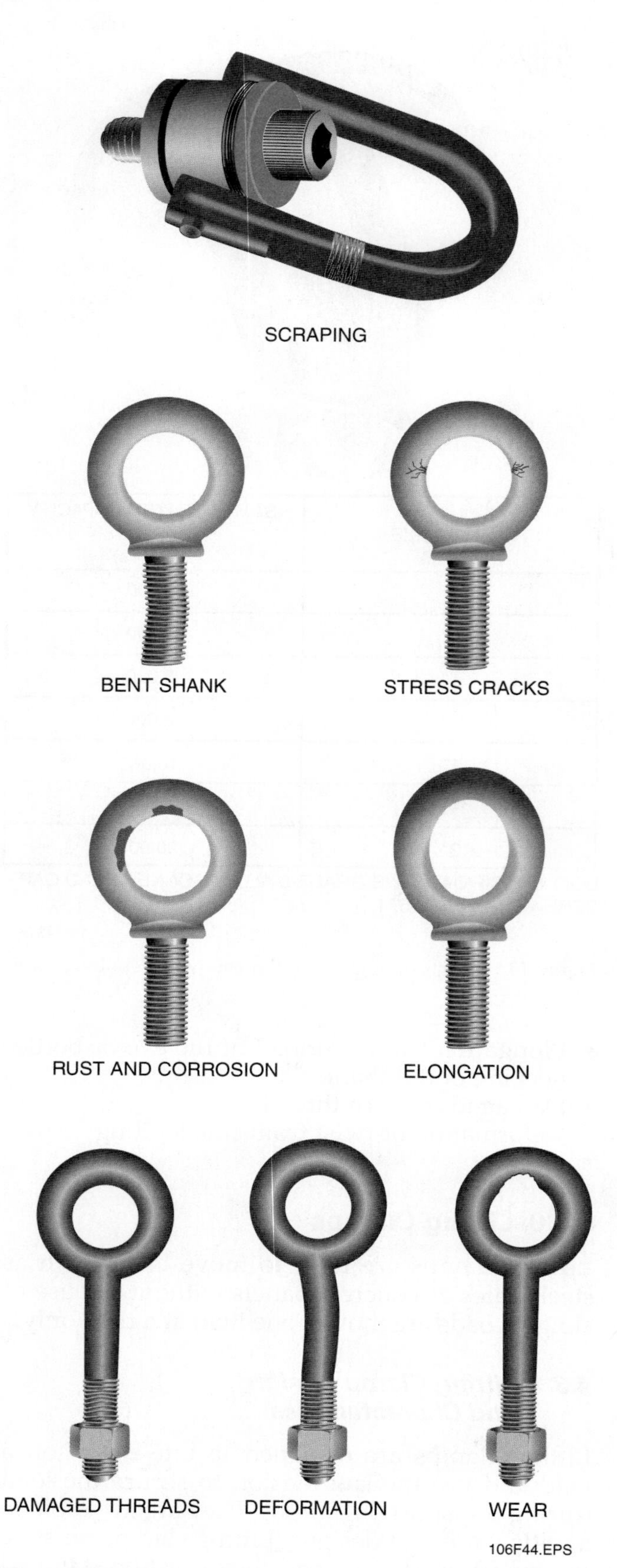

Figure 44 Rejection criteria on eyebolts.

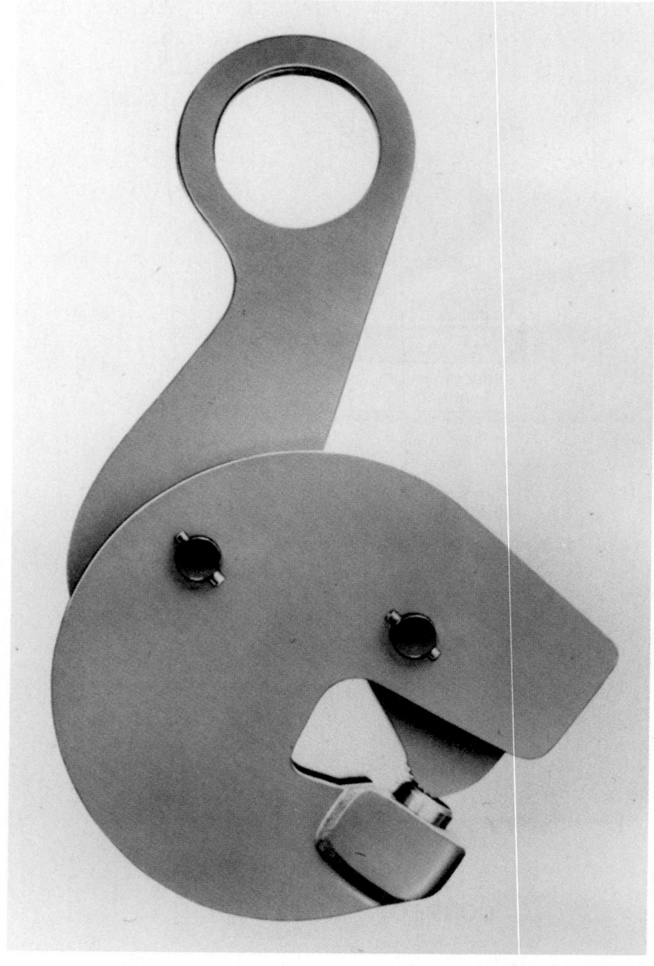

Figure 45 Basic nonlocking clamp.

use an adjustable cam that is set and tightened to maintain a secure grip on the load.

Lifting clamps are designed to carry one item at a time, regardless of the capacity or jaw dimensions of the clamp or the thickness or weight of the item being lifted. Most clamps have a cam, which is adjustable, and a jaw, which is fixed. The item to be lifted is placed in the jaw of the clamp, which bites down onto the item as load stress is applied or as the adjustable cam is tightened.

In order for the clamp to hold an item securely, the cam and the jaw must bite or grip both sides of a single item. Placing more than one plate or sheet into the clamp prevents both the cam and the jaw from securing both sides of the plate or sheet. Clamp designs vary with the application, so it is important to match the design to the intended application. Lifting clamps must function smoothly and adjust with no mechanical difficulty. At least two lifting clamps should always be used when lifting an item. Lifting clamps must be placed to ensure that the load remains balanced.

Loads lifted with both non-marring clamps and plate clamps must be lifted slowly and smoothly. These types of clamps are designed to grip the load gently yet securely. Their cams and jaws do not bite down too forcefully into the load, which might damage it. Therefore, you must be very careful to avoid any sudden movements that could jar the load out of these types of clamps.

4.3.2 Types of Lifting Clamps

Lifting clamps must be carefully selected for the specific lifting application, based on their design. There are four types of specialized lifting clamps and several variations in each type (*Figure 46*): linkage-type cam clamps, locking cam clamps, screw-adjusted cam clamps, and non-marring clamps.

4.3.3 Lifting Clamp Inspection and Rejection Criteria

Lifting clamps, like any other type of rigging hardware, must be inspected by the rigger before every use to make sure there are no defects that would make the clamp unsafe. Each lift may cause some degree of damage or may further reveal existing damage.

If any of the following conditions exists, a clamp must be removed from use (*Figure 47*):

- Cracks
- Abrasion, wear, or scraping
- Any deformation or other impact damage to the shape that is detectable during a visual examination
- Excessive rust or corrosion, meaning rust or corrosion that cannot be removed easily with a wire brush
- Excessive wear of the teeth
- Heat damage
- Loose or damaged screws or rivets
- Worn springs

4.4.0 Rigging Hooks

A **rigging hook** is an item of rigging hardware used to attach a sling to a load. Although there are many classes of rigging hooks used for hoist hooks and rigging, there are only six basic types of rigging hooks (*Figure 48*):

- Eye hooks are the most common type of end fitting hook. These hooks may or may not have safety latches or gates.
- Sorting hooks, also called pipe hooks, are used to lift pipe sections or containers by inserting the hook into the load, thereby avoiding the need for shackles or additional hardware.

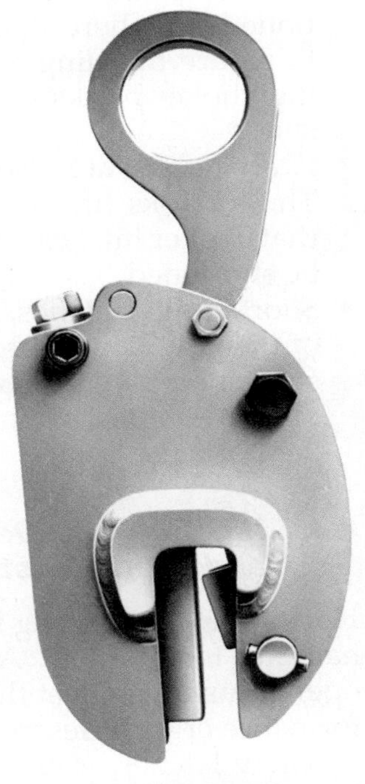

Figure 46 Nonstandard types of lifting clamps.

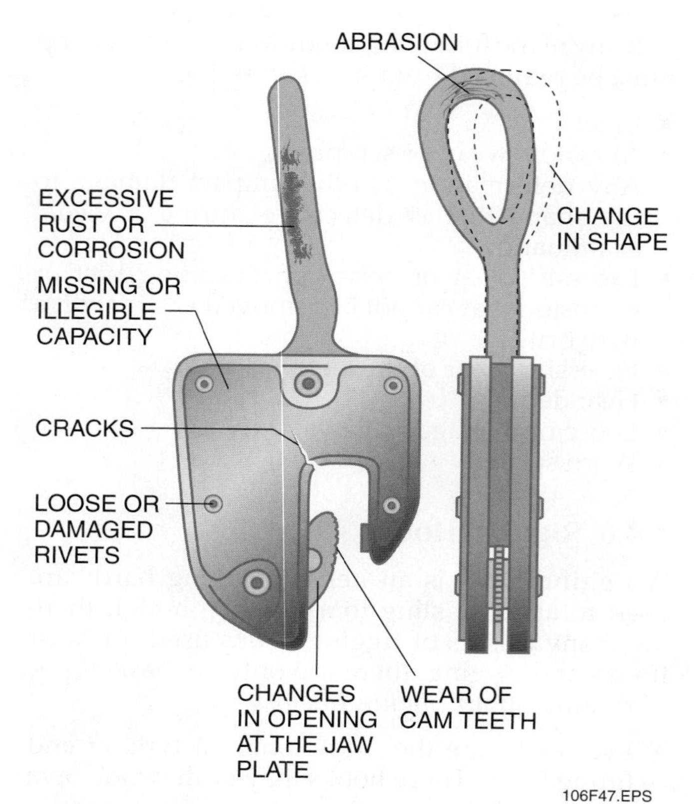

Figure 47 Rejection criteria for lifting clamps.

- Reverse eye hooks position the point of the hook perpendicular to the eye.
- Sliding choker hooks are installed onto the sling when it is made. The hooks, which can be positioned anywhere along the sling body, are used to secure the sling eye in a choker hitch. Sliding choker hooks are available for steel chain slings.
- Grab hooks are used on steel chain slings. These hooks fit securely in the chain link, so that choker hitches can be made and chains can be shortened.
- Shortening clutches, a more efficient version of the grab hooks, provide a secure grab of the shortened sling leg with no reduction in the capacity of the chain because the clutch fully supports the links.

4.4.1 Rigging Hook Design and Characteristics

Hooks used for rigging must be made of drop- or hammer-forged steel. All hooks used in rigging operations must meet the characteristics and performance criteria described in this module.

Safety latches or gates are installed in rigging hooks to prevent a sling from coming out of a hook or load when the sling is slackened. If a safety latch is installed in a rigging hook, the latch must

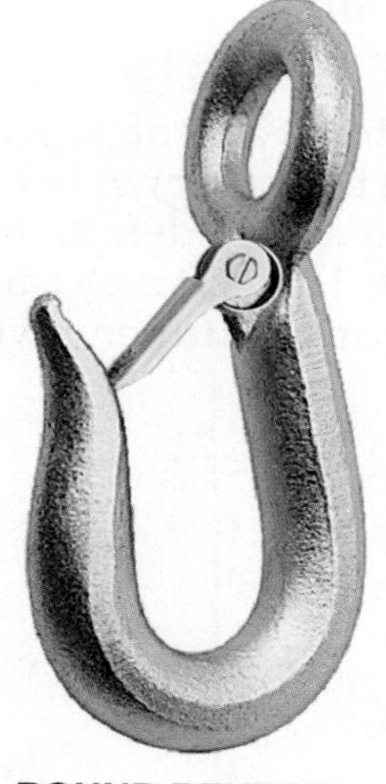

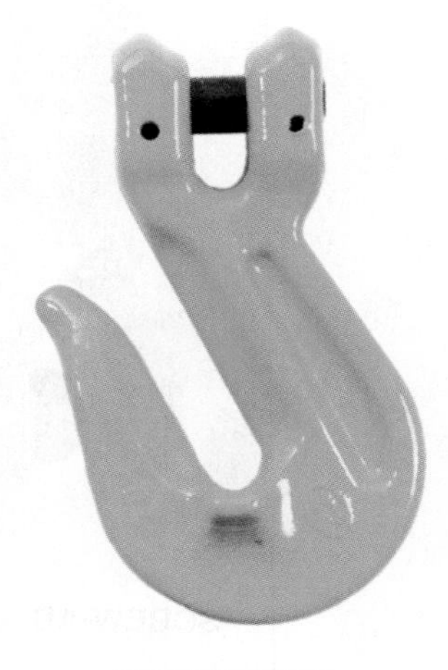

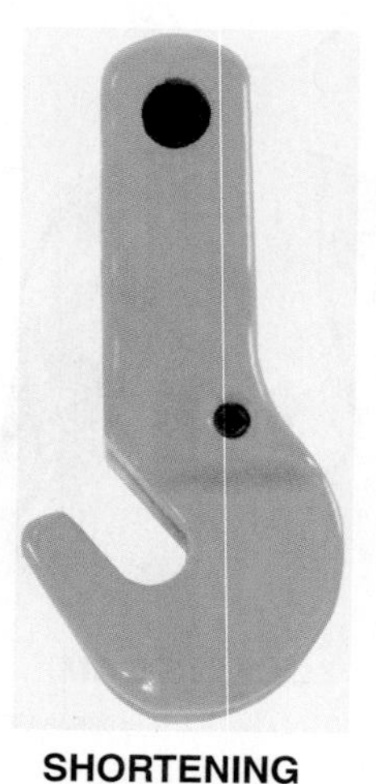

Figure 48 Rigging hooks.

be in good working condition. Damaged safety latches can be easily replaced. Report any damage you detect in a safety hook to your instructor or immediate supervisor.

4.4.2 Rigging Hook Inspection and Rejection Criteria

When hooks are installed as end fittings, they must be inspected along with the rest of the sling before each use. Slings with hook-type end fittings need to be removed from service for any of the following defects (*Figure 49*):

- Wear, scraping, or abrasion
- A broken or missing safety latch
- Cracks
- Cuts, gouges, nicks, or chips
- Excessive rust or corrosion, meaning rust or corrosion that cannot be easily removed with a wire brush
- A twist of 10 degrees or more from the unbent plane of the hook
- An increase in the throat opening of the hook of 15 percent or more—easy to detect if the hook is equipped with a safety latch, because the latch will no longer bridge the throat opening
- An increase of 5 percent or more in the shank or overall elongation of the hook

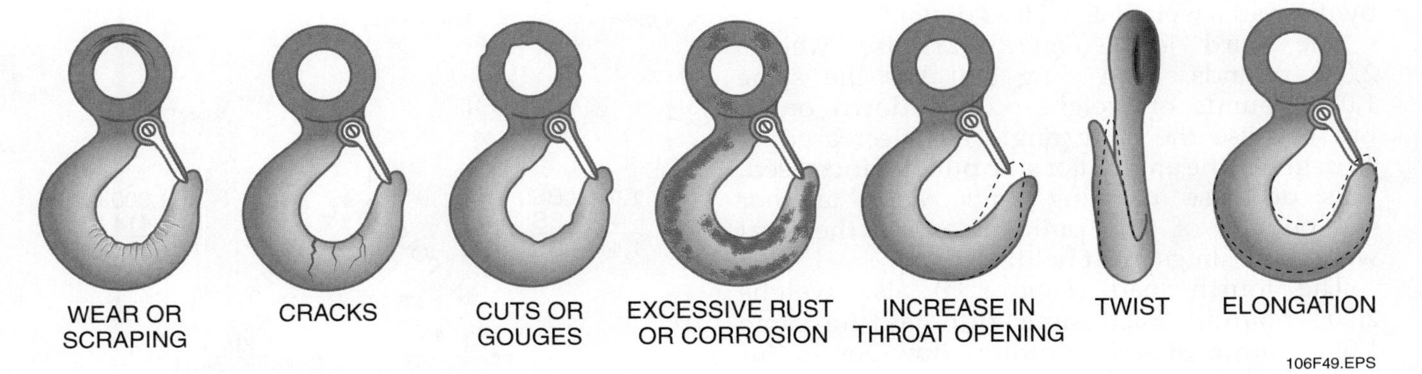

Figure 49 Rejection criteria for rigging hooks.

5.0.0 SLING STRESS

When **sling angle** decreases, **sling stress** increases. This is one of the most important facts you need to know to conduct rigging safely. It is essential that you understand this concept, because not understanding it—or misunderstanding it—could get you or someone else killed.

Here's a good way to understand sling stress. Imagine there are four loads, each weighing 2,000 pounds. If the first load (*Figure 50*) has two slings straight up and down (90 degrees), then each sling is itself holding a weight of 1,000 pounds.

The second load has slings at 60 degrees (*Figure 51*). Each sling still has the original 1,000 pounds of weight pulling down on it. However, because the slings are at 60-degree angles, the **side pull** now adds more stress to the sling. Side pull means the slings are being pulled in, or sideways, by the load's weight as well as down.

The third load (*Figure 52*) also weighs 2,000 pounds. Each sling still has the same 1,000 pounds of weight pulling down on it. But because the sling angle has decreased to 45 degrees, the amount of side pull has increased. This decrease in sling angle adds another 414 pounds of side pull, increasing the total weight the slings must hold.

The fourth load (*Figure 53*) also weighs 2,000 pounds. Each sling still has the same 1,000 pounds of weight pulling down on it. But here the sling angle has decreased to 30 degrees, which causes side pull to increase the sling stress to 1,000 extra pounds. This decrease in sling angle means each sling now has to hold up a total of 2,000 pounds.

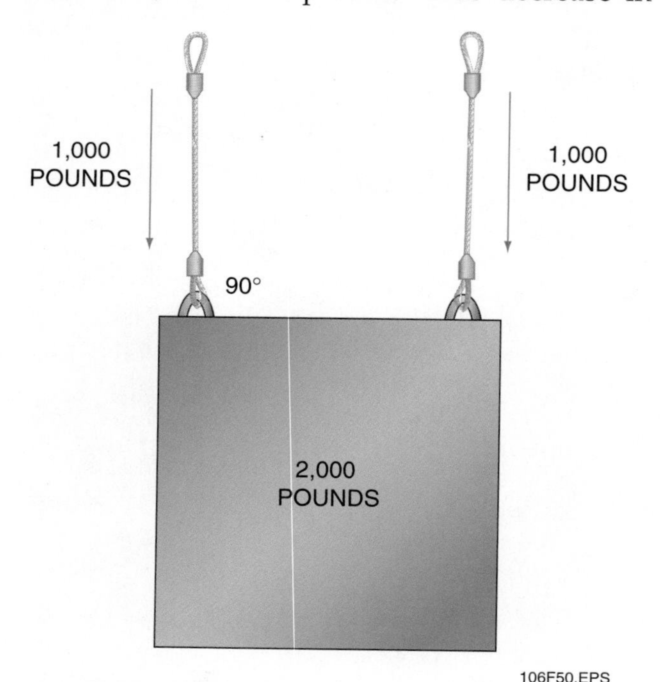

Figure 50 Sling stress example 1.

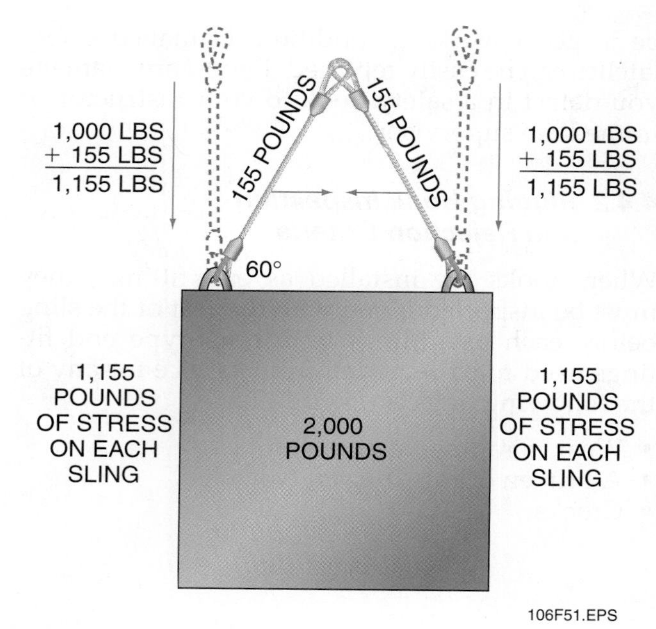

Figure 51 Sling stress example 2.

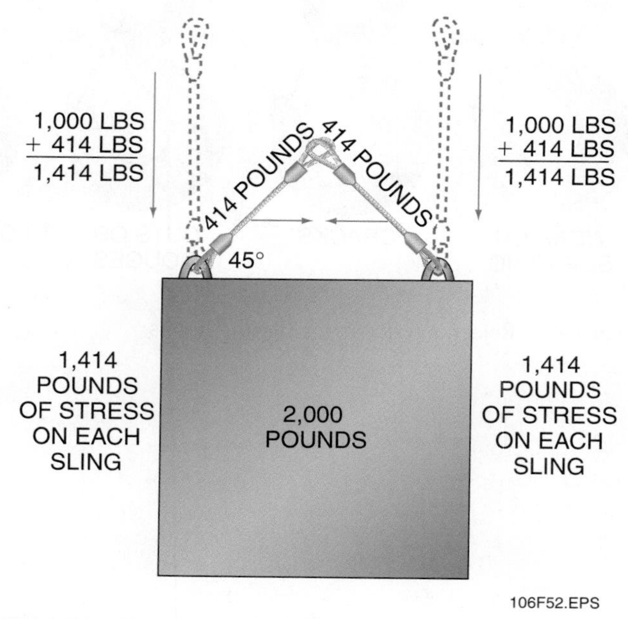

Figure 52 Sling stress example 3.

As you can see from examples 1 through 4, as the sling angle decreases, the sling stress increases. By the time the sling angle gets to 30 degrees, the sling stress has doubled. Although the total weight of the load has stayed the same in each example, the stress on the sling has increased with each reduction in the sling's angle. Below 45 degrees, sling stress increases dramatically. The greater the sling stress, the greater the effect on the lift's safety.

See the *Appendix* in the back of this module for more information on calculating sling stress.

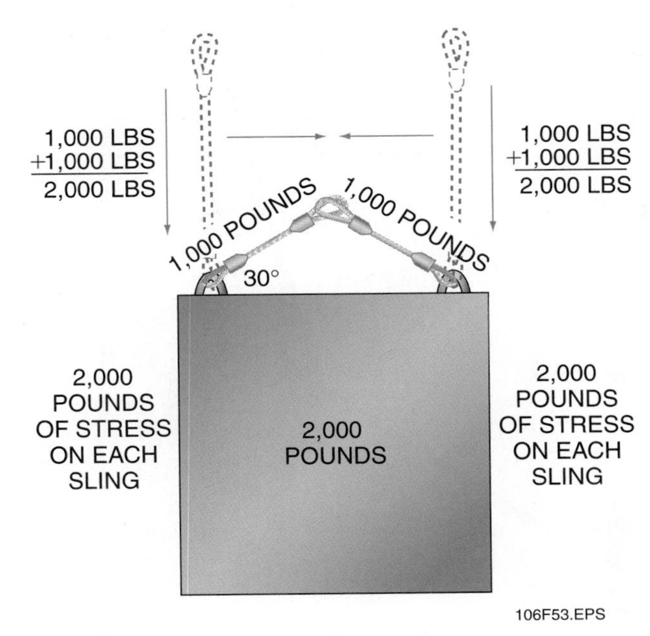

Figure 53 Sling stress example 4.

WARNING!

To rig and move a load safely, the rigger must understand the specific type and amount of stress placed on rigging components such as slings, hooks, and loads. If calculations are not made accurately, the rigging procedure could fail and damage the load and injure or kill the workers.

6.0.0 Hoists

A hoist provides a mechanical advantage for lifting a load, allowing you to move objects that you cannot lift manually. All hoists use a pulley system to transmit power and lift a load. Some hoists are mounted on trolleys and use electricity or compressed air for power. In this section, you will learn about a simple hoisting mechanism called a **block and tackle**, and a more complex hoisting mechanism called a chain hoist.

- *Block and tackle* – A block and tackle is a simple rope-and-pulley system used to lift loads (*Figure 54*). By using fixed pulleys and a wire rope attached to a load, the rigger can raise and lower the load by winding the wire rope around a drum.
- *Chain hoists* – Chain hoists may be operated manually or mechanically. There are three types of chain hoists (*Figure 55*): manual, electric, and pneumatic. Because electric and pneumatic chain hoists use mechanical, not manual, power, they are known as powered chain hoists.

All chain hoists use a gear system to lift heavy loads (*Figure 56*). The gearing is coupled to a sprocket that has a chain with a hook attached to it. The load is hooked onto a chain and the gearing turns the sprocket, causing the chain to travel over the sprocket and moving the load. The hoist can be suspended by a hook connected to an appropriate anchorage point, or it can be suspended from a trolley system (*Figure 57*).

6.1.0 Operation of Chain Hoists

Chain hoists are operated manually (by hand) or by electric or pneumatic power. In this section, you will learn some of the fundamental operating procedures for using both hand chain hoists and powered chain hoists.

- *Hand chain hoist* – To use a hand chain hoist, the rigger suspends the hoist above the load to be lifted, using either the suspension hook or the trolley mount. The rigger then attaches the hook to the load and pulls the hand chain drop to raise the load. The load will either rise or fall, depending on which side of the chain drop is pulled.
- *Powered chain hoist* – To use an electric or a pneumatic powered chain hoist, the rigger positions the chain hoist on the trolley above the load to be lifted, attaches the hook to the load, and uses the control pad to raise the load. Only qualified persons may use powered chain hoists.

6.2.0 Hoist Safety and Maintenance

In addition to the general safety rules you learned in the *Basic Safety* module, there are some specific safety rules for working with hoists. Observe the following guidelines:

- Always use the appropriate personal protective equipment when working with and around any lifting operations.
- Make sure that the load is properly balanced and attached correctly to the hoist before you attempt the lift. Unbalanced loads can slide or shift, causing the hoist to fail.
- Keep gears, chains, and ropes clean. Improper maintenance can shorten the working life of chains and ropes.
- Lubricate gears periodically to keep the wheels from freezing up.
- Never perform a lift of any size without proper supervision.

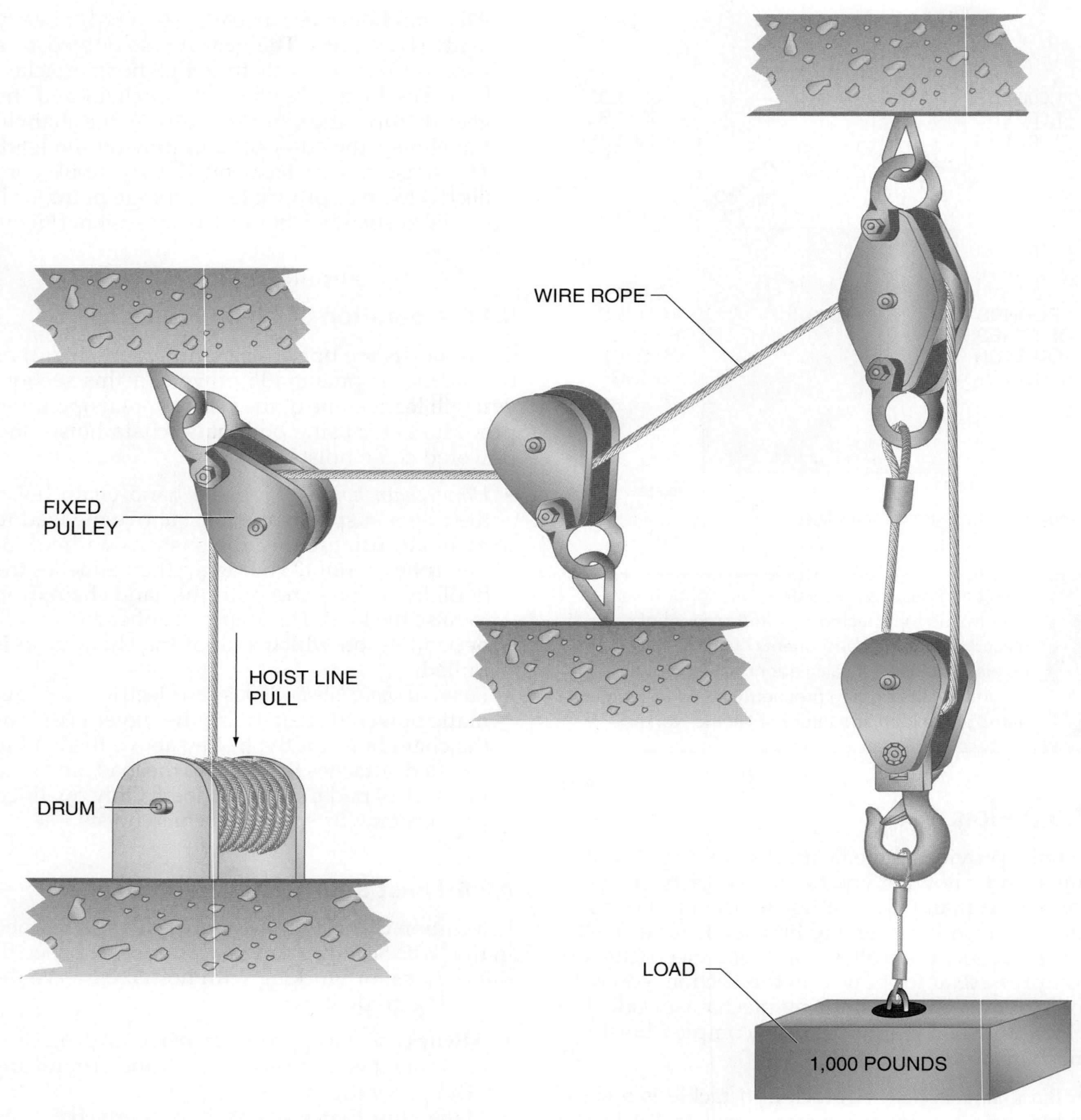

Figure 54 Block and tackle hoist system.

ELECTRIC 106F55A.EPS

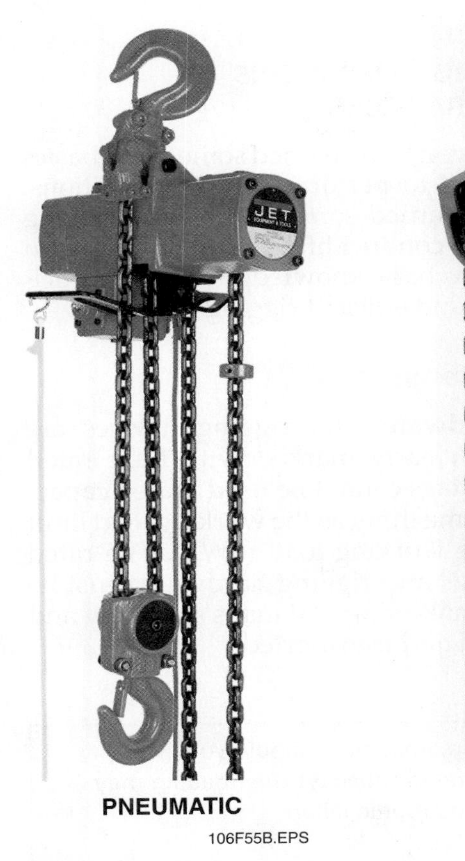

PNEUMATIC 106F55B.EPS

MANUAL 106F55C.EPS

Figure 55 Types of chain hoists.

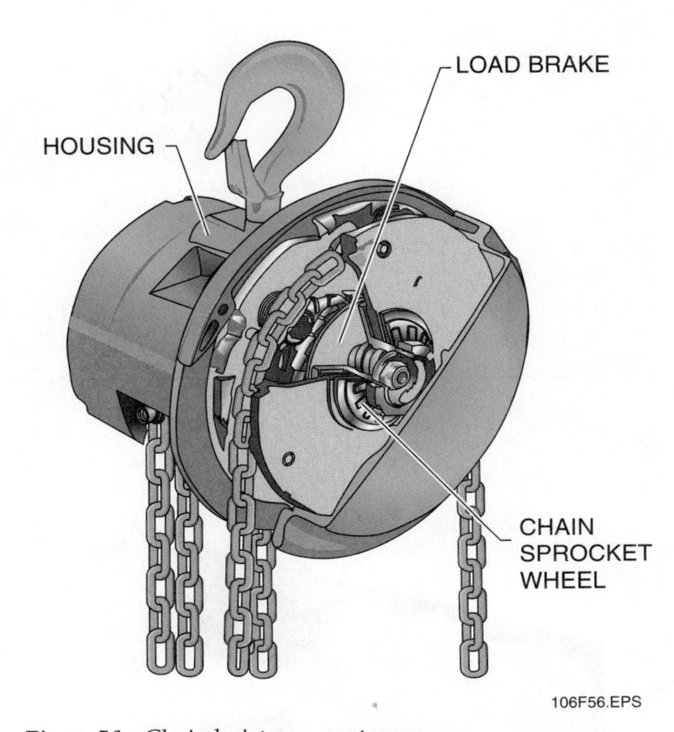

106F56.EPS

Figure 56 Chain hoist gear system.

On-Site

Not for Lifting

Never use a come-along for vertical overhead lifting. Use a come-along only to move loads horizontally over the ground. Be careful not to confuse a come-along with a ratchet lever hoist. Ratchet lever hoists have both a friction-type holding brake and a ratchet-and-pawl load control brake. Come-alongs have only a spring-load ratchet that holds the pawl in place. If the ratchet and pawl fail, the overhead load falls.

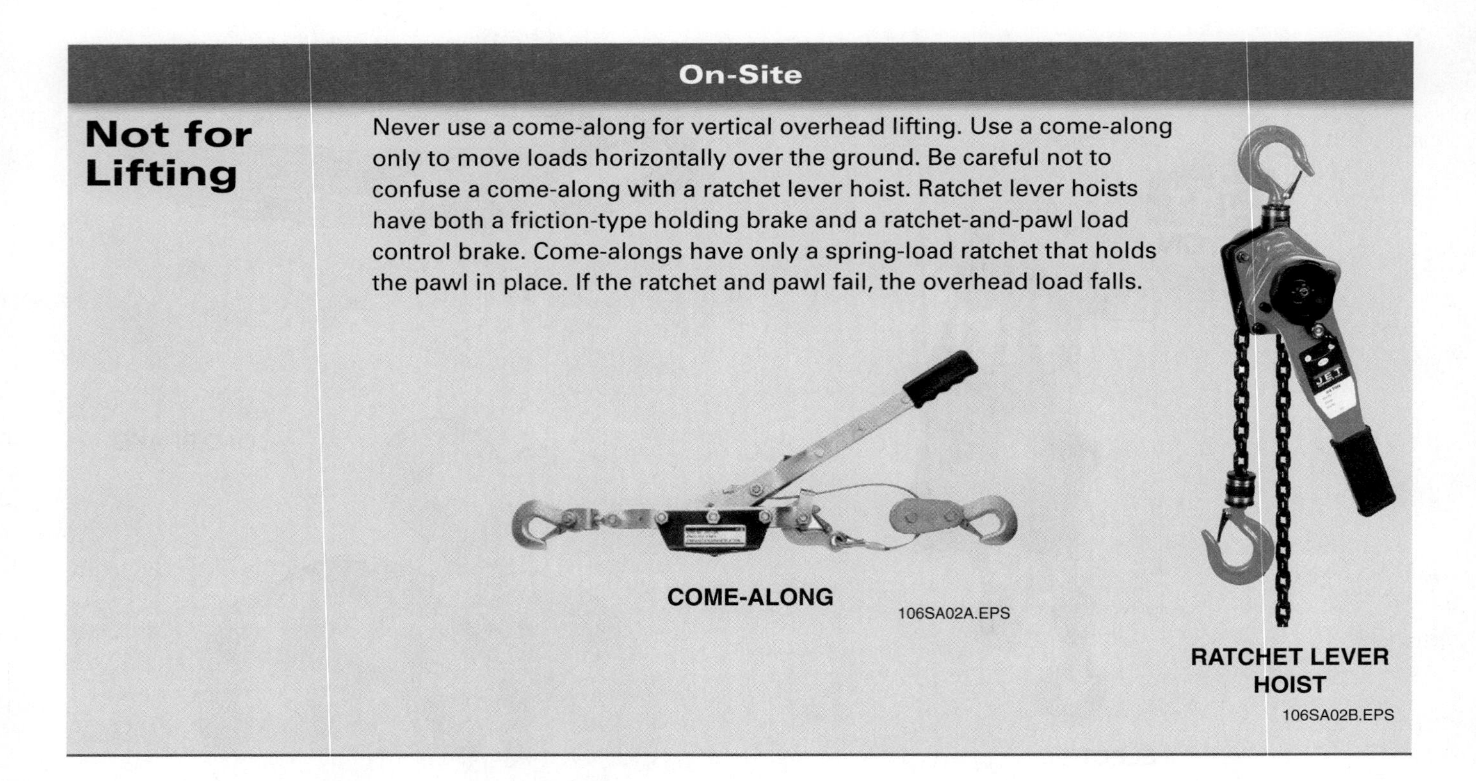

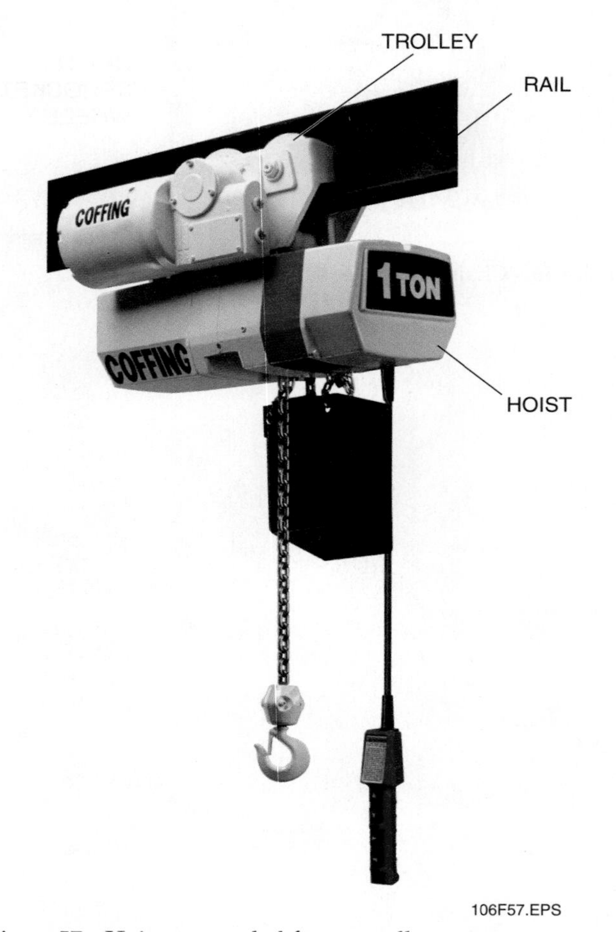

Figure 57 Hoist suspended from a trolley system.

7.0.0 Rigging Operations and Practices

In this module, you have learned some of the basics about tools used to perform rigging operations. You have also learned some of the basic rigging practices used to conduct lifts. Now you will learn how to apply this basic knowledge toward participating in a safe and efficient rigging operation.

7.1.0 Rated Capacity

All slings, hardware, and rigging devices are required to be clearly marked with their rated capacity. Paper tags cannot be used. Rated capacity means the same thing as the working load limit (WLL) or a safe working load (SWL). The rated capacity of slings and rigging hardware must be determined to make sure that loads are safely and effectively lifted and transported.

WARNING! Under no circumstances should you ever exceed the rated capacity! Overloading may result in catastrophic failure.

7.2.0 Sling Attachment

Before attaching slings to a load, riggers must remember several important points. The rigger must select the most appropriate type of sling (synthetic, alloy steel, or wire rope) for the load to be lifted and inspect the sling for damage. Once the type of sling is chosen, the rigger determines the best rigging configuration or hitch (vertical [bridle], choker, or basket) to lift the load. Knowing the type of sling and the best hitch, the rigger then chooses the appropriate rigging hardware to connect the sling to the load. To make the safest selection, the rigger must know the total weight of the load, the angle at which the sling(s) will attach to the load, and the total sling stress applied during the lift. Only then can a sling be rigged safely to a load.

When using slings, follow these safety guidelines:

- Select the correct length of sling for the lift. Never try to shorten a sling by wrapping it around the hoist hook before attaching the eye to the hook.
- Never try to shorten the legs of a sling by knotting, twisting, or wrapping the slings around one another.
- Never try to shorten a chain sling by bolting or wiring the links together.

WARNING!

Use the correct length sling for each lift. Trying to shorten a sling may cause an accident.

- Make sure all personnel are clear of the load before you take full load strain on the slings.
- Never try to adjust the slings while a strain is being taken on the load.
- Make sure all personnel keep their hands away from the slings and the load during hoisting.
- Use sling softeners whenever possible. Sling softeners protect the sling from abrasion, cuts, heat, and chemical damage. Whether the softeners or pads are manufactured or made in the field, using them regularly will considerably extend the life of the sling. If no standard sleeves are available, use rubber belting or sections of old slings. Softeners prevent kinking in wire rope. Corner buffers are specially made for this purpose, but if they are not available, the rigger can use whatever is handy to protect the rope. Wood, old web slings, and factory-made buffers will provide some protection. Softeners for chain slings protect the links contacting the corners of the load and keep the chain links from scraping or crushing the load itself.

7.3.0 Hardware Attachment

Most rigging hitches require some type of hardware to attach the slings to a piece of equipment or crane. You have learned about shackles, eyebolts, and hooks. This section summarizes the basics of how this equipment is attached to the load.

7.3.1 Shackles

Shackles have several installation requirements to prevent damage or failure during the lift. For instance, the shackle must remain in line with the sling so that the shackle does not become side-loaded and get pulled apart by the load stress. Also, when a screw pin type of shackle is installed, the pin should not be overtightened or it will stretch the threads. Overtightening makes the shackle difficult to remove and may damage the shackle so that it has to be removed from service (*Figure 58*).

7.3.2 Eyebolts

Eyebolts must be safely and properly installed. Use the following guidelines:

- Eyebolts must be installed so that the plane of the eye is in direct line with the plane of the sling when the sling is positioned at an angle other than vertical (*Figure 59*). Otherwise, the eyebolt may fold over under the load stress, which could severely deform the eyebolt or even lead to a fracture or a complete failure of the eye.

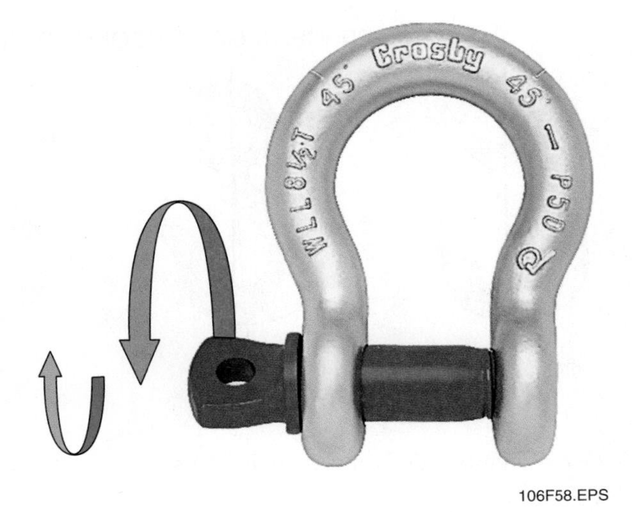

Figure 58 Don't overtighten the pin.

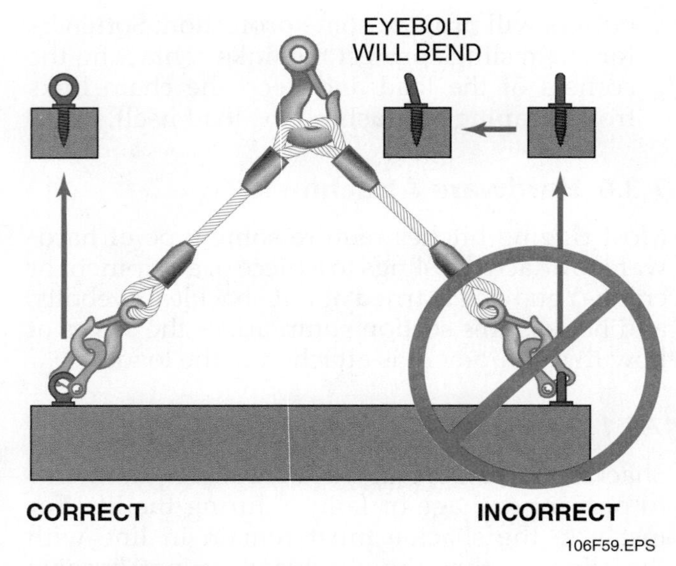

Figure 59 Eyebolt orientation.

- Eyebolts with shoulders or swivel bases must be installed so that the shoulder or base is flush with the surface of the load and makes positive contact around the entire circumference of the shoulder or base.

7.3.3 Hooks

Hooks must be carefully attached to the load to prevent binding of the hook. When a hook becomes bound in a **pad eye** or eyebolt, it can become point-loaded. A point-loaded hook will easily stretch open and slip out of the load. When a hook becomes point-loaded, the capacity of the hook decreases drastically the closer the load moves toward the point of the hook (*Figure 60*). Using a shackle instead of placing the hook directly into the pad eye or eyebolt will prevent this from happening.

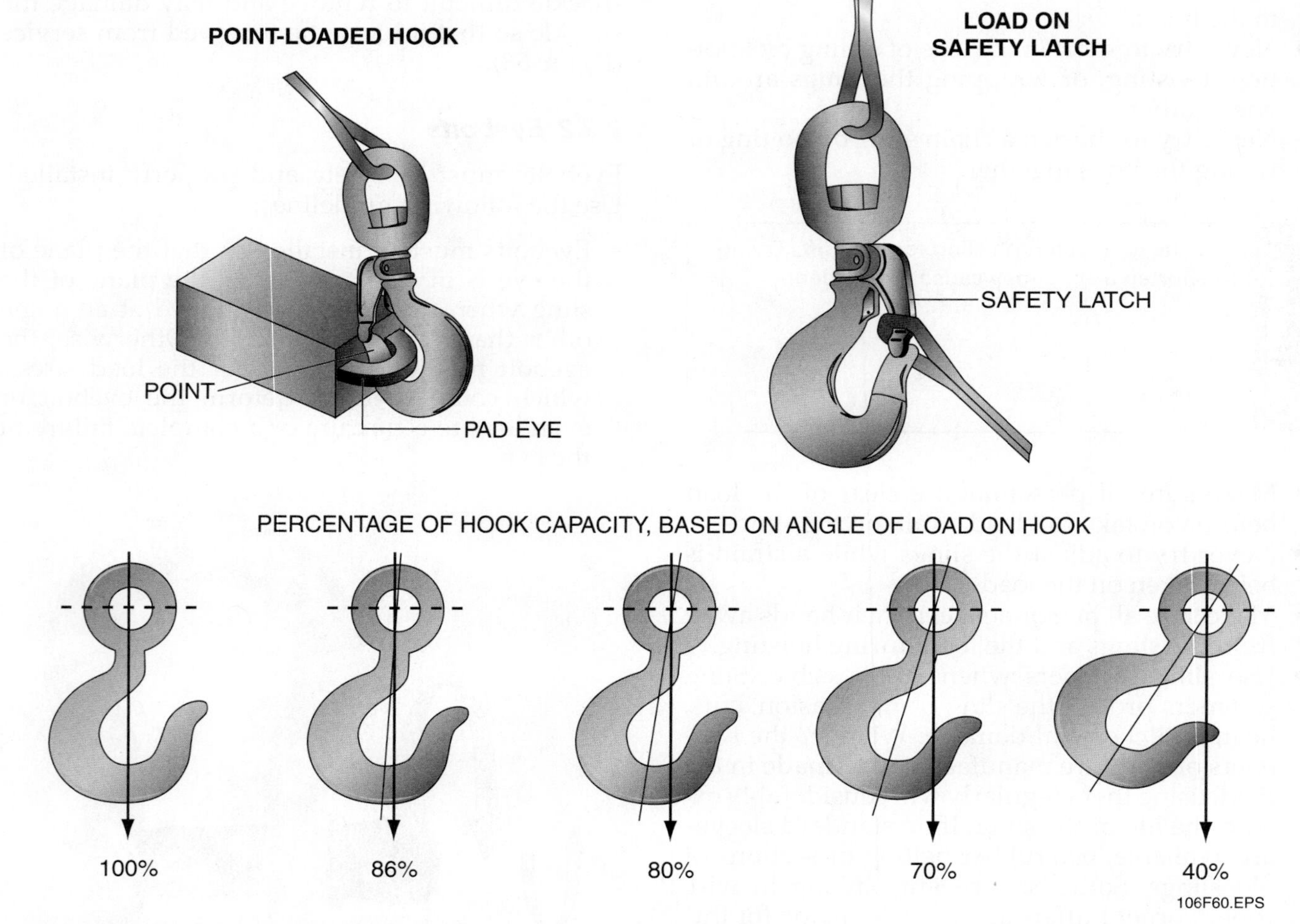

Figure 60 Point-loaded capacity reductions.

On Site

Avoiding Side Buckle

Use a pair of shackles in the eyebolts instead of threading a sling through hardware around a load. Threading the sling through eyebolts as shown and applying stress will cause the eyebolts to pull toward each other and shear off the bolts.

The severe side stress of low sling angles may cause a load to buckle into the center because of side pull. Use extreme caution when attaching a sling to swivel eyebolts, even though the swivel eyebolt does not have restrictions on sling angles.

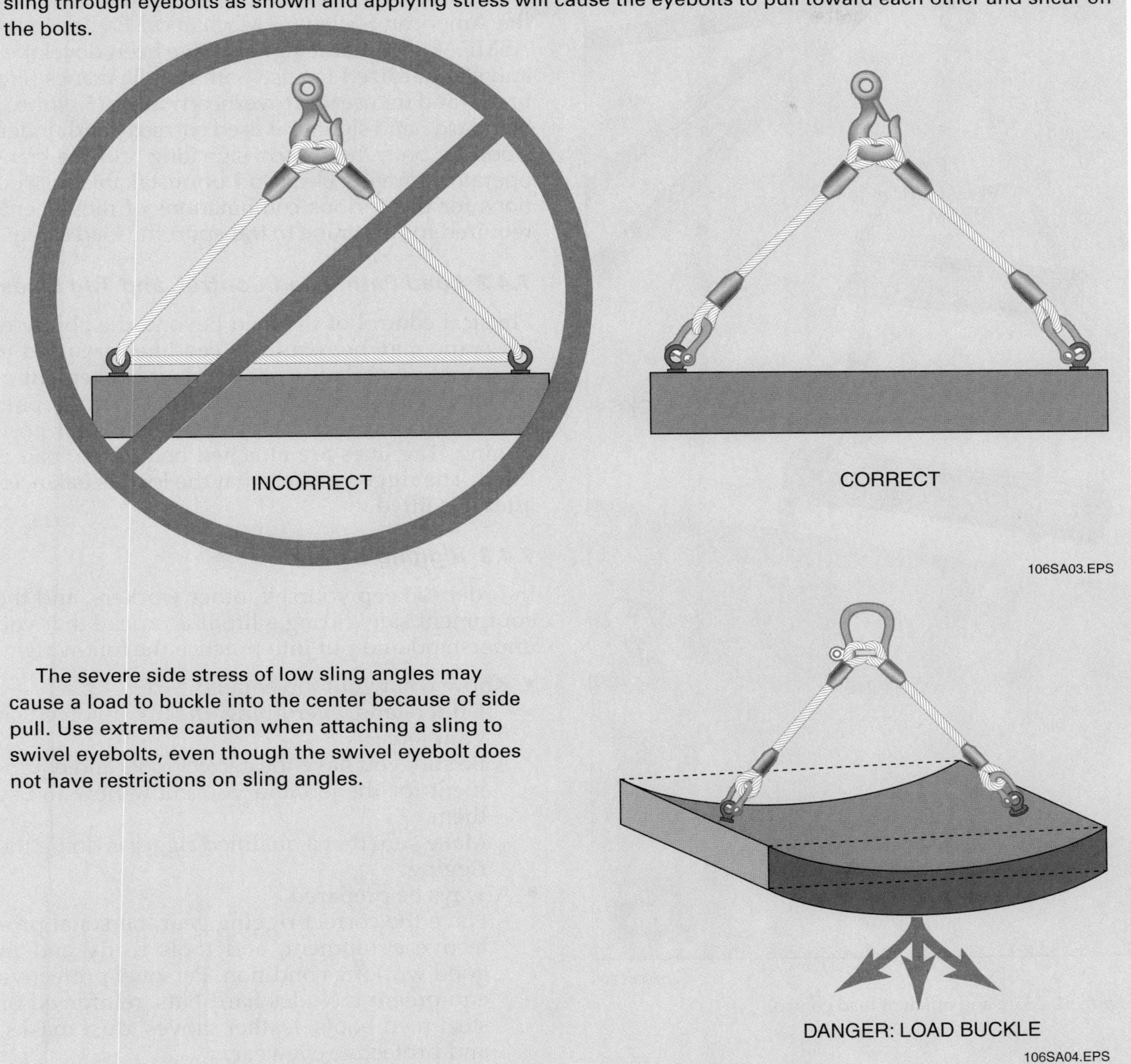

7.4.0 Load Control

Safe and efficient load control (*Figure 61*) involves communication, the use of physical load control techniques, the safe handling of loads, taking landing zone precautions, and following sling disconnection and removal practices.

7.4.1 American Society of Mechanical Engineers (ASME) Hand Signals

Load navigation is accomplished by using common signals—either verbal signals given by radio, or hand signals—to direct a load's movements. Using hand signals to communicate load navigation directions is an integral element in the safe transport and control of all loads. Established hand signals are used for communicating load navigation directions from the signal person to the crane operator. The hand signals used in rigging operations have been developed and standardized by the American Society of Mechanical Engineers, or ASME. **ASME hand signals** have been developed and standardized for use with mobile cranes (*Figure 62*) and for use with overhead cranes (*Figure 63*). Standard hand signals, if used correctly and understood by both the person signaling and the crane operator, provide clear and unmistakable instructions for the various combinations of movements required for the crane to transport the load safely.

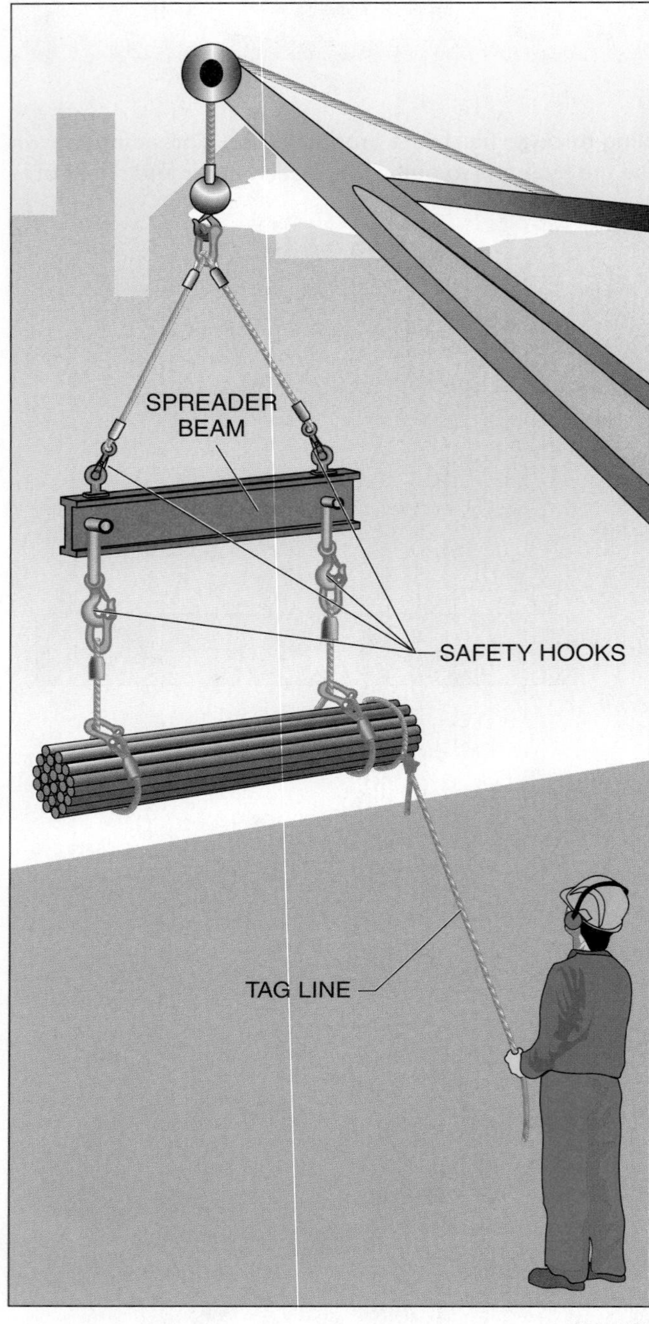

Figure 61 Safe and efficient load control.

7.4.2 Load Path, Load Control, and Tag Lines

Physical control of the load beyond the ability of the crane may be required. Tag lines are used to limit the unwanted or inadvertent movement of the load as it reacts to the motion of the crane, or to allow the controlled rotation of the load for positioning. Tag lines are attached before the load is lifted. The rigger verifies that the load is balanced after it is lifted.

7.4.3 Rigging Safety

In order to keep yourself, other workers, and the equipment safe during a lift, it is crucial that you understand and put into practice the following:

- Know what your job entails.
  - Understand everything that is involved in your job.
  - Be sure you have the correct tools and equipment for the job and you know how to use them.
  - Make sure that a qualified rigger is doing the rigging.
- Always be prepared.
  - Have the correct rigging gear, personal protective equipment, and tools ready and in good working condition. Personal protective equipment includes hard hats, reinforced or steel-toed boots, leather gloves, dust masks, and protective eyewear.

On-Site

Pre-Lift Safety Check

Use these guidelines to ensure your safety and the safety of others on site:

- Use the proper rigging equipment.
- Know how to inspect rigging equipment to ensure that it functions properly.
- Fully understand the principles involved in a lift so that you follow proper rigging procedures.
- Communicate with others effectively before, during, and after a lift.

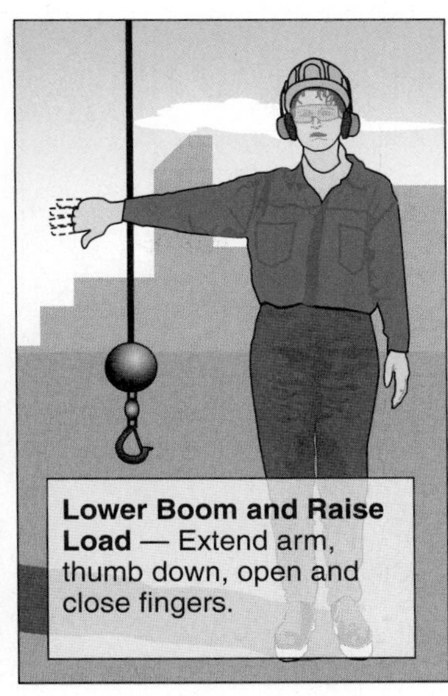

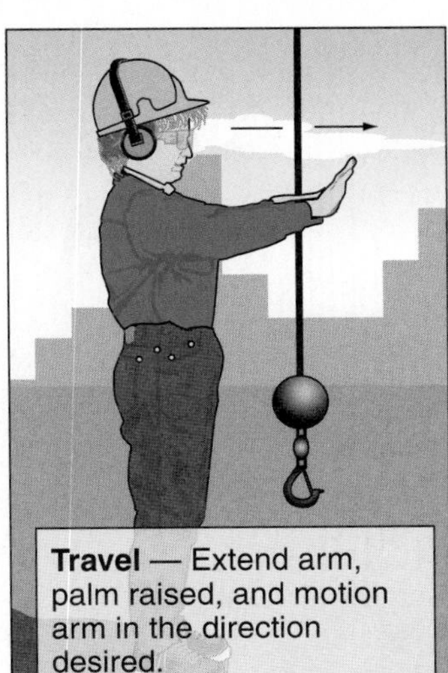

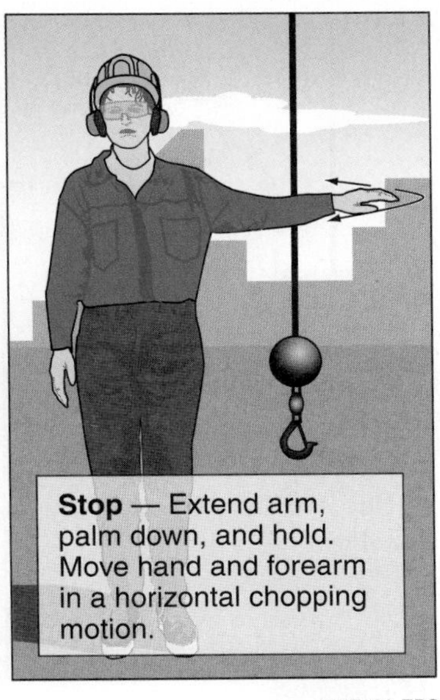

106F62A.EPS

Figure 62 Mobile crane hand signals. (1 of 3)

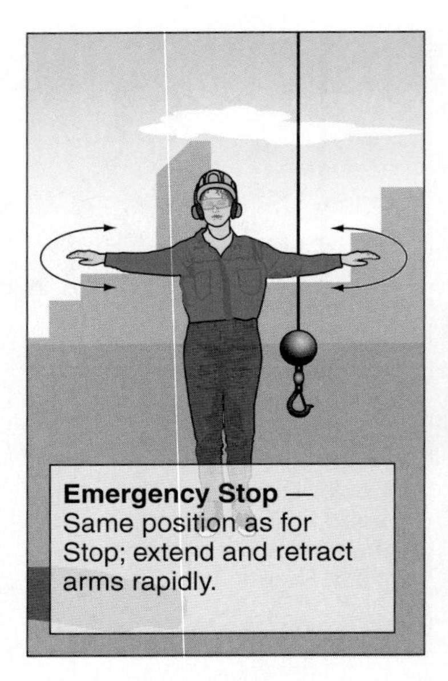

Figure 62 Mobile crane hand signals. (2 of 3)

106F62C.EPS

Figure 62 Mobile crane hand signals. (3 of 3)

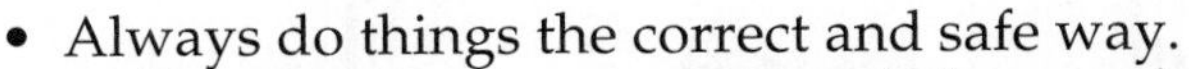

- Always do things the correct and safe way.
  - Always use the recommended procedures for using the lifting equipment and never take shortcuts to save time.
- Understand the importance of good housekeeping.
  - Take care of the equipment and keep it clean and in good working condition.
  - Always properly store tools and equipment.
  - Frequently inspect the rigging equipment. It is the most important thing you can do.
- Understand that safety is everyone's job.
  - Have a safety-oriented attitude.
  - Look out for the safety of not only yourself, but of co-workers as well.

CAUTION

If the lift requires you to work at heights greater than six feet, make sure that you are wearing appropriate fall protection, which includes a safety harness and lanyards.

7.4.4 Risk Management

Risk management is the process of analyzing the work area and the details of the lift prior to beginning the lift. Proper risk management planning can greatly reduce the potential for injuries to workers and pedestrians, and damage to the property, structure, or materials. Risk management allows you to foresee potentially dangerous situations and to develop a plan that will help avoid them.

When considering risk management, keep the following points in mind:

- Measure overhead and width clearances to make sure that the equipment and load can move through the intended route of travel, including aisle dimensions, door widths and heights, and ceiling clearances.
- Check the locations of pipes, power lines, steam lines, or any other structure in the immediate work area.
- Make sure that you will be able to remove the moving equipment once the load has been positioned.
- Control pedestrian traffic and secure the work areas using an appropriate method as approved by the job foreman. Warning tape, rope, warning lights, and a lookout person may all be appropriate.
- If the lift will be made outdoors, make sure that you check the weather forecast several days prior to the lift for wind velocity, extreme temperatures, and factors that may affect visibility, such as rain, snow, fog, and darkness.

CAUTION

Wind can greatly hinder a lift and should be analyzed carefully and accounted for accordingly. Be aware that high winds can create additional forces on a load, and that large dimensional, lightweight loads tend to swing easily.

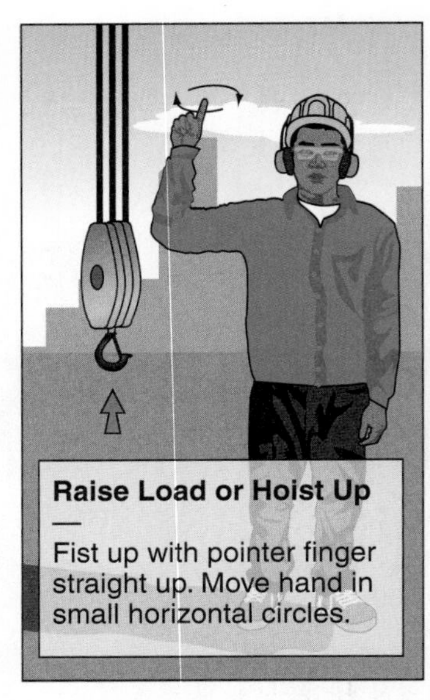

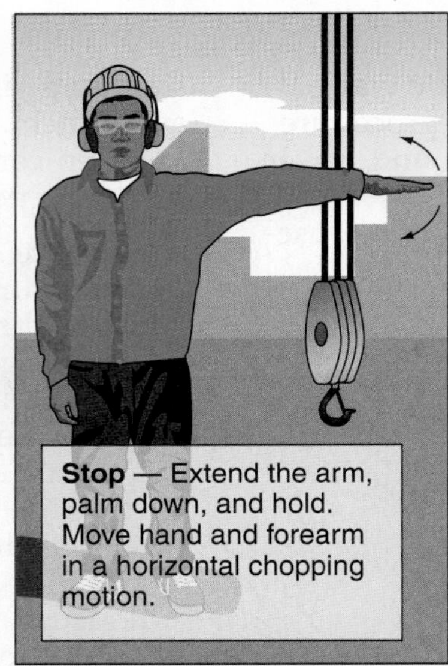

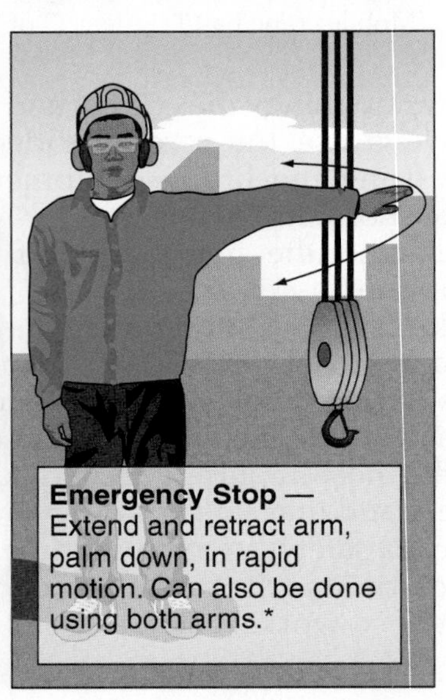

NOTE: The Emergency Stop signal is the only signal that can be given by anyone other than the designated signal person.

106F63.EPS

Figure 63 Overhead crane hand signals.

WARNING!

Never try to lift a load that is frozen to the ground because it can suddenly break loose and shock load the equipment. Shock loading can cause costly damages to the equipment and can create a potentially life threatening situation for the workers.

- Know the **weight capacity** of the structures involved, such as floors, bridges, ramps, elevators, trusses, and beams and columns that the equipment will be attached to. Structural capacities may be posted directly on columns. If they are not, contact the building's engineering department for this information.
- Check the work area for obstacles that may cause you to trip during the lift, such as moveable or projecting objects, slippery surfaces, and holes, dips, or grades along your path of travel.

7.4.5 Safely Handling a Load

Lift safety is the responsibility of everyone involved, including the rigger, equipment operator, and the workers. In order to safely and effectively control a load, you must observe and follow these load-handling safety requirements:

- Keep the front and rear swing paths of the crane clear throughout the lift (*Figure 64*). Most people watch the load when it is in motion, and this prevents them from seeing the back end of the crane coming around. Using safety flags and roping off the area will help prevent accident and injury.
- Keep the landing zone clear of personnel other than the tag line tenders.
- Be sure that the necessary blocking and **cribbing** for the load are in place and set before you position the load for landing. Lowering the load just above the landing zone and then placing the cribbing and blocking is dangerous because the riggers must pass under the load. The layout of the cribbing can be completed in the landing zone before you set the load. Blocking of the load may have to be done after the load is set. In this case, do not take the load stress off the sling until the blocking is set and secured.
- Do not attempt to position the load onto the cribbing by manhandling it.
- Do not move a load if your view of the signal person or load will become restricted at any point.
- Never stand directly under the material being hoisted. Stand as far to one side as possible.
- Never stand between a wall and the load, because a swinging load can crush you.
- Never stand between a load and a roof edge, because a swinging load can knock you off.
- Make sure that no other workers accidentally walk under the load.
- Never reach out over an edge to try to reach the load, because you can fall off.
- Never attempt to receive a load that is spinning out of control.
- Always make sure that there is at least one person on the ground to rig and control the load and one person on the top level to receive it.

CAUTION

Landing the load and disconnecting the rigging are the final steps in any lift. Many rigging accidents and injuries occur at this stage because many riggers mistakenly consider the lift completed just before landing. They do not give their complete attention to the safe landing requirement.

CAUTION

Serious back injuries have resulted from personnel attempting to manually force a load into position or stop even a slight swing of the load. Injuries resulting in crushed hands and other limbs can easily occur should the load shift or swing during the final stages of landing. If the load must be landed with precise control, tag poles, additional tag lines, or hoisting devices should be used.

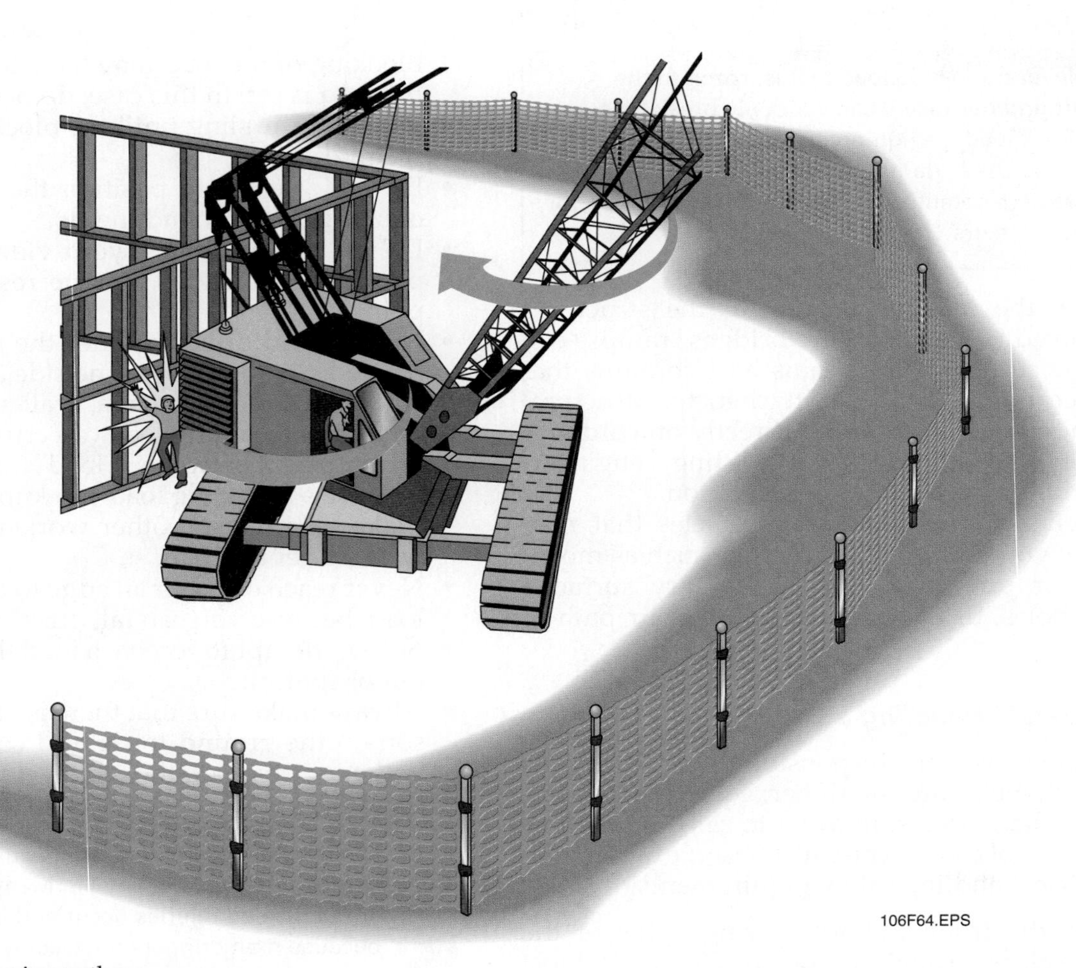

Figure 64 Rear swing path.

On-Site

Keeping Swing Path Clear

Keep the swing path or load path clear of personnel and obstructions. Crane or hoist operators should watch the load when it is in motion. Never work under a suspended load. For more information about swing path safety, refer to OSHA's *CFR 1926, Subpart N and ANSI-ASME B30.2.*

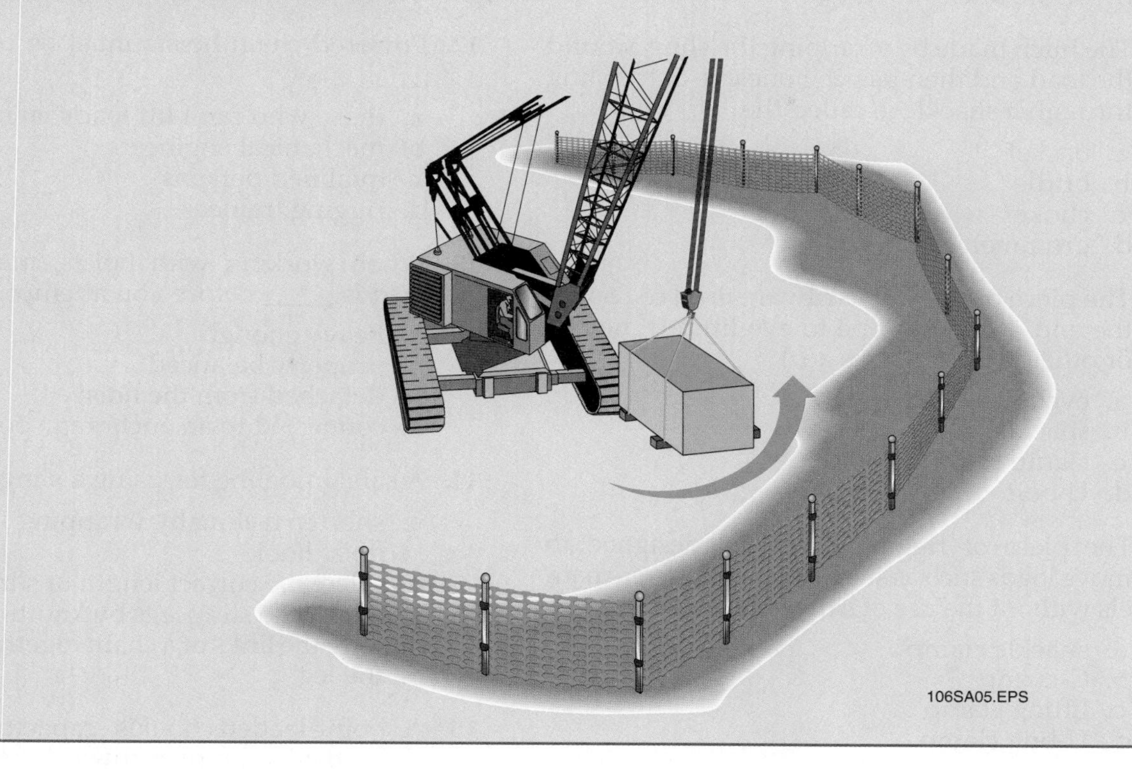

106SA05.EPS

Review Questions

1. Identification tags for slings must include the _____.
   a. type of protective pads to use
   b. type of damage sustained during use
   c. color of the tattle-tail
   d. manufacturer's name or trademark

2. Synthetic slings must be inspected _____.
   a. the first time they are ever used
   b. before every use
   c. every fifth time they are used
   d. by sight only

3. A chain sling that _____ must not be used.
   a. exhibits links stretched from overloading
   b. exhibits paint on more than half of the links
   c. is more than one year old
   d. has an identification tag

4. The type of rope core that is susceptible to heat damage at relatively low temperatures is the _____.
   a. fiber core
   b. strand core
   c. independent wire rope core
   d. supporting core

5. The multiple-leg _____ hitch uses three or four single hitches to increase the stability of the load.
   a. grommet
   b. choker
   c. bridle
   d. basket

6. The _____ allows slings to be connected to the same load without using a spreader beam.
   a. pendulum
   b. choker hitch
   c. basket hitch
   d. bridle hitch

7. The hitch made by wrapping the sling around the load and then passing one eye of the sling through a shackle is called the _____ hitch.
   a. basket
   b. bridle
   c. choker
   d. grommet

8. The piece of rigging hardware used to couple the end of a wire rope to eye fittings, hooks, or other connections is a(n) _____.
   a. eyebolt
   b. shackle
   c. clamp
   d. U-bolt

9. The piece of rigging hardware designed to move loads such as steel plates or concrete panels without the use of slings is called a _____.
   a. shackle clamp
   b. C-clamp
   c. lifting clamp
   d. U-bolt clamp

10. When sling angle decreases, sling stress _____.
    a. stays the same
    b. decreases
    c. increases
    d. becomes zero

11. The greater the sling stress, _____.
    a. the less you have to worry about the lift's safety
    b. the greater the effect on the lift's safety
    c. the safer the lift
    d. the more slings you must use

12. Powered chain hoists must be used only by _____.
    a. those who can't lift loads manually
    b. mechanical engineers
    c. qualified persons
    d. rigging trainees

13. When working with hoists, make sure the load is _____ before you attempt the lift.
    a. heavy enough
    b. properly balanced
    c. detached from the hoist
    d. connected to an anchorage point

14. A safe guideline for using a sling is to _____.
    a. shorten a sling by wrapping it around the hoist hook
    b. select the correct length of sling for the lift
    c. adjust the sling legs by knotting the length
    d. wire the links of a chain together to shorten the leg

15. A point-loaded hook's capacity _____ the closer the load moves toward the point of the hook.
    a. decreases drastically
    b. increases drastically
    c. stays the same
    d. decreases slightly

SUMMARY

Although rigging operations are complex procedures that can present many dangers, a lift executed by fully trained and qualified rigging professionals can be a rewarding operation to watch or participate in. This module has presented many of the basic guidelines you must follow to ensure your safety and the safety of the people you work with during a rigging operation.

The basic approach to rigging involves thorough planning before each lift and executing proper procedures during every lift. The information covered here offers you the groundwork for a safe, productive, and rewarding construction career.

Trade Terms Quiz

Fill in the blank with the correct term that you learned from your study of this module.

1. A simple rope-and-pulley system called a(n) ____________ is used to lift loads.
2. ____________, a type of rigging hardware, are available in three basic designs: unshouldered, shouldered, and swivel.
3. Use a single ring called a(n) ____________ to attach multiple slings to a hoist hook.
4. The ____________ is the distance between the master link of the sling to either the end fitting of the sling or the lowest point on the basket.
5. Also called blocking, ____________ is material used to allow removal of slings after the load is landed.
6. The signal person uses ____________ to communicate load navigation directions to the crane operator.
7. Using a choker hitch and forcing the choke down on loose bundles serves only to drastically increase the ____________ on the choke leg.
8. An endless-loop synthetic web sling, also called a(n) ____________, is made of a single-ply or multiple-ply sling formed into a loop.
9. The way the sling is arranged to hold the load is called the rigging configuration, or ____________.
10. A(n) ____________ hitch uses two or more slings to connect a load to a single hoist hook.
11. To form a choker hitch, wrap the sling around the load and pass one eye of the sling through a(n) ____________ to form a loop around the load.
12. A(n) ____________ uses a pulley system to give you a mechanical advantage for lifting a load.
13. A(n) ____________ is used to move loads such as steel plates or concrete panels without the use of slings.
14. The ____________ is the total amount of what is being lifted.
15. It is important to understand this concept: When ____________ decreases, ____________ increases.
16. The tension applied on the rigging by the weight of the suspended load is called the ____________.
17. The ____________ is the main connection fitting for chain slings.
18. To form an endless-loop web sling, the ends are ____________ together.
19. ____________ equals the lengthwise distance it takes for one strand of wire to make one complete turn around the core.
20. Standard eye-and-eye slings have eyes on the same ____________, whereas twisted eye-and-eye slings have eyes at right angles to each other.
21. The ____________ is the link between the load and the lifting device.
22. The maximum load weight that a sling is designed to carry is called ____________.
23. Examples of ____________ for synthetic slings include a missing identification tag, a puncture, and crushing.
24. Use a(n) ____________ to attach a sling to a load.
25. Often found on a crane, a grooved pulley-wheel for changing the direction of a rope's pull is called a(n) ____________.
26. Be sure to carefully attach hooks to the load to prevent binding of the hook in a(n) ____________, which is a welded structural lifting attachment.
27. When slings are being pulled sideways by the load's weight, this ____________ adds more stress to the sling.
28. The parts of the sling that reach from the attachment device around the object being lifted are called the ____________.
29. A(n) ____________ is a group of wires wound around a center core.
30. Riggers use a(n) ____________ to limit the unwanted movement of the load when the crane begins moving.
31. If the ____________ is showing, the sling is not safe for use.

32. A wire rope sling consists of high-strength steel wires formed into strands wrapped around a supporting ___________.

33. An eyebolt is a piece of rigging hardware with a(n) ___________, which means it has a series of spiral grooves cut into it.

34. To prevent the load from rotating freely, you must use some method of ___________.

35. ___________ slings are made of high-strength steel wires formed into strands wrapped around a core.

36. ___________ is the process of analyzing the work area prior to a lift.

37. The maximum amount of weight a structure can safely support is the ___________.

38. A(n) ___________ is used to determine if an endless sling has been overloaded.

Trade Terms

- ASME hand signals
- Block and tackle
- Bridle
- Bull ring
- Core
- Cribbing
- Eyebolt
- Grommet sling
- Hitch
- Hoist
- Lifting clamp
- Load
- Load control
- Load stress
- Master link
- One rope lay
- Pad eye
- Plane
- Rated capacity
- Rejection criteria
- Rigging hook
- Risk management
- Shackle
- Sheave
- Side pull
- Sling
- Sling angle
- Sling legs
- Sling reach
- Sling stress
- Splice
- Strand
- Stress
- Tag line
- Tattle-tail
- Threaded shank
- Warning yarn
- Weight capacity
- Wire rope

Cornerstone of Craftsmanship

Robert (Bobby) Moffett

Swanson Center for Youth Carpentry Instructor

How did you choose a career in the field?
Construction was natural for me. My dad and older brother were both carpenters. At 12 years old, I had the opportunity to help my dad build a church, and then I was hooked. My dad was a superintendent for a construction company based in Louisiana. When I was not in school, he would take me with him, so I spent a lot of time around construction people. When I turned 13, my brother hired me as a helper, thinking he would discourage me away from carpentry and into college. But I loved it and put myself through high school working summers and holidays.

What types of training have you been through?
After high school I did not receive any formal education. I attended the "University of Hard Knocks."

What kinds of work have you done in your career?
I have done some industrial and commercial carpentry, but mostly custom homes. I moved from Louisiana to Virginia in the 1980s after the economy went sour. With the help of my wife, Peggy, I started Moffett Construction Co. in Manassas, Virginia. It was a small, but successful home improvement company. After six years in Virginia, I got homesick and returned to Louisiana and back to house building. In March 1996, I was offered a job teaching carpentry at Swanson Correctional Center for Youth.

Tell us about your present job and what you like about it.
I love my current job. I teach incarcerated young men ages 16–21. Most of my students have never worked or even been exposed to construction. It gives me great pleasure to see one of my students accomplish his goals. In my present job, instead of building structures, I am building people and I think that is pretty cool.

What factors have contributed most to your success?
Both my dad and brother instilled a strong work ethic in me. Also, in high school some of my teachers said I would never succeed and I was determined to prove them wrong.

What advice would you give to those new to the field?
Remember that to fully succeed, you need formal training along with knowledge gained from experienced professionals. You will learn important fundamentals in the classroom. Out on the job, you will learn from those older craftsmen with well-used tools. If you treat them with respect, they will gladly share their knowledge and experience with you. The more you do, the more you learn; the more you learn, the more valuable you will be on a job site. And never say "that is not my job," or you just may find yourself looking for a new one!

Trade Terms Introduced in this Module

ASME hand signals: Communication signals established by the American Society of Mechanical Engineers (ASME) and used for load navigation for mobile and overhead cranes.

Block and tackle: A simple rope-and-pulley system used to lift loads.

Bridle: A configuration using two or more slings to connect a load to a single hoist hook.

Bull ring: A single ring used to attach multiple slings to a hoist hook.

Core: Center support member of a wire rope around which the strands are laid.

Cribbing: Material used to either support a load or allow removal of slings after the load is landed. Also called blocking.

Eyebolt: An item of rigging hardware used to attach a sling to a load.

Grommet sling: A sling fabricated in an endless loop.

Hitch: The rigging configuration by which a sling connects the load to the hoist hook. The three basic types of hitches are vertical, choker, and basket.

Hoist: A device that applies a mechanical force for lifting or lowering a load.

Lifting clamp: A device used to move loads such as steel plates or concrete panels without the use of slings.

Load: The total amount of what is being lifted, including all slings, hitches, and hardware.

Load control: The safe and efficient practice of load manipulation, using proper communication and handling techniques.

Load stress: The strain or tension applied on the rigging by the weight of the suspended load.

Master link: The main connection fitting for chain slings.

One rope lay: The lengthwise distance it takes for one strand of a wire rope to make one complete turn around the core.

Pad eye: A welded structural lifting attachment.

Plane: A surface in which a straight line joining two points lies wholly within that surface.

Rated capacity: The maximum load weight a sling or piece of hardware or equipment can hold or lift. Also called lifting capacity, working capacity, working load limit (WLL), and safe working load (SWL).

Rejection criteria: Standards, rules, or tests on which a decision can be based to remove an object or device from service because it is no longer safe.

Rigging hook: An item of rigging hardware used to attach a sling to a load.

Risk management: The process of analyzing the work area and the lift prior to the lift being made in order to predict and account for any potential risks.

Shackle: Coupling device used in an appropriate lifting apparatus to connect the rope to eye fittings, hooks, or other connectors.

Sheave: A grooved pulley-wheel for changing the direction of a rope's pull; often found on a crane.

Side pull: The portion of a pull acting horizontally when the slings are not vertical.

Sling: Wire rope, alloy steel chain, metal mesh fabric, synthetic rope, synthetic webbing, or jacketed synthetic continuous loop fibers made into forms, with or without end fittings, used to handle loads.

Sling angle: The angle of an attached sling when pulled in relation to the load.

Sling legs: The parts of the sling that reach from the attachment device around the object being lifted.

Sling reach: A measure taken from the master link of the sling, where it bears weight, to either the end fitting of the sling or the lowest point on the basket.

Sling stress: The total amount of force exerted on a sling. This includes forces added as a result of sling angle.

Splice: To join together.

Strand: A group of wires wound, or laid, around a center wire, or core. Strands are laid around a supporting core to form a rope.

Stress: Intensity of force exerted by one part of an object on another; the action of forces on an object or system that leads to changes in its shape, strain on it, or separation of its parts.

Tag line: Rope that runs from the load to the ground. Riggers hold on to tag lines to keep a load from swinging or spinning during the lift.

Tattle-tail: Cord attached to the strands of an endless loop sling. It protrudes from the jacket. A tattle-tail is used to determine if an endless sling has been stretched or overloaded.

Threaded shank: A connecting end of a fastener, such as a bolt, with a series of spiral grooves cut into it. The grooves are designed to mate with grooves cut into another object in order to join them together.

Warning yarn: A component of the sling that shows the rigger whether the sling has suffered too much damage to be used.

Weight capacity: The maximum amount of weight that a structure can safely support.

Wire rope: A rope made from steel wires that are formed into strands and then laid around a supporting core to form a complete rope; sometimes called cable.

Additional Resources

This module is intended to present thorough resources for task training. The following reference works are suggested for further study. These are optional materials for continued education rather than for task training.

Bob's Rigging and Crane Handbook, Latest Edition. Bob DeBenedictis. Leawood, KS: Pellow Engineering Services, Inc.

High Performance Slings and Fittings for the New Millennium, 1999 Edition. Dennis St. Germain. Aston, PA: I & I Sling, Inc.

Mobile Crane Manual, 1999. Donald E. Dickie, D. H. Campbell. Toronto, Ontario, Canada: Construction Safety Association of Ontario.

Rigging Manual, 1997. Toronto, Ontario, Canada: Construction Safety Association of Ontario.

RATED CAPACITIES

You should never exceed a sling's rated capacity. The rated capacity varies depending upon the type of sling, the size of the sling, and the type of hitch. Rigging operators must know the capacity of the sling they are using. Sling manufacturers usually have charts or tables that contain this information. Sample charts of rated capacities for various types of slings are shown in *Figure A-1*.

Sling Angles

The amount of tension on the sling is directly affected by the angle of the sling. *Figure A-2* shows the effect of sling angles on sling loading. To actually determine the load on a sling, a factor table is used (*Figure A-3*).

In *Figure A-2*, two slings are being used to lift 2,000 pounds. When the slings are at a 45-degree angle, there is 1,414 pounds of tension on each sling. This can be determined mathematically by using the following equation:

$$\text{Sling Tension} = \frac{\text{Load (lbs)} \times \text{Load Angle Factor}}{\text{Number of Legs}}$$

Therefore:

$$\text{Sling Tension} = \frac{2{,}000 \times 1{,}414}{2}$$

$$\text{Sling Tension} = \frac{2{,}828}{2}$$

$$\text{Sling Tension} = 1{,}414$$

Use the weight of the load (2,000 pounds) and multiply it by the corresponding angle factor in *Figure A-3*. Then, divide it by the number of slings (2). The result is 1,414 pounds of tension on each sling.

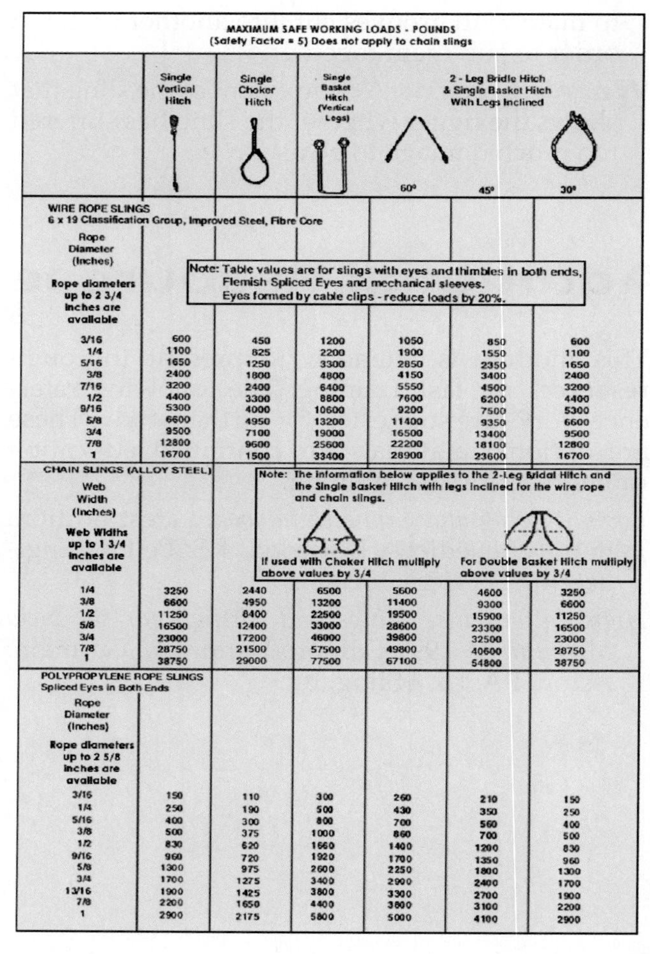

MAXIMUM SAFE WORKING LOADS - POUNDS
(Safety Factor = 5) Does not apply to chain slings

	Single Vertical Hitch	Single Choker Hitch	Single Basket Hitch (Vertical Legs)	2 - Leg Bridle Hitch & Single Basket Hitch With Legs Inclined 60°	45°	30°
WIRE ROPE SLINGS 6 x 19 Classification Group, Improved Steel, Fibre Core — Rope Diameter (Inches). Rope diameters up to 2 3/4 inches are available						
3/16	600	450	1200	1050	850	600
1/4	1100	825	2200	1900	1550	1100
5/16	1650	1250	3300	2850	2350	1650
3/8	2400	1800	4800	4150	3400	2400
7/16	3200	2400	6400	5550	4500	3200
1/2	4400	3300	8800	7600	6200	4400
9/16	5300	4000	10600	9200	7500	5300
5/8	6600	4950	13200	11400	9350	6600
3/4	9500	7100	19000	16500	13400	9500
7/8	12800	9600	25600	22200	18100	12800
1	16700	1500	33400	28900	23600	16700
CHAIN SLINGS (ALLOY STEEL) Web Width (Inches). Web Widths up to 1 3/4 inches are available						
1/4	3250	2440	6500	5600	4600	3250
3/8	6600	4950	13200	11400	9300	6600
1/2	11250	8400	22500	19500	15900	11250
5/8	16500	12400	33000	28600	23300	16500
3/4	23000	17200	46000	39800	32500	23000
7/8	28750	21500	57500	49800	40600	28750
1	38750	29000	77500	67100	54800	38750
POLYPROPYLENE ROPE SLINGS Spliced Eyes in Both Ends — Rope Diameter (Inches). Rope diameters up to 2 5/8 inches are available						
3/16	150	110	300	260	210	150
1/4	250	190	500	430	350	250
5/16	400	300	800	700	560	400
3/8	500	375	1000	860	700	500
1/2	830	620	1660	1400	1200	830
9/16	960	720	1920	1700	1350	960
5/8	1300	975	2600	2250	1800	1300
3/4	1700	1275	3400	2900	2400	1700
13/16	1900	1425	3800	3300	2700	1900
7/8	2200	1650	4400	3800	3100	2200
1	2900	2175	5800	5000	4100	2900

Note: Table values are for slings with eyes and thimbles in both ends, Flemish Spliced Eyes and mechanical sleeves. Eyes formed by cable clips - reduce loads by 20%.

Note: The information below applies to the 2-Leg Bridal Hitch and the Single Basket Hitch with legs inclined for the wire rope and chain slings. If used with Choker Hitch multiply above values by 3/4. For Double Basket Hitch multiply above values by 3/4.

106A01.EPS

Figure A-1 Rated capacities.

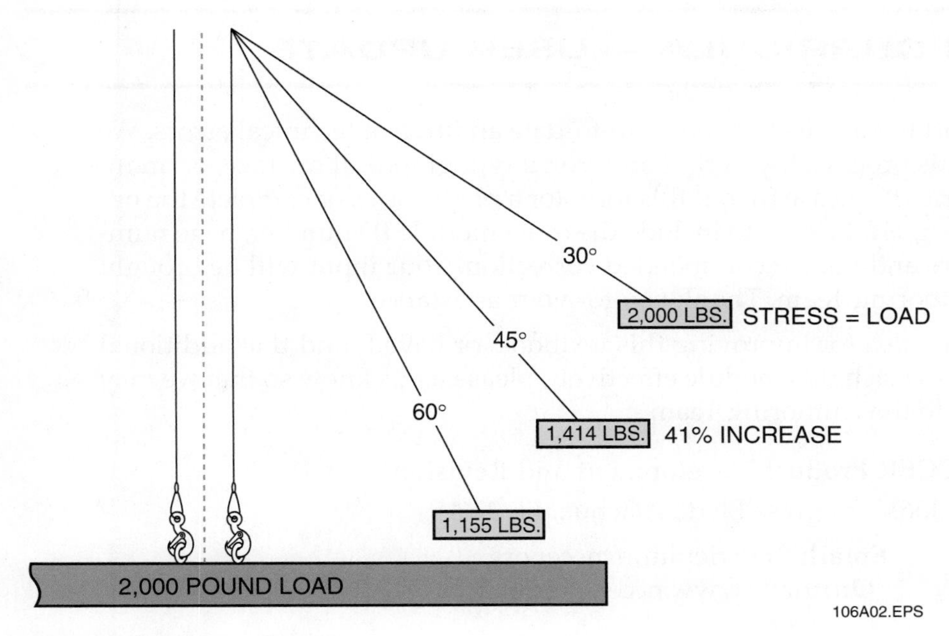

Figure A-2 Sling angles.

SLING ANGLE	LOAD ANGLE FACTOR
30°	2.000
35°	1.742
40°	1.555
45°	1.414
50°	1.305
55°	1.221
60°	1.155
65°	1.104
70°	1.064
75°	1.035
80°	1.015
85°	1.004
90°	1.000

106A03.EPS

Figure A-3 Factor table.

NCCER CURRICULA — USER UPDATE

NCCER makes every effort to keep its textbooks up-to-date and free of technical errors. We appreciate your help in this process. If you find an error, a typographical mistake, or an inaccuracy in NCCER's curricula, please fill out this form (or a photocopy), or complete the online form at **www.nccer.org/olf**. Be sure to include the exact module ID number, page number, a detailed description, and your recommended correction. Your input will be brought to the attention of the Authoring Team. Thank you for your assistance.

Instructors – If you have an idea for improving this textbook, or have found that additional materials were necessary to teach this module effectively, please let us know so that we may present your suggestions to the Authoring Team.

NCCER Product Development and Revision

13614 Progress Blvd., Alachua, FL 32615

Email: curriculum@nccer.org

Online: www.nccer.org/olf

❑ Trainee Guide ❑ AIG ❑ Exam ❑ PowerPoints Other ______________

Craft / Level: Copyright Date:

Module ID Number / Title:

Section Number(s):

Description:

Recommended Correction:

Your Name:

Address:

Email: Phone:

Basic Communication Skills

00107-09

BASIC COMMUNICATION SKILLS

00109-09
Introduction to Materials Handling

00108-09
Basic Employability Skills

00107-09
Basic Communication Skills

00106-09
Basic Rigging

00105-09
Introduction to Construction Drawings

00104-09
Introduction to Power Tools

00103-09
Introduction to Hand Tools

00102-09
Introduction to Construction Math

00101-09
Basic Safety

CORE CURRICULUM

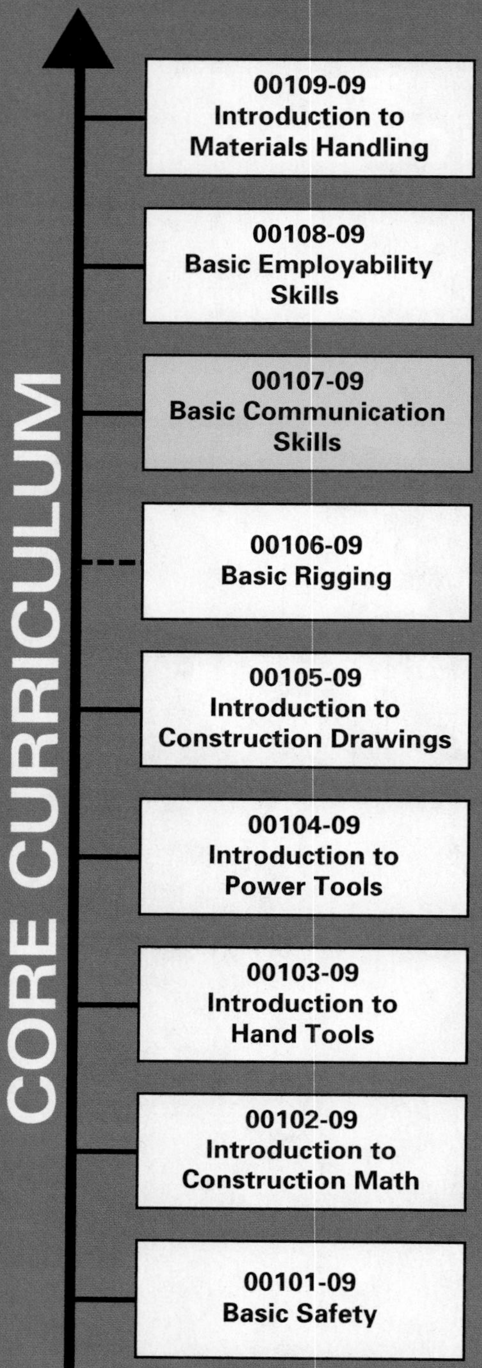

This course map shows all of the modules in the *Core Curriculum: Introductory Craft Skills*. The suggested training order begins at the bottom and proceeds up. Skill levels increase as you advance on the course map. The local Training Program Sponsor may adjust the training order.

Note that Module 00106-09, *Basic Rigging,* is an elective. It is not a requirement for level completion, but it may be included as part of your training program.

Objectives

When you have completed this module, you will be able to do the following:

1. Interpret information and instructions presented in both verbal and written form.
2. Communicate effectively in on-the-job situations using verbal and written skills.
3. Communicate effectively on the job using electronic communication devices.

Trade Terms

Active listening
Appendix
Body language
Bullets
Electronic signature
Font
Glossary
Graph
Index
Italics
Jargon
Memo
Permit
Punch list
Table
Table of contents
Text message

Prerequisites

Before you begin this module, it is recommended that you successfully complete the following: *Core Curriculum,* Modules 00101-09 through 00105-09. Module 00106-09 is an elective and is not a requirement for completion of this course.

Contents

Topics to be presented in this module include:

1.0.0 Introduction . . . 7.1
2.0.0 The Communication Process . . . 7.1
3.0.0 Listening and Speaking Skills . . . 7.2
3.1.0 Active Listening on the Job . . . 7.2
3.1.1 Barriers to Listening . . . 7.5
3.2.0 Speaking on the Job . . . 7.6
3.2.1 Placing Telephone Calls . . . 7.7
3.2.2 Receiving Telephone Calls . . . 7.8
4.0.0 Reading and Writing Skills . . . 7.8
4.1.0 Reading on the Job . . . 7.9
4.2.0 Writing on the Job . . . 7.12
4.2.1 Writing Emails . . . 7.15

1.0.0 INTRODUCTION

Every construction professional learns how to use tools. Depending on your trade, the tools you use could include welding machines and cutting torches, press brakes and plasma cutters, or surveyor's levels and pipe threaders. However, some of the most important tools you will use on the job are not tools you can hold in your hand or put in a toolbox. These tools are your abilities to read, write, listen, and speak.

At first, you might say that these are not really construction tools. They are things you already learned how to do in school, so why do you have to learn them all over again? The types of communication that take place in the construction workplace are very specialized and technical, just like the communications between pilots and air traffic control. Good communications result in a job done safely—a pilot hears and understands the message to change course to avoid a storm, and a construction worker hears and understands the message to install a water heater according to the local code requirements. In a way, you are learning another language—a special language that only trained professionals know how to use.

Here are some specific examples of why these skills are so important in the construction industry:

- *Listening* – Your supervisor tells you where to set up safety barriers but because you did not listen carefully, you missed a spot. As a result, your co-worker falls and is injured.
- *Speaking* – You must train two co-workers to do a task, but you mumble, use words they don't understand, and don't answer their questions clearly. Your co-workers do the task incorrectly, and all of you must work overtime to fix the mistakes.
- *Reading* – Your supervisor tells you to read the manufacturer's basic operating and safety instructions for the new drill press before you use it. You don't really understand the instructions, but you don't want to ask him again. You go ahead with what you think is correct and damage the drill press.
- *Writing* – Your supervisor asks you to write up a material takeoff (supply list) for a project. You rush through the list and don't check what you've written. The supplier delivers 250 feet of PVC piping cut to your specified sizes instead of 25 feet.

As you can see, good communication on the work site has a direct effect on safety, schedules, and budgets. A good communications toolbox is a badge of honor; it lets everyone know that you have important skills and knowledge. And like a physical toolbox, the ability to communicate well verbally and in writing is something that you can take with you to any job. You will find that good communications skills can help you advance your career. This module introduces you to the techniques you will need to read, write, listen, and speak effectively on the job.

2.0.0 THE COMMUNICATION PROCESS

There are two basic steps to clear communication (*Figure 1*). First, a sender sends a spoken or written message through a communication channel to a receiver. (Examples of communication channels include meetings, phones, two-way radios, and email). When the receiver gets the message, he or she figures out what it means by listening or reading carefully. If anything is not clear, the receiver gives the sender feedback by asking the sender for more information.

This process is called two-way communication, and it is the most effective way to make sure that everyone understands what's going on. It sounds simple, doesn't it? So why is good communication so hard to achieve? When we try to communicate, a lot of things—called noise—can get in the

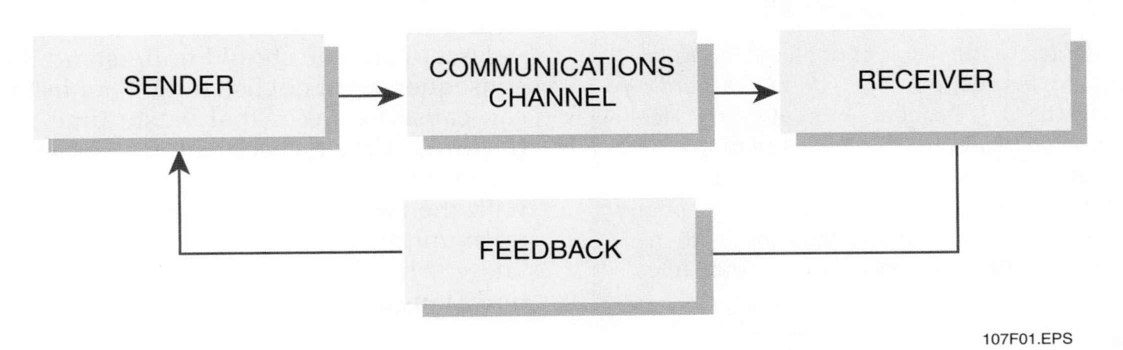

Figure 1 The communication process.

way. Following are some examples of communication noise:

- The sender uses work-related words, or **jargon,** that the receiver does not understand.
- The sender does not speak clearly.
- The sender's written message is disorganized or contains mistakes.
- The sender is not specific.
- The sender does not get to the point.
- The receiver is tired or distracted or just not paying attention.
- The receiver has poor listening or reading skills.
- Actual noise on the construction site makes it physically hard to hear a message.
- There is a mechanical problem with the equipment used to communicate, such as static on a phone or radio line.

3.0.0 Listening and Speaking Skills

Every day on the job can be a learning experience. The more you learn, the more you will be able to help others learn, too (*Figure 2*). An effective method of learning and teaching is through verbal communication—that is, through speaking and listening. As a construction professional, you need to be able to state your ideas clearly. You also need to be able to listen to and understand ideas that other people express. The following are some of the ways that verbal teaching and learning takes place on the job:

- Giving and taking instructions
- Offering and listening to presentations
- Participating in team discussions
- Talking with your co-workers and your supervisor
- Talking with clients

Did You Know?

Paying Attention Prevents Accidents

Many accidents are the result of not listening to, or understanding, instructions. For example, according to a study by the U.S. Occupational Safety and Health Administration (OSHA), over a 10-year period, 39 percent of crane operator deaths resulted from electrocution caused by contact with electrical power lines. This was the single largest cause of death in the study. How many of those accidents could have been prevented if the operator had paid attention to safety instructions before climbing into the cab? Do not become a statistic like that.

107F02.EPS

Figure 2 Teaching and learning are often accomplished by speaking and listening.

Before we discuss some of the ways to become a more effective listener and speaker, you should evaluate your current speaking and listening skills (*Figures 3* and *4*).

At this stage in your career, you will probably do more listening than speaking. You may be wondering why it is so important to be a good listener. The answer is simple: experience. People learn by listening, not by speaking. You are only beginning to learn how the construction industry works, and there is a lot to learn! Teachers, supervisors, and experienced workers can guide you to make sure you are learning what you need to know (*Figure 5*).

3.1.0 Active Listening on the Job

You might think that listening just happens automatically, that someone says something and someone else hears it. However, real listening, the process not only of hearing, but of understanding what is said, is an active process. You have to be involved and paying attention to really listen. Understanding comes from **active listening.** You must develop good listening skills to be able to listen actively. This section presents some tips and suggestions that you can use to develop good listening skills.

First of all, you should understand the possible consequences of not listening. Poor listening skills can cause mistakes that waste time and money (*Figure 6*). Stay focused and do not let your mind wander. One way to do this is to make eye contact with the person who is speaking. Try to keep an open mind; never tune out because you think you know what is being said. Make sure that your **body language,** or your posture and mannerisms, shows that you are paying attention (*Figure 7*). For example, you can nod your head to show that you are listening.

Are You a Good Listener?

Do you have good listening habits? Take the following self–assessment quiz to find out.
Be sure to answer each question honestly.

	Always	Sometimes	Rarely
1. I maintain eye contact when someone is talking to me.	☐	☐	☐
2. I pay attention when someone is talking to me.	☐	☐	☐
3. I ask questions when I don t understand something I hear.	☐	☐	☐
4. I take notes when receiving instructions.	☐	☐	☐
5. I repeat instructions my supervisor has given me to make sure I understand them.	☐	☐	☐
6. I nod my head or say I understand to show others I am listening to them.	☐	☐	☐
7. I let others speak without interrupting.	☐	☐	☐
8. I move to a quieter spot or ask someone to speak up if I am in a noisy location.	☐	☐	☐
9. I put aside what I am doing when someone is speaking to me.	☐	☐	☐
10. I listen with an open mind.	☐	☐	☐

Scoring:
Give yourself 3 points for each "Always" you checked, 2 points for each "Sometimes," and 1 point for each "Rarely." Enter the total for each in the space provided, and then add up your total score.

Always	________	× 3 =	________
Sometimes	________	× 2 =	________
Rarely	________	× 1 =	________
TOTAL			________

Assessment:

25–30 points: You already have excellent listening habits. This section will help you review and practice your listening skills.

18–24 points: You have developed some good listening skills but can benefit from the advice presented in this section.

10–17 points: You have developed some undesirable listening habits. The goal of this section is to help you listen more effectively.

107F03.EPS

Figure 3 Listening skills self-assessment.

Are You a Good Speaker?

How good are your speaking skills? This self-assessment quiz will help you see your speaking strengths and weaknesses.

	Always	Sometimes	Rarely
1. When giving instructions to co-workers, I explain words they might not understand.	☐	☐	☐
2. When giving instructions for a task with several steps, I organize my thoughts first, then give the instructions.	☐	☐	☐
3. I give more details when explaining a task to inexperienced co-workers.	☐	☐	☐
4. When giving instructions to others, I try to keep from sounding like a know-it-all.	☐	☐	☐
5. When giving instructions to others, I encourage them to ask questions about anything they don't understand.	☐	☐	☐
6. I am patient and will explain instructions more than once if necessary.	☐	☐	☐
7. I try to speak more carefully when giving instructions to a co-worker for whom English is a second language.	☐	☐	☐
8. When speaking on the phone or over a two-way radio, I repeat instructions and spell out words when necessary.	☐	☐	☐
9. If someone asks me a question I don't know the answer to, I admit it and then try to find the answer.	☐	☐	☐
10. When I give instructions to others, I am confident, upbeat, and encouraging.	☐	☐	☐

Scoring:
Give yourself 3 points for each "Always" you checked, 2 points for each "Sometimes," and 1 point for each "Rarely." Enter the total for each in the space provided, and then add up your total score.

Always	________	× 3 =	________
Sometimes	________	× 2 =	________
Rarely	________	× 1 =	________
TOTAL			________

Assessment:

25–30 points: You have developed some excellent speaking skills. This section will help you review and practice your speaking skills.

18–24 points: You have developed some good speaking skills but can benefit from the advice presented in this section.

10–17 points: You have developed some undesirable speaking habits. The goal of this section is to help you listen more effectively.

107F04.EPS

Figure 4 Speaking skills self-assessment.

107F05.EPS

Figure 5 Your supervisor can help you learn what you need to know.

107F06.EPS

Figure 6 Use clear and effective statements when speaking on the job.

107F07.EPS

Figure 7 Body language shows whether you are paying attention.

If you do not understand a word or trade term, or if something is not clear, ask questions. Take notes to help you remember better. At the end of the discussion, summarize everything you have heard back to the speaker. That will help you find out immediately if you misunderstood anything.

3.1.1 Barriers to Listening

As you have learned, listening well takes some work on your part. However, even when you have mastered effective listening skills, you will still have to overcome some barriers that will keep the message from getting through. Just like the warning signs on a construction site that indicate hazards to be avoided, there are signs that indicate problems in the process of listening and understanding (*Figure 8*). Following are some of those barriers, along with tips to overcome them:

- *Emotion* – When you're angry or upset, you stop listening. Try counting to 10 or asking the speaker to excuse you for a minute. Go get a drink of water and calm down.
- *Boredom* – Maybe the speaker is dull or overbearing. Maybe you think you know it all already. There is no easy tip for overcoming this barrier. You just have to force yourself to stay focused. Keep in mind that the speaker has important information you need to hear.
- *Distractions* – Anything from too much noise and activity on the site to problems at home can steal your attention. If the problem is noise, ask the speaker to move away from it. If a personal problem is keeping you from listening, concentrate harder on staying focused. In some cases it may help if you explain to your supervisor why you are having trouble concentrating.

107F08.EPS

Figure 8 Learn to read the warning signs of listening problems.

On-Site

Listening in the Classroom

When you are in the classroom, be aware of things that affect your ability to listen well. Is someone on the other side of the room speaking too softly? Is there noise out in the hall? Did your instructor say something you did not understand? Take action to correct these problems. Ask other classmates to face the class when they speak and to speak loudly. Ask permission to close the door against outside noise. Ask your instructor to explain things you don't understand.

- *Your ego* – Do you finish people's sentences for them? Do you interrupt others a lot? Do you think about the things you are going to say instead of listening? That's your ego putting itself squarely between you and effective listening. Be aware of your ego and try to tone it down a bit so you can get the information you need.

3.2.0 Speaking on the Job

Although you can use the skills presented in this module to give a speech if necessary, the term *speaking skills* does not refer to your ability to give a speech or make a presentation to a group of people. Instead, it refers to your ability to communicate effectively, one-on-one, to others on the job every day.

Effective listening depends on effective speaking. After all, you cannot be expected to understand what has not been made clear to you. Look at the following examples of sentences spoken by one worker to another. Which one is the clearest and most effective? Which one would you like to hear if you were the listener?

"Hand me that tool there."
"Hand me the grinder on that bench."
"Hand me the 4-inch angle grinder that's on the bench behind you."

The third example has enough information for you to identify the correct tool and its location. You do not have to ask the speaker, "Which tool? Where is it?" You will not accidentally give the other worker the wrong tool. As a result, time is not wasted trying to clear up confusion. The time it takes for someone to stop what he or she is doing and explain something again because it was not clear the first time is time that the job is not getting done. Time lost this way can add up very quickly.

One of the best ways to learn to speak effectively is to listen to someone who speaks well. Think about what makes that person such an effective speaker. Is it the person's choice of words? Or perhaps their body language? Or their ability to make something complex sound simple? Keep those things in mind when you speak, and they will make a difference for you.

When speaking on the job:

- Think about what you are going to say before you say it.
- As with writing, take time to organize your topic logically.
- Choose an appropriate place and time. For example, if you need to give detailed assembly instructions to your team, pick a quiet place, and do not hold the meeting just before lunch.
- Encourage your listeners to take notes if necessary.

On-Site

Ten Tips for Active Listening

- Maintain eye contact with the speaker.
- Do not allow yourself to be distracted.
- Ask questions.
- Take notes.
- End a conversation by repeating (summarizing) what you heard.
- Nod your head to show that you are listening.
- Recognize that people have valuable contributions to make.
- Recognize the importance of understanding and following instructions on safety.
- Never interrupt other people while they are speaking.
- Put aside your own thoughts, feelings, and opinions when listening.

- Do not over-explain if people are already familiar with the topic.
- Always speak clearly, and maintain eye contact with the person or people you are speaking to.
- Never talk on the phone, **text message**, or listen to music while communicating with the work crew.
- When using jargon be sure that everyone knows what the term means.
- Give your listeners enough time to ask questions, and take the time to answer questions thoroughly.
- When you are finished, make sure that everyone understands what you were saying.

Keep those things in mind when you speak, and they will make a difference for you.

3.2.1 Placing Telephone Calls

You may remember when telephones were anchored to walls and desks. To make or receive a phone call, you had to stop what you were doing and go to the telephone. Today, cell phones allow you to make and receive calls from just about anywhere. A cell phone can be a useful tool on the job site, but keep in mind:

- Cell phones can distract you from your job, so never make or receive personal calls while working.
- Wait until a designated break time to make or receive calls.
- Do not operate cell phones where they would pose a safety hazard, such as while operating a piece of machinery or a power tool or driving a vehicle.

WARNING!

Never make or receive phone calls while driving or operating heavy equipment. Not only is it extremely dangerous, it is also illegal in many states.

When you speak to people face-to-face, you can see them and judge how they react to what you say. When you are on the telephone, you don't have these clues. Effective speaking is all the more important in such cases.

When making a call:

- Start by identifying yourself and ask who you are speaking to.
- Speak clearly and explain the purpose of your call.
- Take notes to help you remember the conversation later (*Figure 9*).

If you leave a message for someone:

- Keep it brief.
- Prepare your message ahead of time so you will know what to say.
- Be sure to leave a number where you can be reached and the best time to reach you.

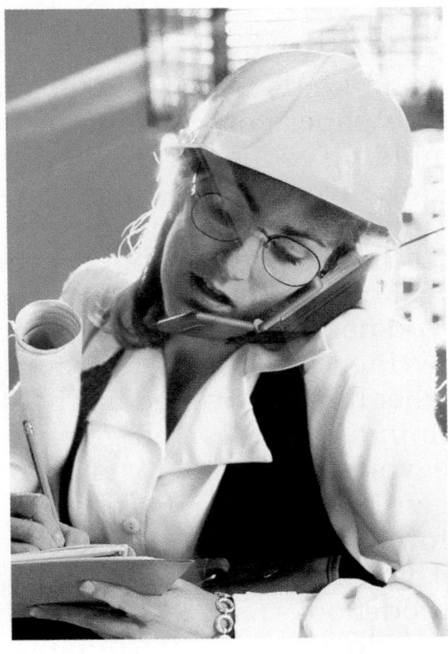

107F09.EPS

Figure 9 Take notes to help you remember important details.

GOING GREEN

Recycling Used Cell Phones

Recycling your mobile devices (cell phones, pagers, PDAs) is a great way to save energy and protect the environment. These mobile devices contain toxic materials. Recycling keeps them out of landfills, cutting down on air and water pollution and greenhouse gas emissions. They contain precious metals, copper, and plastics, which require energy to manufacture and mine.

Over 100 million cell phones are retired each year. The EPA estimates that the energy saved from recycling cell phones alone could power more than 194,000 households for one year.

Your community benefits from cell phone donations. Cell phones can be reconditioned for reuse. They are often donated to charities and law enforcement, and for emergency use by individuals. Recycling is free and some organizations will even pay for your old cell phone.

3.2.2 Receiving Telephone Calls

How you answer a phone call is just as important as how you place a phone call. Remember to be professional and courteous when answering your phone, because you don't know who is going to be on the other end of the line. When you receive a phone call:

- Don't just say "hello." Identify yourself immediately by giving your name and the company name.
- Don't keep people on hold. People resent it. Instead, ask the caller if you can call back at a later time.
- Transfer calls courteously, and introduce the caller to the recipient.
- Keep your calls brief.
- Finally, never talk on the phone in front of coworkers, supervisors, or customers. This is rude and unprofessional.

4.0.0 READING AND WRITING SKILLS

The construction industry depends on written materials of all kinds to carry on business, from routine office paperwork to construction drawings and building codes (*Figure 10*). Written documents allow workers to follow instructions accurately, help project managers ensure that work is on schedule, and enable the company to meet its legal obligations. As a construction professional, you need to be able to read and understand the written documents that apply to your work, whether they are printed in a book or letter, or on a computer. And as you gain professional experience, you will eventually take on responsibilities for writing documents.

107F10.EPS

Figure 10 The written word is extremely important in the construction trade.

Other forms of written construction documents that you are very familiar with are textbooks and codebooks (*Figure 11*). A textbook is an instructional document that contains information that a reader needs to know to carry out a task or a series of tasks. Instruction manuals for tools and installation manuals from manufacturers are other common examples of this type of document. A codebook is a guide that provides the codes and standards for certain areas of construction, such as electrical and building codes. This section reviews some techniques that you may find helpful in reading and writing construction documents.

Did You Know?

Communication is culturally diverse. Our experiences and surroundings, along with context, individual personality, and mood, help to form the ways we learn to speak and give nonverbal messages. For instance, in Europe the correct form for waving hello or goodbye is to keep your hand and arm stationary, palm out, with fingers moving up and down. In America, the common wave is the whole hand in motion moving back and forth. In Europe, the common American hand gesture for a wave means "no," and in Greece it is considered an insult. It is important to remember that not all people communicate in the same way, so be aware of your surroundings when choosing your words and actions.

On-Site

A Sense of Humor

A special tip: Maintain a sense of humor! A good sense of humor will get you through many situations. As a construction professional, you should always take yourself seriously. This means speaking well and conveying the proper professional attitude. But your listeners will always appreciate you more if you show them that you have a sense of humor. Remember, though, never to tell off-color or offensive jokes or play practical jokes.

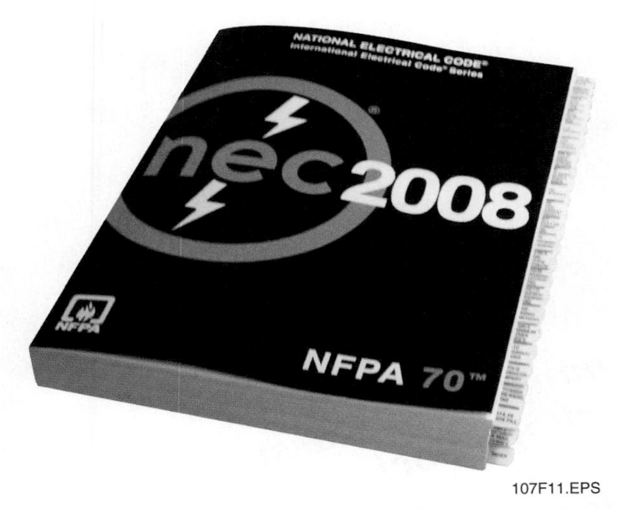

107F11.EPS

Figure 11 Codebooks are critical resources for your job.

4.1.0 Reading on the Job

Say your company has just issued new safety guidelines, or the latest version of the local building code contains new standards that affect how you do your job. You need to be able to understand these documents to perform your job safely and effectively. You should not rely on someone else to tell you how to do your job. Reading is an essential skill for construction professionals.

You may think that you don't have to read on the job, but, in fact, the written word is at the center of the construction trade. The following are some typical examples of things construction workers read on the job:

- Safety instructions
- Construction drawings
- Manufacturer's installation instructions
- Materials lists
- Signs and labels
- Work orders and schedules
- **Permits**
- Specifications
- Change orders
- Industry magazines and company newsletters
- Emails

This section reviews some simple tips and techniques that will help you read faster and more efficiently on the job. And, because someone wrote everything that you read, some of these tips and techniques will help you become a better writer as well. Because you have been reading for some time now, it is probably something you do without even thinking about it. With practice, these guidelines will become second nature, too.

First of all, you should always have a purpose in mind when you read. This will help you find the information faster. For example, say you are looking for a specific installation procedure in a manual. Do not waste time reading other parts of the manual; go straight to the section that deals with that procedure. Read slowly enough to be able to concentrate on what you are reading. Often, technical publications are packed with detailed information, and you do not want to misunderstand it because you rushed.

Most books have special features that can help you locate information, including the following:

- **Table of contents**
- **Index**
- **Glossary**
- **Appendixes**
- **Tables** and **graphs**

Tables of contents are lists of chapters or sections in a book (*Figure 12*). They are usually at the front of a book.

In books, indexes are alphabetical listings of topics with page numbers to show where those topics appear. Indexes are also used to list information in documents other than books—for example, in construction drawings (*Figure 13*). Glossaries are alphabetical lists of terms used in a book, along with definitions for each term.

Appendixes are sources of additional information placed at the end of a section, chapter, or book. Appendixes are separate because the information in them is more detailed or specific than the information in the rest of the book. If the information were placed in the main part of the book, it could distract readers or slow them down too much.

Tables and graphs summarize important facts and figures in a way that lets you understand them at a glance. Tables usually contain text or numbers, and graphs use images or symbols to convey their meaning.

The noise and activity on a construction site or in a shop can easily distract you while you are reading. Try to avoid distractions while you're reading. Radios, cell phones, televisions, conversations, and nearby machinery and power tools can all affect your concentration. Take notes or use a highlighter to help you remember and find important text.

When reading an unfamiliar book, skim or scan the chapter titles and section headings before you start to read. This will help you organize the material in your mind. Look for visual clues that indicate important material. For example, a bold **font** or *italics* indicate important words or information. Bold fonts are letters and numbers that are

107F12.EPS

Figure 12 A table of contents lists the chapters or sections in a book.

heavier and darker than the surrounding text. Italics are letters and numbers that lean to the right, rather than stand straight up.

When reading instructions or a series of steps for performing a task, such as turning on a welding machine, imagine yourself performing the task. You may find the steps easier to remember that way. Be sure to read all the directions through before you begin to follow them. Then you will be able to understand how all of the steps work together.

Always reread what you have just read to make sure you understand it. This is especially important when the reading material is complex or very long. And finally, take a break every now and then. If you read slowly or you find that you are

107F13.EPS

Figure 13 Indexes are often used in construction drawings to identify key information.

getting tired or frustrated, put the material down and do something else to give your mind a break. Reading to understand is hard work!

4.2.0 Writing on the Job

At this stage in your career, you will probably do more reading on the job than writing. But no matter what job you have in the construction industry, you'll eventually have to write something, and writing skills are very important if you want to succeed. Construction workers write work orders, health and safety reports, **punch lists** (written lists that identify deficiencies requiring correction at completion), memoranda (informal office correspondence, or **memos** for short), emails, work orders or change orders, and a whole range of other documents as part of their job (*Figure 14*).

Writing skills will allow you to perform these tasks, too, as you gain experience on the job. They will help you move into positions of greater responsibility. It is never too early to start brushing up on your writing skills, so that when opportunity comes, you will be ready. This section reviews some common writing techniques that you can practice.

Being a good writer is not as hard as you might think. You do not need to be a master of English composition and grammar or have a college degree to write well. A few simple guidelines can make a lot of difference when you are writing on the job. Before you begin to write, organize in your mind what you want to write.

> **Did You Know?**
>
> **The Many Types of Type**
>
> The printed word can be shown in many different ways. Depending on the message that the writer is trying to get across, the font (the shape of the letters), size, and color can all be changed. The shape and style of the letters are often used to show the structure of the text. For example, look at the text used to label the titles, sections, and subsections of this module. You will see that each one has its own style that is used consistently throughout. Some common styles used in text include:
>
> ALL CAPITAL LETTERS
> SMALL CAPITAL LETTERS
> **Bold**
> *Italics*
> Underline

When you start to write, be clear and direct, and use words that people will understand. More words do not mean better writing. Instead of writing "In order to prevent unwanted electrical shocks, workers should make sure they disconnect the main power supply before work commences," write "To avoid being shocked, disconnect the main power supply before starting." Instead of writing "at the present time," simply write "now."

Make your main point easy to find, and include all the necessary details so the reader can understand what you are saying. One way to highlight important information is to list it. Lists can use **bullets** or numbers. Bullets are large dots that line up vertically. When you finish writing, reread what you have written to make sure it is clear and accurate. Check for mistakes and take out words you don't need.

Accuracy is very important in the construction industry. For example, if you are filling out a form to order supplies from a manufacturer's catalog, make sure you use the correct terms and quantities. When writing material takeoffs or other documents that will be used to purchase equipment and assign workers, you need to get it right. Get in the habit of looking over your writing, whether it's a handwritten note, a supply list, a permit, or an email. Errors cost money and time. When writing, ask yourself the following questions:

- Have I identified myself to the reader?
- Have I said why I'm writing this?
- Will the reader know what to do and, if necessary, how to do it?
- Will the reader know when to do it?
- Will the reader know where to do it?
- Will the reader know whom to call with questions?

You should always ask these questions when you are writing something for others to read, even if you are writing something short and simple. Sometimes it is helpful to ask a friend or a co-worker to review something you have written. Another reader can usually spot any mistakes or problems. For example, you might ask a friend to review your job application or resume. You can ask a co-worker to check over a work permit or material takeoff to make sure you haven't left out any information or made any math mistakes. Don't be embarrassed to ask for this type of help, and don't be upset if your friend or co-worker spots a mistake. Professionals are always glad to catch mistakes before those mistakes create bigger problems.

Burning - Welding - Hot Work Permit

Valid from _____ to _____, __________ Master Card No.______________

(am/pm) (am/pm) DATE

1. Work Description

Equipment Location or Area ______________________________

Work to be done:

2. Gas Test	☐ None Required			
	☐ Instrument Check	Test Results	Other Tests	Test Results
	☐ Oxygen 20.8% Min			
	☐ Combustible % LFL			

Gas Tester Signature Date Time

3. Special Instructions ☐ None ☐ Check with issuer before beginning work

4. Hazardous Materials ☐ None What did the line / equipment last contain?

5. Personal Protection ☐ Standard Equipment: welder's hood with long sleeves; cutting goggles

☐ Goggles or Face Shield ☐ Respirator ☐ Forced Air Ventilation

☐ Standby Man ☐ Other, specify: ______________

6. Fire Protection ☐ None Required ☐ Portable Fire Extinguisher

☐ Fire Watch ☐ Fire Blanket ☐ Other, specify: ______________

7. Condition of Area and Equipment

Required Yes	No	THESE KEY POINTS MUST BE CHECKED
		a. Lines disconnected & blanked or if disconnecting is not possible, blinds installed?
		b. Lines steamed, purged, or otherwise properly cleared of combustibles?
		c. Area and equipment satisfactorily clean of oil or combustibles?
		d. Trenches, catch basins & sewer connections properly covered or sealed?
		e. Immediate area and/or area under the work barricaded or roped off?
		f. Adjoining equip. & operations checked to have any effect on the job?
		g. Area fire suppression (fire water and sprinkler system) in service?

Comments

107F14A.EPS

Figure 14 A hot work permit is a typical written product in a construction project. (1 of 2)

Burning - Welding - Hot Work Permit

8. Approval

	Permit Authorization			Permit Acceptance		
	Area Supv.	Date	Time	Maint. Supv./Engineer Contractor Supv.	Date	Time
Issued by						
Endorsed by						
Endorsed by						

9. Individual Review

I have been instructed in the proper Hot Work Procedures

Signed ______________________ Signed ______________________

Persons Authorized to Perform Hot Work ______________________ ______________________

______________________ ______________________

______________________ ______________________

______________________ ______________________

______________________ ______________________

Fire Watch ______________________ ______________________

10. Job Completion

☐ Yes ☐ No Is the work on the equipment completed?

☐ Yes ☐ No Has the worksite been cleaned and made safe?

Worker answering above questions ______________________

Issuer's Acceptance ______________________

Forward to Production Superintendent within 7 days of job completion

107F14B.EPS

Figure 14 A hot work permit is a typical written product in a construction project. (2 of 2)

On-Site

The Five Cs

When you are writing, remember the Five Cs. Your writing should always be:

- Complete
- Clear
- Concise
- Correct
- Considerate of the reader

4.2.1 Writing Emails

With the growing popularity of computers in the construction industry, email (electronic mail) is widely used as a communications tool. You may have used email before; you may have your own personal email account that you use to keep in touch with family and friends. Email is quickly becoming the standard way to send written information, files, and pictures via computer. Like printed letters and memos, email is used to ask and answer questions, and to provide information.

There are advantages and disadvantages to emails. Email delivery is much faster than paper-based delivery, and emails are able to be stored permanently on a hard drive or disk, and are able to be replicated quickly. However, because of their ability to be copied and sent so easily, emails are not as private as paper-based documents. Business emails should not include private, sensitive, or confidential information. They can easily be sent by accident to people who are not authorized to read them.

Another advantage to transmitting information electronically is that it reduces the amount of paper, printing materials, and shipping or processing fees associated with written documents. But in some cases, an email is not considered legally binding. Written documents with handwritten signatures remain the standard for contracts and other formal agreements. However, this situation is changing with the advent of **electronic signatures**.

Because email is now more commonly used in business to communicate with associates, clients, supervisors, and co-workers, it is important to know the proper way to write an email (*Figures 15* and *16*). When writing an email, the nonverbal part of communication (facial expressions, body language, tone of voice) is lost, so it is possible that people may misinterpret the message you are trying to convey if you are not careful. Many of the rules that apply to writing paper-based documents still apply to emails, but there is a certain etiquette involved. Business emailing standards exist so that emails are composed in a professional, efficient, and responsible manner.

The following are some general rules for writing a proper business email. Keep in mind that some rules will vary according to the nature of the business and company culture:

- Write business email the same way you would write a formal business letter or memo.
- Make sure you are sending the email to the correct individual(s) to maintain confidentiality.
- Always start with a clear subject line that indicates the purpose of the message.
- Begin the email by addressing the recipient.

On-Site

He Said *What?*

Being careful about the words you use when writing is especially important when dealing with clients. Always make sure that the person you are communicating with understands the jargon that you are using. To appreciate the humor in this story, you have to know a couple of terms used in the electrical trade. Two types of electrical wires enter a building: drops (overhead wires) and laterals (underground wires).

A power company customer called to report a power outage. The service technician looked at the problem, wrote up the work order, gave the customer a copy, and left. The customer took one look at the work order, got angry, and called the company to complain that its service technician was rude. What made the customer angry? Here is what the service technician wrote on the work order:

INVESTIGATED POWER OUTAGE – DROP DEAD

The technician described the problem accurately, but the customer took it as an insult!

Source: Adapted from *Tools for Success: Soft Skills for the Construction Industry*, Second Edition, 2004, Steven A. Rigolosi. Upper Saddle River, NJ: Pearson.

- Try to keep the body of the email brief and to the point, typically no longer than one screen length, so that the recipient does not have to scroll when reading.
- Use a concise format, such as numbers or bulleted points, to clarify what the email is about and what response or action is required of the recipient.

From: JQSmith@smithcontracting.com Sent: Tue 6/24/2008 1:47 PM
To: WJones@paintersplus.com
Cc:
Subject: Quick Note
Attachments:
View As Web Page

Here are some paint colors and faucets available for your bathroom: Color #1415, Soft Jade; Color #1416, Garden Moss; and Color #1417, Forest Glen. All are available in semi-gloss or eggshell finish. There are also three faucet sets: the Meridian (single handle) $109.88; the Mermaid (dual handles) $83.50; and the Monitor (dual handles) $95.75. All are available in polished brass or polished chrome. I've included paint samples and photos of the faucets. Please tell me your choices by Friday. If you have any questions, call me at 703-555-1212.

107F15.EPS

Figure 15 Bad email example.

From: JQSmith@smithcontracting.com Sent: Tue 6/24/2008 1:47 PM
To: WJones@paintersplus.com
Cc:
Subject: Bathroom Paint and Faucet Options
Attachments:
View As Web Page

Dear Mr. Jones,

The paint colors and faucets available for your bathroom are listed below (photos of faucets and paint colors are attached to this email). Please let me know what you decide by 5:00 pm on Friday, March 22nd. If you have any questions, please do not hesitate to contact me at 703-555-1212.

Paint Colors (Available in semi-gloss or eggshell finish)
- #1415—Soft Jade
- #1416—Garden Moss
- #1417—Forest Glen

Faucet Sets (Available in polished brass or polished chrome)

Model	Price	Handle Style
• Meridian	$109.88	Single
• Mermaid	$83.50	Dual
• Monitor	$95.75	Dual

Regards,
John Q. Smith
Smith Contracting

107F16.EPS

Figure 16 Good email example.

- Write in a positive tone and avoid using negative or blaming statements.
- Do not type in all capital letters, as it gives the impression of shouting.
- Use italicized, bold, or underlined text if you need to stress a point.
- Avoid sarcasm, as it can be easily misunderstood.
- Do not forward junk mail.
- Do not address private issues and concerns.
- When sending an attachment, state the name of the file, and its format.
- Double-check the email for spelling and grammatical mistakes before sending. Remember, once an email is sent it cannot be retrieved.
- Include additional contact information other than your email address, such as a phone number or a mailing address.
- Email is not a substitute for face-to-face interaction. Know when to pick up the phone or schedule a meeting if you cannot express your thoughts or concerns through an email. Never send bad news through email, and don't use email to avoid your responsibilities.

GOING GREEN

Reducing Your Carbon Footprint

Many companies are taking part in the paperless movement. They reduce their environmental impact by reducing the amount of paper they use.

Using email helps to reduce the amount of paper used. There are even postscripts on emails that ask you to reconsider printing that email unless necessary.

Review Questions

1. Good communication on the job site _____.
   a. affects safety, schedules, and budgets
   b. will make you popular
   c. takes too much time
   d. cannot be learned

2. Which of the following is an example of positive verbal communication on the job?
   a. Talking over or interrupting another person
   b. Giving and taking instructions
   c. Tuning out when someone is speaking
   d. Ducking out of team discussions

3. Real listening is _____.
   a. tedious
   b. an active process
   c. unnecessary
   d. an art

4. A supervisor wants an apprentice to insert a plug into a water supply line for a pressure test. The clearest way to instruct the apprentice would be for the supervisor to say, _____.
   a. "Put the plug in"
   b. "Insert that one plug into the pipe"
   c. "Get the water supply line ready for the pressure test"
   d. "Insert the plug into the end of the water supply line"

5. A co-worker asks you to hand her a tool, but does not specify which tool she needs. The most appropriate response would be: _____.
   a. "Get it yourself"
   b. "How am I supposed to figure out which tool you need?"
   c. "I'm sorry; which tool do you need?"
   d. "I am too busy; ask someone else"

6. If you are in the middle of a task and you receive a phone call, _____.
   a. put the caller on hold until you complete the task
   b. ask the caller if you can call back at a later time
   c. tell the caller you don't have time to talk
   d. hand the phone to a co-worker

7. Which is a typical example of an item a construction worker might read for work on a regular basis?
   a. Newspaper
   b. Roadmap
   c. Materials list
   d. Traffic signs

8. When you are looking for a specific installation procedure in a manual, the best way to find the information is to _____.
   a. identify the correct section and go straight there
   b. start reading the book from the beginning
   c. study the terms in the glossary
   d. skim the book and take notes on all important information

9. Many of the rules that apply to writing paper-based documents also apply to _____.
   a. writing emails
   b. leaving voice messages
   c. having a face-to-face conversation
   d. taking orders

10. Which of the following is one disadvantage to using email versus paper-based delivery?
   a. Email is an expensive way to communicate.
   b. The spell checker is unreliable.
   c. Emails cannot easily be replicated.
   d. Emails are not as private as paper-based documents.

SUMMARY

Communications skills—the ability to read, write, listen, and speak effectively—are essential for success in the construction workplace. Effective communication ensures that work is done correctly, safely, and on time. The construction industry relies on a wide variety of written documents. You must be able to read documents that apply to your job. As your responsibilities increase, you will be called on to write some of these documents yourself. This module introduced you to simple tips and techniques that you can use every day to read and write effectively.

Construction professionals state their ideas clearly, and they also listen to the ideas of others. This means not only hearing, but understanding what is said. Real listening is an active process. Listeners are involved with the conversation, and they pay attention to the person who is speaking. Not listening causes mistakes, which waste time and money. Of course, effective listening depends on effective speaking. You should learn how to speak effectively, even though at this stage in your career you will probably be listening more than speaking. No one can understand what is not expressed clearly. When using jargon, be sure that your listeners understand the terms that you are using. Once you master these basic communication skills and add them to your mental toolbox, you will find that you can succeed at more things than you ever thought possible.

Trade Terms Quiz

Fill in the blank with the correct trade term that you learned from your study of this module.

1. Often, books will include additional detailed information in a(n) ____________.
2. In a list, a large dot is called a(n) ____________.
3. Words and numbers can be printed using many different ____________, or type styles.
4. A(n) ____________ includes the definition of terms used in a book.
5. Information can be presented in picture form using a(n) ____________.
6. To find an alphabetical listing of topics in a book, consult the ____________.
7. ____________ are a type style that uses slanted print to emphasize text.
8. Construction workers write ____________ to list deficiencies requiring correction at completion.
9. You can use a(n) ____________ to send an informal written message to someone in your company.
10. You can find a book's chapters and headings listed in a(n) ____________.
11. Numerical or written information can be presented for quick visual scanning in a(n) ____________.
12. A(n) ____________ is another way to sign documents.
13. Providing feedback is an essential part of the ____________ process.
14. Your ____________ silently communicates whether or not you are paying attention to a speaker.
15. As you gain experience, you will learn more of the special ____________ that you can use to communicate with other workers in your trade.
16. A legal document that allows a task, such as constructing a building, to be undertaken is called a(n) ____________.
17. A(n) _____ is used to send a written message from one cell phone to another.

Trade Terms

Active listening
Appendix
Body language
Bullets
Electronic signature
Font
Glossary
Graph
Index
Italics
Jargon
Memo
Permit
Punch list
Table
Table of contents
Text message

Cornerstone of Craftsmanship

Tim Eldridge

North Carolina Department of Public Instruction
Curriculum Consultant

How did you choose a career in the field?
I remember, as a child, I was always taking things apart and putting them back together, attempting to fix anything and everything I could get my hands on. My father is a journeyman electrician, so I worked with him and learned from him every opportunity I could find. I became interested in construction in high school, taking both drafting and carpentry classes.

What types of training have you been through?
I worked at Duke Power as a pipefitter helper, building McGuire Nuclear Station during college. After graduating from Appalachian State University with a bachelor's in industrial arts, I began teaching drafting and electronics at Eastern Wayne High School. The next year I moved to East Gaston High School to teach drafting and continued my education, attending North Carolina Agricultural and Technical State University to complete my master's degree in vocational education. At the same time, I became a licensed general contractor in North Carolina.

I started a small residential construction business to supplement my teaching salary and began designing and building several custom houses each year. We also do major renovations and additions.

What kinds of work have you done in your career?
My career has been progressive and rewarding. I was a pipefitter while attending college. For 16 years, I taught drafting and electronics. I have been a general contractor in North Carolina. I was an apprenticeship manager/trainer at Julius Blum, Inc., a furniture hardware manufacturer. I served as Assistant Bureau Chief for the North Carolina Department of Labor Apprenticeship and Training Bureau and I currently serve as Trade and Industrial Education Consultant for the North Carolina Department of Public Instruction.

Tell us about your present job and what you like about it.
Presently, I am the Trade and Industrial Education Consultant – Construction for the North Carolina Department of Public Instruction. I write curricula, provide technical assistance, and provide professional development for construction teachers in North Carolina.

What factors have contributed most to your success?
I have been fortunate to have worked in so many construction-related areas throughout my career. My background in industrial arts at Appalachian State and my earlier years working with my father helped me to create options and take advantage of opportunities as they arose. My determination to succeed and become successful in my career, and the drive to be the best at whatever I do, are the characteristics that I attribute to my success.

What advice would you give to those new to the field?
Construction and related careers have produced more millionaires than any other field. Those willing to work hard, learn a trade, and take advantage of business opportunities, can be very successful in construction and very satisfied with their accomplishments.

Tell us some interesting career-related facts or accomplishments:
After graduating from Appalachian State in 1980, I received over 50 inquiries from school systems seeking teachers throughout North Carolina, South Carolina, and Virginia. I initiated and implemented the first TV/broadcasting class in North Carolina to produce live daily announcements in every classroom at East Gaston High School. For four consecutive years, I was awarded the North Carolina Apprenticeship Consultant of the Year.

Trade Terms Introduced in This Module

Active listening: A process that involves respecting others, listening to what is being said, and understanding what is being said.

Appendix: A source of detailed or specific information placed at the end of a section, a chapter, or a book.

Body language: A person's physical posture and gestures that reflect how that person is feeling.

Bullets: Large, vertically aligned dots that highlight items in a list.

Electronic signature: A signature that is used to sign electronic documents by capturing handwritten signatures through computer technology and attaching them to the document or file.

Font: The type style used for printed letters and numbers.

Glossary: An alphabetical list of terms and definitions.

Graph: Information shown as a picture or chart. Graphs may be represented in various forms, including line graphs and bar charts.

Index: An alphabetical list of topics, along with the page numbers where each topic appears.

Italics: Letters and numbers that lean to the right rather than stand straight up.

Jargon: Specialized terms used in a specific industry.

Memo: Informal written correspondence. Another term for memorandum (plural: memoranda).

Permit: A legal document that allows a task to be undertaken.

Punch list: A written list that identifies deficiencies requiring correction at completion.

Table: A way to present important text and numbers so they can be read and understood at a glance.

Table of contents: A list of book chapters or sections, usually located at the front of the book.

Text message: A short message (160 characters or fewer) sent from a cell phone.

Additional Resources

This module is intended to present thorough resources for task training. The following references are suggested for further study. These are optional materials for continued education rather than for task training.

Communicating at Work. Tony Alessandra and Phil Hunsaker. New York, NY: Simon and Schuster.

Communicating in the Real World: Developing Communication Skills for Business and the Professions. Terrence G. Wiley and Heide Spruck Wrigley. Englewood Cliffs, NJ: Pearson.

Communication Skills for Business and Professions. Paul R. Timm and James A. Stead. Upper Saddle River, NJ: Pearson.

Elements of Business Writing. Gary Blake and Robert W. Bly. New York, NY: Collier.

Improving Business Communication Skills. Deborah Britt Roebuck. Upper Saddle River, NJ: Pearson.

CONTREN® LEARNING SERIES – USER UPDATE

NCCER makes every effort to keep these textbooks up-to-date and free of technical errors. We appreciate your help in this process. If you have an idea for improving this textbook, or if you find an error, a typographical mistake, or an inaccuracy in NCCER's Contren® textbooks, please write us, using this form or a photocopy. Be sure to include the exact module number, page number, a detailed description, and the correction, if applicable. Your input will be brought to the attention of the Technical Review Committee. Thank you for your assistance.

Instructors – If you found that additional materials were necessary in order to teach this module effectively, please let us know so that we may include them in the Equipment/Materials list in the Annotated Instructor's Guide.

Write: Product Development and Revision
National Center for Construction Education and Research
3600 NW 43rd St., Bldg. G, Gainesville, FL 32606

Fax: 352-334-0932

E-mail: curriculum@nccer.org

Craft ____________ Module Name ____________

Copyright Date ____________ Module Number ____________ Page Number(s) ____________

Description ____________

____

____

____

(Optional) Correction ____________

____

____

(Optional) Your Name and Address ____________

____

____

Basic Employability Skills

00108-09

CORE CURRICULUM

00109-09
Introduction to Materials Handling

00108-09
Basic Employability Skills

00107-09
Basic Communication Skills

00106-09
Basic Rigging

00105-09
Introduction to Construction Drawings

00104-09
Introduction to Power Tools

00103-09
Introduction to Hand Tools

00102-09
Introduction to Construction Math

00101-09
Basic Safety

This course map shows all of the modules in the *Core Curriculum: Introductory Craft Skills.* The suggested training order begins at the bottom and proceeds up. Skill levels increase as you advance on the course map. The local Training Program Sponsor may adjust the training order.

Note that Module 00106-09, *Basic Rigging,* is an elective. It is not a requirement for level completion, but it may be included as part of your training program.

Objectives

When you have completed this module, you will be able to do the following:

1. Explain the role of an employee in the construction industry.
2. Demonstrate critical thinking skills and the ability to solve problems using those skills.
3. Demonstrate knowledge of computer systems and explain common uses for computers in the construction industry.
4. Define effective relationship skills.
5. Recognize workplace issues such as sexual harassment, stress, and substance abuse.

Trade Terms

Absenteeism
Amphetamine
Barbiturate
Browser
CNC
Compromise
Computer literacy
Confidentiality
Constructive criticism
Critical thinking skills
Database
Desktop publishing
Documentation
EDM
Entrepreneur
Goal-oriented
Hallucinogen
Handheld computer
Harassment
Hard drive
Hardware
Initiative
Internet
Leadership
Methamphetamine
Mission statement
Operating system
PDA
Plotter
Processor
Professionalism
Reference
Scanner
Self-presentation
Sexual harassment
Software
Spreadsheet
Tactful
Tardiness
Teamwork
USB flash drive
Wireless
Word processor
Work ethic

Prerequisites

Before you begin this module, it is recommended that you successfully complete the following: *Core Curriculum,* Modules 00101-09 through 00105-09, and Module 00107-09. Module 00106-09 is an elective and is not a requirement for completion of this course.

Contents

Topics to be presented in this module include:

1.0.0 Introduction 8.1
2.0.0 The Construction Business 8.1
2.1.0 Entering the Construction Workforce 8.2
2.2.0 Entrepreneurship 8.4
3.0.0 Critical Thinking Skills 8.7
3.1.0 Barriers to Problem Solving 8.7
3.2.0 Solving Problems Using Critical Thinking Skills 8.8
3.3.0 Problems with Planning and Scheduling 8.10
4.0.0 Computer Skills 8.11
4.1.0 Computer Terms 8.11
4.2.0 Software 8.12
4.3.0 Email 8.12
4.4.0 Computers in the Construction Industry 8.12
5.0.0 Relationship Skills 8.14
5.1.0 Self-Presentation Skills 8.14
5.1.1 Personal Habits 8.14
5.1.2 Work Ethic 8.15
5.1.3 Lateness and Absenteeism 8.15
5.2.0 Conflict Resolution 8.16
5.2.1 Resolving Conflicts with Co-Workers 8.17
5.2.2 Resolving Conflicts with Supervisors 8.17
5.3.0 Giving and Receiving Criticism 8.18
5.3.1 Offering Constructive Criticism 8.18
5.3.2 Receiving Constructive Criticism 8.19
5.4.0 Teamwork Skills and Collaboration 8.19
5.5.0 Leadership Skills 8.20
6.0.0 Workplace Issues 8.23
6.1.0 Harassment 8.23
6.2.0 Stress 8.23
6.3.0 Drug and Alcohol Abuse 8.24

1.0.0 INTRODUCTION

To succeed in the construction industry, you need to know how to do your job well. This means more than just using tools, machines, and equipment with skill. You must also know how to do many things that at first might not seem to have anything to do with being a construction professional. For example, understanding how a construction business works, thinking critically, and being able to solve problems are vital skills. So are working safely, presenting yourself well, getting along with your co-workers and supervisors, and being flexible. Computer skills are also important.

At first, these skills might seem unrelated to construction work, but if you think about it, every good role model exhibits some or all of them. These skills are just as real as your ability to operate a power tool—people can see whether you know how to use it, and they can see the results. The same goes for your behavior on the job. This module introduces the interpersonal skills that you must master to succeed in the construction industry.

2.0.0 THE CONSTRUCTION BUSINESS

The construction industry creates the environment in which we live and work. It is made up of a wide variety of specialized skills that all share a single goal: to make our lives more comfortable (*Figure 1*). The construction industry is the second-largest industry in the United States, larger than the steel and automotive industries combined. The construction industry employs more than five million people. According to the U.S. Census Bureau, over $1 billion was spent on new construction in November of 2008.

The construction industry consists of independent companies of all sizes that specialize in one or more types of work. For example, a company might install heating, ventilating, and air conditioning systems in residences. Another company might oversee the construction of an entire office complex. Sheet metal shops, civil engineering firms, well and septic system installation specialists, welding and cutting specialists—all these and more form the construction industry.

Your company might have a **mission statement**, which explains how it does business. The

GOING GREEN

Green-Collar Jobs

In recent years, there has been a big shift to becoming environmentally friendly. This has created 22 new sectors in the U.S. economy, including everything from recycling to energy retrofits; green building; solar installation/maintenance; water retrofits; and whole house performance (HVAC). These green-collar jobs are in green businesses and all involve work that directly improves environmental quality. Any job that involves the design, manufacture, installation, operation, and/or maintenance of renewable energy and energy-efficient technologies are considered green-collar jobs. Many of these jobs have opened new doors or removed barriers to employment.

In the construction industry, opportunities exist in the following areas:

- Installation, construction, maintenance, and repair of energy retrofits, HVAC, solar panels, and whole home performance
- Installation, construction, maintenance, and repair of water conservation and adaptive grey water reuse
- Construction, carpentry, and demolition in green building and recycling

Did You Know?

How much does it cost your boss to employ you? If your base pay is $18.00 per hour, you are costing your employer $24.30 per hour. The following factors add to the cost of employing you:

- Workers' compensation
- Insurance
- Social Security
- Equipment
- Vacation days
- Sick leave

These factors make hourly costs to your employer 35 percent higher than your salary. The additional 35 percent must come from company profits. Therefore, if your employer does not make a profit, it's unlikely that you will get a raise. You may even lose your job.

You can help your company earn a profit if you do the following:

- Avoid accidents.
- Follow company rules and policies.
- Work quickly and efficiently.
- Use company property appropriately.
- Meet deadlines.
- Show up for work on time.

108F01.EPS

Figure 1 The construction industry offers many career options.

company may describe its philosophy in an employee handbook or other materials you read when you joined the company. However your company describes its role in the construction industry, you should make sure that you understand not only its mission but also your role in the company. In addition, you need to be familiar with your company's policies and procedures. Knowing these things will help you to be a better employee.

2.1.0 Entering the Construction Workforce

When you complete your training, you will need to find a job where you can put your new skills to good use. Perhaps your training is part of your job, or perhaps you have a job waiting for you when you finish your training. Even if you already have a job, brushing up on basic job-search skills is a good idea. You never know when you will want to use them.

When looking for a job, try to find one that is a good match for your skills and experience. For instance, if you are a carpenter with two years' experience, do not apply for a journey-level position. Some employers require that workers are able to legally operate vehicles, so make sure you have a valid driver's license.

You can find jobs listed in newspapers and trade magazines. You can also search for them on the **Internet**. Many companies that announce jobs on the Internet allow you to apply online.

> **NOTE**
> Some employers require applicants to apply online. If you don't have a computer or access to the Internet, check local businesses or the library, or ask if a friend or relative has Internet access.

Ask your current supervisor and your co-workers if they would act as **references** for you. A reference is a person who can vouch for your skills, experience, and work habits. Make sure that your resume is up-to-date, well organized, and easy to read (*Figure 2*). Apply the effective

John Q. Smith

123 Main Street
Anytown, MD 67890
(555) 111-2233 – Home phone
(555) 333-4455 – Cell phone
johnqsmith@email.com

EMPLOYMENT OBJECTIVE

- To obtain a position within a construction company where I can use my carpentry and custom woodworking skills.

SKILLS AND QUALIFICATIONS

- Able to work independently and with a crew.
- Practical knowledge of carpentry hand and power tools, measuring tools, and woodworking tools.
- Trained and certified to use powder-actuated tools.
- Qualified to read residential and commercial construction drawings.
- Experience with energy efficient and conservation building techniques.
- Knowledge of LEED standards.

WORK EXPERIENCE

Journeyman Carpenter *May 2004—present*
LJL Construction, 123 Hammer Heights Road, Fairfax, VA
800-222-2222

- Built custom cabinets and closet systems
- Framed single family homes
- Trained apprentice carpenters

Carpenter *June 1998—May* 2004
Hammer Company, 456 Lathe Lane, Philadelphia, PA
866-555-1234

- Built closets and storage systems for commercial storage units

Carpenter *February 1994 – June 1998*
Blue Ridge Carpenters, 789 Mountain Road, Harrisburg, PA
888-123-4567

- Framed condo units
- Built fences
- Raise wood walkways

CONSTRUCTION EDUCATION

- Carpentry apprenticeship. MK Builders Inc., Gettysburg, PA. Certificate, 1994.
- Professional carpentry certification program. ABC Training Corp., Philadelphia, PA. Certificate, 1992.
- Powder-actuated tools certification program. All States Community College, Marietta, PA. Certificate, 1992.

108F02.EPS

Figure 2 Resume.

writing techniques that you learned in the *Basic Communication Skills* module when writing your resume. You can also find excellent sources in print and online to help you write a resume.

By carefully selecting jobs that you are qualified for and by submitting an accurate, well-written resume, you improve your chances of being called for interviews (*Figure 3*). The effective communication skills that you learned in the *Basic Communication Skills* module will help you present yourself well. In this module, you will learn additional skills that will help you to make a good impression at your interviews.

Good resumes and interviews lead to job offers. You will need to evaluate the offers to select the one that works best for you. The following are some of the questions to consider when selecting a job:

- Is the salary enough to meet my needs?
- Does the company offer a benefits package that covers what I need?
- Will the work be interesting and challenging but not more than I can handle?
- Does the company have a good reputation in the industry?
- Do the people appear to be nice to work with?
- Does the company offer training?
- What is the company's safety record?

108F03.EPS

Figure 3 Job interviews are an important part of the hiring process.

Did You Know?

Hyperstories

Innovative new methods are being developed to train people in key non-technical skills. Many of these methods involve audiovisual techniques. For example, a firm called SUBSTANZ® developed Hyperstories, which are interactive films. Students watch a film of an event, such as a team meeting, on a CD-ROM or over the Internet. By selecting one of the characters, students can watch the same event replayed from that character's point of view. Students can even change the outcome of a story by choosing how the people in the story interact. The ability to see different points of view allows students to understand that situations are often not as simple as they may seem.

2.2.0 Entrepreneurship

Many of the companies in the construction industry are considered small businesses. You may work for a small business right now, or you may

On-Site

Resume Writing Tips

A good resume can make the job of finding a job much easier. The following are some guidelines for resume writing:

- Have someone read your resume to make sure it is free of spelling errors, typos, and poor grammar.
- Use short sentences and bulleted lists to describe your skills.
- Make sure that all contact information is correct and easy to find on the resume.
- Check your resume for inaccurate or missing dates.
- Format your resume chronologically, putting your last job first.
- Use a readable font, such as Times New Roman or Arial, for the text of the resume.
- Remove any personal information not related to the job from your resume.
- Apply only for jobs for which you are qualified.

GOING GREEN

Bank of America Goes Platinum

A new one-billion-dollar building in the city of New York, scheduled for occupancy in 2009, is destined to achieve the world's first LEED Platinum certification for a high-rise office building. The Bank of America Tower at One Bryant Park is a 54-story, 2.2-million-square-foot (204,387 square meters) skyscraper. The tower's architectural spire reaches 1,200 feet (366 meters); in New York City, only the Empire State Building is taller.

The building is being developed with every facet of environmental design taken into consideration. The architectural designers began by considering what natural resources could be applied (sun, air, rain, wind), the technologies to implement, and the determination that it be built largely with recycled and recyclable materials.

The skyscraper will use 50% less energy and generate up to 70% of its annual energy needs using a natural gas co-generation plant. Heat will be contained, in contrast to the typical power plant, which loses 73% of its heat. At night, the co-generation plant will make ice, which will melt the next day to chill liquid in the cooling system.

The building's faces will be designed to work with the site's different solar patterns in winter and summer. Floor-to-ceiling windows made of high-performance, low-emissivity glass filter out infrared rays, while admitting abundant daylight. Because low-iron glass is used, the view to the outside will be exceptionally clear, providing health and productivity benefits to occupants.

A greywater system will capture and reuse rainwater. Waterless urinals will save eight million gallons of water per year and reduce CO_2 emissions by 144,000 pounds per year. Sophisticated air ventilation systems enable each individual workstation to have its own manually-adjusted heating/cooling control. The building's giant air filtration systems mean that air exiting the building will be cleaner than when it came in.

Many other features that enhance the building's efficiency, comfort, and commitment to environmentally responsible design will be implemented. The former governor of New York, George Pataki, noted that the project "is a shining example of how you can create jobs while also protecting the environment." It is expected to create one million jobs by the end of the decade, of which 7,000 are new construction jobs.

108SA01.EPS

even be thinking about starting your own someday. A small business is defined as follows:

- It is independently owned and operated.
- It is not the main or largest company in its field.
- It has fewer than 500 employees.
- It makes less than a certain amount of money per year, depending on the type of work it does (less than $27.5 million for general and heavy construction companies; less than $11.5 million for special trade contractors).

People who start and run their own businesses are called **entrepreneurs**. Before starting a business, an entrepreneur must evaluate the need for such a business. Do other companies already fill that need? If so, will a new company be able to offer something different or better than existing businesses? A new company may not survive long unless it stands out from the competition.

If an entrepreneur believes that there is a need that a new business can fill, the next challenges are to select a location for the business and set up

On-Site

Completing an Application

Completing an application is one of the first steps toward getting a job. Applications require specific information about your work and training history. Be sure to have the following information available when applying for a job:

- Personal identification, such as your driver's license, a passport, or state or military ID.
- Your contact information, including your current phone number and address.
- An up-to-date printed copy of your resume on quality paper.
- The names and locations of the school(s) or training classes you attended.
- The dates you attended the school(s) or classes and the subject(s) you studied.
- The names, addresses, and phone numbers of previous employers.
- The dates of employment for each employer listed.
- The names and contact information for references (non-family members) that can vouch for your personal and professional abilities.

plans for marketing, finances, and management. An organizational chart can help the owner and the employees understand the structure of the company (*Figure 4*). Only when all these elements are in place can the business open. A successful entrepreneur needs to have a lot of energy and be able to adapt and respond to changes in the business situation.

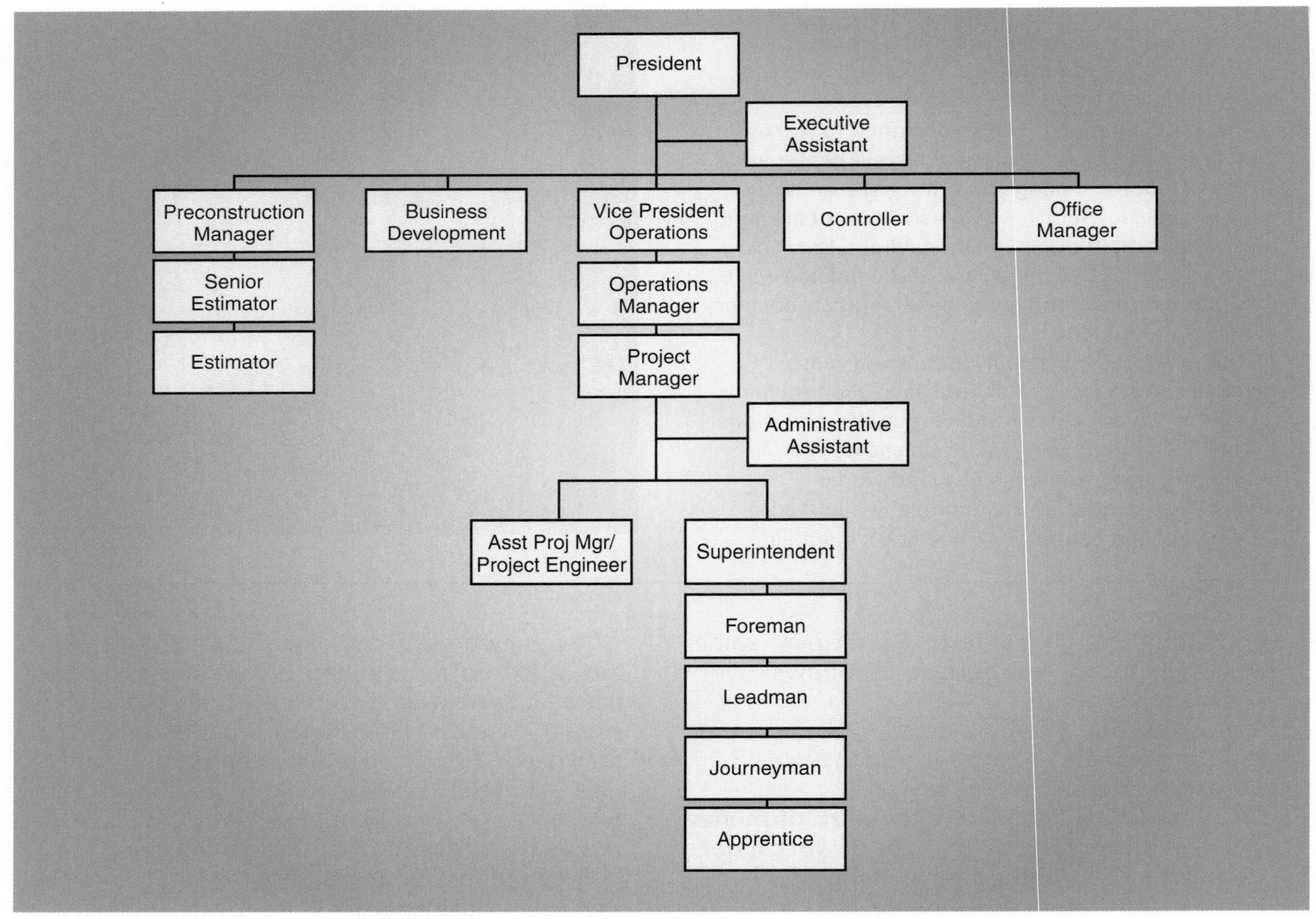

Figure 4 An organizational chart.

3.0.0 Critical Thinking Skills

Suppose that you are the team leader on a construction crew that is building a shopping mall. The job will last approximately 18 months. The parking area for site workers is half a mile from the job site and your team members will have to carry their heavy toolboxes from their cars to the work area every day. Delays caused by late arrivals and team members arriving to work tired from the walk threaten to derail the tight job schedule. As the team leader, you need to solve the problem. What are your options?

Throughout your construction career, you will encounter problems like this one that you will have to solve. **Critical thinking skills** allow you to solve such problems effectively. Critical thinking means evaluating information and then using it to reach a conclusion or to make a decision. Critical thinking allows you to draw sound conclusions and make correct decisions when you use the following approach:

- Evaluate new information with an open mind.
- Identify options and alternatives.
- Weigh the merits of each option and alternative, and justify them.
- Accept or reject options and alternatives based on their merits.
- Discern cause and effect.
- Judge the credibility of your source.

Consider the credibility, or trustworthiness, of a source of information when deciding how to treat it. For example, information provided by your supervisor will be more reliable than gossip you overhear at lunch. Also, think about the expertise or experience of people who give you information. For example, if you need to know about the electrical installation on a project, would you take the carpenter's advice or the electrician's? You'd be better off listening to the electrician. Feel free to ask experts and people you trust for their advice.

Compare information with what you already know. If it does not fit in with what you know, question it. For example, suppose you read in a manual that workers do not need to wear hard hats in the construction area. During your training, you have learned that hard hats should always be worn in a construction area. By comparing these alternatives, you can see that the information presented in the manual is not credible.

Sometimes, however, your personal feelings can get in the way of evaluating information. For example, if you don't like working with computers, you may not want to draft construction drawings with them. But if your shop uses computers to draw construction drawings, then you may disrupt the workflow if you hold on to your bias. Put your personal feelings aside and remain open-minded when evaluating information.

3.1.0 Barriers to Problem Solving

When you are searching for the solution to a problem, you may fall into a trap that prevents you from making the best possible decision. The following are the most common barriers to effective problem solving:

- Closed-mindedness
- Personality conflicts
- Fear of change

To be closed-minded is to distrust any new ideas. Effective problem solving, however, requires you to be open to new ideas. Sometimes the best solution is one that you would have never considered

On-Site

Information on the Internet

Everybody knows that the Internet has information about almost everything. The trick is to know whether the answers you find there are the right ones for you. How do you evaluate sources that you find on the Internet?

One of the easiest and most effective ways is to look at the two- or three-letter extension at the end of the website address. Addresses ending in .gov or .us are official local, state, or federal government Web pages. Use those pages to find accurate information about codes and regulations that apply to your work. Addresses ending in .org are nonprofit organizations. If the organization is affiliated with your industry or sets the standards for it, you can also trust its information.

Manufacturers put a wide variety of information, such as product specifications and catalogs, on their websites. When looking for this type of information, go straight to the manufacturer's website. The websites of companies that sell products from many different manufacturers may not be as complete or as up-to-date as the manufacturer's own site.

on your own. Remember that other people have good ideas, too. You should be willing to listen to them.

Sometimes you may fail to appreciate the value of information or advice simply because you do not get along with the person offering it. Maybe someone has an abusive or insensitive way of talking that offends you. Or maybe someone acts superior or bossy. One of the most important skills you can master is the ability to separate the message from the messenger. Weigh the value of the information separately from your feelings about the individual. This ability will show people that you are a real professional.

People often fear change when they believe it will threaten them, but as the old saying goes, "the only constant is change." Very few changes turn out to be as threatening as people fear. Often, the lack of change is the problem. In the construction industry, new tools, machines, techniques, and materials appear every day. As a construction professional, you need to stay informed about technical advances in your field. If you are open to change, you will never stop learning better ways to solve problems.

3.2.0 Solving Problems Using Critical Thinking Skills

Problems arise when there is a difference between the way something is and the way you would like it to be. You might feel frustrated or intimidated by a problem, or you might feel that you do not have enough time to solve it. Your reaction might be to simply ignore the problem. Instead, try to look at it as an opportunity to demonstrate your skills. By actively seeking solutions to problems, you will demonstrate to your colleagues and supervisors that you are responsible and capable.

To solve problems that arise on the job, use the following five-step procedure (*Figure 5*):

Step 1 Define the problem and identify an acceptable goal.

Step 2 Analyze and explore the alternatives.

Step 3 Choose a solution and plan its implementation.

Step 4 Put the solution into effect and monitor the results.

Step 5 Evaluate the final result.

Reviewing each of these steps in detail will help you understand how to apply them. Before you can solve a problem, you need to know exactly

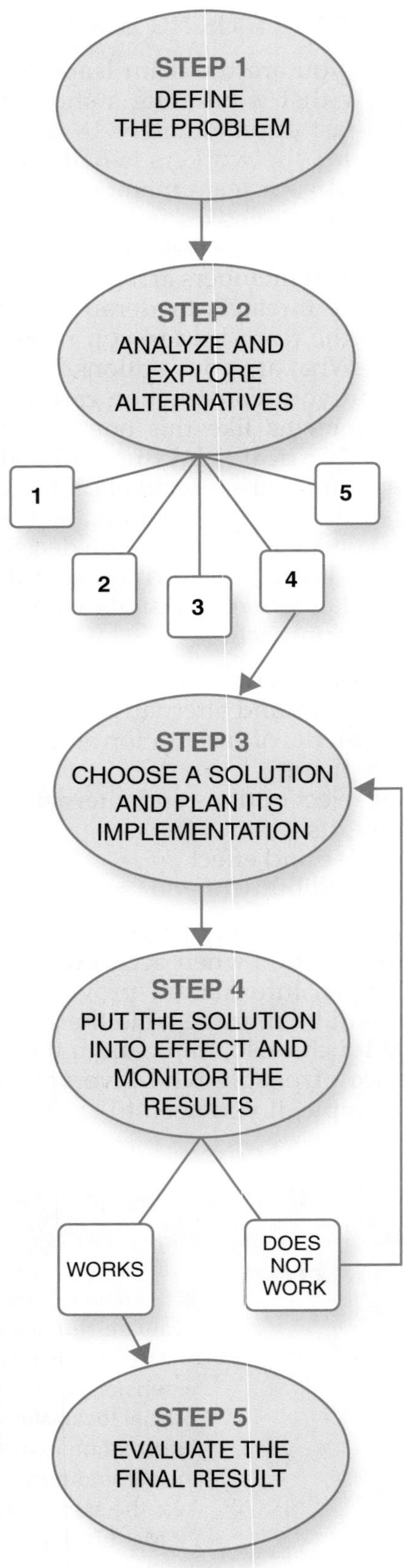

Figure 5 The five-step problem-solving procedure.

what it is. This step might seem obvious, but people often forget it. For example, suppose you are working at a job site when the generator goes out and your team members cannot use their power tools. Is the real problem the fact that the generator quit, or the fact that your team is not able to work? Most likely it's the second issue. You can easily bring in auxiliary generators that will allow your team to keep working. However, your team's inability to work could throw off the entire project's schedule. The generator failure was not the problem but rather the cause of the problem.

Once you've defined the problem, consider different ways to solve it. Collect information from a wide range of sources, and ask people for their opinions. **Teamwork** is an important part of this step. The more suggestions you get from experts and co-workers, the better your chances are of finding the right answer. Identify and compare the alternatives. Look for solutions that will be the most cost-effective, take the least time, and ensure the highest quality. Try to identify the short- and long-term consequences, both good and bad, of each possible solution.

Then, choose the solution that you think will work the best. Develop a plan for carrying it out. The plan should include all necessary tools and materials. It should also specify all the tasks involved and who is responsible for them, and estimate how long each task will take.

When you are ready to put the plan into effect, follow it closely to ensure that it is bringing about the desired results. If it is not, go back to Step 2 and choose another possible solution. Develop and implement a plan suited to the new solution.

When the problem is finally solved, look back over the steps you took and see what lessons you can learn. Perhaps there was something you could have done better. Or perhaps you could have saved time or money by changing part of the plan. When you are satisfied with the solution, remember what worked and what did not. When you face a similar problem in the future, your experience will help you find a solution.

Go back to the beginning of this section and review the hypothetical case discussed there. You are responsible for figuring out a way to help your co-workers carry their toolboxes to and from the work area from a parking lot half a mile away. How would you apply the five-step process to solve that problem? You might approach it in the following way:

Step 1 *Define the problem* – Because the parking area is far from the job site, your team members are forced to waste time walking to and from their cars. They are also forced to waste energy carrying their heavy toolboxes.

Step 2 *Analyze and explore the alternatives* – You could arrange a shuttle service from the parking area to the job site. You could provide lock-up space for tools and toolboxes near the work area. Or you could designate a drop-off point for the tools at the job site. What are some of the benefits and drawbacks of each alternative?

Step 3 *Choose a solution and plan its implementation* – After weighing each option, you decide that the shuttle service is the best one. It allows workers to take their tools home each night. This eliminates the possibility of theft, which was a drawback of the other options. You present your plan to the site superintendent, who agrees with you and announces that every morning at 8:00, the company van will pick up the construction crew in the parking area. Every afternoon at 5:30, the van will make the return trip.

Step 4 *Put the solution into effect and monitor the results* – At first, the new solution seems to work. However, you realize that the van cannot hold everyone. The van will have to make more than one trip. You discuss this with the site superintendent and get permission for three trips in the morning, at 7:45, 8:00, and 8:15, and three trips in the afternoon, at 5:15, 5:30, and 5:45. The site superintendent agrees to schedule the workers' starting times to correspond with the different trips.

Step 5 *Evaluate the final result* – The new plan works well. Everyone gets to the job site on time and ready to work. The schedule is back on track. This cost- and time-effective solution makes everybody happy.

Remember, the word *critical* means important or indispensable. Critical thinking skills are important and indispensable tools in your personal toolkit (*Figure 6*). As you would with any other tool, you must learn how to use these skills correctly and safely. When you do, you will find that one of the most rewarding experiences you can have as a trade professional is to face a problem and solve it yourself.

108F06.EPS

Figure 6 Critical thinking and problem-solving skills are just as important as any of these tools.

3.3.0 Problems with Planning and Scheduling

As a member of a project team, you are a very important link in a chain that stretches from you all the way to the top of your company. Suppose you are a plumber working for a construction company on a new building. At the top of the chain is the company's owner, who is responsible for ensuring that an entire project stays on schedule and within budget and that it meets the client's needs. Underneath the owner is the job superintendent, who is responsible for coordinating the work at the job site. The plumbing foreman is responsible for coordinating the installation of the plumbing, if the project involves more than one trade, and for assigning work to each member of the plumbing crew. And finally, but no less important, the plumber is responsible for performing specific tasks, such as locating waste stacks or installing fixtures. Successful plumbers learn how to undertake and complete their tasks in the right order and on time.

Every task has a logical position in the overall project schedule. Each task must occur before, after, or at the same time as other tasks. Before beginning each task, ensure that you understand what you are expected to do and what is required to accomplish the task. You must know where your work begins and ends and which materials to install. Divide the task into individual steps and decide the sequence in which you will perform the work.

No matter how carefully you plan ahead, unexpected problems can suddenly appear. Usually, extra time is built into project schedules so that simple problems will not cause a delay. However, more complex problems can throw the project off track.

For example, suppose a shipment of bathroom fixtures is delayed by one day. This is a simple problem, and it can be resolved by having the plumbers work on another task until the fixtures arrive. Now suppose that you find out that the bathroom fixtures are no longer being manufactured in the size or color called for in the specifications. This is a more serious problem, because if the plumbers do not get alternative fixtures, they cannot finish the bathroom. As a result, the drywall installers, electricians, and other trades will not be able to complete their tasks in that room, either. The whole project's schedule could be jeopardized.

Foremen and superintendents are responsible for ensuring that a project stays on schedule. Nevertheless, as a member of the team, you may be called on by your supervisor to evaluate and solve some problems. You should be familiar with the types of problems that can disrupt a job. Generally, problems will fall under one or more of the following categories:

- Materials
- Equipment
- Tools
- Labor

The materials required for a job are identified during the planning stages of a project. They are ordered from suppliers and delivered according to a prearranged schedule. Problems with materials can include errors of quantity or type, delays in delivery, and unavailability due to backorders or to business closure. Shortage of materials due to waste, loss of materials due to theft and vandalism, and inability to locate materials in storage are other sources of project delays.

Construction equipment is usually selected and scheduled before the project begins. The site superintendent is responsible for ensuring that equipment is on site at the right time. Often, two pieces of equipment must be scheduled at the same time; for example, backhoes to excavate a foundation and dump trucks to haul away the excavated dirt. Problems with equipment can include unavailability on the scheduled day(s), lack of qualified operators, mechanical breakdown, and extended maintenance.

Supervisors are responsible for ensuring that their workers have the appropriate tools. Workers may need specialized tools that they do not have in their personal toolkits. Or they may be learning how to use such tools on the job, under proper supervision. Common problems related to tools include breakage or damage, loss or theft, or lack of skilled users.

Labor—the men and women who perform the work on the job site—is the most important component of a project. Project planners estimate the number of people needed every day and identify the range of skills required. Supervisors ensure that the right people are available and working on the right tasks. Common problems related to labor include **tardiness** and **absenteeism**, lack of experience or qualifications to perform a given task, and not enough available workers. Idleness due to lack of instructions or because of laziness can also contribute to delays.

Always keep your eyes open for possible delays as you carry out your assigned tasks. If you see a potential cause of delay, notify your supervisor immediately. Always know clearly what your responsibilities are in such situations; be prepared to solve the problem yourself if that is what you are called on to do. Always keep your supervisor up-to-date on your progress. Be sure to let your supervisor know when you have finished a task.

4.0.0 COMPUTER SKILLS

Computers are everywhere. The most familiar types of computers are probably the desktop computers that you see on office desks, and the laptop computers (*Figure 7*) that people use on the road. These are not the only type of computers, however. Computers small enough to hold in your hand, called **handheld computers** or **PDAs** (for personal digital assistants), are now on the market. Small computers are also used in cars, televisions, microwave ovens, and even many alarm clocks and coffee makers.

Computers play an important part in the modern construction industry. Just a few years ago, fast, powerful computers were too expensive for the average business to afford. Now, computers are far more affordable. Industries rely on computers for everything from payroll and billing to design and fabrication. As part of your workplace skills, you need at least a basic understanding of computers and the work they can do. This understanding is called **computer literacy**.

This section introduces some basic computer terms, components, and applications. Because computer technology changes very fast, the topics will be covered generally. Some of the terms and technologies described here may be out-of-date by the time you read about them. Your instructor will be able to suggest sources that contain more specific information.

4.1.0 Computer Terms

Computer systems consist of **hardware**, **software**, and **operating systems**. Together, these three elements allow a computer user to tell the computer to do something, observe the results, make changes, and give new instructions. Hardware is the set of physical components that make up a computer (*Figure 8*). Basic hardware components include the following:

- The **processor**, sometimes called a central processing unit or CPU, which contains the chips and circuits that enable the computer to perform its functions.
- The monitor, a television-like screen that shows information, text, and pictures to the user.
- The keyboard, which allows computer users to type in text and instructions to the processor.
- A mouse or other device that allows users to enter data without having to type on the keyboard.

108F07.EPS

Figure 7 Laptop computer.

108F08.EPS

Figure 8 Computer hardware.

- A **hard drive** that stores software and electronic files, plus a CD/DVD-ROM drive or **USB flash drive** for transferring software and files to and from the computer.
- Printers and **plotters** that print out text, labels, graphics, and drawings on paper, plastic, and other materials.
- **Scanners** that allow printed text or pictures to be copied into an electronic format so they can be manipulated on the computer.

4.2.0 Software

Software tells a computer how to perform one task or a whole series of tasks. Software is usually installed on a computer from one or more CD-ROMs, or may be downloaded from the Internet. CD-ROMs look like music CDs, but they contain electronic files. Operating systems allow the hardware and software to communicate. Popular operating systems include Microsoft® Windows®, Apple® Mac OS®, and Linux.

The following are some of the software packages that are commonly used in office computers in the construction industry:

- **Word processors** are used to write text documents, such as letters, reports, and forms.
- **Spreadsheets** are used to perform math calculations, such as for project budgets and company payrolls.
- **Databases** are used to store large amounts of information, such as employee lists and material inventories, in a way that can be easily retrieved.
- **Desktop publishing** software is used to lay out text and graphics for publications, such as brochures and newsletters.
- Computer-aided design (CAD) systems are used to draw and print civil, architectural, mechanical, structural, plumbing, and electrical drawings (*Figure 9*).
- **Browsers** are used to search for information on the Internet.
- Email programs are used to send and receive electronic mail messages.

Many different companies produce and sell software. Some software is sold separately from hardware so that it can be installed on hardware of the user's choice. Other software is designed to operate on only one model of hardware. Some software is designed for use in many industries; other software is developed specifically to perform a single task in one industry. Sometimes, software is even designed to meet the needs of one company.

Software usually comes with instructional manuals and other information, called **documentation**. In addition, you can buy how-to books for installing and using software at your local bookstore or from vendors on the Internet.

4.3.0 Email

Email is used to send and receive messages electronically over a computer network. Today, it is a common way to communicate, both professionally and personally. Email is typically generated in an email software program on an individual computer and then transmitted to the recipient via a network connection, such as a modem, cable modem, local area network, or the Internet.

4.4.0 Computers in the Construction Industry

Companies use office computers to track time and labor costs, control inventory, and plan and control projects. However, most of the work being done in the construction industry is not done in an office. Work is usually done either on a job site or in a shop. Today, you will find plenty of computers in both of these places.

Field workers can use laptop and handheld computers that are specifically made for the construction industry. Many of them are designed to be operated in the field; they use long-lasting batteries and **wireless** communications technology. Wireless allows you to communicate with other computers without a telephone line or other physical connection. Supervisors can send daily reports, forms, notes, and even photographs to the home office. The project manager can track a job's progress without having to travel back and forth to the job site. Global positioning systems, or GPS, are also useful for field workers. These devices provide reliable location and time data any time of day, in any weather, anywhere in the world. GPS benefits field workers by providing accurate positioning and navigation information, and minimizing travel time to work sites. Many companies that use wireless technology are reporting increased productivity and reduced costs.

If you were to walk into a typical machine shop, you would probably see many different types of machines—lathes, milling machines, and drill presses, for example—that are operated by computers (*Figure 10*). Machine tools that are controlled by computers are often referred to as **CNC**, or computer numerical control, machines. Machine operators enter electronic design drawings into a CNC machine, and the machine creates

AutoCAD™ PROPERTY PALETTE

108F09A.EPS

AutoCAD™ STANDARDS CHECKING

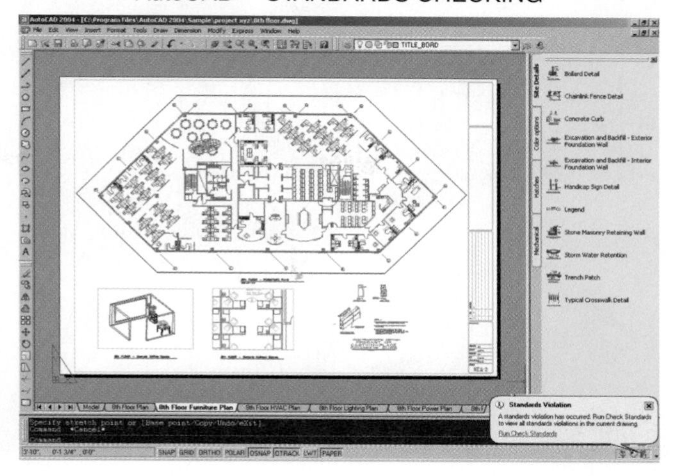

108F09B.EPS

AutoCAD™ TOOL PALETTE

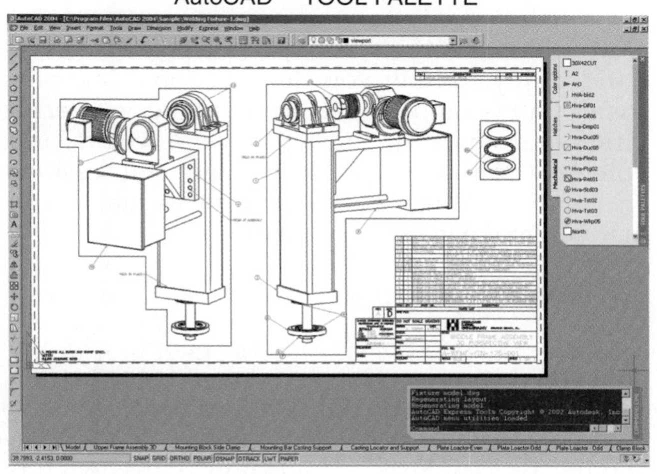

108F09C.EPS

PRIMAVERA EXPEDITION 12.0

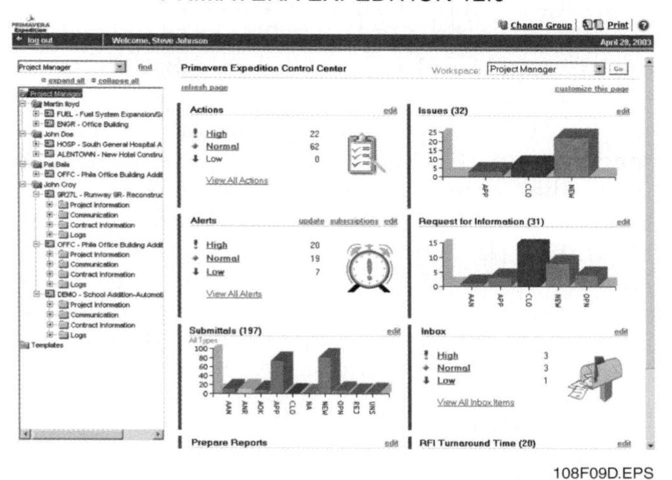

108F09D.EPS

PRIMAVERA P6 FOR CONSTRUCTION

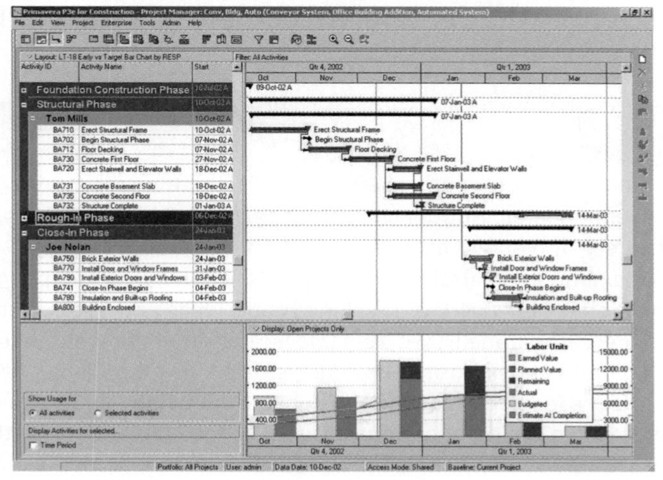

108F09E.EPS

Figure 9 Samples of computer software used in the construction industry, including AutoCAD™ and Primavera.

108F10.EPS

Figure 10 Many modern tools and machines are computer controlled.

a finished product of that exact design. CNC machines can be used to mill materials such as wood, metal, and plastic and to fabricate tools, dies, and fittings.

Computer-controlled electrical discharge machines, or **EDMs**, are used to cut and form parts that cannot be easily handled by other types of machines. EDMs are used on hard metals and on parts that have very complex shapes. Among other things, EDMs are used to make slots and holes in metal, repair damaged dies, fabricate jigs to hold parts, and thread pipe.

5.0.0 RELATIONSHIP SKILLS

A relationship is the process of interacting with another person, or a group of people. Relationships are affected by both actions and perceptions. Every day, you interact with co-workers, supervisors, and members of the public who see you working. No one wants to work with, hire, or spend time with an unprofessional person.

You need to be aware of the appropriate professional conduct for work situations, and you must follow that conduct at all times. Your actions reflect on your own professional status, that of your colleagues, the company you work for, and the image of your profession as seen by the public.

5.1.0 Self-Presentation Skills

Relationships are like equations. They happen between you and another person or between you and a group of people. Because you are half of that equation, proper **self-presentation**—the way you dress, speak, act, and interact—is a vital part of any successful work relationship. Proper self-presentation is not only about respecting others but also about respecting yourself, and it involves developing good personal and work habits. Personal habits apply to your appearance and general behavior. Work habits, called your **work ethic**, apply to how you do your job. Your personal habits and work ethic make powerful impressions on your colleagues, supervisors, and potential employers. Once formed, impressions are hard to change—so you want to make sure that you make good impressions.

5.1.1 Personal Habits

Co-workers and supervisors like people who are dependable. Co-workers know that they can trust dependable people to pull their own weight, take their responsibilities seriously, and look after one another's safety. Supervisors know that a dependable person will do a job correctly and on time. Being dependable means showing up for work on time every day. It also means not stretching out lunch hours and breaks. When dependable

On-Site

Robots on the Job

High-tech computer-controlled machines and equipment are becoming more popular on construction sites. Robots can safely perform dangerous jobs, without putting the health and safety of workers at risk.

For example, the Brokk Company of Sweden has developed a line of remote-controlled demolition robots used for breaking roads, walls and floors, and other pavement or concrete structures. Traditionally, workers use jackhammers or water jets to break up old pavement, putting them at risk from noise, dust, stress injury, and flying debris. In contrast, robot controls are located on a tethered console, allowing workers to operate equipment at a safe distance from the point of impact (up to 20 feet, depending on the robot). Robot concrete breakers are also used in confined spaces or where access is difficult. Computers make possible safer and more efficient tools, and can reduce the risk of injury or death on the job.

workers say they will do a task, they follow through on their promise.

Organizational skills are important as well. Keep your tools and your work areas clean and organized. Remember, craftworkers are judged by their tools, so know which tools you are responsible for, and keep them in good working order. Always know what you need to do each day before you begin. Approach your work in an organized fashion. Follow the schedule that you have been assigned.

Ensure that you are technically qualified to perform your tasks. This means that you know how to use tools, equipment, and machines safely. Take advantage of opportunities to expand your technical knowledge through classes, books and trade periodicals, and mentoring by experienced colleagues.

Offer to pitch in and help whenever you can. The best workers are willing to take on new tasks and learn new ideas. Supervisors notice employees who are willing. Be careful not to take on more tasks than you can reasonably handle—you will not impress anyone if you cannot deliver.

Honesty is one of the most important personal habits you can have. Do not abuse the system, such as by calling in sick just to take a day off, or by leaving early and asking someone else to punch you off the clock. Never steal from the company or from your co-workers. This includes everything from tools and equipment to the simplest of office supplies. If you are struggling with a problem, don't hide it. Speak with your supervisor about it truthfully. Be willing to look actively for a solution to problems you are facing.

Professionalism means that you approach your work with integrity and a professional manner. As you learned in *Basic Safety*, there is no place for horseplay or irresponsible behavior on the construction site. Employers want workers who respect the rules, and who understand that rules exist to keep people safe and projects on schedule. A professional employee always respects company **confidentiality**. This means not sharing with other people any information that belongs to the company.

Most companies have a dress code, and often they have additional special requirements for specific jobs. Always follow these requirements. Do not forget to attend to the basics of good grooming—comb your hair, brush your teeth, and wash your clothes. People who pay attention to their appearance and develop positive personal habits are more likely to be considered for a job over people who do not have good habits. In the field, your personal habits reflect not just your own professionalism, but also the professionalism of your trade.

5.1.2 Work Ethic

Along with good personal habits, a strong work ethic is essential for getting hired and being promoted. Employers look for people who they believe will give them a fair day's work for a fair day's pay. Having a strong work ethic means that you enjoy working and that you always try to do your best on each task. When work is important to you, you believe that you can make a positive contribution to any project you are working on.

Construction work requires people who can work without constant supervision. When there is a problem that you can solve, solve it without waiting for someone to tell you to do it. If you finish your task ahead of time, look for another task that you can do in the time remaining. This type of positive action is called **initiative**. Colleagues and supervisors will respect you more when you demonstrate initiative on the job.

An important part of taking initiative, however, is to know when and how to take it. Suppose your supervisor tells you to perform a task. He shows you the five steps required to complete the task. Later, as you perform the work, you realize that one of the steps is unnecessary. Leaving out the step could save time—but should you?

No. That would not be the right initiative. Instead, tell your supervisor about your discovery, and ask what you should do. Bringing the options to your supervisor is showing initiative. Your supervisor may realize that you are right and allow you to leave out the extra step. After that, every time anyone performs that task, it will take less time and save the company money. Or, your supervisor may explain why the step is important. Then you will have learned something more about the work you are doing. Either way, the result of taking the initiative is a positive one.

5.1.3 Lateness and Absenteeism

The two most common problems supervisors face on the job are lateness and absenteeism. Lateness is when a worker habitually shows up late to work. Absenteeism is when a worker consistently fails to show up for work at all, with or without excuses. People with a strong work ethic are not often late or absent.

To make a profit and stay in business, construction companies operate under tight schedules (*Figure 11*). These schedules are built around workers. If you are late or do not show up regularly, expensive adjustments become

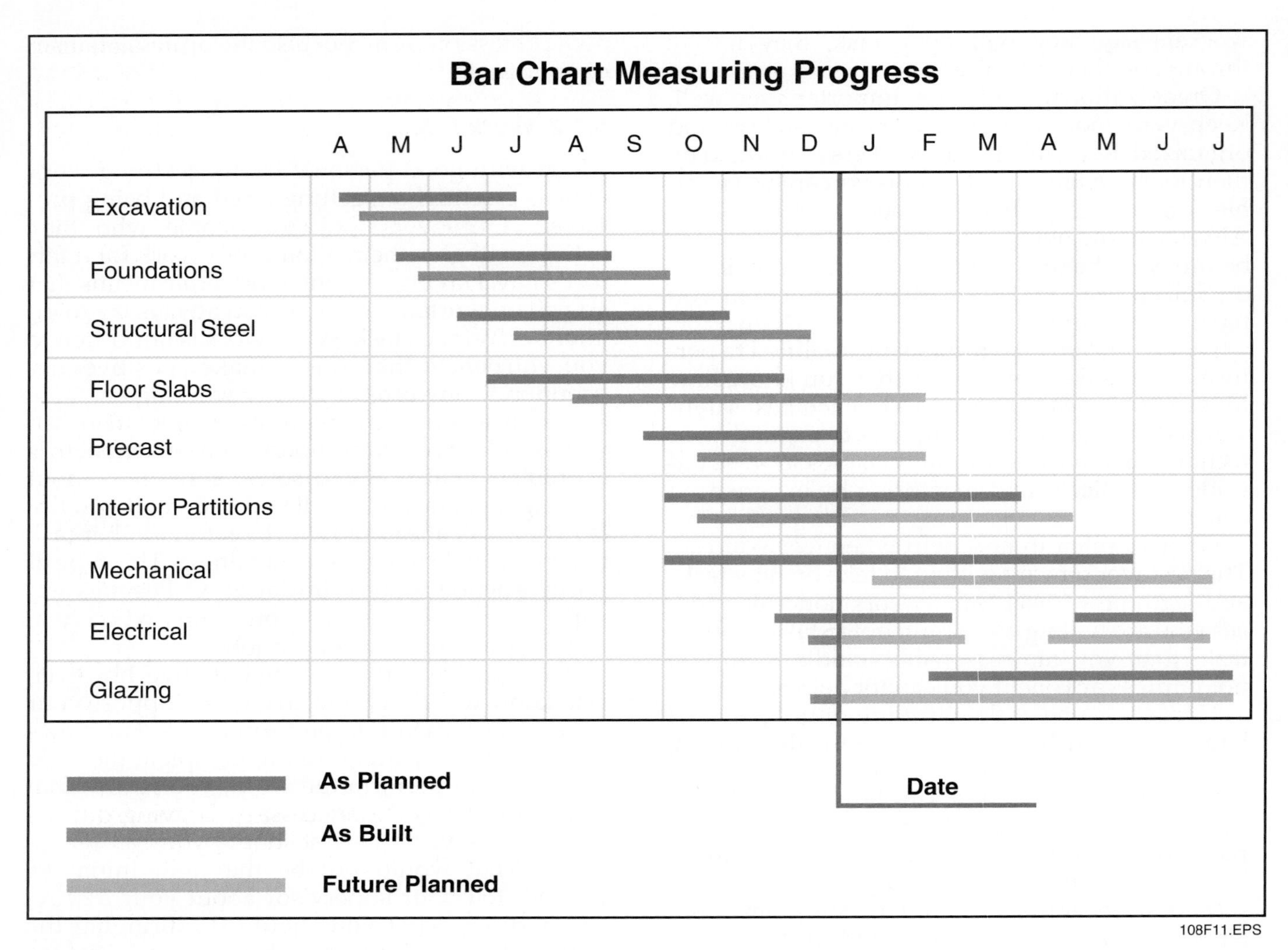

Figure 11 Construction companies operate under tight schedules.

necessary. Your employer may decide that the best way to save money is to stop wasting more of it on you.

Consider the following suggestions for improving or maintaining your record of punctuality and attendance:

- Think about what would happen if everyone on the job were late or absent frequently.
- Think about how being late or absent affects your co-workers.
- Know and follow your company's policy for reporting legitimate absences or lateness.
- Keep your supervisor informed if you need to be out for more than a day.
- Allow yourself enough time to get to work.
- Explain a late arrival to your supervisor as soon as you get to work.
- Do not abuse lunch and break-time privileges.

5.2.0 Conflict Resolution

Suppose you show up to work one day with a new haircut, and one of your co-workers starts to tease you about it. Or you suspect that a co-worker has been stealing tools, but you cannot prove it. These situations are examples of conflicts that you might face on the job. How do you resolve a conflict between you and someone else? You may be the type of person that can resolve a disagreement quickly, or you may prefer to avoid addressing conflicts directly. Some people even become angry and resent the other person.

Conflict resolution is an important relationship skill, because conflict can happen anywhere, anytime. Conflict can happen when people disagree, and people disagree all the time. Conflicts between you and your co-workers can arise because of disagreements over work habits;

different attitudes about the job or the company; differences in personality, appearance, or age; or distractions caused by problems at home. Conflicts between you and your supervisor can happen because of a disagreement over workload, lateness or absenteeism, or criticism of mistakes or inefficiencies.

Most of the time, people are not trying to turn disagreements into conflicts. People are often unaware of the effects of their behavior. Before reacting negatively to a co-worker's behavior, remember to be **tactful**. This means considering how the other person will feel about what you say or do. Do not accuse, embarrass, or threaten the person. Try to have a sense of humor about the situation. Try to avoid calling in your supervisor, except as a last resort.

Never let a disagreement or conflict affect job performance, site safety, team morale, or an individual's well-being. The goal is to keep disagreements from turning into conflicts in the first place. If that is not possible, then the next best thing is to address the conflict quickly and resolve it professionally. If a disagreement with a co-worker is getting out of hand, try one of the following techniques to cool the situation down:

- Think before you react
- Walk away
- Do not take it personally
- Avoid being drawn into others' disagreements

If, despite your best efforts, a disagreement turns into a conflict, then you must take immediate action to resolve it. Do not let a conflict simmer and then boil over before you decide to act. That is not professional. Keep in mind, however, that there are important differences between the way you resolve conflicts with your co-workers and the way you resolve conflicts with your supervisor. The following sections discuss how to handle each type of conflict.

NOTE

Resolve your conflicts quickly. Keep your focus on getting your work done. Avoid prolonged arguments.

5.2.1 Resolving Conflicts with Co-Workers

Remember to have respect for the people you disagree with. After all, they believe they are right, too. Be clear, rational, respectful, and open-minded at all times. Begin by admitting to each other that there is a conflict. Then, analyze and discuss the problem. Allow everyone to describe his or her own perception of the conflict. You may realize that the whole problem was simply miscommunication. Ask the following questions:

- How did the conflict start?
- What is keeping the conflict going?
- Is the conflict based on personality issues or a specific event?
- Has this problem been building up for a while, or did it start suddenly?
- Did the conflict start because of a difference in expectations?
- Could the problem have been prevented?
- Do both sides have the same perception of what's going on?

Once you have analyzed the situation, discuss the possible solutions. You will probably have to **compromise**, or meet in the middle, to find a solution that everyone agrees on. When you agree on the solution, act on it, and see if it works. If it does not, then consult your supervisor for help in resolving the conflict. Notice that this process is similar to the problem-solving techniques discussed earlier.

5.2.2 Resolving Conflicts with Supervisors

You can usually approach co-workers as equals, because you are working on the same job. However, on the job, your supervisor is not your equal. He or she is in charge of you and your work. This means that you must use a different approach to resolve conflicts with your supervisor.

Before going to your supervisor, take some time to think about the cause of the conflict. Consider writing down your thoughts. Organizing the information this way often puts things into perspective. When you approach your supervisor, do so with respect. Wait until your supervisor has a free moment, and then ask if you could arrange a time to talk about something important. Or leave a message or note for your supervisor asking to meet. Be willing to meet at a time that is convenient for your supervisor. Remember, supervisors have many responsibilities. They do not have much free time during the regular workday, and you may have to meet before or after work.

When meeting with your supervisor, speak calmly and clearly. Do not be emotional, sarcastic, or accusatory. Do not confront your supervisor in a threatening or angry way. State only the facts as you see them; never say anything that you cannot prove. Do not mention the names of your co-workers unless they are directly involved. If you want to suggest changes or solutions, explain them clearly and discuss how they will benefit the people involved.

Once you have made your case, allow your supervisor to make a decision. You should accept and respect your supervisor's final decision. It may not be the one you wanted, but it will be the one that is best for all concerned.

5.3.0 Giving and Receiving Criticism

As a construction professional, you will always be learning something new about your job. New technologies, materials, and methods appear all the time. Talking to an experienced construction worker can help you learn a new way to perform a task. As your skills improve with practice, you will be able to use tools that you were not able to use before. All of these are common ways of learning on the job (*Figure 12*).

Another way to learn on the job is through **constructive criticism**. Constructive criticism is advice designed to help you correct a mistake or improve an action. Constructive criticism can improve your job performance and relations with co-workers. You have probably heard the word *criticism* used in a negative way to indicate fault or blame. That is not the type of criticism discussed here. On-the-job criticism does not mean that colleagues and supervisors think little of you. In fact, it means exactly the opposite. Colleagues and supervisors who offer constructive criticism do so because they believe it is worth their time to help you improve your skills.

108F12.EPS

Figure 12 A good employee never stops learning on the job.

As you gain experience on the job, you will be able to give constructive criticism as well as receive it. To make sure that someone does not mistake your constructive criticism for blame, you need to know how to both offer and receive it in the spirit intended. The following sections offer some general advice on offering and receiving constructive criticism.

5.3.1 Offering Constructive Criticism

When you are training a less-experienced person or working with co-workers, you might find yourself offering some constructive criticism. How should you offer it? Before you say anything, think about the rules of effective speaking that you learned in *Basic Communication Tools.* Use positive, supportive words and offer facts, not opinions.

Constructive criticism works best when offered occasionally. Do not constantly comment on other people's work or methods. They may block out your criticism or even become angry. Never criticize people in front of their co-workers or supervisors. They may feel embarrassed and will probably resent you. Criticism should include suggested alternatives. Do not criticize how a co-worker does something unless you can suggest another way. Above all, limit your comments to people's work or methods, not the people themselves.

NOTE

Point out improper behavior or incorrect work techniques the first time they occur. If not addressed, the behavior or technique could become a bad habit. Bad habits, in turn, can lead to accidents. Gently and firmly criticize improper behavior as soon as you see it.

Remember to compliment the person you are criticizing. Compliments have a genuinely positive effect on the person receiving them, and they are easy to offer. You do not have to wait to compliment someone until you offer constructive criticism, either. Try to offer compliments on a regular basis. Your appreciative and respectful approach will help keep team spirits high. Your supervisor will appreciate your professional attitude.

5.3.2 Receiving Constructive Criticism

Not only can criticism make the difference between success and failure, but in cases where workplace safety is involved, it can also make the difference between life and death. Think of criticism as a chance to learn and to improve your skills. Take criticism from your supervisor seriously. Treat criticism from experienced co-workers with the same respect; even though they may not be your supervisor, their experience gives them a different type of authority.

Always take responsibility for your actions. You may want to explain why you did what you did to the person who is offering the criticism. Try not to do this. The person you are talking to may think that you are making excuses or trying to shift responsibility onto someone else. Never be defensive or dispute criticism. If you respond negatively, you will offend the other person. As a result, that person will either lose respect for you or will no longer trust you—or both. If that happens, the quality of your work, and that of everybody who works with you, will suffer.

When someone offers you constructive criticism, demonstrate your positive personal habits and work ethic. If the criticism is vague, ask the person to suggest specific changes that you should try to make. If you do not understand the criticism, ask for more details. Always be respectful to the people you receive criticism from, and take their criticism seriously. The deepest respect you can show them is to improve your performance based on their advice.

You have a right to disagree with criticism that is unacceptable or incorrect. Someone might honestly misunderstand your situation and offer incorrect advice. Or a co-worker might criticize you simply because he thinks he performs a task better than you do. In such cases, clearly and respectfully give your reasons for disagreeing, and explain why you believe that you are correct.

5.4.0 Teamwork Skills and Collaboration

No matter what your job is, chances are that you work with other people. Every day, you interact with members of your work crew and your supervisor. As you have learned, the ability to get along with people is an important skill in the construction industry, but simply working well with others is not enough. The success of any team depends on all of its members doing their parts. Everyone must contribute to the team effort to ensure that it achieves its goals. This cooperation is called teamwork.

From the moment you began your construction career, you've been on different teams for different projects. Teams can be as small as two people and as large as your entire company. Teams are often made up of people from different trades who work together to complete a task. For example, a team that is assigned the task of burying water and electrical lines might include a backhoe operator to dig the trench, plumbers to join and lay the pipe, and electricians to run the electrical cable.

Always show respect for the other members of your team. You can do this by allowing them to share their ideas with you and with the rest of the team. Although you have a specific job to do, always be willing to help a co-worker. Never say that something is not your job—someday, on another project, it might be. Support your co-workers when they need it, and they will support you when you need it.

Good team members are **goal-oriented**. This may sound like management jargon, but it is actually a well-known and accepted concept in the construction industry. Being goal-oriented

On-Site

Destructive Criticism

Constructive criticism has positive effects. However, you may also encounter destructive criticism. Destructive criticism, as its name suggests, is designed to hurt, not help, the person receiving it. The destructive criticism could actually be about something that you could improve, but because it is offered negatively, you would have no desire to hear it.

If you receive destructive criticism, the professional thing to do is to stay calm. Don't get into a fight by replying in a negative tone like the other person's. Find a way to let the criticizer know how the criticism made you feel. If there is a legitimate criticism beneath the destructive tone, ask the person to offer positive suggestions for ways you can improve. By taking a positive approach to negative criticism, you set a good example for others.

means making sure that all activities focus on the team's final objective. Whatever the end result is—fabricating fittings, laying pipe, or framing a house—everyone on the team knows that result in advance and concentrates on achieving it.

As a team member, you should always try to use your skills and strengths, while at the same time accepting your limitations and the limitations of others. To be a good team member, try to do the following:

- Follow your team leader's and/or supervisor's directions.
- Accept that others might be better at some tasks than you.
- Keep a positive attitude when you work with other people.
- Recognize that the work you do is for the benefit of the entire company, not for you personally.
- Learn to work with people who work at different speeds.
- Accept goals that are set by someone else, not by you.
- Trust other members of the team to perform their tasks, just as you perform yours.
- Appreciate the work of others as much as you appreciate your own.

Keep in mind that you are not the only person who is working hard. Everyone on your team is focusing on the goal, too. Offer praise and encouragement to your co-workers, and they will do the same for you. This mutual respect will help you feel confident that your team will reach its goal. Share the credit for good work, and be willing to take blame for your mistakes and errors. These actions will help you earn the respect of your co-workers.

If you practice goal-oriented teamwork, you should be able to meet your deadlines and keep projects on schedule. This translates into time and money saved by your company. There will be times when your co-workers or other trades cannot start their tasks until you have completed yours. Out of respect for them and your company's reputation, always finish your work in a timely manner. You do not like to be kept waiting; do not make others wait.

WARNING!

Distractions caused by personal problems can lead to injury and death. If you are concentrating on a problem instead of paying attention to your job, you can easily make a careless mistake or overlook a vital safety precaution. Stay focused on your job, and take whatever steps are necessary to eliminate distractions.

An important part of teamwork is training. As an apprentice, you received guidance and advice from more experienced colleagues. As you gain your own experience, you will be able to help less-experienced colleagues. Training offers an excellent opportunity to build good relationships. You probably remember at least one teacher who was patient and understanding and made a difference in your life. On a team, you can be that person for someone else.

Being asked to teach someone is an honor. It means that someone believes you do your job well enough to teach it to other people. Such confidence should inspire you to be the best teacher you can be. Remember to be patient with the person you are teaching, and teach by example. Offer encouragement, and give constructive advice as you have learned in this module. Teach people the same way that you would like to be taught.

5.5.0 Leadership Skills

As you gain experience and earn credentials, you will assume positions of greater responsibility in the construction industry. You will earn these

On-Site

Keep Your Problems at Home

Few things are more challenging to deal with than people who take their personal problems to work with them. Being a professional means keeping your personal life and your work life separate. This does not mean that you have to keep your personal life secret from your co-workers. It means that you should not take out your personal frustrations, fears, or hostilities on your colleagues. As you gain more experience on the job, you will discover an appropriate comfort level for discussing personal issues with your co-workers.

Some events—for example, the death of a loved one—can affect you so deeply that it becomes impossible for you to separate your personal life from work. Let your co-workers know that you are having difficulties. Discuss the situation with your supervisor. Your friends and colleagues can help when you are having a problem.

On-Site

How Do Your Co-Workers See You?

Do you exhibit any of the following unpopular behaviors on the job? If you do, they may be keeping you from being an effective member of the team.

If you do exhibit any of these behaviors, develop an action plan to deal with them. Your action plan should be a two- or three-step process that will allow you to correct the behavior. Be honest!

Being a loner
Taking yourself too seriously
Being uptight
Always needing to be the best at everything
Holding a grudge
Arriving late to work
Being inconsiderate
Taking breaks that are too long
Gossiping about others
Sticking your nose into other people's business
Acting as a spy and reporting on the behavior of others to your supervisor
Saying or doing things to create tension or unhappiness
Complaining constantly
Bad-mouthing the company or your boss
Being unable to take a joke
Taking credit for others' work
Bragging
Being sarcastic
Refusing to listen to other people's ideas
Looking down on other people
Being unwilling to pitch in and help
Horsing around when others are trying to work
Thinking you work harder than everyone else
Being stingy with a compliment
Having a chip on your shoulder
Manipulating people

Action Plan for Improvement

Example:

Problem: Gossiping about others

Action Plan: 1. Walk away when people start gossiping or bad-mouthing others.
2. Don't repeat what I hear.
3. Focus on my own work, not on that of others.

Problem: ______________________________

Action Plan: ______________________________

____

Problem: ______________________________

Action Plan: ______________________________

____

Problem: ______________________________

Action Plan: ______________________________

____

108SA02.EPS

positions through your hard work and dedication. Starting as an apprentice, you can work your way up to team leader, foreman, supervisor, and project manager. Someday, with enough hard work and dedication, you might even be able to run your own company.

To progress steadily through your career, you will need to develop **leadership** skills and learn how to use them. Leaders set an example for others to follow. Because of their skills, leaders are trusted not only with the authority to make decisions, but also with the responsibility to carry them out.

The typical image of a leader is that of a boss who has many people working for him or her. But you can be a leader at any stage in your career. As an apprentice, you have the authority to perform the task given to you by your supervisor, and your supervisor expects you to be a responsible worker. By carrying out your task quickly, correctly, and independently, you are setting an example for others to follow. In doing so, you are demonstrating leadership skills.

People with the ability to become leaders often exhibit the following characteristics:

- They lead by example.
- They have a high level of drive, determination, and persistence.
- They are effective communicators.
- They can motivate their team to do its best work.
- They are organized planners.
- They have self-confidence.

The functions of a leader will vary with the environment, the group of workers being led, and the tasks to be performed. However, certain functions are common to all situations. Some of these functions include the following:

- Organizing, planning, staffing, directing, and controlling.
- Empowering team members with authority and responsibility.
- Resolving disagreements before they become problems.
- Enforcing company policies and procedures.
- Accepting responsibility for failures as well as for successes.
- Representing the team to different trades, clients, and others.

Leadership styles vary widely, and they can be classified in many ways. If you classify leadership styles according to the way a leader makes decisions, for example, you end up with three broad categories of leader: autocratic, democratic, and hands-off. An autocratic leader makes all decisions independently, without seeking recommendations or suggestions from the team. A democratic leader involves the team in the decision-making process. Such a leader takes team members' recommendations and suggestions into account before making a decision. A hands-off leader leaves all decision-making to the team members themselves.

Select a leadership style that is appropriate to the situation. You will need to consider your authority, experience, expertise, and personality.

On-Site

Building a Team

In this exercise, you will practice building a team. Consider the following hypothetical situation, and then select an appropriate team. Discuss your selections with your instructor and with the other trainees.

Situation

You are the team leader on a job site where a new single-family home is going to be built. Shortly before construction begins, the architect changes the plans to add a sink to one of the rooms. When you review the plans, you see that an electrical conduit is located where the drain for the new sink should be. In addition, a cabinet needs to be built for the new sink.

You may choose up to five of the following construction professionals to be on your team. Whom do you choose and why?

Mason	Painter
Carpenter	Roofer
Plumber	Landscape designer
Welder	Secretary
Electrician	Cabinetmaker

GOING GREEN

Green Employers

Some employers are doing their part to reduce their environmental footprint. Since 2004, the United Parcel Service (UPS) has revised delivery routes of its drivers to minimize left-hand turns. Doing so has made making the deliveries safer (they no longer have to cross traffic) and quicker (due to the ability to make a right on red and green turn arrows). As a result, the company has shaved 30 million miles off its deliveries in 2007 and thus saved the cost of 3 million gallons of gas. It also reduced UPS truck emissions by 32,000 metric tons (equivalent to the emissions of 5,300 passenger cars).

Leaders need to have the respect of people on the team; otherwise, they will be unable to set an example worth following.

Leaders have to make sure that the decisions they make are ethical. The construction industry demands the highest standards of ethical conduct. The three types of ethics that you will encounter on the job are business or legal ethics, professional ethics, and situational ethics. Business or legal ethics involve adhering to all relevant laws and regulations. Professional ethics involve being fair to everybody. Situational ethics involve appropriate responses to a particular event or situation.

Effective leaders motivate, or inspire, people to do their best. People are motivated by different things at different times. The following are some common ways to motivate people on the job:

- Recognize and praise a job well done.
- Allow people to feel a sense of accomplishment.
- Provide opportunities for advancement.
- Encourage people to feel that their job is important.
- Provide opportunities for change to prevent boredom.
- Reward people for their efforts.

Leaders who can motivate people are more likely to have a team with high morale and a positive work attitude. Morale and attitude are key components of a successful company with satisfied workers, and such a company is more likely to be successful.

6.0.0 Workplace Issues

The modern construction workplace is a cross-section of our society (*Figure 13*). A typical construction project will involve men and women from all walks of life, of many ethnic and racial backgrounds, and from many different countries. Many workers speak more than one language. Workers have grown up in, and currently live in, many different income brackets.

Construction workers face a wide range of mental and physical demands every day on the job. These demands can sometimes feel overwhelming, and people may seek escape through illegal drugs or alcohol abuse. As a construction professional, you need to be aware of these issues. You also need to know what to do if someone is behaving inappropriately.

You may encounter the following issues in the workplace:

- **Harassment** (sexual, age, height, weight, religious, cultural, disability)
- Stress
- Drug and alcohol abuse

6.1.0 Harassment

Harassment is a type of discrimination that can be based on race, age, disabilities, sex, religion, cultural issues, health, or language barriers. Harassment often takes the form of ethnic slurs, racial jokes, offensive or derogatory comments, and other verbal or physical conduct. It can create an intimidating, hostile, or offensive working environment. It can also interfere with an individual's work performance. Harassment can take on many forms. The type most commonly reported and talked about is **sexual harassment**. While other forms of harassment are not talked about as much or as openly, this doesn't mean that they don't happen.

When someone makes unwelcome sexual advances, or requests or exhibits verbal or physical behavior with sexual overtones, he or she is guilty of sexual harassment. Sexual harassment can happen between members of the opposite sex or the same gender. The harasser can threaten to fire the victim to keep him or her quiet. Sexual harassment is illegal, and if you experience it, you should report it to your supervisor immediately. Usually, sexual harassment is a pattern of behavior repeated over a period of time, so if you do not report it, you run the risk of experiencing it again.

6.2.0 Stress

Stress is the tension, anxiety, or strain that you feel when you face unexpected events or things that are outside your control. Stress can cause headaches, irritability, mental and physical exhaustion, and even health problems. Stresses at home can make themselves felt at work, and work stresses can affect home life.

108F13.EPS

Figure 13 Today's construction workplace is a cross-section of our society.

You can take steps to prevent stress from happening and reduce the stresses that you cannot prevent. Plan your work, and try to finish the most difficult tasks first, while you have the energy and attention for it. Always eat properly and get plenty of exercise. Manage your time and money wisely so that you are not surprised by sudden shortages in either resource.

For stresses that you cannot prevent, make a plan to deal with your problems. Control your anger and frustration by relaxing in a way that you enjoy—whether it is taking a hike or getting a massage. Talk with someone who will help you get a better perspective on your situation.

6.3.0 Drug and Alcohol Abuse

Drinking is a common way to avoid or forget about stress. Alcohol is an accepted part of modern social life. Moderate use of alcohol is socially acceptable and is generally believed to cause little or no harm to most people. Alcohol abuse, which means habitually drinking to excess, not only is socially unacceptable, but also poses a serious health and safety risk for yourself, your colleagues, and your family.

Unlike alcohol, illegal drug use is never acceptable. Misusing legal prescription drugs is also not acceptable. Types of drugs that are sometimes used inappropriately by workers include the following:

- **Amphetamines**
- **Methamphetamines**
- **Barbiturates**
- **Hallucinogens**

WARNING!

Working under the influence of drugs or alcohol can lead to fatal accidents on the job site.

Amphetamines and methamphetamine, a highly addictive crystalline drug derived from amphetamines, are both stimulants that affect the central nervous system. Also called uppers, they are used to prolong wakefulness and endurance, and they produce feelings of euphoria, or excitement. They can disturb your vision; cause dizziness and an irregular heartbeat; cause you to lose your coordination; and can even cause you to collapse. Caffeine and tobacco are common legal amphetamines. Cocaine, crack, and crystal meth are illegal amphetamine drugs.

Barbiturates are a sedative, which means they cause you to relax. They are also called downers. They can create slurred speech; slow your reactions; cause mood swings; and cause a loss of inhibition, which means that they make you feel less shy or self-conscious. Alcohol and the prescription drug Valium® are barbiturates.

Hallucinogens distort your perception of reality to the point where you see things that are not there (hallucinations). The effects of hallucinogens can range from ecstasy to terror. When you are hallucinating, you put yourself and others at great risk because you cannot react to situations the right way. Hallucinogens can cause chills, nausea, trembling, and weakness. Mescaline and LSD are two illegal hallucinogens.

If you abuse drugs or alcohol, you will probably not be able to get or keep the job you want. If you are discovered using illegal drugs, you can be fired on the spot because of the safety risks related to drug use. Addiction affects your well-being and state of mind, and you cannot do your best on the job or at home. It can also jeopardize your relationships with friends, family, and co-workers. If people you know are addicted to alcohol or drugs, seek help for them or encourage them to get help for themselves. The Appendix contains a list of organizations that can help people cope with addictions.

Did You Know?

Several federal laws prohibit job discrimination:

- *Title VII of the Civil Rights Act of 1964* – Prohibits employment discrimination based on race, color, religion, sex, or national origin.
- *Equal Pay Act of 1963* – Protects men and women who perform substantially equal work in the same establishment from sex-based wage discrimination.
- *Age Discrimination in Employment Act of 1967* – Protects individuals who are 40 years of age or older.
- *Title I and Title V of the Americans with Disabilities Act of 1990* – Prohibits employment discrimination against qualified individuals with disabilities in the private sector and in state and local governments.
- *Sections 501 and 505 of the Rehabilitation Act of 1973* – Prohibits discrimination against qualified individuals with disabilities who work in the federal government.
- *Civil Rights Act of 1991* – Provides monetary damages in cases of intentional employment discrimination.

Source: U.S. Equal Employment Opportunity Commission

Review Questions

1. To succeed in the construction industry, you must _____ as well as know your technical skills.

a. own a cell phone
b. have a driver's license
c. master interpersonal skills
d. have an email account

2. A(n) _____ can vouch for your skills, experience, and work habits.

a. mission statement
b. entrepreneur
c. interviewer
d. reference

3. A(n) _____ is a person who starts and runs his or her own business.

a. entrepreneur
b. visionary
c. competent person
d. qualified person

4. A common barrier to effective problem solving includes _____.

a. overwork
b. fear of change
c. inadequate supervision
d. procrastination

5. After putting a solution to a problem into effect, it is important to _____.
   a. stop considering other solutions
   b. insist it become a company procedure
   c. monitor the results
   d. develop a business plan for it

6. Who is responsible for ensuring that equipment is on site at the right time?
   a. The owner of the company.
   b. The manufacturer.
   c. The supply company.
   d. The site superintendent.

7. PDA stands for _____.
   a. personal drafting aid
   b. personality disorder analysis
   c. portable document assistance
   d. personal digital assistant

8. Browsers and word processors are examples of _____.
   a. software
   b. hardware
   c. hard drives
   d. operating systems

9. Actions and perceptions affect _____.
   a. construction drawings
   b. tool use
   c. relationships
   d. work schedules

10. A person who works without constant supervision is showing _____.
    a. initiative
    b. fortitude
    c. respect
    d. self-presentation

11. The process of solving disagreements between co-workers is called _____.
    a. tactfulness
    b. conflict resolution
    c. relationship resolution
    d. addressing conflict

12. To progress steadily through your career, you will need to develop _____ skills and learn how to use them.
    a. leadership
    b. classroom
    c. study
    d. driving

13. The most commonly reported and talked-about form of harassment is _____.
    a. age discrimination
    b. racism
    c. religious discrimination
    d. sexual harassment

14. You can help prevent or reduce stress by _____.
    a. leaving your most difficult tasks until the end of the day
    b. eating properly and exercising often
    c. not speaking with anyone about what may be bothering you
    d. exhibiting anger and frustration openly and often

15. Caffeine and tobacco are legal forms of _____.
    a. methamphetamines
    b. prescription drugs
    c. amphetamines
    d. hallucinogens

Summary

This module reviewed some of the important nontechnical skills that you must learn to be successful as a construction professional. Whatever positions you may achieve throughout your career, these skills are vital.

Whether you are just starting out, or moving on to new opportunities and challenges, you should be able to write a top-notch resume, prepare for interviews, and select a job for which you are qualified. Eventually, many construction professionals become entrepreneurs and start their own businesses.

The ability to solve problems using critical thinking skills is important for any employer and employee. Critical thinking involves evaluating and using information to reach conclusions or make decisions.

Computer literacy is also important for construction professionals. As part of your workplace skills, you need to have at least a basic understanding of computers and the work they perform. This module introduced you to hardware, software, and operating systems.

Good self-presentation skills, including personal habits, work ethic, and dependability, are characteristics of professionalism. Developing your skills in conflict resolution, teamwork, and leadership will help you to advance in your career.

The construction industry demands the best from its people. You are choosing a career that will bring out the best in you, but only if you are willing to give it your best to begin with.

Trade Terms Quiz

Fill in the blank with the correct trade term that you learned from your study of this module.

1. A company uses a(n) ____________ to state how it does business.
2. Someone who can vouch for your skills, experience, and work habits is a(n) ____________.
3. A(n) ____________ is someone who starts and runs a business.
4. Problems can be effectively resolved using ____________.
5. ____________ are small enough to hold in your hand.
6. The basic understanding of how computers work is called ____________.
7. ____________, ____________, and ____________ are the three elements that make it possible for a computer to operate.
8. A(n) ____________ is sometimes called a CPU.
9. The ____________ on a computer stores software and electronic files.
10. A(n) ____________ is used to transfer text, labels, graphics, and drawings onto paper and other materials.
11. Printed text or pictures can be changed into electronic files by using a(n) ____________.
12. Letters, reports, and forms can be written using a(n) ____________.
13. Math problems can be calculated using a software program that utilizes ____________.
14. Employee lists and material inventories are often organized and stored in a(n) ____________.
15. ____________ software is used to lay out text and graphics.
16. The Internet is typically searched using a(n) ____________.
17. The instructional manuals that come with software are called ____________.
18. ____________ communication allows computers to communicate without being physically connected.
19. Machine tools controlled by computers are called ____________ machines.
20. Slots and holes in metal can be made using a(n) ____________.
21. ____________ is the way you act, speak, and dress.
22. Another term for work habit is ____________.
23. ____________ is closely related to self presentation, but it relies more on your integrity and manner.
24. When you break ____________, you share with other people information that belongs to the company.
25. You will gain the respect of your supervisor if you regularly take the ____________ to complete additional tasks when your work is done.
26. A(n) ____________ approach should be used when approaching a co-worker about negative behavior.
27. ____________ and ____________ are the two most common problems supervisors are faced with workers.
28. ____________ is often necessary to resolve some situations.
29. Advice that is given to point out or correct a mistake is called ____________.
30. ____________ is often called cooperation.
31. Workers who focus on reaching final outcomes are considered ____________.
32. A person who sets an example for other people to follow is generally demonstrating good ____________ skills.
33. Racial jokes are considered a form of ____________.
34. Unwelcome sexual advances or requests are considered ____________.
35. Uppers are another word for ____________.
36. ____________ cause you to feel sedated.
37. The drug LSD is considered a(n) ____________.

38. The electronic communications network that connects computer networks and organizational computer facilities around the world is known as the ___________.

39. A(n) ___________ is a lightweight, portable data storage device.

40. The crystalline derivative of amphetamines is called ___________.

Trade Terms

Absenteeism	Desktop publishing	Internet	Self-presentation
Amphetamine	Documentation	Leadership	Sexual harassment
Barbiturate	EDM	Methamphetamine	Software
Browser	Entrepreneur	Mission statement	Spreadsheet
CNC	Goal-oriented	Operating system	Tactful
Compromise	Hallucinogen	PDA	Tardiness
Computer literacy	Handheld computer	Plotter	Teamwork
Confidentiality	Harassment	Processor	USB flash drive
Constructive criticism	Hard drive	Professionalism	Wireless
Critical thinking skills	Hardware	Reference	Word processor
Database	Initiative	Scanner	Work ethic

Drug and Alcohol Abuse Resources

Many organizations throughout the United States are devoted to helping people and families facing problems with alcohol or drugs. Contact any of the organizations listed below, and ask them to send you an information packet. Share the information in your packet with the class.

Al-Anon Family Group Headquarters, Inc.
http://www.al-anon.alateen.org/
1600 Corporate Landing Parkway
Virginia Beach, VA 23454-5617
Tel: (757) 563-1600
email: wso@al-anon.org

Alcoholics Anonymous World Services, Inc.
http://www.aa.org
P.O. Box 459
New York, NY 10163
Tel: (212) 870-3400
email: regionalforums@aa.org

Cocaine Anonymous World Service Office
http://www.ca.org
3740 Overland Avenue, Suite C
Los Angeles, CA 90034
Tel: 800-347-8998
email: cawso@ca.org

Marijuana Anonymous World Services
http://www.marijuana-anonymous.org
P.O. Box 2912
Van Nuys, CA 91404
Tel: 1-800-766-6779
email: office@marijuana-anonymous.org

Mothers Against Drunk Driving (MADD)
http://www.madd.org
511 E. John Carpenter Freeway, Suite 700
Irving, TX 75062
Tel: (800-438-6233)
Victim Services 24-hour hotline: 877-623-3435

Narcotics Anonymous World Service Office
http://www.na.org
P.O. Box 9999
Van Nuys, CA 91409
Tel: 818-773-999
email: fsmail@na.org

National Association for Children of Alcoholics
http://www.nacoa.org
11426 Rockville Pike, Suite 301
Rockville, MD 20852
Tel: 888-55-4COAS
email: nacoa@nacoa.org

National Clearinghouse for Alcohol and Drug Information
http://ncadi.samhsa.gov
P.O. Box 2345
Rockville, MD 20847-2345
Tel: 800-729-6686

National Council on Alcoholism and Drug Dependence
http://www.ncadd.org
244 East 58th Street, 4th Floor
New York, NY 10022
Tel: 800-NCA-CALL
email: national@ncadd.org

National Highway Traffic Safety Information
http://www.nhtsa.dot.gov
1200 New Jersey Avenue, SE West Building
Washington, DC 20590
Tel: 888-327-4236

Cornerstone of Craftsmanship

Erin M. Hunter

River Valley Technical Center
Carpentry Instructor
Springfield, Vermont

How did you become interested in carpentry?
I have always loved working with wood, building things, and being outside. My father has been in concrete construction as long as I can remember. He also did a lot of work renovating the house I grew up in. There was always a project to help with, so I guess you could say I grew up with it. My love of working with wood led me to jobs building furniture. My career in construction started on a historical restoration project as a finish carpenter/project manager. I later began taking on jobs of my own and learning a variety of rough carpentry skills. This led to the start of my own company called E.M. Hunter Construction, where I did everything from roofing to renovations.

What types of training have you been through and how did it help you get to where you are now?
I had no formal training in carpentry. I had to learn the hard way and from anyone willing to take the time to show me. I wish the opportunities that are available to today's young people were available when I started.

What kinds of work have you done in your career?
I have done custom furniture, historical restoration, finish carpentry, cabinetry, built-ins, tiling, decks, gazebos, timber-framing, roofing, additions; you name it.

What are some of the things you do in your job?
I currently teach a two-year carpentry program to juniors and seniors in a high school career and technical center.

What do you think it takes to be a success in your trade?
Carpentry is a craft that requires integrity, attention to detail, good people skills, adaptability, problem-solving skills, and — most importantly — the willingness to work. To be successful, carpenters need to care about the quality of their work and take pride in what they do.

What do you like about the work you do?
I really enjoy teaching 16- to 18-year-old students. I love helping them learn skills that are relevant and real-life. Carpentry is a great springboard into learning other life skills and disciplines like math. My students get to see themselves as successful in a school setting where they may not have been able to in other classes.

What advice would you give someone just starting out?
Stay open and be willing to learn. Carpentry is an ever-changing field. Sometimes I see young people cut themselves short because they feel that they should already know it all. One of the joys of carpentry is that there is always something you can learn or get better at. I encourage my students to be lifelong learners. Our class motto is "no fear." No fear of learning and no fear of trying something new.

Trade Terms Introduced in This Module

Absenteeism: Consistent failure to show up for work.

Amphetamine: A class of drugs that causes mental stimulation and feelings of euphoria.

Barbiturate: A class of drugs that induces relaxation.

Browser: Software that allows users to search the Internet.

CNC: Abbreviation for computer numerical control. A general term used to describe computer-controlled machine tools.

Compromise: When people involved in a disagreement make concessions to reach a solution that everyone agrees on.

Computer literacy: An understanding of how computers work and what they are used for.

Confidentiality: Privacy of information.

Constructive criticism: A positive offer of advice intended to help someone correct mistakes or improve actions.

Critical thinking skills: The skills required to evaluate and use information to make decisions or reach conclusions.

Database: Software that stores, organizes, and retrieves information.

Desktop publishing: Software used to lay out text and graphics for publication.

Documentation: Instruction manuals and other information for software.

EDM: Abbreviation for electrical discharge machines. Computer-controlled machine tools that cut and form parts that cannot be easily fabricated otherwise.

Entrepreneur: A person who starts and runs his or her own business.

Goal-oriented: To be focused on an objective.

Hallucinogen: A class of drugs that distort the perception of reality and cause hallucinations.

Handheld computer: A computer that is designed to be small and light enough to be carried.

Harassment: A type of discrimination that can be based on race, age, disabilities, sex, religion, cultural issues, health, or language barriers.

Hard drive: Hardware that stores software and electronic files.

Hardware: The physical components that make up a computer.

Initiative: The ability to work without constant supervision and solve problems independently.

Internet: Electronic communications network that connects computer networks and organizational computer facilities around the world.

Leadership: The ability to set an example for others to follow by exercising authority and responsibility.

Methamphetamine: A highly addictive crystalline drug, derived from amphetamines, that affects the central nervous system.

Mission statement: A statement of how a company does business.

Operating system: A complex set of commands that enables hardware and software to communicate.

PDA: Abbreviation for personal digital assistant. Another name for a handheld computer.

Plotter: Hardware that prints large architectural and construction drawings and images.

Processor: The part of a computer that contains the chips and circuits that allow it to perform its functions.

Professionalism: Integrity and work-appropriate manners.

Reference: A person who can confirm to a potential employer that you have the skills, experience, and work habits that are listed in your resume.

Scanner: Hardware that converts printed text or pictures into an electronic format.

Self-presentation: The way a person dresses, speaks, acts, and interacts with others.

Sexual harassment: A type of discrimination that results from unwelcome sexual advances, requests, or other verbal or physical behavior with sexual overtones.

Software: A large set of commands and instructions that direct a computer to perform certain tasks.

Spreadsheet: Software that performs mathematical calculations.

Tactful: Aware of the effects of your statements and actions on others.

Tardiness: Habitually showing up late for work.

Teamwork: The cooperation of co-workers to achieve one or more goals.

USB flash drive: A lightweight, portable data storage device.

Wireless: A technology that allows computers to communicate without physical connections.

Word processor: Software that is used for writing text documents.

Work ethic: Work habits that are the foundation of a person's ability to do his or her job.

Additional Resources

This module is intended to present thorough resources for task training. The following reference works are suggested for further study. These are optional materials for continued education rather than for task training.

Art and Science of Leadership. Afsaneh Nahavandi. Upper Saddle River, NJ: Prentice Hall.

Computer Numerical Control. John S. Stenerson. Upper Saddle River, NJ: Prentice Hall.

Introduction to Computer Numerical Control. James Valentino. Upper Saddle River, NJ: Prentice Hall.

Tools for Teams: Building Effective Teams in the Workplace. Craig Swenson, ed. Leigh Thompson, Eileen Aranda, Stephen P. Robbins. Boston, MA: Pearson Custom Publishing.

Your Attitude Is Showing. Elwood M. Chapman. Upper Saddle River, NJ: Prentice Hall.

CONTREN® LEARNING SERIES – USER UPDATE

NCCER makes every effort to keep these textbooks up-to-date and free of technical errors. We appreciate your help in this process. If you have an idea for improving this textbook, or if you find an error, a typographical mistake, or an inaccuracy in NCCER's Contren® textbooks, please write us, using this form or a photocopy. Be sure to include the exact module number, page number, a detailed description, and the correction, if applicable. Your input will be brought to the attention of the Technical Review Committee. Thank you for your assistance.

Instructors – If you found that additional materials were necessary in order to teach this module effectively, please let us know so that we may include them in the Equipment/Materials list in the Annotated Instructor's Guide.

Write: Product Development and Revision
National Center for Construction Education and Research
3600 NW 43rd St., Bldg. G, Gainesville, FL 32606

Fax: 352-334-0932

E-mail: curriculum@nccer.org

Craft ______________ Module Name ______________

Copyright Date ______________ Module Number ______________ Page Number(s) ______________

Description

____

____

____

____

(Optional) Correction

____

____

____

(Optional) Your Name and Address

____

____

____

Introduction to Materials Handling

00109-09

00109-09 Introduction to Materials Handling

CORE CURRICULUM

00109-09
Introduction to
Materials Handling

00108-09
Basic Employability
Skills

00107-09
Basic Communication
Skills

00106-09
Basic Rigging

00105-09
Introduction to
Construction Drawings

00104-09
Introduction to
Power Tools

00103-09
Introduction to
Hand Tools

00102-09
Introduction to
Construction Math

00101-09
Basic Safety

This course map shows all of the modules in the *Core Curriculum: Introductory Craft Skills.* The suggested training order begins at the bottom and proceeds up. Skill levels increase as you advance on the course map. The local Training Program Sponsor may adjust the training order.

Note that Module 00106-09, *Basic Rigging,* is an elective. It is not a requirement for level completion, but it may be included as part of your training program.

Objectives

When you have completed this module, you will be able to do the following:

1. Define a load.
2. Establish a pre-task plan prior to moving a load.
3. Use proper materials-handling techniques.
4. Choose appropriate materials-handling equipment for the task.
5. Recognize hazards and follow safety procedures required for materials handling.

Trade Terms

Concrete mule
Fall zone
Freight elevator
Hand truck
Industrial forklift
Jack
Load
Material cart
Pallet jack
Pipe mule
Pipe transport
Powered wheelbarrow
Roller skids
Rough terrain forklift
Spotter
Wheelbarrow
Work zone

Prerequisites

Before you begin this module, it is recommended that you successfully complete the following: *Core Curriculum,* Modules 00101-09 through 00105-09, 00107-09, and 00108-09. Module 00106-09 is an elective and is not a requirement for completion of this course.

Contents

Topics to be presented in this module include:

1.0.0 Introduction . . . 9.1
2.0.0 Materials-Handling Basics . . . 9.1
2.1.0 Pre-Task Planning . . . 9.1
2.2.0 Personal Protective Equipment . . . 9.1
2.3.0 Proper Lifting Procedures . . . 9.1
2.3.1 Lowering Loads from Overhead . . . 9.2
2.3.2 Back Injury Prevention . . . 9.3
3.0.0 Materials-Handling Safety . . . 9.3
3.1.0 Stacking and Storing Materials . . . 9.3
3.2.0 Working from Heights . . . 9.5
3.3.0 Working with Cables . . . 9.5
4.0.0 Materials-Handling Equipment . . . 9.6
4.1.0 Non-Motorized Materials-Handling Equipment . . . 9.6
4.1.1 Material Carts . . . 9.6
4.1.2 Hand Trucks . . . 9.7
4.1.3 Roller Skids . . . 9.7
4.1.4 Wheelbarrows . . . 9.7
4.1.5 Pipe Mule . . . 9.8
4.1.6 Pipe Transport . . . 9.8
4.1.7 Jack . . . 9.8
4.1.8 Pallet Jack . . . 9.9
4.2.0 Motorized Materials-Handling Equipment . . . 9.10
4.2.1 Powered Wheelbarrow . . . 9.10
4.2.2 Concrete Mule . . . 9.11
4.2.3 Industrial Forklift . . . 9.11
4.2.4 Rough Terrain Forklift . . . 9.12
4.2.5 Freight Elevator . . . 9.12
4.3.0 Hand Signals . . . 9.13

1.0.0 INTRODUCTION

Manual materials handling is a common task on most construction sites, and is one of the leading causes of non-fatal injuries in the construction industry. According to the Bureau of Labor Statistics, each year over one million workers suffer from back injuries, and one of every five workplace injuries or illnesses is associated with back injuries.

Most tasks performed in construction involve the handling of some type of material or **load**, such as wood, bricks, lumber, pipes, or other supplies, on a daily basis. A load is defined as the quantity of materials able to be carried, transported, or relocated at one time by a machine, vehicle, piece of equipment, or person. The risk of injury is present each time you handle materials, so always be conscious of what you're doing and follow basic materials-handling safety rules.

2.0.0 MATERIALS-HANDLING BASICS

To reduce the risk of injury when manually handling material, plan your task before doing it, wear the appropriate personal protective equipment (PPE), and follow proper lifting procedures. You must also be aware of hazards when working from heights or working near suspended loads.

2.1.0 Pre-Task Planning

When it comes to materials handling, it is just as important to be mentally fit as it is to be physically fit. Most materials-handling accidents occur when workers are new and relatively inexperienced at a job, or when very experienced workers think that accidents cannot happen to them. Before handling any material, consciously think about what you are doing. Always assess the situation before you attempt to lift any material by doing the following:

- Check to make sure the load is not too big, too heavy, or hard to grasp.
- Make sure the load does not have protruding nails, wires, or sharp edges.
- Make sure the material is something you are able to lift by yourself. If not, ask a co-worker for assistance.
- Inspect your path of travel. Look out for hazards that could make you slip, trip, or fall. If there are hazards in your path of travel, move them or go around them.
- Always read the warning labels or instructions on materials before they are moved, and be aware of the potential dangers if you mishandle a particular product.

2.2.0 Personal Protective Equipment

It's important to wear the proper clothing and personal protective equipment (PPE) when moving or handling materials. Remember the following safety guidelines when dressing to perform materials-handling operations:

- Do not wear loose clothing that can get caught in moving parts.
- Be sure to button shirt sleeves and tuck in shirt tails.
- Remove all rings and jewelry.
- Tie back and secure long hair underneath your hard hat.
- If you wear a wristwatch, wear one that will easily break away if it gets caught in machinery.
- Wear gloves whenever cuts, splinters, blisters, or other hand injuries are possible. Select gloves that fit properly. Tight gloves may increase hand fatigue. Loose gloves reduce grip strength and may get caught on moving objects or machinery.
- Gloves should be removed when working with rotating machinery and equipment with exposed moving parts.

2.3.0 Proper Lifting Procedures

To reduce the risk of back injuries, you must use proper lifting techniques. Determine the weight of the load prior to lifting and plan your lift. Know where it is to be unloaded and remove any slipping or tripping hazards from your path of travel.

When lifting objects, it is important to avoid unnecessary physical stress and strain. Know your limits and what you are able to physically handle when lifting a load. If the load is too heavy, ask a co-worker for help.

As you lift an object, make sure you have firm footing. Bend your knees and get a good grip. Be sure to lift with your legs, keep your back straight, and keep your head up. Keep the load close to your body (*Figure 1*). Never turn or twist until you are standing straight, then pivot your feet and body.

When unloading materials, use your leg muscles to set them down. Space your feet far enough apart to maintain good balance and control of the object. When placing the load down, move your fingers out of the way to avoid pinching or crushing them. When possible, attach handles to loads to reduce the chances of getting your fingers smashed.

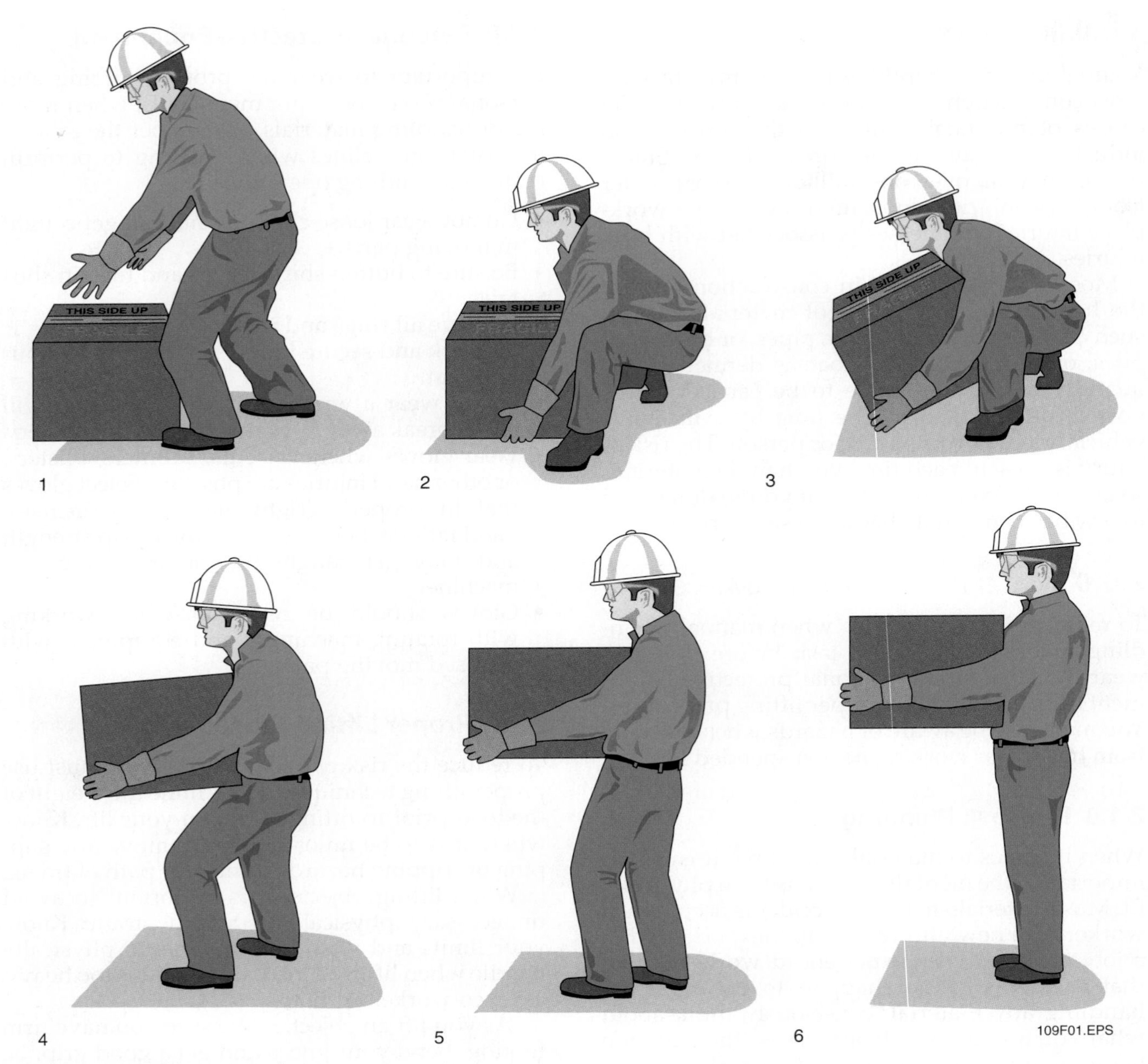

Figure 1 Proper lifting procedures.

Ask a co-worker for assistance when:

- The load is too heavy (weighing 50 pounds or more) to carry alone.
- Handling objects longer than 10 feet, such as lumber, conduit, pipes, and scaffolding poles.
- Handling objects that can be affected by wind gusts, such as plywood and tarps.

Sometimes it may not be the weight of a load that gives you trouble, but its configuration. Secure long or cumbersome objects. Floppy rods, pipes, or lumber should be tied together in several places before you try to carry them, and never carry more pieces than you can control.

When you are handling long objects, such as scaffold poles or planks, be aware of objects and persons that may be struck by them as you turn your body. Never swing around quickly or without checking your surroundings. Never pass any material to another worker unless the person is looking directly at you and expecting to receive it.

2.3.1 Lowering Loads from Overhead

Most workers have had training about the dangers of trying to lift heavy loads. Likewise, it is important to talk about lowering heavy loads from overhead. Reaching and lowering heavy loads from overhead can be very dangerous. When you are lifting something, you can always stop and put it down if you find that it is too heavy to handle alone. However, if you are lowering something from overhead, it is usually too late by the time you realize it is too heavy. Also, reaching above shoulder height can cause stress to the shoulders and back, and often results in awkward hand and wrist postures.

Before you attempt to lower overhead loads, consider the following:

- Size up the load. If it looks too heavy to have been lifted by one person to its current location, then it is probably too heavy for one person to take down.
- Ask yourself, "How did it get up there?" Was it put there by a lift truck or by more than one person? The way it got up is probably the best way to get it back down.
- Lower objects down the same way you would lift them up. Keep your knees bent and your back straight. If you have to place it to one side or another, move your feet instead of twisting your body.

2.3.2 Back Injury Prevention

Back injuries can easily occur when lifting and lowering materials. Keep in mind that exercise, good posture, and a healthy weight and diet can help prevent such injuries. You can also ask your doctor about muscle-strengthening techniques for your back.

3.0.0 Materials-Handling Safety

When working on a job site, you must ensure that you, your materials, and your equipment are safe from unexpected movement such as falling, slipping, tipping, rolling, blowing, or any other uncontrolled motion.

3.1.0 Stacking and Storing Materials

In order to work efficiently and effectively, workers should properly stack and secure materials. Taking the time to stack and secure materials correctly can save space and time, and helps to avoid accidents and injuries. Stacking materials keeps them from interfering with other activities and makes them more organized and readily accessible for use. Stacking materials also helps to create safe storage without danger to others and allows more room in the work area or on the job site.

You must always store materials and equipment properly when they are not in use. Learn your company- and site-specific rules, as well as those laid out by OSHA. Failing to store materials properly can lead to accidents, injuries, wasted materials, increased job costs, and project delays. For example, failing to remove nails from reusable lumber prior to storing it can cause puncture wounds when the lumber is taken out again. Stacking materials on the ground when they belong on a pallet can mean having to move them again in order to transport them safely and can also damage the materials. Leaving tools out when they should be stored can create an untidy work site, which can lead to an accident, tool damage, and misplaced items. Take the time to learn to store your materials properly. If you are unsure of how or where to store something, ask your supervisor.

WARNING!

Remember to remove all nails from reusable lumber to prevent accidental injuries.

Observe the following guidelines when stacking and storing materials:

- Keep aisles and passageways clear in the storage area.
- Materials stored in cartons should be kept away from rain and moisture.
- Never stack cartons higher than the height listed on the carton (*Figure 2*).
- Stack lumber on level, solidly supported sills, and remember to remove all nails.
- Stack pipes neatly and chock them so that they cannot roll.
- Stack pipes and fittings according to size so that you do not have to dig to find the piece required for the job (*Figure 3*).
- Chock all material and equipment such as pipes, drums, tanks, reels, trailers, and wagons, as necessary to prevent shifting or rolling.
- Tie down or band all light, large surface area materials that might be moved by the wind.
- Do not stack bagged material the same width all the way to the top of the pile—stack bagged material by stepping back the layers and cross-keying the bags at least every 10 bags high (*Figure 4*).
- Stack bricks no higher than seven feet—taper brick stacks back two inches for every foot above four feet (*Figure 5*).
- Taper masonry blocks back one-half block per tier above six feet (*Figure 6*).
- Store flammables, such as gasoline, in a well-ventilated area away from danger of ignition or in an approved flammable storage cabinet.

Figure 2 Properly stacked cartons.

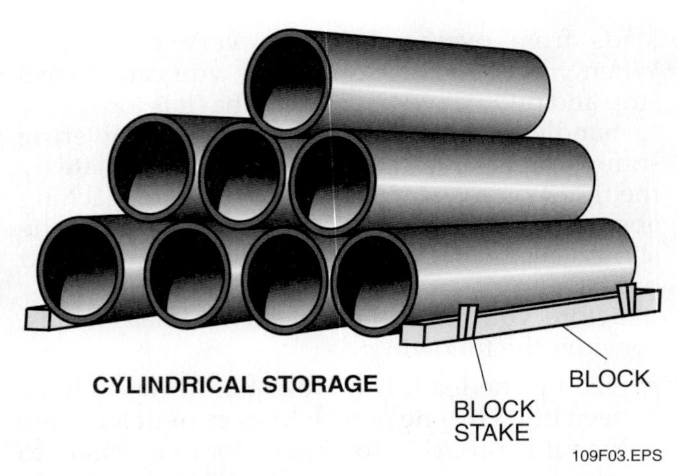

Figure 3 Properly stacked pipes.

Figure 4 Properly stacked bags.

GOING GREEN

When cleaning up materials around the job site, check to see if they can be reused or recycled rather than putting them in the trash.

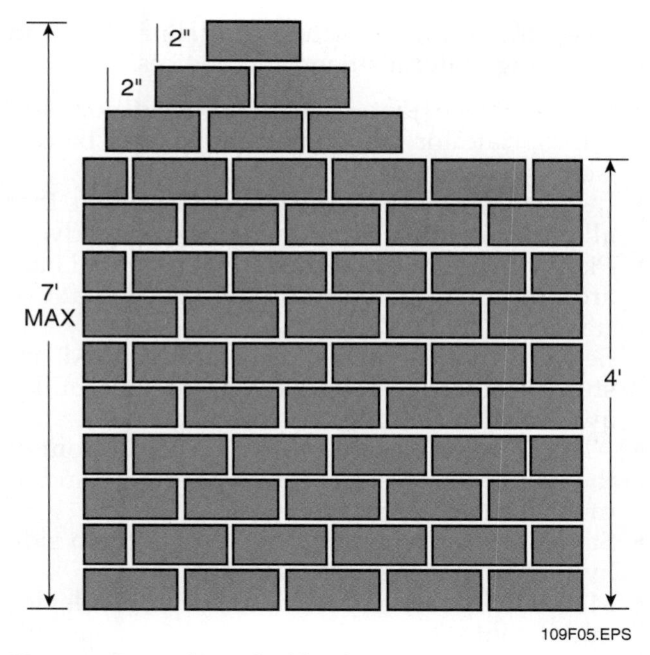

Figure 5 Properly stacked bricks.

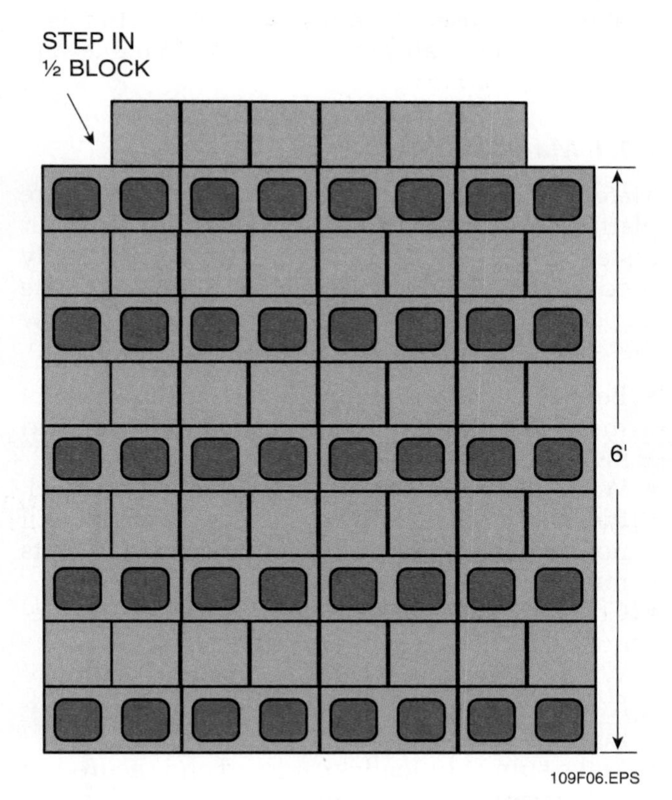

Figure 6 Properly stacked masonry blocks.

3.2.0 Working from Heights

Moving material to a higher location requires good planning and safety awareness. Use a safety harness or positioning belt with a lanyard when working from heights over six feet. Never throw tools or materials up to a worker on a higher level, or down to a worker on a lower level. Instead, use a rope to raise or lower tools and materials to and from an elevated work area. Once material is lifted to an elevated level, tie off the rope and secure the material to keep it from falling.

When climbing up or down a ladder, check your pockets and tool belt to ensure that no objects will fall out of them or get caught on something along the way. Always keep your hands free when climbing a ladder. Never carry tools or materials by hand up a ladder, or throw material up to or down from a work platform. Place objects in a bucket and hoist the materials to the platform.

Never stack or store materials on scaffolds or runways. Stacking materials on scaffolds or storing materials in high-traffic areas such as runways can increase the chances of falling object hazards.

3.3.0 Working with Cables

Often in construction, you may be required to pull various types of cable through conduit and wireways, or run it through walls, over ceilings, and under floors in order to install it at desired locations for a particular job. In some instances, you may be required to select the proper type of cabling to be used for the job. The following are several important safety precautions that will help to reduce the chance of being injured while pulling cable:

- Read and understand both the operating and safety instructions for the pull system before pulling cable.
- When moving reels of cable, avoid back strain by using the proper lifting techniques discussed earlier. Also, when manually pulling wire, spread your legs to maintain your balance and do not stretch.
- Select a rope that has a pulling load rating greater than the estimated forces required for the pull.
- Use only low-stretch rope, such as multi-ply and double-braided polyester, for cable pulling. High-stretch ropes store energy much like a stretched rubber band. If the rope, pulling grip, conductors, or other component in the pulling system fails, this potential energy will suddenly be unleashed. The whipping action of a rope can cause considerable damage, serious injury, or death.

> **WARNING!** Do not use high-stretch ropes to pull a load. If any component fails, the stored energy could sling-shot like a rubber band and injure someone or cause damage.

- Inspect the rope thoroughly before use. Make sure there are no cuts or frays in the rope. Remember, the rope is only as strong as its weakest point.
- When designing the pull, keep the rope confined in conduit wherever possible. Should the rope break or any other part of the pulling system fail, releasing the stored energy in the rope, the conduit will confine the whipping action of the rope.
- Wrap up the pulling rope after use to prevent others from tripping over it.

4.0.0 Materials-Handling Equipment

Instead of physically lifting heavy loads at work, use materials-handling devices when they are available to reduce your risk of back injury. Materials-handling devices help you to work smarter instead of harder, easing stress on your muscles and joints. These work-saving devices also increase productivity and lessen the chance of dropping and damaging equipment or supplies. There are two types of materials-handling equipment, non-motorized and motorized.

4.1.0 Non-Motorized Materials-Handling Equipment

When using non-motorized, or manual, materials-handling equipment, you use the force and strength of your body to move the equipment. Examples of manual materials-handling equipment include the following:

- **Material carts**
- **Hand trucks**
- **Roller skids**
- **Wheelbarrows**
- **Pipe mules**
- **Pipe transports**
- **Jacks**
- **Pallet jacks**

Keep the following safety guidelines in mind when using materials-handling devices:

- Use a device that is in good condition and appropriate for the job and the load to be carried.
- Inspect the device before using it to ensure that all parts are intact and functioning properly.
- Plan your route, and be aware of potential hazards that may be encountered on the path of travel.
- Be sure that all items to be transported are sturdy enough to be moved. Secure any bulky, awkward, or delicate objects.
- Always put the heaviest load on the bottom of the device to lower the center of gravity and to make it easier to handle.
- Stagger boxes when you must stack them side by side so that they are locked in.
- Maintain a safe speed and keep the device under control.
- Keep your hands and feet from underneath work-saving devices at all times.
- Try not to stack items higher than your line of sight. If your view is obstructed, a **spotter** should be used to assist you. A spotter is a worker who walks in front of you to make sure your path is clear.

4.1.1 Material Carts

Material carts, also known as platform trucks, are platforms or boards laid horizontally on four caster wheels. These types of carts are typically used to transport materials around a job site (*Figure 7*). When using a material cart, follow these safety guidelines:

- Before using carts with caster wheels, inspect them to ensure that the casters will roll and swivel freely during transport.
- When using a cart with pneumatic tires, check the tires' air pressure before use. Tires that will not hold pressure or those that have flat spots must be replaced.
- Check to make sure that the surface has adequate traction to avoid slips.
- When moving a cart, keep your hands away from the edges of the handles to avoid pinching, crushing, or cutting your hands.
- Make sure your load is centered and secured so it does not roll or fall off.
- If materials are protruding out in front of the cart, have a spotter walk ahead of you to help avoid contact with objects or other workers.

109F07.EPS

Figure 7 Material cart.

- Use caution when moving a cart on an inclined or declined surface, and never load a cart past its labeled weight capacity. If the load is too heavy on a decline, it may drag you down with it. If the load is too heavy on an incline, it may back over you. The weight capacity must be marked in plain view on the side of the cart.

4.1.2 Hand Trucks

Hand trucks, also known as dollies, are two-wheeled carts that are used to transport large, heavy loads, such as gas cylinders or drums. Hand trucks are vertical materials-handling devices that have a metal blade at the bottom that is inserted beneath the load. The entire assembly tilts backward so that it may be easily pulled or pushed (*Figure 8*).

Before using a hand truck, inspect the framework for stress fractures, and check the tires. When loading a hand truck, tilt the object to be moved forward and slide the cart under the object. Hold the object against the cart as you tilt both back on the wheels.

109F08.EPS

Figure 8 Hand truck.

4.1.3 Roller Skids

Roller skids move materials by pushing them on a table surface that is placed on top of two, three, or four roller skids. These devices are available in different capacities, depending on their intended usage. Roller skids used for moving heavy equipment use steel rollers or steel chain rollers, while those used for moving lighter equipment may use polyurethane or nylon rollers. The table surface of roller skids may also differ from case to case. Some come equipped with the rotating table surface, some have spikes on the table surface for better grip, and some simply have a plain surface just like any other table. Many roller skids have a pull-push handle attachment for operators to move them easily (*Figure 9*). As with other materials-handling equipment, inspect the roller skids before use and only use roller skids for their intended purpose and weight capacity.

4.1.4 Wheelbarrows

A wheelbarrow is a one- or two-wheeled vehicle with handles at the rear, used to carry small loads (*Figure 10*). On job sites, it is recommended that you use a two-wheeled wheelbarrow. Remember

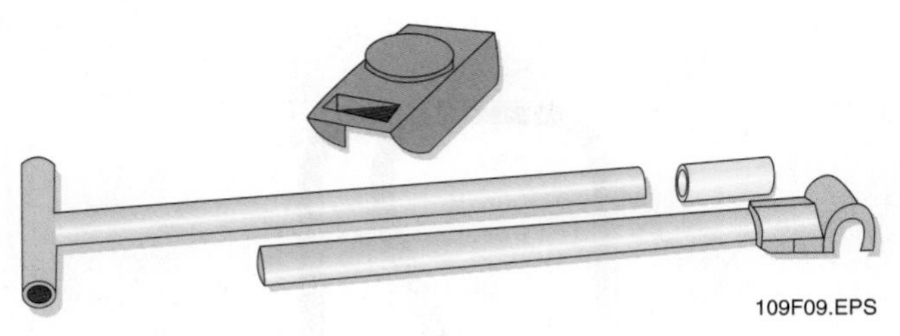

Figure 9 Roller skids.

Figure 10 Wheelbarrow.

to check the air pressure in both tires, and check the body of the wheelbarrow for cracks, breaks, or punctures before using it. Use proper lifting techniques when moving a wheelbarrow. Keep your head up, your back straight, bend your knees, and lift with your legs.

4.1.5 Pipe Mule

A pipe mule, sometimes referred to as a tunnel buggy, is a two-wheeled device used to transport medium-length pieces of pipe, tubing, or scaffolding (*Figure 11*). When loading or unloading a pipe mule, use proper lifting techniques. The proper loading method is to have one worker lift one end of the load while another worker rolls the pipe mule under the load. Place the wheels of the pipe mule underneath the center of the load. Use a spotter when using a pipe mule, as you would with any long load.

4.1.6 Pipe Transport

The pipe transport is similar to a pipe mule, but it is used to move larger pieces of pipe. The pipe is slung underneath the frame that is attached to two rubber wheels (*Figure 12*). This device requires two people to operate. Use caution when moving the pipe into position.

Figure 11 Pipe mule.

Figure 12 Pipe transport.

> **WARNING!**
> When using a pipe transport, beware of pinch points when securing chains around the pipe. Do not overload the equipment. Follow the manufacturer's recommendations.

4.1.7 Jack

A jack is a portable device used for raising heavy objects by means of force applied with a lever, screw, or hydraulic press (*Figure 13*). Jacks can be either motorized or non-motorized. Before using a jack, make sure you have the proper equipment and that all equipment has been inspected to ensure there are no oil leaks or cracks in the hardware. Also, check the area prior to jacking a load to ensure that the load will clear all obstacles.

Proper positioning of the jack is critical to safe lifting. Position the jack so that the load is centered. When using more than one jack, position the jacks so that the weight is uniformly distributed to avoid unexpected load shifts. When raising a load with multiple jacks, brace the load laterally with struts to ensure that the load cannot

109F13.EPS

Figure 13 Jack.

move laterally, causing the jacks to lean and ultimately to fail. When performing a multiple-jack lift, use extreme caution in controlling the lowering speed of the jacks. All jacks must be lowered at the same rate to ensure stability of the load as it is lowered. When using more than one ratchet-type jack, use a matched set of jacks to ensure a uniform lift.

Before jacking a load, make sure the base of the jack is placed on a solid, even, and level surface. Never place the base of the jack on bare soil or any other surface that could compact or shift under the load. If there is a possibility that the load could move while jacking, make sure it is chocked and restrained. Never jack metal against metal. Instead, use wood softeners as a buffer between metal surfaces.

If you are qualified to operate a jack, never stand under or place any part of your body under the load of a lifted jack, and never operate a jack while another worker is standing under or has any part of his or her body under the load. Operate the jack by applying pressure with your hands only. Never step on a jack handle to get additional force or leverage. Never use jack handle extensions or cheater bars. Check the jack's maximum rated capacity. If hand pressure is not enough to operate the jack, then stop, block up the load, and get a bigger jack.

Sometimes, when making a jacked lift, the supervisor may choose to use a hydraulic ram, also referred to as a Porta-Power®. A Porta-Power® is especially useful for horizontal jacking. It is a good practice to have the body of the cylinder blocked from beneath when jacking horizontally. As with any piece of equipment, always inspect the Porta-Power® before use. See that all hose fittings are tight and that there are no leaks. Also inspect the hose to ensure that it has not been damaged.

Never leave a jack under a load without blocking it first. When it becomes necessary to use blocking for elevation adjustments, the blocking should always be placed directly under the jack base. Never place blocking material between the jack ram and the object being lifted. Blocking material must be of sufficient size and solidarity to provide for a stable base for the jack.

To avoid accidentally disengaging the jack, and to reduce tripping hazards, always remove the operating lever when the jack is not being used. When storing hydraulic jacks, always store them upright to prevent the loss of fluid.

4.1.8 Pallet Jack

A pallet jack, also known as a pallet truck, is a device that typically uses hydraulics to lift and move heavy or stacked pallets (*Figure 14*). Remember the following safety guidelines when using a pallet jack:

- Inspect pallet jacks before use, looking carefully for malfunctions or missing parts.
- Check for oil leaks around fittings or hoses of hydraulic pallet jacks, and repair any leaks before the jack is put into service.
- Never use a pallet jack to lift or transport a load that exceeds the jack's rated capacity. The load rating should be clearly posted on the body of the jack.
- Inspect the wood pallet to ensure it is in good condition to adequately handle the load, and make sure the load is centered between the pallet jack forks.

109F14.EPS

Figure 14 Pallet jack.

4.2.0 Motorized Materials-Handling Equipment

Motorized materials-handling equipment is powered by gasoline or electric motors. You must be trained, certified, and authorized to operate motorized materials-handling equipment. Training courses must consist of a combination of formal instruction, hands-on training, and evaluation of the operator's performance in the workplace for each type of materials-handling equipment being used. All operator training and evaluation must be conducted by a person who has knowledge, training, and experience.

When lifting materials mechanically, you must carefully secure the materials you are loading. You must also ensure that you are aware of any obstructions, such as people or equipment, that are in your path so that you avoid potential accidents.

When handling materials with a machine, such as a pallet jack or forklift, always follow these guidelines:

- Know the weight of the object to be handled/ moved.
- Know the capacity of the handling device that you intend to use and never exceed it.
- Ensure that your handling equipment is in good working order and free of damage.

Some of the motorized materials-handling equipment discussed in this section include the following:

- **Powered wheelbarrow**
- **Concrete mule**
- **Industrial forklift**
- **Rough terrain forklift**
- **Freight elevator**

> **NOTE**
> Nameplates must be posted on each materials-handling device. The nameplate must indicate the capacity of the device, the approximate weight, and any instructional information.

GOING GREEN: Fuel Cell Technology

It is anticipated that within the next few years, fuel cell technology will be applied to industrial materials-handling equipment. A fuel cell is an electrochemical device that combines hydrogen and oxygen to produce electricity. Since the conversion of the fuel to energy takes place using an electrochemical process, and not combustion, the process is clean, quiet, and highly efficient. Fuel cell technology is estimated to be two to three times more efficient than fuel burning.

4.2.1 Powered Wheelbarrow

A powered wheelbarrow, also known as a power buggy, is similar to a manual wheelbarrow, but is powered by an electric or gas motor (*Figure 15*). Before using a powered wheelbarrow, conduct a thorough inspection of the machine. Check the brakes and brake linkages for proper operation and adjustment. Inspect tires and wheels for damage and proper tire pressure. Test the directional controls, dumping controls, and speed controls to verify they are all working properly. Be sure to read safety labels and decals for proper safety instructions. Safety labels should be clearly visible and readable on the side of the machine.

If you are certified to operate a powered wheelbarrow, you must have a thorough understanding of all operating controls. Keep the following safety guidelines in mind when operating a powered wheelbarrow:

- Avoid pinch points on dumping mechanisms.
- Be sure the dump bucket is securely down at all times when not dumping.
- Do not move the buggy with the bucket in the dumping position.
- Do not mount or dismount the buggy while it is moving.
- Shut off the engine and lock the parking brake when you are finished using the machine.
- Do not allow riders.
- Avoid any condition of slope and/or grade which could cause the buggy to tip.
- Do not exceed load limits in weight or height.
- Follow all procedures outlined on the safety labels and decals.

109F15.EPS

Figure 15 Powered wheelbarrow.

4.2.2 Concrete Mule

A concrete mule, sometimes referred to as a Georgia buggy, is a wheeled device used when a concrete pour is in a place that a concrete delivery truck or pump cannot reach (*Figure 16*). Concrete mules are designed to carry concrete, sand, and gravel, or materials of that nature, and should not be used to transport undesignated materials.

Before using a concrete mule, you must conduct a thorough inspection of the following:

- Tires
- Fluids
- Mechanical hinges and joints
- Throttle cables and steering mechanisms

When refueling a concrete mule, or any other gas-powered materials-handling machine, always be sure that the engine is turned off and has cooled down before refueling. Only use fuel and oil types that are specified by the manufacturer.

When discharging concrete mix from a concrete mule near an open pit, use a tire stop board or chocks to prevent the mule from rolling forward into the pit during the emptying process.

When operating a concrete mule, make sure there are no obstacles or other workers in the turning radius of the rear platform. Always look behind you before operating in reverse, and never allow anyone other than the operator to ride on the machine.

4.2.3 Industrial Forklift

An industrial forklift is a vehicle with a power-operated pronged platform that can be raised and lowered for insertion under a load to be lifted and moved (*Figure 17*). Forklifts are typically used to lift, lower, and transport large or heavy loads in areas with a smooth terrain, such as warehouses or shops.

Figure 16 Concrete mule.

Figure 17 Industrial forklift.

When working in the vicinity of a forklift, remember the following safety guidelines:

- All workers must keep a safe distance from the machine. This perimeter is known as the **work zone**. It is the area in which the machine may come in contact with objects or people either with the rear end or the front forks.
- Workers must also stay clear of the **fall zone**. This area includes a diameter twice the height of the object being lifted. For instance, if an object is lifted 10 feet, then the fall zone is a diameter of 20 feet.
- Stay in designated walkways and make eye contact with the forklift operator to ensure that he or she sees you. Always avoid the driver's blind spot. Be aware that forklifts pivot on the front wheels and turn with the back wheels, which could cause a potential pinch point.
- Plan your route of escape. Never get caught in a position where you are trapped in a corner or on the side of a truck.

If you are trained and certified to operate a forklift, there are certain traffic safety rules and regulations you must observe while driving.

Before operating a forklift, put on the proper PPE and fasten your seat belt. If you are not wearing a seat belt and the forklift tips, you can easily be pitched under it, or have some of its load land on you. Falling objects are common when stacking and lifting with a forklift, so be sure to wear a hardhat. As with most machinery, when working with a forklift do not wear baggy clothes that can catch on controls, cargo, or stacked objects.

Never horse around on a forklift. Accidents often happen because workers forget to respect equipment. A forklift is not a toy and should not be treated like one.

Before operating the forklift each day, inspect the following:

- Tires
- Fluid levels
- Battery/fuel level
- Fire extinguisher
- Brakes
- Deadman control
- Warning lights
- Horn
- Backup alarm

Use a checklist, if one is available. If you find any defects, report them to your supervisor.

Forklift operators with the fewest accidents drive with slow deliberation, take their time, and do not make mistakes. Do not attempt to cut corners to speed up your work. Drive slowly and keep your arms, hands, and legs inside the cage of the forklift. Forklifts are top-heavy and, when driven recklessly, can tip over on a curve, drive off the edge of a dock, or fail to stop in time. Drive at a speed that is appropriate and safe for the location and surface condition and use extra caution on hills, corners, and ramps. Keep the forklift to the right on roadways and in wide aisles, and stop at all designated stop signs. Always start and stop smoothly. This helps to preserve the equipment by putting less wear and tear on the forklift. It also keeps loads from shifting, objects from falling, and helps prevent damage to cargo, equipment, and operators.

Remember that pedestrians have the right-of-way. Forklifts are often so quiet that they are difficult to hear. Pedestrians may not be aware that you are there, so be careful when you are driving near people, especially if they have their back to you. Make sure that the backup alarm is functional. Driving in reverse presents special dangers to pedestrians, especially if you are watching the load in front as you back up.

Workers sometimes attempt to make modifications to forklifts so that they can transport heavier or wider loads than the equipment is designed to lift. This presents the hazards of broken masts or forks, dropped loads, and tipovers. Never modify a forklift without obtaining the consent from the manufacturer or from a mechanical engineer familiar with the equipment.

4.2.4 Rough Terrain Forklift

Rough terrain forklifts are designed to be used on rough surfaces. They are typically used for jobs where there are no paved surfaces. Rough terrain forklifts are characterized by large pneumatic tires, usually with deep treads that allow the vehicle to grab onto the roughest of roads or ground cover without sliding or slipping (*Figure 18*). Be sure to apply the same safety guidelines to rough terrain forklifts as you would to industrial forklifts.

One of the most common accidents involving rough terrain forklifts is tipovers. This is because these types of forklifts are often operated on unlevel ground. To prevent tipovers with rough terrain forklifts, observe the following safety guidelines:

- Study load charts carefully and only operate the forklift within its stability limits.
- Before placing a load, apply the brakes, shift to neutral, level the frame, and engage the stabilizers, if so equipped.
- Carry the load low. Avoid driving with an elevated load except for short distances.
- Move carefully on slopes and only operate within specific grade limits.
- Retract the boom and lower the forks before moving a forklift.

4.2.5 Freight Elevator

On multi-level sites, you may use a freight elevator to transport materials from floor to floor. Typically, when freight elevator doors close they travel from the top and bottom simultaneously (*Figure 19*). These doors do not have the safety bumpers that you would find on a commercial

109F18.EPS

Figure 18 Rough terrain forklift.

109F19.EPS

Figure 19 Freight elevator.

passenger elevator. Rather, as you pull the top door down, by way of a strap or handle, the bottom door rises to meet the top door. When these doors close, it is normally metal-against-metal and this can cause injury. Make sure that you have all body parts such as arms, hands, and fingers clear of the door before it closes. Check to make sure that tool belts, body harnesses, and all materials are also clear of the door. Always check the weight capacity shown on the elevator wall before placing material on an elevator.

4.3.0 Hand Signals

When working with motorized lifting equipment, such as forklifts, the noise generated by the equipment and the distance that workers are set apart can make verbal communication difficult. Standard hand signals (*Figure 20*) allow both the workers and the operator to go through the lifting, loading, moving, and unloading procedures at a more controlled setting, and help to avoid potential misunderstandings and accidents.

For instance, it is important for the operator to know when to raise or lower the tines (the L-shaped parts that support the material load) to avoid obstructions that may be in their path. Another very important signal for the operator to know is "dog everything." This signal means to pause, and can be used when potentially risky situations arise, such as when it starts raining, when the load doesn't fit the space for which it was planned, or when a bystander gets too close to the action.

When using hand signals, make sure you keep eye contact with the operator at all times. There may be special operations that require adaptations of the basic hand signals, so be sure to review the signals for your work site from time to time. If there is a misunderstanding regarding hand signals, go over the signals with the operator before proceeding. One wrong signal could cause a serious injury or death.

Figure 20 Common forklift hand signals.

Review Questions

1. When performing materials-handling tasks, it is important to be _____.
   a. mentally and physically fit
   b. able to carry large or heavy loads
   c. nimble enough to move around obstacles
   d. able to work from heights

2. Loose gloves reduce _____ and may get caught on moving objects and machinery.
   a. friction
   b. rigidity
   c. torque
   d. grip strength

3. To reduce the risk of back injuries, you must use proper _____ techniques.
   a. machine-operating
   b. electrical wiring
   c. lifting
   d. conversational

4. When working with motorized lifting equipment and verbal communication is difficult, workers should rely on _____ to communicate.
   a. written notes
   b. shouting
   c. hand signals
   d. a relay person

5. Reaching above _____ height can cause stress to the shoulders and back.
   a. knee
   b. hip
   c. shoulder
   d. head

6. The proper way to get tools to a worker on a higher level is to _____.
   a. toss them carefully
   b. carry them up a ladder
   c. place them in your pocket
   d. use a rope and bucket

7. Never leave a jack under a load without _____.
   a. checking it
   b. weighing it
   c. lowering it
   d. blocking it

8. When handling materials with a machine, know the _____ of the object to be moved.
   a. height
   b. volume
   c. weight
   d. girth

9. An example of a motorized materials-handling device is a _____.
   a. concrete mule
   b. jack
   c. Porta-Power®
   d. pipe mule

10. If your view is obstructed when handling materials, use a _____ to assist you and make sure your path is clear.
    a. mirror
    b. spotter
    c. tow rope
    d. step ladder

SUMMARY

Materials handling is one of the most common tasks on a job site, and it is also one that can cause accidents or injuries if not done properly. Workers must follow safety procedures for lifting, carrying, and transporting materials, whether doing so manually or using a piece of materials-moving equipment.

This module has presented many of the basic guidelines you must follow to ensure your safety and the safety of your co-workers. These guidelines fall into the following categories:

- Using proper PPE for materials-handling procedures
- Planning your route before handling materials
- Using proper lifting procedures
- Inspecting materials-handling equipment before use
- Following proper materials-handling equipment procedures

Trade Terms Quiz

Fill in the blank with the correct key term that you learned from your study of this module.

1. The area whose diameter is twice the height of the object being lifted is known as the ____________.
2. A(n) ____________ is used to move large pieces of pipe, while a(n) ____________ is used to transport medium-length pieces of pipe, tubing, or scaffolding.
3. A(n) ____________ is a one- or two-wheeled vehicle with handles at the rear, used to carry small loads.
4. A(n) ____________ is a vehicle with a power-operated pronged platform that can be raised and lowered for insertion under a load to be lifted and moved.
5. Sometimes referred to as a Georgia buggy, a(n) ____________ is a wheeled device used when a delivery truck or pump cannot reach.
6. A(n) ____________ is similar to its industrial counterpart, except that it's designed to be used on rough surfaces.
7. Also known as dollies, ____________ are two-wheeled carts that are used to transport large, heavy loads, such as gas cylinders or drums.
8. The quantity of materials able to be carried, transported, or relocated at one time by a machine, vehicle, piece of equipment, or person is known as a(n) ____________.
9. Also known as a power buggy, a(n) ____________ is powered by an electric or gas motor.
10. The area in which a forklift may come in contact with objects or people either with the rear end or the front forks is known as the ____________.
11. A(n) ____________ is a device that includes a surface table and two, three, or four skids.
12. Also known as a pallet truck, a(n) ____________ is a device used to lift and move heavy or stacked pallets.
13. A(n) ____________ is a four-wheeled device used to transport supplies around a job site.
14. A machine used to transport materials from one floor to another is called a(n) ____________.
15. A(n) ____________ is a portable device used for raising heavy objects by means of force applied with a lever, screw, or hydraulic press.
16. A(n) ____________ is a person who walks in front of you when you are transporting a long load to ensure you have a clear, unobstructed path.

Trade Terms

Concrete mule
Fall zone
Freight elevator
Hand truck
Industrial forklift
Jack
Load
Material cart
Pallet jack
Pipe mule
Pipe transport
Powered wheelbarrow
Roller skids
Rough terrain forklift
Spotter
Wheelbarrow
Work zone

Cornerstone of Craftsmanship

Tim Grattan

Corbins Electric, LLC
Superintendent

How did you choose a career in the field?
I chose the electrical field because I was looking for a well-paying construction trade with growth potential.

What types of training have you been through?
I have been able to take advantage of a wealth of training. The list is long; however, I can highlight a few. I have taken NCCER's Master Trainer course and attended the Supervisors Academy at Clemson University, which uses NCCER's curriculum. I am a master graduate of Rapport Leadership International. I have undertaken numerous electrical trade training courses, including High Voltage Terminations, NFPA70® Energized Electrical Work certification, NEC Code courses, OSHA 10, Cadweld, Boom Lift, Reach Fork, Backhoe, 3M Fire Seal, and more. And I took a course on BIM Preparing Interview Team for Success at the Del Webb School of Construction at Arizona State University.

What kinds of work have you done in your career?
I have performed electrical installations for a variety of diverse construction projects, large and small. Some of the interesting large projects include installations in a prison facility; German Air Force hangars at Holloman Air Force Base; a University of Texas at El Paso (UTEP) classroom building; barracks buildings and dining facilities at Fort Bliss, Texas; a Joint Use Sports Training Facility at Holloman Air Force Base; an underground Primary Loop at Fort Bliss, Texas; the Arizona State University Foundation Building; Intel Corporation's Ocotillo manufacturing and development campus in Chandler, Arizona; a Kohl's Department Store; and the Westin Kierland Resort.

Tell us about your present job and what you like about it.
I am a superintendent for Corbins Electric and really love what I do. My main duties are to distribute and maintain appropriate manpower throughout the state of Arizona. I teach foreman's development classes along with apprenticeship courses through the NCCER. I am involved and chair several committees in our organization that deal with everything from organizational strategic planning to bonus programs and training oversight. It is challenging and very time consuming, but also very rewarding.

What factors have contributed most to your success?
Dedication and always being willing to go the extra mile without complaining or making excuses. Being focused on success and having a plan and goals with realistic milestones has always paid off for me. Another thing that has always helped is self-analysis and seeking feedback for improvement, always working on improving my performance regardless of the level I had achieved at the time.

What advice would you give to those new to the field?
Decide early on what you want to do with your career and make a plan as to how to achieve your goals. Learn all you can. Always keep an open mind. New products and methods are always becoming available.

Tell us some interesting career-related facts or accomplishments:
I never knew until I had been in the trade for a while, and started seeing comparisons on wages between college graduates and construction professionals, how much money we really make in this industry. It's a great profession for compensation and job satisfaction.

Trade Terms Introduced in This Module

Concrete mule: Sometimes referred to as a Georgia buggy, a wheeled device used when a concrete pour is in a place that a concrete delivery truck or pump cannot reach.

Fall zone: When lifting, the area having a diameter twice the height of the object being lifted.

Freight elevator: An elevator used to transport materials from floor to floor.

Hand truck: Also known as dollies, hand trucks are two-wheeled carts that are used to transport large, heavy loads, such as gas cylinders or drums.

Industrial forklift: A vehicle with a power-operated pronged platform that can be raised and lowered for insertion under a load to be lifted and moved.

Jack: Portable device used for raising heavy objects by means of force applied with a lever, screw, or hydraulic press.

Load: The quantity of materials able to be carried, transported, or relocated at one time by a machine, vehicle, piece of equipment, or person.

Material cart: Four-wheeled device used to transport materials around a job site.

Pallet jack: Also known as a pallet truck, a device used to lift and move heavy or stacked pallets.

Pipe mule: Sometimes referred to as a tunnel buggy, a two-wheeled device used to transport medium-length pieces of pipe, tubing, or scaffolding.

Pipe transport: Similar to a pipe mule, but used to move larger pieces of pipe.

Powered wheelbarrow: Also known as a power buggy, similar to a manual wheelbarrow, but powered by an electric or gas motor.

Roller skids: A device that includes a surface table and two, three, or four roller skids. Materials that are to be moved are placed on the table surface and then pushed on the skids.

Rough terrain forklift: Similar to an industrial forklift, but designed to be used on rough surfaces. Rough terrain forklifts are characterized by large pneumatic tires, usually with deep treads that allow the vehicle to grab onto the roughest of roads or ground cover without sliding or slipping.

Spotter: A person who walks in front of another worker who is carrying or transporting a long load to ensure there is a clear, unobstructed path.

Wheelbarrow: A one- or two-wheeled vehicle with handles at the rear, used to carry small loads.

Work zone: The area in which a forklift may come in contact with objects or people, either with the rear end or the front forks.

Additional Resources

This module is intended to present thorough resources for task training. The following reference work is suggested for further study. This is optional material for continued education rather than for task training.

Make More Money with Construction Machine Control—A How To Manual for Site-Prep Contractors. First Edition. 2008. TrenchSafety. Little Rock, AK: TrenchSafety.

CONTREN® LEARNING SERIES – USER UPDATE

NCCER makes every effort to keep these textbooks up-to-date and free of technical errors. We appreciate your help in this process. If you have an idea for improving this textbook, or if you find an error, a typographical mistake, or an inaccuracy in NCCER's Contren® textbooks, please write us, using this form or a photocopy. Be sure to include the exact module number, page number, a detailed description, and the correction, if applicable. Your input will be brought to the attention of the Technical Review Committee. Thank you for your assistance.

Instructors – If you found that additional materials were necessary in order to teach this module effectively, please let us know so that we may include them in the Equipment/Materials list in the Annotated Instructor's Guide.

Write: Product Development and Revision
National Center for Construction Education and Research
3600 NW 43rd St., Bldg. G, Gainesville, FL 32606

Fax: 352-334-0932

E-mail: curriculum@nccer.org

Craft ____________ Module Name ____________

Copyright Date ____________ Module Number ____________ Page Number(s) ____________

Description ____________

____

____

____

(Optional) Correction ____________

____

____

(Optional) Your Name and Address ____________

____

____

Core Curriculum Glossary

Abrasive: A substance, such as sandpaper, that is used to wear away material.

Absenteeism: Consistent failure to show up for work.

AC (alternating current): An electrical current that reverses its direction at regularly recurring intervals; the current delivered through wall plugs.

Active listening: A process that involves respecting others, listening to what is being said, and understanding what is being said.

Acute angle: Any angle between 0 degrees and 90 degrees.

Adjacent angles: Angles that have the same vertex and one side in common.

Adjustable end wrench: A smooth-jawed adjustable wrench used for turning nuts, bolts, and pipe fittings. Often referred to as a Crescent® wrench.

Amphetamine: A class of drugs that causes mental stimulation and feelings of euphoria.

Angle: The shape made by two straight lines coming together at a point. The space between those two lines is measured in degrees.

ANSI hand signals: Communication signals established by the American National Standards Institute (ANSI) and used for load navigation for mobile and overhead cranes.

Apparatus: An assembly of machines used together to do a particular job.

Appendix: A source of detailed or specific information placed at the end of a section, a chapter, or a book.

Arc: The flow of electrical current through a gas (such as air) from one pole to another pole.

Arc welding: The joining of metal parts by fusion, in which the necessary heat is produced by means of an electric arc.

Architect: A qualified, licensed person who creates and designs drawings for a construction project.

Architect's scale: A specialized ruler used in making or measuring reduced scale drawings. It is marked with a range of calibrated ratios used for laying out distances, with scales indicating feet, inches, and fractions of inches. Used on drawings other than site plans.

Architectural plans: Drawings that show the design of the project. Also called architectural drawings.

Area: The surface or amount of space occupied by a two-dimensional object such as a rectangle, circle, or square. To calculate the area for rectangles and squares, multiply the length and width. To calculate the area for circles, multiply the radius squared and pi.

Auger: A tool with a spiral cutting edge for boring holes in wood and other materials.

Ball peen hammer: A hammer with a flat face that is used to strike cold chisels and punches. The rounded end—the peen—is used to bend and shape soft metal.

Barbiturate: A class of drugs that induces relaxation.

Beam: A large, horizontal structural member made of concrete, steel, stone, wood, or other structural material to provide support above a large opening.

Bell-faced hammer: A claw hammer with a slightly rounded, or convex, face.

Bevel: To cut on a slant at an angle that is not a right angle (90 degrees). The angle or inclination of a line or surface that meets another at any angle but 90 degrees.

Bisect: To divide into equal parts.

Block and tackle: A simple rope-and-pulley system used to lift loads.

Blueprints: The traditional name used to describe construction drawings.

Body language: A person's physical posture and gestures.

Booster: Gunpowder cartridge used to power powder-actuated fastening tools.

Borrow: To move numbers from one value column (such as the tens column) to another value column (such as units) to perform subtraction problems.

Box-end wrench: A wrench, usually double-ended, that has a closed socket that fits over the head of a bolt.

Bridle: A configuration using two or more slings to connect a load to a single hoist hook.

Browser: Software that allows users to search the Internet.

Bull ring: A single ring used to attach multiple slings to a hoist hook.

Bullets: Large, vertically aligned dots that highlight items in a list.

Carbide: A very hard material made of carbon and one or more heavy metals. Commonly used in one type of saw blade.

Carpenter's square: A flat, steel square commonly used in carpentry.

Carry: To transfer an amount from one column to another column.

Cat's paw: A straight steel rod with a curved claw at one end that is used to pull nails that have been driven flush with the surface of the wood or slightly below it.

Chisel: A metal tool with a sharpened, beveled edge used to cut and shape wood, stone, or metal.

Chisel bar: A tool with a claw at each end, commonly used to pull nails.

Chuck: A clamping device that holds an attachment; for example, the chuck of the drill holds the drill bit.

Chuck key: A small, T-shaped steel piece used to open and close the chuck on power drills.

Circle: A closed curved line around a central point. A circle measures 360 degrees.

Circumference: The distance around the curved line that forms a circle.

Civil plans: Drawings that show the location of the building on the site from an aerial view, including contours, trees, construction features, and dimensions.

Claw hammer: A hammer with a flat striking face. The other end of the head is curved and divided into two claws to remove nails.

CNC: Abbreviation for computer numerical control. A general term used to describe computer-controlled machine tools.

Combination square: An adjustable carpenter's tool consisting of a steel rule that slides through an adjustable head.

Combination wrench: A wrench with an open end and a closed end.

Combustible: Capable of easily igniting and rapidly burning; used to describe a fuel with a flash point at or above 100°F.

Competent person: A person who is capable of identifying existing and predictable hazards in the surroundings or working conditions which are unsanitary, hazardous, or dangerous to employees, and who has authorization to take prompt corrective measures to eliminate them.

Compromise: When people involved in a disagreement make concessions to reach a solution that everyone agrees on.

Computer-aided drafting (CAD): The making of a set of construction drawings with the aid of a computer.

Computer literacy: An understanding of how computers work and what they are used for.

Concealed receptacle: The electrical outlet that is placed inside the structural elements of a building, such as inside the walls. The face of the receptacle is flush with the finished wall surface and covered with a plate.

Concrete mule: Sometimes referred to as a Georgia buggy, a wheeled device used when a concrete pour is in a place that a concrete delivery truck or pump cannot reach.

Confidentiality: Privacy of information.

Confined space: A work area large enough for a person to work, but arranged in such a way that an employee must physically enter the space to perform work. A confined space has a limited or restricted means of entry and exit. It is not designed for continuous work. Tanks, vessels, silos, pits, vaults, and hoppers are examples of confined spaces. See also *permit-required confined space.*

Construction drawings: Architectural or working drawings used to represent a structure or system.

Constructive criticism: A positive offer of advice intended to help someone correct mistakes or improve actions.

Contour lines: Solid or dashed lines showing the elevation of the earth on a civil drawing.

Convert: To change from one unit of expression to another. For example, to convert a decimal to a percentage: 0.25 to 25%; or to convert a fraction to an equivalent: ¾ to ⅝.

Core: Center support member of a wire rope around which the strands are laid.

Countersink: A bit or drill used to set the head of a screw at or below the surface of the material.

Cribbing: Material used to either support a load or allow removal of slings after the load is landed. Also called blocking.

Critical thinking skills: The skills required to evaluate and use information to make decisions or reach conclusions.

Cross-bracing: Braces (metal or wood) placed diagonally from the bottom of one rail to the top of another rail that add support to a structure.

Cubic: Measurement found by multiplying a number by itself three times; it describes volume measurement.

Database: Software that stores, organizes, and retrieves information.

DC (direct current): Electrical current that flows in one direction, from the negative (2) to the positive (1) terminal of the source, such as a battery.

Decimal: Part of a number represented by digits to the right of a point, called a decimal point. For example, in the number 1.25, .25 is the decimal part of the number.

Degree: A unit of measurement for angles. For example, a right angle is 90 degrees, an acute angle is between 0 and 90 degrees, and an obtuse angle is between 90 and 180 degrees.

Denominator: The part of a fraction below the dividing line. For example, the 2 in ½ is the denominator.

Desktop publishing: Software used to lay out text and graphics for publication.

Detail drawings: Enlarged views of part of a drawing used to show an area more clearly.

Diagonal: Line drawn from one corner of a rectangle or square to the farthest opposite corner.

Diameter: The length of a straight line that crosses from one side of a circle, through the center point, to a point on the opposite side. The diameter is the longest straight line you can draw inside a circle.

Difference: The result you get when you subtract one number from another. For example, in the problem 8 – 3 = 5, the number 5 is the difference.

Digit: Any of the numerical symbols 0 to 9.

Dimension line: A line on a drawing with a measurement indicating length.

Dimensions: Measurements such as length, width, and height shown on a drawing.

Disk drive: Hardware that allows software and files to be transferred between computers.

Documentation: Instruction manuals and other information for software.

Dowel: A pin, usually round, that fits into a corresponding hole to fasten or align two pieces.

Dross: Waste material resulting from cutting using a thermal process.

EDM: Abbreviation for electrical discharge machines. Computer-controlled machine tools that cut and form parts that cannot be easily fabricated otherwise.

Electric tools: Tools powered by electricity. The electricity is supplied by either an AC source (wall plug) or a DC source (battery).

Electrical distribution panel: Part of the electrical distribution system that brings electricity from the street source (power poles and transformers) through the service lines to the electrical meter mounted on the outside of the building and to the panel inside the building. The panel houses the circuits that distribute electricity throughout the structure.

Electrical plans: Engineered drawings that show all electrical supply and distribution.

Electronic signature: A signature that is used to sign electronic documents by capturing handwritten signatures through computer technology and attaching them to the document or file.

Elevation (EL): Height above sea level, or other defined surface, usually expressed in feet.

Elevation drawing: Side view of a building or object, showing height and width.

Engineer: A person who applies scientific principles in design and construction.

Engineer's scale: A tool for measuring distances and transferring measurements at a fixed ration of length. It is used to make drawings to scale. Usually used for land measurements on site plans. A straightedge measuring device divided uniformly into multiples of 10 divisions per inch so drawings can be made with decimal values.

English ruler: Instrument that measures English measurements; also called the standard ruler. Units of English measure include inches, feet, and yards.

Entrepreneur: A person who starts and runs his or her own business.

Equilateral triangle: A triangle that has three equal sides and three equal angles.

Equivalent fractions: Fractions having different numerators and denominators, but equal values, such as ½ and ¾.

Excavation: Any man-made cut, cavity, trench, or depression in an earth surface, formed by removing earth. It can be made for anything from basements to highways. Also see *trench*.

Experience modification rate (EMR): A rate computation to determine surcharge or credit to workers' compensation premium based on a company's previous accident experience.

Extension ladder: A ladder made of two straight ladders that are connected so that the overall length can be adjusted.

Eyebolt: An item of rigging hardware used to attach a sling to a load.

Fall zone: When lifting, the area having a diameter twice the height of the object being lifted.

Fastener: A device such as a bolt, clasp, hook, or lock used to attach or secure one material to another.

Ferromagnetic: Having magnetic properties. Substances such as iron, nickel, cobalt, and various alloys are ferromagnetic.

Fire protection plan: A drawing that shows the details of the building's sprinkler system.

Flammable: Capable of easily igniting and rapidly burning; used to describe a fuel with a flash point below 100°F.

Flash: A sudden bright light associated with starting up a welding torch.

Flashback: A welding flame that flares up and chars the hose at or near the torch connection. It is caused by improperly mixed fuel.

Flash burn: The damage that can be done to eyes after even brief exposure to ultraviolet light from arc welding. A flash burn requires medical attention.

Flash goggles: Eye protective equipment worn during welding operations.

Flash point: The temperature at which fuel gives off enough gases (vapors) to burn.

Flat bar: A prying tool with a nail slot at the end to pull nails out in tightly enclosed areas. It can also be used as a small pry bar.

Flats: The straight sides or jaws of a wrench opening. Also, the sides on a nut or bolt head.

Floor plan: A drawing that provides an aerial view of the layout of each room.

Font: The type style used for printed letters and numbers.

Foot-candle: A unit of measure of the intensity of light falling on a surface, equal to one lumen per square foot and originally defined with reference to a standardized candle burning at one foot from a given surface.

Foot-pounds: Unit of measure used to describe the amount of pressure exerted (torque) to tighten a large object.

Formula: A mathematical process used to solve a problem. For example, the formula for finding the area of a rectangle is side A times side B = Area, or $A \times B = \text{Area}$.

Foundation plan: A drawing that shows the layout and elevation of the building foundation.

Fraction: A number represented by a numerator and a denominator, such as ½.

Freight elevator: An elevator used to transport materials from floor to floor.

Glossary: An alphabetical list of terms and definitions.

Goal-oriented: To be focused on an objective.

Graph: Information shown as a picture or chart. Graphs may be represented in various forms, including line graphs, bar charts, and pie charts.

Grit: A granular abrasive used to make sandpaper or applied to the surface of a grinding wheel to give it a nonslip finish. Grit is graded according to its texture. The grit number indicates the number of abrasive granules in a standard size (per inch or per cm). The higher the grit number, the finer the abrasive material.

Grommet sling: A sling fabricated in an endless loop.

Ground: The conducting connection between electrical equipment or an electrical circuit and the earth.

Ground fault circuit interrupter (GFCI): A circuit breaker device that de-energizes an electrical circuit. Used to protect people from electric shock and protect equipment from damage by interrupting the flow of electricity if a circuit fault occurs.

Ground fault protection: Protection against short circuits; a safety device cuts power off as soon as it senses any imbalance between incoming and outgoing current.

Guarded: Enclosed, fenced, covered, or otherwise protected by barriers, rails, covers, or platforms to prevent dangerous contact.

Hallucinogen: A class of drugs that distort the perception of reality and cause hallucinations.

Hand line: A line attached to a tool or object so a worker can pull it up after climbing a ladder or scaffold.

Hand truck: Also known as dollies, hand trucks are two-wheeled carts that are used to transport large, heavy loads, such as gas cylinders or drums.

Handheld computer: A computer that is designed to be small and light enough to be carried.

Harassment: A type of discrimination that can be based on race, age, disabilities, sex, religion, cultural issues, health, or language barriers.

Hard drive: Hardware that stores software and electronic files.

Hardware: The physical components that make up a computer.

Hazard Communication Standard (HazCom): The Occupational Safety and Health Administration standard that requires contractors to educate employees about hazardous chemicals on the job site and how to work with them safely.

Hazardous materials: Materials (such as chemicals) that must be transported, stored, applied, handled, and identified according to federal, state, or local regulations. Hazardous materials must be accompanied by material safety data sheets (MSDSs).

Heating, ventilating, and air conditioning (HVAC): Heating, ventilating, and air conditioning.

Hex key wrench: A hexagonal steel bar that is bent to form a right angle. Often referred to as an Allen® wrench.

Hidden line: A dashed line showing an object obstructed from view by another object.

Hitch: The rigging configuration by which a sling connects the load to the hoist hook. The three basic types of hitches are vertical, choker, and basket.

Hoist: A device that applies a mechanical force for lifting or lowering a load.

Hydraulic tools: Tools powered by fluid pressure. The pressure is produced by hand pumps or electric pumps.

Improper fraction: A fraction whose numerator is larger than its denominator. For example, $\frac{5}{4}$ and $\frac{6}{5}$ are improper fractions.

Inch-pounds: Unit of measure used to describe the amount of pressure exerted (torque) to tighten a small object.

Index: An alphabetical list of topics, along with the page numbers where each topic appears.

Industrial forklift: A vehicle with a power-operated pronged platform that can be raised and lowered for insertion under a load to be lifted and moved.

Initiative: The ability to work without constant supervision and solve problems independently.

Internet: Electronic communications network that connects computer networks and organizational computer facilities around the world.

Invert: To reverse the order or position of numbers. In fractions, to turn upside down, such as $\frac{3}{4}$ to $\frac{4}{3}$. When you are dividing by fractions, one fraction is inverted.

Isometric drawing: A three-dimensional drawing of an object.

Isosceles triangle: A triangle that has two equal sides and two equal angles.

Italics: Letters and numbers that lean to the right rather than stand straight up.

Jack: Portable device used for raising heavy objects by means of force applied with a lever, screw, or hydraulic press.

Jargon: Specialized terms used in a specific industry.

Joint: The point where members or the edges of members are joined. The types of welding joints are butt joint, corner joint, and T-joint.

Kerf: A cut or channel made by a saw.

Lanyard: A short section of rope or strap, one end of which is attached to a worker's safety harness and the other to a strong anchor point above the work area.

Leader: In drafting, the line on which an arrowhead is placed and used to identify a component.

Leadership: The ability to set an example for others to follow by exercising authority and responsibility.

Legend: A description of the symbols and abbreviations used in a set of drawings.

Level: Perfectly horizontal; completely flat; also, a tool used to determine if an object is level.

Lifting clamp: A device used to move loads such as steel plates or concrete panels without the use of slings.

Load: The quantity of materials able to be carried, transported, or relocated at one time by a machine, vehicle, piece of equipment, or person; the total amount of what is being lifted, including all slings, hitches, and hardware.

Load control: The safe and efficient practice of load manipulation, using proper communication and handling techniques.

Load stress: The strain or tension applied on the rigging by the weight of the suspended load.

Lockout/tagout: A formal procedure for taking equipment out of service and ensuring that it cannot be operated until a qualified person has removed the lockout or tagout device (such as a lock or warning tag).

Machinist's rule: A ruler that is marked so that the inches are divided into 10 equal parts, or tenths.

Management system: The organization of a company's management, including reporting procedures, supervisory responsibility, and administration.

Masonry: Building material such as stone, brick, or concrete block.

Master link: The main connection fitting for chain slings.

Material cart: Four-wheeled device used to transport materials around a job site.

Material safety data sheet (MSDS): A document that must accompany any hazardous substance. The MSDS identifies the substance and gives the exposure limits, the physical and chemical characteristics, the kind of hazard it presents, precautions for safe handling and use, and specific control measures.

Maximum allowable slope: The steepest incline of an excavation face that is acceptable for the most favorable site conditions as protection against cave-ins, expressed as the ratio of horizontal distance to vertical rise.

Maximum intended load: The total weight of all people, equipment, tools, materials, and loads that a ladder can hold at one time.

Mechanical plans: Engineered drawings that show the mechanical systems, such as motors and piping.

Memo: Informal written correspondence. Another term for memorandum (plural: memoranda).

Meter: The base unit of length in the metric system; approximately 39.37 inches.

Methamphetamine: A highly addictive crystalline drug, derived from amphetamines, that affects the central nervous system.

Metric ruler: Instrument that measures metric lengths. Units of measure include millimeters, centimeters, and meters.

Metric scale: A straightedge measuring device divided into centimeters, with each centimeter divided into 10 millimeters. Usually used for architectural drawings.

Mid-rail: Mid-level, horizontal board required on all open sides of scaffolding and platforms that are more than 14 inches from the face of the structure and more than 10 feet above the ground. It is placed halfway between the toeboard and the top rail.

Mission statement: A statement of how a company does business.

Miter joint: A joint made by fastening together usually perpendicular parts with the ends cut at an angle.

Mixed number: A combination of a whole number with a fraction or decimal. For example, mixed numbers are $3\frac{7}{16}$, 5.75, and $1\frac{1}{4}$.

Nail puller: A tool used to remove nails.

Negative numbers: Numbers less than zero. For example, −1, −2, and −3 are negative numbers.

Not to scale (NTS): Describes drawings that show relative positions and sizes.

Numerator: The part of a fraction above the dividing line. For example, the 1 in ½ is the numerator.

Obtuse angle: Any angle between 90 degrees and 180 degrees.

Occupational Safety and Health Administration (OSHA): An agency of the U.S. Department of Labor. Also refers to the Occupational Safety and Health Act of 1970, a law that applies to more than 111 million workers and 7 million job sites in the country.

One rope lay: The lengthwise distance it takes for one strand of a wire rope to make one complete turn around the core.

Open-end wrench: A nonadjustable wrench with an opening at each end that determines the size of the wrench.

Operating system: A complex set of commands that enables hardware and software to communicate.

Opposite angles: Two angles that are formed by two straight lines crossing. They are always equal.

Pad eye: A welded structural lifting attachment.

Pallet jack: Also known as a pallet truck, a device used to lift and move heavy or stacked pallets.

PDA: Abbreviation for personal digital assistant. Another name for a handheld computer.

Peening: The process of bending, shaping, or cutting material by striking it with a tool.

Percent: Of or out of one hundred. For example, 8 is 8 percent (%) of 100.

Perimeter: The distance around the outside of any closed shape, such as a rectangle, circle, or square.

Permit: A legal document that allows a task to be undertaken.

Permit-required confined space: A confined space that has been evaluated and found to have actual or potential hazards, such as a toxic atmosphere or other serious safety or health hazard. Workers need written authorization to enter a permit-required confined space. Also see *confined space.*

Personal protective equipment (PPE): Equipment or clothing designed to prevent or reduce injuries.

Pi: A mathematical value of approximately 3.14 (or $\frac{22}{7}$) used to determine the area and circumference of circles. It is sometimes symbolized by π.

Pipe mule: Sometimes referred to as a tunnel buggy, a two-wheeled device used to transport medium-length pieces of pipe, tubing, or scaffolding.

Pipe transport: Similar to a pipe mule, but used to move larger pieces of pipe.

Pipe wrench: A wrench for gripping and turning a pipe or pipe-shaped object; it tightens when turned in one direction.

Piping and instrumentation drawings (P&IDs): Schematic diagrams of a complete piping system.

Place value: The exact quantity of a digit, determined by its place within the whole number or by its relationship to the decimal point.

Plane: A surface in which a straight line joining two points lies wholly within that surface.

Planed: Describing a surface made smooth by using a tool called a plane.

Planked: Having pieces of material 2 or more inches thick and 6 or more inches wide used as flooring, decking, or scaffolding.

Pliers: A scissor-shaped type of adjustable wrench equipped with jaws and teeth to grip objects.

Plotter: Hardware that prints large architectural and construction drawings and images.

Plumb: Perfectly vertical; the surface is at a right angle (90 degrees) to the horizon or floor and does not bow out at the top or bottom.

Plumbing: A general term used for both water supply and all liquid waste disposal.

Plumbing plans: Engineered drawings that show the layout for the plumbing system.

Pneumatic tools: Air-powered tools. The power is produced by electric or fuel-powered compressors.

Points: Teeth on the gripping part of a wrench. Also refers to the number of teeth per inch on a handsaw.

Positive numbers: Numbers greater than zero. For example, 1, 2, and 3 are positive numbers.

Powered wheelbarrow: Also known as a power buggy; similar to a manual wheelbarrow, but powered by an electric or gas motor.

Processor: The part of a computer that contains the chips and circuits that allow it to perform its functions.

Product: The answer to a multiplication problem. For example, the product of 6×6 is 36.

Professionalism: Integrity and work-appropriate manners.

Proximity work: Work done near a hazard but not actually in contact with it.

Punch: A steel tool used to indent metal.

Punch list: A written list that identifies deficiencies requiring correction at completion.

Qualified person: A person who, by possession of a recognized degree, certificate, or professional standing, or by extensive knowledge, training, and experience, has demonstrated the ability to solve or prevent problems relating to a certain subject, work, or project.

Quotient: The result of a division. For example, when dividing 6 by 2, the quotient is 3.

Radius: The distance from a center point of a circle to any point on the curved line, or half the width (diameter) of a circle.

Rafter angle square: A type of carpenter's square made of cast aluminum that combines a protractor, try square, and framing square.

Rated capacity: The maximum load weight a sling or piece of hardware or equipment can hold or lift. Also called lifting capacity, working capacity, working load limit (WLL), and safe working load (SWL).

Reciprocating: Moving back and forth.

Rectangle: A four-sided shape with four 90-degree angles. Opposite sides of a rectangle are always parallel and the same length. Adjacent sides are perpendicular and are not equal in length.

Reference: A person who can confirm to a potential employer that you have the skills, experience, and work habits that are listed in your resume.

Rejection criteria: Standards, rules, or tests on which a decision can be based to remove an object or device from service because it is no longer safe.

Remainder: The leftover amount in a division problem. For example, in the problem $34 \div 8$, 8 goes into 34 four times ($8 \times 4 = 32$) and 2 is left over; in other words, it is the remainder.

Request for information (RFI): A means of clarifying a discrepancy in the construction drawings.

Respirator: A device that provides clean, filtered air for breathing, no matter what is in the surrounding air.

Revolutions per minute (rpm): The number of times (or rate) a motor component or accessory (drill bit) completes one full rotation every minute.

Rigging hook: An item of rigging hardware used to attach a sling to a load.

Right angle: An angle that measures 90 degrees. The two lines that form a right angle are perpendicular to each other. This is the angle used most in the trades.

Right triangle: A triangle that includes one 90-degree angle.

Ring test: A method of testing the condition of a grinding wheel. The wheel is mounted on a rod and tapped. A clear ring means the wheel is in good condition; a dull thud means the wheel is in poor condition and should be disposed of.

Ripping bar: A tool used for heavy-duty dismantling of woodwork, such as tearing apart building frames or concrete forms.

Risk management: The process of analyzing the work area and the lift prior to the lift being made in order to predict and account for any potential risks.

Roller skids: A device that includes a surface table and two, three, or four roller skids. Materials that are to be moved are placed on the table surface and then pushed on the skids.

Roof plan: A drawing of the view of the roof from above the building.

Rough terrain forklift: Similar to an industrial forklift, but designed to be used on rough surfaces. Rough terrain forklifts are characterized by large pneumatic tires, usually with deep treads that allow the vehicle to grab onto the roughest of roads or ground cover without sliding or slipping.

Round off: To smooth out threads or edges on a screw or nut.

Safety culture: The culture created when the whole company sees the value of a safe work environment.

Scaffold: An elevated platform for workers and materials.

Scale: The ratio between the size of a drawing of an object and the size of the actual object.

Scalene triangle: A triangle with sides of unequal lengths.

Scanner: Hardware that converts printed text or pictures into an electronic format.

Schematic: A one-line drawing showing the flow path for electrical circuitry.

Section drawing: A cross-sectional view of a specific location, showing the inside of an object or building.

Self-presentation: The way a person dresses, speaks, acts, and interacts with others.

Sexual harassment: A type of discrimination that results from unwelcome sexual advances, requests, or other verbal or physical behavior with sexual overtones.

Shackle: Coupling device used in an appropriate lifting apparatus to connect the rope to eye fittings, hooks, or other connectors.

Shank: The smooth part of a drill bit that fits into the chuck.

Sheave: A grooved pulley-wheel for changing the direction of a rope's pull; often found on a crane.

Shoring: Using pieces of timber, usually in a diagonal position, to hold a wall in place temporarily.

Side pull: The portion of a pull acting horizontally when the slings are not vertical.

Signaler: A person who is responsible for directing a vehicle when the driver's vision is blocked in any way.

Six-foot rule: A rule stating that platforms or work surfaces with unprotected sides or edges that are six feet or higher than the ground or level below it require fall protection.

Slag: Waste material from welding operations.

Sling: Wire rope, alloy steel chain, metal mesh fabric, synthetic rope, synthetic webbing, or jacketed synthetic continuous loop fibers made into forms, with or without end fittings, used to handle loads.

Sling angle: The angle of an attached sling when pulled in relation to the load.

Sling legs: The parts of the sling that reach from the attachment device around the object being lifted.

Sling reach: A measure taken from the master link of the sling, where it bears weight, to either the end fitting of the sling or the lowest point on the basket.

Sling stress: The total amount of force exerted on a sling. This includes forces added as a result of sling angle.

Software: A large set of commands and instructions that direct a computer to perform certain tasks.

Spall: A chip or fragment of rock or soil that has broken off from the main mass.

Specifications: Precise written presentation of the details of a plan.

Spliced: Having been joined together.

Spotter: A person who walks in front of another worker who is carrying or transporting a long load to ensure there is a clear, unobstructed path.

Spreadsheet: Software that performs mathematical calculations.

Square: (1) A special type of rectangle with four equal sides and four 90-degree angles. (2) The product of a number multiplied by itself. For example, 25 is the square of 5; 16 is the square of 4. (3) Exactly adjusted; any piece of material sawed or cut to be rectangular with equal dimensions on all sides. (4) A tool used to check angles.

Standard ruler: An instrument that measures English lengths (inches, feet, and yards). See *English ruler*.

Stepladder: A self-supporting ladder consisting of two elements hinged at the top.

Straight angle: A 180-degree angle or flat line.

Straight ladder: A nonadjustable ladder.

Strand: A group of wires wound, or laid, around a center wire, or core. Strands are laid around a supporting core to form a rope.

Stress: Intensity of force exerted by one part of an object on another; the action of forces on an object or system that leads to changes in its shape, strain on it, or separation of its parts.

Striking (or slugging) wrench: A nonadjustable wrench with an enclosed, circular opening designed to lock on to the fastener when the wrench is struck.

Strip: To damage the threads on a nut or bolt.

Structural plans: A set of engineered drawings used to support the architectural design.

Sum: The total in an addition problem. For example, in the problem 7 + 8 = 15, 15 is the sum.

Switch enclosure: A box that houses electrical switches used to regulate and distribute electricity in a building.

Symbol: A drawing that represents a material or component on a plan.

Table: A way to present important text and numbers so they can be read and understood at a glance.

Table of contents: A list of book chapters or sections, usually located at the front of the book.

Tactful: Aware of the effects of your statements and actions on others.

Tag line: Rope that runs from the load to the ground. Riggers hold on to tag lines to keep a load from swinging or spinning during the lift.

Tang: Metal handle-end of a file. The tang fits into a wooden or plastic file handle.

Tardiness: Habitually showing up late for work.

Tattle-tail: Cord attached to the strands of an endless loop sling. It protrudes from the jacket. A tattle-tail is used to determine if an endless sling has been stretched or overloaded.

Teamwork: The cooperation of co-workers to achieve one or more goals.

Tempered: Treated with heat to create or restore hardness in steel.

Tenon: A piece that projects out of wood or another material for the purpose of being placed into a hole or groove to form a joint.

Text message: A short message (160 characters or fewer) sent from a mobile phone.

Threaded shank: A connecting end of a fastener, such as a bolt, with a series of spiral grooves cut into it. The grooves are designed to mate with grooves cut into another object in order to join them together.

Title block: A part of a drawing sheet that includes some general information about the project.

Toeboard: A vertical barrier at floor level attached along exposed edges of a platform, runway, or ramp to prevent materials and people from falling.

Top rail: A top-level, horizontal board required on all open sides of scaffolding and platforms that are more than 14 inches from the face of the structure and more than 10 feet above the ground.

Torque: The turning or twisting force applied to an object, such as a nut, bolt, or screw, using a socket wrench or screwdriver to tighten it. Torque is measured in inch-pounds or foot-pounds.

Trench: A narrow excavation made below the surface of the ground that is generally deeper than it is wide, with a maximum width of 15 feet. Also see *excavation*.

Triangle: A closed shape that has three sides and three angles.

Trigger lock: A small lever, switch, or part that you push or pull to activate a locking catch or spring. Activating the trigger lock causes the trigger to stay in the operating mode even without your finger on the trigger.

Try square: A square whose legs are fixed at a right angle.

USB flash drive: A lightweight, portable data storage device.

Vertex: A point at which two or more lines or curves come together.

Volume: The amount of space occupied in three dimensions (length, width, and height/depth/thickness).

Warning yarn: A component of the sling that shows the rigger whether the sling has suffered too much damage to be used.

Weight capacity: The maximum amount of weight that a structure can safely support.

Weld: To heat or fuse two or more pieces of metal so that the finished piece is as strong as the original; a welded joint.

Welding shield: (1) A protective screen set up around a welding operation designed to safeguard workers not directly involved in that operation. (2) A shield that provides eye and face protection for welders by either connecting to helmet-like headgear or attaching directly to a hard hat; also called a welding helmet.

Wheelbarrow: A one- or two-wheeled vehicle with handles at the rear, used to carry small loads.

Whole numbers: Complete units without fractions or decimals.

Wind sock: A cloth cone open at both ends mounted in a high place to show which direction the wind is blowing.

Wire rope: A rope made from steel wires that are formed into strands and then laid around a supporting core to form a complete rope; sometimes called cable.

Wireless: A technology that allows computers to communicate without physical connections.

Word processor: Software that is used for writing text documents.

Work ethic: Work habits that are the foundation of a person's ability to do his or her job.

Work zone: The area in which a forklift may come in contact with objects or people, either with the rear end or the front forks.

Figure Credits

Module 1: Basic Safety

Accuform Signs, 101F03
DBI/SALA & PROTECTA, 101F13
North Safety Products USA, 101F14, 101F44, 101F45 (glasses, goggles), 101F46, 101F49, 101F51A, 101F51B, 101F51C
Emerson Tool Company, Ridge Tool Company/RIDGID®, 101F22, 101F23, 101F24
Werner Co., 101F25
www.trenchsafety.com, 101F35, 101F36
Makita USA, Inc., 101F39B
Grainger, 101F40B, 101F40C
Mid-State Manufacturing Corporation, 101F42
DeWALT Power Tools, 101F50
WD-40 Company, 101F52A, 101F52B, 101F52C, 101F52D
NAFED, 101F63
Badger Fire Protection, 101F64A, 101F64B, 101F64C, 101F64D, 101F64E

Module 2: Introduction to Construction Math

Mary Fudge & Steve Anderson (Van Buren Intermediate School District), 102F04, 102F30, 102F31, 102F32, 102F34, 102F35, 102F36, 102F39, 102F53, 102F54, 102SA01, 02SA04, 102SA05, 102SA07, 102SA09
Cooper Hand Tools, 102F06 (standard ruler)
Courtesy of Oak Ridge National Laboratory, 102F43
Calculated Industries, Inc., 102A01

Module 3: Introduction to Hand Tools

The Stanley Works, 103F01, 103F02, 103F03A, 103F03B, 103F03C, 103F06B, 103F09A, 103F09B, 103F09C, 103F16D, 103F21, 103F25, 103F27, 103F28B, 103F32A, 103F32B, 103F32C, 103F32D, 103F36A, 103F36B, 103F38A, 103F38B, 103F40, 103F47C, 103F47G
Cooper Hand Tools, 103F05A, 103F05B, 103F05C, 103F05D, 103F13A, 103F13B, 103F18C, 103F24, 103F41A, 103F41B, 103F41C, 103F41F, 103F43A, 103F43B, 103F43C, 103F43D, 103F43E, 103F43F, 103F46, 103F47B
Emerson Tool Company, Ridge Tool Company/RIDGID®, 103F06A, 103F11, 103F13C, 103F18A, 103F18B, 103F26, 103F28A, 103F41E, 103F51
Channellock, Inc., 103F13D
IRWIN Industrial Tool Co., 103F13E, 103F41D, 103F47D
Klein Tools, Inc., 103F16A, 103F16C, 103F22A
S-K Hand Tool Corporations, 103F16B
JET Brand of WMH Tool Group, 103F20, 103F47A, 103F49, 103F50A, 103F50B
Futek Advanced Sensor Technology, 103F22B
Holloway Engineering, 103F22C
M-D Building Products, Inc., 103F30
Porter Cable Robo Toolz, 103F31A, 103F31B
American Clamping Corp./Bessy, 103F47F
Northern Tool and Equipment, 103F47H
Estwing Manufacturing Co., 103SA01
Chicago Brand Industrial, 103SA02

Module 4: Introduction to Power Tools

DeWALT Power Tools, 104F01, 104F02A, 104F02E, 104F03, 104F04A, 104F04B, 104F04C, 104F05A, 104F06A, 104F06B, 104F07, 104F08, 104F09, 104F10A, 104F10B, 104F13, 104F14A, 104F14B, 104F15A, 104F15B, 104F16, 104F17, 104F18, 104F19, 104F20A, 104F20B, 104F21A, 104F23A, 104F23B, 104F24, 104F25A, 104F26, 104F27A, 104F27B, 104F29
Courtesy of Milwaukee Electric Tool Corp., 104F02B, 104F02C, 104F22, 104F25B, 104F25C, 104F32A, 104F32B, 104F32D, 104F32E
Bosch Power Tools & Accessories, 104F05B, 104F32C
JET Brand of WMH Tool Group, 104F11, 104F31A, 104F31B
Makita USA, Inc., 104F21B
Northern Tool and Equipment, 104F21C
Stanley Fastening Systems, LP/Stanley Bostitch, 104F28, 104SA02
Shinn Fu Company of America, 104F33
Torin Jacks, Inc., 104F34
Robert Bosch Tool Corp., 104SA01
SENCO Products, Inc., 104SA03

Module 5: Introduction to Construction Drawings

Croxton Collaborative/Gould Evans Associates, 105F02, 105F05, 105F06, 105F07, 105F08, 105F09, 105F10, 105F11, 105F12, 105F13, 105F14, 105F19, 105F20, 105F21, 105F22, 105F24, 105F25, 105F26, 105F27, 105F28A, 105F28A, 105F29, 105F30, 105F31, 105F32A, 105F32B, 105SA03
Ritterbush-Ellig-Hulsing, PC, 105F29
Printed by permission of Pearson, 105F30, 105F32, 105F33, 105F34, 105F35
Colonial Webb, 105SA01

Module 6: Basic Rigging

North American Industries, Inc., 106F01
Mammoet USA, Inc., 106F02A, 106F02B, 106SA01
Lift-It Manufacturing Co., Inc., 106F04, 106F12, 106F13, 106F14, 106F15
Lift-All Company, Inc., 106F05, 106F06, 106F07, 106F08A, 106F08B, 106F08C, 106F09A, 106F09B, 106F10, 106F11, 106F16C, 106F16I, 106F16M, 106F22, 106F25A, 106F25B, 106F25C, 106F25D, 106F25E
The Crosby Group, Inc., 106F17, 106F20A, 106F20B, 106F20C, 106F36A, 106F36B, 106F36C, 106F36D, 106F36E, 106F36F, 106F37, 106F38, 106F39A, 106F39B, 106F39C, 106F48A, 106F48B, 106F48C, 106F48D, 106F54, 106F58, 106F60 (top), 106SA04
J.C. Renfroe & Sons, Inc., 106F45, 106F46A, 106F46B, 106F46C
Gunnebo Johnson Corporation, 106F48E, 106F48F
JET Brand of WMH Tool Group, 106F55A, 106F55B, 106F55C, 106SA02A, 106SA02B
Columbus McKinnon Corporation, 106F56
Coffing Hoists, 106F57

Module 7: Basic Communication Skills

Courtesy of Oak Ridge National Laboratory, 107F02, 107F08
M.C. Dean, Inc., 107F05
Ritterbush-Ellig-Hulsing PC, 107F12, 107F13

Module 8: Basic Employability Skills

Courtesy of Oak Ridge National Laboratory, 108F01 (background)
Autodesk, Inc., 108F09A, 108F09B, 108F09C
Primavera Systems, 108F09D, 108F09E
League Manufacturing, 108F10

Module 9: Introduction to Materials Handling

Northern Tool and Equipment, 109F08, 109F14
GAR-BRO Manufacturing Co., Herber Springs, AR, 109F10
BE&K, 109F11, 109F19
Sumner, 109F12
Arizona Tools, 109F13
The Power Barrow Company, LLC, 109F15
Multiquip, Inc., 109F16
Taylor Machine Works, 109F17, 109F18

Index

Note: The letter *f* denotes pages containing figures.

A

Abbreviations
- drawings, 5.34–5.38, 5.35*f*–5.39*f*
- electrical, 5.29*f*
- mechanical plans, 5.11, 5.17*f*, 5.18*f*

Abrasive, defined, 4.36
Abrasive cutoff saw, 4.19, 4.19*f*
Absenteeism, defined, 8.32
AC (alternating current), 4.1, 4.36
Accidents
- alcohol and drug use, 1.6–1.7
- causes, 1.3–1.9
- compliance, policies, 1.11–1.12
- costs, 1.2–1.3
- employee rights and responsibilities, 1.10
- evacuation procedures, 1.13
- four high-hazard areas, 1.13
- general duty clause, 1.10
- housekeeping and, 1.9
- inspections, 1.11
- intentional acts, 1.8
- lack of skill and, 1.7–1.8
- ladders and stairs, 1.21–1.28, 1.21*f*–1.27*f*
- management system failures, 1.9
- overview, 1.1–1.2
- policies and regulations, 1.9–1.12
- poor work habits, 1.5
- rationalizing risk, 1.8
- record keeping, 1.12
- unsafe acts, 1.8
- unsafe conditions, 1.8

Acetylene, 1.62, 1.64
Acids, eye and face protection, 1.45–1.46, 1.46*f*
Active listening, 7.1, 7.5, 7.6, 7.21
Acute angles, 2.41–2.42, 2.42*f*, 2.63
Addition
- calculators and, 2.69–2.70
- decimals, 2.29–2.30
- fractions, 2.23
- whole numbers, 2.2

Adjacent angles, 2.42*f*, 2.63
Adjustable wrenches, 3.15–3.16, 3.15*f*, 3.16*f*, 3.44
Air conditioning
- Going Green, Bank of America, 8.5
- mechanical plans, 5.11, 5.16*f*–5.21*f*

Air impact wrench, 4.27–4.28, 4.27*f*
Air masks, 1.49–1.52, 1.50*f*
Alcohol, accidents and, 1.6–1.7, 8.24–8.25, 8.30
Allen (hex) screwdrivers, 3.9–3.11, 3.9*f*
Allen® wrench, 3.14–3.15, 3.14*f*
Alloy steel chain slings, 6.10–6.11, 6.10*f*–6.12*f*
Alphabet of Lines, 5.25*f*, 5.34
Alternating current (AC), 4.1, 4.36
Aluminum, cutting, 4.11
Aluminum ladders, 1.22
American National Standards Institute (ANSI)
- eye and face protection standards, 1.45–1.46, 1.46*f*
- hand signals, 6.35–6.36, 6.36*f*–6.39*f*, 6.47

Anchors, fastening systems, 4.26–4.27, 4.26*f*
Angle grinders, 4.19–4.21, 4.20*f*
Angle iron, 4.19
Angles
- carpenter's squares, 3.23, 3.23*f*
- types of, 2.41–2.42, 2.42*f*

Angular pull, eyebolts, 6.22, 6.22*f*, 6.23*f*
Apparatus, defined, 1.82
Appendix, defined, 7.21
Applications, employment, 8.2–8.4, 8.3*f*, 8.6
Arc, defined, 1.82
Architect, defined, 5.47
Architect's scale
- construction drawings, 5.16, 5.18
- defined, 2.63
- description of, 2.9*f*, 2.14–2.16, 2.15*f*, 2.16*f*
- study problems, 2.17

Architectural plans
- defined, 5.47
- overview, 5.1, 5.2*f*, 5.4, 5.7–5.8, 5.9*f*–5.12*f*
- symbols, 5.7, 5.35, 5.37*f*

Arc welding, 1.45–1.46, 1.46*f*, 1.62, 1.82
Area
- calculating, 2.45–2.47, 2.68
- defined, 2.63
- formulas for, 2.68
- practical application, 2.43
- On Site, cylinders, 2.53
- unit conversions, 2.50

Arrowheads, drawings, 5.25*f*, 5.34
Asbestos, 1.57
Assured equipment grounding, 1.42–1.43
Auger, defined, 4.36
Auger drill bits, 4.2, 4.2*f*

B

Back injuries, ergonomics, 1.67–1.69, 1.69*f*, 9.1–9.3, 9.2*f*
Backsaws, 3.27–3.29, 3.28*f*
Bagged materials, storing, 9.3
Ball peen hammers, 3.1*f*, 3.2, 3.44
Band clamp, 3.33*f*, 3.34–3.35
Bandsaws, handheld, 4.16–4.17, 4.17*f*
Bar clamps, 3.32–3.35, 3.33*f*
Barricades, 1.17–1.18
Barriers, 1.17–1.18
Basket hitch, 6.18, 6.19*f*
Basket slings, 6.10, 6.11*f*
Batteries, rechargeable, 4.5
Beams
- defined, 5.47
- foundation plans, 5.11, 5.14*f*

Bell-faced hammer, 3.1, 3.44
Bench grinders, 4.21–4.23, 4.22*f*, 4.23*f*
Benching, trenches, 1.35
Benefits, employee, 8.1
Bevel, defined, 3.44
BIM (Building Information Modeling), 5.7
Biodiesel, 6.16

Biological hazards, 1.56–1.58
Birdcaging, wire rope, 6.14, 6.14*f*
Bisect, defined, 2.63
Bits, drill, 4.2, 4.2*f*
Blades, saws
  abrasive cutoff saws, 4.19
  circular saw, 4.11
  handheld bandsaw, 4.16–4.17, 4.17*f*
  power miter saws, 4.13
  saber saws (jig saws), 4.14
  safety and, 4.16
  types of, 4.11
  worm drive, 4.13
Block and tackle hoists, 6.29–6.32, 6.30*f*–6.32*f*, 6.47
Blocking and cribbing loads, 6.40–6.42, 6.41*f*, 6.42*f*, 6.47
Blocks, concrete, 2.7
Blood borne pathogens, 1.57–1.58
Blueprints, 5.2, 5.47. *See also* Construction drawings
Boards, lumber sizes, 2.8
Body harnesses, 1.19–1.21, 1.19*f*, 1.20*f*
Body language, defined, 7.21
Bolts, air impact wrench, 4.27–4.28, 4.27*f*
Boosters, 4.26–4.27, 4.26*f*, 4.36
Boots, safety, 1.48
Borers, core, 4.28–4.30, 4.29*f*
Borrow, subtraction, 2.63
Box-end wrench, 3.14–3.15, 3.14*f*, 3.44
Brass, cutting, 4.11
Breakers, electrical plans, 5.16, 5.24*f*, 5.25*f*, 5.27*f*–5.29*f*
Break lines, drawings, 5.25*f*, 5.34
Bricklayers, 3.22
Bricks
  architectural symbols, 5.7
  stacking and storing, 9.3
Bridle, defined, 6.47
Bridle hitch, 6.15, 6.16*f*
Builder's saw. *See* Circular saws
Building Information Modeling (BIM), 5.7
Bullets, defined, 7.21
Bull ring, defined, 6.47
Business, construction overview, 8.1–8.6

C

Cable pullers, 3.35–3.36, 3.35*f*, 3.36*f*
Cables, handling, 9.5–9.6
CAD (computer-aided drafting), 5.1, 5.47, 8.12, 8.13*f*. *See also* Construction drawings
Cadmium, 4.5
Calculators
  decimal numbers, 2.34–2.35
  use instructions, 2.69–2.70
Calipers, 3.20
Carbide, defined, 4.36
Carbon footprint, reducing, 7.17
Carpenter's square, 3.23, 3.23*f*, 3.44
Carry, addition, 2.63
Carts, 9.6–9.7, 9.7*f*
Catastrophe inspections, 1.11
Cat's paw, 3.5*f*, 3.6, 3.44
Caught-in accidents, 1.13
Caught-in-between accidents, 1.33–1.37, 1.35*f*–1.37*f*
Caution signs, 1.4, 1.4*f*
Cave-ins, trenches, 1.33–1.34
C-clamps, 3.32–3.35, 3.32*f*
Ceiling joists, 5.11, 5.14*f*
Cell phones, 7.7–7.8
Centerlines, drawings, 5.25*f*, 5.34
Center punch, 3.8, 3.9*f*
Centi (c), metric units, 2.12, 2.14
Centimeters (cm)
  cubic, 2.48
  defined, 2.12, 2.14
  metric rulers, 2.10–2.11, 2.11*f*
  square centimeters, 2.46
  unit conversion, 2.67
CFR *(Code of Federal Regulations)*, 1.9–1.10
Chain bridle slings, 6.10, 6.11*f*
Chain falls, 3.35–3.36, 3.35*f*, 3.36*f*
Chain hoists, 6.29–6.32, 6.30*f*–6.32*f*
Chain slings, alloy steel, 6.10–6.11, 6.10*f*–6.12*f*
Chalk lines, 3.26, 3.26*f*
Channel, cutting, 4.19
Chemicals
  eye and face protection, 1.45–1.46, 1.46*f*
  hazardous, 1.52–1.58, 1.53*f*–1.56*f*
  spills, evacuation procedures, 1.13
  splashes, 1.58
Chimneys, bricklayers, 3.22
Chipping, eye and face protection, 1.45–1.46, 1.46*f*
Chisel bar, 3.5*f*, 3.6, 3.44
Chisels, 3.6–3.8, 3.7*f*, 3.44
Choker hitches, 6.5, 6.6*f*, 6.16–6.18, 6.17*f*, 6.18*f*
Chucks/keys, 4.2, 4.3, 4.3*f*, 4.36
Circles
  area, 2.47, 2.68
  defined, 2.45–2.46, 2.63
  volume, 2.50, 2.68
Circular saws, 4.11–4.14, 4.12*f*
Circumference, circles, 2.45, 2.63
Citations, safety, 1.10
Civil engineering symbols, 5.35, 5.37*f*
Civil plans, 5.1, 5.2*f*, 5.47
Civil scale. *See* Engineer's scale
Clamps, 3.32–3.35, 3.33*f*, 6.23–6.25, 6.24*f*–6.26*f*
Claw hammer, 3.1–3.4, 3.1*f*, 3.44
Clay spades, 4.28–4.30, 4.29*f*
Cleancutting, 4.14
Clothing
  dress codes, 8.15
  safety and, 1.45, 1.48, 1.66, 9.1
Clutch-drive screwdrivers, 3.9–3.11, 3.9*f*
CNC (computer numerical control) machines, 8.12, 8.14, 8.32
Codebook, 7.8, 7.9*f*
Code of Federal Regulations (CFR), 1.9–1.10
Cold chisels, 3.7*f*, 3.8
Cold stress, human, 1.60–1.62
Columns, foundation plans, 5.11, 5.14*f*
Combination pliers, 3.11, 3.12*f*
Combination saw blades, 4.11
Combination squares, 3.23–3.24, 3.23*f*, 3.24*f*, 3.44
Combination wrench, 3.14–3.15, 3.14*f*, 3.44
Combustible, defined, 1.82
Combustible materials, 1.9, 1.62, 1.70–1.71, 1.82, 9.3
Come-alongs, chains, 3.35–3.36, 3.35*f*, 3.36*f*
Communication skills
  communication process, 7.1–7.2, 7.1*f*
  listening, 7.1–7.6, 7.3*f*
  overview, 7.1
  reading, 7.8–7.12
  safety and, 1.3–1.5, 1.4*f*, 1.5*f*
  speaking skills, 7.1, 7.4*f*, 7.6–7.8
  writing skills, 7.8, 7.12–7.17
Communication standard, hazards, 1.52–1.56, 1.53*f*–1.56*f*
Compass (keyhole) saws, 3.27–3.29, 3.28*f*
Competent person, defined, 1.11–1.12, 1.82

Compound miter saws, 4.18, 4.18*f*
Compressed air, 1.64
Compressed gases, 1.62, 1.64, 1.71
Compressors
  drills, pneumatic, 4.10–4.11, 4.10*f*
  nailers, pneumatic, 4.23–4.25, 4.23*f*, 4.24*f*, 4.25*f*
  pneumatic tools, 4.1
Computer-aided drafting (CAD), 5.1, 5.47, 8.12, 8.13*f*. *See also* Construction drawings
Computer numerical control (CNC) machines, 8.12, 8.14, 8.32
Computer skills, 8.11–8.14
Concealed receptacle, 1.82
Concrete
  architectural symbols, 5.7
  beams, fastening systems, 4.26–4.27, 4.26*f*
  blocks, dimensions of, 2.7, 2.9
  footings, 5.11, 5.14*f*
  forms, 1.17
  silica, 1.57
  skin protection, 1.48
Concrete mules, 9.11, 9.11*f*, 9.19
Conduit, 9.5–9.6
Confidential information, 7.15, 8.15, 8.32
Confined spaces, 1.66–1.67, 1.67*f*, 1.68*f*, 1.82
Conflict resolution, 8.16–8.18
Construction business, 8.1–8.6
Construction drawings
  abbreviations, symbols and keynotes, 5.34–5.38, 5.35*f*–5.39*f*
  architectural plans, 5.4, 5.7–5.8, 5.9*f*–5.12*f*
  basic components, 5.1–5.4, 5.2*f*, 5.3*f*, 5.4*f*
  care of, 5.35
  civil plans, 5.4, 5.5*f*, 5.6*f*
  defined, 5.47
  dimensions, 5.39, 5.41*f*
  electrical plans, 5.16, 5.24*f*, 5.25*f*, 5.27*f*–5.29*f*
  fire protection plans, 5.16
  gridlines and plan locations, 5.39, 5.41*f*
  lines of construction, 5.25*f*, 5.34
  mechanical plans, 5.11, 5.16*f*–5.21*f*
  plumbing/piping plans, 5.11, 5.22*f*, 5.23*f*
  request for information (RFI), 5.16, 5.33*f*
  scale of, 5.16, 5.18
  specifications, 5.16, 5.30*f*, 5.31*f*
  structural plans, 5.8, 5.11, 5.13*f*–5.15*f*
Contact lenses, 1.51, 1.64
Contour lines, 5.4, 5.5*f*, 5.6*f*, 5.47
Conversion, units
  defined, 2.63
  inches to feet, 2.65
Cooling. *See also* Air conditioning
  conservation, Rinker Hall, 5.40
  Going Green, Bank of America, 8.5
Coping saws, 3.27–3.29, 3.28*f*
Copper, cutting, 4.11
Cordless tools, 4.5–4.6, 4.6*f*, 4.7*f*
Core, defined, 6.47
Core borers, 4.28–4.30, 4.29*f*
Corrosive liquids, eye and face protection, 1.45–1.46, 1.46*f*
Costs
  calculating, 2.39
  employee benefits, 8.1
Countersink, defined, 4.36
Cranes, 6.1, 6.2*f*. *See also* Rigging
Cribbing, 6.40–6.42, 6.41*f*, 6.42*f*, 6.47
Critical thinking skills, 8.7–8.11
Criticism, giving and receiving, 8.18–8.19, 8.32
Cross-bracing, scaffolds, 1.30, 1.82
Crosscut saws, 3.27–3.29, 3.28*f*, 4.11
Crosspeen sledgehammer, 3.3*f*
Crushing, wire rope, 6.14, 6.14*f*
Cubes, volume of, 2.50, 2.68
Cubic units, volume, 2.46, 2.48–2.52, 2.63
Culture of safety, 1.1
Cut lines, 5.25*f*, 5.34
Cutting tools
  flame cutting, 1.64
  guards, 1.38
  hazards, 1.62–1.66, 1.63*f*, 1.65*f*
Cylinder
  area, 2.53
  volume, 2.50–2.51, 2.51*f*, 2.68
Cylinders, compressed gas, 1.62, 1.71

D

Danger, signs for, 1.4, 1.4*f*
Database, defined, 8.32
DC (direct current), 4.1, 4.36
Deci (d), metric units, 2.12, 2.14
Decimals
  addition and subtraction, 2.29–2.30
  comparing decimals, 2.28–2.29
  comparing to whole numbers, 2.26–2.27
  converting inches to feet, 2.39–2.40, 2.65
  converting to percentages, 2.37–2.38
  defined, 2.63
  division, 2.31–2.33
  fraction conversions, 2.38–2.39, 2.40
  multiplication, 2.30–2.31
  practical application, 2.28, 2.35–2.36, 2.38
  rounding, 2.33–2.34
  using calculators, 2.34–2.35
Decking, 1.17
Degrees, angles, 2.63
Deka (da), metric units, 2.12, 2.14
Demolition
  pavement breakers, 4.28–4.30, 4.29*f*
  sledgehammers, 3.3–3.4
Denominator, fractions, 2.20, 2.21–2.23, 2.63
Depth, practical application, 2.43
Desktop publishing, defined, 8.32
Detail drawings, 5.7, 5.11*f*, 5.47
Detail grinders, 4.19–4.21, 4.21*f*
Diagonals, 2.44, 2.44*f*, 2.45, 2.63
Diameter, 2.45, 2.63
Difference (subtraction), 2.63
Digital levels, 3.21–3.22, 3.22*f*
Digits, numerical, 2.1, 2.63
Dimension lines, 5.25*f*, 5.34, 5.47
Dimensions, drawings, 5.39, 5.41*f*, 5.47
Direct costs, 2.39
Direct current (DC), 4.1, 4.36
Discrimination, 8.25
Distribution panel, electrical plans, 5.16, 5.24*f*, 5.25*f*, 5.27*f*–5.29*f*
Division
  calculators and, 2.70
  decimals, 2.31–2.33
  fractions, 2.25–2.26
  whole numbers, 2.6–2.8
Door schedules, 5.7, 5.12*f*
Double-basket slings, 6.10, 6.11*f*
Double-faced sledgehammer, 3.3*f*
Double-insulated extension cords, 1.39, 1.40*f*
Dowel, 3.44

Drainage, 5.8, 5.8*f*
Dressing the stone, 3.5
Drill bits, 4.2, 4.2*f*
Drills, power
  cordless, 4.5–4.6, 4.6*f*, 4.7*f*
  electromagnetic drills, 4.7–4.9, 4.9*f*
  hammer drills, 4.6–4.7, 4.7*f*, 4.8*f*
  overview, 4.1–4.4, 4.2*f*
  pneumatic drills, 4.10–4.11, 4.10*f*
  rock drills, 4.28–4.30, 4.29*f*
Dross, 1.64, 1.82
Drug abuse, accidents and, 1.6–1.7, 8.24–8.25, 8.30
Drywall compounds, 1.57
Drywall saws, 3.27–3.29, 3.28*f*
Duty ratings, ladders, 1.21

E

EDMs (electrical discharge machines), 8.14, 8.32
Egress, means of, 1.9–1.10, 1.37
Electrical discharge machines (EDMs), 8.14, 8.32
Electrical distribution panels, 1.44, 1.82
Electrical plans
  defined, 5.47
  HVAC, 5.11, 5.20*f*, 5.21*f*
  meters, 5.16, 5.24*f*, 5.25*f*, 5.27*f*–5.29*f*
  overview, 5.1, 5.2*f*, 5.16, 5.24*f*, 5.25*f*, 5.27*f*–5.29*f*
  schematic drawings, 5.26, 5.26*f*
Electric handsaw. *See* Circular saws
Electric hoists, 6.31*f*
Electricians, 3.6
Electricity. *See also* Electrical plans
  accidents involving, 1.13
  current, 4.1
  drills and, 4.4
  ground fault protection, 4.4
  hazards, 1.38–1.45, 1.40*f*–1.42*f*, 1.44*f*
  ladders and, 1.22
  power adapters, 4.1
  proximity work, 1.58–1.59, 1.58*f*, 1.59*f*
  scaffolds and, 1.31
  utility knives, 3.27
Electric pumps, 4.1
Electric tools, 4.36. *See also* Power tools
Electromagnetic drills, 4.7–4.9, 4.9*f*
Electronic levels, 3.21–3.22, 3.22*f*
Electronic signatures, 7.15, 7.21
EL (elevation) drawings, 5.7, 5.10*f*
Elevated work, safety, 1.17–1.21, 1.18*f*, 1.19*f*, 1.20*f*
Elevation (EL), defined, 5.47
Elevation (EL) drawings, 5.7, 5.10*f*, 5.47
Emails, 7.15–7.17, 7.16*f*
Emergency action plans
  requirements for, 1.9–1.10
  trenches and excavations, 1.37
Emery cloth, 3.29
Employee medical records
  exposure to hazards, 1.56–1.57
  regulations about, 1.9–1.10, 1.11
Employees, rights and responsibilities, 1.10
Employment, skills for
  applications, 8.2–8.4, 8.3*f*
  computer skills, 8.11–8.14
  conflict resolution, 8.16–8.18
  construction business, overview, 8.1–8.6
  critical thinking skills, 8.7–8.11
  criticism, giving and receiving, 8.18–8.19
  lateness and absenteeism, 8.15–8.16
  leadership skills, 8.20–8.23
  planning and scheduling, 8.10–8.11
  self-presentation skills, 8.14–8.16
  teamwork skills, 8.19–8.20
EMR (experience modification rate), 1.1, 1.82
End grinders, 4.19–4.21, 4.20*f*
Endless web slings, 6.5, 6.5*f*
Energy
  biodiesel fuels, 6.16
  fuel cells, 9.10
  Going Green, 2.41, 5.25, 5.40, 8.5
  green-collar jobs, 8.1
Engineer, defined, 5.47
Engineer's hammers. *See* Ball peen hammers
Engineer's scale
  construction drawings, 5.16, 5.18
  defined, 2.63
  description of, 2.8, 2.9*f*, 2.16–2.17, 2.17*f*
English ruler, 2.8–2.10, 2.9*f*, 2.63
English units, conversion of, 2.67
Entrepreneurship, 8.4–8.6, 8.6*f*, 8.32
Equilateral triangles, 2.45, 2.45*f*, 2.63
Equipment
  concrete mule, 9.11, 9.11*f*
  freight elevators, 9.12–9.13, 9.13*f*
  hand signals, 9.13, 9.14*f*
  hand trucks, 9.7, 9.7*f*
  industrial forklift, 9.11–9.12, 9.11*f*
  jacks, 9.8–9.9, 9.9*f*
  material carts, 9.6–9.7, 9.7*f*
  pallet jacks, 9.9, 9.9*f*
  pipe mule, 9.8, 9.8*f*
  pipe transport, 9.8, 9.8*f*
  roller skids, 9.7, 9.8*f*
  rough terrain forklift, 9.12, 9.12*f*
  scheduling, 8.10–8.11
  stacking and storing, 9.3
  tool and machine guards, 1.37–1.38, 1.38*f*
  vehicle and roadway hazards, 1.31–1.33
  wheelbarrows, 9.7–9.8, 9.8*f*, 9.10, 9.10*f*
Equivalent fractions, 2.20–2.21, 2.63
Ergonomics, 1.67–1.69, 1.69*f*, 9.1–9.3, 9.2*f*
Evacuation procedures, 1.13
Excavations, 1.17, 1.33–1.37, 1.35*f*–1.37*f*, 1.82
Experience modification rate (EMR), 1.1, 1.82
Exponential notation, numbers, 2.34
Extension cords, 1.39
Extension ladders, 1.24–1.26, 1.25*f*, 1.26*f*, 1.82
Exteriors
  architectural symbols, 5.7
  elevation drawings, 5.10*f*
  stonemasons, 3.5
Eye-and-eye slings, 6.5, 6.6*f*
Eyebolts, 6.21–6.23, 6.22*f*–6.24*f*, 6.32–6.34, 6.34*f*, 6.47
Eye hooks, 6.10, 6.11*f*, 6.25–6.27, 6.26*f*
Eye protection, 1.45–1.46, 1.46*f*
Eyes, contact lenses, 1.51, 1.64

F

Face protection, 1.45–1.46, 1.46*f*, 1.62
Falling objects, 1.32
Falls, 1.13, 1.17–1.21, 1.18*f*, 1.19*f*, 1.20*f*
Fall zone, 9.11, 9.19
Fastener, defined, 3.44
Fastening systems, powder-actuated, 4.26–4.27, 4.26*f*
Fatalities, OSHA inspections, 1.11
Ferromagnetic metal, 4.7–4.8, 4.36

Fiberglass ladders, 1.22
Fiber-optic inspection cable, 6.6, 6.7*f*
Files, 3.30–3.32, 3.30*f*
Filter masks, 1.49–1.52, 1.50*f*
Finish carpentry, 4.19–4.23
Finish schedules, 5.7, 5.12*f*
Fire extinguishers, 1.71, 1.72*f*, 1.73*f*, 1.74*f*
Fire hazards, 1.69–1.74, 1.70*f*–1.74*f*
Fireplaces, bricklayers, 3.22
Fire protection plans, 5.1, 5.2*f*, 5.16, 5.47
Fires, evacuation procedures, 1.13
First Aid
  Blood borne pathogens, 1.57–1.58
  cold related illness, 1.60–1.62
  electrical shocks, 1.44–1.45
  heat related illness, 1.59–1.60
Fittings, storing, 9.3
Fixtures, schedules, 5.8
Flame cutting, 1.64
Flammable materials, 1.9, 1.62, 1.70–1.71, 1.82, 9.3
Flash, defined, 1.82
Flashback, defined, 1.82
Flash burn, 1.63, 1.82
Flashdrives, 8.33
Flash goggles, 1.63, 1.63*f*, 1.82
Flash point, 1.70
Flat bar, 3.5*f*, 3.6, 3.44, 4.19
Flat file, 3.30–3.32, 3.30*f*
Flats, 3.44
Floor plans, 5.4, 5.7, 5.16, 5.18, 5.47
Floors
  bricklayers, 3.22
  holes in, 1.17–1.19
  stonemasons, 3.5
Floor tiles, 1.57
Flying objects, 1.32–1.33
Folding rule, 3.19, 3.19*f*
Follow-up inspection, 1.11
Font, defined, 7.21
Foot, measurement
  converting from inches, 2.39–2.40, 2.65
  cubic, 2.48
  origin of, 2.13
  square foot, 2.46
  unit conversion, 2.67
Foot-candles, 1.28, 1.82
Footings, foundation plans, 5.11, 5.14*f*
Foot-pounds, defined, 3.44
Foot protection, 1.48
Force, hammer strike, 3.2
Forklifts, 9.11–9.12, 9.11*f*, 9.12*f*, 9.19
Formulas, defined, 2.63
Forstner drill bits, 4.2, 4.2*f*
Foundation plans, 5.11, 5.14*f*, 5.47
Fractions
  addition, 2.23
  decimal conversions, 2.38–2.39, 2.40
  defined, 2.20, 2.63
  dividing, 2.25–2.26
  equivalent fractions, 2.20–2.21
  lowest common denominator, 2.22–2.23
  multiplying, 2.25
  practical applications, 2.26
  reducing, 2.22
  subtracting, 2.24
Framing, foundation plans, 5.11, 5.14*f*
Framing squares, 3.23, 3.23*f*
Freight elevators, 9.12–9.13, 9.13*f*, 9.19
Frostbite, 1.60–1.61
Fuel cells, 9.10
Fuels, biodiesel, 6.16
Furring, architectural symbols, 5.7

G

Gallons, unit conversion, 2.67
Galvanized metal, 1.64
Gas leaks, evacuation procedures, 1.13
Gas masks, 1.49–1.52, 1.50*f*
General duty clause, OSHA standards, 1.10
Geometry
  angles, 2.41–2.42, 2.42*f*
  combination square, 3.24
  practical applications, 2.51–2.52
Georgia buggy, 9.11, 9.11*f*
GFCI (ground fault circuit interrupter), 1.40, 1.41*f*, 1.43, 1.82, 4.4
Glasses, safety, 1.46*f*
Global positioning systems (GPS), 8.12
Glossary, defined, 7.21
Gloves, safety, 1.46–1.47, 1.47*f*, 9.1
Glue, clamps and, 3.35
Goggles, safety, 1.46*f*, 1.63, 1.63*f*, 1.82
Going Green
  biodiesel, 6.16
  carbon footprint, reducing, 7.17
  cell phones, recycling, 7.7
  city projects, 2.41
  fuel cells, 9.10
  green-collar jobs, 8.1
  green employers, 8.23
  LEED program, 5.25, 5.40, 8.5
  power adapters, 4.1
  rechargeable batteries, 4.5
  recycling, Rinker Hall, University of Florida, 5.40
  recycling equipment, 3.11
  recycling lumber, 2.11
GPS (global positioning systems), 8.12
Grab hooks, 6.10, 6.11*f*, 6.26, 6.26*f*
Grains, unit conversion, 2.67
Grams (g), 2.12, 2.14, 2.67
Graph, defined, 7.21
Green Building Council, 5.25
Green construction. *See* Going Green
Greenhouse-gas emissions, 2.39
Gridlines, drawings, 5.39, 5.41*f*
Grilles, mechanical plans, 5.11, 5.16*f*–5.21*f*
Grinding hand tools
  eye and face protection, 1.45–1.46, 1.46*f*
  guards, 1.38
  On-Site, 3.8
Grinding power tools
  bench grinders, 4.21–4.23, 4.22*f*, 4.23*f*
  handheld, 4.19–4.21, 4.20*f*, 4.21*f*
Grit, defined, 4.36
Grommet slings, 6.5, 6.5*f*, 6.47
Ground, defined, 1.82
Ground fault circuit interrupter (GFCI), 1.40, 1.41*f*, 1.43, 1.82, 4.4, 4.36
Guarded, defined, 1.82

H

Hacksaws, 3.27–3.29, 3.28*f*
Hair, safety concerns, 9.1
Hammer drills, 4.6–4.7, 4.7*f*, 4.8*f*

Hammers
ball peen, 3.1*f*, 3.2, 3.44
claw hammers, 3.1–3.4, 3.1*f*, 3.44
grinding, 3.8
sledgehammers, 3.3–3.4, 3.3*f*
Hand chain hoists, 6.29–6.32, 6.30*f*–6.32*f*
Handheld computers, 8.11–8.14, 8.32
Hand line, 1.82
Hand pumps, 4.1
Handsaws, 3.27–3.29, 3.28*f*
Hand-screw clamp, 3.33*f*, 3.34–3.35
Hand signals, lifting equipment, 9.13, 9.14*f*
Hand signals, load navigation, 6.35–6.36, 6.36*f*–6.39*f*, 6.47
Hand tools
chain falls and come alongs, 3.35–3.36, 3.35*f*, 3.36*f*
chalk lines, 3.26, 3.26*f*
chisels, 3.6–3.8, 3.7*f*
clamps, 3.32–3.35, 3.33*f*
files and rasps, 3.30–3.32, 3.30*f*
grinding, 3.8
hammers, ball peen, 3.1*f*, 3.2, 3.44
hammers, claw, 3.1–3.4, 3.1*f*, 3.44
hammers, sledgehammers, 3.3–3.4, 3.3*f*
levels, 3.20–3.22, 3.21*f*, 3.22*f*
metrics and, 3.16
nail pullers, 3.5–3.6, 3.5*f*
picks, 3.38, 3.38*f*
pliers, 3.11–3.14, 3.44
plumb bobs, 3.25–3.26, 3.25*f*
precision measuring tools, 3.20
ripping bars, 3.5–3.6, 3.5*f*, 3.44
rules and measures, 3.18–3.20
safety, 3.1
saws, 3.27–3.29, 3.28*f*
screwdrivers, 3.8, 3.9–3.11, 3.9*f*
shovels, 3.36–3.37, 3.36*f*
sockets and ratchets, 3.17, 3.17*f*
squares, 3.23–3.25, 3.23*f*, 3.24*f*, 3.44
torque wrenches, 3.17–3.18, 3.17*f*, 3.18*f*
utility knives, 3.27, 3.27*f*
wire cutters, 3.11–3.14
wrenches, 3.14–3.16, 3.14*f*, 3.15*f*, 3.16*f*
Hand trucks, 9.7, 9.7*f*, 9.19
Harassment, 8.23, 8.32
Hard hats, 1.45, 1.45*f*
Hardware, computer, 8.11–8.12, 8.32
Hardware, schedules, 5.8
Hazard Communication Standard (HazCom), 1.52–1.58, 1.53*f*–1.56*f*, 1.82
Hazardous materials
communication standards, 1.52–1.56, 1.53*f*–1.56*f*
defined, 4.36
spoil pile, 1.36
Hazards
communication standards, 1.52–1.56, 1.53*f*–1.56*f*
four high-hazard areas, 1.13
recognition and control, 1.13–1.16, 1.14*f*, 1.16*f*
signs for, 1.4, 1.4*f*
spoil pile, 1.36
struck-by hazards, 1.31–1.33
HazCom, 1.52–1.58, 1.53*f*–1.56*f*, 1.82
Heads, sprinkler system, 5.16
Hearing protection, 1.48–1.49, 1.49*f*
Heating, ventilating, and air conditioning (HVAC)
defined, 5.47
Going Green, 5.40, 8.5
green-collar jobs, 8.1
mechanical plans, 5.11, 5.16*f*–5.21*f*
Heat-resistant gloves, 1.46–1.47, 1.47*f*
Heat stress, human, 1.59–1.60
Heavy metals, 4.5
Hecto (h), metric units, 2.12, 2.14
Height, measuring, 2.46, 2.48–2.51
Hex key wrench, 3.14–3.15, 3.14*f*, 3.44
Hex screwdrivers, 3.9–3.11, 3.9*f*
Hidden lines, drawings, 5.25*f*, 5.34, 5.47
High-temperature systems, 1.59
Hitches, 6.15–6.18, 6.15*f*–6.19*f*, 6.47
Hoists, 6.29–6.32, 6.30*f*–6.32*f*, 6.47
Hole, hazards of, 1.17, 1.18–1.19
Hooks, rigging, 6.10, 6.11*f*, 6.25–6.27, 6.26*f*, 6.27*f*, 6.34
Horizontal pencil grinders, 4.19–4.21, 4.20*f*
Horizontal surfaces, level measurement, 3.20, 3.21*f*
Hoses
compressors, 4.11, 4.28
safety concerns, 1.64–1.66, 1.65*f*
Hot work permit, 7.13*f*, 7.14*f*
Housekeeping, 1.9–1.10, 1.28
HVAC (heating, ventilation, and air conditioning)
defined, 5.47
Going Green, 5.40, 8.5
green-collar jobs, 8.1
mechanical plans, 5.11, 5.16*f*–5.21*f*
Hydraulic tools. *See also* Power tools
defined, 4.1, 4.36
jacks, 4.30, 4.30*f*, 9.8–9.9, 9.9*f*
Hyperstories, 8.4
Hypothermia, 1.60–1.62

I

Illumination, regulations, 1.9–1.10
Imminent danger inspections, 1.11
Improper fractions, 2.23, 2.63
Inches (in)
converting to feet, 2.39–2.40, 2.65
cubic, 2.48
square inches, 2.46
unit conversion, 2.67
Inch-pounds (torque), 3.44
Index, defined, 7.21
Indirect costs, 2.39
Indoor environment, 5.25
Industrial forklift, 9.11–9.12, 9.11*f*, 9.19
Initiative, 8.15, 8.32
Injuries, reporting and recording, 1.9–1.10, 1.12
Inspections, safety
extension ladders, 1.24–1.25
hoses and regulators, 1.64–1.66, 1.65*f*
ladders, 1.23
respirators, 1.52
scaffolds, 1.28, 1.30, 1.30*f*
stepladders, 1.25
trenches and excavations, 1.34
types of, 1.11
Insulation
architectural symbols, 5.7
asbestos exposure, 1.57
Insurance, worker's compensation, 1.3
Invert, fractions, 2.63
Iron, cutting, 4.19
Isometric drawings, 5.11, 5.22*f*, 5.23*f*, 5.26, 5.26*f*, 5.47
Isosceles triangle, 2.45, 2.45*f*, 2.63
Italics, defined, 7.21

J

Jackhammers, 4.28–4.30, 4.29*f*
Jacks, 4.30, 4.30*f*, 9.8–9.9, 9.9*f*, 9.19
Jargon, defined, 7.21

Jewelry, safety and, 1.45, 9.1
JHA (job hazard analysis), 1.14–1.15, 1.14*f*
Jig saws, 4.14–4.15, 4.14*f*
Job hazard analysis (JHA), 1.14–1.15, 1.14*f*
Jobs, green-collar, 8.1
Job safety analysis (JSA), 1.14–1.15, 1.14*f*
Job skills. *See* Employment, skills for
Joint, defined, 3.44
JSA (job safety analysis), 1.14–1.15, 1.14*f*

K

Kerf, defined, 3.44
Keyhole saws, 3.27–3.29, 3.28*f*
Keynotes, drawings, 5.34–5.38, 5.35*f*–5.39*f*
Keys, chuck, 4.3, 4.3*f*, 4.36
Kilograms (kg), 2.67
Kilo (k), metric units, 2.12, 2.14
Kilometers (km), 2.12, 2.14, 2.67
Kinks, wire rope, 6.14, 6.14*f*

L

Ladders, 1.17, 1.21–1.28, 1.21*f*–1.27*f*, 9.5
Landscape plans, 5.4, 5.5*f*, 5.6*f*
Land use, topographic maps, 5.8, 5.8*f*
Lanyards, 1.19–1.21, 1.19*f*, 1.20*f*, 1.82
Laser measuring tools
  length measurement, 3.19–3.20, 3.20*f*
  levels, 3.22, 3.22*f*
Lead, 1.57, 4.5, 4.11
Leaders and arrowheads, drawings, 5.25*f*, 5.34, 5.47
Leadership skills, 8.20–8.23, 8.32
LEED program, 5.25, 5.40, 8.5
Legal concerns, construction drawings, 5.3
Legend, construction drawings
  defined, 5.3, 5.4*f*, 5.47
  electrical symbols, 5.16, 5.24*f*, 5.25*f*, 5.27*f*–5.29*f*
  mechanical plans, 5.11, 5.17*f*, 5.18*f*
Leg protection, 1.48
Length, measuring
  metric units, 2.12, 2.15–2.16
  ruler types, 2.8–2.11, 2.9*f*
  unit conversion, 2.67
  volume of shapes, 2.48, 2.50–2.52
Level, defined, 3.44
Levels
  digital levels, 3.21–3.22, 3.22*f*
  laser levels, 3.22, 3.22*f*
  spirit levels, 3.20–3.21, 3.21*f*
Lifelines, 1.19–1.21, 1.19*f*, 1.20*f*
Lifting, ergonomics, 1.67–1.69, 1.69*f*, 9.1–9.2, 9.2*f*
Lifting capacity, 6.3
Lifting clamps, 6.23–6.25, 6.24*f*–6.26*f*, 6.47
Lighting
  plans for, 5.16, 5.24*f*, 5.25*f*, 5.27*f*–5.29*f*
  stairways, 1.28
Lineman pliers, 3.12, 3.12*f*
Lines of construction, 5.25*f*, 5.34
Listening skills, 7.1–7.6, 7.3*f*
Liters (l), 2.67
Load capacities, ladders, 1.21
Loads
  control, 6.35–6.41, 6.36*f*–6.39*f*, 6.41*f*, 6.47
  defined, 6.47, 9.19
  lowering, 9.2–9.3
  oscillation, 6.19
  overloading, 1.21
Load stress, defined, 6.47
Locking cam, lifting clamp, 6.25, 6.25*f*
Locking C-clamps, 3.32–3.35, 3.33*f*
Locking pliers, 3.12*f*, 3.13, 3.13*f*
Lockout/tagout systems, 1.43–1.44, 1.44*f*, 1.82
Long-nose pliers, 3.12, 3.12*f*
Lubricants. *See* Maintenance
Lumber
  board sizes, 2.9
  recycling, 2.11
  stacking and storing, 9.3
  warping, 3.23
Lung cancer, 1.57

M

Machinist's hammers. *See* Ball peen hammers
Machinist's rule, 2.26, 2.27*f*, 2.63
Maintenance
  air impact wrenches, 4.28
  chain pulls and come alongs, 3.36
  chisels, 3.8
  clamps, 3.35
  drills, 4.4–4.9, 4.11
  emery cloth, 3.29
  files and rasps, 3.32
  grinders and sanders, 4.19–4.23
  hammers, 3.2–3.3, 3.4
  hoists, 6.29
  hydraulic jacks, 4.30
  levels, 3.21
  measuring tools, 3.20
  pavement breakers, 4.28, 4.30
  picks, 3.38
  pliers, 3.14
  pneumatic nailers, 4.25
  powder-actuated fastening systems, 4.27
  punches, 3.8
  respirators, 1.52
  ripping bars and nail pullers, 3.6
  saws, 3.29, 4.13–4.19
  shovels, 3.37
  sockets and ratchets, 3.17
  squares, 3.25
  stairways, 1.28
  torque wrenches, 3.18
  utility knives, 3.27
  wrenches, 3.16
Management system, defined, 1.83
Management system failures, 1.9
Maps, topographic, 5.8, 5.8*f*
Masks, air, 1.49–1.52, 1.50*f*
Masonry
  architectural symbols, 5.7
  bricklayers, 3.22
  defined, 4.36
  drill bits, 4.2, 4.2*f*
  silica, 1.57
  stacking and storing, 9.3
  stonemasons, 3.5
Master link, slings, 6.10, 6.11*f*, 6.47
Material cart, 9.19
Material Safety Data Sheets (MSDS), 1.52–1.56, 1.53*f*–1.56*f*, 1.83
Materials handling
  cables, working with, 9.5–9.6
  carts, 9.6–9.7, 9.7*f*
  concrete mule, 9.11, 9.11*f*
  freight elevators, 9.12–9.13, 9.13*f*
  hand signals, 9.13, 9.14*f*
  hand trucks, 9.7, 9.7*f*
  industrial forklifts, 9.11–9.12, 9.11*f*
  jacks, 9.8–9.9, 9.9*f*
  lifting procedures, 9.1–9.2, 9.2*f*

Materials handling, *continued*
loads, lowering, 9.2–9.3
pallet jacks, 9.9, 9.9*f*
personal protective equipment (PPE), 9.1
pipe mule, 9.8, 9.8*f*
pipe transport, 9.8, 9.8*f*
pre-task planning, 9.1
roller skids, 9.7, 9.8*f*
rough terrain forklift, 9.12, 9.12*f*
stacking and storing, 9.3, 9.4*f*, 9.5*f*
wheelbarrows, 9.7–9.8, 9.8*f*, 9.10, 9.10*f*
working from heights, 9.5
Math. *See also* Measurement
angles, 2.41–2.42
area of shapes, 2.46–2.47
circles, 2.45–2.46
converting decimals and percentages, 2.37–2.39
converting fractions to decimals, 2.38–2.39
converting inches to feet, 2.39–2.40
decimals, 2.26–2.35
division, 2.6–2.8, 2.25–2.26, 2.31–2.32, 2.70
fractions, 2.20–2.25
geometry, 2.41, 2.51–2.52
multiplication, 2.4–2.8, 2.25, 2.30–2.31, 2.64, 2.70
rectangles, 2.44
squares, 2.44
triangles, 2.43–2.45, 2.48
volume of shapes, 2.48, 2.50–2.52
whole numbers, 2.1–2.2
Maximum allowable slope (MAS), 1.35, 1.35*f*, 1.83
Maximum intended load, 1.21, 1.83
Means of egress, 1.9–1.10
Measurement
architect's scale, 2.14–2.16, 2.15*f*, 2.16*f*
converting units, 2.18–2.19, 2.39–2.40
engineer's scale, 2.16–2.17, 2.17*f*
laser measuring tools, 3.19–3.20, 3.20*f*
level, 3.20–3.22, 3.21*f*, 3.22*f*
measuring tape, 2.11–2.13, 2.13*f*, 2.15, 3.19, 3.19*f*
precision measuring tools, 3.20
torque, 3.44
wooden folding rule, 3.19, 3.19*f*
Measuring tape, 2.11–2.13, 2.13*f*, 2.15, 3.19, 3.19*f*
Mechanical plans
defined, 5.47
overview, 5.1, 5.2*f*, 5.11, 5.16*f*–5.21*f*
symbols, 5.35, 5.38*f*
Medical examinations, 1.56–1.57
Medical records
hazardous material exposure, 1.56–1.57
regulations about, 1.9–1.10, 1.11
Megameters, defined, 2.12, 2.14
Memo, defined, 7.21
Mercury, 4.5
Mesothelioma, 1.57
Metal, cutting, 4.11
Metal joists, foundation plans, 5.11, 5.14*f*
Metal ladders, 1.22–1.24, 1.23*f*, 1.24*f*
Metal lath furring, 5.7
Meters (m)
cubic, 2.48
defined, 2.12, 2.14, 2.63
square meters, 2.46
unit conversion, 2.67
Metric ruler scales
architect's scale, 2.15–2.16, 2.15*f*, 2.16*f*
construction drawings, 5.18
defined, 2.10–2.11, 2.10*f*–2.11*f*, 2.64
study problems, 2.17
Metric system
defined, 2.12, 2.16
tools and, 3.16
unit conversion, 2.67
Micrometers, 3.20
Mid-rail, defined, 1.83
Miles, unit conversion, 2.67
Milli (m), metric units, 2.12, 2.14
Millimeters (mm)
defined, 2.12, 2.14
unit conversion, 2.67
Millwrights, 2.44
Mission statements, 8.1–8.2, 8.32
Miter joint, 3.44
Miter saws, power, 4.18, 4.18*f*
Mixed numbers, defined, 2.64
Mobile cranes, 6.1, 6.2*f*. *See also* Rigging
Molten materials, safety, 1.45–1.46, 1.46*f*
Monitoring inspection, 1.11
Monkey wrenches, 3.15–3.16, 3.15*f*, 3.16*f*
Motor control centers, plans for, 5.16, 5.24*f*, 5.25*f*, 5.27*f*–5.29*f*
Motors, mechanical plans, 5.11, 5.16*f*–5.21*f*
MSDS (Material Safety Data Sheets), 1.52–1.56, 1.53*f*–1.56*f*, 1.83
Multiple-leg bridle hitch, 6.15, 6.16*f*
Multiplication
calculators and, 2.70
decimals, 2.30–2.31
fractions, 2.25
table of, 2.64
whole numbers, 2.4–2.8

N

Nail cutter saw blades, 4.11
Nail guns, 4.23–4.25, 4.23*f*–4.25*f*
Nail pullers, 3.5–3.6, 3.5*f*, 3.44
Nails, 4.11. *See also* Hammers; Nail pullers
Needle-nose pliers, 3.12, 3.12*f*
Negative numbers, 2.1, 2.64
Nickel, 4.5
Noise hazards, 1.13, 1.48–1.49, 1.49*f*
Nonadjustable wrenches, 3.14–3.15, 3.14*f*
Nonferrous metal cutter saw blades, 4.11
Non-marring lifting clamps, 6.25, 6.25*f*
North arrow, construction drawings, 5.11
Not to scale (NTS), 5.18, 5.47
Numeral systems, 2.2, 2.27, 2.30
Numerator, fractions, 2.20, 2.64
Nuts, air impact wrench, 4.27–4.28, 4.27*f*

O

Object lines, drawings, 5.25*f*, 5.34
Obtuse angles, 2.42*f*, 2.64
Occupational Safety and Health Administration (OSHA)
confined spaces, 1.67
contact information, 1.12
defined, 1.83
hazard communication, 1.52–1.56, 1.53*f*–1.56*f*
inspections, 1.11
ladder restrictions, 1.23
overview, 1.9–1.12
respiratory protection, 1.51
scaffolds, 1.28, 1.30, 1.30*f*
stairway lighting, 1.28
One rope lay, 6.14, 6.14*f*, 6.47
On-Site
addition and subtraction, 2.3, 2.4
architectural drawings, 5.8
architectural symbols, 5.7

blocks, number of courses, 2.9
bricklayers, 3.22
Building Information Modeling (BIM), 5.7
coating thickness, calculating, 2.31
Code of Federal Regulations, 1.10
computer-aided drafting (CAD), 5.1
construction drawings, care of, 5.35
contact lenses, 1.51
criticism, destructive, 8.19
cutting, circular saws, 4.13
cutting tools, sharpening, 1.5
cylinders, area of, 2.53
decimals, use of, 2.28, 2.32, 2.33, 2.40
diagonals, 2.45
drilling metal, 4.4
electrical cord safety, 1.42
electricians, 3.6
falls from scaffold and ladders, 1.31
fire safety, 1.64, 1.66, 1.71, 1.73
fractions, use of, 2.20, 2.21, 2.40
geometry, use of, 2.43
grinding hand tools, 3.8
hammers, physics of, 3.2
harnesses, 1.8
hoists, 6.32
humor, use of, 7.8
Internet, information on, 8.7
isometric drawings, 5.26, 5.26*f*
job applications, 8.6
ladders, three-point contact, 1.24
listening in the classroom, 7.6
load oscillation, 6.19
measuring tools, precision, 3.20
metrics and tools, 3.16
metric system, 2.18
mobile cranes, 6.1, 6.1*f*
multiplication and division, 2.5, 2.6
orthographic drawings, 5.32, 5.32*f*
OSHA contact information, 1.12
percentage-decimal conversions, 2.38
personal evaluation, 8.21
personal problems, 8.20
plans, checking, 5.34
pneumatic nailers, 4.24, 4.25
power screwdrivers, 4.24
pre-lift safety check, 6.36
Pythagorean Theorem, 2.49
resumes, 8.4
riggings, 6.3
robots on the job, 8.14
saw blades, 4.11, 4.16
scales on drawings, 2.18
schematic drawings, 5.26, 5.26*f*
screw pin shackles, 6.21
screws, threads, 3.10
shackles, 6.35
sling jackets, 6.9
stonemasons, 3.5
team building, 8.22
topographic maps, 5.8, 5.8*f*
triangles, types of, 2.46
unit conversion, area and volume, 2.50
visualizing plans, 5.15
worm-drive saw, 4.13
writing skills, 7.15
Open-end wrench, 3.14–3.15, 3.14*f*, 3.44
Operating systems, computer, 8.11–8.12
Opposite angles, 2.42*f*, 2.64
Organizational chart, 8.6
Organizational skills, 8.15
Orthographic drawings, 5.32, 5.32*f*
OSHA (Occupational Safety and Health Administration)
confined spaces, 1.67
contact information, 1.12
defined, 1.83
hazard communication, 1.52–1.56, 1.53*f*–1.56*f*
inspections, 1.11
ladder restrictions, 1.23
overview, 1.9–1.12
respiratory protection, 1.51
scaffolds, 1.28, 1.30, 1.30*f*
stairway lighting, 1.28
Ounces (oz), unit conversion, 2.67
Outlets, electrical plans, 5.16, 5.24*f*, 5.25*f*, 5.27*f*–5.29*f*
Overhead costs, 2.39
Overhead cranes, 6.1, 6.2*f*. *See also* Rigging
Overloading, 1.21
Over-the-counter medications, use of, 1.6–1.7, 8.24–8.25
Oxyacetylene equipment, 1.64
Oxygen, 1.62, 1.64

P

Paddle drill bits, 4.2, 4.2*f*
Pad eye, defined, 6.47
Paint, calculating amount, 2.50
Pallet jacks, 9.9, 9.9*f*, 9.19
Pallet trucks, 9.9, 9.9*f*
Panel schedules, electrical plans, 5.16, 5.24*f*, 5.25*f*, 5.27*f*–5.29*f*
Pavement breakers, 4.28–4.30, 4.29*f*
PDAs (personal digital assistants), 8.11, 8.32
Peening, 3.2, 3.44
Pencil grinders, 4.19–4.21, 4.20*f*
Percentages, 2.37–2.38, 2.64
Perimeter, defined, 2.64
Permissible exposure limits (PELs), 1.56–1.57
Permit-required confined space, 1.66–1.67, 1.67*f*, 1.68*f*, 1.83
Permits
defined, 7.21
hot work permits, 7.13*f*, 7.14*f*
Perpendicular pull, eyebolts, 6.22, 6.22*f*
Personal digital assistants (PDAs), 8.11, 8.32
Personal fall arrest systems (PFAS), 1.19–1.21, 1.19*f*, 1.20*f*
Personal protective equipment (PPE)
defined, 1.83
materials handling, 9.1
regulations, 1.9–1.10
types of, 1.45–1.52, 1.45*f*–1.50*f*
welding, 1.62–1.63, 1.63*f*
Phillips screwdrivers, 3.9–3.11, 3.9*f*
Pi, 2.45, 2.64
Picks, 3.38, 3.38*f*
P&IDs (piping and instrumentation drawings), 5.11, 5.19*f*, 5.47
Pints (pts), unit conversion, 2.67
Pipe clamp, 3.33*f*, 3.34–3.35
Pipe hooks, 6.10, 6.11*f*, 6.25–6.27, 6.26*f*
Pipe mule, 9.8, 9.8*f*, 9.19
Pipes and piping
fire protection plans, 5.16
grinders, 4.19–4.21, 4.20*f*
mechanical plans, 5.11, 5.16*f*–5.21*f*
piping and instrumentation drawings (P&IDs), 5.11, 5.19*f*, 5.47
plumbing plans, 5.11, 5.22*f*, 5.23*f*
stacking and storing, 9.3
Pipe transport, 9.8, 9.8*f*, 9.19
Pipe wrenches, 3.15–3.16, 3.15*f*, 3.16*f*, 3.44

Place value, numbers, 2.1, 2.29, 2.64
Plane, 3.44, 6.47
Planked, defined, 1.83
Plan locations, drawings, 5.39, 5.41*f*
Planning and scheduling, 8.10–8.11
Plan view. *See* Floor plans
Plaster, 1.57
Plate hooks, 6.10, 6.11*f*
Plates, grinders, 4.19–4.21, 4.20*f*
Platforms, 9.5
Platform Twinring cranes, 6.1, 6.1*f*
Pliers, 3.11–3.14, 3.44
Plot plans, 5.4, 5.5*f*, 5.6*f*
Plotter, defined, 8.32
Plumb, 3.20, 3.21*f*, 3.44
Plumb bobs, 3.25–3.26, 3.25*f*
Plumbing. *See also* Pipes and piping
  defined, 5.47
  drills and, 4.4
  plumbing plans, 5.1, 5.2*f*, 5.11, 5.22*f*, 5.23*f*, 5.47
  schematic drawings, 5.26, 5.26*f*
  symbols, drawings, 5.38, 5.38*f*
Pneumatic tools. *See also* Power tools
  air impact wrench, 4.27–4.28, 4.27*f*
  defined, 4.36
  drills, 4.10–4.11, 4.10*f*
  hoists, 6.31*f*
  power nailers, 4.23–4.25, 4.23*f*, 4.24*f*, 4.25*f*
Point-loaded hook, 6.34
Points, wrenches, 3.44
Polishing, grinders, 4.19–4.21, 4.20*f*
Portable handheld bandsaw, 4.16–4.17, 4.17*f*
Porta Power®, 4.30, 4.30*f*
Positive numbers, 2.1, 2.64
Pounds (lb), unit conversion, 2.67
Powder-actuated fastening systems, 4.26–4.27, 4.26*f*
Power adapters, 4.1
Powered chain hoists, 6.29–6.32, 6.30*f*–6.32*f*
Powered wheelbarrow, 9.10, 9.10*f*, 9.19
Power miter saws, 4.18, 4.18*f*
Power plans, 5.16, 5.24*f*, 5.25*f*, 5.27*f*–5.29*f*
Power tools
  abrasive cutoff saw, 4.19, 4.19*f*
  air impact wrench, 4.27–4.28, 4.27*f*
  circular saws, 4.11–4.14, 4.12*f*
  cordless drills, 4.5–4.6, 4.6*f*, 4.7*f*
  drills, 4.1–4.4, 4.2*f*
  electromagnetic drills, 4.7–4.9, 4.9*f*
  eye and face protection, 1.45–1.46, 1.46*f*
  grinders, 4.19–4.23
  hammer drills, 4.6–4.7, 4.7*f*, 4.8*f*
  hydraulic jack, 4.30, 4.30*f*
  machine guards, 1.37–1.38, 1.38*f*
  overview, 4.1
  pavement breakers, 4.28–4.30, 4.29*f*
  pneumatically powered nailers, 4.23–4.25, 4.23*f*, 4.24*f*, 4.25*f*
  pneumatic drills, 4.10–4.11, 4.10*f*
  portable handheld bandsaw, 4.16–4.17, 4.17*f*
  powder-actuated fastening systems, 4.26–4.27, 4.26*f*
  power miter saw, 4.18, 4.18*f*
  reciprocating saws, 4.15–4.16, 4.15*f*, 4.16*f*
  rock drills, 4.28–4.30, 4.29*f*
  saber saws (jig saws), 4.14–4.15, 4.14*f*
  sanders, 4.19–4.23
  screwdrivers, 4.2, 4.2*f*, 4.24, 4.24*f*
  worm-drive saw, 4.13
PPE (personal protective equipment)
  defined, 1.83
  materials handling, 9.1
  regulations, 1.9–1.10
  types of, 1.45–1.52, 1.45*f*–1.50*f*
  welding, 1.62–1.63, 1.63*f*
Prescription medications, use of, 1.6–1.7, 8.24–8.25
Presentation skills, self, 8.14–8.16
Presses, guards, 1.38
Pressurized air. *See* Compressors
Pressurized systems, 1.59
Prick punch, 3.8, 3.9*f*
Problem solving skills, 8.7–8.11
Product (multiplication), 2.64
Professionalism, 8.15
Programmed inspection, 1.11
Project teams, 8.10–8.11
Property lines, 5.25*f*, 5.34
Proximity work
  defined, 1.83
  electrical equipment, 1.44
  hazards, 1.58–1.59, 1.58*f*, 1.59*f*
Pumps, mechanical plans, 5.11, 5.16*f*–5.21*f*
Punches, 1.38, 3.8, 3.9*f*, 3.44
Punch lists, 7.12, 7.21
Purple, warning signs, 1.17
Pythagorean Theorem, 2.48, 2.50

Q

Qualified persons, defined, 1.12, 1.83
Quarts (qt), unit conversion, 2.67
Quick Grip® bar clamp, 3.33*f*, 3.34–3.35
Quotient, 2.5, 2.64

R

Radiation warnings, 1.17
Radius, circles, 2.45, 2.64
Rafters, roof, 5.11, 5.14*f*
Rafters angle square, 3.23, 3.23*f*, 3.44
Railings, 1.17–1.18, 1.30
Rasps, 3.30–3.32, 3.30*f*
Ratchets and sockets, 3.17, 3.17*f*
Rated capacity, defined, 6.47, 6.49–6.50, 6.50*f*
Rat-tail file, 3.30–3.32, 3.30*f*
Reading skills, 7.8–7.12
Receptacles, electrical, 1.82, 5.16, 5.24*f*, 5.25*f*, 5.27*f*–5.29*f*
Rechargeable batteries, recycling, 4.5
Reciprocating, defined, 4.36
Rectangles
  area, 2.46, 2.68
  defined, 2.44, 2.64
  volume, 2.48, 2.68
Recycling
  cell phones, 7.7
  equipment, 3.11
  green collar jobs, 8.1
  green construction, 5.25, 5.40
  lumber, 2.11
  rechargeable batteries, 4.5
Red, warning signs, 1.17
Red core yarns, 6.4, 6.5*f*
Reducing fractions, 2.22
References, jobs, 8.2, 8.32
Registers, mechanical plans, 5.11, 5.16*f*–5.21*f*
Regulators, safety concerns, 1.64–1.66, 1.65*f*
Rejection criteria, defined, 6.47
Remainder (division), 2.64

Repetitive motion hazards, 1.13
Reporting injuries, accidents, 1.12
Request for information (RFI), 5.16, 5.33*f*, 5.47
Respirator, defined, 1.83
Respiratory protection, 1.49–1.52, 1.50*f*, 1.64
Resumes, 8.2, 8.3*f*, 8.4
Reverse eye hooks, 6.26, 6.26*f*
Revision blocks, construction drawings, 5.2, 5.3, 5.4*f*
Revolutions per minute (rpm), defined, 4.36
RFI (request for information), 5.16, 5.33*f*, 5.47
Rigging
  eyebolts, 6.21–6.23, 6.22*f*–6.24*f*, 6.32–6.34, 6.34*f*
  hardware attachment, 6.32–6.34, 6.34*f*
  hitches, 6.15–6.18, 6.15*f*–6.19*f*
  hoists, 6.29–6.32, 6.30*f*–6.32*f*
  lifting clamps, 6.23–6.25, 6.24*f*–6.26*f*
  load control, 6.35–6.41, 6.36*f*–6.39*f*, 6.41*f*, 6.47
  overview, 6.1, 6.2*f*
  rated capacity, 6.32
  rigging hooks, 6.25–6.27, 6.26*f*, 6.27*f*
  risk management, 6.40, 6.47
  shackles, 6.19–6.21, 6.20*f*, 6.21*f*, 6.32, 6.33*f*, 6.35, 6.35*f*
  slings, attachments, 6.32
  slings, steel alloy chain, 6.10–6.11, 6.10*f*–6.12*f*
  slings, synthetic, 6.4–6.9, 6.4*f*–6.9*f*
  slings, tagging requirements, 6.3, 6.3*f*
  slings, wire rope, 6.11–6.14, 6.12*f*–6.14*f*
  sling stress, 6.28, 6.28*f*–6.29*f*
Right angles
  carpenter's squares, 3.23, 3.23*f*
  defined, 2.42, 2.42*f*, 2.64
  3/4/5 Rule, 2.47
Right triangles
  angles of, 2.42, 2.42*f*
  defined, 2.45, 2.45*f*, 2.64
  Pythagorean Theorem, 2.48, 2.50
  3/4/5 Rule, 2.47
Ring test, 4.36
Rinker Hall, University of Florida, 5.40
Ripping bars, 3.5–3.6, 3.5*f*, 3.44
Ripsaw blades, 4.11
Ripsaws, 3.27–3.29, 3.28*f*
Risk management, 1.15, 6.40, 6.47
Rivets, peening, 3.2
Roadway hazards, 1.31–1.33
Robertson® (square) screwdrivers, 3.9–3.11, 3.9*f*
Robots, use of, 8.14
Rock drills, 4.28–4.30, 4.29*f*
Roller skids, 9.7, 9.8*f*, 9.19
Rolling machines, guards, 1.38
Rolling scaffolds, 1.28–1.31, 1.29*f*, 1.30*f*
Roofs
  rafters, 5.11, 5.14*f*
  roof-framing plans, 5.11, 5.14*f*
  roofing materials, 1.57
  roof plan, 5.7, 5.9*f*, 5.47
Rooftop gardens, 2.41
Rough terrain forklifts, 9.12, 9.12*f*, 9.19
Round-bladed shovel, 3.36–3.37, 3.36*f*
Round chain shackles, 6.19–6.21, 6.20*f*
Rounding decimal numbers, 2.33–2.34
Round off, defined, 3.45
Round pin anchor shackle, 6.19–6.21, 6.20*f*
Round reverse eye hook, 6.26, 6.26*f*
Round slings, 6.5–6.7, 6.5*f*, 6.6*f*, 6.7*f*
RPM (revolutions per minute), defined, 4.36
Rubber-insulated gloves, 1.46–1.47, 1.47*f*
3/4/5 Rule, 2.47
Rulers
  architect's scale, 2.8, 2.9*f*, 2.14–2.16, 2.15*f*, 2.16*f*
  engineer's (civil) scale, 2.8, 2.9*f*, 2.16–2.17, 2.17*f*
  flat steel rules, 3.18–3.19, 3.19*f*
  laser measuring tools, 3.19–3.20, 3.20*f*
  machinist's rule, 2.26, 2.27*f*, 2.63
  measuring tape, 3.19, 3.19*f*
  standard (English), 2.8–2.10, 2.9*f*, 2.64
  wooden folding rule, 3.19, 3.19*f*
Rung locks, 1.25, 1.25*f*

S

Saber saws, 4.14–4.15, 4.14*f*
Safety. *See also* Materials handling
  access and egress, 1.37
  accidents, causes and results, 1.1–1.13
  air impact wrenches, 4.28
  alloy steel chains, 6.10, 6.11*f*
  asbestos, 1.57
  Blood borne pathogens, 1.57–1.58
  caught-in-between hazards, 1.33–1.37, 1.35*f*–1.37*f*
  cell phones, 7.7–7.8
  chain pulls and come alongs, 3.36
  chemical splashes, 1.58
  chisels, 3.8
  clamps, 3.35
  cold stress, human, 1.60–1.62
  compressor hoses, 4.11
  confined spaces, 1.66–1.67, 1.67*f*, 1.68*f*
  culture of, 1.1
  cutting hazards, 1.62–1.66, 1.63*f*, 1.65*f*
  drills, 4.3, 4.4–4.9, 4.11
  electrical hazards, 1.38–1.45, 1.40*f*–1.42*f*, 1.44*f*
  elevated work, fall protection, 1.17–1.21, 1.18*f*, 1.19*f*, 1.20*f*
  ergonomics, 1.67–1.69, 1.69*f*
  exposure limits, 1.56–1.57
  eyebolts, 6.23
  files and rasps, 3.32
  fire hazards, 1.69–1.74, 1.70*f*–1.74*f*
  fire protection plans, 5.16
  grinders and sanders, 4.19–4.23
  hammers, 3.1, 3.2–3.4, 3.4*f*
  hazard communication, 1.52–1.56, 1.53*f*–1.56*f*
  hazards, recognition and control, 1.13–1.16, 1.14*f*, 1.16*f*
  heat stress, human, 1.59–1.60
  high-temperature systems, 1.59
  hoists, 6.29
  hoses and regulators, 1.64–1.66, 1.65*f*
  hydraulic jacks, 4.30
  inspections, 1.11
  job safety analysis (JSA), 1.14–1.15, 1.14*f*
  ladders and stairs, 1.21–1.28, 1.21*f*–1.27*f*
  laser tools, 3.22
  lead, 1.57
  levels, 3.21
  lifting clamps, 6.25, 6.26*f*
  lifting procedures, 9.1–9.2, 9.2*f*
  loads, lowering, 9.2–9.3
  logout/tagout, 1.43–1.44, 1.44*f*
  measuring tools, 3.20
  pavement breakers, 4.28, 4.30
  personal fall arrest systems, 1.19–1.21, 1.19*f*, 1.20*f*
  personal protective equipment (PPE), 1.45–1.52, 1.45*f*–1.50*f* 9.1
  picks, 3.38
  pliers, 3.14

Safety, *continued*
pneumatic nailers, 4.24, 4.25
policies and regulations, 1.9–1.12
powder-actuated fastening systems, 4.27
power tools, 4.1
pressurized systems, 1.59
proximity work, hazards, 1.58–1.59, 1.58*f*, 1.59*f*
punches, 3.8
rigging hooks, 6.27, 6.27*f*
riggings, 6.32–6.42, 6.33*f*–6.42*f*
ripping bars and nail pullers, 3.5, 3.6
risk management, 1.15
saws, 3.29, 4.11, 4.13–4.19
shackles, 6.21
shovels, 3.37
signs for, 1.4–1.5, 1.5*f*
silica, 1.57
sledgehammers, 3.3
slings, 6.3, 6.7–6.9, 6.8*f*, 6.9*f*
sling stress, 6.28, 6.28*f*–6.29*f*
sockets and ratchets, 3.17
squares, 3.25
struck-by hazards, 1.31–1.33
task safety analysis, 1.14–1.15, 1.16*f*
tool and machine guards, 1.37–1.38, 1.38*f*
toolbox talks, 1.3
torque wrenches, 3.18
utility knives, 3.27
vertical lifting, 6.15
welding hazards, 1.62–1.66, 1.63*f*, 1.65*f*
wire rope slings, 6.13–6.14, 6.14*f*
work area, 1.66
wrenches, 3.16
Safety anchor shackle, 6.19–6.21, 6.20*f*
Safety chain shackle, 6.19–6.21, 6.20*f*
Safety culture, defined, 1.83
Safety glasses, 1.46*f*
Safety harnesses, 1.19–1.21, 1.19*f*, 1.20*f*
Safety tags, 1.4–1.5, 1.5*f*
Safe working load (SWL), 6.3, 6.32
Sanders, 4.19–4.23
Sanitary riser system, 5.11, 5.22*f*, 5.23*f*
Sanitation, work site, 1.9–1.10
Saws
abrasive cutoff saw, 4.19, 4.19*f*
circular saws, 4.11–4.14, 4.12*f*
eye and face protection, 1.45–1.46, 1.46*f*
guards, 1.38
portable handheld bandsaw, 4.16–4.17, 4.17*f*
power miter saw, 4.18, 4.18*f*
reciprocating, 4.15–4.16, 4.15*f*, 4.16*f*
saber saws (jig saws), 4.14–4.15, 4.14*f*
types of, 3.27–3.29, 3.28*f*
worm-drive saw, 4.13
SawZall®, 4.15–4.16, 4.15*f*, 4.16*f*
Scaffolds, 1.28–1.31, 1.29*f*, 1.30*f*, 1.83, 9.5
Scale
construction drawings, 5.2, 5.16, 5.18
defined, 5.47
Scalene triangles, 2.45, 2.45*f*, 2.64
Schedules
door and window, 5.7, 5.12*f*
finish, 5.7, 5.12*f*
problems with, 8.10–8.11
Schematic drawings, 5.26, 5.26*f*, 5.47
Scientific notation, numbers, 2.34
Screw-adjusted lifting clamps, 6.25, 6.25*f*
Screwdrivers
drill bits, 4.2, 4.2*f*
grinding blades, 3.8
power, 4.24, 4.24*f*
types of, 3.9–3.11, 3.9*f*
Screw pin anchor shackles, 6.19–6.21, 6.20*f*
Screw pin chain shackle, 6.19–6.21, 6.20*f*
Screws. *See also* Screwdrivers
head types, 3.9*f*, 3.11
threads, 3.10
Section cuts, 5.25*f*, 5.34
Section drawings, 5.7, 5.10*f*
Self-contained breathing apparatus (SCBA), 1.49–1.52, 1.50*f*
Self-presentation, 8.32
Self-retracting lifelines, 1.19–1.21, 1.19*f*, 1.20*f*
Sewage disposal, plans, 5.11, 5.22*f*, 5.23*f*
Sexual harassment, 8.23, 8.32
Shackles, 6.19–6.21, 6.20*f*, 6.21*f*, 6.32, 6.33*f*, 6.35, 6.35*f*, 6.47
Shank, defined, 4.36
Shapes
area, 2.46–2.47
circles, 2.45–2.46
cylinders, 2.50
rectangles, 2.44
squares, 2.44
triangles, 2.44–2.45, 2.45*f*
unit conversions, 2.50
volume, 2.48, 2.50–2.52
Shearing tools, 1.38
Sheave, defined, 6.47
Sheet metal workers, 2.46
Shielding, trenches, 1.35
Shields, welding, 1.63, 1.63*f*
Shock loading
personal fall arrest systems, 1.21
rigging, 6.40
Shoes, safety, 1.48
Shoring, defined, 1.83
Shoring, trenches, 1.35
Shortening clutches, 6.26, 6.26*f*
Shouldered eyebolts, 6.22, 6.22*f*, 6.23*f*
Shovels, 3.36–3.37, 3.36*f*
Side cutter pliers, 3.12, 3.12*f*
Side grinders, 4.19–4.21, 4.20*f*
Side pull, defined, 6.47
Signaler, 1.83
Signals, lifting equipment, 9.13, 9.14*f*
Signals, load navigation, 6.35–6.36, 6.36*f*–6.39*f*, 6.47
Signs, hazards, 1.4, 1.4*f*
Silica, 1.57
Single-basket slings, 6.10, 6.11*f*
Site plans, 5.4, 5.5*f*, 5.6*f*, 5.16, 5.18
Six-foot rule, 1.83
Skids, roller, 9.7, 9.8*f*, 9.19
Skilsaw. *See* Circular saws
Skin protection, 1.48
Slabs, foundation plans, 5.11, 5.14*f*
Slag, 1.66, 1.83
Sledgehammers, 3.3–3.4, 3.3*f*
Sliding choker hooks, 6.26, 6.26*f*
Sling angle, defined, 6.47, 6.49–6.50, 6.50*f*
Sling legs, defined, 6.47
Sling reach, defined, 6.47

Slings
  alloy steel chain, 6.10–6.11, 6.10*f*–6.12*f*
  defined, 6.47
  rated capacity, 6.49–6.50, 6.50*f*
  sling jackets, 6.9
  stress, 6.28, 6.28*f*–6.29*f*, 6.47
  synthetic, 6.4–6.9, 6.4*f*–6.9*f*
  tagging requirements, 6.3, 6.3*f*
  wire rope, 6.11–6.14, 6.12*f*–6.14*f*
Slip-Joint (combination) pliers, 3.11, 3.12*f*
Slipping hazards, 1.17
Sloping, trenches, 1.35
Slotted screwdrivers, 3.9–3.11, 3.9*f*
Slugging wrench, 3.14–3.15, 3.14*f*, 3.45
Sockets and ratchets, 3.17, 3.17*f*
Software, computer, 8.11–8.12
Soil types, 1.34, 1.35
Solar panel installations, 8.1
Sorting hooks, 6.10, 6.11*f*, 6.25–6.27, 6.26*f*
Spade drill bits, 4.2, 4.2*f*
Spades, 3.36–3.37, 3.36*f*
Spades, clay, 4.28–4.30, 4.29*f*
Spall, 1.83
Speaking skills, 7.1, 7.4*f*, 7.6–7.8
Specifications, drawings, 5.16, 5.30*f*, 5.31*f*, 5.47
Spliced, defined, 6.47
Spoil pile, 1.36
Spotter, defined, 9.19
Spring clamps, 3.32–3.35, 3.33*f*
Sprinkler system, plans, 5.16
Spud wrenches, 3.15–3.16, 3.15*f*, 3.16*f*
Square, angles, 3.45
Square-bladed shovel, 3.36–3.37, 3.36*f*
Square centimeters, 2.46
Square feet, 2.46
Square file, 3.30–3.32, 3.30*f*
Square inches, 2.46
Square meters, 2.46
Squares
  area, 2.47, 2.68
  defined, 2.44, 2.64
  hand tools, 3.23–3.25, 3.23*f*, 3.24*f*, 3.44
  volume, 2.48, 2.50, 2.68
Square screwdrivers, 3.9–3.11, 3.9*f*
Square units, 2.46
Square yards, 2.46
Stacking materials, 9.3, 9.4*f*, 9.5*f*
Stairs
  hazards, 1.17
  maintenance and housekeeping, 1.28
  overview, 1.28
Standard eye-and-eye slings, 6.5, 6.6*f*
Standard ruler, 2.8–2.10, 2.9*f*, 2.64
Standard tip screwdrivers, 3.9–3.11, 3.9*f*
Steel
  architectural symbols, 5.7
  chain slings, 6.10–6.11, 6.10*f*–6.12*f*
  fastening systems, 4.26–4.27, 4.26*f*
  girders, 5.11, 5.14*f*
Stepladders, 1.25, 1.26*f*, 1.83
Stonemasons, 3.5
Storing materials, 9.3, 9.4*f*, 9.5*f*
Straight angles, 2.42*f*, 2.64
Straight-blade screwdrivers, 3.9–3.11, 3.9*f*
Straight ladders, 1.22–1.24, 1.23*f*, 1.24*f*, 1.83
Strand, defined, 6.47
Strap clamp, 3.33*f*, 3.34–3.35
Stress, accidents and, 1.6, 8.23–8.24
Stress, defined, 6.47
Striking wrench, 3.14–3.15, 3.14*f*, 3.45
Strip, 3.45
Struck-by accidents, 1.13, 1.31–1.33
Structural engineering symbols, 5.35, 5.37*f*
Structural plans, 5.1, 5.2*f*, 5.8, 5.11, 5.13*f*–5.15*f*, 5.47
Structural section drawings, 5.11, 5.15*f*
Substance abuse, accidents and, 1.6–1.7
Subtraction
  calculators and, 2.70
  decimals, 2.29–2.30
  fractions, 2.24
  whole numbers, 2.3–2.4
Sum, defined, 2.64
Supplied air mask, 1.49–1.52, 1.50*f*
Surface area, units of, 2.46
Surface water, 1.34
Survey plans, 5.4, 5.5*f*, 5.6*f*
Swing paths, 6.40–6.42, 6.41*f*, 6.42*f*
Switch enclosure, 1.83
Switchgear, plans for, 5.16, 5.24*f*, 5.25*f*, 5.27*f*–5.29*f*
Swivel eyebolts, 6.22, 6.22*f*
SWL (safe working load), 6.3, 6.32
Symbols
  architectural plans, 5.7
  defined, 5.47
  drawings, 5.34–5.38, 5.35*f*–5.39*f*
  electrical plans, 5.16, 5.24*f*, 5.25*f*, 5.27*f*–5.29*f*
  mechanical plans, 5.11, 5.17*f*, 5.18*f*
Synthetic web shackle, 6.20, 6.21*f*

T

Table, information, 7.21
Table of contents, defined, 7.21
Tag lines, 6.36, 6.48
Tagout systems, 1.43–1.44, 1.44*f*
Talc, 1.64
Tang, 3.45
Tapered punch, 3.8, 3.9*f*
Tardiness, 8.15–8.16, 8.32
Task hazard analysis (THA), 1.14–1.15, 1.16*f*
Task safety analysis (TSA), 1.14–1.15, 1.16*f*
Tattle-tail yarns, 6.6, 6.7*f*, 6.48
Teamwork skills, 8.19–8.20
Telephone calls, 7.7–7.8
Temperature, metric units, 2.14
Tempered, defined, 3.45
Tenon, 3.45
Tensiometer, 3.17*f*
Tension, slings, 6.49–6.50, 6.50*f*
Text messages, defined, 7.21
THA (task hazard analysis), 1.14–1.15, 1.16*f*
Threaded shank, defined, 6.48
Three-wire extension cords, 1.39
Title block, 5.2, 5.3*f*, 5.47
Toeboard, defined, 1.83
Tongue-and-groove pliers, 3.12*f*, 3.13, 3.13*f*
Tons (metric), unit conversion, 2.67
Tons (short), unit conversion, 2.67
Toolbox talks, 1.3
Tools. *See* Equipment; Hand tools; Power tools
Topographic maps, 5.8, 5.8*f*

Top rail, defined, 1.83
Torpedo level, 3.21*f*
Torque, defined, 3.45
Torque wrenches, 3.17–3.18, 3.17*f*, 3.18*f*
Torx® screwdrivers, 3.9–3.11, 3.9*f*
Toxic gases, trenches, 1.33
Toxins, 1.52–1.56, 1.53*f*–1.56*f*
Training and education, 1.9–1.12
Transformers, electrical plans, 5.16, 5.24*f*, 5.25*f*, 5.27*f*–5.29*f*
Trenches
  defined, 1.83
  hazards, 1.33–1.37, 1.35*f*–1.37*f*
Triangle file, 3.30–3.32, 3.30*f*
Triangles
  area, 2.47, 2.68
  defined, 2.64
  equilateral, 2.45, 2.45*f*, 2.63
  isosceles, 2.45, 2.45*f*, 2.63
  On-Site, 2.46
  Pythagorean Theorem, 2.48, 2.50
  right triangles, 2.42, 2.42*f*, 2.45, 2.45*f*, 2.47, 2.64
  3/4/5 Rule, 2.47
  scalene, 2.45, 2.45*f*, 2.64
  types of, 2.44–2.45, 2.45*f*
  volume, 2.51, 2.68
Trigger lock, power tools, 4.1, 4.36
Tripping hazards, 1.17
Trolley systems, hoists and, 6.32*f*
Trusses, 5.11, 5.14*f*
Try square, 3.23, 3.23*f*, 3.45
TSA (task safety analysis), 1.14–1.15, 1.16*f*
Twin-Path® slings, 6.5–6.7, 6.6*f*, 6.7*f*
Twist drill bits, 4.2, 4.2*f*
Twisted eye-and-eye slings, 6.5, 6.6*f*
Two-foot level, 3.21*f*

U

U.S. Green Building Council (USGBC), 5.25
Units, measurement
  area of shapes, 2.46–2.47, 2.53
  converting, 2.18–2.19, 2.50, 2.67
  cubic, volume, 2.12, 2.14, 2.48, 2.50–2.52, 2.67
  foot-pounds (torque), 3.44
  inch-pounds (torque), 3.44
  length, 2.8–2.14, 2.9*f*, 2.13*f*, 2.48, 2.50–2.52, 2.67
  metric, 2.12, 2.14, 2.67
  volume, 2.48, 2.50–2.52
  weight, 2.12, 2.14, 2.67
University of Florida, Rinker Hall, 5.40
Unprotected openings, 1.17–1.18
Unsafe conditions, 1.8
Unshouldered eyebolts, 6.21, 6.22*f*
Up-cutting, 4.14
USB flashdrive, 8.33
USGBC (U.S. Green Building Council), 5.25
Utility knives, 3.27, 3.27*f*
Utility saw, 4.11–4.14, 4.12*f*

V

Valves, sprinkler system, 5.16
Vapors, 1.70
Vehicle hazards, 1.31–1.33
Veneer knife file, 3.30–3.32, 3.30*f*
Ventilation
  conservation, Rinker Hall, 5.40
  Going Green, Bank of America, 8.5
  mechanical plans, 5.11, 5.16*f*–5.21*f*
Vertex, defined, 2.64
Vertical hitches, 6.15–6.16, 6.15*f*, 6.16*f*
Vertical pull, eyebolts, 6.22, 6.22*f*
Vertical surfaces, level measurement, 3.20, 3.21*f*
Vibration hazards, 1.13
Volume
  defined, 2.64
  formulas for, 2.68
  metric units, 2.12, 2.14
  practical application, 2.43
  shapes, 2.48, 2.50–2.52
  unit conversions, 2.50, 2.67

W

Walking surfaces, 1.17
Wall openings, 1.17–1.18
Walls
  blocks, number required, 2.7, 2.9
  bricklayers, 3.22
  fastening systems, 4.26–4.27, 4.26*f*
  stonemasons, 3.5
Warnings
  color codes, 1.17
  signs for, 1.4, 1.4*f*
  trench cave-ins, 1.34
Warning yarns, slings, 6.4, 6.5*f*, 6.48
Water
  conservation, Rinker Hall, 5.40
  Going Green, Bank of America, 8.5
  green-collar jobs, 8.1
  green construction, 5.25
  plumbing plans, 5.11, 5.22*f*, 5.23*f*
  in trenches, 1.33
Weather conditions
  cold stress, human, 1.60–1.62
  heat stress, human, 1.59–1.60
  trenches, 1.34
Web clamp, 3.33*f*, 3.34–3.35
Web slings, 6.4–6.5, 6.4*f*, 6.5*f*
Weight
  metric units, 2.12, 2.14
  unit conversion, 2.67
Weight capacity, defined, 6.48
Weight-forward hammers, 3.2, 3.2*f*
Welding
  defined, 3.45
  eye and face protection, 1.45–1.46, 1.46*f*
  grinders, 4.20–4.21, 4.20*f*
  hazards, 1.62–1.66, 1.63*f*, 1.65*f*
  protective equipment, 1.62–1.63, 1.63*f*
Welding shield, defined, 1.83
Weld joints, peening, 3.2
Wheelbarrows, 9.7–9.8, 9.8*f*, 9.10, 9.10*f*, 9.19
Whip check, 4.11
Whole numbers
  addition and subtraction, 2.2–2.4
  calculators, 2.69–2.70
  decimals and, 2.27, 2.28
  defined, 2.1–2.2, 2.64
  dividing, 2.5–2.8
  multiplying, 2.4–2.6
Wide-body shackle, 6.20, 6.21*f*
Width, measuring, 2.48, 2.50–2.52
Windows, green construction, 8.5
Window schedules, 5.7, 5.12*f*
Wind sock, 1.83
Wind turbines, 2.41

Wire cutters, 3.11–3.14
Wireless communications technology, 8.12, 8.14
Wire rope, defined, 6.48
Wire rope slings, 6.11–6.14, 6.12*f*–6.14*f*
Wires, electrical. *See* Electricity
WLL (working load limit), 6.3, 6.32
Wood, cutting, 4.11
Wood chisels, 3.7*f*
Wooden folding rule, 3.19, 3.19*f*
Wood joists, foundation plans, 5.11, 5.14*f*
Wood ladders, 1.22
Worker complaint and referral inspections, 1.11
Worker's compensation insurance, 1.3
Work ethic, 8.14, 8.15, 8.33
Work habits, accidents and, 1.5
Working capacity, 6.3
Working load limit (WLL), 6.3, 6.32
Working surfaces, 1.17
Workplace issues, 8.23–8.25
Work platforms, 1.17
Work site
  cleanliness, 1.9
  safety citations, 1.10
Work zone, 9.11, 9.19
Worm-drive saw, 4.13
Wrecking bar, 3.5*f*
Wrenches
  adjustable, 3.15–3.16, 3.15*f*, 3.16*f*, 3.44
  air impact wrench, 4.27–4.28, 4.27*f*
  nonadjustable, 3.14–3.15, 3.14*f*
  torque wrenches, 3.17–3.18, 3.17*f*, 3.18*f*

Y

Yard, measurement
  cubic, 2.48
  history of, 2.13
  square yard, 2.46
  unit conversion, 2.67
Yellow, warning signs, 1.17

Z

Zinc oxide, 1.64